Modified Atmosphere and Active Packaging Technologies

Edited by
Ioannis S. Arvanitoyannis

Contemporary Food Engineering

Series Editor

Professor Da-Wen Sun, Director
Food Refrigeration & Computerized Food Technology
National University of Ireland, Dublin
(University College Dublin)
Dublin, Ireland
http://www.ucd.ie/sun/

Modified Atmosphere and Active Packaging Technologies, *edited by Ioannis Arvanitoyannis* (2012)
Engineering Aspects of Membrane Separation and Application in Food Processing, *edited by Gyula Vatai, Erika Bekassy-Molnar, Robert W. Field, and Frank Lipnizki* (2012)
Advances in Fruit Processing Technologies, *edited by Sueli Rodrigues and Fabiano Andre Narciso Fernandes* (2012)
Biopolymer Engineering in Food Processing, *edited byVânia Regina Nicoletti Telis* (2012)
Operations in Food Refrigeration, *edited by Rodolfo H. Mascheroni* (2012)
Thermal Food Processing: New Technologies and Quality Issues, Second Edition, *edited by Da-Wen Sun* (2012)
Physical Properties of Foods: Novel Measurement Techniques and Applications, *edited by Ignacio Arana* (2012)
Handbook of Frozen Food Processing and Packaging, Second Edition, *edited by Da-Wen Sun* (2011)
Advances in Food Extrusion Technology, *edited by Medeni Maskan and Aylin Altan* (2011)
Enhancing Extraction Processes in the Food Industry, *edited by Nikolai Lebovka, Eugene Vorobiev, and Farid Chemat* (2011)
Emerging Technologies for Food Quality and Food Safety Evaluation, *edited by Yong-Jin Cho and Sukwon Kang* (2011)
Food Process Engineering Operations, *edited by George D. Saravacos and Zacharias B. Maroulis* (2011)
Biosensors in Food Processing, Safety, and Quality Control, *edited by Mehmet Mutlu* (2011)
Physicochemical Aspects of Food Engineering and Processing, *edited by Sakamon Devahastin* (2010)
Infrared Heating for Food and Agricultural Processing, *edited by Zhongli Pan and Griffiths Gregory Atungulu* (2010)
Mathematical Modeling of Food Processing, *edited by Mohammed M. Farid* (2009)
Engineering Aspects of Milk and Dairy Products, *edited by Jane Sélia dos Reis Coimbra and José A. Teixeira* (2009)
Innovation in Food Engineering: New Techniques and Products, *edited by Maria Laura Passos and Claudio P. Ribeiro* (2009)
Processing Effects on Safety and Quality of Foods, *edited by Enrique Ortega-Rivas* (2009)
Engineering Aspects of Thermal Food Processing, *edited by Ricardo Simpson* (2009)
Ultraviolet Light in Food Technology: Principles and Applications, *Tatiana N. Koutchma, Larry J. Forney, and Carmen I. Moraru* (2009)
Advances in Deep-Fat Frying of Foods, *edited by Serpil Sahin and Servet Gülüm Sumnu* (2009)
Extracting Bioactive Compounds for Food Products: Theory and Applications, *edited by M. Angela A. Meireles* (2009)
Advances in Food Dehydration, *edited by Cristina Ratti* (2009)
Optimization in Food Engineering, *edited by Ferruh Erdoğdu* (2009)
Optical Monitoring of Fresh and Processed Agricultural Crops, *edited by Manuela Zude* (2009)
Food Engineering Aspects of Baking Sweet Goods, *edited by Servet Gülüm Sumnu and Serpil Sahin* (2008)
Computational Fluid Dynamics in Food Processing, *edited by Da-Wen Sun* (2007)

Contemporary Food
Engineering Series
Da-Wen Sun, Series Editor

Modified Atmosphere and Active Packaging Technologies

Edited by
Ioannis S. Arvanitoyannis

CRC Press
Taylor & Francis Group
Boca Raton London New York

CRC Press is an imprint of the
Taylor & Francis Group, an **informa** business

CRC Press
Taylor & Francis Group
6000 Broken Sound Parkway NW, Suite 300
Boca Raton, FL 33487-2742

First issued in paperback 2016

ISBN 13: 978-1-138-19902-6 (pbk)
ISBN 13: 978-1-4398-0044-7 (hbk)

Library of Congress Cataloging-in-Publication Data

Modified atmosphere and active packaging technologies / editor, Ioannis Arvanitoyannis.
p. cm. -- (Contemporary food engineering)
Summary: "While other packaging books focus on individual types of packaging, this volume takes an applied engineering approach by matching packaging types to specific food types. The material provides unique insight into Modified/Controlled Atmosphere Packaging/Storage (MAP/CAS) and Active Packaging (AP). Depending on the kind of food, packaging material and the corresponding technique employed can both vary considerably. In appreciation of consumer focus on shelf life and safety, the book addresses a range of aseptic, smart, and modified atmosphere packaging. With the increased expense of active packaging, this resource will help product developers make cost-effective decisions early in product development"-- Provided by publisher.
Includes bibliographical references and index.
ISBN 978-1-4398-0044-7 (hardback)
1. Food--Packaging. 2. Protective atmospheres. I. Arvanitoyannis, Ioannis.

TP374.M59 2012
664--dc23 2012006572

Visit the Taylor & Francis Web site at
http://www.taylorandfrancis.com

and the CRC Press Web site at
http://www.crcpress.com

To

my beloved and patient wife Nicole for her continuous and unfailing support over the last 13 years and to our three children

- *Nefelli-Kallisti (the youngest and most communicative)*
- *Artemis-Eleni (the thoughtful)*
- *Iasson (the most sentimental with the fighting spirit)*

who work hard toward conveying to us a new, more interesting and promising perspective of life.

And

to the memory of my grandparents (Dimitrios and Evangelia) for their love and support in my first steps.

Contents

Series Preface

CONTEMPORARY FOOD ENGINEERING

Food engineering is the multidisciplinary field of applied physical sciences combined with the knowledge of product properties. Food engineers provide the technological knowledge transfer essential to the cost-effective production and commercialization of food products and services. In particular, food engineers develop and design processes and equipment to convert raw agricultural materials and ingredients into safe, convenient, and nutritious consumer food products. However, food engineering topics are continuously undergoing changes to meet diverse consumer demands, and the subject is being rapidly developed to reflect market needs.

In the development of food engineering, one of the many challenges is to employ modern tools and knowledge, such as computational materials science and nanotechnology, to develop new products and processes. Simultaneously, improving food quality, safety, and security continues to be a critical issue in food engineering study. New packaging materials and techniques are being developed to provide more protection to foods, and novel preservation technologies are emerging to enhance food security and defense. Additionally, process control and automation regularly appear among the top priorities identified in food engineering. Advanced monitoring and control systems are developed to facilitate automation and flexible food manufacturing. Furthermore, energy saving and minimization of environmental problems continue to be important food engineering issues, and significant progress is being made in waste management, efficient utilization of energy, and reduction of effluents and emissions in food production.

The Contemporary Food Engineering series, consisting of edited books, attempts to address some of the recent developments in food engineering. The series covers advances in classical unit operations in engineering applied to food manufacturing as well as such topics as progress in the transport and storage of liquid and solid foods; heating, chilling, and freezing of foods; mass transfer in foods; chemical and biochemical aspects of food engineering and the use of kinetic analysis; dehydration, thermal processing, nonthermal processing, extrusion, liquid food concentration, membrane processes, and applications of membranes in food processing; shelf life and electronic indicators in inventory management; sustainable technologies in food processing; and packaging, cleaning, and sanitation. These books are aimed at professional food scientists, academics researching food engineering problems, and graduate-level students.

The editors of these books are leading engineers and scientists from many parts of the world. All the editors were asked to present their books to address market needs and pinpoint cutting-edge technologies in food engineering.

All contributions have been written by internationally renowned experts who have both academic and professional credentials. All the authors have attempted to provide critical, comprehensive, and readily accessible information on the art and science of a relevant topic in each chapter, with reference lists for further information. Therefore, each book can serve as an essential reference source to students and researchers in universities and research institutions.

Da-Wen Sun
Series Editor

Series Editor

Professor Da-Wen Sun, PhD, is a world authority on food engineering research and education. He is a member of the Royal Irish Academy, which is the highest academic honor in Ireland; he is also a member of Academia Europaea. His main research activities include cooling, drying, and refrigeration processes and systems; quality and safety of food products; bioprocess simulation and optimization; and computer vision technology. His innovative studies on vacuum cooling of cooked meat, pizza quality inspection using computer vision, and edible films for shelf life extension of fruits and vegetables, in particular, have been widely reported in the national and international media. Results of his work have been published in about 600 papers, including about 250 peer-reviewed journal papers (h-index = 36). He has also edited 13 authoritative books. According to Thomson Scientific's *Essential Science Indicators*SM updated as of July 1, 2010, based on data derived over a period of 10 years and 4 months (January 1, 2000–April 30, 2010) from ISI Web of Science, a total of 2554 scientists are among the top 1% of the most-cited scientists in the category of agriculture sciences, and Professor Sun is listed at the top with a ranking of 31.

Dr. Sun received first class BSc honors and his MSc in mechanical engineering and PhD in chemical engineering from China before working at various universities in Europe. He became the first Chinese national to be permanently employed in an Irish university, when he was appointed as a college lecturer at University College Dublin (UCD)—National University of Ireland, Dublin, in 1995. He was then continuously promoted in the shortest possible time to the position of senior lecturer, associate professor, and full professor. Dr. Sun is now professor of food and biosystems engineering and director of the Food Refrigeration and Computerized Food Technology Research Group at UCD.

As a leading educator in food engineering, Dr. Sun has contributed significantly to the field of food engineering. He has guided many PhD students who have made their own contributions to the industry and academia. He has also, on a regular basis, given lectures on the advances in food engineering at international academic institutions and delivered keynote speeches at international conferences. As a recognized authority in food engineering, he has been conferred adjunct/visiting/consulting professorships by over 10 top universities in China, including Zhejiang University, Shanghai Jiaotong University, Harbin Institute of Technology, China Agricultural University, South China University of Technology, and Jiangnan University. In recognition of his significant contribution to food engineering worldwide and for his outstanding leadership in the field, the International Commission of Agricultural and Biosystems Engineering (CIGR) awarded him the CIGR Merit Award in 2000 and again in 2006; the U.K.-based Institution of Mechanical Engineers named him Food

Engineer of the Year 2004; in 2008, he was awarded the CIGR Recognition Award in recognition of his distinguished achievements as the top 1% of agricultural engineering scientists in the world; in 2007, he was presented the only AFST(I) Fellow Award in that year by the Association of Food Scientists and Technologists (India); and in 2010, he was presented the CIGR Fellow Award (the title of "Fellow" is the highest honor in CIGR and is conferred upon individuals who have made sustained, outstanding contributions worldwide).

Dr. Sun is a fellow of the Institution of Agricultural Engineers and a fellow of Engineers Ireland (the Institution of Engineers of Ireland). He has also received numerous awards for teaching and research excellence, including the President's Research Fellowship, and has received the President's Research Award from the UCD on two occasions. He is the editor in chief of *Food and Bioprocess Technology—An International Journal* (Springer) (2010 Impact Factor = 3.576, ranked at the fourth position among 126 ISI-listed food science and technology journals); series editor of Contemporary Food Engineering series (CRC Press/Taylor & Francis); former editor of *Journal of Food Engineering* (Elsevier); and an editorial board member of *Journal of Food Engineering* (Elsevier), *Journal of Food Process Engineering* (Blackwell), *Sensing and Instrumentation for Food Quality and Safety* (Springer), and *Czech Journal of Food Sciences*. Dr. Sun is also a chartered engineer.

On May 28, 2010, Dr. Sun was awarded membership to the Royal Irish Academy (RIA), which is the highest honor that can be attained by scholars and scientists working in Ireland. At the 51st CIGR General Assembly held during the CIGR World Congress in Quebec City, Canada, in June 2010, he was elected as incoming president of the CIGR. He will become the president of the CIGR in 2013–2014. The term of the presidency is six years, two years each for serving as incoming president, president, and past president.

On September 20, 2011, he was elected to Academia Europaea, which is functioning as European Academy of Humanities, Letters and Sciences and is one of the most prestigious academies in the world; election to the Academia Europaea represents the highest academic distinction.

Preface

The first human beings survived mainly by hunting wild animals, and only later did they learn to harvest vegetables, fruits, and cereals. Once they had gathered surplus food, they tried to find ways of preserving this. Apart from heating, drying, smoking, and freezing food, they used packaging materials abundant in nature such as leaves, wood, and bamboo. Later on, animal skin and woven baskets were also used for transporting water and food respectively. The canning process was introduced in 1810 by N. Appert (the method was initially known as Appertization) as a promising technique for preserving food.

Over the last two centuries, there have been lots of advances in the process of packaging. To make the proper choice of the packaging material, several parameters have to be taken into account, such as imparting the protective properties for the anticipated shelf life and availability of proper size and weight.

Although "food packaging" has been defined in several ways, some of the most representative ones are as follows: "a system for (i) preparing goods for transport, distribution, storage, (ii) ensuring safe delivery to consumer, and (iii) minimizing costs in conjunction with maximizing profits." Another factor that has recently been introduced is the biodegradability of the packaging material, or its being environment friendly.

Apart from the so-called classical or traditional preservation techniques such as heating, drying, smoking, cooling, and freezing, there is a plethora of novel processing and preservation techniques that have gradually gained ground. Some of these techniques are the application of high pressure, microwave technology, irradiation (X-rays, γ-rays, e-beam), and ohmic heating, on the one hand, and the packaging-based techniques such as aseptic packaging, smart/intelligent packaging, active packaging (AP), packaging under vacuum, and modified atmosphere packaging (MAP) and controlled atmosphere packaging, on the other. Although MAP first appeared in the 1960s, its application at the commercial level was rather restricted. It is only in the last 10–15 years that this technique has seen increased usage. The same was valid for active packaging where the inclusion of a small bag (containing the active component due to be released or react with a non-desirable one) on several occasions was not highly appreciated by the consumers. Another factor limiting the extensive usage of MAP and AP was their higher cost compared to conventional packages. However, recently there has been an increasing demand for large amounts of fresh food (in view of globalization), which has favored the upgradation of MAP and AP. Since both techniques can be applied to fresh food stored in the fridge, their application has been enhanced considerably.

This book consists of 19 chapters and has been divided into seven parts: Part I (Principles, Materials, Gases, and Machinery for MAP) consists of Chapters 1 and 2. Part II (Safety and Quality Control of MAP Produces) consists of Chapter 3. Part III (Applications of MAP in Foods of Animal Origin) consists of Chapters 4 through 7, which cover fish, meat, poultry, and dairy products. Part IV (Applications of MAP in Foods of Plant Origin) consists of Chapters 8 through 11, which deal with cereals, minimally processed vegetables, fruits, and bakery products. Part V (Other Applications of MAP) consists of Chapters 12 and 13, which describe RTE food and other miscellaneous types of foods. Part VI (Active Packaging and Its New Trends) consists of Chapters 14 through 16, which deal with active packaging, nanotechnology, and bioactive packaging. Part VII (Consumer Behavior/Sensory Analysis and Legislation) consists of Chapters 17 through 19, which cover the issues of sensory analysis and consumer search; EU, U.S., and Canadian legislation; and, finally, conclusions and new trends.

The aim of this book is to convey, both to the average and specialized reader, an overview of the current status quo of MA and AP in terms of applied techniques and methodologies in conjunction with a large number of applications on food of animal and plant (both raw and processed) origin and updated legislation for packaging coming in contact with foods. Emphasis was also given to novel technologies such as nanotechnology and bioactive packaging (chitosan).

The uniqueness of this book is that it covers practically all issues related to packaging under modified atmosphere and vacuum, controlled atmosphere, and active packaging, starting with the very basics (films, gases, techniques and applications, legislation) up to the latest advances (nanotechnology and bioactive compounds), and it is supported in this endeavor by a large number of sources (more than 2000), which will be of use to the reader.

It is anticipated that this book will be of interest and use to scientists and technologists coming from different backgrounds (veterinary doctors, agriculturists, chemists, chemical engineers, food scientists, and technologists) and levels, that is, from undergraduate students up to graduates, postgraduates, instructors, professors (research institutes and universities or polytechnics), and professionals well versed in this topic.

Dr. Ioannis S. Arvanitoyannis PhD
Associate Professor
University of Thessaly
Volos, Greece

Editor

Ioannis S. Arvanitoyannis graduated from the Department of Chemistry at Aristotle University of Thessaloniki (AUTH), Hellas. He did his first PhD in polymer science, Department of Chemistry, AUTH, Hellas, and his second PhD in physical chemistry of foods, Department of Food Science and Applied Microbiology. He has worked as a postdoctoral researcher at the Research Center of Plastic Materials of Ciba-Geigy, Marly, Fribourg, Switzerland; in the Department of Chemistry at Loughborough University of Technology, United Kingdom, for two years; and in Osaka National Research Institute of Japan for two and a half years.

Since 2005, Dr. Arvanitoyannis has been serving as associate professor in the Department of Agriculture, Ichthyology and Aquatic Environment in the School of Agricultural Sciences at the University of Thessaly. He has published more than 180 research and review articles in well-known peer review journals and is the author of 35 invited chapters (32 in English). He is also author and coauthor of 16 books on topics related to food technology; food packaging (MAP, AP); food safety (HACCP), ISO 22000:2005; genetically modified foods; implementation of TQM, ISO 9001 to the food industry; quality assurance and safety guide for the food and drink industry; irradiation of food commodities; waste management for the food industries; and ISO 14000 and a laboratory guide on food quality control. He has been invited to give lectures in Kyoto University, Japan, the University of Bangkok, Thailand, the University of Chester, United Kingdom, the University of Helsinki, Finland, Mediterranean Agronomic Institute of Chania (MAICh), Greece, and Mediterranean Agronomic Institute of Bari, Italy.

Dr. Arvanitoyannis enjoys international recognition, with more than 3000 citations (cross-references) to his credit, is associate editor of the *International Journal of Food Science and Technology*, and is a member of the editorial board of six journals. He has also served as reviewer for more than 25 international journals.

Contributors

Maria Andreou
Department of Ichthyology & Aquatic Resources
University of Thessaly
Volos, Greece

Ioannis S. Arvanitoyannis
Department of Ichthyology & Aquatic Resources
University of Thessaly
Volos, Greece

Konstantinos Bosinas
Department of Dietetics
Technological Institute of Larissa
Larissa, Greece

Achilleas Bouletis
Department of Ichthyology & Aquatic Resources
University of Thessaly
Volos, Greece

Stéphane Desobry
Laboratory of Biomolecules Engineering
Engineering School in Food and Agricultural Sciences
University of Lorraine
Vandoeuvre, France

Nikoletta K. Dionisopoulou
Department of Ichthyology & Aquatic Resources
University of Thessaly
Volos, Greece

Vasiliki I. Giatrakou
Department of Chemistry
University of Ioannina
Ioannina, Greece

Muhammad Imran
Laboratory of Biomolecules Engineering
Engineering School in Food and Agricultural Sciences
University of Lorraine
Vandoeuvre, France

Konstantinos Kotsanopoulos
Department of Ichthyology & Aquatic Resources
University of Thessaly
Volos, Greece

Nikoletta Manti
Department of Ichthyology & Aquatic Resources
University of Thessaly
Volos, Greece

Georgios Oikonomou
Department of Ichthyology & Aquatic Resources
University of Thessaly
Volos, Greece

Anne-Marie Revol-Junelles
Laboratory of Biomolecules Engineering
Engineering School in Food and Agricultural Sciences
University of Lorraine
Vandoeuvre, France

Maria Savva
Department of Ichthyology & Aquatic Resources
University of Thessaly
Volos, Greece

Ioannis N. Savvaidis
Department of Chemistry
University of Ioannina
Ioannina, Greece

Alexandros Ch. Stratakos
Department of Ichthyology & Aquatic Resources
University of Thessaly
Volos, Greece

Persephoni Tserkezou
Department of Ichthyology & Aquatic Resources
University of Thessaly
Volos, Greece

Georgios Tziatzios
Department of Ichthyology & Aquatic Resources
University of Thessaly
Volos, Greece

Part I

Principles, Materials, Gases, and Machinery for MAP

1 Principles of MAP and Definitions of MAP, CA, and AP

Ioannis S. Arvanitoyannis

CONTENTS

1.1 INTRODUCTION

1.1.1 MODIFIED ATMOSPHERE PACKAGING

Temperature control and modification of atmosphere are two important factors in extending the shelf life of perishable food. Modified Atmosphere Packaging (MAP) of fresh produce relies on modification of the gas composition inside the package, based on the interactions between two processes, the respiration of product and the transfer of gases through the packaging, thereby resulting in gas composition richer in CO_2 and poorer in O_2 (Fonseca et al., 2000). In contrast to controlled atmosphere (CA) systems, modified atmosphere (MA) technology has higher flexibility in prolonging the CA benefits for improving shelf life of a larger number of fresh produce during distribution and storage. MA conditions can be effected via packaging a passive system, by balancing produce respiration and gas exchange through package materials. Such systems, called MAP, can be visualized as bulk packaging containers, as unit retail packages, and as individual produce coatings (Lee et al., 1996). Bulk packaging systems are pallet bags and paperboard containers employed for several processes such as transportation, handling, and storage. They have the advantage of being able to handle mixed loads in the same storage place. The prepackaging system is based on a plastic film, such as low-density polyethylene (LDPE), polyvinylchloride (PVC), polyethylene terephthalate (PET), and polypropylene (PP). The use of these materials is advantageous because of their transparency, their action as barriers to water vapor transmission, and their selectivity in gas permeability (Lee et al., 1996).

MA packages must be carefully designed, because, otherwise, the results will be adverse, such as system ineffectiveness or even product shelf life shortening (Fonseca et al., 2000). The design should take into account both the steady-state conditions and the dynamic process. In fact, if the product is exposed for a long time to unsuitable gas composition before reaching appropriate gas composition, the package is not likely to be suitable to the produce enclosed. The MA package design depends on various factors such as the temperature, the gas composition that is recommended, the characteristics of the product and its mass, and the permeability of the packaging material. The known tendency for the package volume to shrink during storage of produce is due to the gas exchange (through a semipermeable membrane) and can be easily eliminated by tiny holes that allow pressure equalization at constant volume by convective flow of gas. There is always the possibility that the number and size of the holes may be greater than what they should be, thereby altering the selectivity of gas exchange between the air and the package (Paul and Clarke, 2002). Except for the permeability of metabolic gases, permeability for water vapor, ethylene, and volatiles can be of crucial importance as well. Low permeability to water vapor can augment the risk for condensation. Condensation should be avoided at any cost since it provides an ideal environment for microbial growth. Furthermore, discoloration of the product could be due to condensation.

The main gases employed in MAP are oxygen, nitrogen, and carbon dioxide used in different combinations and proportions depending on (a) the product, (b) the anticipated product shelf life, and (c) the needs of the processor and the consumer. The final choice is greatly influenced by the microbiological flora growing on the product, the sensitivity of the product to oxygen and carbon dioxide, and its color stability requirements (Church, 1994, Phillips, 1996). Retail MAP of freshly prepared produce could be successfully accomplished by using a packaging film of proper permeability so as to establish optimal EMAs of typically 3%–10% O_2 and 3%–10% CO_2.

To maintain headspace O_2 levels at 40% and CO_2 levels in the range of 10%–25% over the chilled shelf life of the product, it is desirable to minimize the produce volume or gas volume ratio of fresh prepared produce MA packs. This can be achieved by either diminishing the pack fill weight or augmenting the pack headspace volume. Reducing the pack fill weight of fresh prepared produce will have the effect of lowering the overall respiratory load or activity within MA packs, and thus, the rate of O_2 depletion will be decreased. Augmentation of the pack headspace volume is bound to have the effect of enhancing the reservoir of O_2 for respiratory purposes, and thus, the rate of O_2 depletion will be decreased as well. Consequently, low produce volume or gas volume ratios are conducive to maintaining headspace O_2 levels >40% and CO_2 levels in the range of 10%–25%.

Even after harvesting, fresh fruits continue to be metabolically active for long periods because of their endogenous activity (respiration) and other external factors, such as physical injury, microbial flora, water loss, and storage temperature (Kader et al., 1989). Concerns have been occasionally expressed that the extension in shelf life of MAP products could unfortunately give adequate time to human pathogens to grow to contents that render the food unsafe while still edible (Jay, 1992). It should be noted at this point that pathogens capable of growing at chill temperatures such as *Listeria monocytogenes* and *Yersinia enterocolitica* and those growing at minimal oxygen contents like psychotropic *Clostridium botulinum* must be taken into account.

The introduction of "smart," "intelligent," or "active" packaging systems to prolong fruit safety during packaging and storage is maybe the most important field of development in MAP techniques and methods (Summers, 1992). Such techniques are an indispensable part of the packaging system, thereby imparting intelligence appropriate to the function and use of the product itself, and have the ability to perceive any changes occurring and to inform the consumer about them (Church, 1994). A general classification of AP techniques depending on the fruit or packaging affected is oxygen absorbers or scavengers, carbon dioxide absorbers or scavengers, flavor removers or releasers, ethylene removers, ethanol releasers or emitters, moisture absorbers or scavengers, time–temperature indicators (TTIs), and antimicrobial-containing films (Kader, 1980, Robertson, 1991).

The factors that can affect the shelf life of MAP foods can be distinguished as internal and external factors. The internal factors are the water activity (a_w), pH, microbial flora, nutrient availability, content and type of preservative agent, redox potential, and the presence of naturally occurring antimicrobial compounds and spores. The external factors are the temperature control, the hygienic processing (HACCP application), the raw material quality, the blending or mixing of ingredients in package, the period of time prior to packaging, the initial and final gas purity or composition, the relative permeability of packaging film to gases, the gas product ratio, and the pack design (Church, 1993).

The loss of vitamin C after harvest can also be decreased with storage of fruits and vegetables in atmospheres of reduced O_2 and up to 10% CO_2 as Lee and Kader (2000) have reported. CA conditions do not have a positive effect on vitamin C in case enhanced CO_2 concentrations are applied. However, the contents above which CO_2 influences the loss of AA should be calculated per commodity (Kader, 2001).

1.2 SYNERGISTIC ACTION OF MAP

1.2.1 Active Packaging

Active packaging (AP) technologies opened a new horizon of new opportunities for food preservation. AP is defined as an intelligent or smart system involving interactions between package or package components and food or internal gas atmosphere. AP aims at meeting the consumer demands for high quality and safety in conjunction with fresh (not strongly processed) foods. AP prolongs the shelf life of foods, keeping at the same time their nutritional quality, suppressing or limiting the growth of pathogenic and spoilage microorganisms, preventing and indicating the migration of contaminants, and exhibiting any package leaks present, thereby ensuring food safety. Active packaging has been used with many food products and is being tested in conjunction with many others. It should be noted that all food products have a unique deterioration mechanism that must be comprehended before implementing this technology. The shelf life of packed food depends on many parameters such as the intrinsic nature of the food (e.g., pH, a_w, nutrient content, occurrence of antimicrobial compounds, redox potential, respiration rate, and biological structure) and extrinsic factors (e.g., storage temperature, relative humidity [RH], and the gaseous composition). These factors will directly affect the chemical, biochemical, physical, and microbiological spoilage mechanisms of individual food products and their obtained shelf lives. Should one take into consideration all these factors, it is possible to assess both current and developing AP technologies and apply them for keeping the quality and prolonging the shelf life of different food products. Recently, a number of new "intelligent" concepts have been introduced, the functions of which include more than just scrubbing or emitting compounds. These types of packages are due to be "activated" when a specific presupposition has been met. A great deal of these packages focus on prevention of problems linked to anaerobic conditions.

According to Kader (2001), "the aim of application of MAP to fleshy fruit is to annihilate or minimize phenomena related to the ripening of fleshy fruit such as: seed maturation, color change, abscission, change in respiration rate, change in ethylene production rate, change in tissue permeability, softening, changes in pectic composition, change in carbohydrate composition, protein changes, production of flavor volatiles, development of wax on skin, and organic acid changes."

REFERENCES

Church, P.N. (1993). Meat products. In: *Principles and Applications of Modified Atmosphere Packaging of Food*, ed. Parry, R.T., Blackie Academic & Proffessional, Glasgow, U.K., pp. 229–268.

Church, N. (1994). Developments in modified-atmosphere packaging and related technologies. *Trends in Food Science and Technology*, 5(11): 345–352.

Fonseca, S.C., Oliveira, F.A.R., Lino, I.B.M., Brecht, J.K., and Chau, K.V. (2000). Modelling O_2 and CO_2 exchange for development of perforation-mediated modified atmosphere packaging. *Journal of Food Engineering*, 43(1): 9–15.

Jay, J.M. (1992). *Modern Food Microbiology*, 4th edn. New York: Chapman & Hall, pp. 18–32.

Kader, A.A. (1980). Prevention for ripening in fruits by use of controlled atmospheres. *Food Technology*, 34(3): 51.

Kader, A.A., ed. (2001). CA Bibliography (1981–2000) and CA Recommendations CD. University of California, Post harvest Technology Center, Post harvest Horticulture Series No. 22 (The CA Recommendations, 2001 portion is also available in printed format as Post harvest Horticulture Series No. 22A). For more information, go to: http://postharvest.ucdavis.edu

Kader, A.A, Zagory, D., and Kerbel, E.L. (1989). Modified atmosphere packaging of fruits and vegetables. *Critical Reviews in Food Science and Nutrition*, 28(1): 1–10.

Lee, L.Z., Arult, J., Lencki, R., and Castaignet, F. (1996). Methodology for determining the appropriate selectivity of mass transfer devices for modified atmosphere packaging of fresh produce. *Packaging Technology and Science*, 9: 55–72.

Lee, S.K. and Kader, A.A. (2000). Pre harvest and post harvest factors influencing vitamin C content of horticultural crops. *Postharvest Biology and Technology*, 20: 207–220.

Paul, D.R. and Clarke, R. (2002). Modelling of modified atmosphere packaging based on designs with a membrane and perforations. *Journal of Membrane Science*, 208: 269–283.

Phillips, C.A. (1996). Review: Modified atmosphere packaging and its effects on the microbiological quality and safety of produce. *International Journal of Food Science and Technology*, 31(6): 463–479.

Robertson, G.L. (1991). In: *Proceedings of CAP 91, Sixth International Conference on Controlled/Modified Atmosphere/Vacuum Packaging*, Scotland Business Research, Princeton, NJ, pp. 163–181.

Summers, L. (1992). *Intelligent Packaging*. Centre for Exploitation of Science and Technology, London, U.K.

2 Materials (Films), Gases, and Machinery (Techniques) for MAP

Ioannis S. Arvanitoyannis and Achilleas Bouletis

CONTENTS

2.1 MACHINERY USED IN MAP

In the late 1950s, the first efforts were made to develop the principles of atmosphere modification on a larger scale. The experimental work in this field resulted in the development of equipment for industrial use. During the same decade, another useful technique called vacuum packaging was presented. In this phase, the production of efficient, safe, and low-cost machinery was very important.

Modified atmosphere packaging (MAP) equipment is divided into two main categories: pillow wrap and chamber. Pillow wrap can be divided into two further categories: horizontal and vertical processing machines. Chamber machines can use two different techniques. The first technique is the thermoforming technique. In the second technique, ready containers can be used for the packaging of products (Hastings, 1999).

2.2 CHAMBER MACHINES

2.2.1 Thermoforming Machines

Thermoforming machines demand the use of a rigid or semirigid material. This material is used as a substrate of the packaging tray. In a heating station, it could have a place in the main steps of the process. In this station, the material is heated to the point that can be treated. Then, the treated material is transferred to a forming station where it takes its final shape. This process can be supported by the use of vacuum air pressure.

The next step in this procedure is the cooling of the material and its transfer to the product loading area. The completion of this phase can occur either automatically or manually. The nature of the product plays the most important role in the chosen method. Atmosphere modification in the packages can be carried out by air extraction and gas flush of the trays. The gas flush takes place in a sealed chamber. After passing through the sealing station, the packages are stored or further processed (e.g., labeling) (Hastings, 1993).

2.2.2 Preformed Container Machines

Performed container machines are usually known as tray sealers. Tray sealing and thermoforming machines seem to be similar to each other. The preformed trays used are fed either manually or automatically, and the product is loaded. The tray and the material of the top lid are then transferred to the atmosphere modification chamber. In the chamber, vacuum and gas flush create the best environment for the maintenance of the product. It should be added that the top lid is heat-sealed. In the next stations, the material of lids is removed from the web and the packages get rid of the carriers (Hastings, 1999).

The advantages of the preformed tray system under the thermoforming process are as follows:

- Preformed trays are very flexible in regard to the tray shape and construction used.
- Changes in the tray material do not involve modifications in the packaging system.
- Preformed trays stabilize product appearance and presentation.
- Different colored trays can be used without altering the production method.
- A separate area can be used for tray loading without having impact on the packaging process.

The low packaging cost is the main advantage of thermoforming machines. The process cost could be reduced about 30%–50%, if the cost of transportation, base material storage, and tray is also reduced (Mullan and McDowell, 2003).

2.3 PILLOW WRAP MACHINES

2.3.1 Horizontal Form-Fill-Seal Machines

During this procedure, a roll of flexible packaging material is used. The roll passes through a forming tool, and then it is transformed into a tube. The foodstuff is then passed through the tube in different ways depending on its nature. Gas flush is used for atmosphere modification. This technique is applied by a lance that is entered in the tube when the horizontal form-fill-seal machine is operated.

Package integrity and leak avoidance must ensure the success of MAP operation. Is should be added that the implementation of hygiene practices is very important in the food industry. Machinery, especially parts that are in contact with foodstuffs, should be cleaned and disinfected regularly (Hastings, 1999). A typical horizontal form-fill-seal machine is shown in Figure 2.1.

2.3.2 Vertical Form-Fill-Seal Machines

Foodstuffs such as coffee, nuts, and cereals use vertical form-fill-seal machines for their packaging. The packaging material passes though a forming tube that is raised in the vertical plane. Then the foodstuff is placed in the packaging material. The vertical tube consists of two layers when atmosphere modification is in motion.

FIGURE 2.1 A horizontal form-fill-seal machine. (The photo was reproduced with permission from CFS, Bakel, the Netherlands.)

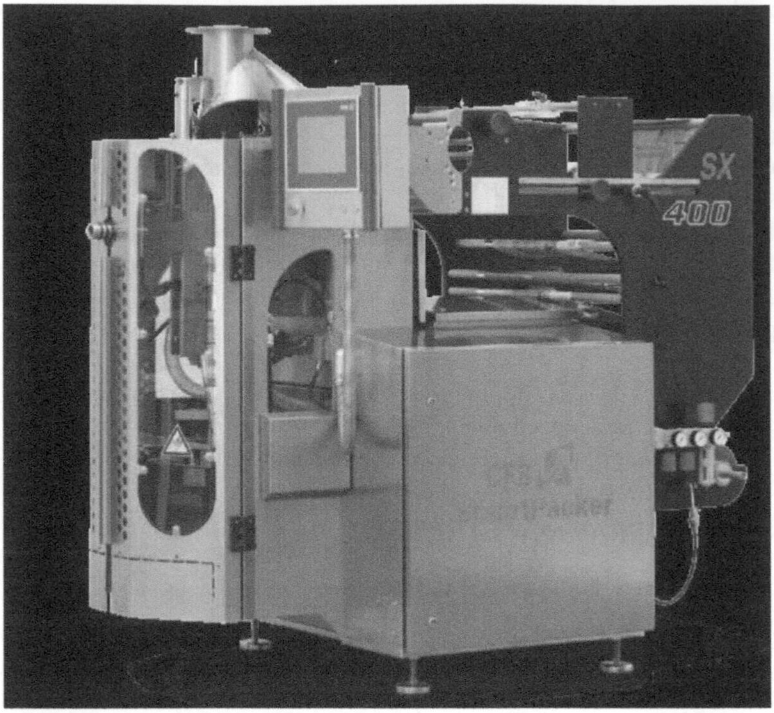

FIGURE 2.2 Vertical packaging machines offer a convenient and inexpensive means of packaging a wide variety of convenience goods, mainly protein, fresh shredded cheese, frozen vegetables, French fries, confectionery, and snacks. (The photo was reproduced with permission from CFS, Bakel, the Netherlands.)

The foodstuffs are entered in their packages through the inner tube, while the gap between the inner and the outer tubes is filled by the gas mix (Hastings, 1993).

A representative vertical packaging machine offering a convenient and inexpensive means of packaging and applicable for a wide variety of foods is exhibited in Figure 2.2.

2.4 SNORKEL MACHINES

Snorkel machines do not need chamber presence during atmosphere modification. A pouch is used for the foodstuff, and probes or snorkels are entered. Through these machinery parts, the air is replaced with the desired gas mix. The package is sealed and heated and is removed from the chain. The snorkel machines are used to pack bulks. Master packs also use the same machinery (Robertson, 2006).

2.5 GASES USED IN MAP

O_2, CO_2, and N_2 are the main gases used in MAP. The type of packaged foodstuffs determines the percentage of the gas that will be used. The created atmosphere aims

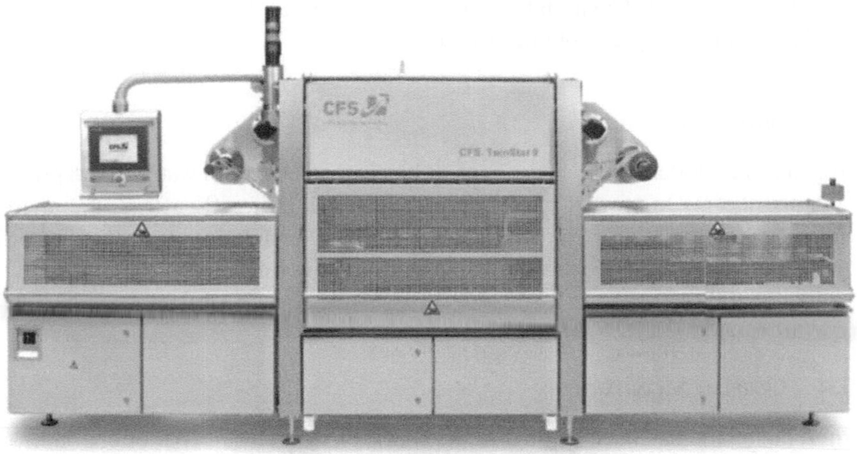

FIGURE 2.3 Auto tray sealer for MAP, skin, and normal air packages and a wide variety of trays. (The photo was reproduced with permission from CFS, Bakel, the Netherlands.)

to balance the shelf-life extension and the optimal organoleptic quality of the packaged foodstuffs. In several experiments, carbon monoxide (CO) and sulfur dioxide (SO_2) were used, but noble gases are also commercially used in a wide variety of products.

An auto tray sealer for MAP is displayed in Figure 2.3.

2.5.1 Oxygen

The reduction of the surrounding oxygen can control the oxidative and browning reactions. Oxidative reactions occur in meat and fish and are known as lipid rancidity. On the other hand, the browning reactions are presented on the surfaces of fruits and vegetables. In modified atmosphere packages, low-level oxygen is preferred to avoid both fermentation and the growth of aerobic spoilage microorganisms. Some microorganisms such as *Clostridium botulinum* can grow in complete oxygen absence. This microbiological hazard could be eliminated, if complete oxygen absence is avoided (Blakistone, 1999). The fruit and vegetable respiration and the retention of color in red meat need high-level oxygen (Parry, 1993).

2.5.2 Carbon Dioxide

Carbon dioxide is a colorless gas and has important bacteriostatic and fungistatic properties. It has a slight odor and can cause asphyxia in high concentrations. The carbon dioxide is water soluble and creates carbonic acid, causing pH reduction and increasing solution acidity, which has further preservative effects. Carbon dioxide also dissolves in lipids and in other organic compounds. When temperature is increased, the gas solubility also increases. Therefore, in low temperatures, the antimicrobial activity of CO_2 increases as a result of the increased concentration (Mullan and McDowell, 2003). The high solubility of CO_2 can result in pack collapse due to the reduction of headspace volume (Sandhya, 2010).

2.5.3 Nitrogen

Nitrogen is an odorless and tasteless gas. It has low solubility in water (0.018 g/kg at 20°C), and it aims to prevent pack collapse. Nitrogen indirectly affects the microorganism growth in foods, which is achieved with the avoidance of aerobic bacteria growth. Therefore, aerobic spoilage is prevented but not anaerobic bacteria growth (Robertson, 2006).

2.5.4 Carbon Monoxide

Carbon monoxide has been studied in the MAP of meat. The MAP effect on myoglobin was indicated, which leads to the formation of carboxymyoglobin, a bright red pigment. The commercial use of carbon monoxide is still under consideration because of its high toxicity and flammability. These characteristics pose health hazard to packaging machine operators. Another disadvantage of carbon monoxide is its poor bacteriostatic properties (Parry, 1993).

2.5.5 Sulfur Dioxide

Sulfur dioxide is antimicrobial in its nonionized molecular form. This is achieved in low pH environments (pH < 4). Its fungicidal and bacteriostatic properties affect Gram (–) rods (e.g., *Escherichia coli*, *Pseudomonas*) and Gram (+) rods (e.g., lactobacilli). In many foodstuff packages, such as fruits, sausages, juices, wines, and shrimps packages, sulfur dioxide is used (Blakistone, 1999).

2.5.6 Noble Gases

Noble gases are a group of chemical elements that are odorless and colorless and have very low chemical reactivity. Helium (He), neon (Ne), argon (Ar), and xenon (Xe) are members of this group. Ar can replace N_2 to fill in the wine bottle neck before corking. Many studies and experiments were conducted using noble gases as they are chemically steady elements. As a result, biological interactions and benefits were established (Robertson, 2006). Noble gases can extend the shelf life of fruits and vegetables. This property appears in the clathrate hydrate formation, which is diluted in water and has limited water molecule activity. The main gas that is used in modified atmosphere packages is Ar. It has the ability to prevent the growth of microorganisms in foodstuffs such as broccoli and lettuce (Jamie and Saltveit, 2002). A survey by Zhang et al. (2008) indicated that a compressed mix of Ar and Xe in asparagus packages could keep them in a satisfactory condition for consumption up to for 12 days at 4°C.

A MAP tray lidding and sealing machine is shown in Figure 2.4.

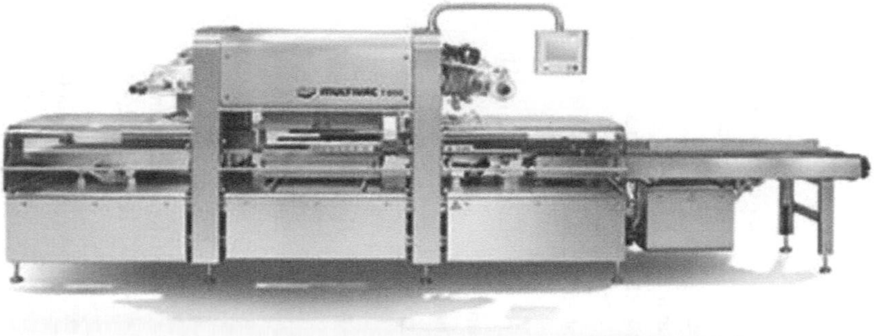

FIGURE 2.4 MAP tray lidding and sealing machines with a capacity between 80 and 100 packs per minute. (The photo was reproduced with permission from Multivac, Kansas, MO.)

2.6 FILMS USED IN MAP

The selection of packaging film is the key process for a valuable and efficient use of MAP. It will offer the exact method for the best maintenance of foodstuffs. As the number of foodstuffs stored under modified atmosphere conditions is increasing, the need for new packaging materials with new characteristics is growing. Material selection is the main step in good packaging, and the first target of packaging material is to preserve the foodstuffs from external conditions with all the essential characteristics. Some of the commonly used plastics in MAP applications are mentioned in the following section.

Table 2.1 summarizes the combination of food packaging film and gas composition for various foods.

A semiautomatic tray lidding and sealing equipment is displayed in Figure 2.5.

2.7 POLYOLEFINS

Olefin is a common term in the plastics industry, and refers to materials that are derivatives of ethylene or propylene. Polyethylene (PE), low-density polyethylene (LDPE), high-density polyethylene (HDPE), linear LDPE, and polypropylene (PP) are included in polyolefins.

Guevara et al. (2001) used a polyolefin bag to store prickly pear cactus stems. The polyolefin bag in combination with active modification of the atmosphere in the package extended the storage life and reduced texture, weight, and chlorophyll content losses, and crude fiber and color deterioration.

A survey by Zakrys et al. (2009) showed that beef steaks stored in PS/EVOH/PE trays with a polyolefin lid under 40% O_2 modified atmosphere are more acceptable by the consumers because of their reduction in toughness and increase in juiciness.

Another survey by McConnell et al. (2005) showed that the quality of sweet potatoes was maintained for 14 days when they were stored in a polyolefin bag under modified atmosphere conditions. The advantages of this packaging are the reduction

TABLE 2.1
Combination of Food Packaging Film and Gas Composition for Various Foods

Product	Initial Atmosphere	Film	Permeability	References
Bread	Air			Lavermicocca et al. (2003)
Sponge cake	1. 100% N_2 2. 30% CO_2–70% N_2	Polyamide and polyethylene coextrusion mix film with a thickness of 90 μm	OTR: 19.913 cm^3/m^2 day atm CO_2TR: 164.903 cm^3/m^2 day atm Water vapor: 2.60 cm^3/m^2 day atm	Guynot et al. (2003)
Spanish sponge cake	1. Air for sponge cakes with a_w 0.75–0.80 2. 100% CO_2 for a_w 0.85–0.90 3. 50% CO_2 for a_w 0.85	OPP20/ (PELD/EVOH/PELD) 45/PELLD30 with a total thickness of 95 μm	OTR: 2 cm^3/m^2 day atm (at 50% RH and 23°C) Water vapor: 1 g/m^2 day atm (at 75% RH and 25°C)	Abellana et al. (2000)
Sponge cake	1. 100% N_2 2. 50% CO_2–50% N_2 3. 70% CO_2–30% N_2	Polyamide–polyethylene co-extrusion mix film (total thickness 90 μm)	OTR: 19.913 cm^3/m^2 day atm CO_2TR: 164.903 cm^3/m^2 day atm Water vapor: 2.60 cm^3/m^2 day atm	Guynot et al. (2004)
High moisture, high pH bakery products	Air	1. Cryovac bags 2. Metallized bags		Koukoutsis et al. (2004)
Pizza crust	1. 60% CO_2–40% N_2 2. 80% CO_2–20% N_2	High gas barrier Cryovac bags	OTR: 3–6 cm^3/m^2 day atm at 4.4°C, 0% RH	Hasan (1997)
Crusty rolls	1. 60% CO_2–40% N_2 2. 100% N_2	PE coated nylon film	OTR: 40 cm^3/m^2 day atm CO_2TR: 155 cm^3/m^2 day atm NTR: 14 cm^3/m^2 day atm Water vapor: 11 g/m^2 day atm	Smith et al. (1986)
Brown rice muffin	60% CO_2–40% N_2	Co-extrusion film (OPP and cast polypropylene [CPP]), 45 μm thickness		Keawmanee and Haruenkit (2007)
Wheat bread	Air	Laminate with ethylene–vinyl alcohol as barrier layer with 95 μm thickness	OTR: <2 cm^3/m^2 day atm CO_2TR: <2.3 cm^3/m^2 day atm Water vapor: 0.1 g/24 h	Rasmussen and Hansen (2001)

Whole wheat flour bread, white breads, and biscuits	100% CO_2	Flexible package		Knorr and Tomlins (1985)
Cheese cake	1. 30% CO_2–70% N_2 2. 80% CO_2–20% N_2 3. O_2 absorbers	Multistrate (EVOH/OPET/PE) gas and water barrier film with a thickness of 54 µm		Sanguinetti et al. (2009)
Bread	1. 100% CO_2 2. 50% N_2, 50% CO_2	High-barrier plastic bag (laminate of OPP20/EVOH/PELD45/PELLD30)		Nielsen and Rios (2000)
Crumpets	60% CO_2–40% N_2	12/75 polyester/PE top web and 100/100 nylon/PE bottom web	OTR: 40 cm^3/m^2 day atm CO_2TR: 155 cm^3/m^2 day atm NTR: 14 cm^3/m^2 day atm Water vapor: 11 g/m^2 day atm	Smith et al. (1983)
Crumpets	100% CO_2	High gas barrier bags	Ethanol transmission rate (ETR) 0.21 g/m^2/day at 25°C	Daifas et al. (1999)
Vegetables				
Chicory endive	1. 5% O_2/5% CO_2 2. 10% O_2/10% CO_2	Laminated foil/plastic barrier bag		Niemira et al. (2005)
Chicory endive	3% O_2 and 2%–5% CO_2	Plastic film	OTR: 3,704 mL O_2/m^2 24 h atm	Jacxsens et al. (2003)
Chicory endive	1. 3% O_2 and 5% CO_2 2. HOA (70%, 80% and 95% O_2)	Plastic film	OTR: 1,446 mL O_2/m^2 24 h atm	Jacxsens et al. (2001)
Tomato slices	1. 1% O_2/4% CO_2 2. 20% O_2/4% CO_2 3. 1% O_2/8% CO_2 4. 20% O_2/8% CO_2 5. 1% O_2/12% CO_2 6. 20% O_2/12% CO_2	Films A and B	A: OTR: 87.4 mL/h/m^2/atm at 5°C and 119.3 at 10°C B: OTR: 60 mL/h/m^2/atm at 5°C and 77.8 at 10°C	Hong and Gross (2001)

(continued)

TABLE 2.1 (continued)
Combination of Food Packaging Film and Gas Composition for Various Foods

Product	Initial Atmosphere	Film	Permeability	References
Tomato slices	1. 3 kPa O_2 2. 3 kPa O_2/4 kPa CO_2	OPP film of 35 μm thickness	OTR: 5,500 mL/m^2 day atm CO_2TR: 10,000 mL/m^2 day atm	Aguayo et al. (2004)
Tomato slices	12–14 kPa O_2/0 kPa CO_2	1. Vascolan with 80 μm thickness 2. Bioriented polypropylene	1. OTR: 2.4×10^{-14} mol/s/m^2/Pa CO_2TR: 6.1×10^{-14} mol/s/m^2/Pa 2. OTR: 3.3×10^{-12} mol/s/m^2/Pa CO_2TR: 3.1×10^{-9} mol/s/m^2/Pa	Gil et al. (2002)
Tomato	5 kPa O_2/5 kPa CO_2	ILPRA	OTR: 110 mL/m^2 day bar CO_2TR: 500 mL/m^2 day bar	Odriozola-Serrano et al. (2008)
Grated carrots	60% O_2/30% CO_2	0.5 mm metallized polyester/2 mm EVA copolymer sterile bag	OTR: 0.7 cm^3/m^2 day atm Water vapor: 1.2 g/m^2 day atm	Lacroix and Lafortune (2004)
Carrots	1. 5% O_2/10% CO_2 2. 80% O_2/10% CO_2	CPP–OPP	OTR: 1,296 mL/m^2 day atm CO_2TR: 3,877 mL/m^2 day atm	Ayhan et al. (2008)
Grated carrots	2.1% O_2/4.9% CO_2	PE bags	OTR: 1,000 mL/m^2 day atm CO_2TR: 5,450 mL/m^2 day atm	Tassou and Boziaris (2002)
Orange and purple carrots	1. 5% O_2/5% CO_2 2. 95% O_2/5% CO_2	PE bags		Alasalvar et al. (2005)
Shredded carrots	3% O_2/97% CO_2	L bags (Cryovac)	OTR: 3,000 mL/m^2 day atm	Beuchat and Brakett (1990)
Grated carrots	4.5% O_2/8.9% CO_2	Plastic film	OTR: 3,529 mL/kg h	Gomez-Lopez et al. (2007)
Nantes carrots	2% O_2/10% CO_2	BOPP/LDPE plastic bags		Pilon et al. (2006)
Carrots	1. 2.1% O_2/4.9% CO_2 2. 5.2% O_2/5% CO_2	PE bags 60 μm	OTR: 995.9 mL/m^2 day bar CO_2TR: 5,452.5 mL/m^2 day bar	Kakiomenou et al. (1998)

Broccoli heads		PD-961	OTR: 6,000–8,000 mL/m^2 day atm. CO_2TR: 19,000–22,000 mL/m^2 day atm Water vapor: 1.1 g/m^2 day atm	De Ell et al. (2006)
Broccoli heads	Air	1. Macroperforated polypropylene 2. Microperforated polypropylene 3. Nonperforated polypropylene	1. OTR: 1,600 mL/m^2 day atm CO_2TR: 3,600 mL/m^2 day atm 2. OTR: 2,500 mL/m^2 day atm CO_2TR: 25,000 mL/m^2 day atm	Serrano et al. (2006)
Broccoli	Air	1. OPP 35 μm thickness 2. LDPE	1. OTR: 1,300 mL/m^2 day atm CO_2TR: 4,900 mL/m^2 day atm Water vapor: 1.6 g/m^2 day atm 2. OTR: 4,290 mL/m^2 day atm CO_2TR: 18,600 mL/m^2 day atm	Jacobsson et al. (2004a) Jacobsson et al. (2004b)
Broccoli florets	Air	2. LDPE 15 μm thickness		Artes et al. (2001)
Broccoli		26 L plastic containers	OTR: 21.76.10^{12} mol/s Pa CDTR: 61.52.10^{-12} mol/s Pa at 3°C	Tano et al. (2007)
Kohlrabi sticks		1. OPP 20 μm thickness 2. Amide-PE 40 μm thickness 3. OPP 40 μm thickness	1. OTR: 5,500 mL/m^2 day atm CO_2TR: 10,000 mL/m^2 day atm	Escalona et al. (2007a) Escalona et al. (2007b)
White mushrooms	5% ± 1% O_2/3% ± 1% CO_2	LDPE 25 μm thickness		Tao et al. (2006)
Mushrooms	Air	1. Polyolefin pouches 2. Biopolymer film (wheat-gluten-based material) 3. Hydrophilic polyether polyamide (PA) film 4. PVC 12 μm thickness 5. OPP 35 μm thickness 6. OPP_2 35 μm thickness	1. OTR: 13,800 mL/m^2 day atm 2. OTR: 4,160 mL/m^2 day atm CO_2TR: 65,380 mL/m^2 day atm 3. OTR: 6,500 mL/m^2 day atm CO_2TR: 56,000 mL/m^2 day atm 4. OTR: 25,000 mL/m^2 day atm CO_2TR: 75,000 mL/m^2 day atm 5. OTR: 45,000 mL/m^2 day atm CO_2TR: 45,000 mL/m^2 day atm	(1) Roy et al. (1996) (2, 3) Barron et al. (2001) (4, 5, 6) Simon et al. (2005)

(continued)

TABLE 2.1 (continued)
Combination of Food Packaging Film and Gas Composition for Various Foods

Product	Initial Atmosphere	Film	Permeability	References
			6. OTR: 2,400 mL/m^2 day atm CO_2TR: 2,400 mL/m^2 day atm	
Pleurotus Mushrooms	1. Air	1. PVC 11 µm thickness 2. LDPE 11 µm thickness 3. Microperforated PP 35 µm thickness 4. Macroperforated PP 35 µm thickness	1. OTR: 0.28 × 10^{-10} mol/s m^2 Pa CO_2TR: 2.52 × 10^{-10} mol/s m^2 Pa 2. OTR: 0.48 × 10^{-10} mol/s m^2 Pa CO_2TR: 2.4 × 10^{-10} mol/s m^2 Pa 3. OTR: 9.34 × 10^{-10} mol/s m^2 Pa CO_2TR: 9.34 × 10^{-10} mol/s m^2 Pa 4. OTR: 3.36 × 10^{-10} mol/s m^2 Pa CO_2TR: 3.36 × 10^{-10} mol/s m^2 Pa	Villaescusa and Gil (2003)
Shiitake mushrooms	1. 15% O_2 2. 25% O_2 3. 5% O_2/2.5% CO_2 for second and third film	1. LDPE 30 µm thickness 2. LDPE 60 µm thickness 3. PE 40 µm thickness	1. OTR: 6.8–8 × 10^{-7} mL/m^2 s Pa CO_2TR: 1.8–3.4 × 10^{-7} mL/m^2 s Pa Water vapor: 2.8–6.5 × 10^{-5} g/m^2 s	(1) Antmann et al. (2008) (2, 3) Parentelli et al. (2007)
Mushrooms	1. Air 2. HOA (70%, 80%, and 95% O_2) for the fifth film	1. 4 L plastic containers 2. Bioriented PP	1. OTR: 5.58 × 10^{-12} mol/s m^2 Pa CO_2TR: 13.55 × 10^{-12} mol/s m^2 Pa 2. OTR: 914 mL/m^2 day atm	(1) Tano et al. (2007) (2) Jacxsens et al. (2001)
Asparagus spears	Air	Stretch film	OTR: 583 mL/m^2 day atm CO_2TR: 1,750 mL/m^2 day atm	Siomos et al. (2000)
White asparagus	Air	OPP P-Plus 160 35 µm thickness		Simon et al. (2004)
Asparagus	1. 5% O_2/5% CO_2 2. 10% O_2/5% CO_2 3. Air	1. PVC 2. LDPE 25 µm thickness 3. OPP P plus	1. OTR: 7.8 × 10^{-12} mol/s mm^2 kPa CO_2TR: 3.6 × 10^{-12} mol/s mm^2 kPa 2. OTR: 2.1 × 10^{-12} mol/s mm^2 kPa CO_2TR: 6.5 × 10^{-12} mol/s mm^2 kPa 3. OTR: 14,000 mL/m^2 day atm	(1) Zhang et al. (2008) (2) An et al. (2007) (3) Villanueva et al. (2005)

Mixture of green, red, and yellow shredded bell peppers	3% O_2 and 2%–5% CO_2	1. Prototype film 2. Plastic bags	1. OTR: 2,897 mL O_2/m^2 24 h atm 2. OTR: 1.57 × 10^{-11} mol O_2/m^2 s Pa	(1) Jacxsens et al. (2003) (2) Jacxsens et al. (2002)
Green peppers	1. Air 2. 2% O_2/10% CO_2	1. PLA 2. LDPE 3. BOPP/LDPE for active MAP	1. OTR: 7.17 × 10^{-10} mol cm/cm^2/h/atm CO_2TR: 1.1 × 10^{-10} mol cm/cm^2/h/atm 2. OTR: 3.43 × 10^{-8} mol cm/cm^2/h/atm CO_2TR: 5.26 × 10^{-8} mol cm/cm^2/h/atm	(1, 2) Koide and Shi (2007) (3) Pilon et al. (2006)
Green bell peppers	Air	1. PE 2. PD-961	1. OTR: 4,000 mL O_2/m^2 24 h atm 2. OTR: 6,000–8,000 mL O_2/m^2 24 h atm	(1) Toivonen and Stan (2004) (2) Gonzalez-Aguilar et al. (2004)
Romaine lettuce	1. Flushing with 0, 1, 2 5, 10, or 21 kPa O_2 2. Evacuation until O_2 is 1% and 4% CO_2 3. 30% CO_2 4. 50% CO_2 5. 100% CO_2	1. PP_1 2. PP_2 3. Packaging film 4. Barrier plastic bags for MAP 3, 4, and 5	1. OTR: 8 pmol/s m^2 Pa 2. OTR: 16.6 pmol/s m^2 Pa 3. OTR: 16.6 pmol/s m^2 Pa 4. 0.46–0.93 mL/100 mL day atm	(1, 2) Kim et al. (2005) (3) Chua et al. (2008) (4) Bidawid et al. (2001)
Iceberg lettuce	1. Air 2. 4% O_2/12% CO_2 3. 4% O_2 and 0% CO_2 with gas flush with N_2 4. Nitrogen flush	1. E-300 Cryovac 2. PET-PP film 3. Polyolefin laminate film for MAP 2 4. OPP 35 μm thickness for MAP 3	1. OTR: 4,000 mL O_2/m^2 24 h atm 2. OTR: 4.2 × 10^{-13} mol/s/m^2/Pa 3. OTR: 3,800 mL/m^2 day atm CO_2TR: 13,000 mL/m^2 day atm	(1) Fan et al. (2003) (2) Beltran et al. (2005) (3) Hagenmaier and Baker (1997) (4) Francis and O'Beirne (1997)

(continued)

TABLE 2.1 (continued)
Combination of Food Packaging Film and Gas Composition for Various Foods

Product	Initial Atmosphere	Film	Permeability	References
		5. OPP 35 μm thickness for MAP 4 6. Packaging film for PMAP	5. OTR: 1,200 mL/m^2 day atm CO_2TR: 4,000 mL/m^2 day atm 6. OTR: 3,529 mL O_2/m^2 24 h. atm	(5) Barry-Ryan and O'Beirne (1999) (6) Gomez-Lopez et al. (2005)
Red pigmented lettuce "Lollo Rosso"	1. 3% O_2/5% CO_2 2. Air	1. PP 35 μm thickness 2. BOPP	1. OTR: 1,000 mL/m^2 day atm CO_2TR: 3,400 mL/m^2 day atm 2. OTR: 1,800 mL/m^2 day atm	(1) Allende et al. (2004a) (2) Allende and Artes (2003)
Butterhead lettuce	1. Air 2. 5% O_2/2.5% CO_2 3. Nitrogen flush	1. BOPP 2. Polyolefin PD-961 3. PP 40 μm thickness for MAP 2 4. OPP 35 μm thickness for MAP 3	1. OTR: 2,000–3,000 mL/m^2 day atm CO_2TR: 6,000–7,000 mL/m^2 day atm 2. OTR: 6,000–8,000 mL/m^2 day atm CO_2TR: 19,000–22,000 mL/m^2 day atm	(1, 2) Martinez et al. (2008) (3) Ares et al. (2008) (4) Gleeson and O'Beirne (2005)
"Red Oak Leaf" lettuce	Air	BOPP	OTR: 1,800 mL/m^2 day atm	Allende and Artes (2003)
Mixed lettuce (20% endive, 20% curled endive, 20% radicchio lettuce, 20% lollo rosso, and 20% lollo bionta lettuces—red and green variety)	1. 3% O_2/2%–5% CO_2 2. Air 3. 3% O_2/2%–5% CO_2	1. BOPP film (30 μm), PVC coated 2. High permeability film 3. Plastic bags	1. OTR: 15 mL O_2/m^2 24 h atm 2. OTR: 2,270 mL O_2/m^2 24 h atm 3. OTR: 1.04.10^{-11} mol O_2/m^2 s Pa	(1, 2) Jacxsens et al. (2003) (3) Jacxsens et al. (2002)
Lettuce	1. 2.1% O_2/4.9% CO_2 2. 5.2% O_2/5% CO_2	PE 60 μm thickness	OTR: 996 mL O_2/m^2 24 h bar CO_2TR: 5,452.5 mL O_2/m^2 24 h bar	Kakiomenou et al. (1998)

Ready to eat food				
Precooked chicken	1. 10.7% O_2/76% CO_2 2. 0% O_2/80% CO_2 3. 0% O_2/30% CO_2 4. 0% O_2/60% CO_2 5. 0% O_2/90% CO_2	1. Nylon-PE barrier bag for MAP 1 and 2 2. LDPE/PA/LDPE 75 μm thickness for MAP 3, 4, and 5	1. OTR: 9 mL O_2/m^2 24 h bar 2. OTR: 52.2 mL O_2/m^2 24 h bar Water vapor: 2.4 g/m^2 day	(1) Marshall et al. (1992) (2) Patsias et al. (2006)
Cooked poultry	0% O_2/40% CO_2	Plastic film	OTR: 5.5 mL O_2/m^2 24 h bar Water vapor: 13.85 g/m^2 day	Barakat and Harris (1999)
Cooked minced pork meat	1. 5% O_2/95% N_2 2. 3% O_2/97% N_2 3. 1% O_2/99% N_2 4. 100% N_2	PE	OTR: 3,800 mL O_2/m^2 24 h bar	Huisman et al. (1994)
Minced turkey	21% O_2/79% N_2	High-barrier bag	OTR: <2 mL O_2/m^2 24 h bar	Bruun-Jensen et al. (1994)
Smoked turkey	1. 0% O_2/30% CO_2 2. 0% O_2/50% CO_2	PET//LDPE/EVOH/LDPE	OTR: <2.32 mL O_2/m^2 24 h bar	Ntzimani et al. (2008)
Cooked frankfurter-type sausages and sliced, cooked, cured, pork shoulder	0% O_2/80% CO_2	Cryovac-type bags	OTR: 35 mL O_2/m^2 24 h bar	Metaxopoulos et al. (2002)
Sliced turkey breast fillets Piroshki pork sausage	1. 0% O_2/80% CO_2 2. 20% O_2/60% CO_2 3. 0.4% CO/80% CO_2 4. 1% CO/80% CO_2 5. 24% O_2/50% CO_2/0.5 CO 6. 100% N_2	Water vapor impermeable film	OTR: <35 mL O_2/m^2 24 h bar	Pexara et al. (2002)
Pastirma	1. 50% CO_2 2. 80% CO_2 3. 50% CO_2	1.80% PE–20% PA for MAP 1 and 2 2. OPAEVOH/PE	1. OTR: 40–50 mL O_2/m^2 24 h atm Water vapor: 100 g/m^2 day 2. OTR: 5 mL O_2/m^2 24 h atm Water vapor: 15 g/m^2 day	(1) Gok et al. (2008) (2) Aksu and Kaya (2005)

(continued)

TABLE 2.1 (continued)
Combination of Food Packaging Film and Gas Composition for Various Foods

Product	Initial Atmosphere	Film	Permeability	References
Cottage cheese	100% CO_2	Polystyrene containers	OTR: 1.7 mL/m^2 day atm CO_2TR: 8.5 mL/m^2 day atm	Mannheim and Soffer (1996)
Greek cheese *Anthotyros*	1. 30% CO_2 2. 70% CO_2	LDPE/PA/LDPE 75 μm thickness	OTR: 52.2 mL O_2/m^2 24 h atm Water vapor: 2.4 g/m^2 day	Papaioannou et al. (2007)
Cheddar cheese	73% CO_2	Nylon/LLDPE/LDPE/LLDPE	OTR: <20 mL O_2/m^2 24 h atm	Oyugi and Buys (2007)
Cheese *Myzithra Kalathaki*	1. 20% CO_2 2. 40% CO_2 3. 60% CO_2	LDPA/PE/LDPE 75 μm thickness	OTR: 52.2 mL O_2/m^2 24 h atm Water vapor: 2.4 g/m^2 day	Dermiki et al. (2008)
Graviera hard cheese	1. 100% CO_2 2. 100% N_2 3. 50% CO_2	LDPE/PA/LDPE 75 μm thickness	OTR: 52.2 mL O_2/m^2 24 h atm Water vapor: 1.29 g/m^2 day	Trobetas et al. (2008)
Fruits				
Strawberry *Fragaria Ananassa*	1. 100% CO_2 2. 100% N_2 3. Air 4. 3% O_2/5% CO_2 5. 95% O_2 6. 2.5% O_2/10% CO_2 7. 15% CO_2 8. Air and 10.9% CO_2 9. Air and 5% CO_2	1. Microperforated PP 30 μm thickness for MAP 1 and 2 2. 4 L containers 3. WA7217-3 with 30 μm thickness 4. PVC 48 μm thickness 5. LDPE 30 μm thickness 6. Microperforated OPP_1 for MAP 8 and 9 7. Microperforated OPP_2 for MAP 8 and 9	2. OTR: 5.58×10^{-12} mol/s/Pa CO_2TR: 13.55×10^{-12} mol/s/Pa 3. OTR: 4,679 mL O_2/m^2 24 h atm C_2H_4TR: 5,390 mL O_2/m^2 24 h atm 4. OTR: 1–5 mL O_2/m^2 24 h atm Water vapor: 5–13 g/m^2 day 6. OTR: 1,560 mL O_2/bag 24 h atm 7. OTR: 6,750 mL O_2/bag 24 h atm	(1) Renault et al. (1994) (2) Tano et al. (2008) (3) Siro et al. (2006) (4) Zhang et al. (2006) (5) Picon et al. (1993) (6, 7) Nielsen and Leufven (2008)

Strawberry "Fragaria Elvira"	1. 3% O_2/5% CO_2 2. 95% O_2	1. Low-barrier film for MAP 1 2. High-barrier film for MAP 2	1. OTR: 2.54 × 10^{-11} mol/m²/s/Pa Water vapor: 1.16 × 10^{-4} g/m²/s 2. OTR: 1.083 × 10^{-14} mol/m²/s/Pa. Water vapor: 2.89 × 10^{-5} g/m²/s	Van der Steen et al. (2002)
Pineapple	1. 38%–40% O_2 2. 10%–12% O_2/1% CO_2 3. 4% O_2/10% CO_2 4. Air	1. PP film for MAP 1 and 2 2. Polystyrene containers for MAP 3 3. Oriented polylactide containers for MAP 4 4. PET containers for MAP 4 5. Oriented polystyrene containers For MAP 4	1. OTR: 110 mL/m² bar day CO_2TR: 110 mL/m² bar day 2. OTR: 1.7 × 10^{-7} mL/cm² Pa day 3. OTR: 56.33 mL/m² bar day 4. OTR: 9.44 mL/m² bar day 5. OTR: 531.58 mL/m² bar day	(1) Montero-Calderon et al. (2008) (2) Liu et al. (2007) (3, 4, 5) Chonhenchob et al. (2007)
Conference pear	1. Air 2. 21% O_2/10% CO_2 3. 2% O_2 4. 2% O_2/10% CO_2 5. 2.5% O_2/7.5% CO_2	1. LDPE 30 μm thickness 2. 25 L boxes for MAP 2, 3, and 4 3. PE bags for MAP 5	3. OTR: 15 and 30 mL/m² bar day	(1) Geeson et al. (1991b) (2) Arias et al. (2008) (3) Soliva-Fortuny et al. (2007)
Pear "Flor de Invierno"	1. 2.5% O_2/7.5% CO_2 2. 70% O_2	PP	OTR: 5.24 × 10^{-13} mol/m² s Pa CO_2TR: 2.38 × 10^{-13} mol/m² s Pa	Oms-Oliu et al. (2008)
Comice pear	Air	LDPE 25 μm thickness		Geeson et al. (1991a)
Pear "Bartlett"	CA: 0%, 1%, and 2% O_2	Glass jars		Rosen and Kader (1989)
Apple "Bravo de esmolfe"	Air	PP 100 μm thickness	OTR: 227 mL/m² bar day CO_2TR: 711 mL/m² bar day	Rocha et al. (2004)
Discovery apple	Air	EVA 30 μm thickness		Smith et al. (1988)
Apple Golden delicious	1. 0% O_2 2. 2.5% O_2/7% CO_2	Plastic bags	OTR: 15 and 30 mL/m² bar day	Soliva-Fortuny et al. (2005)

(continued)

TABLE 2.1 (continued)
Combination of Food Packaging Film and Gas Composition for Various Foods

Product	Initial Atmosphere	Film	Permeability	References
Banana "Musa cavendishii"	1. Air 2. 3% O_2/5% CO_2	1. PVC 12.5 μm thickness 2. PE bags 11 μm thickness 3. PE pouches for MAP 2	1. OTR: 0.031 nmol/s m^2 Pa CO_2TR: 0.066 nmol/s m^2 Pa Water vapor: 1.33 nmol/s m^2 Pa	(1) Choehom et al. (2004) (2) Nguyen et al. (2004) (3) Chauhan et al. (2006)
Miscellaneous				
Almonds	1. Air and 100% N_2 2. N_2	1. Flexible plastic pouches 2. Metal cans	1. Low permeability to O_2, CO_2, N_2, and H_2O vapors	(1) Garcia-Pascual et al. (2003) (2) Kazantzis et al. (2003)
Almond paste pastries	Nitrogen or air with oxygen scavengers for both films	1. PE 40 μm/EVOH 5 μm/PE 40 μm/PET 12 μm 2. PP 75 μm/Nylon 6.6 15 μm		Baiano and Del Nobile (2005)

FIGURE 2.5 Semiautomatic tray lidding and sealing equipment. (The photo was reproduced with permission from Multivac, Kansas, MO.)

in the presence of aerobic bacteria and the restriction of tissue firmness, dry matter, and starch. Polyolefin bags were also used in packaging *enoki* mushrooms and proved advantageous for freshness maintenance of samples (Kang et al., 2000). An organoleptic analysis indicated that the use of polyolefin as packaging material created the ideal environment for maintenance of butterhead lettuce quality (Martinez et al., 2008). De Ell et al. (2006) showed that broccoli samples packed in polyolefin bags with sorbitol and $KMnO_4$ maintained quality and marketability more than the control samples. This happened because of the elimination of fermentative volatiles from the inside of the bag. Moreover, semiprocessed potatoes stored in polyolefin bags under active modified atmosphere conditions can be maintained in low temperature for about 3 weeks (Gunes and Lee, 1997). If fresh-cut green bell peppers were packaged in polyolefin bags, they could be maintained for 21 days under refrigeration (Gonzalez-Aguilar et al., 2004).

Sea bass fillet samples had higher organoleptic scores, lower microbial counts, and better pH, when they were stored in polystyrene trays with a polyolefin lid under modified atmosphere conditions, than those samples stored either in air or in ice (Poli et al., 2006).

2.7.1 Polyethylene

PE is a thermoplastic that is chemically stable and has many similarities to wax. In addition, PE is a material that has the following characteristics: high elasticity, easy heat processing material, good cold resistance, and sufficient water vapor barrier properties. The mechanical properties of PE are strongly associated with the chain branching degree and polymer molecular weight. PE presents resistance to acid and alkalis but it can be oxidized by nitric acid. Finally, halogens can react with PE through exchange mechanisms (Piringer and Baner, 2000).

In banana packages, PE is used in combination with ethylene absorbers and CO_2 scrubbers. These materials could achieve the reduction of chilling injury effect that causes browning in banana (Nguyen et al., 2004). Banana shelf life is extended when they were stored in PE bags either under active MAP or under passive MA conditions combined with ethylene absorbent, moisture traps, and CO_2 scrubbers (Chauhan et al., 2006). In sweet cherry fruits, PE bags under passive MAP decreased brown rot (Spotts et al., 1998, 2002). Sweet cherries were stored under passive MAP (two microperforated PP films were used with 0.55 [MAP .55] and 0.30 [MAP .30] μmol cm/cm^2 day atm permeability) for eight days at 4°C followed by four days at 20°C. The use of MAP .30 film extended shelf life and retarded the quality, color, and decay occurrence losses (Alique et al., 2003). Tian et al. (2005) showed that when litchi flesh was packaged in PE bags under high O_2 modified atmosphere, the ethylene production decreased in the early storage period. Conference pear cubes prepared with ascorbic acid and calcium chloride and packaged in PE bags under MA preserved their properties without significant alternations in comparison to the fresh samples (Soliva-Fortuny et al., 2007).

According to Viuda-Martos et al.' (2010) experiment, it was indicated that the organoleptic acceptance of Bologna sausages with orange dietary fiber and oregano essential oil stored in PE/PA laminate bags under MA was higher than that of the control sample.

Tomatoes heat treated and stored in PE bags under low oxygen modified atmosphere presented a greater delay in color development than untreated samples (Ali et al., 2004). The shelf life of white and violet savoy salad stored in PE bags was prolonged for 25 days with acceptable quality (Kim et al., 2004). Purple carrots stored in PE bags under low oxygen atmosphere modification had better organoleptic quality and shelf-life extension (Alasalvar et al., 2005). Kohlrabi stems stored in amide/PE film under MAP preserved their quality and extended the time of leaves' wilting (Escalona et al., 2007c). Another survey for kohlrabi sticks indicated that when they were stored in amide/PE bags, they maintained edibility and organoleptic acceptance for 14 days (Escalona et al., 2007a). Jia et al. (2009) found that broccoli stored in PE bags with no microholes maintained the external appearance and the indole and aliphatic glucosinolate concentrations similar to the initial ones for 13 days under refrigeration. Peppers washed three times and stored in PE bags under modified atmosphere conditions maintained their firmness because of the removal of stress-related compounds (Toivonen and Stan, 2004). Baby spinach leaves stored in PE bags with the addition of superatmospheric oxygen decreased microorganism growth and preserved the foodstuff quality (Allende et al., 2004b).

The use of nisin and sodium hexametaphosphate in PE bags under CO_2 rich atmosphere had a positive impact on refrigerated blue whiting storage (Cabo et al., 2005). The counts of psychrotrophic bacteria in marinated Pacific saury fillets are minor in samples stored in PE bags under vacuum than in samples treated with other techniques (Sallam et al., 2007). Spotted wolf fish stored aerobically had higher microbial counts, worse odor and flavor scores, and reduced shelf life than those of samples packaged in PE/PA film under modified atmosphere conditions (Rosnes et al., 2006). Pastoriza et al. (1998) showed that hake slices packaged in a multilayer PE/Saran film under MA doubled the shelf-life period of fish. This happened because of microbial growth prevention and total volatiles and trimethylamine content reduction. Aquacultured mussels stored in LDPE/PA/LDPE bags under modified atmosphere (80% CO_2) prolonged their shelf life about 5 or 6 days. This extension depended on biochemical and sensory parameters during storage (Goulas et al., 2005). Pournis et al. (2005) studied the packaging of open sea red mullet in PE/PA/LDPE films. They resulted in the aerobic microorganisms' growth eliminated under MAP in contrast to the H_2S-producing bacteria and pseudomonades that were partially reduced under these conditions. It should be added that the trimethylamine nitrogen and total volatile nitrogen behaved similar to aerobic microorganisms during MAP storage. During storage of prerigor filleted farmed cod fillets in HDPE trays with an oriented PA/linear medium-density PE lid, it was found that the best packaging practice was a high oxygen atmosphere (HOA) (63% O_2 and 37% CO_2) (Sivertsvik, 2007).

A thermoforming machine for small production volumes is exhibited in Figure 2.6.

2.7.2 Low-Density Polyethylene

LDPE, both film and blow-molded, is used in the food industry. It is the largest volume single polymer, and its availability in many forms is very important. This

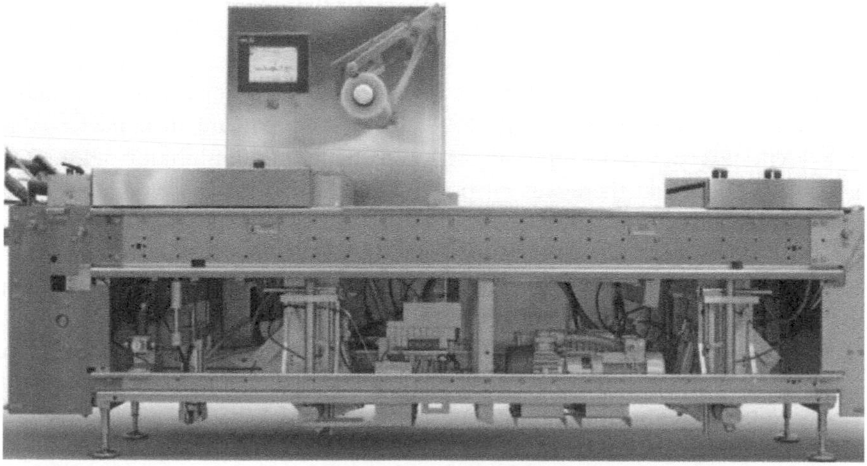

FIGURE 2.6 A thermoforming machine especially designed for small output requirements and small production volumes. (The photo was reproduced with permission from CFS, Bakel, the Netherlands.)

thermoplastic is produced by the polymerization of ethylene. Ethylene is a very ordinary petroleum refinement by-product. The most important role in polymer's properties is played by CH_2 chain branching. This formation is achieved by applying high pressures. The percentage of crystallinity varies from 50% to 70%, and the softening point is influenced by the chain branching level (Robertson, 2006).

LDPE is a chemically static material. It is less penetrated by vapor, whereas it is highly penetrated by gases that have a poor odor barrier. LDPE could easily be passed by essential oils (Greengrass, 1999). LDPE is a strong material. Some of its positive properties are its good tensile, burst, and tear strength, and its impact resistance. It is extremely resistant to alkalis, acids, and inorganic solutions. On the other hand, it is sensitive to the influence of hydrocarbons, halogenated hydrocarbons, oils, and greases (Robertson, 2006). LDPE can be used for lids, bulk films, base webs, and trays. When it is mixed with other films, it can be extrusion coated, laminated, and in some cases coextruded (Greengrass, 1993). The form of LDPE mostly used is films. Film thicknesses vary from 15 to 250 µm (Brandsch and Piringer, 2000). LDPE alone or combined with other plastics is used widely as a packaging material in the food industry.

Anthotyros cheese samples stored in LDPE/PA/LDPE barrier film under active atmosphere modification reduced microbial growth in comparison to vacuum-packaged samples (Papaioannou et al., 2007). Precooked chicken stored in LDPE/PA/LDPE film under active modified atmosphere maintained attractive color and taste quality for 20 days (Patsias et al., 2006). *Myzithra kalathaki* cheese stored in LDPE/PA/LDPE film under active MA eliminated the growth of lactic acid bacteria, enterobacteria, molds, and yeasts for at least 33 days during storage (Dermiki et al., 2008). *Graviera* cheese stored in LDPE/PA/LDPE bags under modified atmosphere indicated less lipid oxidation than aerobically packaged samples (Trobetas et al., 2008). Fresh chicken meat antimicrobially treated with nisin, stored in LDPE/PA/LDPE bags under active MAP, maintained its acceptable odor for at least 20 days of cold storage (Economou et al., 2009). Pate is a foodstuff made with fried chicken liver, onions, and hard-boiled eggs. It is stored in PE/EVOH/PE bags under active atmosphere modification conditions, and it was maintained for about 14 days (Soffer et al., 1994). Souvlaki-type lamb meat packaged in LDPE/PA/LDPE bags under high CO_2 MA had an important inhibition of all TVC. It should be noted that there was no positive effect with *Pseudomonas* spp. and yeasts. The lipid oxidation and the color stability were also maintained (Soldatou et al., 2009). Breast chicken meat samples stored in LDPE/PA/LDPE bags under MA conditions had lower levels of putrescine and cadaverine than aerobically stored samples (Balamatsia et al., 2006).

Bamboo shoots stored in LDPE under active MAP reduced the peroxidase and phenylalanine ammonia lyase activity, checking the browning and lignification occurrence (Shen et al., 2006). Fresh ginseng stored in LDPE packages under passive MAP decreased the rot rate at 1.3% after 210 days (Hu et al., 2005). When hot water–treated tomatoes are packaged in LDPE bags, the weight loss and decay were decreased (Suparlan and Itoh, 2003). Inoculated cherry tomatoes in LDPE bags presented higher *Salmonella enteritidis* death rate than those stored either in CA or in air (Das et al., 2006). Celery sticks packaged in LDPE bags presented better sensorial attributes, maintained color, and delayed the growth of microorganism (Gomez and Artes, 2005). Gil-Izquierdo et al. (2002) proved that with the health claims and nutritional properties of phenolics and vitamin C, LDPE is recommended as a film that retained artichoke quality. Endives stored in LDPE bags combined with oxygen scavengers could reduce the transient period of MAP about 50%. Also, a remarkably significant delay in browning was noticed (Charles et al., 2008). A mix salad dish consisting of cucumber, carrot, garlic, and green pepper was packaged in LDPE bags. The result was the preservation of better quality (Lee et al., 1996). Carrots and green peppers packaged in biaxially oriented polypropylene (BOPP)/LDPE bags maintained vitamin C and low microbial contamination (Pilon et al., 2006). LDPE bags used for the storage of pumpkin pieces and other foodstuff maintained not only its edibility for 25 days but also all the quality properties at adequate levels (Habibunnisa et al., 2001). Bunched onions stored in LDPE packages under moderate vacuum conditions showed that these conditions were ideal for microbial decay prevention and visual organoleptic properties maintenance (Hong and Kim, 2004). White mushrooms stored in LDPE under passive MAP and vacuum cooling prolonged their shelf life. They also presented a higher enzymatic antioxidant system than control samples (Tao et al., 2006, 2007). Broccoli heads stored in LDPE packages under MAP did not present glucoraphanin reduction over 10 days of storage (Jones et al., 2006). Another survey of broccoli conducted by Jacobsson et al. (2004a,c) showed that when stored in LDPE bags, the results are shelf-life extension and the maintenance of dimethylsulfide, dimethyldisulfide, and dimethyltrisulfide concentrations. The use of LDPE combined with ethylene absorbers maintained the organoleptic characteristics of broccoli such as fresh ones (Jacobsson et al., 2004b). "Shogun" broccoli packaged in LDPE bags had 60 times lower weight loss than the samples packaged in the other tested films (Artes et al., 2001). Fresh-cut spinach stored in LDPE bags obtained CO_2 and O_2 at the same levels as the recommended ones (Piagentini et al., 2003). Rangkadilok et al. (2002) showed that broccoli stored in LDPE bags with no microholes and under refrigeration did not change the glucoraphanin concentration. Asparagus wrapped in LDPE under MAP reduced the phenylalanine ammonia lyase activity (An et al., 2007). Cucumbers wrapped in LDPE film under modified atmosphere decreased chilling injury in comparison to unwrapped samples (Wang and Qi, 1997).

Bramley's apples packaged in LDPE bags under MA conditions maintained better firmness and color than those stored in CA. The maintenance duration could be more than 22 weeks (Geeson et al., 1987). *Embul* bananas could be preserved up to 21 days in cold stores, if they contained cinnamon oil combined with LDPE packages under MAP (Ranasinghe et al., 2005). Burlat cherries stored in LDPE bags under refrigeration at 2°C had a perfect appearance and were commercial for 3 weeks (Remon et al., 1999). Conference pears stored in LDPE film under MAP completely prevented the chlorophyll degradation (Geeson et al., 1991b).

Salted sea bream extended its shelf life for 7–8 days when it was stored in LDPE/PA/LDPE bags under MAP. MAP combined with oregano oil addition could extend shelf life for 33 days (Goulas and Kontominas, 2007a).

High-pressure processing (HPP) packaging targets eliminating foodborne microorganisms Figure 2.7.

FIGURE 2.7 HPP packaging line has a primary aim to reduce or eliminate the relevant foodborne microorganisms, thereby extending the shelf life and enhancing the food safety. (The photo was reproduced with permission from Multivac, Kansas, MO.)

2.7.3 Linear Low-Density Polyethylene

Linear low-density polyethylene (LLDPE) film is known as low-pressure LDPE. LLDPE has greater stiffness, better impact strength and tear resistance, and higher tensile strength than LDPE. This material presents environmental stress resistance. The higher the temperatures are the better the resistances presented in LLDPE. It also has good elongation potential. On the other hand, there are some disadvantages in this material such as higher heat seal temperature use and inferior gloss and transparency characteristics in comparison with those used in LDPE (Greengrass, 1993).

Cooked and peeled shrimps stored in LLDPE packages under refrigerated temperatures (2°C) maintained freshness for 20–21 days (Mejlholm and Dalgaard, 2005). Cheddar cheese wrapped in Nylon/LLDPE/LDPE/LLDPE film under active MA conditions decreased yeast counts in comparison to control samples (Oyugi and Buys, 2007).

Mushrooms pretreated with calcium chloride and packaged in LLDPE bags containing 3 g magnesium oxide extended shelf life twofold from 6 to 12 days (Jayathunge and Illeperuma, 2005).

2.7.4 High-Density Polyethylene

HDPE has a higher softening point, which gives better hurdle properties than LDPE and HDPE. This material cannot be used as a sealing layer. This characteristic is present because it is not part of thermoformable base webs. However, this packaging material is in a coextruded form at the top or lidding of webs (Greengrass, 1999).

Filleted wolf fish stored in HDPE trays combined with PE/PA lidding film under MA decreased aerobic and psychrotrophic bacteria counts more than samples stored in air (Rosnes et al., 2006). Microbial growth of ready-to-eat sushi rolls could be minimized if they were stored in HDPE bags combined with bacteriostatic agents such as limonene and ethanol vapor under high CO_2 MA (Chen et al., 2003).

Dry-salted olives wrapped in HDPE bags prevented fungal contamination and yeast growth for a period of 180 days (Panagou, 2006). Grafted eggplants packaged in HDPE bags under MAP preserved sweetness and acceptance and maintained freshness better than control samples (Arvanitoyannis et al., 2005).

Halibut stored in HDPE trays combined with PE lids under an oxygen-enriched atmosphere prolonged shelf life from 13 to 20 days because of the reduction of microbial growth. These samples also had high organoleptic scores (Hovda et al., 2007).

2.7.5 Polypropylene-Oriented Polypropylene

Polypropylene is a thermoplastic and linear polymer with little or no unsaturation. Most commercial PP is isotactic, has an intermediate level of crystallinity, and can resist high temperatures without softening. PP has medium gas hurdle properties and high temperature stability. Its advantages focus on its low water vapor transmission, good resistance to greases and chemicals, good resistance to abrasions, good gloss, and high clarity (Robertson, 2006).

Oriented polypropylene is one of the most popular high-growth films. It is produced by stretching, usually biaxially, under melting temperatures. Oriented polypropylene (OPP) has enhanced physical properties because of orientation. The main physical properties are strength, heat resistance, and cold stability (Piringer and Baner, 2000). OPP can be used well as a moisture vapor hurdle with a low penetrability to gases. It is seven to ten times higher than that of PE. It is also resistant to grease (Blakistone, 1999).

The shelf life of cottage cheese prolonged about 150% when PP/EVOH/PP barrier trays with an aluminum/PP laminate cover were used. This shelf-life extension did not alter the organoleptic properties of the cheese (Mannheim and Soffer, 1996). Cheese packaged in PP/EVOH/PP film under high CO_2 atmospheres decreased fungi growth by 80% and aflatoxin formation (Taniwaki et al., 2001). Mandarin segments wrapped in microperforated PP/EVOH/PP packages under passive MA maintained the quality indexes when the rate of O_2/CO_2 reached 19.8%/1.2% (Del-Valle et al., 2009). Precooked red claw crayfish tails packed in PP/PE film under active MAP eliminated the aerobic bacteria growth, pH rise, and texture toughening (Chen and Xiong, 2008).

Apples maintained better color quality, firmness, and less weight loss than the control samples when packed in PP films under passive MAP (Rocha et al., 2004). Cherry fruits maintained quality and decreased the occurrence of decay when they were packed in OPP bags combined with essential oils such as eugenol, thymol, and menthol under passive MA conditions (Serrano et al., 2005). BOPP film under atmosphere modification conditions was used in litchi fruit packaging and resulted in weight and subsequent quality loss prevention (Sivakumar and Korsten, 2006). Sivakumar et al. (2008) also indicated that BOPP combined with microbial antagonists such as *Bacillus subtilis* under MAP controlled decay and maintained the total litchi quality when they were stored for 29 days. Fresh-cut pears packaged in PP bags under passive or active MAP decreased microorganism growth and increased shelf life up to 14 days (Oms-Oliu et al., 2008). Active and passive modified atmosphere conditions in combination with PP bags maintained pineapples well. Using this type of packaging, symptoms of fermentation and deterioration could be prevented during

2 weeks of storage (Montero-Calderon et al., 2008). Honeoye strawberries packaged in perforated PP bags retained their initial weight during storage and their aroma profile (Nielsen and Leufven, 2008).

Kozova et al. (2009) indicated that chicken breasts stored in PP trays with a PP/PE lid under active MAP decreased the existing spermine in the samples. A study by Babic et al. (2009) demonstrated that freeze drying of "broiler" chicken breasts stored in PP/EVOH/PP films under MA conditions can lead to products with similar sensory properties to fresh chicken meat.

The shelf life of bean sprouts could be prolonged for about 4–5 days if they were stored in an OPP film, which created an atmosphere of 5% O_2 and 15% CO_2 (Varoquax et al., 1996). Tomatoes stored in PP packages reduced color development and had lower weight loss and higher soluble solids than untreated samples (Batu and Thompson, 1998). Moreover, *Durinta* tomatoes packaged in a PP film under passive and active MA conditions maintained aroma and had better appearance and overall quality than control samples (Aguayo et al., 2004). Bailen et al. (2006) showed that tomatoes packaged in OPP bags had significantly reduced volatile compounds compared with control samples.

Cauliflower and broccoli florets packaged in microperforated BOPP bags under MAP conditions maintained aliphatic and indole glucosinolates (Schreiner et al., 2007). Dry coleslaw mix treated with chlorine combined with OPP films under MA conditions developed organoleptic scores and decreased microbial loads (Cliffe-Byrnes and O'Beirne, 2005). Mini broccoli packaged in microperforated BOPP films under MA conditions retained their aliphatic and indole glucosinolates for 7 days (Schreiner et al., 2006). Coleslaw mix stored in perforated OPP bags had better sensory properties and reduced membrane damage during storage (Cliffe-Byrnes et al., 2003). Carrots stored in PP trays with a CPP/OPP lid under passive and high-oxygen MA presented better quality characteristics compared with samples under low-oxygen MA (Ayhan et al., 2008). Carrot disks packaged in microperforated OPP films under atmosphere modification maintained good texture, aroma, and flavor (Cliffe-Byrnes and O'Beirne, 2007). Cauliflower presented better organoleptic properties for about 20 storage days when it was stored in PP film (Simon et al., 2008). Cauliflowers stored in a low-penetrability PP film under passive MAP reduced alterations in color (Sanz et al., 2007). Fresh sliced mushrooms washed, pretreated, and stored in OPP film under MAP maintained organoleptic characteristics and prolonged shelf life (Cliffe-Byrnes and O'Beirne, 2008). Kohlrabi stems packaged in PP film retained high quality and marketability (Escalona et al., 2007b). *Pleurotus* mushrooms stored in PP bags under MAP retained quality for a long period (Villaescusa and Gil, 2003). When fennel was stored in an OPP film able to create a modified atmosphere of 6–7 kPa O_2 and 10–12 kPa CO_2, the browning decreased at the end of its storage (Escalona et al., 2004). Another survey showed that fennel packaged in an OPP bag under MAP retained its green color and firmness (Artes et al., 2002). Fresh processed lettuce was packaged in a bioriented polypropylene film combined with UV-C. This package was valuable in inhibiting the growth of psychrotrophic bacteria, coliform, and yeast (Allende and Artes, 2003). Winter-harvested lettuce packaged either in a thicker PP film under passive MAP or in a more penetrable PP film under low-oxygen MA preserved quality. Both packages were the best treatments for lettuce preservation (Martinez and Artes, 1999). Ares et al. (2008) indicated that butterhead lettuce leaves stored in PP bags under active MAP combined with low temperatures held organoleptic attributes and prolonged shelf life. Minimally processed broccoli packaged using microperforated and nonperforated PP prolonged the storage period for 28 days without significant changes in the organoleptic and the nutritional quality (Serrano et al., 2006). Irradiated potatoes cubes packaged in PP bags under vacuum had good color and total sugar parameters at the end of the 4-week storage period (Baskaran et al., 2007). Packed celery sticks retained their best quality whey they were packaged in OPP films under MAP (Gomez and Artes, 2005). Asparagus extended their shelf life for 4 weeks and maintained color characteristics when they were packaged in PP film (Tenorio et al., 2004). Peeled white asparagus stored in PP bags prevented dehydration and retained their fresh appearance (Simon et al., 2004). Villanueva et al. (2005) showed the use of OPP packages under atmosphere modification conditions. This type of package combined with refrigerated storage managed to maintain the organoleptic and nutritional properties of green asparagus.

2.8 POLYAMIDE

PAs are also known as nylons. Polycondensation of amino carboxylic acids from diamines and dicarboxylic acids or functional derivatives of them (e.g., lactams) is the process that leads to the mass production of a large variety of PA. In 1997, 1.6×10^6 ton of PA with the use of 75% of casting processed materials was produced. PA is a resistant material in high temperatures and has good hurdle properties. These advantages occur because of the strong hydrogen bonds formed between neighboring macromolecules. Therefore, this formation makes it a hard material (Piringer and Baner, 2000).

PAs are extremely penetrable to water vapor because of the amide group polarity. PAs that have large carbon chains, more than six, present lower melting points and penetrability to water vapor. The amides catch the water molecules and give a plasticizing effect. This happens when the tensile strength decreases and the impact strength increases. If the amide molecules do not have water, the oxygen and gas penetrability is quite low. PA films are very thermally stable and flexible in low temperatures and have resistance to dilute acids and alkalis (Robertson, 2006).

Reduced pH, total volatile basic nitrogen, trimethylamine nitrogen, and thiobarbituric acid values are observed in smoked whole and gutted mullets stored in PA/PE packages under active MAP (Ibrahim et al., 2008). Sliced pastirma packed in OPA/EVOH/PE film under high CO_2 MA can be retained about 150 days in stores and maintained marketability (Aksu and Kaya, 2005). Sivertsvik and Birkeland (2006) showed that ready-to-eat shrimp packaged in an OPA/linear medium-density PE film under active MAP combined with soluble gas stabilization developed organoleptic and microbiological properties. Rye bread packed in PA/EVOH/PA film combined with allyl isothiocyanate achieved the preferred shelf life (Nielsen and Rios, 2000). Whole or decapitated Chinese shrimps submerged in a bactericide compound and packaged in nylon/PE film under active MAP extended their shelf life for 13 and 17 days, respectively (Lu, 2009). Korean braised green peppers with anchovies prolonged their shelf life particularly at low temperatures when they were stored in a nylon/PE/nylon/EVOH/nylon/PE/LLDPE film under active MA (Lee et al., 2008). Cooked sirloin steaks could maintain their red color for 21 days and did not demonstrate any premature browning signs and rancidity when they were packaged in nylon/PE bags under a rich oxygen atmosphere and CO presence (John et al., 2005). Bovine muscles injected with phosphate and potassium L- or D-lactate solutions maintained better in a nylon/EVA/metallocenece PE barrier film under a rich oxygen atmosphere modification. Color deterioration reduced and NADH concentration increased when lactate enhancement and MAP were used (Kim et al., 2009).

Meat of lambs fed with quebracho tannins when stored under rich oxygen atmosphere modification in PA/PE packages developed color stability during storage (Luciano et al., 2009a). Meat of lamps fed with herbs when stored under a rich oxygen modified atmosphere in PA/PE films had better oxidative stability than meat of lambs fed a concentrate-based diet (Luciano et al., 2009b). Poultry patties packed in nylon/PE films under active MAP had high reduction in *Pseudomonas* spp. during the whole storage period (Mastromatteo et al., 2009).

Celeriac flakes packed in OPA/PE laminate film under low O_2 and CO_2 atmospheres (>10%) maintained better quality and low mesophilic, psychrophilic, and coliform numbers until the end of storage (Radziejewska-Kubzdela et al., 2007). Mung bean sprouts treated with ClO_2 and packed in Nylon/PE bags under CO_2 modified atmosphere had lower microbial load and retained their quality during storage (Jin and Lee, 2007). Commercial spinach stored in Nylon/PE packages under a CO_2 or N_2 atmosphere combined with ClO_2 as a decontamination agent drastically reduced the risk of *E. coli* O157:H7 (Lee and Baek, 2008).

Cold-smoked salmons packed in polystyrene/PA films under vacuum presented shelf-life reduction when the storage temperature increased (Dondero et al., 2004). Chub mackerel packaged in PA/LDPE bags under MAP of 70% CO_2 and 30% N_2 reduced the production rate of total volatile basic nitrogen and trimethylamine nitrogen and finally the spoilage rate (Goulas and Kontominas, 2007b). Frozen shrimps packaged in an OPA/CPA/LLDPE film under a 100% N_2 atmosphere maintained higher quality in color, flavor, and texture (Bak et al., 1999). Gutted farmed bass packed in PA/EVOH/PE films under active MAP (30% O_2 and 50% CO_2) showed the best quality during storage (Torrieri et al., 2006). Filleted rainbow trout preserved better under modified atmosphere consisting of 10% O_2, 50% CO_2, and 40% N_2/Ar while the fish were stored in PA/PE bags (Gimenez et al., 2002). Deepwater pink shrimp was preserved for up to 9 days when it was stored in PA/PE bags under MAP compared with samples stored under air that were maintained 4–7 days (Goncalves et al., 2003). Sardines stored in nylon–PE packages under MA had a lower concentration of biogenic amines than samples stored under air and VP (Ozogul and Ozogul, 2006). Blue fish burgers continued being microbial acceptable up to 28 days of storage when they were packed in nylon/PE film under a high CO_2 atmosphere modification combined with a natural preservative addition such as thymol, grapefruit seed extract, and lemon extract (Del Nobile et al., 2009). Ozogul and Ozogul (2004) indicated that rainbow trouts slaughtered by percussive stunning and stored in Nylon/PE packages under MAP reduced bacterial counts compared with samples stored in ice.

Cod fillets presented a longer shelf life when they were packed in nylon/PE films under a low O_2 atmosphere. This happened because of bacterial growth reduction compared with the other used atmospheres (Corbo et al., 2005).

2.9 ETHYLENE–VINYL ACETATE

Ethylene and vinyl acetate polymerization gives ethylene–vinyl acetate (EVA). There are many similarities with PE especially in its characteristics and properties. In many cases, EVA and PE become a blend in a variety of concentrations. When VA increases in the polymer, reduction in sealing temperature and impact strength stress resistance is observed while clarity increases. EVA can be used for vacuum packaging when it is combined with polyvinylidene chloride (PVdC). This blend gives a polymer with high-barrier film properties (Kirwan and Strawbridge, 2003).

Early climacteric and immediately preclimacteric Discovery apples packed in EVA films under MAP maintained their texture and skin color for more time than control samples (Smith et al., 1988).

Ground beef stored in EVA film under low-oxygen atmospheres for 2 days can cause the production of a purple pigment mainly consisting of deoxymyoglobin (Sorheim et al., 2009). Steaks treated with L-lactate and stored in PP/EVA trays with a nylon/EVA/metallocene PE film under HOA modification had the lowest L* values, higher a*/b* ratios, and lower hue angles than other examined samples (Kim et al., 2010). Fish and pork fillets stored in EVA film, which was encapsulated with microcapsules containing horseradish extract, successfully delayed oxidative discoloration and rancidization (Jung et al., 2009).

Grated carrots stored in polyester/EVA packages combined with irradiation at doses more than 0.3 kGy reduced the *E. coli* growth during the whole storage period (Lacroix and Lafortune, 2004).

2.10 ETHYLENE–VINYL ALCOHOL

Ethylene–vinyl alcohol (EVOH) copolymers can be treated without difficulty and can be an efficient hurdle to odors, gases, and solvents. It should be added that when the ethylene in the copolymer increases, the gas penetrability of the film also increases and the water hurdle properties develop. EVOH resins present high mechanical strength elasticity and surface hardness and have thermal stability. EVOH can replace glass and metal containers in the food packaging industry (Robertson, 2006). Pita bread packed in EVOH film under MA delayed staling and extended shelf life for 14 days (Avital and Mannheim, 1988).

2.11 POLYESTERS

Polyesters are polymers that are produced by esters. The basis of their formation is their carbon–oxygen–carbon links and condensation polymerization. During this process, the two molecules are unified and a smaller molecule, usually H_2O, is removed. Fiber structured polyesters were examined in aromatic materials and aliphatic polyesters. This resulted in the creation of polyethylene terephthalate (PET) (Robertson, 2006).

Murcia et al. (2003) showed that many ready-to-eat food (Spanish omelet, tuna and macaroni, spaghetti and lean pork, etc.) stored in polyester film under active MAP have lower counts of microbial load in all treatments than the conventional packaged samples.

Rainbow trout and Baltic herring fillets packed in polypropylene trays with polyester/PE laminate lids under MA resulted in slower growth of microorganisms and better organoleptic attributes. Both fillets were better preserved (Randell et al., 1997). Pearl spot packed polyester/LDPE bags under active MAP of 60% CO_2 and 40% O_2 prolonged the shelf life for 21 days, whereas samples stored under air have a 12–14 day shelf life (Lalitha et al., 2005).

2.11.1 Polyethylene Terephthalate

The polymerization of terephthalic acid and ethylene glycol molecule gives PET. PET is resistant to heat and presents high mechanical strength when oriented. One of the most important PET properties is its reactive surface to ink. PET can be applied both in high and low temperatures and has medium oxygen penetrability, which reduces significantly when metallized with aluminum (Coles et al., 2003).

Smoked turkey breast fillets packed in PET//LDPE/EVOH/LDPE film under active atmosphere modification delayed the biogenic amine formation and kept counts of *Pseudomonas* spp. and *Enterobacteriaceae* under the detection limit (Ntzimani et al., 2008). Fresh strawberries stored in PET/EVOH-LAF film under MA retained the overall fruit quality (Caner and Aday, 2008). Fresh-cut mangoes, pineapples, and mixes stored in PET packages under MA conditions extended shelf life (Chonhenchob et al., 2007).

Chicken breasts stored in PET film under vacuum combined with UV-C light radiation reduced foodborne pathogen population when they were inoculated with *Campylobacter jejuni*, *Listeria monocytogenes*, and *Salmonella enterica* (Chun et al., 2010).

Lettuce leaves disinfected with ozonated water and packed in PET/PP films under active MAP had a 2–3.5 fold reduction of coliform numbers compared with water-washed samples (Beltran et al., 2005).

2.12 POLYVINYL CHLORIDE

Polyvinyl chloride (PVC) is a polymer that has effective hurdle properties for gases and is a medium hurdle for water vapor. It is extremely resistant to oil and grease. PVC can be formed in any shapes like shallow or deep trays. Physical and hurdle properties of this film differ and change when the gauge thickness changes. Printing and coloring are allowed in PVC film (Parry, 1993). Senescent spotting of banana peel was avoided using a PVC film under passive modified atmosphere (Choehom et al., 2004).

Mushrooms stored in PVC packages under modified atmospheres of 2.5% CO_2 and 10%–20% O_2 decreased microbial growth and retained the organoleptic properties compared to control samples (Simon et al., 2005). Asparagus stored in PVC bags under passive MAP inhibited weight loss during the whole storage period (Zhang et al., 2008).

2.13 POLYVINYLIDENE CHLORIDE

PVdC is a copolymer of vinyl and vinylidene chloride. It is not easily penetrated by gases and water vapor and is highly resistant to oil and fat. These beneficial properties led to PVdC film's use in the food industry. Several foodstuffs such as cured meats, cheese, snack foods, tea, coffee, and confectionary can be packed in PVdC film. It can also be used as a monolayer film or as a coating in coextrusion (Coles et al., 2003).

Smoked sausages stored in PVdC/Nylon/PE packages under active MA conditions reduced microbial growth and prolonged shelf life (FRDC, 2000).

Table 2.2 shows the gas permeability and water vapor transmission rate (WVTR) and other properties of most usually applied polymeric films for food packaging applications.

ABBREVIATIONS

BOPP	bioriented polypropylene
CA	controlled atmosphere
CPA	cast polyamide
CPP	cast polypropylene
EVA	ethylene–vinyl acetate
EVOH	ethylene–vinyl alcohol
HDPE	high-density polyethylene
LAF	polyethylene-low-acetyl fractions
LDPE	low-density polyethylene
LLDPE	linear low density polyethylene
MA	modified atmosphere
MAP	modified atmosphere packaging
NADH	nicotinamide adenine dinucleotide
OPA	oriented polyamide
OPP	oriented polypropylene
PA	polyamide
PE	polyethylene
PET	polyethylene terephthalate
PP	polypropylene
PVC	poly vinyl chloride
PVdC	polyvinylidene chloride
TVC	total viable counts
UV-C	ultra violet-C
VP	vacuum packed

TABLE 2.2
Gas (O_2, CO_2, N_2) and Water Vapor Transmission Rate and Other Processing and Mechanical Properties of Most Usually Applied Polymeric Films for Food Packaging Applications

Film	O_2 Permeability	CO_2 Permeability	N_2 Permeability	Water Vapor Transmission Rate	Properties
Rigid PVC	150–350 mL/m^2 day atm	450–1000 mL/m^2 day atm	60–150 mL/m^2 day atm	30–40 g/m^2 day atm	Good processing properties, sensitive to organic solvents
LDPE	7800 mL/m^2 day atm	2800 mL/m^2 day atm	2800 mL/m^2 day atm	18 g/m^2 day atm	High gas permeability, low odor barrier, versatile
HDPE	2600 mL/m^2 day atm	7600 mL/m^2 day atm	7600 mL/m^2 day atm	7–10 g/m^2 day atm	Superior barrier properties, hard film, poor clarity
PVdC	2–4 mL/m^2 day atm	20–30 mL/m^2 day atm	35–50 mL/m^2 day atm	0.5–1 g/m^2 day atm	Resistance to oil, grease, and organic solvents
EVOH	0.5 mL/m^2 day atm	—	—	1000 g/m^2 day atm	High gas barrier, expensive, good processing properties
PP	2000–4500 mL/m^2 day atm	10,000 mL/m^2 day atm	680 mL/m^2 day atm	5–12 g/m^2 day atm	Clear, readily processed
LLDPE	200 mL/m^2 day atm	—	—	15.5–18.5 g/m^2 day atm	Tear resistance, higher elongation potential, higher tear resistance
Ionomer	6000 mL/m^2 day atm	—	—	25–35 g/m^2 day atm	Metallic salts of acid copolymers of PE; broad heat sealant range
EVA	12,500 mL/m^2 day atm	50,000 mL/m^2 day atm	4900 mL/m^2 day atm	40–60 g/m^2 day atm	High flexibility, high permeability to gases and vapor
PA	50–75 mL/m^2 day atm	150–190 mL/m^2 day atm	14 mL/m^2 day atm	300–400 g/m^2 day atm	High tensile strength, good puncture, and abrasion resistance
PET	100–150 mL/m^2 day atm	—	—	15–20 g/m^2 day atm	Good clarity, temperature resistant
PS	4500–6000 mL/m^2 day atm	18,000 mL/m^2 day atm	—	70–150 g/m^2 day atm	High tensile strength, poor barrier
OPP	2000 mL/m^2 day atm	8000 mL/m^2 day atm	400 mL/m^2 day atm	6–7 g/m^2 day atm	Grease resistance, higher vapor barrier than of PE
OPP–PVdC coated	10–20 mL/m^2 day atm	35–50 mL/m^2 day atm	8–13 mL/m^2 day atm	4–5 g/m^2 day atm	High barrier to moisture vapor and gases

Plasticized PVC	500–30,000 mL/m² day atm	1500–46,000 mL/m² day atm	300–10,000 mL/m² day atm	15–40 g/m² day atm	Resistant to nonpolar and strongly polar substances
PS oriented	5000 mL/m² day atm	18,000 mL/m² day atm	800 mL/m² day atm	100–125 g/m² day atm	
Polyurethane	800–1500 mL/m² day atm	7000–25,000 mL/m² day atm	600–1200 mL/m² day atm	400–600 g/m² day atm	Cross-linked polyurethanes are used as glue layers in food packaging
PVdC–PVC copolymer (Saran)	8–25 mL/m² day atm	50–150 mL/m² day atm	2–2.6 mL/m² day atm	1.5–5 g/m² day atm	Low permeability and good environmental stress, crack resistance to many agents
Polyester Oriented	50–130	180–390	15–18 mL/m² day atm	25–30	
Oriented Polyester PVdC coated	9–15	20–30	—	1–2	Good gas barrier and heat-sealing capability
Nylon 6	40	150–190	14 mL/m² day atm	84–3100	PA made from a polymer of ε-caprolactam
Nylon 6,6	78	140	6 mL/m² day atm	45–90	Contains six carbon atoms
Nylon 11	500	2000	52 mL/m² day atm	5–13	Nylon 11 is made from a polymer of ω-undecanolactam
Polyacrylonitrile	12	17	3 mL/m² day atm	78	Resistance to wide range of chemicals, inability to be melt processed
Polybutylene	5000	—	—	8–10	Impact resistant, smooth, and with good wear resistance

Sources: Data taken from Parry, T.R. Introduction, in *Principles and Applications of Modified Atmosphere Packaging of Food*, 1st edn., Parry, T.R. (ed.), Blackie Academic and Professional, London, U.K., ISBN: 0-7514-0084-X, 1993; McMillin, K.W., *Meat Sci.*, 80, 43, 2008; Blakistone B.A., Introduction, in *Principles and Applications of Modified Atmosphere Packaging of Food*, 2nd edn., Blakistone, B.A. (ed.), Aspen publications, Gaithersburg, MD, ISBN: 0-8342-1682-5, 1999; FDA/CFSAN, Microbiological safety of controlled and modified atmosphere packaging of fresh and fresh-cut produce, 2001, http://www.fda.gov/Food/ScienceResearch/ResearchAreas/SafePracticesforFoodProcesses/ucm091368.htm; Coles, R. et al., *Food Packaging Technology*, Blackwell Publishing, Oxford, U.K., 2003; Robertson, G.L., *Food Packaging: Principles and Practice*. Food Science and Technology, Taylor & Francis, Boca Raton, FL, ISBN: 0-8493-3775-5, 2006; Piringer, O.G. and Baner, A.L., *Plastic Packaging Materials for Food*, Wiley VCH, Weinheim, Germany, 2000, ISBN: 3-527-28868-6.

REFERENCES

Abellana, M., Sanchis, V., Ramos, A.J., and Nielsen, P.V. (2000). Effect of modified atmosphere packaging and water activity on growth of *Eurotium amstelodami*, *E. chevalieri* and *E. herbariorum* on a sponge cake analogue. *Journal of Applied Microbiology*, 88(4): 606–616.

Aguayo, E., Escalona, V., and Artes, F. (2004). Quality of fresh-cut tomato as affected by type of cut, packaging, temperature and storage time. *European Food Research and Technology*, 219: 492–499.

Aksu, M.I. and Kaya, M.K. (2005). Effect of storage temperatures and time on shelf-life of sliced and modified atmosphere packaged pastırma, a dried meat product, produced from beef. *Journal of the Science of Food and Agriculture*, 85: 1305–1312.

Alasalvar, C., Al-Farsi, M., Quantick, P.C., Shahidi, F., and Wiktorowicz, R. (2005). Effect of chill storage and modified atmosphere packaging (MAP) on antioxidant activity, anthocyanins, carotenoids, phenolics and sensory quality of ready-to-eat shredded orange and purple carrots. *Food Chemistry*, 89: 69–76.

Ali, S., Nakano, K., and Maezawa, S. (2004). Combined effect of heat treatment and modified atmosphere packaging on the color development of cherry tomato. *Postharvest Biology and Technology*, 34: 113–116.

Alique, R., Martinez, M.A., and Alonso, J. (2003). Influence of the modified atmosphere packaging on shelf life and quality of Navalinda sweet cherry. *European Food Research and Technology*, 217: 416–420.

Allende, A., Aguayo, E., and Artes, F. (2004a). Microbial and sensory quality of commercial fresh processed red lettuce throughout the production chain and shelf life. *International Journal of Food Microbiology*, 91: 109–117.

Allende, A. and Artes, F. (2003). Combined ultraviolet-C and modified atmosphere packaging treatments for reducing microbial growth of fresh processed lettuce. *LWT—Food Science and Technology*, 36: 779–786.

Allende, A., Luo, Y., McEvoy, J.L., Artés, F., and Wang, C.Y. (2004b). Microbial and quality changes in minimally processed baby spinach leaves stored under super atmospheric oxygen and modified atmosphere conditions. *Postharvest Biology and Technology*, 33: 51–59.

An, J., Zhang, M., and Lu, Q. (2007). Changes in some quality indexes in fresh-cut green asparagus pretreated with aqueous ozone and subsequent modified atmosphere packaging. *Journal of Food Engineering*, 78: 340–344.

Antmann, G., Ares, G., Lema, P., and Lareo, C. (2008). Influence of modified atmosphere packaging on sensory quality of *shiitake* mushrooms. *Postharvest Biology and Technology*, 49: 164–170.

Ares, G., Lareo, C., and Lema, P. (2008). Sensory shelf life of butterhead lettuce leaves in active and passive modified atmosphere packages. *International Journal of Food Science and Technology*, 43: 1671–1677.

Arias, E., Gonzalez, J., Lopez-Buesa, P., and Oria, R. (2008). Optimization of processing of fresh-cut pear. *Journal of the Science of Food and Agriculture*, 88: 1755–1763.

Artes, F., Escalona, V.H., and Artes-Hdez, F. (2002). Modified atmosphere packaging of fennel. *Journal of Food Science*, 67(4): 1550–1554.

Artes, F., Vallejo, F., and Martinez, J.A. (2001). Quality of broccoli as influenced by film wrapping during shipment. *European Food Research and Technology*, 213: 480–483.

Arvanitoyannis, I.S., Khah, E.M., Christakou, E.C., and Bletsos, F.A. (2005). Effect of grafting and modified atmosphere packaging on eggplant quality parameters during storage. *International Journal of Food Science and Technology*, 40: 311–322.

Avital, Y. and Mannheim, C.H. (1988). Modified atmosphere packaging of pita (*Pocket*) bread. *Packaging Technology and Science*, 1: 7–23.

Ayhan, Z., Esturk, O., and Tas, E. (2008). Effect of modified atmosphere packaging on the quality and shelf life of minimally processed carrots. *Turkish Journal of Agriculture and Forestry*, 32: 57–64.

Babic, J., Cantalejo, M.J., and Arroqui, C. (2009). The effects of freeze-drying process parameters on Broiler chicken breast meat. *LWT—Food Science and Technology*, 42: 1325–1334.

Baiano, A. and Del Nobile, M.A. (2005). Shelf life extension of almond paste pastries. *Journal of Food Engineering*, 66: 487–495.

Bailen, G., Gullen, F., Castillo, S., Serrano, M., Valero, D., and Martinez-Romero, D. (2006). Use of activated carbon inside modified atmosphere packages to maintain tomato fruit quality during cold storage. *Journal of Agricultural and Food Chemistry*, 54: 2229–2235.

Bak, L.S., Andersen, A.B., Andersen, E.M., and Bertelsen, G. (1999). Effect of modified atmosphere packaging on oxidative changes in frozen stored cold water shrimp (*Pandalus borealis*). *Food Chemistry*, 64: 169–175.

Balamatsia, C.C., Paleologos, E.K., Kontominas, M.G., and Savvaidis, I.N. (2006). Correlation between microbial flora, sensory changes and biogenic amines formation in fresh chicken meat stored aerobically or under modified atmosphere packaging at 4°C: Possible role of biogenic amines as spoilage indicators. *Antonie van Leeuwenhoek*, 89: 9–17.

Barakat, R.K. and Harris, J. (1999). Growth of *Listeria monocytogenes* and *Yersinia enterocolitica* on cooked modified-atmosphere-packaged poultry in the presence and absence of a naturally occurring microbiota. *Applied and Environmental Microbiology*, 65(1): 342–345.

Barron, C., Varoquaux, P., Guilvert, S., Gontard, N., and Gouble, B. (2001). Modified atmosphere packaging of cultivated mushroom (*Agaricus bisporus* L.) with hydrophilic films. *Journal of Food Science*, 66(8): 251–255.

Barry-Ryan, C. and O'Beirne, D. (1999). Ascorbic acid retention in shredded iceberg lettuce as affected by minimal processing. *Journal of Food Science*, 64(3): 498–500.

Baskaran, R., Usha Devi, A., Nayak, C.A., Kudachikar, V.B., Prakash, M.N.K., Prakash, M., Ramana, K.V.R., and Rastogi, N.K. (2007). Effect of low-dose γ-irradiation on the shelf life and quality characteristics of minimally processed potato cubes under modified atmosphere packaging. *Radiation Physics and Chemistry*, 76: 1042–1049.

Batu, A. and Thompson, A.K. (1998). Effects of modified atmosphere packaging on post harvest qualities of pink tomatoes. *Turkish Journal of Agriculture and Forestry*, 22: 365–372.

Beltran, D., Selma, M.V., Marian, A., and Gil, M.I. (2005). Ozonated water extends the shelf life of fresh-cut lettuce. *Journal of Agricultural and Food Chemistry*, 53: 5654–5663.

Beuchat, L.R. and Brakett, R.E. (1990). Inhibitory effects of raw carrots on *Listeria monocytogenes*. *Applied and Environmental Microbiology*, 56(6): 1734–1742.

Bidawid, S., Farber, J.M., and Sattar, S.A. (2001). Survival of *hepatitis A* virus on modified atmosphere-packaged (MAP) lettuce. *Food Microbiology*, 18: 95–102.

Blakistone, B.A. (1999). Introduction, in *Principles and Applications of Modified Atmosphere Packaging of Food*, 2nd edn., Blakistone, B.A. (ed.), Aspen Publications, Gaithersburg, MD. ISBN: 0-8342-1682-5.

Brandsch, J. and Piringer, O.G. (2000). Characteristics of plastic materials, in *Plastic Packaging Materials for Food*, Piringer, O.G. and Baner, A.L. (eds.), Wiley VCH, Weinheim, Germany. ISBN: 3-527-28868-6.

Bruun-Jensen, L., Skovgaard, M., Skibsted, L.H., and Bertelsen, G. (1994). Antioxidant synergism between tocopherols and ascorbyl palmitate in cooked, minced turkey. *Zeitschrift für Lebensmittel-Untersuchung und Forschung*, 199: 210–213.

Cabo, M.L., Herrera, J., Sampedro, G., and Pastoriza, L. (2005). Application of nisin, CO_2 and a permeabilizing agent in the preservation of refrigerated blue whiting. *Journal of Science of Food and Agriculture*, 85: 1733–1740.

Caner, C. and Aday, M.S. (2008). Extending the quality of fresh strawberries by equilibrium modified atmosphere packaging. *European Food Research and Technology*, 227: 1575–1583.

Charles, F., Guillaume, C., and Gontard, N. (2008). Effect of passive and active modified atmosphere packaging on quality changes of fresh endives. *Postharvest Biology and Technology*, 48: 22–29.

Chauhan, O.P., Raju, P.S., Dasgupta, D.K., and Bawa, A.S. (2006). Instrumental textural changes in banana (var. *Pachbale*) during ripening under active and passive modified atmosphere. *International Journal of Food Properties*, 9: 237–253.

Chen, S.C., Lin, C.A., Fu, A.H., and Chuo, Y.W. (2003). Inhibition of microbial growth in ready-to-eat food stored at ambient temperature by modified atmosphere packaging. *Packaging Technology and Science*, 16: 239–247.

Chen, G. and Xiong, Y.L. (2008). Shelf-stability enhancement of precooked red claw crayfish (*Cherax quadricarinatus*) tails by modified $CO_2/O_2/N_2$ gas packaging. *LWT*, 41: 1431–1436.

Choehom, R., Ketsa, S., and Van Doorn, W.G. (2004). Senescent spotting of banana peel is inhibited by modified atmosphere packaging. *Postharvest Biology and Technology*, 31: 167–175.

Chonhenchob, V., Chantarasomboon, Y., and Singh, S.P. (2007). Quality changes of treated fresh-cut tropical fruits in rigid modified atmosphere packaging containers. *Packaging Technology and Science*, 20: 27–37.

Chua, D., Goh, K., Saftner, R.A., and Bhagwat, A.A. (2008). Fresh-cut lettuce in modified atmosphere packages stored at improper temperatures supports enterohemorrhagic *E. coli* isolates to survive gastric acid challenge. *Journal of Food Science*, 73(9): M148–M153.

Chun, H.H., Kim, J.Y., Lee, B.D., Yu, D.J., and Song, K.B. (2010). Effect of UV-C irradiation on the inactivation of inoculated pathogens and quality of chicken breasts during storage. *Food Control*, 21(3): 276–280.

Cliffe-Byrnes, V., Mc Laughlin, C.P., and O'Beirne, D. (2003). The effects of packaging film and storage temperature on the quality of a dry coleslaw mix packaged in a modified atmosphere. *International Journal of Food Science and Technology*, 38: 187–199.

Cliffe-Byrnes, V. and O'Beirne, D. (2005). Effects of chlorine treatment and packaging on the quality and shelf-life of modified atmosphere (MA) packaged coleslaw mix. *Food Control*, 16: 707–716.

Cliffe-Byrnes, V. and O'Beirne, D. (2007). The effects of modified atmospheres, edible coating and storage temperatures on the sensory quality of carrot discs. *International Journal of Food Science and Technology*, 42: 1338–1349.

Cliffe-Byrnes, V. and O'Beirne, D. (2008). Effects of washing treatment on microbial and sensory quality of modified atmosphere (MA) packaged fresh sliced mushroom (*Agaricus bisporus*). *Postharvest Biology and Technology*, 48: 283–294.

Coles, R., McDowell, D., and Kirwan, M.J. (2003). *Food Packaging Technology*, Blackwell Publishing, Oxford, U.K.

Corbo, M.R., Altieri, C., Bevilacqua, A., Campaniello, D., Amato, D.D., and Sinigaglia, M. (2005). Estimating packaging atmosphere–temperature effects on the shelf life of cod fillets. *European Food Research and Technology*, 220: 509–513.

Daifas, D.P., Smith, J.P., Blanchfield, B., and Austin, J.W. (1999). Effect of pH and CO_2 on growth and toxin production by *Clostridium botulinum* in English-style crumpets packaged under modified atmospheres. *Journal of Food Protection*, 62(10): 1157–1161.

Das, E., Gurakan, G.C., and Bayındirli, A. (2006). Effect of controlled atmosphere storage, modified atmosphere packaging and gaseous ozone treatment on the survival of *Salmonella Enteritidis* on cherry tomatoes. *Food Microbiology*, 23: 430–438.

De Ell, J.R., Toivonen, P.M.A., Cornut, F., Roger, C., and Vigneault, C. (2006). Addition of sorbitol with $KMnO_4$ improves broccoli quality retention in modified atmosphere packages. *Journal of Food Quality*, 29: 65–75.

Del Nobile, M.A., Corbo, M.R., Speranza, B., Sinigaglia, M., Conte, A., and Caroprese, M. (2009). Combined effect of MAP and active compounds on fresh blue fish burger. *International Journal of Food Microbiology*, 135: 281–287.

Del-Valle, V., Hernandez-Munoz, P., Catala, R., and Gavara, R. (2009). Optimization of an equilibrium modified atmosphere packaging (EMAP) for minimally processed mandarin segments. *Journal of Food Engineering*, 91(3): 474–481.

Dermiki, M., Ntzimani, A., Badeka, A., Savvaidis, I.N., and Kontominas, M.G. (2008). Shelf-life extension and quality attributes of the whey cheese *Myzithra Kalathaki* using modified atmosphere packaging. *LWT*, 41: 284–294.

Dondero, M., Cisternas, F., Carvajal, L., and Simpson, S. (2004). Changes in quality of vacuum-packed cold-smoked salmon (*Salmo salar*) as a function of storage temperature. *Food Chemistry*, 87: 543–550.

Economou, T., Pournis, N., Ntzimani, A., and Savvaidis, I.N. (2009). Nisin–EDTA treatments and modified atmosphere packaging to increase fresh chicken meat shelf-life. *Food Chemistry*, 114: 1470–1476.

Escalona, V.H., Aguayo, E., and Artes, F. (2007a). Quality changes of fresh-cut kohlrabi sticks under modified atmosphere packaging. *Journal of Food Science*, 72(5): 303–307.

Escalona, V.H., Aguayo, E., and Artes, F. (2007b). Extending the shelf life of kohlrabi stems by modified atmosphere packaging. *Journal of Food Science*, 72(5): 308–313.

Escalona, V.H., Aguayo, E., and Artes, F. (2007c). Modified atmosphere packaging improved quality of kohlrabi stems. *Lebensmittel-Wissenschaft und Technologie*, 40: 397–403.

Escalona, V.H., Aguayo, E., Gomez, P., and Artes, F. (2004). Modified atmosphere packaging inhibits browning in fennel. *Lebensmittel-Wissenschaft und Technologie*, 37: 115–121.

Fan, X., Toivonen, P.M.A., Rajkowski, K.T., and Sokorai, K.J.B. (2003). Warm water treatment in combination with modified atmosphere packaging reduces undesirable effects of irradiation on the quality of fresh-cut iceberg lettuce. *Journal of Agricultural and Food Chemistry*, 51: 1231–1236.

FDA/CFSAN. (2001). Microbiological safety of controlled and modified atmosphere packaging of fresh and fresh-cut produce. http://www.fda.gov/Food/ScienceResearch/ResearchAreas/SafePracticesforFoodProcesses/ucm091368.htm (accessed on 24 October 2007).

Food Research and Development Centre (FRDC). (2000). The third research co-ordination meeting of the co-ordinated research programme on the development of safe, shelf-stable and ready-to-eat food through high-dose radiation processing In: *Proceedings of a Final Research Co-Ordination Meeting*, Montreal, Quebec, Canada, July 10–14, 2000, IAEA-TECDOC-1337. http://wwwpub.iaea.org/MTCD/publications/PDF/te_1337_web.pdf (accessed on July 12, 2011).

Francis, G.A. and O'Beirne, D. (1997). Effects of gas atmosphere, antimicrobial dip and temperature on the fate of *Listeria innocua* and *Listeria monocytogenes* on minimally processed lettuce. *International Journal of Food Science and Technology*, 32: 141–151.

Garcia-Pascual, P., Mateos, M., Carbonell, V., and Salazar, D.M. (2003). Influence of storage conditions on the quality of shelled and roasted almonds. *Biosystems Engineering*, 84(2): 201–209.

Geeson, J.D., Genge, P.M., Sharples, R.O., and Smith, S.M. (1991a). Limitations to modified atmosphere packaging for extending the shelf-life of partly ripened Doyenne du Comice pears. *International Journal of Food Science and Technology*, 26: 225–231.

Geeson, J.D., Genge, P.M., Smith, S.M., and Sherples, S.O. (1991b) The response of unripe Conference pears to modified atmosphere retail packaging. *International Journal of Food Science and Technology*, 26: 215–223.

Geeson, J.D., Smith, S.M., Everson, H.P., Gengret, P.M., and Browne, K.M. (1987). Responses of CA-stored Bramley's Seedling and Cox's Orange Pippin apples to modified atmosphere retail packaging. *International Journal of Food Science and Technology*, 22: 659–668.

Gil, M.I., Conesa, M.A., and Artes, F. (2002). Quality changes in fresh cut tomato as affected by modified atmosphere packaging. *Postharvest Biology and Technology*, 25: 199–207.

Gil-Izquierdo, A., Conesa, M.A., Ferreres, F., and Gil, M.A. (2002). Influence of modified atmosphere packaging on quality, vitamin C and phenolic content of artichokes (*Cynara scolymus* L.). *European Food Research and Technology*, 215: 21–27.

Gimenez, B., Roncales, P., and Beltran, J.A. (2002). Modified atmosphere packaging of filleted rainbow trout. *Journal of the Science of Food and Agriculture*, 82: 1154–1159.

Gleeson, E. and O'Beirne, D. (2005). Effects of process severity on survival and growth of *Escherichia coli* and *Listeria innocua* on minimally processed vegetables. *Food Control*, 16: 677–685.

Gok, V., Obuz, E., and Akkaya, L. (2008). Effects of packaging method and storage time on the chemical, microbiological, and sensory properties of Turkish pastirma—A dry cured beef product. *Meat Science*, 80: 335–344.

Gomez, P.A. and Artes, F. (2005). Improved keeping quality of minimally fresh processed celery sticks by modified atmosphere packaging. *Lebensmittel-Wissenschaft und Technologie*, 38: 323–329.

Gomez-Lopez, V.M., Devlieghere, F., Bonduelle, V., and Debevere, J. (2005). Intense light pulses decontamination of minimally processed vegetables and their shelf-life. *International Journal of Food Microbiology*, 103: 79–89.

Gomez-Lopez, V.M., Devlieghere, F., Ragaert, P., and Debevere, J. (2007). Shelf-life extension of minimally processed carrots by gaseous chlorine dioxide. *International Journal of Food Microbiology*, 116: 221–227.

Goncalves, A.C., Lopez-Caballero, M.E., and Nunes, M.I. (2003). Quality changes of deepwater pink shrimp (*Parapenaeus longirostris*) packed in modified atmosphere. *Journal of Food Science*, 68(8): 2586–2590.

Gonzalez-Aguilar, G.A., Ayala-Zavala, J.F., Ruiz-Cruz, S., Acedo-Felix, E., and Diaz-Cinco, M.E. (2004). Effect of temperature and modified atmosphere packaging on overall quality of fresh-cut bell peppers. *Lebensmittel-Wissenschaft und Technologie*, 37: 817–826.

Goulas, A.E., Chouliara, I., Nessi, E., Kontominas, M.G., and Savvaidis, I.N. (2005). Microbiological, biochemical and sensory assessment of mussels (*Mytilus galloprovincialis*) stored under modified atmosphere packaging. *Journal of Applied Microbiology*, 98: 752–760.

Goulas, A.E. and Kontominas, M.G. (2007a). Combined effect of light salting, modified atmosphere packaging and oregano essential oil on the shelf-life of sea bream (*Sparus aurata*): Biochemical and sensory attributes. *Food Chemistry*, 100: 287–296.

Goulas, A.E. and Kontominas, M.G. (2007b). Effect of modified atmosphere packaging and vacuum packaging on the shelf-life of refrigerated chub mackerel (*Scomber japonicus*): Biochemical and sensory attributes. *European Food Research and Technology*, 224: 545–553.

Greengrass, J. (1993). Films for MAP of foods, in *Principles and Applications of Modified Atmosphere Packaging of Food*, 1st edn., Parry, T.R. (ed.), Blackie Academic and Professional, London, U.K. ISBN: 0-7514-0084-X.

Greengrass, J. (1999). Packaging materials for MAP of foods, in *Principles and Applications of Modified Atmosphere Packaging of Food*, 2nd edn., Blakistone, B.A. (ed.), Aspen Publications, Gaithersburg, MD. ISBN: 0-8342-1682-5.

Guevara, J.C., Yahia, E.M., and Brito de la Fuente, E. (2001). Modified atmosphere packaging of prickly pear cactus stems (*Opuntia spp.*). *Lebensmittel-Wissenschaft und Technologie*, 34: 445–451.

Gunes, G. and Lee, C.Y. (1997). Color of minimally processed potatoes as affected by modified atmosphere packaging and antibrowning agents. *Journal of Food Science*, 62(3): 572–575.

Guynot, M.E., Marın, S., Sanchis, V., and Ramos, A.J. (2003). Modified atmosphere packaging for prevention of mold spoilage of bakery products with different pH and water activity levels. *Journal of Food Protection*, 66(10): 1864–1872.

Guynot, M.E., Marın, S., Sanchis, V., and Ramos, A.J. (2004). An attempt to minimize potassium sorbate concentration in sponge cakes by modified atmosphere packaging combination to prevent fungal spoilage. *Food Microbiology*, 21: 449–457.

Habibunnisa, Baskaran, R., Prasad, R., and Shivaiah, K.M. (2001). Storage behaviour of minimally processed pumpkin (*Cucurbita maxima*) under modified atmosphere packaging conditions. *European Food Research and Technology*, 212: 165–169.

Hagenmaier, R.D. and Baker, R.A. (1997). Low-dose irradiation of cut iceberg lettuce in modified atmosphere packaging. *Journal of Agricultural and Food Chemistry*, 45: 2864–2868.

Hasan, S. (1997). Methods to extend the mold free shelf life of pizza crusts. MSc thesis, McGill University, Montreal, Quebec, Canada.

Hastings, M.J. (1993). MAP machinery, in *Principles and Applications of Modified Atmosphere Packaging of Food*, 1st edn., Parry, T.R. (ed.), Blackie Academic and Professional, London, U.K.

Hastings, M.J. (1999). MAP machinery, in *Principles and Applications of Modified Atmosphere Packaging of Food*, 2nd edn., Blakistone, B.A. (ed.), Aspen Publications, Gaithersburg, MD.

Hong, J.H. and Gross, K.C. (2001). Maintaining quality of fresh-cut tomato slices through modified atmosphere packaging and low temperature storage. *Journal of Food Science*, 66(7): 960–965.

Hong, S.I. and Kim, D. (2004). The effect of packaging treatment on the storage quality of minimally processed bunched onions. *International Journal of Food Science and Technology*, 39: 1033–1041.

Hovda, M.B., Sivertsvik, M., Lunestad, B.T., Lorentzen, G., and Rosnes, J.T. (2007). Characterisation of the dominant bacterial population in modified atmosphere packaged farmed halibut (*Hippoglossus hippoglossus*) based on 16S rDNA-DGGE. *Food Microbiology*, 24: 362–371.

Hu, W., Xu, P., and Uchino, T. (2005). Extending storage life of fresh ginseng by modified atmosphere packaging. *Journal of the Science of Food and Agriculture*, 85: 2475–2481.

Huisman, M., Madsen, H.L., Skibsted, L.H., and Bertelsen, G. (1994). The combined effect of rosemary (*Rosmarinus officinalis L.*) and modified atmosphere packaging as protection against warmed over flavor in cooked minced pork meat. *Zournal von Lebensmittel Untersuchung und Forschung*, 198: 57–59.

Ibrahim, S.M., Nassar, A.G., and El-Badry, N. (2008). Effect of modified atmosphere packaging and vacuum packaging methods on some quality aspects of smoked mullet (*Mugil cephalus*). *Global Veterinaria*, 2(6): 296–300.

Jacobsson, A., Nielsen, T., and Sjoholm, I. (2004a). Influence of temperature, modified atmosphere packaging, and heat treatment on aroma compounds in broccoli. *Journal of Agricultural and Food Chemistry*, 52: 1607–1614.

Jacobsson, A., Nielsen, T., and Sjoholm, I. (2004b). Effects of type of packaging material on shelf-life of fresh broccoli by means of changes in weight color and texture. *European Food Research and Technology*, 218: 157–163.

Jacobsson, A., Nielsen, T., Sjoholm, I., and Werdin, K. (2004c). Influence of packaging material and storage condition on the sensory quality of broccoli. *Food Quality and Preference*, 15: 301–310.

Jamie, P. and Saltveit, M.E. (2002). Postharvest changes in broccoli and lettuce during storage in argon, helium, and nitrogen atmospheres containing 2% oxygen. *Postharvest Biology and Technology*, 26: 113–116.

Jacxsens, L., Devlieghere, F., and Debevere, J. (2002). Temperature dependence of shelf-life as affected by microbial proliferation and sensory quality of equilibrium modified atmosphere packaged fresh produce. *Postharvest Biology and Technology*, 26: 59–73.

Jacxsens, L., Devlieghere, F., Ragaert, P., Vanneste, E., and Debevere, J. (2003). Relation between microbiological quality, metabolite production and sensory quality of equilibrium modified atmosphere packaged fresh cut produce. *International Journal of Food Microbiology*, 83: 263–280.

Jacxsens, L., Delvieghere, F., Van Der Steen, C., and Debevere, J. (2001). Effect of high oxygen modified atmosphere packaging on microbial growth and sensorial qualities of fresh-cut produce. *International Journal of Food Microbiology*, 71: 197–210.

Jayathunge, L. and Illeperuma, C. (2005). Extension of postharvest life of oyster mushroom by modified atmosphere packaging technique. *Journal of Food Science*, 70(9): 573–578.

Jia, C.G., Wei, C.J., Wei, J., Yan, G.F., Wang, B.L., and Wang, Q.M. (2009). Effect of modified atmosphere packaging on visual quality and glucosinolates of broccoli florets. *Food Chemistry*, 114(1): 28–37.

Jin, H.H. and Lee, S.Y. (2007). Combined effect of aqueous chlorine dioxide and modified atmosphere packaging on inhibiting *Salmonella typhimurium* and *Listeria monocytogenes* in mungbean sprouts. *Journal of Food Science*, 72(9): 441–445.

John, L., Cornforth, D., Carpenter, C.E., Sorheim, O., Pettee, B.C., and Dick, R. (2005). Whittier color and thiobarbituric acid values of cooked top sirloin steaks packaged in modified atmospheres of 80% oxygen, or 0.4% carbon monoxide, or vacuum. *Meat Science*, 69: 441–449.

Jones, R.B., Faragher, J.D., and Winkler, S. (2006). A review of the influence of postharvest treatments on quality and glucosinolate content in broccoli (*Brassica oleracea* var. *italica*) heads. *Postharvest Biology and Technology*, 41: 1–8.

Jung, D.C., Lee, S.Y., Yoon, J.H., Hong, K.P., Kang, Y.S., Park, S.R., Park, S.K., Ha, S.D., Kim, G.H., and Bae, D.H. (2009). Inhibition of pork and fish oxidation by a novel plastic film coated with horseradish extract. *LWT—Food Science and Technology*, 42: 856–861.

Kakiomenou, K., Tassou, C., and Nychas, G.J. (1998). Survival of *Salmonella enteritidis* and *Listeria monocytogenes* on salad vegetables. *World Journal of Microbiology and Biotechnology*, 14: 383–387.

Kang, J.S., Park, W.P., and Lee, D.S. (2000). Quality of enoki mushrooms as affected by packaging conditions. *Journal of the Science of Food and Agriculture*, 81: 109–114.

Kazantzis, I., Nanos, G.D., and Stavroulakis, G.G. (2003). Effect of harvest time and storage conditions on almond kernel oil and sugar composition. *Journal of the Science of Food and Agriculture*, 83: 354–359.

Keawmanee and Haruenkit. (2007). Shelf life of brown rice muffin under modified atmosphere packaging. http://iat.sut.ac.th/food/FIA2007/FIA2007/paper/P2-11-NC.pdf (accessed on 10 September 2007).

Kim, J.G., Luo, Y., and Gross, K.C. (2004). Effect of package film on the quality of fresh-cut salad savoy. *Postharvest Biology and Technology*, 32: 99–107.

Kim, J.G., Luo, Y., Tao, Y., Saftner, R.A., and Gross, K.C. (2005). Effect of initial oxygen concentration and film oxygen transmission rate on the quality of fresh-cut romaine lettuce. *Journal of the Science of Food and Agriculture*, 85: 1622–1630.

Kim, Y.H., Keeton, J.T., Hunt, M.C., and Savell, J.W. (2010). Effects of L- or D-lactate enhancement on the internal cooked colour development and biochemical characteristics of beef steaks in high oxygen modified atmosphere. *Food Chemistry*, 119(3): 918–922.

Kim, Y.H., Keeton, J.T., Yang, H.S., Smith, S.B., Sawyer, J.E., and Savell, J.W. (2009). Color stability and biochemical characteristics of bovine muscles when enhanced with L- or D-potassium lactate in high-oxygen modified atmospheres. *Meat Science*, 82: 234–240.

Kirwan, M.J. and Strawbridge, J.W. (2003). Plastics in food packaging. *Food Packaging Technology*, Coles, R., McDowell D., Kirwan, M.J. (eds.), Blackwell Publishing, Oxford, U.K.

Knorr, D. and Tomlins, R. (1985). Effect of carbon dioxide modified atmosphere on the compressibility of stored baked goods. *Journal of Food Science*, 50: 1172.

Koide, S. and Shi, J. (2007). Microbial and quality evaluation of green peppers stored in biodegradable film packaging. *Food Control*, 18: 1121–1125.

Koukoutsis, J., Smith, J.P., Daifas, D.P., Yayalan, V., Kayouette, B., Ngadi, M., and El Khoury, W. (2004). In vitro studies to control the growth of microorganisms of spoilage and safety concern in high-moisture, high-pH bakery products. *Journal of Food Safety*, 24(3): 211–230.

Kozova, M., Kalac, P., and Pelikanova, T. (2009). Contents of biologically active polyamines in chicken meat, liver, heart and skin after slaughter and their changes during meat storage. *Food Chemistry*, 116: 419–425.

Lacroix, M. and Lafortune, R. (2004). Combined effects of gamma irradiation and modified atmosphere packaging on bacterial resistance in grated carrots (*Daucus carota*). *Radiation Physics and Chemistry*, 71: 77–80.

Lalitha, K.V., Sonaji, E.R., Manju, S., Jose, L., Gopal, T.K.S., and Ravisankar, C.N. (2005). Microbiological and biochemical changes in pearl spot (*Etroplus suratensis* Bloch) stored under modified atmospheres. *Journal of Applied Microbiology*, 99: 1222–1228.

Lavermicocca, P., Valerio, F., and Visconti, A. (2003). Antifungal activity of phenyllactic acid against molds isolated from bakery products. *Applied and Environmental Microbiology*, 69: 634–640.

Lee, S.Y. and Baek, S.Y. (2008). Effect of chemical sanitizer combined with modified atmosphere packaging on inhibiting *Escherichia coli O157:H7* in commercial spinach. *Food Microbiology*, 25: 582–587.

Lee, K.E., Kim, H.J., An, D.S., Lyu, E.S., and Lee, D.S. (2008). Effectiveness of modified atmosphere packaging in preserving a prepared ready-to-eat food. *Packaging Technology and Science*, 21: 417–423.

Lee, K.S., Park, I.S., and Lee, D.S. (1996). Modified atmosphere packaging of a mixed prepared vegetable salad dish. *International Journal of Food Science and Technology*, 31: 7–13.

Liu, C.L., Hsu, C.K., and Hsu, M.M. (2007). Improving the quality of fresh-cut pineapples with ascorbic acid/sucrose pretreatment and modified atmosphere packaging. *Packaging Technology and Science*, 20: 337–343.

Lu, S. (2009). Effects of bactericides and modified atmosphere packaging on shelf-life of Chinese shrimp (*Fenneropenaeus chinensis*). *LWT—Food Science and Technology*, 42(1): 286–291.

Luciano, G., Monahan, F.J., Vasta, V., Biondi, L., Lanza, M., and Priolo, A. (2009a). Dietary tannins improve lamb meat colour stability. *Meat Science*, 81: 120–125.

Luciano, G., Monahan, F.J., Vasta, V., Pennisi, P., Bella, M., and Priolo, A. (2009b). Lipid and colour stability of meat from lambs fed fresh herbage or concentrate. *Meat Science*, 82: 193–199.

Mannheim, C.H. and Soffer, T. (1996). Shelf-life extension of cottage cheese by modified atmosphere packaging. *Lebensmittel-Wissenschaft und Technologie*, 29: 767–771.

Marshall, D.L., Andrews, L.S., Wells, J.H., and Farr, A.J. (1992). Influence of modified atmosphere packaging on the competitive growth of *Listeria monocytogenes* and *Pseudomonas fluorescens* on precooked chicken. *Food Microbiology*, 9: 303–309.

Martinez, J.A. and Artes, F. (1999). Effect of packaging treatments and vacuum-cooling on quality of winter harvested iceberg lettuce. *Food Research International*, 32: 621–627.

Martinez, I., Ares, G., and Lema, P. (2008). Influence of cut and packaging film on sensory quality of fresh-cut butterhead lettuce (*Lactuca sativa L., cv. Wang*). *Journal of Food Quality*, 31: 48–66.

Mastromatteo, M., Lucera, A., Sinigaglia, M., and Corbo, M.R. (2009). Combined effects of thymol, carvacrol and temperature on the quality of non conventional poultry patties. *Meat Science*, 83: 246–254.

McConnell, R.Y., Truong, V.D., Walter, W.M. Jr., and McFeeters, R.F. (2005). Physical, chemical and microbial changes in shredded sweet potatoes. *Journal of Food Processing and Preservation*, 29: 246–267.

McMillin, K.W. (2008). Where is MAP going? A review and future potential of modified atmosphere packaging for meat. *Meat Science*, 80: 43–65.

Mejlholm, N.B. and Dalgaard, P. (2005). Shelf life and safety aspects of chilled cooked and peeled shrimps (*Pandalus borealis*) in modified atmosphere packaging. *Journal of Applied Microbiology*, 99: 66–76.

Metaxopoulos, J., Mataragas, M., and Drosinos, E.H. (2002). Microbial interaction in cooked cured meat products under vacuum or modified atmosphere at 4°C. *Journal of Applied Microbiology*, 93: 363–373.

Montero-Calderon, A., Rojas-Graó, M.A., and Martin-Belloso, O. (2008). Effect of packaging conditions on quality and shelf-life of fresh-cut pineapple (*Ananas comosus*). *Postharvest Biology and Technology*, 50: 182–189.

Mullan, M. and McDowell, D. (2003). Modified atmosphere packaging, in *Food Packaging Technology*, Coles, R., McDowell, D., Kirwan, M.J. (eds.), Blackwell Publishing, Oxford, U.K., pp. 312–319. ISBN: 1-84127-221-3.

Murcia, M.A., Martınez-Tome, M., Carmen, M.N., and Vera Ana, M. (2003). Extending the shelf-life and proximate composition stability of ready to eat foods in vacuum or modified atmosphere packaging. *Food Microbiology*, 20: 671–679.

Nguyen, T.B.T., Ketsa, S., and Van Doorn, W.G. (2004). Effect of modified atmosphere packaging on chilling-induced peel browning in banana. *Postharvest Biology and Technology*, 31: 313–317.

Nielsen, T. and Leufven, A. (2008). The effect of modified atmosphere packaging on the quality of *Honeoye* and *Korona* strawberries. *Food Chemistry*, 107: 1053–1063.

Nielsen, P.V. and Rios, R. (2000). Inhibition of fungal growth on bread by volatile components from spices and herbs, and the possible application in active packaging, with special emphasis on mustard essential oil. *International Journal of Food Microbiology*, 60(2–3): 219–229.

Niemira, B.A., Fan, X., and Sokorai, K.J.B. (2005). Irradiation and modified atmosphere packaging of endive influences survival and regrowth of *Listeria monocytogenes* and product sensory qualities. *Radiation Physics and Chemistry*, 72: 41–48.

Ntzimani, A.G., Paleologos, E.K., Savvaidis, I.N., and Kontominas, M.G. (2008). Formation of biogenic amines and relation to microbial flora and sensory changes in smoked turkey breast fillets stored under various packaging conditions at 4°C. *Food Microbiology*, 25: 509–517.

Odriozola-Serrano, I., Soliva-Fortuny, R., and Martin-Belloso, O. (2008). Effect of minimal processing on bioactive compounds and color attributes of fresh-cut tomatoes. *LWT*, 41: 217–226.

Oms-Oliu, G., Soliva-Fortuny, R., and Martın-Belloso, O. (2008). Physiological and microbiological changes in fresh-cut pears stored in high oxygen active packages compared with low oxygen active and passive modified atmosphere packaging. *Postharvest Biology and Technology*, 48: 295–301.

Oyugi, E. and Buys, E.M. (2007). Microbiological quality of shredded Cheddar cheese packaged in modified atmospheres. *International Journal of Dairy Technology*, 60(2): 89–95.

Ozogul, Y. and Ozogul, F. (2004). Effects of slaughtering methods on sensory, chemical and microbiological quality of rainbow trout (*Onchorynchus mykiss*) stored in ice and MAP. *European Food Research and Technology*, 219: 211–216.

Ozogul, F. and Ozogul, Y. (2006). Biogenic amine content and biogenic amine quality indices of sardines (*Sardina pilchardus*) stored in modified atmosphere packaging and vacuum packaging. *Food Chemistry*, 99: 574–578.

Panagou, E.Z. (2006). Greek dry-salted olives: Monitoring the dry-salting process and subsequent physico-chemical and microbiological profile during storage under different packing conditions at 4 and 20°C. *LWT*, 39: 322–329.

Papaioannou, G., Chouliara, I., Karatapanis, A.E., Kontominas, M.G., and Savvaidis, I.N. (2007). Shelf-life of a Greek whey cheese under modified atmosphere packaging. *International Dairy Journal*, 17: 358–364.

Parentelli, C., Ares, G., Corona, M., Lareo, C., Gambaro, A., Soubes, M., and Lema, P. (2007). Sensory and microbiological quality of *shiitake* mushrooms in modified-atmosphere packages. *Journal of the Science of Food and Agriculture*, 87: 1645–1652.

Parry, T.R. (1993). Introduction, in *Principles and Applications of Modified Atmosphere Packaging of Food*, 1st edn., Parry, T.R. (ed.), Blackie Academic and Professional, London, U.K. ISBN: 0-7514-0084-X.

Pastoriza, L., Sampedro, G., Herrera, J.J., and Cabo, M.L. (1998). Influence of sodium chloride and modified atmosphere packaging on microbiological, chemical and sensorial properties in ice storage of slices of hake (*Merluccius merluccius*). *Food Chemistry*, 61(1/2): 23–28.

Patsias, A., Chouliara, I., Badeka, A., Savvaidis, I.N., and Kontominas, M.G. (2006). Shelf-life of a chilled precooked chicken product stored in air and under modified atmospheres: Microbiological, chemical, sensory attributes. *Food Microbiology*, 23: 423–429.

Pexara, S., Metaxopoulos, J., and Drosinos, E.H. (2002). Evaluation of shelf life of cured, cooked, sliced turkey fillets and cooked pork sausages '*piroski*'—Stored under vacuum and modified atmospheres at +4 and +10°C. *Meat Science*, 62: 33–43.

Piagentini, A.M., Guemes, D.R., and Pirovani, M.E. (2003). Mesophilic aerobic population of fresh-cut spinach as affected by chemical treatment and type of packaging film. *Journal of Food Science*, 68(2): 602–606.

Picon, A., Martinez-Jaivega, J.M., Cuquerella, J., Del Rio, M.A., and Navarro, P. (1993). Effects of precooling, packaging film, modified atmosphere and ethylene absorber on the quality of refrigerated Chandler and Douglas strawberries. *Food Chemistry*, 48: 189–193.

Pilon, L., Oetterer, M., Gallo, C.R., and Spoto, M.H.F. (2006). Shelf life of minimally processed carrot and green pepper. *Ciência e Tecnologia de Alimentos Campinas*, 26(1): 150–158.

Piringer, O.G. and Baner, A.L. (2000). *Plastic Packaging Materials for Food*, Wiley VCH, Weinheim, Germany. ISBN: 3-527-28868-6.

Poli, B.M., Messini, A., Parisi, G., Scappini, F., Vigiani, V., Giorgi, G., and Vincenzini, M. (2006). Sensory, physical, chemical and microbiological changes in European sea bass (*Dicentrarchus labrax*) fillets packed under modified atmosphere/air or prepared from whole fish stored in ice. *International Journal of Food Science and Technology*, 41: 444–454.

Pournis, N., Papavergou, A., Badeka, A., Kontominas, M.G., and Savvaidis, I.N. (2005). Shelf-life extension of refrigerated Mediterranean mullet (*Mullus surmuletus*) using modified atmosphere packaging. *Journal of Food Protection*, 68(10): 2201–2207.

Radziejewska-Kubzdela, E., Czapski, J., and Czaczyk, K. (2007). The effect of packaging conditions on the quality of minimally processed celeriac flakes. *Food Control*, 18: 1191–1197.

Ranasinghe, L., Jayawardena, B., and Abeywickrama, K. (2005). An integrated strategy to control post-harvest decay of Embul banana by combining essential oils with modified atmosphere packaging. *International Journal of Food Science and Technology*, 40: 97–103.

Randell, K., Hattula, T., and Ahvenainen, R. (1997). Effect of packaging method on the quality of rainbow trout and Baltic herring fillets. *Lebensmittel-Wissenschaft und Technologie*, 30: 56–61.

Rangkadilok, N., Tomkins, B., Nicolas, M.E., Premier, R.R., Bennett, R.N., Eagling, D.R., and Taylor, P.W.J. (2002). The effect of post-harvest and packaging treatments on glucoraphanin concentration in broccoli (*Brassica oleracea var. italica*). *Journal of Agricultural and Food Chemistry*, 50: 7386–7391.

Rasmussen, P.H. and Hansen, A. (2001). Staling of wheat bread stored in modified atmosphere. *Lebensmittel-Wissenschaft und Technologie*, 34: 487–491.

Remon, S., Ferrer, A., Marquina, P., Burgos, J., and Oria, R. (1999). Use of modified atmospheres to prolong the postharvest life of Burlat cherries at two different degrees of ripeness. *Journal of the Science of Food and Agriculture*, 80: 1545–1552.

Renault, P., Houal, L., Jacquemin, G., and Chambroy, Y. (1994). Gas exchange in modified atmosphere packaging. 2: Experimental results with strawberries. *International Journal of Food Science and Technology*, 29: 379–394.

Robertson, G.L. (2006). *Food Packaging: Principles and Practice*. Food Science and Technology, Taylor & Francis, Boca Raton, FL. ISBN: 0-8493-3775-5.

Rocha, A.M.C.N., Barreiro, M.G., and Morais, A.M.M.B. (2004). Modified atmosphere package for apple *'Bravo de Esmolfe'*. *Food Control*, 15: 61–64.

Rosen, J.C. and Kader, A.A. (1989). Postharvest physiology and quality maintenance of sliced pear and strawberry fruits. *Journal of Food Science*, 54(3): 656–659.

Rosnes, J.T., Kleiberg, G.H., Sivertsvik, M., Lunestad, B.T., and Lorentzen, G. (2006). Effect of modified atmosphere packaging and superchilled storage on the shelf-life of farmed ready-to-cook spotted wolf-fish (*Anarhichas minor*). *Packaging Technology and Science*, 19: 325–333.

Roy, S., Anantheswaran, R.C., and Beelman, R.B. (1996). Modified atmosphere and modified humidity packaging of fresh mushrooms. *Journal of Food Science*, 61(2): 391–397.

Sallam, K.I., Ahmed, A.M., Elgazzar, M.M., and Eldaly, E.A. (2007). Chemical quality and sensory attributes of marinated Pacific saury (*Cololabis saira*) during vacuum-packaged storage at 4°C. *Food Chemistry*, 102: 1061–1070.

Sandhya (2010). Modified atmosphere packaging of fresh produce: Current status and future needs. *LWT - Food Science and Technology*, 43: 381–392.

Sanguinetti, A.M., Secchi, N., Del Caro, A., Stara, G., Roggio, T., and Piga, A. (2009). Effectiveness of active and modified atmosphere packaging on shelf life extension of a cheese tart. *International Journal of Food Science and Technology*, 44(6): 1192–1198.

Sanz, S., Olarte, C., Echavarri, J.F., and Ayala, F. (2007). Evaluation of different varieties of cauliflower for minimal processing. *Journal of the Science of Food and Agriculture*, 87: 266–273.

Schreiner, M., Peters, P., and Krumbein, A. (2006). Glucosinolates in mixed-packaged mini broccoli and mini cauliflower under modified atmosphere. *Journal of Agricultural and Food Chemistry*, 54: 2218–2222.

Schreiner, M., Peters, P., and Krumbein, A. (2007). Changes of glucosinolates in mixed fresh-cut broccoli and cauliflower florets in modified atmosphere packaging. *Journal of Food Science*, 72(8): 585–589.

Serrano, M., Martınez-Romero, D., Castillo, S., Guillen, F., and Valero, D. (2005). The use of natural antifungal compounds improves the beneficial effect of MAP in sweet cherry storage. *Innovative Food Science and Emerging Technologies*, 6: 115–123.

Serrano, M., Martinez-Romero, D., Guillen, F., Castillo, S., and Valero, D. (2006). Maintenance of broccoli quality and functional properties during cold storage as affected by modified atmosphere packaging. *Postharvest Biology and Technology*, 39: 61–68.

Shen, Q., Kong, F., and Wang, Q. (2006). Effect of modified atmosphere packaging on the browning and lignification of bamboo shoots. *Journal of Food Engineering*, 77: 348–354.

Simon, A., Gonzales-Fandos, E., and Rodriguez, D. (2008). Effect of film and temperature on the sensory, microbiological and nutritional quality of minimally processed cauliflower. *International Journal of Food Science and Technology*, 43: 1628–1636.

Simon, A., Gonzalez-Fandos, E., and Tobar, V. (2004). Influence of washing and packaging on the sensory and microbiological quality of fresh peeled white asparagus. *Journal of Food Science*, 69(1): 6–12.

Simon, A., Gonzalez-Fandos, E., and Tobar, V. (2005). The sensory and microbiological quality of fresh sliced mushroom (*Agaricus bisporus* L.) packaged in modified atmospheres. *International Journal of Food Science and Technology*, 40: 943–952.

Siomos, A.S., Sfakiotakis, E.M., and Dogras, C.C. (2000). Modified atmosphere packaging of white asparagus-spears: Composition, color and textural quality responses to temperature and light. *Scientia Horticulturae*, 84: 1–13.

Siro, I., Devlieghere, F., Jacxsens, L., Uyttendaele, M., and Debevere, J. (2006). The microbial safety of strawberry and raspberry fruits packaged in high-oxygen and equilibrium-modified atmospheres compared to air storage. *International Journal of Food Science and Technology*, 41: 93–103.

Sivakumar, D., Arrebola, E., and Korsten, L. (2008). Postharvest decay control and quality retention in litchi (*cv. McLean's Red*) by combined application of modified atmosphere packaging and antimicrobial agents. *Crop Protection*, 27: 1208–1214.

Sivakumar, D. and Korsten, L. (2006). Influence of modified atmosphere packaging and postharvest treatments on quality retention of litchi *'cv. Mauritius. Postharvest Biology and Technology*, 41: 135–142.

Sivertsvik, M. (2007). The optimized modified atmosphere for packaging of pre-rigor filleted farmed cod (*Gadus Morhua*) in 63 ml/100 ml oxygen and 37 ml/100 ml carbon dioxide. *LWT*, 40: 430–438.

Sivertsvik, M. and Birkeland, S. (2006). Effects of soluble gas stabilisation, modified atmosphere, gas to product volume ratio and storage on the microbiological and sensory characteristics of ready-to-eat shrimp *(Pandalus borealis). Food Science and Technology International*, 12(5): 445–454.

Smith, S.M., Geeson, J.D., and Genge, P.M. (1988). The effect of harvest date on the responses of Discovery apples to modified atmosphere retail packaging. *International Journal of Food Science and Technology*, 23: 81–90.

Smith, J.P., Jackson, E.D., and Ooraikul, B. (1983). Storage study of a gas-packaged bakery product. *Journal of Food Science*, 48: 1370.

Smith, J.P., Ooraikul, B., Koersen, W.J., Jacksont, E.D., and Lawrence, R.A. (1986). Novel approach to oxygen control in modified atmosphere packaging of bakery products. *Food Microbiology*, 3: 315–320.

Soffer, T., Margalith, P., and Mannheim, C.H. (1994). Shelf-life of chicken liver/egg pate in modified atmosphere packages. *International Journal of Food Science and Technology*, 29: 161–166.

Soldatou, N., Nerantzaki, A., Kontominas, M.G., and Savvaidis, I.N. (2009). Physicochemical and microbiological changes of "Souvlaki"—A Greek delicacy lamb meat product: Evaluation of shelf-life using microbial, colour and lipid oxidation parameters. *Food Chemistry*, 113: 36–42.

Soliva-Fortuny, R., Ricart-Coll, M., Elez-Martınez, P., and Martın-Belloso, O. (2007). Internal atmosphere, quality attributes and sensory evaluation of MAP packaged fresh-cut Conference pears. *International Journal of Food Science and Technology*, 42: 208–213.

Soliva-Fortuny, R.C., Ricart-Coll, M., and Martın-Belloso, O. (2005). Sensory quality and internal atmosphere of fresh-cut Golden Delicious apples. *International Journal of Food Science and Technology*, 40: 369–375.

Sorheim, O., Westad, F., Larsen, H., and Alvseike, O. (2009). Colour of ground beef as influenced by raw materials, addition of sodium chloride and low oxygen packaging. *Meat Science*, 81: 467–473.

Spotts, R.A., Cervantes, L.A., and Facteau, T.J. (2002). Integrated control of brown rot of sweet cherry fruit with a preharvest fungicide, a postharvest yeast, modified atmosphere packaging, and cold storage temperature. *Postharvest Biology and Technology*, 24: 251–257.

Spotts, R.A., Cervantes, L.A., Facteau, T.J., and Chand-Goyal, T. (1998). Control of brown rot and blue mold of sweet cherry with preharvest iprodione, postharvest *Cryptococcus infirmo-miniatus*, and modified atmosphere packaging. *Plant Disease*, 82(10): 1158–1160.

Suparlan, and Itoh, K. (2003). Combined effects of hot water treatment (HWT) and modified atmosphere packaging (MAP) on quality of tomatoes. *Packaging Technology and Science*, 16: 171–178.

Taniwaki, M.H., Hocking, A.D., Pitt, J.I., and Fleet, G.H. (2001). Growth of fungi and mycotoxin production on cheese under modified atmospheres. *International Journal of Food Microbiology*, 68: 125–133.

Tano, K., Kouamé, F.A., Nevry, R.K., and Oulé, M.K. (2008). Modified atmosphere packaging of strawberries (*Fragaria X Ananassa Duch.*) stored under temperature fluctuation conditions. *European Journal of Scientific Research*, 21(2): 353–364.

Tano, K., Oule, M.K., Doyon, G., Lencki, R.W., and Arul, J. (2007). Comparative evaluation of the effect of storage temperature fluctuation on modified atmosphere packages of selected fruit and vegetables. *Postharvest Biology and Technology*, 46: 212–221.

Tao, F., Zhang, M., Hangqing, Y., and Jincai, S. (2006). Effects of different storage conditions on chemical and physical properties of white mushrooms after vacuum cooling. *Journal of Food Engineering*, 77: 545–549.

Tao, F., Zhang, M., and Yu, H. (2007). Effect of vacuum cooling on physiological changes in the antioxidant system of mushroom under different storage conditions. *Journal of Food Engineering*, 79: 1302–1309.

Tassou, C.C. and Boziaris, J.S. (2002). Survival of *Salmonella enteritidis* and changes in pH and organic acids in grated carrots inoculated or not with *Lactobacillus sp.* and stored under different atmospheres at 4°C. *Journal of the Science of Food and Agriculture*, 82: 1122–1127.

Tenorio, M.D., Villanueva, M.J., and Sagardoy, M. (2004). Changes in carotenoids and chlorophylls in fresh green asparagus (*Asparagus officinalis L.*) stored under modified atmosphere packaging. *Journal of the Science of Food and Agriculture*, 84: 357–365.

Tian, S.P., Li, B.Q., and Xu, Y. (2005). Effects of O_2 and CO_2 concentrations on physiology and quality of litchi fruit in storage. *Food Chemistry*, 91: 659–663.

Toivonen, P.M.A. and Stan, S. (2004). The effect of washing on physicochemical changes in packaged, sliced green peppers. *International Journal of Food Science and Technology*, 39: 43–51.

Torrieri, E., Cavella, S., Villani, F., and Masi, P. (2006). Influence of modified atmosphere packaging on the chilled shelf life of gutted farmed bass. *Journal of Food Engineering*, 77: 1078–1086.

Trobetas, A., Badeka, A., and Kontominas, M.G. (2008). Light-induced changes in grated *Graviera* hard cheese packaged under modified atmospheres. *International Dairy Journal*, 18: 1133–1139.

Van der Steen, C., Jacxsens, L., Devlieghere, F., and Debevere, J. (2002). Combining high oxygen atmospheres with low oxygen modified atmosphere packaging to improve the keeping quality of strawberries and raspberries. *Postharvest Biology and Technology*, 26: 49–58.

Varoquax, P., Albagnac, G., Nguyen, C., and Varoquax, F. (1996). Modified atmosphere packaging of fresh beansprouts. *Journal of the Science of Food and Agriculture*, 70: 224–230.

Villaescusa, R. and Gil, M.I. (2003). Quality improvement of *Pleurotus* mushrooms by modified atmosphere packaging and moisture absorbers. *Postharvest Biology and Technology*, 28: 169–179.

Villanueva, M.J., Tenorio, M.D., Sagardoy, M., Redondo, A., and Saco, M.D. (2005). Physical, chemical, histological and microbiological changes in fresh green asparagus (*Asparagus officinalis L.*) stored in modified atmosphere packaging. *Food Chemistry*, 91: 609–619.

Viuda-Martos, M., Ruiz-Navajas, Y., Fernαndez-Lσpez, J., and Pιrez-Alvarez, J.A. (2010). Effect of orange dietary fibre, oregano essential oil and packaging conditions on shelf-life of bologna sausages. *Food Control*, 21(4): 436–443.

Wang, C.Y. and Qi, L. (1997). Modified atmosphere packaging alleviates chilling injury in cucumbers. *Postharvest Biology and Technology*, 10: 195–200.

Zakrys, P.I., O'Sullivan, M.G., Allen, P., and Kerry, J.P. (2009). Consumer acceptability and physiochemical characteristics of modified atmosphere packed beef steaks. *Meat Science*, 81: 720–725.

Zhang, M., Xiao, G., and Salokhe, V.M. (2006). Preservation of strawberries by modified atmosphere packages with other treatments. *Packaging Technology and Science*, 19: 183–191.

Zhang, M., Zhan, Z.G., Wang, S.J., and Tang, J.M. (2008). Extending the shelf-life of asparagus spears with a compressed mix of argon and xenon gases. *LWT*, 41: 686–691.

Part II

Safety and Quality Control of MAP Produces

3 Safety and Quality Control of Modified Atmosphere Packaging Products

Ioannis S. Arvanitoyannis and Konstantinos Kotsanopoulos

CONTENTS

3.1 INTRODUCTION

In Europe, the 1990s was a very important decade because the implementation of the Single European Market began and the "mad cow disease" (BSE) outbreak hit the European food sector. Because of the introduction of the Single Market, several novel issues regarding integration and harmonization of safety and quality procedures in food production and processing arose. BSE—one of the worst outbreaks mainly related to erroneous management—helped to upgrade the role of food safety and placed it at the very top of the political agenda in Europe (Halkier and Holm, 2006). Nowadays, consumers are faced with dilemmas on a daily basis in view of the food or food ingredients that may have been produced in distant countries or continents and not necessarily with the most transparent food supply. This globalization has resulted in enhanced food quality and safety concerns. Although some of these concerns can be true, they can also act as arguments for restricting trade (van Veen, 2005). Therefore, for the past decade, the marked increase recorded in foodborne infections has become a universal public health concern. Among the many factors that have played an important role in changing the epidemiology of microbial foodborne illnesses were (i) greater susceptibility of human population to diseases; (ii) changing life styles, adopting more adventurous eating habits (foreign cuisine); (iii) more convenience foods and less time devoted to food preparation; and (iv) emergence of recently recognized microbial pathogens and ever-evolving technologies for food production, processing, and distribution (Meng and Doyle, 2002). A hazard analysis critical control point (HACCP) system can be effective only if it is based on sound good manufacturing and hygienic practices (GMP/GHP). Consequently, it is the responsibility of the government agencies to ensure that these prerequisite programs are properly implemented before assessing HACCP implementation. HACCP aims to indentify problems before their occurrence and, if possible, establish preventive measures for their control at all stages (starting with receipt of raw materials) in productions that are critical to ensure the safety of the food. Control is primarily based on preventive actions or proactive measures, and there are corrective actions to be undertaken in case the former proves to be unsuccessful (Notermans and Veld, 1994; Arvanitoyannis, 2001).

3.1.1 Food Safety

The principles of food safety objectives (FSOs) as suggested by the International Commission on Microbiological Specifications for Foods (ICMSF [2002]) and

Codex Committee on Food Hygiene (CCFH [2004a,b]) are very comprehensive and this is their strong point. Should one integrate the changes in a hazard from the initial level (Ho) minus the sum of the reductions (R) plus the sum of growth (G) and (re)contamination (C), one reaches a concentration/prevalence, the consumption level of which must be lower with regard to the established FSO. The latter is closely related to the so-called appropriate level of protection (ALOP) (Zwietering, 2005).

Safe food is produced by implementing the GHP, GMP, good agricultural practices (GAP), etc., and application of food safety risk management systems such as HACCP. However, the level of safety that these food safety systems are anticipated to deliver has seldom been defined in quantitative terms. Establishment of FSOs and performance objectives (POs) provides the industry with quantitative targets to be met. When required, industry may have to exhibit that their food safety system is capable of controlling the hazard of concern, by displaying evidence of applied control measures. Moreover, industry must periodically prove that their control or preventive measures are functioning as intended (Van Schothorst et al., 2009). Although food safety regulation was the field of food technologists and government regulators where economic efficiency played no role in the design of most regulations focusing on food safety, this has changed over the last 10 years thereby affecting the way food regulations are designed. Consumer concerns have gradually shifted from the availability of food-to-food quality and safety (Antle, 1999). In China, for example, the economic growth and development of populated urbanized centers made Chinese consumers more interested in food quality and safety. Similarly, the frequent outbreaks of foodborne illnesses in China made clear the urgent requirement for more effective food safety assurance systems (Bai et al., 2007). Viruses are increasingly recognized as an important cause of foodborne outbreaks. In the United States, noroviruses (NoV), hepatitis A virus (HAV), rotavirus, astrovirus, and enteric adenoviruses are described as foodborne-associated viruses, out of which NoV was by far the most causative agent. Application of modified atmosphere packaging (MAP) can effectively inhibit spoilage due to bacterial and fungal microorganisms to prolong the shelf life of foods (Baert et al., 2009).

The results of consumer studies on food safety knowledge and practices have revealed that consumers are aware of and are thinking about food safety, although there are several gaps in food safety knowledge and practices that may potentially lead to foodborne diseases. The reports of epidemiologic surveillance of foodborne diseases exhibited that consumer behavior regarding ingestion of raw/undercooked foods and poor GHPs are important parameters to outbreaks of foodborne diseases. It was reported that although people of all ages seem to think they know how to handle food safely, their self-reported food-handling behaviors do not fall in line with their confidence (Jevŝnik et al., 2008). In the case of China, the senior officer of Food and Agriculture Organization of UN argued that effective food control systems are feasible through a combination and harmonic cooperation of food legislation, national food control strategy, and food control agencies. The development and implementation of Food Safety Control Systems (FSCS) for the food service sector in Taiwan, set a good example for the country-level adaptation to food-safety system, particularly helpful for the factories exporting their agricultural products (Jeng and Fang, 2003). In Taiwan, the Department of Health was the major food control agency involved actively in HACCP expansion in Taiwan's food service sector in an attempt to enhance the competitive power of Taiwan food industry in the international market (Bai et al., 2007). Formal assessment of the economic impact of regulation is part of the governments' desire for evidence-based policy making. In fact, both consideration and quantification (if possible) of economic issues and publication of respective findings and their assumptions are vital for establishing transparent and accountable policy making. As a result, guidelines have been publicized both at EU and Member State levels (Traill and Koenig, 2009).

A good tool toward a structured approach to the management of food safety is the risk analysis described by Codex Alimentarius (CA). According to ICMSF and CA, the establishment of an FSO is a basic tool to meet public health requirements such as ALOP. An FAO/WHO expert consultation reemphasized the original definition for ALOP that was part of the Sanitary and Phytosanitary (SPS) Measures Agreement, namely that it is the "expression of the level of protection in relation to food safety that is currently achieved" (Van Schothorst et al., 2009). According to Egan et al. (2007), food safety is still a critical issue with outbreaks of foodborne illness leading to substantial costs to individuals, the food industry, and the economy. The number of food poisoning notifications in England and Wales rose steadily from approximately 15,000 cases in the early 1980s to a peak of over 60,000 cases in 1996. This could be partly attributed to enhanced surveillance but may equally reflect greater global trade and travel, changes in modern food production and consumption, the effect of modern lifestyles, and the emergence of new pathogens.

3.1.2 Food Quality

Typical factors that are part of the evaluation of food quality are food producers, government officials, marketing people, and consumers (Wandel and Bugge, 1997). The scientific strategy most appropriate for sensory quality evaluation considers the relation from two types of data: tests with consumers (affective and hedonic type) and trained analytical panels (descriptive and analytic type). The relationship between them determines the sensory profiles to be adapted to the concept of the product quality in the target market. Large companies are thereby induced to establish proper control activities, ameliorate quality, and develop novel products. It is noteworthy, however, that this particular approach has certain limitations in terms of its being implemented by small producers, such as those with protected designation of origin (PDO) (Elortondo et al., 2007). The producers' preference to technical use-attributes, such as enhanced yield, suitability for mechanical harvesting, as well as resistance against insects and diseases. Government officials are involved in regulations related to health aspects, that is, dosimetry of contaminants and types of additives allowed in food. Consumers are interested in food quality aspects and food safety (Wandel and Bugge, 1997).

Consumer requirements generally include the following:

1. Safety requirements, expressed as the absence of "risk factors." Any failure to respect safety requirements represents a risk for consumer health and is against the current legislation.
2. Commodity requirements, the conformance of a product to its definition.
3. Nutritional requirements are obviously extremely important because the main purpose of eating is to satisfy nutritional needs (Peri, 2006).

Investigations in several countries have shown that consumers are getting increasingly interested in food qualities related to the nutritional content and safety issues, food additives, and the presence and levels of agrochemical residues and contaminants from environmental pollution (Wandel and Bugge, 1997).

4. Sensory requirements (S′R). The way S′R are perceived makes them critical means of interaction between products and consumers.
5. Requirements concerning the production context.
6. Ethical requirements (Peri, 2006). Consumers are getting interested in foods produced and manufactured in compliance with ethical aspects of animal rearing (Wandel and Bugge, 1997).
7. Guarantee requirements.
8. The requirements of the product/packaging system facilitate product recognition, marketing, and use.
9. Requirements of the product/market system (Peri, 2006).

Computer vision systems have been increasingly used in the food industry for quality evaluation purposes in view of their low cost and consistent and rapid performance. The food industry is ranked among the top 10 industries applying computer vision technology, which has the advantages of being objective and nondestructive vis á vis food products. Quality has become a key factor for the modern food industry because the high quality of products is the basis for success in today's highly competitive market (Du and Sun, 2006).

3.2 EFFECT OF MAP ON SAFETY OF FOODS OF ANIMAL ORIGIN

3.2.1 Meat/Poultry

Meat is a very complex ecosystem endowed with physical and chemical characteristics that can allow the colonization and the development of a great variety and number of organisms. Several studies on the microbial spoilage ecology identified *Brochothrix thermosphacta*, *Pseudomonas* spp., *Carnobacterium* spp., *Enterobacteriaceae*, *Lactobacillus* spp., *Leuconostoc* spp., and *Shewanella putrefaciens* as the predominant members of spoilage microflora in refrigerated meat (beef and pork) and meat products (Pennacchia et al., 2009). The role of food packaging in the food industry is being increasingly recognized in view of its numerous functions and its great importance with regard to enhancing product shelf life by delaying food quality degradation. Moreover, packaging of fresh red meat is carried out to avoid contamination, retard spoilage, allow some enzymatic activity to enhance tenderness, reduce weight loss, and wherever possible, to ensure a cherry-red color in red meats at retail or consumer level (Zakrys et al., 2009). The consumption of cooked meat products such as ham, turkey, and chicken breast displays a steadily increasing trend because of the growing consumers' interest in low-calorie meat products. The majority of these products are sold as sliced VP or MAP products. The exclusion or reduction of oxygen in MAP products by applying a barrier film extends the shelf life of meat by lowering down oxidative rancidity and microbial growth (Audenaert et al., 2009).

Overall appearance of retail meat cuts is the main factor consumers take into account when considering meat freshness and making purchasing decisions. It has been reported that the U.S. beef industry could avoid $520 million losses in annual revenue from retail sales by delaying surface discoloration of meats during retail display (Kim et al., 2009a). The level of O_2 in MAP has a direct impact on the color of the muscle pigment myoglobin. High concentrations of O_2 enhance the bright red oxygenated oxymyoglobin (OMB), low concentrations of O_2 the brown oxidized metmyoglobin (MMB), and anaerobic conditions the reduced purple deoxymyoglobin (DMB) (Sørheim et al., 2009). MAP systems of a high oxygen level (80%) are extensively used in retail meat markets to maintain the bright-red color of meat, which is appealing to the consumers. However, the inclusion of high oxygen levels will most probably enhance the incidence of oxidative changes in meat and consequently accelerate muscle surface discoloration, resulting in decrease in the desirable flavor and tenderness of meat. Moreover, meat with higher concentrations of OMB or MMB can develop brown color faster at a relatively lower cooking temperature leading to "premature browning" of cooked meat (Kim et al., 2009c). CO became highly relevant since its approval for use at 0.4% in MAP systems for red meats. Despite the fact that the color-stabilizing effect of CO on meat color has been extensively investigated, many fundamental concepts of carboxymyoglobin (COMb) redox chemistry are not entirely comprehended (Joseph et al., 2009).

The use of technologies that modify food environment for preservation purposes has been thoroughly examined in the past two decades. Despite the already accomplished technological advances, most of the MAP development has been based on empirical observations (Simpson et al., 2009).

3.2.2 Effect of MAP on the Microflora of Turkey

Recently, attention has been directed to the use of sucuk (dry, uncooked, cured, and fermented turkey sausage) as food carrier for probiotics because its production includes no heating and at the same time harbors high numbers of lactic acid bacteria (LAB). Identifying a single or a mixture of probiotic bacteria that inhibit the growth of spoilage and pathogenic bacteria is of increasing interest for research to improve the shelf life and safety of the meat products. Therefore, it is important to look into probiotics for potential applications in fermented meat products. The application of probiotics (*Lactobacillus acidophilus* and *Bifidobacterium lactis*) in sucuk manufacture decreases lipid oxidation (LO), total aerobic bacteria, LAB, and *Micrococcus/Staphylococcus* counts in both vacuum and MA (50% N_2+50% CO_2) packed and sliced sucuk samples. Probiotic and bioprotective *Lactobacillus rhamnosus* strains GG, LC-705, and E-97800 can result in high-quality dry sausage with low risk for *Listeria monocytogenes* or *Escherichia coli* O157:H7 (Kilic, 2009).

3.2.3 Effect of MAP on the Quality of Chicken

The combined effect of thymol (0–300 ppm), carvacrol (0–300 ppm), and temperature (0°C–18°C) on the quality of nonconventional poultry patties packed in air and MA (MAP: 40% CO_2; 30% O_2; 30% N_2) was studied using a simplex centroid mixture design. The patties were monitored for microbiological (TVC, *Enterobacteriaceae*, LAB, *Pseudomonas* spp.) physicochemical (pH, color), and sensory attributes. For the poultry patties packaged in MAP, the greatest log reduction was reported for *Pseudomonas* spp. over the entire storage period (Masdtromatteo et al., 2009). Polyamine contents were determined in chilled chicken meat and giblets (n=20) and skin (n=10) 24h after slaughter. Mean spermidine (SPD) values were 4.8, 10.2, 11.4, 48.7, and 12.1mg/kg and spermine (SPM) values were 36.8, 38.0, 24.3, 133, and 82.7mg/kg in breast, thigh, skin, liver, and heart, respectively. Significant statistical correlations between SPD and SPM contents were observed in breast, thigh, skin, and liver. A considerable enhancement of decrease in SPM to about 60% of the initial contents was observed in both VP- and in MA (20% CO_2 and 80% O_2)-stored breasts on day 21 at 2°C (Kozová et al., 2009b). In the design of novel raw products, three supplementary barriers were considered (ozonization, freeze-drying, and MAP) to obtain a new food product from chicken by using ozone to get a very hygienic product endowed with a high nutritional value. Afterward, freeze-drying would be applied. It was shown that it is possible to obtain freeze-dried poultry meat that looks and tastes similarly to fresh poultry meat. As for further research, the optimization of ozonization conditions and MAP is being carried out, to apply the barrier technology toward developing a long shelf-life lyophilized product (Babić et al., 2009).

3.2.4 Effect of MAP on the Microflora of Chicken

Campylobacteriosis in humans is caused by thermotolerant *Campylobacter* spp. This pathogen is one of the typical causes of zoonotic enteric infections in most developed and developing nations worldwide. Although it is generally accepted that there are numerous sources of *Campylobacter*, campylobacteriosis is predominantly believed to be associated with the consumption of poultry meat and, in particular, fresh broiler meat (Nauta et al., 2009).

There is rather limited information regarding the level of contamination of foods with *L. monocytogenes* at the point of consumption. Predictive microbiology was used to estimate the microorganism level at consumption based on known contamination frequencies and levels at production or retail, product formulation (e.g., salt/water activity, pH, other additives), times and temperatures between production and consumption, and the ecology of *L. monocytogenes* in foods, including lag times and the effects of LAB, in VP or MAP product (Ross et al., 2009). Zhang et al. (2009) investigated the antimicrobial activity of 14 spice extracts against 4 meat spoilage and pathogenic bacteria (*L. monocytogenes*). A rising trend was observed over the past 5 years in listeriosis incidence at the EU level. Most of the cases are sporadic and reported in the age group of 65 years and older (Uyttendaele et al., 2009). The results revealed that individual extracts of clove, rosemary, cassia bark, and liquorice contained strong antimicrobial activity, but the mixture of rosemary and liquorice extracts was the best inhibitor against most types of microbes.

An evaluation of the inactivation of foodborne pathogens inoculated on chicken breasts by UV-C treatment occurred when chicken breasts were inoculated with *Campylobacter jejuni*, *L. monocytogenes*, and *Salmonella enterica serovar*

Typhimurium at 6–7 log CFU/g. The inoculated chicken breasts were subjected to irradiation with UV-C light of dose 0, 0.5, 1, 3, and 5 kJ/m^2. Microbiological data indicated that the populations of the foodborne pathogens dropped considerably with rising UV-C irradiation. UV-C irradiation at 5 kJ/m^2 decreased the initial populations of the microorganisms by 1.26, 1.29, and 1.19 log CFU/g, respectively. After UV-C irradiation, the samples were individually packed using polyethylene terephthalate (PET) containers and stored at 4°C ± 1°C for 6 days. It was found that UV-C irradiation can be effective in improving the microbial safety of chicken breasts during storage, without affecting quality (Chun et al., 2009). Economou et al. (2009) investigated the effect of nisin and EDTA treatments on the shelf life of fresh chicken meat stored under MAP at 4°C. Chicken meat was subjected to several antimicrobial treatment combinations of Nisin–EDTA. N3, N4, N5, N6, and N7 affected populations of mesophilic bacteria, *Pseudomonas* sp., *B. thermosphacta*, LAB, and *Enterobacteriaceae*. The use of MAP in combination with antimicrobial treatments resulted in an organoleptic (acceptable odor attributes) extension of refrigerated, fresh chicken meat even up to 24 and 20 days of storage, respectively.

In chilled, sliced, cooked ham stored under vacuum or MAP, the contaminating microbiota is generally composed of psychrotrophic LAB, *B. thermosphacta*, and, occasionally, *Enterobacteriaceae*. The shelf life of such products is often restricted from 3 to 6 weeks. Metabolite production attributed to bacterial outgrowth and oxidation phenomena result in spoilage development of cooked ham. Cold chain variations related to distribution and consumer habits have a negative effect on the shelf life. Sliced, MAP artisan-type cooked ham was stored at different temperatures (4°C, 7°C, 12°C, and 26°C). It was found that application of SH-GC-MS effectively analyzed the volatile profiles developing in spoiled artisan-type cooked ham as a function of temperature and time (Leroy et al., 2009). The effect of different MAPs on the growth of *L. monocytogenes* on ham slices and fresh pork at 4°C is shown in Figure 3.1.

3.2.5 Effect of MAP on the Quality of Pork

The effect of irradiation (0, 5, and 10 kGy) on the oxidative and color stability of VP Iberian dry cured loin slices from pigs fed on concentrate feed (CON) or free-range reared (FRG) stored under refrigerated storage was examined by Cava et al. (2009b). Irradiation treatment augmented LO, measured as TBARS values and hexanal content of dry-cured loins and redness and lightness. Refrigerated storage diminished the differences because of irradiation treatment of instrumental color values like lightness. Storage enhanced the differences in TBARS values between irradiated and nonirradiated FRG dry-cured loin, whereas the opposite trend was reported for CON dry-cured loins. Moreover, no differences in the hexanal content were reported after 30 days of refrigerated storage. It was concluded that storage of Iberian dry-cured loin in the absence of oxygen by applying a VP could be an adequate method to minimize any changes related to irradiation treatment in Iberian dry-cured loin.

The evaluation of the effect of γ-irradiation and packaging on the lipolytic and oxidative processes in lipid fraction of Bulgarian fermented salami (30% brisket

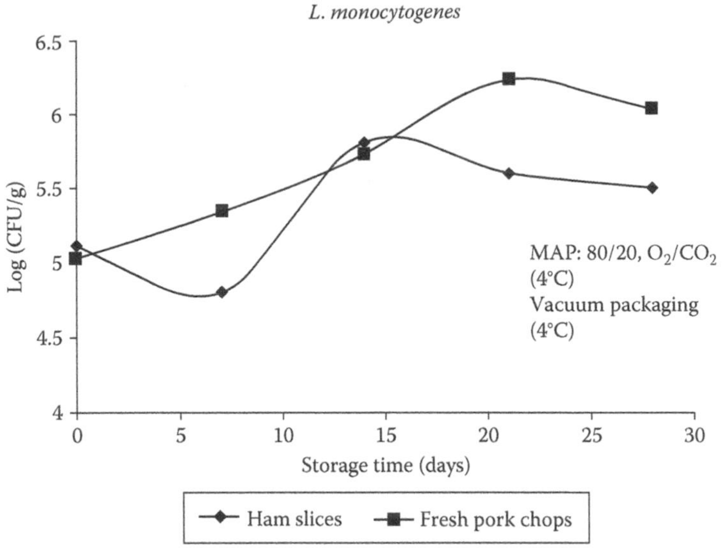

FIGURE 3.1 Effect of different MAPs on *L. monocytogene* growth of ham slices stored at 4°C and fresh pork chops stored at 4°C. (From Zhang, H. et al., *Meat Sci.*, 81, 686, 2009.)

and chuck beef mixed with 70% pork belly, collar and shoulder, filled in pig's small guts) during storage at 5°C (1st, 15th, 30th day) demonstrated no significant differences in the amounts of total lipids (TL), total phospholipids (TPL), and acid number (AN) for the VP samples of salami treated with 4 and 9 kGy during storage. The changes in TBA greatly depended on the irradiation dose applied and did not exceed 1.37 mg/kg in all groups. The most intensive lipolytic and oxidative processes and the lowest sensory assessment were recorded in the nonpacked irradiated (9 kGy) samples (Bakalivanova et al., 2009). Viuda-Martos et al. (2009) investigated the effect of orange dietary fiber (1%) (ODF), oregano essential oil (0.02%) (OEO), and the storage conditions (VP, air, and MAP) on the shelf life of bologna sausage (50% lean pork meat and 50% pork backfat; 15% water [ice, w/w], 3% potato starch [w/w], 2.5% sodium chloride [w/w], 300 mg/kg sodium tripolyphosphate, 500 mg/kg sodium ascorbate, 150 mg/kg sodium nitrite, spices [0.01% black pepper, 0.005% nutmeg, and 0.2% garlic powder]). Samples with ODF + OEO stored in VP revealed the lowest TBA values. ODF + OEO samples stored in VP displayed the lowest aerobic and LAB counts. The sensory evaluation scores were almost the same for samples with ODF + OEO and stored either in air or VP. ODF and OEO could be well applied in the food industry to prolong the meat products' shelf life. Cava et al. (2009a) assessed the effect of high-pressure (HP) treatments (200 MPa 15 min, 200 MPa 30 min, 300 MPa 15 min, 300 MPa 30 min) on color, lipid, and protein oxidation in sliced VP dry-cured Iberian ham and loin during refrigerated storage (90 days, +4°C). Both pressure level and holding time enhanced the extent of LO in both products. Dry-cured ham displayed a higher susceptibility to LO than dry-cured loin since HP treatment enhanced TBA-RS values in dry-cured ham samples whereas HP treatment diminished TBA-RS values in dry-cured loin samples. However, HP treatment did not affect protein oxidation in both meat products. HP treatment had a strong effect on instrumental color since nonpressurized dry-cured meat products displayed higher redness than pressurized ones. As regards the changes under storage, after 90 days of refrigerated storage lipid and protein oxidation augmented while redness dropped in both HP-treated and nontreated dry-cured meat products.

Morcilla de Burgos is the most famous blood sausage in Spain. Producers are interested in prolonging its shelf life, while the consumer demand for natural food displays a steady increase. This situation has resulted in the current search for new and mild preservation technologies. Two batches of four different products, control without any treatment, control with organic acid salts (CnOAS; a 3% mixture of potassium/sodium L-lactate), control with high hydrostatic pressure (HHP) processing (CnHPP; 600 MPa—10 min), and a combination of both treatments (OAS + HPP), were evaluated for any synergistic effect that may occur when combining OAS and HPP and the effect of different preservative treatments on the spoilage bacterial population and their evolution. Since HPP (with or without addition of OAS) cannot produce any negative changes on sensory properties, it would be considered the most suitable method for preserving morcilla de Burgos (Diez et al., 2009b).

Product samples of a Danish lightly fermented heat-processed cold cut pork product called "rullepølse" were stored under MAP (30% CO_2/70% N_2) (0, 28, and 34 days) and with subsequent aerobic storage (4 days) (MAP–OPEN) (4°C and 8°C). LAB and their corresponding metabolites were examined for sensory shelf-life indexing potential for the "rullepølse". Storage temperature was found to have a strong impact on the sensory characterized shelf life of "rullepølse" stored under MAP and MAP–OPEN conditions. The MAP stored "rullepølse" with subsequent 4 days storage in air (MAP–OPEN) could be effectively stored for at least 28 days at 4°C without displaying any apparent change in the sensory quality when opened. However, comparison of MAP-stored "rullepølse" at 8°C with subsequent open storage (MAP–OPEN) to the lower temperature revealed a decreased shelf life of less than 28 days should sensory quality of the "rullepølse" be maintained (Stolzenbach et al., 2009). Heterofermentative LAB contributes actively to the spoilage of morcilla de Burgos whenever identified as the main microbial spoilage group involved in the spoilage, especially in VP and MAP. Several previous studies concluded that *Weissella viridescens*, *Leuconostoc mesenteroides*, *Leuconostoc carnosum*, and *Weissella confusa* are the main LAB species in morcilla de Burgos. Some representative sensory changes taking place in VP morcilla are swelling of the packs, development of drip, milky exudates, slime formation and souring, and discoloration (Diez et al., 2009a).

3.2.6 Effect of MAP on the Microflora of Pork

MAP has been applied to obstruct the growth of a psychrotolerant toxin-producing *Bacillus* spp. during chill storage at 8°C and minimize the risk of emetic food poisoning at abuse condition. A model agar system mimicking a cooked meat product was applied in initial experiments. Incubation at refrigeration temperature of 8°C for 5 weeks of 26 *Bacillus weihenstephanensis* including two emetic toxin (cereulide) producing strains revealed that *B. weihenstephanensis* is sensitive to MAP containing CO_2. The susceptibility to 20% CO_2 depended strongly on strain and oxygen level, being enhanced in the absence of oxygen from the MAP. Results were validated in a cooked meat sausage model for two nonemetic and one emetic *B. weihenstephanensis* strain. The packaging film and oxygen transfer rate (OTR) were 1.3 and 40 mL/m^2/24 h, and the atmospheres were 2% O_2/20% CO_2 and "0%" O_2/20% CO_2. Oxygen availability had a large impact on the spore growth in MAP meat sausage; only the most oxygen restricted condition (OTR of 1.3 mL/m^2/24 h and "0"% O_2/20% CO_2) limited the growth of the strains during 4 weeks storage at 8°C (Thorsen et al., 2009). Cabeza et al. (2009) investigated the inactivation kinetics in the death of *Listeria innocua* NTC 11288 and *Salmonella enterica serovar Enteritidis* and *Salmonella enterica serovar Typhimurium* with e-beam irradiation in two types of VP dry fermented sausages to optimize the sanitation treatment of these products. A treatment of 1.29 kGy was considered adequate to reach the FSO based on the "zero tolerance" criterion. No irradiation treatment was necessary to meet the 10^2 CFU/g microbiological criterion for *L. monocytogenes*. No sensory changes were reported when the dry fermented sausages were treated with 62 kGy.

3.2.7 Effect of MAP on the Quality of Beef

Vaikousi et al. (2009) investigated the applicability of a microbial time–temperature indicator (TTI), based on the growth and metabolic activity of a *Lactobacillus sakei* strain in monitoring quality of MAP minced beef at conditions similar to that of the chill chain. At storage temperatures examined (0°C, 5°C, 10°C, 15°C), the results revealed that LAB were the dominant bacteria and can be well applied as a promising spoilage index of MAP minced beef. The results displayed that the end point of TTI, after storage at those fluctuating temperatures, was very close to the end of product's sensorial shelf life. This finding corroborates the applicability potential of the developed microbial TTI as a valuable tool for monitoring the quality status during distribution and storage of chilled meat products, spoiled by LAB or other bacteria showing similar kinetic responses and spoilage potential. The studies of Boselli et al. (2009) clearly corroborated the effect of the fluorescent light exposure and type of packaging (normal atmosphere and oxygen-rich atmosphere) on the oxidation parameters of raw beef slices in packed and refrigerated vessels. The concentration of COPs in meat treated under MAP ranged from 0.15 to 0.52 mg/100 g meat (average value of 0.27 mg COPs/100 g meat), twice as much as the average COP content (0.14 mg/100 g) of meat packed under air (0.04–0.27 mg COPs/100 g meat). The main cholesterol oxide was 7k, representing about one-third of the total cholesterol oxides, followed by 7b-OH (20%–25% of total COPs), 7a-OH (about 20%), and b-epoxy (12%–18%).

Two different bovine muscles—*Musculus longissimus lumborum* (LL) and *Musculus psoas major* (PM)—were injection-enhanced (n = 10, respectively) with solutions containing phosphate and potassium L- or D-lactate, cut into steaks, packaged with a high-oxygen (80% O_2) MAP, stored 9 days at 2°C and then exhibited for 5 days at 1°C. The parameters determined instrumentally were color, total reducing activity (TRA), lactate dehydrogenase (LDH) activity, and NADH. An increase in L-lactate resulted in less color deterioration and higher chroma values ($P<0.05$) than nonenhanced control of the bovine muscles. L-lactate rise significantly augmented NADH concentration and TRA of LL and PM than the nonenhanced control through increased LDH-B flux at 14 days. L-lactate increase can be effectively used for improving muscles with lower color stability in high-oxygen MA (Kim et al., 2009c). To study the effect of MAP on the internal cooked color of beef steaks, LL and PM muscles from 16 (n = 16) beef carcasses (USDA Select) were investigated on four enhancement treatments (noninjected control, distilled water–enhanced control, 1.25% and 2.5% lactate) and formed into 2.54 cm steaks. Steaks were packaged in high oxygen MAP (HIOX; 80% O_2 + 20% CO_2) and stored for 0, 5, or 9 days at 1°C. The interior cooked redness decreased over storage for steaks in VP and HIOX, whereas it was stable for steaks in CO. The findings indicated that the beef industry could use a combination of lactate enhancement and CO MAP to minimize premature browning in whole-muscle beef steaks (Suman et al., 2009). Sørheim et al (2009) studied the effects of freezing of raw materials, holding time for fresh raw materials *post mortem*, and addition of 0.5%–1.0% NaCl on the color of ground beef under low-oxygen (O_2) MA storage. The samples were subjected to 0.1%–3.0% O_2 at 4°C for up to 10 days and analyzed for O_2 concentrations and instrumental and visual color. Residual O_2 in the headspace of the packages oxidizes myoglobin and discolors the meat. Meat may be capable of scavenging residual O_2, and ground beef differs from intact muscles because of its much higher capacity for O_2 consumption. Using raw materials from 2 days rather than 7 days *post mortem* greatly enhanced the O_2 removal rate and improved redness. In the case of low-O_2 packaging, ground beef preferably should be stored 2 days in an atmosphere with 0.1% residual O_2 to produce a purple pigment predominantly consisting of DMB. Zakrys et al. (2009) studied the physiochemical changes of beefsteaks packed under various gas compositions and the relationship between consumer perception of flavor and acceptability of MA-packed beef steaks during retail display. Experimental gas atmospheres included 40%, 50%, 60%, 70%, and 80% O_2, with all packs containing 20% CO_2 and the rest N_2 (4°C, 12 days) and the samples tested for lipid and protein oxidation, heme iron, color, OMb concentration, Warner–Bratzler shear force (WBSF), and consumer acceptability of the resulting cooked meat. The results from 134 consumers revealed a directional preference for the steaks stored in packs containing 40% and 80% O_2.

Fifteen USDA Select beef strip loins were divided individually into four equal width sections, and one of six treatments containing phosphate and/or calcium lactate (CAL) enhancement solutions were attributed randomly to each loin section (n = 10). Steaks from each loin section were packed with high-oxygen (80% O_2) MAP and/or irradiated at 2.4 kGy, stored 10 days, and then displayed for 5 days at 1°C. Loins with CAL and phosphate maintained the most stable red color, enhanced NADH, and were the least oxidized. Among irradiated steaks, CAL with phosphate treatment significantly minimized LO, increased NADH and TRA, and consequently had a higher a* value. These results clearly displayed that lactate inclusion improves the color stability of fresh beef by providing superior antioxidant capacity (Kim et al., 2009b).

Ramamoorthi et al. (2009) evaluated the combined effects of irradiation and carbon monoxide in MAP (CO-MAP) on TPC, *E. coli* K12, color, and odor of fresh beef during refrigerated storage. Beef was packed aerobically or in CO-MAP, irradiated at 0, 0.5, 1.0, 1.5, or 2.0 kGy, and then held at 4°C for 28 days. Raw beef odor decreased and acid/sour, rancid, and grassy odors became more pronounced starting on day 14. However, initially, no difference was recorded for visual green color scores due to gas atmosphere. After 14 days of storage, aerobically packaged beef was greener and less red than CO-MAP packaged beef. On day 0 and thereafter, no coliforms were detected after irradiation (1.5 or 2.0 kGy). These findings suggest that CO-MAP could be used to preserve color of beef irradiated at doses adequate to decrease microbial loads to safe levels during 28 days of storage.

Fourier transform infrared (FTIR) spectroscopy was used to determine the biochemical changes within fresh minced beef in an attempt to rapidly monitor beef spoilage. Minced beef was packed either aerobically, under MAP, and using an active packaging. Qualitative interpretation of spectral data was conducted and used to corroborate the obtained sensory data and to accurately assess sample freshness and packaging. Partial least-squares (PLS) regressions provided estimates of bacterial loads and pH values from the spectral data with a fit of $R_2=0.80$ for total viable counts (TVC) and fit of $R_2=0.92$ for the pH. The collected data confirmed that a FTIR spectrum could be seen as a metabolic fingerprint and the method in tandem with chemometrics is a very powerful, promising, rapid, economical, and noninvasive method for monitoring minced beef freshness with respect to the storage conditions (Ammor et al., 2009).

3.2.8 Effect of MAP on the Microflora of Beef

Dietary polyamines such as putrescine (PUT), SPD, and SPM participate in numerous human physiological processes, including tumor growth. Reliable information on their contents in foods is therefore urgently needed. Nine experiments with beef loin (LL) were conducted. Loin cuts were stored at −18°C for 178 days or beef was stored aerobically, VP, and packaged in an MA; 70% N_2 and 30% CO_2, at +2°C for 9, 21, and 21 days, respectively. In loins stored under MA, the mean decrease in SPM content was slightly more extensive than in the VP loins and levels reached approximately 80% of the initial value on final day 21. Although the differences among the cooking treatments were not significant, differences were observed among the loins used (Kozová et al., 2009a). Buffalo meat packaged under MAP was analyzed with regard to its structural and physical parameters. Variations in cooking time and temperature were investigated for collagen solubility of *Semimembranosus* muscle (SM) in beef. Following the comparison of physicochemical and functional properties of meat obtained from three different groups of buffaloes, it was shown that young male buffalo meat is more suitable for processing in chunks while spent male and female buffalo meat is more appropriate for processing smaller particles (Kandeepan et al., 2009).

3.2.9 Effect of MAP on the Quality of Lamb

The effect of different stunning methods (using two different CO_2 concentrations and exposure times) on lamb meat quality was investigated on 49 Manchenga breed male lambs. The lambs were allocated to five stunning treatments including four CO_2 treatments (80% CO_2 for 90 s [G1]; 90% CO_2 for 90 s [G2]; 90% CO_2 for 60 s [G3]; 80% CO_2 for 60 s [G4]) and an electrically stunned control group (G5). Meat quality was assessed by testing pH, color (L*, a*, b*, chroma, hue values), water holding capacity (WHC), cooking loss (CL), shear force (SF), drip loss (DL), and total aerobic bacteria. Both G2 as G3 could be recommended as suitable for stunning suckling lambs since the highest stability with aging time on meat quality was found using 90% CO_2 (Bórnez et al., 2009a). Figure 3.2 displays the effect of MAP on the LAB growth of bolognas and Manchega lamb at two different temperatures. Vergara et al. (2009) investigated the impact of different gas stunning methods (concentration of CO_2/time of exposure (G1: 80% 90 s; G2: 90% 90 s; G3: 90% 60 s; G4: 80% 60 s) on the meat quality of Manchego breed light lambs (25 kg live weight) assessed by pH, colorimetric parameters, WHC, CL, DL, SF, and LO. An electrically stunned control group (G5) was also used. Vergara's et al. (2009) stunning method had a strong impact on pH values. The lowest pH was determined at 24 h postslaughter in G1 and the highest one on G5. The greatest drop in pH (pH 0–pH 24) was found in G1 and G5 while the smallest in G3. In general, values of color coordinates, WHC, and DL were similar in all groups. The stunning method affected CL at 7 days postslaughter, with the lowest values being recorded in G1. Significant differences among groups were reported for SF values at both postmortem times, with less tender meat in groups stunned with 80% CO_2, especially in G1. A significant effect depending on the type of stunning was recorded at 24 h on LO, with the highest value in G5. The lowest value for this parameter was recorded in G1 and G4.

The effect of the following stunning methods (80% CO_2 for 90 s [G1]; 90% CO_2 for 90 s [G2]; 90% CO_2 for 60 s [G3]; 80% CO_2 for 60 s [G4]) (TS) plus an electrically

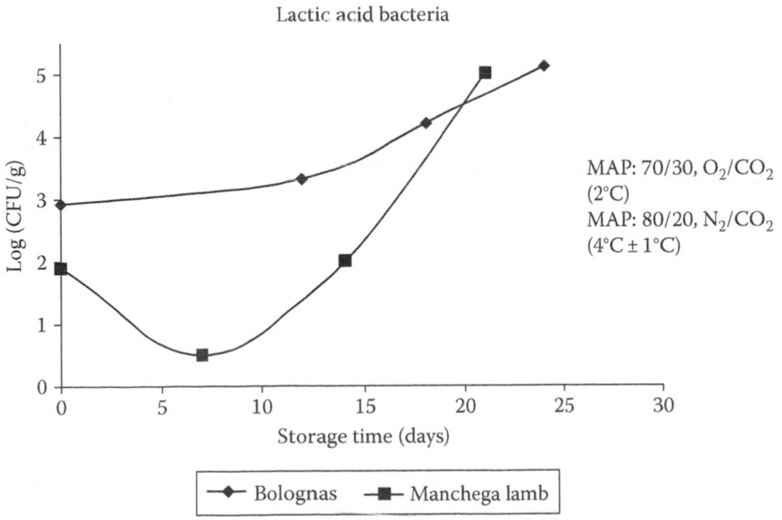

FIGURE 3.2 Effect of different MAPs on LAB and growth of bolognas stored at 4°C ± 1°C (From Viuda-Martos, M. et al., *Food Control*, 21, 436, 2009.) and Manchega lamb stored at 2°C. (From Bórnez, R. et al., *Meat Sci.*, 83, 383, 2009b.)

stunned control group (G5) on both LO and microbial counts were assessed in samples packed under two different types of modified atmospheres (MA: MA-A: 70% O_2+30% CO_2; MA-B: 69.3% N_2+30% CO_2+0.7% CO) at 7, 14, and 21 days post-packaging. Both factors (TS and MA) significantly affected LO, which was highest in the samples from the MA-A/G4 group. No significant differences in microbial quality between MA were reported. However, it was found that the applied method of stunning affected microbial count at all analysis times. G4 and G5 showed the highest level in all microorganisms assessed, while the rest of the gas-stunning groups showed more stability with aging (Bórnez et al., 2009b). Over 105 days, seven lambs were fed a concentrate-based diet (C), whereas the remaining animals received vetch (*Vicia sativa*; H) harvested daily and given fresh. LO was determined in both minced cooked meat (SM) over 4 days of aerobic refrigerated storage and on minced raw meat stored over 14 days in a high oxygen atmosphere. It was concluded that under conditions that promote oxidative stress in meat, a herbage-based diet can improve the oxidative stability of meat compared with a concentrate-based diet (Luciano et al., 2009).

3.2.10 Effect of MAP on the Lamb Microflora

Fresh Souvlaki-type lamb meat was packaged in VP and MAs and stored under refrigeration (4°C) for a period of 13 days. The following gas mixtures were used: M1: 30%/70% (CO_2/N_2) and M2: 70%/30% (CO_2/N_2). Identical samples were aerobically packaged and used as control samples. Of the two MAs and VP used, gas mixture M2 and VP were the most effective treatments for the inhibition of TVC, *Pseudomonas* spp., yeasts, and *B. thermosphacta* in Souvlaki meat. LAB and *Enterobacteriaceae* were also determined in the microbial flora of Souvlaki and enhanced during storage under all packaging conditions applied. Based on microbiological analysis data and on the proposed a* values, the use of VP and MAP (M2: 70% CO_2/30 N_2) extended the shelf life of "Souvlaki" meat stored at 4°C by approximately 4–5 days compared with aerobic packaging (Soldatou et al., 2009).

The effect of MAP on sensory assessment and microflora of various meat products is given in Table 3.1.

3.3 DAIRY

Milk is one of the few foodstuffs consumed without further processing. Nutritionists worldwide agree that it is of great value in not only promoting growth and development of children and young animals but continues to be essential in our diets right throughout our adult life (Chandan, 2008). The most important proteins in milk are the whey proteins such as serum albumin, immunoglobulin, α-lactalbumin, and β-lactoglobulin. Denaturation of the latter (due to heat treatment) is strongly related to many functional properties of dairy products (De Jong, 2008). The extent of denaturation is about 10% during pasteurization, 70% during UHT treatment, and 75% during in-bottle sterilization. However, this denaturation hardly affects the nutritive value of the whey proteins for the human infant, and UHT milk has been extensively used in infant feeding (Rolls and Porter, 1973). Separation technologies are the basis

TABLE 3.1
The Effect of MAP on Shelf Life, Microflora, and Sensory Assessment of Meat, Poultry, and Meat Products

Food Type	MAP	Other Technologies	Storage Temperature	Shelf Life	Sensory Assessment	Microflora	References
Chicken meat	65% CO_2/30% N_2/5% O_2	—	4°C	—			Economou et al. (2009)
		500 IU/g Nisin		1–2 days extension			
		1500 IU/g Nisin		3–4 days extension		Mesophilic bacteria, *Pseudomonas* sp., *B. thermosphacta* LAB, and *Enterobacteriaceae*	
		500 IU/g Nisin–10 mM EDTA					
		1500 IU/g Nisin–10 mM EDTA		7–8 days extension			
		500 IU/g Nisin–50 mM EDTA		13–14 days extension	Acceptable odor for more than 24 days		
		1500 IU/g Nisin–50 mM EDTA		9–10 days extension	Acceptable odor for more than 20 days		
Dry-cured Iberian hum	Vacuum 60% N_2/40% CO_2 70% N_2/30% CO_2 80% N_2/20% CO_2 70% argon/30% CO_2		4°C ± 1°C	120 days	Loss of intensity of red color; slices at MAP 60/40 and 70/30 showed lower a* values than rest of batches after 60 days	*Enterobacteriaceae*, aerobic bacterial counts, *E. coli*, molds, yeasts	Parra et al. (2010)

(continued)

TABLE 3.1 (continued)
The Effect of MAP on Shelf Life, Microflora, and Sensory Assessment of Meat, Poultry, and Meat Products

Food Type	MAP	Other Technologies	Storage Temperature	Shelf Life	Sensory Assessment	Microflora	References
Ground beef patties	Vacuum 80% O_2/20% CO_2 0.4% CO_2/30% CO_2 69.6% N_2	25% lactate w/w	2°C	0.2 or 4 days (time of storage)	Lactate improved color stability of PVC, high-oxygen, and vacuum samples. Patties packaged in high O_2 showed premature browning		Mancini et al. (2010)
Dry-cured hum	$[O_2]<4.5\%$	—	2°C–4°C	275 or 289 days	The use of MAP increased the L^* color parameter in the subcutaneous fat and proteolysis index and decreased b^* in the external part of subcutaneous fat and cholesterol oxide concentration. MAP with low RH retarded microbial growth and prevented mite growth	Bacteria, fungi, *Tyrophagus putrescentiae*, *Penicillium*, *Aspergillus*, LAB, *B. thermosphata*, *semitendinosus*, *Rectus fermoris*	Sánchez-Molinero et al. (2010)

for adding value to milk through the production of proteins for the food industry. There has been a great progress in the development of functional and nutritional attributes for milk lipids, calcium delivery, and nutraceuticals (Huffman and Harper, 1999). According to De Jong (2008), "Different species of mammals produce milk with widely varying composition. Obviously, cows' milk accounts for the vast bulk of dairy production, but goats, sheep, and even buffalo are milked commercially. The last two species provide milk with higher solids than cows' milk."

Four fifths of milk protein consists of casein, actually a mixture of four proteins: αS1-, αS2-, β-, and κ-casein. Moreover, milk contains numerous minor proteins, including a wide range of enzymes (Walstra et al., 2006). Dairy products are often exposed to light during retail storage and display. Light initiates oxidation processes in foods, resulting in losses of valuable nutrients, discoloration, and formation of off-flavors from compounds such as aldehydes, ketones, methional, and dimethyl disulfide. Light-induced oxidation requires both oxygen, the presence of a light source, and a photosensitizer (riboflavin [vitamin B_2]) in order to occur. Dairy products are well known for being very sensitive to light, due to a high amount of the strong photosensitizer, riboflavin (vitamin B_2) (Pettersen et al., 2005). Quality changes are apparent in milk due to the presence of light and residual oxygen levels even as low as 0.5% in headspace. Spectral distribution and photon flux of the light source determine the extent of quality changes, since photochemical processes have limited temperature dependence, in contrast to the consecutive lipid auto-oxidation process (Mortensen et al., 2004). Solid-sample fluorescence has been used as a nondestructive and rapid tool to determine the degree of light-induced degradation of riboflavin in dairy products. Images of fluorescence can be used to visualize the intensity and propagation of this process. Solid-sample fluorescence can be used as a nondestructive and rapid tool to determine the sensory properties related to the storage of dairy products (Wold et al., 2002). Concentrations of retinol, TH, and a tocopherolquinone have been determined in dairy products by HPLC according to a published procedure (Bergamo et al., 2003). Some individuals display allergic responses to one or more proteins. It has been reported that symptoms vary widely, from fairly mild ones like rhinitis or diarrhea, to more serious ones like dermatitis or asthma. It has been calculated that about 2% of babies display a cows' milk allergy, the intensity of which becomes less sensitive later on. Although both caseins and serum proteins can cause allergic reactions, β-lactoglobulin is considered responsible for most incidents in milk. Heat denaturation may downsize the allergenic properties; however, substantial hydrolysis of the protein is more effective. It was also shown that a potential substitution of cows' milk with goats' milk hardly has any effect on the allergic reaction (Walstra et al., 2006). In view of the fact that limited unbiased research was carried out to provide evidence and decrease discrimination against goats and substantiate the medical benefits from goat milk consumption, further research is urgently required (Haenlein, 2004).

An important feature of quality assurance in modern dairy processing plants is the ability to predict rapidly the potential shelf life of the finished products. Methods designed to estimate the food shelf life include bioluminescence, impedance microbiology, limulus amoebocyte lysate (LAL), direct reflectance colorimetry, Virginia Tech shelf-life procedure, and the Moseley keeping quality test (White, 1993). Dairy processing involves chemical, microbiological, physical, and engineering principles, and it is imperative to understand them for effective management of a dairy plant. Moreover, meeting consumer expectations by controlling the processes to deliver quality, safety, and shelf life of the products is of utmost importance to successful dairy processing operation. Major advances in dairy processing have resulted in considerable improvement in the safety and quality of products (Kailasapathy, 2008). The consumption of fermented milk products (i.e., yogurts) dates back to several thousand years. However, it is only in recent years that scientific support for these beliefs has begun to build. Fermented milk products are made and are rich in protein, vitamins, and minerals. Further to these purely nutritional properties, there is increasing support for a number of other health advantages. For example, while the support for improvement in tolerance of lactose by maldigesters of this disaccharide is now very robust, evidence to support the suggested ability of fermented milk products comprising particular bacterial cultures to reduce cancer risk is still at a very premature storage (Buttriss, 2007).

Yogurt's high nutritious capability is due to the high level of milk solids in addition to nutrients developed during the fermentation process. The various forms of yogurt currently available in the market are known as stirred, set, frozen, and liquid yogurt. Yogurt can be stored for up to 4 weeks. Over this time, the product gradually undergoes physicochemical and rheological changes that may severely affect its organoleptic quality (Fadela et al., 2009). During storage of yogurt powder, many chemical changes occur and starter culture counts decrease. TBARS is frequently determined to characterize oxidative processes in milk powder while hydroxymethyl furfural (HMF) has been used to evaluate the nonenzymatic browning in nonfat dried milk under normal or accelerated storage conditions (Kumar and Mishra, 2004).

The shelf life of nonsterile dairy products, including pasteurized milk, cottage cheese, and some types of yogurt and fermented milk products, is often limited to 1–3 weeks. Spoilage results primarily from the growth of organisms surviving pasteurization and postprocessing microbial contamination and degradative enzymes surviving the high temperature short time (HTST) process (Hotchkiss et al., 2006). Among the methods applied to enhance the shelf life of the product are addition of gas, application of preservatives, heat treatment after incubation, and carbonation. The latter is of low cost and safe and does not impart any negative effects to dairy products (Karagül-Yüceer et al., 2001). In dairying, several microorganisms are considered to be harmful, for example, spoilage organisms and pathogens whereas others are beneficial, for example, cheese and yogurt starters and yeasts and molds used in controlled fermentations and cheesemaking. The main microorganisms occurring in the dairy industry are bacteria, yeasts, molds, and viruses (O'Connor, 1995). Milk and dairy products constitute a suitable environment for the growth of various gram-negative rod-shaped bacteria (e.g., *Pseudomonas* spp., coliforms), gram-positive spore-forming bacteria (e.g., *Bacillus* spp., *Clostridium* spp.), lactic-acid-producing bacteria (e.g., *Lactococcus* spp.), yeasts, and molds (Van Asselt and Te Giffel, 2009). It has been found that many foodborne diseases are caused by the harmful bacteria *L. monocytogenes* Scott A. Even if *L. monocytogenes* has been found at relatively low rates in raw milk (2.2%), it can be generally observed at a wide range of rates

(0%–45.3%). There is no evidence that the outbreaks of human listeriosis are directly associated with raw milk intake, even if it seems that two of the most principal outbreaks causing many deaths have occurred due to consumption of contaminated raw milk (Mussa et al., 1999).

A photosensitizer along with the presence of oxygen is usually applied to oxidize dairy components. Photosensitivity and photodegradation of yogurt, milk, and other dairy products are closely associated with riboflavin. This vitamin can play the role of a photosensitizer in its capability to absorb light. Oxygen can be thereby activated, inducing irreversible changes to nutrients (i.e., proteins and lipids can be degraded). Several studies have shown that both light and oxygen favor riboflavin degradation. These chemical reactions lead to several organoleptic changes in dairy products, such as discoloration and reduction of nutrient value (Becker et al., 2003).

Four different categories (processed cheese, processed cheese food, processed cheese spread, processed cheese analogues) can be used to characterize cheese products after processing. Because of the great variety of these products, the latter can be divided in appropriate categories. It is noteworthy that ingredients of many cheese products have anticarcinogenic properties. A typical example is the conjugated linoleic acid (CLA) the concentration of which in foods is not affected by any industrial processing (Henning et al., 2006).

Modern yogurt packaging industries have initially focused on materials such as HDPE (high-density polyethylene) and recently on PP (polypropylene). Despite the significant advantages that these materials offer, they can contribute to the pollution of the environment as, even though they can be recycled, they are not accepted by recycling facilities in the United States (Keoleian et al., 2004).

Products stored under MAP have a significantly extended shelf life even when perishable products (meat, fishes, milk, etc.) are treated. During MAP application on food products, it is essential to take into account not only the interactions of the packaging material with the environment but also any possible migration from packages to food. In addition, product respiration can dramatically alter the percentages of gases contained in the package, leading to the creation of inadequate storage conditions (Rodriguez-Aguilera et al., 2009).

3.3.1 Raw Milk

A very important factor during storage of foods at low temperatures is the operation of mechanisms that protect products from psychrotrophic microorganisms. MAP is an effective way to protect products at low or higher temperatures. It consists of package air substitution by gases like CO_2 and N_2 thereby creating unfavorable conditions for the growth of harmful microorganisms (Munsch-Alatossava et al., 2010).

Dechemi et al. (2005) examined the effect of five different MAPs on effective shelf-life extension of raw milk. The samples remained at 7°C for 10 days. Data extracted after the experiment showed that the combination of CO_2 and N_2 was able to effectively inhibit psychrotrophic microorganisms. Specifically, best results were observed at the concentrations of 50% CO_2 and 50% N_2. Under these conditions, no enzymatic activity, as concerns protease and lipase, was observed, rendering it clear that this MAP is the most appropriate for the safe preservation of raw milk.

3.3.2 Milk and Cream Powder

Shelf life, uniformity, and quality of milk powders are the most important factors that play an important role in the price of these products. Nowadays, milk powder manufacturing companies are trying to achieve the highest possible prices. A few variations between the batches of powders can be one of the main obstructive factors in this target. These variations can occur even after adherence to the protocol that can include control of raw materials, CCPs during the whole process, and systematic chemical analysis (Nielsen et al., 1997).

Treatment of milk powders using spray-drying procedures is widely applied. Such procedures (direct-fired heating and increased rates of nitrogen oxides) and their effects, in combination with various packages (PE, crimp-sealed vials, oxygen absorbers application) were examined. After 6 months of storage, it was concluded that O_2 absorbers contributed to the effective reduction of cholesterol oxidation. The use of low nitrogen-oxide drying processes, in combination with oxygen impermeable packages, was able to significantly enhance the stability of the products (Chan et al., 1993).

According to Gallagher et al. (2003), a bread formulation without gluten was enriched with 3% milk protein isolate and 3% novel starch to study the effect of these two ingredients on the product. The bread was stored under MAP (80% CO_2/20% N_2), and samples were taken for examination on the 8th and 43rd day of storage. It was concluded that the two additives enhanced loaf volume, while significant differentiations in the hardness of the crumb were observed during the first days of the testing periods for the control bread without gluten in both examinations after 8 and 43 days.

CO_2 is widely used to protect food from harmful microorganisms. A prerequisite for the use of MAP is the use of containers that are capable of holding the gas under storage conditions. CO_2 sensory threshold has been identified at >2.8 mM and <9.1 mM in pasteurized milk. The product was inoculated with several microorganisms and stored at 6.1°C for approximately 28 days. For the packaging of milk, various barrier film pouches were used to evaluate the appropriateness of each package. Increasing levels of CO_2 effectively delayed the microorganisms' growth, in comparison with control milk. Although control milk curdled in approximately 17 days, when high-barrier packages were used, it had not curdled at 28 days. Scientists concluded that the shelf life of the product can be extended up to 200%. The most important role as concerns the extension of milk shelf life is the CO_2 amount and the properties of packaging material (Hotchkiss et al., 1999).

Light-induced oxidation and auto-oxidation phenomena, were examined after volatile profiles observation of pasteurized milk. The product was stored under fluorescent light and in different packages for 7 days at 4°C. It was observed that in samples exposed to light, light-induced oxidation rate was significantly higher than that of auto-oxidation, whereas the opposite occurred in products stored in darkness. Dimethyl disulphide, pentanal, hexanal, and heptanal were used as markers to measure the milk quality. The shelf life of the product was identified at 5 days (Karatapanis et al., 2006).

3.3.3 Fresh Cheese

When converting milk into cheese, various reactions take place mainly due to bacterial growth. These bacteria can be separated into two categories: the starter bacteria like *Lactococcus* and different adventitious microbes. The first category consists of bacteria added to raw material to produce the final product. Bacteria of the second category may exist in the processing environment and can affect the product during the procedure. Examples of such bacteria can be found in *Lactobacillus*, *Pediococcus*, *Enterococcus*, *Leuconostoc*, *Micrococcus*, and *Staphylococcus* (Ogier et al., 2002).

3.3.4 Cameros Cheese

The province of La Rioja, in Spain, is the area of Cameros cheese production. This cheese is made from pasteurized goat's milk and its production is increasing because of various programs that many governments in Europe have established to promote agricultural production (Olarte et al., 2001).

The shelf life of this cheese, stored under five different MAPs, was examined. CO_2 and N_2 gases were used in various proportions to preserve the product while the control samples remained in air. The cheese was stored at 3°C–4°C, and various microbiological and physicochemical controls as well as quality controls took place. The acidity and the weight loss were similar as concerns the packaging in vacuum and air, but when a MAP of 100% CO_2 was applied, the highest weight losses and the lowest pH occurred while it terminated the growth of different microorganisms, protecting the product effectively and extending its shelf life. It is important to mention that cheese packaged in air had a shelf life of just 7 days. It was concluded that MAPs consisted of 50% CO_2/50% N_2 and 40% CO_2/60% N_2 presented the best results, preserving the sensory quality and extending the shelf life of the product (Gonzales-Fandos et al., 2000).

Olarte et al. (2002) examined the growth of the bacterium *L. monocytogenes* in inoculated and noninoculated Cameros cheese. Samples were packaged under three different MAPs (20% CO_2/80% N_2, 40% CO_2/60 % N_2, and 100% CO_2) and in air. The storage temperature remained at 4°C, and various controls took place systematically. It was clear that samples stored under MAP presented a significantly longer shelf life due to CO_2, which effectively protected products from mesophiles, psychrotrophs, and anaerobes. By increasing the level of CO_2, a reduction in microbial population was observed, and in 100% CO_2, *L. monocytogenes* presented the lowest counts. Nevertheless, after 28 days of storage *L. monocytogene* counts were significantly lower in inoculated samples stored under MAP of 100% CO_2 in comparison with those stored in air conditions. The use of MAP therefore is not capable of protecting Cameros cheese against *L. monocytogenes*.

3.3.5 Cottage Cheese

Cottage cheese, though being a minor product, is a product with a significantly high added value. During the manufacturing process, a curd is formed, which is annealed, and finally a cream dressing is used to coat it. The curd formation takes place with the help of lactic starter bacteria, which acidify skim milk. After its formation, curd has to be washed and cooked, and then a cream dressing with or without fruits, herbs, or even spices has to be added (Muir, 1996).

Chen and Hotchkiss (1991) investigated how CO_2 affects the life cycle of psychrotrophic spoilage bacteria in 2% fat creamed cottage cheese. For the packaging of the product, sealed glass cans were used and the cheese inoculated with psychrotrophic spoilage bacteria remained at 4°C and 7°C for 80 days. An acceptable product appearance remained for 80 days at 4°C or 60 days at 7°C. Ever since the first days of the storage, samples stored in CO_2 had more than 100-fold more colony-forming units than samples packaged in air. It was proved that dissolved CO_2 inhibited very effectively the Gram-negative bacteria, indicating that this gas can effectively preserve cottage cheese from these microorganisms when high-barrier cans are used.

Chen and Hotchkiss (1992, 1993) studied the effect of CO_2 and different packages on the growth rate of various microorganisms in stored cottage cheese. Specifically, they studied three strains of *L. monocytogenes* and *Clostridium sporogenes* ATCC 3584, while the product was packed in polystyrene tubs overwrapped with or without high-barrier heat shrink film. MAP (CO_2 35%) was applied in some of the samples, and the rest of them were packaged without CO_2. A significant reduction (by 1/3) in the CO_2 concentration over the 63 days of storage at 4°C was observed. This method inhibited the growth of *C. sporogenes*. It was concluded that cottage cheese preservation with CO_2 does not pose risks for listeriosis or botulism, but the product itself may function as a carrier of *L. monocytogenes*.

The quality of cottage cheese, stored under three different MAP conditions (100% CO_2, 75% CO_2:25% N_2, 100% N_2) at 4°C for 28 days, was examined. Barrier cans were used, and tests included several sensory, microbiological, and chemical examinations. MAP implementation inhibited the growth of the population of psychrotrophic and LAB while retaining the color of the cheese. An increase in the acidity was observed, but this phenomenon did not take place due to microorganisms. Finally, it was concluded that by implementing a MAP of 100% CO_2, organoleptic characteristics of the product remained at a satisfactory level, rendering the product acceptable even after 28 days (Maniar et al., 1994). Mannheim and Soffer (1996) studied the effect of flushing the headspace with pure CO_2 over the shelf life of cottage cheese. The most effective headspace (25% [v/v] in 250 mL packages) was able to extend the shelf life of the product by about 150% (at 8°C) while its quality characteristics remained unalterable. Adding CO_2 to the headspace of the packages protected cheese from harmful microorganisms more effectively. High-barrier packages were used to ensure the appropriate amount of CO_2 in the headspace as well as its dissolution inside the cheese.

3.3.6 Requeijão Cremoso

Pintado and Malcata (2000) evaluated the extent to which the use of MAP affects the organoleptic characteristics of Portuguese whey cheese, Requeijão. The response surface approach was applied to conduct the study, taking into account storage time, temperature, and fraction of CO_2 in the flushing gas. The sensorial optima were monitored regularly proving that the optimal storage temperature was at 4°C, as under these conditions no significant lipolytic reactions occurred. Application of plain CO_2 may play a role in stabilizing the product and protecting it from extensive lipolysis until the 15th day of storage. It was concluded that packaging is essential for the proper maintenance of Requeijão. The temperature of 4°C is best suited for optimizing the quality of the product.

Packaging commonly used for retail sales of cheese Requeijão cremoso is transparent and, as a consequence, the product is exposed to light. As a result, many photochemical reactions take place altering the environment inside the package. The purpose of the study of Alves et al. (2007) was to examine the effect of light (1000 lx) on cheese. The product was packaged in various packages like glass cup with a tinplate cap, glass, PP cups heat-sealed with aluminum foil, PE squeeze tubes with O_2 barrier, and PE squeeze tubes without O_2 barrier. From a microbiological point of view, the product remained unchanged while significant changes were observed in O_2 content due to its consumption by photochemical reactions. It was concluded that the quality of the product decreased significantly when stored in transparent containers and was maintained satisfactorily when stored in dark. The major factors that affected the quality degradation were the packages and the rate of O_2 inside the package.

Pintado and Malcata (2000) examined the population variations of different microorganisms after the application of three MAPs (100% CO_2, 50% N_2/50% CO_2, 100% N_2) to Requeijão cheese. The method of surface response was used to carry out the research. The product was stored for 15 days at 4°C, 12°C, and 18°C, and samples were taken for examination at the 2nd, 6th, 10th, and 15th day. At elevated CO_2 concentrations, no significant changes in bacterial population were observed, and this method did not inhibit the growth of *Enterococci*, *Bacillus*, *Pseudomonads*, *Lactobacilli*, and *Streptococci*. It is important to mention that a MAP of 100% N_2 at 4°C effectively protected the product from *Staphylococci*, *Lactobacilli, and Bacillus* during the first 2 days of the experiment. It was concluded that the shelf life of cheese can be extended until 15 days at 4°C and 100% CO_2 without significant deterioration of product quality.

3.3.7 Giuncata and Primosale Cheese

Gammariello et al. (2009) examined the sensory properties of Giuncata and Primosale cheeses and the microorganisms that grow on products during packaging under vacuum, various MAPs, or in air. These cheeses are Italian products made from cow and goat milk, respectively. It was proved that the best preservation of Giuncata cheese was achieved using MAP (75% CO_2/25% N_2). Under these conditions, the growth of harmful microorganisms was delayed without having any effect on dairy microflora and quality of products. Primosale cheese could not be maintained effectively by any of the methods tested.

3.3.8 Fior di latte

According to Del Nobile et al. (2009c), Fior di latte cheese can be preserved for a longer period of time with the addition of chitosan and active coating, under MAP

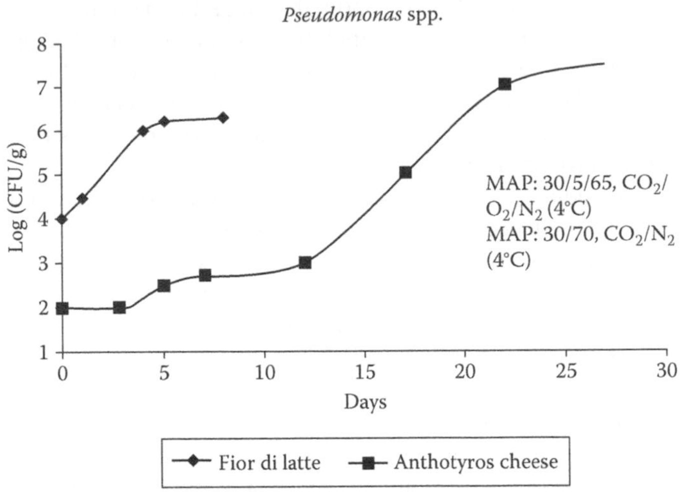

FIGURE 3.3 Effect of different MAPs on *Pseudomonas* spp. growth of Fior di latte stored at 4°C (From Del Nobile, M.A. et al., *Carbohydr. Polym.*, 78, 151, 2009c) and Anthotyros cheese stored at 4°C. (From Papaioannou, G. et al., *Int. Dairy J.*, 17, 358, 2007.)

conditions. Scientists examined if chitosan and coating or active coating can effectively preserve the product extending its shelf life. The product was stored at 4°C and various controls took place for more than a week. These controls included any variations of sensorial properties, pH, gas composition, and microbiological tests. Compared with traditional methods of preservation, a significant expansion of cheese lifetime was achieved while maintaining quality at high levels. Thus, it is clear that this preservation method can be used to extend the storage time of this vulnerable product. In Figure 3.3, the effect of different MAP conditions on *Pseudomonas* spp. growth of Fior di latte and Anthotyros cheese at 4°C is demonstrated.

In another investigation, Conte et al. (2009) used active coating under MAP (30% CO_2/5% O_2/65% N_2) conditions but without the addition of chitosan to achieve a proper conservation of Fior di latte cheese for a longer period of time. The active coating consisted mainly of sodium alginate, the enzyme lysozyme, and EDTA. The product was stored at 10°C and samples were taken regularly to examine the same parameters as in the investigation of Del Nobile et al. (2009c). The method has proven effective in extending the lifetime of the product to more than 3 days maintaining the qualitative value. Furthermore, unlike brine, coating allows for easier transportation due to the reduced weight of the final product.

3.3.9 Cheddar Cheese

O'Mahony et al. (2006) evaluated the application of MAP (70% N_2:30% CO_2) as a method for the effective preservation of Cheddar cheese. Several samples of this cheese were stored under the conditions mentioned earlier, and an optical O_2-analyzer was chosen to obtain the results so that no damages occur to the product. During the experiment, and specifically on the 1st, 4th, and 11th day, three damaged packages were found, while in the examination of the rest of them for residual oxygen, only a small portion (0.5%–2.0%) presented such phenomena. The method seemed to present as satisfactory results as the conventional methods while maintaining the cost at low levels. The use of an oxygen sensor system provided sufficient information as concerns the packaging procedure, the conditions in which the product is exposed during storage, and the quality of the cheese. Cheese products were packaged in containers characterized by different O_2-transmission rates and light limiting properties. Cheddar cheese remained exposed to fluorescent light for a period of 14 days. By increasing the amount of O_2 in the package, significant changes in yellowness and redness were presented. Nevertheless, only the differences in yellowness were associated with changes to package oxygen transmission rate. However, an enhancement of yellowness and a decrease in redness were presented when product remained under vacuum and in light. During exposure to light, the highest stability was observed when the product was packaged in aluminum foil packages. The growth of oxygen transmission rate was accompanied by the formation of thiobarbituric acid due to LO (Hong et al., 1995). A MAP of 73% CO_2/27% N_2 was evaluated for the effective protection of cheddar cheese against harmful molds. The samples consisted of shredded cheese, and the experiment included the effect of O_2-scavengers in the microbiological load of the products. Although containers with O_2-scavengers presented very admirable results in comparison with bags without O_2-scavengers, it was proved that MAP was the most effective method of preservation. The application of MAP, as well as the packages, was instrumental in the development of the different species of molds, as species that dominated the initial samples differed from those isolated at the end of the experimental procedure (after 16 weeks). It was concluded that O_2-scavengers effectively limited the product damages caused by molds (Oyugi and Buys, 2007).

Scannell et al. (2000) studied the absorbance capacity of bacteriocins nisin and laticin 3147 from packaging materials. The packages used for conducting the experiment consisted of cellulose-based bioactive inserts and polyethylene/polyamide pouches with antimicrobial properties. All packages with bacteriocin absorbance capacity were able to protect, to some extent, the products from microorganisms such as *Lactococcus lactis* subsp. *lactis* HP, *L. innocua* DPC 1770, and *Staphylococcus aureus* MMPR3. It was concluded that even though lacticin 3147 was not absorbed effectively, neither under room conditions nor by lowering the temperature, the bacteriocin-adsorbed packages limited the growth of bacteria. The most expanded shelf life occurred while storing the products under refrigeration conditions. Almost the same results were observed in Cheddar cheese stored under VP in nisin-adsorbed packages.

It is commonly accepted that MAP may contribute to the protection of packaged products from undesirable microorganisms. Taniwaki et al. (2001) studied the impact of MAP on the growth of different species of fungi (*Mucor plumbeus*, *Fusarium oxysporum*, *Byssochlamys fulva*, *B. nivea*, *Penicillium commune*, *P. roqueforti*, *Aspergillus flavus*, and *Eurotium chevalieri*) that are usually found in foods. Cheddar cheese was used for the inoculation of the microorganisms, and the product was stored under MAP (5%–<0.5% O_2 or 20%–40% CO_2). The characteristics of the fungal populations examined were colony diameter and ergosterol content. The species of the microorganisms studied grew under conditions of 20% and 40% CO_2

with 1% or 5% O_2, while the growth rate decreased at higher concentrations of CO_2 (20%–80%) in comparison with growth in air. At such high concentrations of this gas, the production of aflatoxins (B1 and B2, roquefortine C, and cyclopiazonic acid) was significantly reduced, but was not completely discontinued so that the product can be rendered acceptable for consumption.

3.3.10 Pasta Flata

3.3.10.1 Mozzarella Cheese

Mozzarella cheese is an Italian product made from water buffalo milk. It has a high moisture (55%–62%) and fat (>45%) concentration. According to Laurenzio et al. (2006), "Mozzarella cheese is characterized by a soft body and a juicy appearance, and by a pleasant, fresh, sour, and slightly nutty flavor." The maintenance of the quality of Mozzarella cheese under different combinations of gases, mainly CO_2 and N_2, was evaluated by Eliot et al. (1998). VP and air were also applied during a 2 month period at 10°C. Products were stored in barrier containers and systematically examined both for microbiological growth and changes in the rates of the gases contained inside the package. Reduction in O_2 percentage and increase in CO_2 levels were observed in several packages due to biochemical reactions. MAPs with elevated levels of CO_2 hindered the growth of lactic and mesophilic bacteria and restrained the growth of microorganisms while completely inhibiting *Staphylococci*, molds, and yeasts. It was proved that MAP with 75% CO_2 was the most effective technique for the protection of the product inhibiting or restricting the growth of harmful microorganisms and stabilizing the combination of gases inside the package. Figure 3.4 depicts the effect of different MAPs on LAB growth of Mozzarella cheese at two different temperatures. Pluta et al. (2005) used several

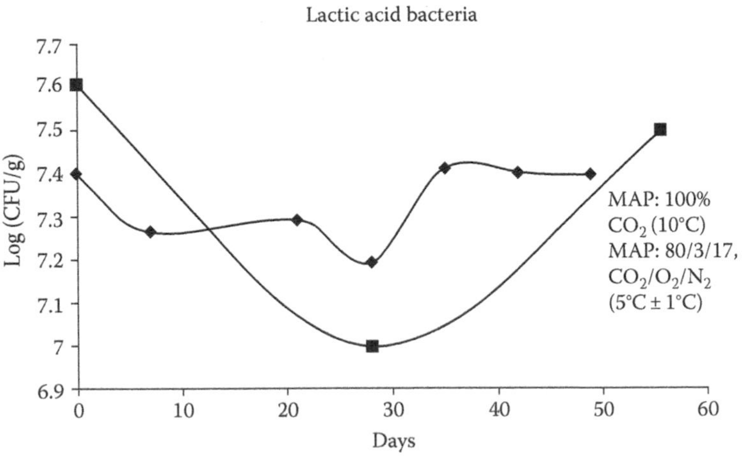

FIGURE 3.4 Impact of different MAPs on LAB growth of Mozzarella cheese stored at 10°C (◆) (From Eliot, S.C. et al., J. Food Sci., 63(6), 1075, 1998) and Mozzarella cheese stored at 5°C ± 1°C (■). (From Oyugi, E. and Buys, E.M., *Int. J. Dairy Technol.*, 60(2), 89, 2007.)

MAPs to examine their effects on grated Mozzarella cheese quality and preservation. The initial products were taken from an industrial unit so that the initial microflora and their physicochemical properties correspond to the products intended for consumption. The samples were incubated with a mixture of microorganisms and stored into bags and under MAPs consisting of CO_2 and N_2 in different combinations as well as lowered pressure conditions (400 mbar). It was proved that CO_2 was the most effective gas for the restriction of undesirable microorganisms (coliforms, yeast, molds). The microbial load of the products stored under MAP consisting mainly of CO_2 was significantly lower in comparison to the products stored under MAP mainly consisting of N_2 or low pressure. Nevertheless, low pressure showed the best results regarding quality maintenance of Mozzarella cheese.

Alves et al. (1996) evaluated the sensory degradation of mozzarella cheese under various MAPs. Expanded polystyrene trays placed in gas-barrier bags and three MAPs (100% N_2, 100% CO_2, and 50% CO_2/50% N_2) were used for packaging 12 samples of the product at 7°C ± 1°C. The samples did not show any differentiation between various preservation methods regarding the physicochemical characteristics. However, a significant extension of the shelf life was observed in products stored under CO_2 (63 days) and 50% CO_2/50% N_2 (45 days) in comparison with storage in MAP consisting mainly of N_2 or in air. Increasing the levels of CO_2, a reduction in the growth of various microorganisms was observed, but full fungi inhibition was achieved only under 100% CO_2. Figure 3.5 demonstrates the effect of MAP on psychrotrophic bacterial growth in Mozzarella and Myzithra cheese at two different temperatures.

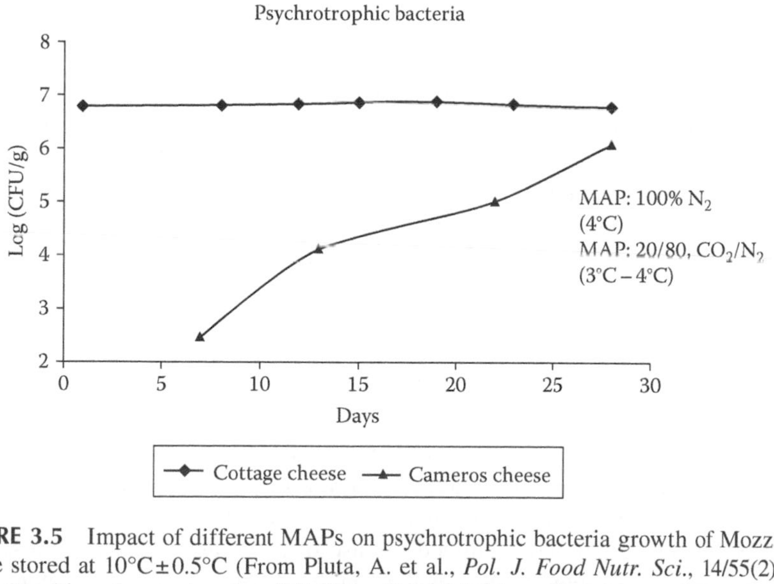

FIGURE 3.5 Impact of different MAPs on psychrotrophic bacteria growth of Mozzarella cheese stored at 10°C ± 0.5°C (From Pluta, A. et al., *Pol. J. Food Nutr. Sci.*, 14/55(2), 117, 2005), Myzithra cheese stored at 4°C (From Dermiki, M. et al., *Lebensm. Wiss. Technol.*, 41, 284, 2008), and Mozzarella cheese stored at 10°C. (From Eliot, S.C. et al., *J. Food Sci.*, 63(6), 1075, 1998.)

3.3.11 Whey Cheese

3.3.11.1 Anthotyros

Anthotyros is a type of low-fat cheese produced from a mixture of sheep and goat milk. Anthotyros cheese in Crete is produced from the whey of hard cheese such as kefalotyri and graviera. Anthotyros is a cheese that comes from myzithra cheese (whey) dehydration. The dehydrated anthotyros' physicochemical properties are moisture content of 35%, fat content in dry matter of 55%, and low concentration of salt. The purpose of this study was to identify which of the three gas compositions applied can lead to greater shelf life prolongation of Anthotyros: MAP_1: 40% CO_2, 55% N_2, 5% O_2; MAP_2: 60% CO_2, 40% N_2; and MAP_3: 50% CO_2, 50% N_2. The control samples were packaged in air. All cheese samples were kept under refrigeration (4°C ± 0.5°C) for 55 days. MAP_2 and MAP_3 mixtures proved to be the most effective for inhibiting total mesophilic microorganisms and *E. coli*. Neither *S. aureus* nor *L. monocytogenes* were detected over the duration (56 days) of the experiment. Similar to sensory analysis, good overall acceptability of the control samples was reported until the 7th day and of MAP samples until 24th day. The most effective gas mixtures in regard to sensory analysis were the MAP_1 and MAP_2. An important observation was that the anthotyros cheese with 60% CO_2/40% N_2 strongly inhibited the growth of *E. coli* and TVC (Arvanitoyannis et al., 2011a).

MAPs (30%/70% CO_2/N_2 or 70%/30% CO_2/N_2) and VP at 4°C or 12°C were applied to investigate the most suitable method for maintaining *Anthotyros* cheese. The experiment showed that the growth of microorganisms was more limited in MAP conditions rather than VP. As a result, by applying MAP the product shelf life was increased significantly, while the organoleptic characteristics remained at a satisfactory level (Papaioannou et al., 2007).

3.3.11.2 Myzithra Kalamaki

Different MAPs were applied on *Myzithra Kalathaki* cheese at 4°C ± 0.5°C for 45 days to examine the shelf life and course of degradation of product quality under these conditions. Four MAPs (VP, 20% CO_2/80% N_2, 40% CO_2/60% N_2, and 60% CO_2/40% N_2) were applied while control samples were stored in air. By increasing the rate of CO_2 better retention of the sensory properties and lengthening of the shelf life from approximately 10 days in air to 30 days under MAP were achieved (Dermiki et al., 2008). Figure 3.6 depicts the impact of different MAPs on yeasts-mold growth of Mozzarella and Myzithra cheese at 4°C and 10°C.

3.3.11.3 Ricotta and Stracchino Cheese

According to Paris et al. (2004), "Stracchino cheese is a soft fat cheese-spread without rind, produced from cow's whole milk only and characterized by fast lactic maturation and a delicate taste." It contains high levels of protein and minerals, but also fat and it is not commercially competitive. It is originated from Lombardia in Italy.

Del Nobile et al. (2009a) examined the use of MAP as a means of preserving Ricotta cheese. The deterioration evaluation of product quality was conducted using microbiological and physicochemical tests for 8 days. It was found that MAP can effectively protect the product from undesirable microorganisms. Specifically,

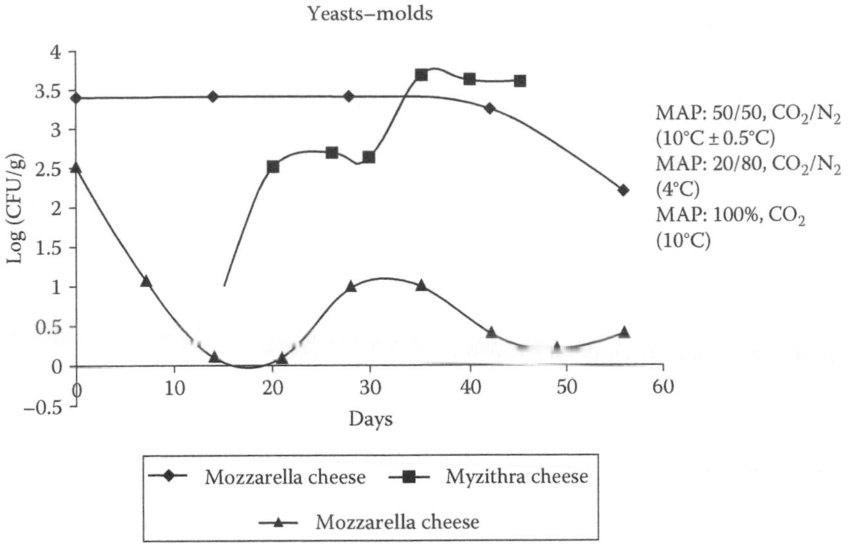

FIGURE 3.6 Impact of different MAPs on yeast–mold growth of Mozzarella cheese stored at 10°C ± 0.5°C (From Pluta, A. et al., *Pol. J. Food Nutr. Sci.*, 14/55(2), 117, 2005), Myzithra cheese stored at 4°C (From Dermiki, M. et al., *Lebensm. Wiss. Technol.*, 41, 284, 2008), and Mozzarella cheese stored at 10°C. (From Eliot, S.C. et al., *J. Food Sci.*, 63(6), 1075, 1998.)

increase in CO_2 levels leads not only to growth suspension of harmful microorganisms, without affecting LAB but also to better preservation of the cheese color. By applying MAP of 95% CO_2, greater expansion of the product shelf life was achieved.

3.3.12 Semi-Hard Cheese

Holm et al. (2006) examined the effect of a poly-lactic acid (PLA) package on qualitative characteristics of semi-hard cheese under light or dark conditions and with or without O_2-scavengers. The samples were stored under MAP (30% CO_2/70% N_2). The migration degree from the package materials to the product was minimal and could not affect the marketability of the product. The LO rate was higher in products stored in PLA because of the higher O_2 transmission rate of PLA in comparison with other packages, but it was decreased significantly with the use of O_2-scavengers. It was proved that better product quality (decrease in LO and moisture retention) can be achieved with storage under dark and reduction of water vapor transmission degree.

As CO_2 is widely used for the preservation of food and is also the main gas used in MAP formation, it was necessary to develop a computational model for predicting the absorbance of this gas by the food, taking into account not only the initial CO_2 rate of the product but also all the other factors affecting the absorbance. Jakobsen and Risbo (2009) proceeded to the development of such a model using a packaging experiment with semi-hard cheese (10.7% fat; 49.5% water, and 34.5% protein) as a means of its validation.

3.3.13 Sliced Samso Cheese

Sliced Samso cheese, packaged in three different MAPs (0% CO_2 [100% N_2], 20% CO_2 [80% N_2], or 100% CO_2) under light conditions for 21 days, showed significant loss of yellowness and increase in redness. Samples stored in 100% CO_2 under light conditions were notably darker than products stored in 0% CO_2. The different MAPs, light, and storage period greatly affected the flavor and the smell of the product. Cheeses stored in 20% CO_2 or 0% CO_2 without light were characterized as "buttermilk-like," while those stored in 100% CO_2 and exposed to light were assessed as rancid and distinguished by a "dry/crumbly texture." Significant light exposure effects during storage were noted for 1-pentanol, 1-octanol, 2-ethyl-1-butanol, 2-butanol, 2-heptanol, 2-pentanol, 2-nonanol, benzaldehyde, 2-butanone, 2-nonanone, dimethyl trisulfide, and tetrahydrofuran (Juric et al., 2003).

3.3.14 Provolone

Provolone is a pasta filata cheese (stretched like Mozzarella) originating from Italy. It can be distinguished by loss of crumbles and a nice mellow, piquant flavor. Its color is light and its shape is ovoid (Cakir and Clark, 2009). Favati et al. (2007) examined the effect of several MAPs consisting of CO_2 and N_2 on the life of this cheese. Four gas mixtures (10% CO_2/90% N_2, 20% CO_2/80% N_2, 30% CO_2/70% N_2, and 100% CO_2/0% N_2) were applied, while control samples were stored under vacuum. All products were maintained at 4°C and 8°C. The maximum shelf life (280 days) was achieved after storage under MAP with 30% CO_2/70% N_2 as in these conditions proteolytic and lipolytic reactions were slower and the quality of the product maintained at high levels.

3.3.15 Sliced Havarti Cheese

The effect of MAP on sliced Havarti cheese quality and shelf life and various light storage parameters were examined. It was proved that product quality is directly affected by light or dark storage, color of light, intensity, and O_2 fluctuations. After analyzing the data using PCA and 50-50 MANOVA, it was shown that exposure of the product to white light may reduce the levels of riboflavin, whereas loss of porphyrins can occur after exposure to both white and yellow light (Andersen et al., 2006).

Kristensen et al. (2000) evaluated the color changes and other sensory characteristics of sliced Havarti cheese during a period of 21 days at 5°C and under MAP (25% CO_2/75% N_2), with or without product's exposure to light (1000 lx). All tested samples showed an enhancement in redness and a reduction in yellowness. As in the study of Andersen et al. (2006), it was proved that storing the product under light, a reduction in the levels of riboflavin was observed. The increased production of free radicals showed a gradual reduction during storage, with the fastest reduction occurring with exposure to light. After examinations of odor and taste, the oxidation caused by exposure to light became clear. Mortensen et al. (2003) evaluated the effect of several MAPs and other factors such as light, residual oxygen, and temperature, on photooxidation of sliced Havarti cheese. Products stored during a period of 168 h and color alterations as well as oxidation products (1-pentanol and 1-hexanol) were used as oxidation indicators. It was proved that in light-exposed cheeses, increasing the residual oxygen caused greater oxidation. Factors such as temperature and light intensity had no effect on oxidation reactions. However, light color affected riboflavin degradation, as yellow light caused relatively little degradation in comparison with common light source. All in all, the most critical factor in determining the degree of oxidation, and as a result the quality of the product, is the residual oxygen, which should remain at low levels (<0.6%).

3.3.16 Hard Cheese

3.3.16.1 Grated Graviera

Graviera is a full fat cheese, of rather high added value, the preservation of which is of great importance for its producers. Graviera cheese was stored at 4°C for up to 90 days in polyamide packages under three different modified atmosphere compositions. Control cheeses were packaged in air whereas MAP mixtures were packaged in MAP_1: 40% CO_2/55% N_2/5% O_2, MAP_2: 60% CO_2/40% N_2, and MAP_3: 50% CO_2/50% N_2. Sampling of product was carried out every 10 days to investigate its sensory quality and microbiological characteristics. Ten trained panelists participated in the sensory panel to evaluate the cheeses for external appearance (color, texture), taste, and flavor on a scale from 1 to 10 (1 very poor, 10 very good). The microbiological analysis revealed that there were no colonies of *Staphylococcus aureus* and *L. monocytogenes*, whereas both *E. coli* and TVC increased strongly in control samples but were inhibited under all MAP compositions (Arvanitoyannis et al., 2011b). The mean mesophile counts of the samples stored in air (control) enhanced rapidly and were higher than 5.2 log CFU/g after 50 days of storage at 4°C. However, hard cheeses packaged under modified atmospheres reached populations above 4.4 log CFU/g on day 20 depending on the amount of oxygen contained in MAP_1 gas mixture, apart from cheeses packaged under 60% CO_2 which gave a value of 4.7 log CFU/g. Furthermore, the mesophilic counts of control and MAP cheeses increased progressively with storage time in grated Graviera cheese in agreement with Mexis et al. (2011).

Grated Graviera cheese organoleptic characteristics, LO, and discoloration were examined under different MAP conditions (100% CO_2, 100% N_2, 50% CO_2/50% N_2) and in air. The samples remained for 9 weeks under light conditions (fluorescent light) or in dark. It was clear that products stored in light not only were more discolored than those stored in dark but also LO reactions were more extensive. It was proved that light, gas mixture, and storage period had a great impact on the taste and odor properties of the product. It was concluded that the best results were observed at 100% N_2 or 50% N_2/50% CO_2 in the absence of light, as the quality of the product remained high for 9 weeks (only 2 weeks in light conditions). The quality of the products stored in air remained at acceptable levels for 2.5 and 2 weeks, with storage in dark and light, respectively (Trobetas et al., 2008).

3.3.17 White Cheeses, Cheese, and Cheese Cakes

Sanguinetti et al. (2009) examined the effect of two different MAPs (70% N_2/30% CO_2 and 20% N_2/80% CO_2) and an AP on cheese cake shelf life. The AP was formed using an iron oxide–based O_2-absorber. Samples remained at 20°C during a period of 48 days. Gas mixture physicochemical parameters as well as alterations in proliferation of various microorganisms were recorded regularly. AP effectively protected product against molds during the whole period. As concerns MAPs, it was proved that enhancement at CO_2 level of 30% and 80% results in an extension of product's shelf life up to 14 and 34 days, respectively. Products stored in MAP presented a significant increase in hardness, especially 14 days after packaging, whereas AP preserved the quality of the product during the whole 48 days of the storage. In comparison, control samples stored in air had a shelf life of just 7 days.

3.3.18 Yogurt

According to Saint-Eve et al. (2008), "yogurt is a very popular fermented milk product, consumable all over the world." Packaging technology has been extensively used to preserve the quality and organoleptic properties of this product.

During the fermentation of milk into yogurt, the production of lactic acid and several metabolites from LAB leads to pH reduction, creating adverse conditions for the growth of harmful microorganisms. Because of that, yogurt products produced in well-equipped production units can very rarely cause foodborne diseases. However, any change in the pH of the product can eventually result in contamination and proliferation of harmful microorganisms (Birollo et al., 2001).

The effect of MAP storage on sensory assessment and shelf life (extension) in conjunction with packaging material and storage temperature on different dairy products packaging is given in Table 3.2.

3.4 FISH

MAP technology has extensively been used to protect fisheries from harmful microorganisms and oxidation reactions, extending the shelf life of these products (Sivertsvik et al., 2002). This technology has been successfully used in preservation of refrigerated fisheries inhibiting the growth of Gram-negative bacteria (Cabo et al., 2005). As the shelf life of fresh fisheries is not very extended, several problems have to be faced during the transport of these products and their storage in retail and wholesale areas (Mejlholm and Dalgaard, 2002). Many factors have to be taken into account when the shelf life of fisheries is calculated. Some of the most critical are the type of fish catch, the fat content, and the combination of gases used (if a MAP is applied), but mostly the storage temperature. Microorganisms are mainly responsible for the extensive quality deterioration and reduced shelf life of seafood. MAP as well as VP can be applied in order to significantly extend the shelf life of fisheries (Sivertsvik et al., 2002).

TABLE 3.2
The Effect of MAP (Gas Mix, Packaging Material) and Storage Temperature on Shelf Life and Sensory Assessment of Dairy Products

Dairy Product	MAP (Gas Mix)	Sensory Assessment	Shelf Life (Days)	Shelf-Life Extension (Days)	Packaging Material	Storage Temperature	References
Cheddar cheese	60% N_2:40% CO_2			24	PS/EVOH/PE/cellulose based paper, plastic (PE:PA = 70:30) film	4°C	Scannell et al. (2000)
Cheese	20% CO_2:1% O_2 40% CO_2:5% O_2 40% CO_2:1% O_2		14	15	Plastic bags (PP:EVOH:PP), PE plastic bags	25°C	Taniwaki et al. (2001)
Cream cheese		Better protection			Amorphous PE + PET + PE (750 and 550 mm), PS + EVOH copolymer + PE (750 mm), PP + PE (600 mm)	4°C	Pettersen et al. (2005)
Giuncata cheese	50% CO_2, 50% N_2	No significant changes	180		Nylon (PA) + Polyolefin (PO) bags (95 μm)	8°C	Gammariello et al. (2009)
Fior di latte cheese	30% CO_2, 5% O_2, 65% N_2	No color changes		1	PA/PO bags	4°C	Del Nobile et al. (2009c)
Mozzarella cheese	50% N_2:50% CO_2 70% CO_2:30% N_2	No statistical changes	13		Laminated PE/PA bags	10°C ± 0.5°C	Pluta et al. (2005)

3.4.1 Fish Stored under MAP

According to Dalgaard et al. (1997), a mathematical model can be used to predict the expected shelf life of fish products packaged under MAP conditions. It was proved that using this model, the shelf life of fresh cod can be predicted with high accuracy. The model was tested in cod fillets contaminated with the bacterium *Photobacterium phosphoreum*. Figure 3.7 demonstrates the effect of different MAPs on *P. phosphoreum* growth on cod fillets and *Belone belone* at 0°C. Özogul and Özogul (2006) studied the formation of biogenic amines during storage of *Sardina pilchardus* samples packaged under MAP, VP, and in air. Samples remained at 4°C while several tests were taking place systematically, to record the quality deterioration of the product. It was proved that MAP was the most effective way of preservation. Using this method, the quality of the product remained at high levels even 12 days after packaging. In comparison, fishes were preserved for 9 days using VP, while in air the shelf life of the products was just 2 days. Although increased production of biogenic amines was observed in all products, the amine creation reactions were taking place much slower when MAP and VP were applied. PUT and cadaverine rates of sardines that remained in air hit a peak at 12 (12.2 mg/100 g) and 15 (10.0 mg/100 g) days, respectively, while significant differences at the levels of these substances were observed when MAP and VP were applied. The formation of amides and, as a consequence, the qualitative deterioration of the products were greater in products that remained in air, while MAP was proved to be the most effective method for the preservation of sardines.

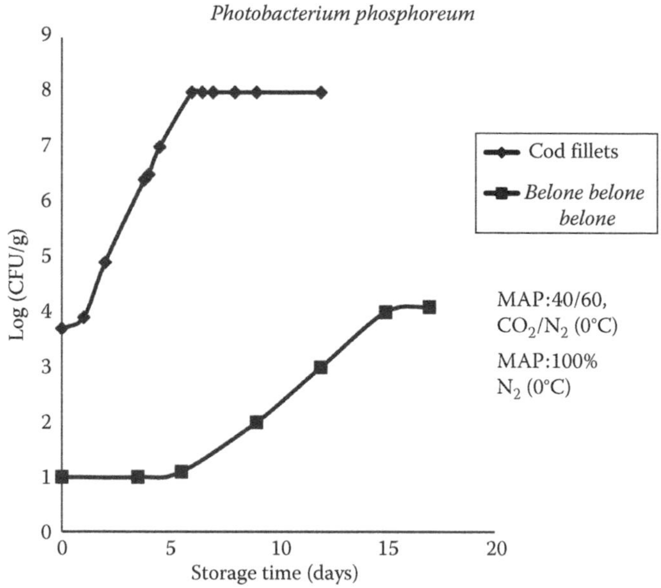

FIGURE 3.7 Effect of different MAPs on *P. phoshoreum* growth of cod fillets stored at 0°C (From Dalgaard, P. et al., *Int. J. Food Microbiol.*, 38, 169, 1997) and *B. belone* stored at 0°C. (From Dalgaard, P. et al., *J. Appl. Microbiol.*, 101, 80, 2006.)

Corbo et al. (2005) studied the life cycle of microorganisms naturally found on fresh cod fillets. The samples were stored in air, under VP, or MAP with or without high oxygen concentration (80% O_2 or 5% O_2, respectively). Microbial tests took place at different temperatures from 4°C to 12°C, and the shelf life of all products was recorded. It was clear that reducing the temperature and the O_2 level and using VP or MAP a more extensive shelf life can be achieved. As a result, the quality of products stored under VP or MAP, with low O_2 level, was acceptable even after 6 and 5.42 days, respectively.

Bøknæs et al. (2002a) studied the NIR spectroscopy capability to calculate the optimum freezing temperature of cod mince, the period that the product can be frozen but also the period that thawed-chilled MAP Barents Sea cod fillets can be effectively preserved in chill storage. A total of 105 samples of the product were used to conduct the experiment, and several microbial and physicochemical tests were carried out to receive the requested data. It was proved that there was a relatively good correlation between the measured and the expected shelf life of the product.

MAPs mainly consisting of CO_2 and a low rate of either O_2 or N_2 were applied by Masniyom et al. (2002) to maintain the quality of sea bass slices. It was shown that at 4°C MAPs preserved the product from spoilage and pathogen microorganisms. Furthermore, CO_2 limited considerably the total volatile base, trimethylamine, ammonia, and formaldehyde formation in comparison to samples that remained in air. The shelf life of products stored under MAP with more than 80% of CO_2 can be extended to more than 20 days, while the qualitative characteristics remain at acceptable levels.

Several samples of chilled swordfish (*Xiphias gladius*), caught in the Mediterranean sea, were used to carry out an experiment in which the shelf life of this product after its storing under MAP (40%/30%/30%, CO_2/N_2/O_2), VP, or in air was examined. Samples remained at 4°C for a period of 16 days and were systematically analyzed for monitoring the growth of the products' microorganisms. It was proved that MAP and VP showed the best results adequately protecting the product from harmful microorganisms and maintaining the quality of the product at high levels for 9–10 days. Under VP and MAP storage, the growth of several bacteria (*Pseudomonas* spp., *S. putrefaciens*) commonly found in swordfish flesh limited significantly whilst populations of LAB and *Enterobacteriaceae* were observed in all samples. It was demonstrated that the shelf life of the products stored in air and under VP and MAP conditions were 7, 9, and 11–12 days, respectively (Pantazi et al., 2008). Figure 3.8 depicts the impact of different MAPs on *Pseudomonas* spp. growth of chilled swordfish, fresh tuna, and fish burgers at different temperatures.

The formation of histamine is a serious problem that many professionals dealing with seafood have to be faced with. Histamine concentration increases steadily after the killing of fish to consumption and reduces product shelf life, as increased concentrations of this substance can even make the product dangerous for consumption. MAP consisting of 4% CO_2/60% O_2 effectively preserved seafood by inhibiting the growth of histamine-producing bacteria (*Morganella morganii* and *P. phosphoreum*) for 28 days at 1°C (Emborg et al., 2005). Sivertsvik (2007) examined several MAPs to find the best preservation method for fillets of farmed Atlantic cod (*Gadus morhua*). The samples were stored for 14 days at 0°C under MAPs consisting of CO_2, O_2, and N_2, and many

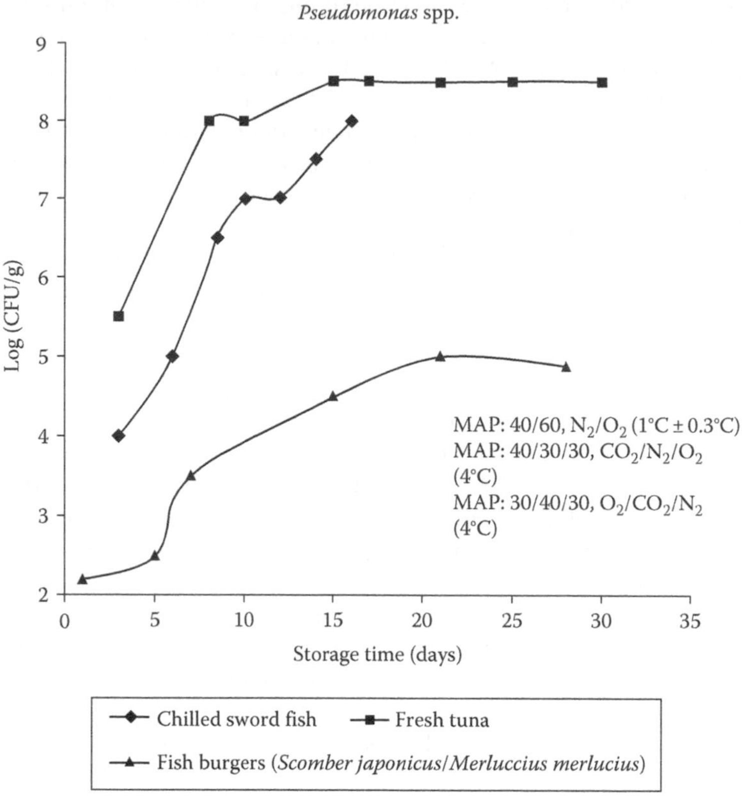

FIGURE 3.8 Effect of different MAPs on *Pseudomonas* spp. growth of chilled sword fish stored at 4°C (From Pantazi, D. et al., *Food Microbiol.*, 25, 136, 2008), fresh tuna stored at 1°C±0.3°C (From Emborg, J. et al., *Int. J. Food Microbiol.*, 101, 263, 2005), and fish burgers (*S. japonicus/M. merluccius*) stored at 4°C. (From Del Nobile, M.A. et al., *Int. J. Food Microbiol.*, 135, 281, 2009b.)

physicochemical and organoleptic tests were carried out at the 7th, 10th, and 14th day. Enhanced concentrations of CO_2 were shown to limit the growth of spoilage bacteria and completely inhibit the growth of H_2S-producing bacteria. When MAPs low in CO_2 and O_2 were used, trimethylamine oxide rate was reduced while the trimethylamine content of the flesh increased considerably. To be more specific, 63 mL/100 mL O_2 and 37 mL/100 mL CO_2 appeared to be the optimum gas combination.

3.4.2 Fish Stored with a Combination of MAP and Other Technologies

Although nisin can be used for the preservation of many food products, like milk and meat, it is impossible to inhibit Gram-negative bacteria with it. As a result, this substance can only be used in combination with other antibacterial compounds. Cabo et al. (2005) examined the application of nisin in conjunction with other factors like sodium hexametaphosphate (SMP) and ethylenediaminetetraacetic acid (EDTA) that favor the introduction of antimicrobials in packaging. It was shown that nisin can act synergistically with SMP and CO_2, thereby effectively protecting seafood from harmful microorganisms.

Goulas and Kontominas (2007b) examined how MAP in combination with oregano oil can decrease total volatile basic nitrogen (TVBN) and trimethylamine nitrogen (TMAN) formation of salted sea bream (*Sparus aurata*) fillets. Samples of the product were refrigerated and remained under MAP (40% CO_2/30% O_2/30% N_2) with or without oregano oil (0.4% [v/w] or 0.8% [v/w]). It was demonstrated that high concentrations of oregano oil (0.8% [v/w]) in combination with MAP showed the best results proving that the oregano oil can be used as an additive for sea bream fillets to extend their shelf life. By using oregano oil, a significant reduction in oxidative reactions was observed. Combined use of salting, oregano oil, and MAP effectively preserved the products during a period of 27–28 days.

Del Nobile et al. (2009b) examined the effects of thymol, lemon extract, and grapefruit seed extract in combination with MAP (air, 30% O_2:40%:CO_2 30% N_2, 50% O_2:50% CO_2 or 5% O_2:95% CO_2) on preservation of blue fish burgers. All samples were stored for 28 days at 4°C. It was proved that MAP, mainly consisting of CO_2, in combination with any of these three essential oils, can extend the shelf life of the products up to 24 days at 4°C maintaining an acceptable quality level. Very small quantities of these three components (110 ppm of thymol, 100 ppm of GFSE, and 120 ppm of lemon extract) are needed as additives for the formation of the best conditions to effectively preserve the products.

Pastoriza et al. (2002) demonstrated that lauric acid in combination with MAP (50% CO_2:50% N_2) was the most appropriate method for the shelf-life extension of refrigerated minced cod or shrimp tails. Samples of these seafood were stored with or without lauric acid, and it was proved that the best results were obtained when there was synergistic use of both methods. Under these conditions, products maintained a high quality even 30 days after packaging.

Rosnes et al. (2006) examined the implementation of MAP (60% CO_2:40% N_2) in combination with superchilled conditions (−1°C) as a method for the protection of farmed spotted wolf-fish (*Anarhichas minor*) products. Samples remained in air or under MAP conditions at −1°C and 4°C. It was proved that, at both temperatures, MAP application was able to restrict the growth of different harmful microorganisms such as aerobic plate counts (APC) and psychrotrophic bacteria maintaining smell and flavor characteristics to acceptable levels. Under MAP and superchilled conditions, products were preserved for a period of 15 days whilst at 4°C, MAP effectively preserved samples for 13 days. Samples stored in air at −1°C and 4°C had a relatively short shelf life of just 8–10 and 6–8 days, respectively.

According to Franzetti et al. (2001), two different bags made of foam plastic, capable of absorbing volatile amines and liquids, were examined in combination with MAP (40% CO_2:60% N_2) as a potential method for the shelf-life extension of several seafoods. Specifically, samples of sole (*Solea solea*), cod (*Merluccius merluccius*), and cuttlefish (*Sepia fillouxi*) remained in these bags at 3°C and were examined regularly for 10 days. It was demonstrated that the qualitative characteristics of the products were effectively preserved but an increase in the growth of microorganisms, for example, *Moraxella phenylpiruvica* was observed.

The combination of sodium chloride and MAP application was studied by Pastoriza et al. (1998) to find a preservation method for extending the shelf life of hake pieces (*M. merluccius*). Before packing in MAP, samples were infused for 5 min in 5% NaCl solution. Several qualitative characteristics as well as the microbial growth were studied to determine if the combination of the two methods can lead to better results in maintaining and extending the shelf life of the product. It was found that combined use of MAP and NaCl effectively protected the product from physicochemical and microbiological changes, extending the shelf life of the fish for 8 days. The quality of samples stored under MAP, without prior treatment with NaCl, remained at high levels only for 2 days.

3.4.3 Quality of Fish Stored under MAP

Two different MAPs (MAP_1: 70% CO_2/30% N_2 and MAP_2: 50% CO_2/30% N_2/20% O_2) as well as VP and storage in air were applied to fillets of *Scomber japonicus* to examine the effectiveness of each atmosphere in the preservation of the product. All samples remained under refrigeration during the whole experiment. The deterioration of qualitative characteristics was measured through various sensory and physicochemical tests. It was proved that the first MAP followed by the second and the VP were able to ensure the preservation of product quality for the longest period of time reducing significantly the total volatile basic nitrogen (TVBN) and TMAN formation. While the products stored in air were acceptable for consumption only for 11 days, products stored under the first MAP were acceptable for 20–21 days, indicating that this method is the most appropriate for the extension of products' shelf life. The second MAP and VP imparted to the product a shelf life of about 15–16 days (Goulas and Kontominas, 2007a).

Pastoriza et al. (1996) examined the effectiveness of four different MAPs (MAP_1: 40% CO_2, 50% N_2, 10% O_2; MAP_2: 60% CO_2, 30% N_2, 10% O_2; MAP_3: 40% CO_2, 30% N_2, 30% O_2; MAP_4: 60% CO_2, 10% N_2, 30% O_2) in the preservation of hake slices (*M. merluccius*). Products stored at 2°C ± 1°C for 3 weeks and different qualitative characteristics were recorded through determination of pH, TVB, TMA content, microbiological tests, and several physicochemical and sensory tests. After the experiment, no significant deterioration of product quality was observed. The most important factors affecting the shelf life of the product were the time elapsed between harvesting and packaging, the initial microbiological quality, and the correct or incorrect handling throughout the whole process.

Pieces of *Gadus morhua* were packaged under MAP conditions and two different temperatures (20°C and 30°C) for 3, 6, 9, and 12 months and for 21 days at 2°C. Several physicochemical, microbiological, and sensorial tests were carried out during storage, and it was found that 12 months after packaging at 20°C and 30°C, there were only minor changes in the qualitative characteristics of the product. At 2°C the shelf life of the product was determined at 14 days (Bøknæs et al., 2002b).

Qualitative characteristics of gutted farmed bass stored in several MAPs (0% O_2–70% CO_2; 20% O_2–70% CO_2; 30% O_2–60% CO_2; 40% O_2–60% CO_2; 30% O_2–50% CO_2; 21% O_2–0% CO_2) were determined by Torrieri et al. (2006). Samples remained at 3°C for 9 days and several tests, including microbiological tests and headspace gas composition determination, a_w, sensorial and organoleptic examinations, were carried out not only at the beginning of the experimental procedure but also on the 2nd, 5th, 7th, and 9th day. It was demonstrated that the fifth MAP (30% O_2/50% CO_2) showed better results in maintaining product quality.

Five different MAPs (40% CO_2/60% O_2, 50% CO_2/50% O_2, 60% CO_2/40% O_2, 70% CO_2/30% O_2, and 40% CO_2/30% O_2/30% N_2) as well as storage in air were used to determine the effectiveness of MAP on the preservation of fresh pearl spot (*Etroplus suratensis* Bloch) and its protection from harmful microorganisms. All samples were maintained at 0°C, and the measurement of product qualitative deterioration indicated that the MAP of 60% CO_2:40% O_2 showed the best results as concerns the limited increase in TVDN level. It is important to mention that no difference in microbial growth was observed between samples stored in air and in the MAP of 40% CO_2/30% O_2/30% N_2 (Lalitha et al., 2005).

Giménez et al. (2002) compared six MAPs (10% O_2+50% CO_2+40% N_2, 10% O_2+50% CO_2+40% Ar, 20% O_2+50% CO_2+30% N_2, 20% O_2+50% CO_2+30% Ar, 30% O_2+50% CO_2+20% N_2, and 30% O_2+50% CO_2+20% Ar), VP, and an overwrap package to determine the most appropriate method for the preservation of *Oncorhynchus mykiss* fillets. The samples were stored at 1°C±1°C and the quality degradation was evaluated through systematic measurements of various parameters such as pH, TVB, hypoxanthine concentration, LO, microbial growth, and sensorial characteristics. It was proved that MAP application effectively protected the product from harmful microorganisms and undesirable oxidation reactions, extending significantly its shelf life and maintaining the quality at high levels.

According to Metin et al. (2002), MAP can be used to extend the shelf life of fish salads up to 50%. Under MAP conditions (O_2/CO_2/N_2:5/35/60 and CO_2/N_2:30/70) adverse conditions are created for the growth of various microorganisms, delaying the deterioration of the product quality and facilitating its transportation and storage.

3.4.4 Microflora of Fish Stored under MAP

Schirmer et al. (2009) investigated the microbial populations of fresh salmon stored under MAP of 100% CO_2 and a brine solution consisting of citric acid (3% w/w, pH 5), acetic acid (1% w/w, pH 5), and cinnamaldehyde (200 μg/mL). The largest bacterial populations belonged to the species *P. phosphoreum*, *Carnobacterium maltaromaticum*, and LAB, whilst *Yersinia aldovae*, *Aeromonas salmonicida*, and *S. putrefaciens* were found in markedly smaller populations. It is important to note that conditions created in the food environment limited the population of *P. phosphoreum*.

According to Sivertsvik et al. (2002), maintaining seafood at low-temperature conditions is essential for the proper action of MAP and the effective protection of temperate region fishes against the bacterium *P. phosphoreum* known as the specific spoilage organism (SSO) of MA. Low storage temperature allows higher CO_2 gas solubility and thus a more effective inhibition of harmful microorganisms and reduction of enzymatic activity. Mejlholm and Dalgaard (2002) examined several essential oils for their effectiveness in extending the shelf life of fresh fish fillets stored under MAP, limiting the growth of *P. phosphoreum*. When essential oils are sprayed on fish flesh, they can contribute to the preservation of the product's organoleptic characteristics and limitation of qualitative degradation. Their advantages include low cost (0.05% v/w costs approximately 1% of fish raw material like cod fillets) and ease of implementation in comparison to other techniques like smoking, HP treatment, etc.

Dalgaard et al. (2006) studied the biogenic amine formation as well as the bacterial growth in fresh and thawed chilled garfish. All samples were stored under MAP (40% CO_2 and 60% N_2) at 0°C or 5°C. More than 1000 ppm of histamine detected in chilled fresh garfish were the product of the metabolic activity of the bacterium *P. phosphoreum*, while some isolates produced up to 2080–4490 ppm at 5°C. Histamine production was limited significantly by storing thawed garfish under freezing conditions.

Hovda et al. (2007) examined the microbial content of farmed Atlantic halibut packaged in two different MAPs (50% CO_2:50% N_2 and 50% CO_2:50% O_2). Products were maintained for 23 days and remained at 4°C during the whole experiment. Microbial analysis carried out during the experiment indicated that the largest bacterial populations belonged to the species *P. phosphoreum* and *Pseudomonas* spp., while *B. thermosphacta* and *S. putrefaciens* were found in smaller populations.

Debevere and Boskou (1996) used several MAPs to find the best preservation method for cod fillets (*Gadus morhua*). All samples remained at 6°C while microbiological analysis took place on the 3rd, 4th, 5th, 6th, and 7th day of storage. It was proved that MAPs severely limited the growth of microorganisms (aerobic and anaerobic bacteria, LAB, H_2S-producing bacteria, and *Enterobacteriaceae*) but did not affect significantly the formation of TVB and TMA. High concentrations of TVB and TMA made the product unacceptable for consumption just 4 days after packaging.

MAP of fresh fish is used to market high-quality products in some European countries. Although MAP is an effective method for the inhibition or reduction of harmful microorganisms, it does not offer sufficient protection against the bacterium *Clostridium botulinum*. Because of the extremely dangerous toxins produced by this bacterium and the difficulty in detecting them, an in-depth study of this microorganism and the accurate record of the condition under which it releases its toxins were necessary. Baker et al. (1990) used data from rockfish, salmon, and sole muscle tissues vaccinated with a strain of *C. botulinum* and maintained under VP and MAP (100% CO_2) at 4°C–30°C for approximately 60 days. The purpose of the experiment was not only to investigate the bacterium lag phase, during which it starts the formation of toxins, but also the effect of storage conditions on toxin production. The predictive formulae designed during the experiment can be used to protect consumers' health against *C. botulinum*.

Table 3.3 provides the MAP characteristics when applied to different fisheries.

3.5 EFFECT OF MAP ON SAFETY OF FOODS OF PLANT ORIGIN

3.5.1 Fruits

The use of proper packaging materials and methods to minimize food losses and provide safe and wholesome food products has always been the focus of food packaging.

TABLE 3.3
Effect of MAP (Gas Mix and Storage Temperature) on Shelf Life and Sensory Assessment of Fish and Seafood

Species	MAP (Gas Mix)	Sensory Assessment	Shelf Life (Days)	Storage Temperature	References
Fresh garfish spring	Air	Dark color, sour and oxidized flavor, dry texture	15	0°C	Dalgaard et al. (2006)
	Air		17		
	40% CO_2/60% N_2		20		
	40% CO_2/60% N_2		38		
Salmo salar	25% CO_2:75% N_2 40% CO_2:60% N_2		17, 17, 18–19	−1.5°C	Nilsson et al. (1997)
	75% CO_2:25% N_2				
	60% CO_2:40% N_2 75% CO_2:25% N_2 90% CO_2:10% N_2		20–22, 22, 22–23		
Pangasius hypophthalmus	60% N_2/35% CO_2/5% O_2	Acceptable	>15	4°C	Maqsood and Benjakul (2010)
Scomberomorus commerson	<0.01% O_2		>25	1°C–2°C	Mohan et al. (2010)
Fresh cod fillets	Air		1.96	4°C	Corbo et al. (2005)
	Vacuum		6		
	65% N_2/30% CO_2/5% O_2		5.42		
	20% CO_2/80% O_2		2.62		
Fish salads (Rainbow trout *Oncorhynchus mykiss*)	Air	Higher sensory quality	7	4°C	Metin et al. (2002)
	O_2:CO_2:N_2 = 1:7:12		14		
	CO_2:N_2 = 3:7				
European sea bass (*Dicentrarchus labrax* fillets)	40% CO_2/60% N_2	Higher sensory quality	8	2°C ± 1°C	Poli et al. (2006)
	Air		7		

Many packaging materials have been used in an attempt to identify containers suitable for storing food products. The ever-increasing consumer demands for freshness, higher quality, safety, and environmental protection have led to the development of active packaging technology (Ozdemir and Floros, 2004).

Respiration activity is a characteristic of fruits and vegetables. MAP technology has widely been used for the preservation of these products. Packages of proper permeability can be used to bind the CO_2 produced by products' respiration and form MAP conditions, extending the shelf life and protecting the qualitative characteristics of the products. At high CO_2 concentrations, plant respiration reduces, resulting in the preservation of the products for a longer period of time (Sandhya, 2010). Fresh or lightly processed fruit and vegetable products are affected by a wide range of microorganisms during harvesting or processing (Rico et al., 2007). Furthermore, several microorganisms such as *Lactobacillus*, *Leuconostoc*, thermophilic *Bacillus*, *E. coli*, and *Saccharomyces cerevisiae* can be found in fruit juices such as orange and apple juice (Keyser et al., 2008). MAP rich in CO_2 can effectively protect fruits and vegetables against harmful microorganisms and also reduce the natural process of ripening (Hintlian and Hotchkiss, 1986). Fruit ripening is a complicated process that involves many reactions and physicochemical and biochemical changes (Kader, 1980).

Although the usual shelf life of fruits can be up to several weeks, after mechanical procedure, (peeling, grating, shredding), they usually get a very limited shelf life of about 1–3 days at chilling conditions. Many physicochemical changes, occurring during storage of fruits and vegetables, result in changes in the products' organoleptic characteristics. MAP is a commonly applied method used in the packaging of these products (Ahvenainen, 1996). Fruits contain large portions of nutrients including several antioxidants such as vitamins C and E and β-carotene (Prior et al., 1998; Alasalvar et al., 2005). In general, atmospheric modification reduces the physiological and chemical changes in fruits and vegetables during storage. It is generally accepted that fruit storage under MAP conditions can preserve products from qualitative degradation for a long period of time, but increased CO_2 concentrations can lead to vitamin C degradation. This phenomenon has been observed in fresh kiwifruit slices by Lee and Kader (2000).

3.6 APPLICATION OF MAP TO FRUITS

3.6.1 Strawberries

According to Zhang et al. (2006), strawberries (*Fragaria ananassa* Duch) are among the fruits with the shortest shelf life (1–2 days when stored in air conditions) due to the increased respiration rate of the product as well as its easy microbiological contamination mainly due to fungi. The most common preservation method of these products is the application of MAP with high CO_2 concentrations and storage at chill temperatures. Under these conditions, the production of C_2H_4 is inhibited due to inactivation of C_2H_4 formation enzymes. Strawberries can be effectively preserved under MAP with at least 10% CO_2 at 10°C, protecting fruits from fungi and improving its appearance and texture. However, very high concentrations of this gas favor the formation of off-flavors degrading the quality of the product (Renault et al., 1994). Four MAPs ((1) 2.5 kPa O_2 + 7 kPa CO_2, (2) 10 kPa O_2 + 5 kPa CO_2, (3) 21 kPa O_2, 60 kPa O_2, and (4) 80 kPa O_2) were applied by Odriozola-Serrano et al. (2010) to examine these atmospheres as potential preservation methods of fresh-cut strawberries, able to preserve the antioxidants and the quality of the fruits. The strawberries were stored under MAP conditions for 21 days at 4°C, and several tests were carried out to detect changes in the concentrations of phenolic derivatives, flavonoids (anthocyanins and flavonols), vitamin C, and antioxidants. It was proved that an increased proportion of O_2 in food environment results in increased losses in phenolic acids and vitamin C. Anthocyanin rates were markedly increased after application of the first three MAPs but remained stable at higher O_2 concentrations. It was demonstrated that low O_2 concentrations and passive atmospheres in combination with storage under chilling conditions preserved the product effectively, maintaining its antioxidant properties at high levels. Figure 3.9 demonstrates the growth of psychrotrophic bacteria of minimally processed litchis and strawberries under MAP conditions.

3.6.2 Cherries

Although preservation of cherries with MAP is quite common, there are only a few reports in the literature on such applications. It was demonstrated that hydrocooling can extend the shelf life of Lambert cherries up to 30 days. Products were stored in plastic bags and fungicides were used to protect them against harmful fungi.

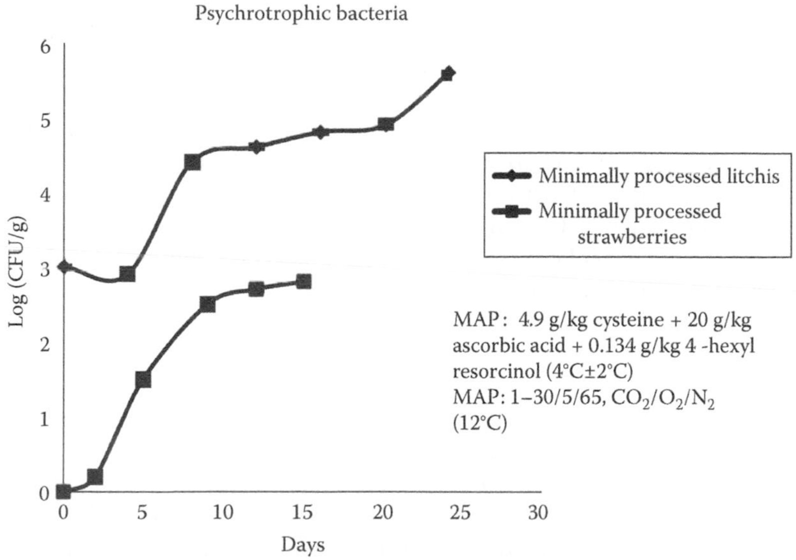

FIGURE 3.9 Impact of different MAPs on psychrotrophic bacteria growth of minimally processed litchis stored at 4°C ± 2°C (From Shah, N.S. and Nath, N., *Lebensm. Wiss. Technol.*, 41, 660, 2008) and minimally processed strawberries stored at 12°C. (From Campaniello, D. et al., *Food Microbiol.*, 25, 992, 2008.)

Furthermore, there are a few reports claiming that MAP is able to effectively preserve strawberries for 2±4 weeks facilitating transport of products over long distances (Remón et al., 2000). According to Alique et al. (2003), application of MAP (8% CO_2/5% O_2 and 10% CO_2/5% O_2) can protect cherry products against fungi contamination, maintaining the quality of the products at high levels. Specifically, MAP protected cherries from rotting, browning of peduncles, color changes, and loss of firmness and acidity, extending the shelf life of these products. It was proved that MAPs of 3%–10% of O_2 and 10%–12% of CO_2 can create the most suitable conditions for extending the shelf life of cherries. The main problem over storage of cherries under MAP rich in CO_2 is the formation of off-odors and off-flavors due to anaerobic respiration. MAP can be used for the preservation of these products provided that the right combination of gases and pressure is used so as to inhibit the action of harmful microorganisms without enhancing the anaerobic respiration (Petracek et al., 2002).

3.6.3 Pineapple

Pineapple (*Ananas comosus* L. Merrill) is a fruit originated from the tropical zone and is characterized by relatively low rates of respiration and C_2H_4 formation. Although these fruits can be found all over the world, very few reports can be found related to their preservation methods (Marrero and Kader, 2006). Rocculi et al. (2009) studied the impact of three different MAPs (86.13 kPa N_2O, 10.13 kPa O_2, and 5.07 kPa CO_2) on qualitative characteristics and microorganism growth in pineapples. Fruits were stored under MAP conditions or in air for 10 days at 4°C. Changes of gases ratio in package internal, organoleptic characteristics, and in the process of ripening were observed and recorded during the experiment. It was proved that MAP rich in N_2O markedly reduced respiration rates and ethylene formation, maintaining the quality of the product at high levels. Under the same MAP, a shelf-life extension of 3–4 days was observed. According to Chonhenchob et al. (2007), passive MAP in combination with a PET package can effectively preserve the qualitative characteristics and prolong the shelf life of fresh-cut pineapples up to 7 days. Products stored under MAP of 6% O_2 and 14% CO_2 were effectively preserved for 13 days. Figure 3.10 depicts the growth of mold–yeasts on fresh-cut pineapples and litchis under MAP conditions at two different temperatures.

3.6.4 Apple

"Bravo de Esmolfe" (BE) is a type of apple originated from Portugal and is well known for its intense aroma. According to Reis et al. (2009) it was characterized as a "Protected Designation of Origin" fruit. It has been found that storage of product under MAP conditions can maintain its quality for a longer period of time, enhancing the stability of cell wall structure. Further investigation is needed to find the best preservation atmospheres for prolonging the shelf life of this apple product. According to Lakakul et al. (1999), a MAP was applied with the aim of recording the respiration rates of apple Malus × domestica Borkh. Products were sliced and stored at 0°C, 5°C, 10°C, and 15°C. It was demonstrated that an increase in temperature

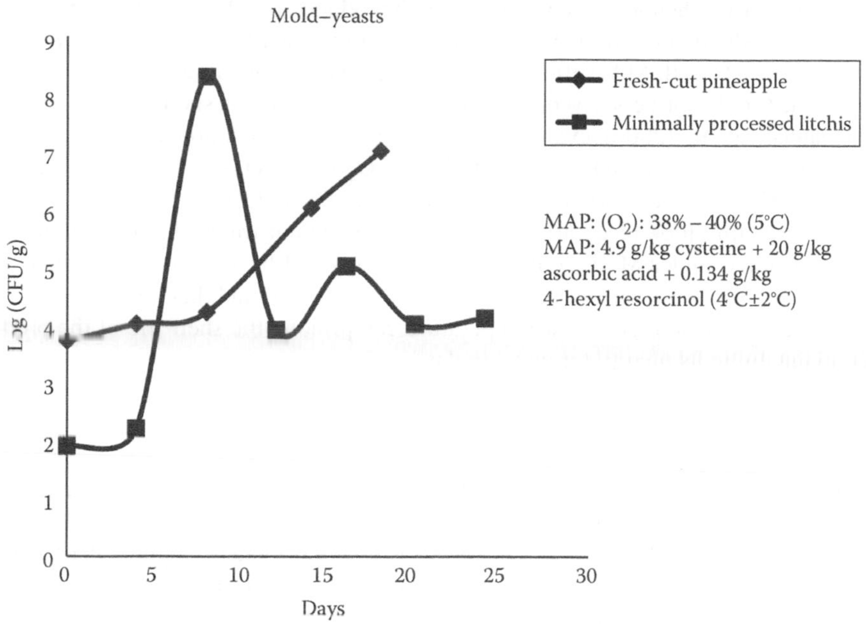

FIGURE 3.10 Effect of different MAPs on mold–yeasts growth of fresh-cut apples stored at 5°C (From Calderón-Montero, M. et al., *Postharv. Biol. Technol.*, 50, 182, 2008) and minimally processed litchis stored at 4°C ± 2°C. (From Shah, N.S. and Nath, N., *Lebensm. Wiss. Technol.*, 41, 660, 2008.)

resulted in enhancement of O_2 consumption and increase in the O_2 partial pressure that leads to fermentation procedures. Several parameters like film permeability to O_2, activation energy of O_2 permeation, temperature, and the film type were described using a computational model. The data can be used for the implementation of appropriate apple packaging.

Soliva-Fortuny et al. (2005) evaluated the quality of Golden Delicious fresh-cut apples after storage under MAP and refrigerated temperatures. The products were immersed in ascorbic acid and $CaCl_2$ and packaged in bags with an O_2 permeability of 15 cm^3/m^2/bar for 3 weeks. It was proved that the method preserved the organoleptic characteristics of the product, maintaining the quality of the fruits at high levels especially during the first 2 weeks of storage. During the third week, the product quality was deteriorated due to increased rates of respiration, although ethylene formation and CO_2 release delayed the ripening progress.

3.6.5 Banana

Nguyen et al. (2004) evaluated the quality of Sucrier bananas (Kluai Khai) in an attempt to clarify whether MAP can be used to effectively preserve this product during a longer period of time. Samples were stored under MAP (12% O_2, 4% CO_2) or in air at 10°C. Products stored under MAP showed less chilling injury involving minimization of browning reactions and, after a period, they were characterized by higher rates of total free phenolics in comparison with products stored in air. Oxidative reactions and production of ammonia derivatives occurred in a larger degree in samples stored in air. Overall, MAP effectively preserved the organoleptic characteristics of the product (pulp softness, sweetness, and flavor) increasing its shelf life.

Chauhan et al. (2006) evaluated the qualitative degradation of banana (var. Pachbale) under active and passive MAP (3% O_2 + 5% CO_2 + 92% N_2) and partial vacuum (52.63 kPa) conditions. Samples were stored at 13°C ± 1°C before ethrel-induced ripening took place at 30°C ± 1°C. It was demonstrated that the quality of products stored under MAP conditions was determined by various factors such as penetration, SF, force relaxation, and TPA factors. It was found that VP and MAP in combination with storage at low temperature can prolong the shelf life of the product, maintaining its quality for up to 36 days.

3.6.6 Litchi Cultivars

Somboonkaew and Terry (2010) evaluated the changes in weight and color and the quantities of pigments, carbohydrates, and organic acids of litchi samples. Samples were packaged in different packages made of microperforated polypropylene (PP), PropaFreshTM PFAM (PF), NatureFlexTM NVS (NVS), and CellophaneTM WS (WS) and maintained at 13°C for 9 days. The highest rates of CO_2 were observed in the insides of packages made of CellophaneTM WS, while lower rates were observed in NVS, PF, and PP packages. It was demonstrated that the MAP created inside PropaFreshTM PFAM packages showed the best results, effectively protecting the quality of the product and maintaining, to a large extent, its initial organoleptic characteristics.

Shah and Nath (2008) evaluated the quality of processed litchi samples stored for 24 days. After peeling, destoning, and processing of produces with antibrowning agents (4.9 g/kg cysteine, 20 g/kg ascorbic acid, and 0.134 g/kg 4-hexyl resorcinol) and osmo-vacuum dehydration, several tests were carried out to assess the nutrient content, the qualitative deterioration of the product, and the microbial contamination. Twenty-four days after treatment, the loss of nutrients (vitamins, carbohydrates, etc.) and organoleptic characteristics was evident, and it is important to mention that microorganisms contributed significantly to this quality degradation. The combination of the two processing procedures prolonged the shelf life of the fruits, making them marketable for up to 24 days.

3.6.7 Pears

The concentration reduction and increase in ascorbic acid after application of various MAPs (1% ± 0.1% or 3% ± 0.1% O_2 + <0.7% or 3% ± 0.1% CO_2, 3% O_2 and <0.7% CO_2, 21% O_2 and <0.7% CO_2 at −0.5°C) and during browning effect were evaluated in 'Conference' and 'Rocha' pears (*Pyrus communis* L.). It was demonstrated that browning reactions and MAP application caused a reduction in ascorbic acid concentration up to 50% of the initial concentration. Specifically, as concerns the Conference cultivar, the greatest losses of ascorbic acid occurred during the transfer to CA, while most Rocha pear losses were caused due to long storage under AP (Veltman et al., 2000).

According to Chavez-Franco and Kader (1993), several Mature-green "Bartlett" pear (*P. communis* L.) fruit disks of given diameter and thickness (8 and 10 mm, respectively) were stored under a continuous flow of air (control) or MAPs with 0.1%, 0.5%, 1%, 5%, 10%, or 20% CO_2 for 10 days at 20°C. Samples stored in atmospheres with 5%–20% CO_2 yielded smaller quantities of ethylene, aminocyclopropane-L-carboxylic acid (ACC), and ACC oxidase (ethylene-forming enzyme) retaining their color to a greater extent. On the other hand, samples stored in an environment of low CO_2 concentrations were characterized by visible quality degradation and a shorter shelf life. In atmospheres with 1% CO_2, ethylene formation was reduced.

Table 3.4 summarizes MAP conditions, packaging material, shelf-life parameters, and usage of other parameters in different kinds of fruit packaging.

3.7 VEGETABLES

The consumption of fresh vegetables is a common practice in all countries. Although raw products maintain all nutritive elements, many of which would be destroyed after processing (e.g., thermal processing), they increase the danger of foodborne diseases (Sengun and Karapinar, 2005). Vegetables are characterized by a high content of nutrients like carbohydrates, proteins, organic acids, vitamins, minerals, and water, making them key elements in a healthy diet. Different practices are applied to preserve the quality of vegetables for longer periods and to extend their shelf life. The basic protection methods of these products are chilling and refrigeration, as temperature is the main parameter causing qualitative degradation (Irtwange, 2006).

Consumption of all carotenoids, and especially of β-carotene, protects human organisms because of their antioxidant and anticancer properties. Chlorophylls are pigments responsible for the color of plant origin foods. A reduction in the quantity of these substances causes discoloration and may reduce consumer acceptance (Tenorio et al., 2004). A few mammalian species, including primates, humans, and guinea pigs, have lost the capability of synthesizing vitamin C (L-ascorbic acid, ascorbate) and therefore require vitamin C as an essential component of their diet. The human being has to take sufficient quantities of vitamin C through diet, as the human body cannot synthesize it. Vegetables are a rich source of vitamin C, and therefore, it is necessary to consume them daily and in large quantities. The vitamin C content may change due to various factors and, for this reason, it can be used for the evaluation of qualitative deterioration (Kim and Ishii, 2007).

According to Tano et al. (2007), temperature can drastically affect the quality of plant origin foods, as increased temperatures enhance oxidative reactions, browning, microbial spoilage, etc., leading to quality degradation. Respiration, transpiration, and ripening rates are directly connected with temperature, and a temperature increase of 10°C can double or triple the rate of these processes.

3.7.1 Physiological Factors Affecting Shelf Life of Minimally Processed Vegetables

Respiration is the process through which energy stored in organic molecules is released to be used in metabolic procedures. It is controlled by enzymes and in the

TABLE 3.4
Impact of MAP (Gas Mix, Packaging Material) in Conjunction with Other Treatments on the Shelf Life of Fruits

Fruit	MAP (Gas Mix)	Packaging Material	Parameters Determined Related to Shelf Life	Usage of Other Parameters	Shelf Life	References
Fragaria × Ananassa, "Camarosa" cv	EMA: 3% O_2, 5% CO_2, balance N_2 HOA: 95% O_2, 5% N_2	Macroperforated bioriental PP film (30 μm)	Gas composition, microbial infection, quality assessment		Control: 3 days MAP: 5 days	Siro et al. (2006)
Fragaria Ananassa Duch	2.5% O_2 15% CO_2	PVC film (48 μm)	Soluble sugars, acidity, titratable acidity, anthocyanin	Two ozone treatments	Control: 2 days MAP: 6–7 days	Zhang et al. (2006)
Fragaria × Ananassa	High 80% CO_2 and low 5% O_2	Plastic	Microbial quality, color, shelf life of fresh-cut strawberries	Chitosan	Control: <5 days MAP: 11–15 days	Campaniello et al. (2008)
Prunus avium	10% O_2 20% CO_2	Nonperforated OPP film (20 μm)	Weight loss, color, firmness, sweetness, microbial infection	Antifungal treatments (eugenol, thymol, menthol, eucalyptol)	Control: 6 days MAP: 16 days	Serrano et al. (2005)
Prunus avium "Lapis" cv	18% O_2 4% CO_2	PE film (40 μm)	Acidity, pH, color, firmness, ethanol and ethylene productivity, vitamin C		Control: 14 days MAP: 20 days	Tian et al. (2004)
Navalinda sweet cherry	3% O_2 12% CO_2	Macroperforated and microperforated PP film (35 μm)	Acidity, color, firmness, rotting	Precooling Hydrocooling	Control: 10 days MAP: 16 days	Alique et al. (2003)
Ananas comosus Gold pineapple	HO: 38%–40% O_2 LO: 10%–12% O_2, 1% CO_2 AIR: 20.9% O_2	PP film (64 μm)	Gas composition, titratable acidity, pH, SS, color, texture, microbial growth, juice leakage	Alginate coating, ascorbic acid, citric acid	Control: 2 days MAP: 30 days	Calderón-Montero et al. (2008)

Ananas comosus Gold pineapple	4% O_2 10% CO_2 86% N_2	PS container	Respiration rate, ethylene production, textural and color deterioration, sensory quality	Ascorbic acid, sucrose, sodium hypochlorite (NaClO)	Control: 2 days MAP: 30 days	Liu et al. (2007)
Ananas comosus Gold pineapple	8% O_2 10% CO_2	Oriented polystyrene cups	Conservation conditions, pH, firmness, quality evaluation	Ascorbic acid, sodium hypochlorite (NaClO)	Control: 2 days MAP: 30 days	Marrero and Kader (2006)
Ananas cosmosus L	86.13 kPa N_2O, 10.13 kPa O_2, and 5.07 kPa CO_2	Plastic bag (40 µm)	Ripening index, firmness, color	Mesophilic bacteria, yeasts, molds	Control: 4–8 days MAP: 12 days	Rocculi et al. (2009)
Malus domestica "Bravo de Esmolfe" cv	13.9% O_2 2.5% CO_2	Plastic 2-PP film (100 µm)	Weight loss, color, pH, SS, firmness		Control: 2 months MAP: 6.5 months	Rocha et al. (2004)
Malus domestica "Borkh" cv	2.5% O_2 7% CO_2		SS, color, pH, sensory analysis, firmness, acidity	Ascorbic acid, calcium chloride, ethanol, acetaldehyde	Control: 6 days MAP: 21 days	Soliva-Fortuny et al. (2005)
Musa acuminate Embul cv	8% O_2 11% CO_2	LDPE film (75 µm)	Organoleptic properties, pathological properties, physicochemical parameters	Cinnamon leaf or bark, clove oils, fungicide	Control: 14 days MAP: 21 days	Ranasinghe et al. (2005)
Pachbale variety	3% O_2 5% CO_2 92% N_2	PE pouches, 100 gouge	Penetration, shear, force relaxation, sensory analysis	Ethylene scrubber	Control: 8 days MAP: 21 days	Chauhan et al. (2006)

presence of O_2, CO_2 and H_2O are produced ($C_6H_{12}O_6 + 6O_2 \rightarrow 6CO_2 + 6H_2O + \text{Energy}$ (heat). The reactions occurring during respiration constitute a serious problem in the preservation of vegetables because of their high energy cost. Furthermore, the heat produced by the plant's metabolic activity can alter storage conditions with adverse effects on the expected shelf life of vegetable products. As a result, increased rates of respiration can cause qualitative deterioration of the products, leading to reduced consumer acceptability (Irtwange, 2006).

Changes in concentrations of CO_2 and O_2 are used to measure the rate of respiration. According to Fonseca et al. (2002), "The usual methods of respiration rate determination are:

(i) the closed or static system,
(ii) the flowing or flushed system and
(iii) the permeable system."

According to Murcia et al. (2009), "high CO_2 stimulates the oxidation of ascorbic acid due to an increase of ascorbate peroxidase activity in response to ethylene production." The loss of cellular compartmentation, due to peeling and cutting, causes mixing of previously sequestered metabolites of the ethylene-generating system, stimulating ethylene production (Rocculi et al., 2005). It was shown that ethylene production can increase after peeling and/or cutting of vegetables. The quality degradation caused by ethylene formation can be avoided by storage in CO_2-enriched atmosphere. According to Beaudry (1999), the partial pressure of oxygen at which the response of ethylene is inhibited by 50% (Ks) is approximately 2.8 kPa.

3.7.2 Chemical Hazards

Vegetables can often be contaminated with various chemicals that may prove hazardous for the health of consumers. Such substances can be various organochlorides (e.g., DDT), aldrin, endrin, heptachlor, hexachlorobenzene, mirex, and toxaphene, known as persistent organic pollutants (POPs), which can threaten human health and damage ecosystems. Furthermore, various chemical substances contained in industrial wastes that can contaminate vegetables include polychlorinated biphenyls (PCBs), dioxins, and furans (Schafer and Kegley, 2002).

3.7.3 Microbiological Factors Affecting Shelf Life

Various types of microorganisms can be found in fresh or lightly processed vegetables. Because of the minimal processing of these plant products (they are commonly consumed when raw), microorganisms can grow, endangering the marketability of the products and the health of the consumers (Rico et al., 2007). The shift of people to the consumption of minimally processed vegetables inevitably leads to an increased risk of outbreaks of foodborne diseases. The need to preserve the heat-sensitive food components requires little or no thermal processing, thereby increasing the risk of pathogen microorganism growth. Consequently, it is necessary to improve the detection systems of various food hazards (Slifko et al., 2000).

Open-field vegetable crops, like carrots, grow in an uncontrolled environment, and the plants are exposed to various microbiological hazards. Several incidents of diseases caused by the consumption of raw vegetables have been recorded (Rudi et al., 2002). Various forms of spoilage may be caused by different microorganisms. Several spoilage microorganisms like pectinolytic bacteria infect a wide range of products causing tissue breakdown, while others infect only specific vegetables. An example is the bacterium *L. mesenteroides* found in carrots after the second week of storage at 10°C (Zagory, 1999).

3.7.4 Pathogenic Organisms

There is a huge variety of microorganisms found in raw vegetables. Species found in each food of plant origin depend on several factors such as the type of the plant, plant growing area, possible treatment, and pesticide use. Because of this complexity, the detection of potential microbiological hazards in each product and the establishment of prerequisite preventive measures can be quite a difficult process (Burnett and Beuchat, 2000).

It has been shown that many enteric pathogen microorganisms found in foods are directly linked to raw fruits and vegetables. Specifically, *E. coli* O157:H7 is often found in tomatoes and lettuce, and hepatitis A is directly connected to the consumption of spring onion (Heaton and Jones, 2008). The MAP preservation method of vegetables has been extensively studied. Some of the most common microorganisms occurring in food products that can cause quality degradation are *Pseudomonas*, *Aeromonas*, *Enterobacteriaceae*, *L. monocytogenes*, *Clostridium*, *Campylobacter* spp., *Cryptococcus*, *Rhodotorula*, and *Candida* spp. The latter three organisms are yeasts, which usually occupy a large proportion (~10%) of the total microbial load of vegetables stored under MAP conditions at 4°C (Edgar and Aidoo, 2001).

3.8 INTERVENTION METHODS

In the last decades, many cases of foodborne diseases associated with consumption of raw or minimally processed vegetables have been reported. This is a result of several factors including increased consumption of these products but also wrong postharvest handling, transportation over long distances, etc. Many measures applied for the inhibition of microorganisms have failed as the exact growth of the population and the physiology of many microorganisms are not well known yet. It is necessary for bactericidal substances to reach all possible places of the produces where microorganisms can grow, but this presupposes very deep knowledge of their behavior (Beuchat, 2002).

According to Sengun and Karapinar (2005) and Rudi et al. (2002), many sanitizers have been examined for their ability to inhibit harmful microorganisms. Chlorinated water can be mentioned as a major sanitizer. Application of chlorine (2 and 10 mg/L) can cause contraction of the bacterial population by 10 times. According to Akbas and Ölmez (2007), many organic acids, which are components of fruits and vegetables, have antimicrobial properties. These include acetic acid, which inhibits the growth of *E. coli*, *L. monocytogenes*, *Salmonella typhimurium*, and *Yersinia enterocolitica*, and lactic acid and citric acid that create an unfavorable environment for the growth of *Aeromonas* and *S. typhimurium*, respectively.

The major organic acids in vegetables are malic and citric acids, malic acid being the most abundant. The sugar and organic acid contents of vegetables may be affected by various factors. Furthermore, vegetables and fruits contain plenty of other organic acids that may provide protection to products against spoilage microorganisms. Some of them are succinic, fumaric, quinic, tartaric, malonic, shikimic, isocitric, and taconitic acids (Haila et al., 1992).

Improving existing sanitizers and finding new more effective ones is now imperative. Ukuku (2004) found that hydrogen peroxide can be used to preserve raw foods of plant origin, inhibiting the microbial populations. The potency of each substance is related with the production of hydroxyl radicals, which can cause extensive damages to the genetic materials of microorganisms. Hydrogen peroxide (H_2O_2) may be involved in the formation of hydroxyl radicals, which are highly reactive, destructive, and result in direct DNA damage (Arranz et al., 2007). According to Ukuku (2004), application of 2.5% and 5% hydrogen peroxide for 5 min can reduce the bacterial population of *Salmonella*. This procedure is indicated for the removal of *Salmonella* from the melon rind before cutting and can prevent possible contamination of edible products from this bacterium.

Gaseous and aqueous ozone, at a low dose and with short contact time, is effective against numerous bacteria, molds, yeasts, parasites, and viruses. The efficacy of ozone as a sanitizer, however, depends on the target microorganism and treatment conditions. Microorganisms inherently vary in sensitivity to ozone. Ozone can be used for the inhibition of various microorganisms met in vegetables. However, different organisms react differently in the presence of ozone. A typical example is bacteria. Although their cells are effectively destroyed in the presence of ozone, the bacterial spores can be maintained even at high concentrations of bactericidal substances (Kim et al., 2003).

Chitosan has recently gained more interest due to its applications in food and pharmaceutics. Among others, the antimicrobial activity of chitosan has been pointed out as one of its most interesting properties. There are many chitosan derivatives that have proven antimicrobial action and can be used for the protection of food products. Some of these are acid-free-water-soluble chitosan, quaternary N-alkylchitosan, sulfonated chitosan, and N-carboxybutyl chitosan. The potential applications of chitosan are numerous, and the use of the method is becoming very common (Devlieghere et al., 2004).

3.9 PHYSICAL TREATMENT FOR MINIMALLY PROCESSED VEGETABLES

According to Suparlan and Itoh (2003), thermal processing is widely used not only to protect vegetable products from spoilage and harmful microorganisms but also to make them more resistant to potential chilling damages when stored under refrigeration. Chlorine is applied by many fresh vegetable processing industries and has the greatest impact among the methods of killing and inactivation of microorganisms, due to the low cost and effectiveness of the method. However, due to the fact that there is evidence of formation of harmful chlorine by-products and environmental pollution, research for finding other methods is carried out (warm water, etc.) (Klaiber et al., 2005).

A variety of heat processes including hot air, vapor heat, dips, sprays, and hot water brushing are used to protect fresh vegetable products from harmful organisms. Although in general these methods can be applied in a wide range of products, different problems are often presented. In peaches treated with hot water (55°C) for 90–300 s, increased vulnerability to *Molina fructicola*–induced deterioration of the quality was observed (Bai et al., 2006). UC-V has been used to inhibit or destroy microorganisms met in different vegetables and fruits like lettuce, pomegranate, and fresh broccoli (Lemoine et al., 2008). Moreover, Nam (2008) used UV-spectrophotometers to evaluate the nitrate NO_3^- concentration of vegetables. It is a very inexpensive method, easy to carry out measurements and distinguished by low maintenance.

3.10 USE OF MAP IN MINIMALLY PROCESSED VEGETABLES

MAP has been proven beneficial in extending the postharvest life span of a wide variety of fresh horticultural commodities.

According to this method, appropriate packaging materials are used and the correct combination of gases is created by CO_2 generated by plants and the appropriate gas diffusion degree through the packaging material (Hu et al., 2005). Fresh or minimally processed vegetables are usually packed in packages of a certain size and weight and using proper packaging materials. The respiration level of every product can be influenced by many factors, and thus the use of predetermined packaging materials and quantity of product in each package is essential to create appropriate conditions for the effective protection of the product (Kim et al., 2004). Moreover, active MAP is often applied for the preservation of vegetables during the last 30 years. Active MAP can be referred to as a package that is not a simple barrier between product and environment but is engaged in a process of product maintenance (Charles et al., 2006).

The concentration of O_2 in air is around 21%. MAP is usually applied to reduce this O_2 percentage in the territory of the package and replace it with other gases (MAP) or vacuum (VP). However, O_2 reduction level should be controlled carefully as O_2 concentration lower than 2%–4% enhances the anaerobic metabolism, resulting in the formation of off-flavors, off-odors, and by-products and finally the degradation of the qualitative characteristics of the product. Furthermore, increased levels of CO_2 cause increase in the rate of ripening due to ethylene. For vegetables that can be stored at high levels of CO_2, the formation of an atmosphere with more than 10% CO_2 is recommended to effectively inhibit most spoilage and pathogen microorganisms (Zagory, 1998).

3.10.1 Lettuce

Kim et al. (2005) used OTRs film of 8.0 and 16.6 pmol/s/m²/Pa to extend the shelf life of Romaine lettuce. The O_2 of the different packages was 0, 1, 2.5, 10, and 21 kPa. It was demonstrated that the qualitative characteristics of the product remained at the highest levels by applying 21 kPa of O_2 and 8.0 pmols/m²/Pa.

Heads of winter-harvested iceberg lettuce (*Lactuca sativa* L. cv. Coolguard) were stored under MAP conditions using perforated (22 mm thickness) or unperforated (25, 30, or 40 mm thickness) polypropylene packages for 14 days at 2°C and during a 2.5 days period at 12°C to evaluate the effectiveness of MAP in preserving this product. All changes and qualitative characteristics were evaluated throughout the experiment. It was proved that MAP in combination with low temperature was able to preserve the quality and extend the shelf life of lettuces. Passive MAP in 40 mm PP and active (initial 5% O_2 and 0% CO_2) MAP in 30 mm PP were the most effective methods of preservation (Martínez and Artés, 1999). Figure 3.11 demonstrates the effect of different MAP conditions on yeast growth of lettuce and fresh-cut pepper stored at 7°C and 5°C, respectively.

3.10.2 Chicory Endive

Charles et al. (2008) evaluated the application of MAP (3 kPa of O_2 and 4.5 kPa of CO_2) on quality degradation of fresh endives. Products were stored either under

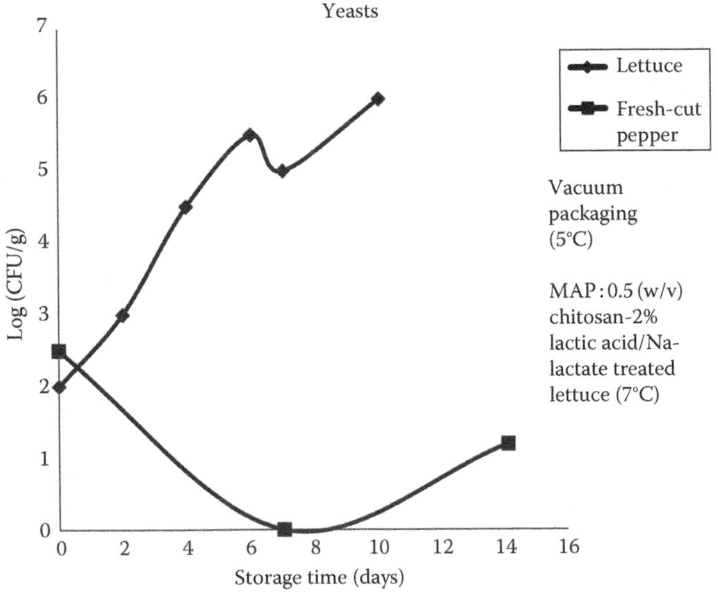

FIGURE 3.11 Impact of different MAPs on yeasts growth of lettuce stored at 7°C (From Devlieghere, F. et al., *Food Microbiol.*, 21, 703, 2004) and fresh-cut pepper stored at 5°C. (From González-Aguilar, G.A. et al., *Lebensm. Wiss. Technol.*, 37, 817, 2004.)

MAP conditions in low-density polyethylene (LDPE) or in air. The effect of O_2 scavengers was also examined. It was demonstrated that although changes in color of the product did not differ between different storage conditions, the use of O_2-scavengers reduced greening and browning of products effectively.

Bennik et al. (1996) examined the application of CA for the preservation of fresh-cut chicory endive. It was demonstrated that MAP, in combination with storage at low temperatures, maintained the quality of the product at high levels protecting it from spoilage and pathogen microorganisms. After treatment, populations of the bacterium *L. monocytogenes* were detected. The size of the populations was affected by various factors such as the initial infection and other species of microorganisms. Although product quality was improved by reducing the total microbial load, the growth of *L. monocytogenes* was favored. It was proved that the providence to avoid contamination is always very important.

3.10.3 Tomato

Artés et al. (1999) examined the effectiveness of MAP (active and passive) and calcium chloride on the preservation of fresh-cut tomato. Samples were stored in plastic trays under MAP or in air at 2°C and 10°C. Visual and sensorial characteristics were evaluated before packaging and after 7 and 10 days under MAP or in air. Under passive MAP conditions, an increase in CO_2 levels of around 6% at 2°C and 20% on the 10th day at 10°C was observed. In contrast, the O_2 level was limited to 14% and 2.5%. Under active MAP conditions (7.5% O_2+0% CO_2), the O_2 level was limited to 6% and 1.5%, while the level of CO_2 was increased to 6% and 14% on the 10th day at 2°C and 10°C, respectively. It was proved that the qualitative characteristics of the product were better preserved at low temperatures regardless of which MAP (active or passive) was applied. Long-life "Calibra" tomatoes (*Lycopersicum esculentum* Mill.) were cut and maintained in PP trays under passive or active (3kPa O_2+0kPa CO_2 and 3kPa O_2+4kPa CO_2) MAP at 0°C and 5°C. After 1 week at 0°C, samples showed a better organoleptic quality in comparison with storage at 5°C. It was obvious that low temperatures enhanced MAP activity, better preserving the quality of the products. At 0°C, MAP maintained the quality of the tomatoes at high levels 2 weeks after storage, but at 5°C, a significant degradation of product quality was observed. MAP in combination with storage at low temperatures limited the growth of harmful microorganisms, extending the shelf life of the product (Aguayo et al., 2004).

Suparlan and Itoh (2003) examined the effectiveness of MAP and hot water application in preserving the quality of tomatoes. Samples were dipped in hot water (42.5°C) for half a minute and then maintained in LDPE film. Control samples were not processed and maintained unpacked for 14 days at 10°C and then for 3 days at 20°C. It was demonstrated that 7 days after packaging, a significant deterioration in the quality of control samples and extensive fungal growth were observed, making the product inappropriate for consumption. Hot water in combination with MAP preserved the qualitative characteristics of the samples extending their shelf life and inhibiting the growth of harmful microorganisms. Figure 3.12 demonstrates the effect of different MAPs on *P. phosphoreum* growth of tomatoes and sterile grated carrots stored at 7°C and 4°C, respectively.

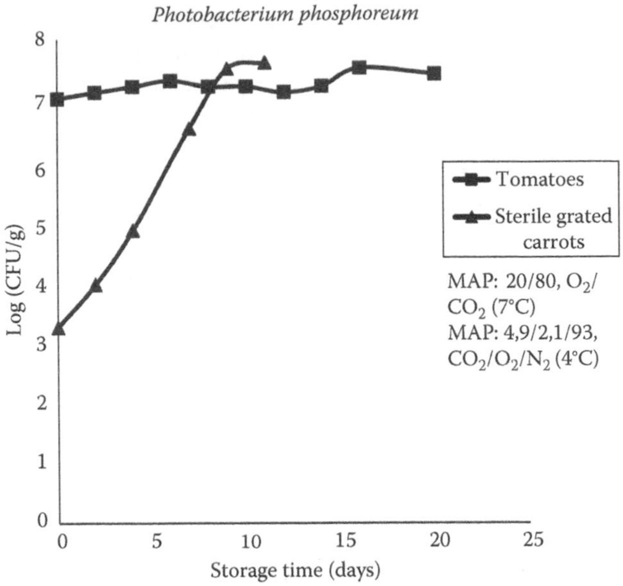

FIGURE 3.12 Effect of different MAPs on *P. phosphoreum* growth of tomatoes stored at 7°C (From Daş, E. et al., *Food Microbiol*., 23, 430, 2006) and sterile grated carrots stored at 4°C. (From Tassou, C.C. and Boziaris, J.S., *J. Sci. Food Agric*., 82, 1122, 2002.)

3.10.4 Carrot

Peeled and aromatic sliced carrots were stored under MAP (active and passive) conditions at low (5% O_2, 10% CO_2, 85% N_2) and high oxygen levels (80% O_2, 10% CO_2, 10% N_2) at 4°C. PP trays were used for the adequate preservation of the samples. Several examinations were carried out during the 21 days of the experiment. It was proved that although the total microbial growth was effectively limited, an increase in the population of mesophilic aerobic bacteria was observed. Because of this growth, samples stored under MAP with relatively high concentrations of O_2 protected the qualitative characteristics of the product, inhibiting the growth of anaerobic microorganisms. Using this method, the shelf life of the carrots was prolonged up to 7 days while MAP characterized by decreased O_2 level preserved the quality of the product just for 2 days (Ayhan et al., 2008).

Tassou and Boziaris (2002) examined the development course of the bacterium *Salmonella enteritidis* in grated carrots maintained in air, under MAP and under VP conditions at 4°C. By the end of the experiment, populations of this bacterium were observed in all samples. Furthermore, the growth of LAB caused a pH reduction. The formation of lactate and other organic acids limited the growth of total microbial counts in all samples because of their action as hurdles.

3.10.5 Broccoli

Three different forms of polypropylene films (macro-perforated, microperforated, and nonperforated) were used for the storage of broccoli (*Brassica oleracea* L. var.

Italica) heads under MAP, for almost 1 month at 1°C. It was demonstrated that the quality of the samples stored under MAP conditions was maintained at significantly higher levels in comparison to control samples stored in air. All organoleptic characteristics of the product were better preserved and the shelf life of broccoli was effectively prolonged (Serrano et al., 2006).

The parameters for maintaining visual and nutritional quality in broccoli heads after harvest are well understood, with low-temperature maintenance being of paramount importance. Relatively little is known of the effects of commonly used postharvest handling procedures designed to maintain broccoli quality on glucosinolate content. Although most factors for preserving the qualitative characteristics of broccoli heads after harvest have been extensively examined, the impact of usually applied postharvest handling to preserve the glucosinolate content of broccoli is now of great importance. Low temperature is the most common method for broccoli preservation. Jones et al. (2006) examined the effectiveness of storage temperature, a_w, application of CA and MAP conditions, and processing on glucosinolate content in broccoli samples. It was demonstrated that the products were effectively preserved by applying MAP (1%–2% O_2 and 5%–10% CO_2) and low-temperature conditions (<4°C) in combination with increased a_w. Carotenoids, vitamins, and other nutrients were also effectively preserved under the above MAP conditions.

3.10.6 Cabbage

Irradiation combined with MAP was used for the preservation of minimally processed Chinese cabbage. Products were stored for 3 weeks at refrigeration temperatures, and several tests were carried out to evaluate their organoleptic characteristics and microbial quality. By application of irradiation, there was an obvious limitation to the growth of all harmful microorganisms. The use of MAP enhanced the irradiation effects, effectively protecting the product from total aerobic and coliform bacteria. Furthermore, the irradiation method limited the pH changes while antiradical, antioxidant activity and the phenolic concentration were barely increased by irradiation at 0.5 kGy. The phenolic contents were limited by irradiation over 1 kGy. It was proved that irradiation is a very effective method for the preservation of cabbage and the maintenance of its quality at high levels (Ahn et al., 2005).

Pirovani et al. (1997) examined the effectiveness of three different packaging materials (monooriented polypropylene film and polyethylene trays overwrapped with a multilayer polyolefin or with a plasticized PVC film) and MAP application in combination with low temperature (3°C) in the preservation of cut cabbage. It was proved that weight decrease was effectively limited regardless of the package used. Although all packaging conditions maintained product at high microbiological quality, maintaining microbial load at low levels, monooriented polypropylene film presented the best results preserving the organoleptic characteristics and maintaining low rates of wilting and browning. The only disadvantage of the method was the formation of an off-odor.

3.10.7 Mung Bean

The effectiveness of ClO_2 treatment in combination with MAP and refrigeration conditions in the preservation of mung bean was examined by Jin and Lee (2007). MAP effectively protected products from harmful microorganisms like total mesophilic microorganisms, *Salmonella Typhimurium*, and *L. monocytogenes*. Samples were stored under four different storage conditions (in air, under vacuum, CO_2 gas, and N_2 gas) and then water or 100 ppm ClO_2 was applied for 5 min. All products were stored at 5°C ± 2°C, and it was proved that ClO_2 significantly limited the growth of *S. Typhimurium* and *L. monocytogenes* (by 3.0- and 1.5-log CFU/g, respectively) and MAs maintained the microbial loads at low levels. It was demonstrated that MAP combined with ClO_2 can effectively protect mung bean sprouts from spoilage microorganisms and maintain the quality of the product at high levels.

Varoquaux et al. (1996) investigated the respiration rate of fresh bean sprouts in order to store them in an appropriate package (permeabilities ranging from 950 to 200,000 mL O_2/m^2/day/atm, and stored at 8°C) achieving the longest shelf life. After determination of the respiration rate at 1 mmol O_2/kg/h at 10°C, it was found that by storing products at 8°C and using 0.24 m^2 of film per kg of sprouts and, as a consequence, the formation of MAP (5% O_2, 15% CO_2), the slowest change in pH and the smallest degradation of product quality took place.

3.10.8 Artichoke

Gil-Izquierdo et al. (2002) examined the effectiveness of MAP in the preservation of artichoke nutrients like phenolics concentration and vitamin C. Samples were stored in packages made of perforated polypropylene (Control), polyvinylchloride (PVC), LDPE, and three microperforated polypropylene films (PP1, PP2, and PP3). The MA achieved was 14.4 kPa + 5.2 kPa for PVC, 7.7 kPa + 9.8 kPa for LDPE, 8.6 kPa + 15.3 kPa for PP1, 9.7 kPa + 13.8 kPa for PP2, and 13.7 kPa + 9.5 kPa for PP3. It was demonstrated that 8 days after packaging at 5°C there was a significant deterioration in qualitative characteristics of the products mainly due to a_w decrease. MAP limited the qualitative deterioration by maintaining humidity at high levels preserving vitamin C concentration.

3.10.9 Cucumber

Wang and Qi (1997) used perforated or sealed 31.75 pm (1.25 mil) LDPE bags to reduce the qualitative degradation from chilling storage. Samples were stored at 5°C and 90–95X1 relative humidity, while control samples remained unpacked. An increase in the level of CO_2 (3%) and decrease in the level of O_2 (16%) were observed. Products stored in sealed bags maintained their weight almost at initial levels (1% loss of weight), while control samples reduced their weight up to 9% in 18 days. Polyamine formation may be a very important factor in the maintenance of quality under chilled temperatures.

According to Karakaş and Yıldız (2007), lipid peroxidation in slightly processed cucumbers maintained under MAP or in covered Petri dishes was evaluated by the FOX2 and TBARS assays. The level of lipid hydroperoxides and TBARS was recorded at 1.44–2.00 and 0.11–0.20 nmol/g, respectively. It was demonstrated that samples stored under MAP with increased concentration of O_2 significantly limited the formation of lipid hydroperoxides and TBARS. The effectiveness of $CaCl_2$ and/or ascorbic acid on peroxidation was studied in the same investigation. Samples stored under MAP in sealed packages presented a significant tissue hardness from the third day of storage. However, qualitative characteristics of samples stored under MAP in combination with chilling conditions were prolonged up to 6 days.

3.10.10 Sweet Potato

Erturk and Picha (2007) examined the low (PD900), medium (PD961), and high permeability (PD941) film packages as potential methods for the preservation of qualitative characteristics of minimally processed sweet potatoes. Samples were stored at 2°C or 8°C for a 2 week period. It was proved that the weight of samples was better preserved at 2°C while there were significant weight losses by storing the products in PD941 film bags and at 8°C. None of the methods of packaging presented significant quality degradation at 2°C. Dry matter and alcohol-insoluble solids (AIS) and mineral concentrations were limited, whereas nutrients like fructose, sucrose, total sugar, and total carotenoid content increased during storage. These changes were more notable at 2°C than at 8°C. Preservation methods have no impact on glucose and crude protein concentrations.

McConnell et al. (2005) examined the effectiveness of MAP (5% O_2, 4% CO_2, 91% N_2) and low or medium O_2 permeability packages in quality preservation of minimally processed sweet potatoes. It was shown that samples stored in air were preserved just for 1 week at 4°C. Under MAP conditions, the shelf life of samples was extended up to 2 weeks at the same temperature. MAP in combination with moderately O_2-permeable film (7000 $cm^3/atm/m^2/24\,h$) effectively preserved the qualitative characteristics of the product, maintaining nutrient content at high levels. Moreover, MAP protected potatoes from harmful microorganisms by limiting the microbial growth.

3.10.11 Cauliflower

Simón et al. (2008) used three packaging films (PVC, nonperforated PVC, and PP films) to evaluate the preservation of quality in cauliflower. Samples were maintained at 4°C or 8°C for up 20 days. Several tests were carried out throughout the experimental procedure to record changes in the organoleptic properties of the samples. It was demonstrated that the atmosphere formed in PVC films was more stable in comparison to the other two packaging materials. Yellowing was significantly affected by the conditions created in each package. Mesophilic bacteria and *Pseudomonas* remained at low levels (<7 log CFU/g), especially by using nonperforated PVC and PP films. Except for a significant loss of carbohydrates (27% in 20 days of storage), no other degradation of nutrients was observed.

Sanz et al. (2007) evaluated the effect of four different films (one PVC and three P-Plus) in the quality preservation and shelf-life extension of cauliflower. Samples were stored for 25 days at 4°C. It was proved that the microbial load initially existing in products played a key role in the microbial degradation of the final product. MAP formed in P-Plus 120 film reduced the concentration of O_2 in less than 10%, whereas it increased the concentration of CO_2 over 10%, effectively preserving the qualitative characteristics of the final product for more than 25 days.

3.11 EGGPLANT

The physicochemical parameters (pH, mechanical firmness, and vitamin C) and sensory parameters of grafted and ungrafted eggplant plants were studied in relation to storage time (up to 17 days at 10°C). Eggplant plants of cultivar 'Tsakoniki' were grafted on *Solanum torvum* and *Solanum sisymbriifolium* rootstocks to avoid the soilborne diseases caused by *Verticillium dahliae*. The fruits were stored under MAP. Vitamin C was negatively affected by grafting over storage, while MAP prolonged the shelf life. pH was not affected by grafting but positively affected by MAP. Flesh firmness was negatively affected by grafting and reduced over storage but positively affected by MAP. Sensory analysis showed higher ratings for fruits from ungrafted plants for sweetness, acceptance, and hardness, whereas no difference was detected for overall acceptance. Fruits stored under MAP were better maintained compared with those stored in air (Arvanitoyannis et al., 2005).

3.11.1 Potato

Beltrán et al. (2005) used different sanitizers to preserve the organoleptic characteristics and inhibit the growth of harmful microorganisms in minimally processed potatoes. Samples were stored under MAP and VP conditions, and washing treatments containing water, sodium sulfite, sodium hypochlorite, Tsunami, ozone, and the combination of ozone–Tsunami were applied and examined. It was demonstrated that VP showed better results compared with MAP. Although MAP protected the product from extensive browning, it resulted in the formation of off-odors, degrading its quality. Ozonated water or ozone–Tsunami and storage under VP effectively protected the product from browning for up to 2 weeks, maintaining the quality of the samples at high levels.

Tudela et al. (2002) examined the effectiveness of MAP and deep-freezing (DF) on the preservation of L-ascorbic acid (AA) concentration of long-term stored and fresh potato strips. Fresh, minimally processed, and deep frozen samples were maintained for 6 days at 4°C and 5 weeks at −22°C, respectively. MAP storage caused a degradation in AA content (by 14%–34%) in comparison to products stored in air. Samples stored at −22°C presented a reduction in AA concentration (23%) of "Spunta" potato strips 5 weeks after packaging, whereas "Agria" tubers did not present any change. It was demonstrated that samples stored in air at 4°C maintained their vitamin content for 6 days while those stored under MAP and DF conditions had a limited shelf life.

3.11.2 Celery

Gómez and Artés (2005) examined the effectiveness of MAP formed in two films made of polymers on physicochemical, organoleptic, and microbial characteristics of "Trinova" cv. celery sticks stored for 2 weeks at 4°C. Samples were stored in hermetically sealed plastic bags made of LDPE, oriented polypropylene (OPP), and perforated polyethylene. It was shown that MAP maintained the qualitative characteristics of the product, reducing color losses and inhibiting the growth of harmful microorganisms. The quality of samples stored in OPP bags was preserved for 15 days at 4°C.

Several MAPs (0% CO_2/100% N_2, 5% CO_2/95% N_2, 10% CO_2/90% N_2, 20% CO_2/80% N_2, 30% CO_2/70% N_2, 50% CO_2/50% N_2 plus 2% O_2/98% N_2) were applied to evaluate the best preservation method of minimally processed celeriac samples. The color (recorded in the CIE L*a*b* system), organoleptic characteristics, and microbial quality of the product were determined throughout the experiment. After 12 days of storage at 4°C, samples stored in MAP of 5% or 10% CO_2, 2% O_2 maintained their qualitative characteristics at high levels in comparison to products stored under the other MAPs or in air. The same combination of gases effectively inhibited the growth of all harmful microorganisms (Radziejewska-Kubzdela et al., 2007).

3.11.3 Pepper

Lee et al. (1994) evaluated the effectiveness of MAP and VP on the shelf-life extension of minimally processed bell pepper stored at 5°C and 10°C. Different factors of qualitative degradation were determined and evaluated throughout the experiment. It was demonstrated that microbial spoilage limited the shelf life of products to 14 days at 10°C and 21 days at 5°C, indicating the importance of low-temperature conditions in the effectiveness of MAP. Packaging materials made of 25 pm LDPE and 30 pm cast polypropylene could form a MAP of 3% O_2 and 5% CO_2, achieving the best quality preservation. Figure 3.13 demonstrates the effect of MAP conditions on LAB growth of tomatoes and peppers at two different temperatures.

Table 3.5 summarizes the gas mix, materials, other characteristics, and microflora on different vegetables packaging.

3.12 CONCLUSIONS

The MAP preservation technique can be described as the replacement of the air content of the package with a combination of gases, creating an adverse environment for the growth of harmful microorganisms and several oxidative and browning reactions. This balance between internal and external package atmosphere, packaging materials, and different products must be maintained to effectively preserve food (Rodriguez-Aguilera et al., 2009). It has been proved that increased concentration of CO_2 or N_2 and decreased O_2 levels protect food products against harmful microorganisms and extend their shelf life, facilitating their transportation and storage (Özogul et al., 2004).

Many perishable dairy products are packaged and maintained under MAP conditions. The nature of these products and the increased proportion of CO_2 produced during maturation permit the storage of dairy products at high concentrations of CO_2. This enables the effective protection of products against harmful microorganisms (Jakobsen and Risbo, 2009).

The shelf life of fresh or minimally processed fish products is very limited because the microbial activity causes rapidly extended damages to these products

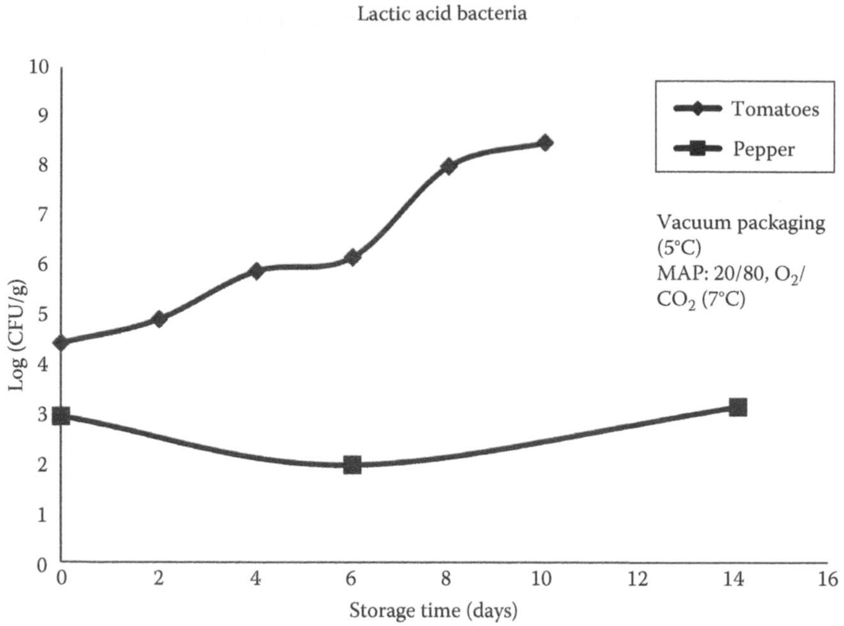

FIGURE 3.13 Effect of different MAPs on LAB growth of tomatoes stored at 7°C (From Daş, E. et al., *Food Microbiol.*, 23, 430, 2006) and pepper stored at 5°C. (From González-Aguilar, G.A. et al., *Lebensm. Wiss. Technol.*, 37, 817, 2004.)

making them inappropriate for consumption. The activity of different microorganisms and enzymes that cause spoilage of fish products is affected by various factors including temperature and storage conditions (Altieri et al., 2005).

MAP has been widely applied to protect fish and fisheries against harmful microorganisms and extend their shelf life, maintaining their quality at high levels. This method is usually used to preserve products like shrimps for long periods of time, and many studies have been carried out describing these applications (Lu, 2009). The main problem during the application of MAP in fisheries is the production of toxins by the nonproteolytic clostridia. These microorganisms can grow under refrigeration conditions, in air, 100% O_2, or even under MAP, making products dangerous and inappropriate for consumption (Huss, 1997). Nevertheless, MAP is widely applied to protect fish products against harmful microorganisms (Nilsson et al., 1997).

The trend of consumers switching to raw fruits and vegetables has led to the development of research around food packaging with MAP application. It has been demonstrated by many studies that MAP limits respiration levels and browning while increasing the preservation of the products' qualitative characteristics (Chonhenchob et al., 2007). According to Irtwange (2006), "Certain climacteric fruits, e.g., apples, avocados pears and tomatoes, experience a marked and transient increase in respiration during their ripening which is associated with increased production of and sensitivity to ethylene." Respiration rates can be altered under MAP conditions resulting in the reduction of ripening procedure.

TABLE 3.5
Impact of MAP (Initial Gas Mix, Packaging Materials) on Physical (Color, Texture-Weight Loss), Microbiological, and Sensory Analysis and Shelf-Life Extension

Species and Food Type	MAP (Initial Gas Mix)	Packaging Material	Storage Temperature (°C) and Storage Period (Days)	Color	Microflora	Texture-Weight Loss	Sensory Analysis	Shelf-Life Extension (Days)	References
Chicory endive *Cichorium endivia* L.	(1) Air (2) 1.5% O_2/20% CO_2 (3) 0% O_2/20% CO_2	Airtight containers with continuous flow of gas mixture (flow rate 200 mL/min)	8°C—9 days		No significant differences in Total Aerobic Count (TAC) *Pseudomonas* counts were lower under 0% O_2. *Enterobacteria* were more abundant under 0% O_2 *L. Monocytogene* growth rate higher in low O_2 and high CO_2 conditions			Under air: 2 days MAP: 4 days	Bennik et al. (1996)

(*continued*)

TABLE 3.5 (continued)
Impact of MAP (Initial Gas Mix, Packaging Materials) on Physical (Color, Texture-Weight Loss), Microbiological, and Sensory Analysis and Shelf-Life Extension

Species and Food Type	MAP (Initial Gas Mix)	Packaging Material	Storage Temperature (°C) and Storage Period (Days)	Color	Microflora	Texture-Weight Loss	Sensory Analysis	Shelf-Life Extension (Days)	References
Endive *Cichorium intybus* L.	Passive MAP: 3 kPa O_2/5 kPa CO_2 with 50% reduction of transient period for active MAP (with O_2 scavengers)	LDPE film (50 μm), macroperforated OPP	20°C—7 days	Color changes of endives delayed with active MAP			Passive MAP: the browning was delayed Active MAP: acceptable appearance maintained	Active MAP: prolonged acceptable appearance of endives from 3 to 6 days	Charles et al. (2008)
Tomato slices (*Lycopersicum esculentum Mill.*)	(1) Passive MAP (2) 3 kPa O_2 (3) 3 kPa O_2/4 kPa CO_2	OPP film of 35 μm thickness	0°C and 5°C–14 days	At 5°C, a slight ripening occurred	Temperature more effective than MAP in reducing microbial growth Fungi were detected only in control samples at 5°C		Passive MAP: the C_2H_4 level was higher at 5°C than at 0°C. Effect of temperature inhibited by	Combined use of 0°C and passive or active MAP allowed a shelf life of	Aguayo et al. (2004)

					Yeast counts were higher in control than in MAP samples		active MAP. Significant reduction in sensorial scores at 0°C and 5°C	14 days	
Tomato (*Lycopersicum esculentum Mill.*) cv *Rambo*, *Durinta*, *Bodar*, *Pitenza*, *Cencara*, and *Bola*	(1) 5 kPa O_2/5 kPa CO_2	(1) ILPRA with O_2 and CO_2 permeabilities 110 and 500 cm^3/m^2 day. bar at 23°C and 0% RH	4°C–21 days	Neither CIE a* nor CIE b* shifted significantly with time			Vitamin C concentration revealed no substantial variations Phenolics enhanced after 14 days of storage at 4°C	Fresh-cut tomatoes maintained the main antioxidant compounds and color parameters for 21 days at 4°C ± 1°C	Odriozola-Serrano et al. (2008)
Cherry tomatoes	(1) Passive MAP: 6% O_2/4% CO_2 (2) CA with 5% CO_2	LDPE film (50 µm), macroperforated OPP	7°C for 20 days and 22°C for 10 days		Death rate of *S. Enteritidis* in MA greater than that stored in CA at 7°C. LAB and TVC grew regardless of the atmosphere and storage temperature	At the last 3–4 days of limited softening and darkening recorded at 22°C			Daş et al. (2006)

(*continued*)

TABLE 3.5 (continued)
Impact of MAP (Initial Gas Mix, Packaging Materials) on Physical (Color, Texture-Weight Loss), Microbiological, and Sensory Analysis and Shelf-Life Extension

Species and Food Type	MAP (Initial Gas Mix)	Packaging Material	Storage Temperature (°C) and Storage Period (Days)	Color	Microflora	Texture-Weight Loss	Sensory Analysis	Shelf-Life Extension (Days)	References
Grated carrots	(1) Passive MAP (2) 2.1% O_2/4.9% CO_2	(1) PE bags with O_2 and CO_2 permeability 1000 and 5450 mL/m^2 day bar	4°C—12 days		*S. enteritidis*		Samples inoculated with *Lactobacillus* or *Salmonella* showed a marked pH decrease		Tassou and Boziaris (2002)
Orange and purple carrots	(1) Air (2) 5% O_2/5% CO_2 (3) 95% O_2/5% CO_2	PE bags	5°C ± 2°C–13 days				Loss of total carotenoids (in the 95% O_2). 5% O_2 + 5% CO_2 reduced phenols	Shredded purple carrot can be stored under 5% O_2/5% CO_2 (up to 10 days)	Alasalvar et al. (2005)

Asparagus (*Asparagus officinalis* L.)	Passive MA: (1) 15% O_2/5% CO_2 at 2°C (2) 5% O_2/13% CO_2 at 10°C	(1) 2°C for 26–33 days (2) 2°C for 5 days and then at 10°C for 20 days		Total oxygenated carotenoid loss was lower, whereas higher retention of chlorophylls, lower increases in pheophytins	MAP was more effective in prolonging the shelf life for up to 4 weeks at 2°C	Tenorio et al. (2004)

Several methods have been used for the preservation of meat products and the extension of their shelf life. VP and MAP have attracted the interest of researchers because of their efficiency and the reduced impact on the sensory components of these products (Parra et al., 2010).

During the last decades, a large number of industries have used MAP technology to extend the shelf life of perishable products (O'Mahony et al., 2006). MA has proved very effective for the protection of products from spoilage and pathogen microorganisms (Baert et al., 2009).

ABBREVIATIONS

AIS	alcohol-insoluble solids
ALOP	appropriate level of protection
AN	acid number
AP	active packaging
APC	aerobic plate counts
BE	Bravo de Esmolfe
BSE	bovine spongiform encephalopathy
CA	Codex Alimentarius
CAL	calcium lactate
CCFH	Codex Committee on Food Hygiene
CCP	critical control point
CFU	colony forming unit
CL	cooking loss
CLA	conjugated linoleic acid
COMb	carboxymyoglobin
DDT	dichlorodiphenyltrichloroethane
DL	drip loss
DMB	deoxymyoglobin
EDTA	ethylene diaminetetraacetic acid
FAO	Food and Agriculture Organization
FTIR	Fourier transform infrared
GAP	good agricultural practices
GMP	good manufacturing practices
HACCP	hazard analysis critical control point
HAV	hepatitis A virus
HDPE	high-density polypropylene
HHP	high hydrostatic pressure
HMF	hydroxymethyl furfural
HP	high pressure
HPLC	high-performance liquid chromatography
HTST	high temperature short time
ICMSF	International Commission on Microbiological Specifications for Foods
LAB	lactic acid bacteria
LAL	limulus amoebocyte lysate
LDPE	low-density polyethylene

MA	modified atmosphere
MANOVA	multivariate analysis of variance
MAP	modified atmosphere packaging
MMb	metmyoglobin
MMB	oxidized metmyoglobin
NADH	nicotinamide adenine dinucleotide
ODF	orange dietary fiber
OEO	Oregano essential oil
OMb	oxymyoglobin
OMB	oxygenated oxymyoglobin
OPP	oriented polypropylene
PCA	principal component analysis
PCBs	polychlorinated biphenyls
PDO	protected designation of origin
PE	polyethylene
PET	polyethylene terephthalate
PLA	poly-lactic acid
PLS	partial least-squares
PO	performance objective
PP	polypropylene
PUT	putrescine
PVC	polyvinyl chloride
SF	shear force
SM	semimembranosus muscle
SMP	hexamethaphosphate
SPD	spermidine
SPM	spermine
SPS	sanitary and phytosanitary
SR	sensory requirements
TBARS	thiobarbituric acid reactive substances
TL	total lipids
TMA	trimethylamine
TMAN	trimethylamine nitrogen
TPL	total phospholipids
TRA	total reducing activity
TTI	time temperature indicator
TVBN	trimethylamine nitrogen
VP	vacuum packaging
WHC	water-holding capacity
WHO	World Health Organization

REFERENCES

Aguayo, E., Escalone, V., and Artés, F. (2004). Quality of fresh-cut tomato as affected by type of cut packaging, temperature and storage time. *European Food Research and Technology*, 219: 492–499.

Ahn, H.J., Kim, J.H., Kim, J.K., Kim, D.H., Yook, H.S., and Byun, M.W. (2005). Combined effects of irradiation and modified atmosphere packaging on minimally processed Chinese cabbage (*Brassica rapa* L.). *Food Chemistry*, 89: 589–597.

Ahvenainen, R. (1996). New approaches in improving the shelf life of minimally processed fruit and vegetables. *Trends in Food Science and Technology*, 7: 179–187.

Akbas, M.Y. and Ölmez, H. (2007). Inactivation of *Escherichia coli* and *Listeria monocytogenes* on iceberg lettuce by dip wash treatments with organic acids. *The Society of Applied Microbiology*, 44: 619–624.

Alasalvar, C., Al-Farsi, M., Quantick, P.C., Shahidi, F., and Wiktorowicz, R. (2005). Effect of chill storage and modified atmosphere packaging (MAP) on antioxidant activity, anthrocyanins, carotenoids, phenolics and sensory quality of ready-to-eat shredded orange and purple carrots. *Food Chemistry*, 89: 69–76.

Alique, R., Martinez, M.A., and Alonso, J. (2003). Influence of the modified atmosphere packaging on shelf life and quality of Navalinda sweet cherry. *European Food Research and Technology*, 217: 416–420.

Altieri, C., Speranza, B., Del Nobile, M.A., and Sinigaglia, M. (2005). Suitability of bifidobacteria and thymol as biopreservatives in extending the shelf life of fresh packed plaice fillets. *Journal of Applied Microbiology*, 99: 1294–1302.

Alves, R.M.V., De Luca Sarantópoulos, C.I.G., Van Dender, A.G.F., and Faria, J.D.A.F. (1996). Stability of sliced Mozzarella cheese in modified-atmosphere packaging. *Journal of Food Protection*, 59(8): 838–844.

Alves, R.M.V., Van Dender, A.G.F., Jaime, S.B.M., Moreno, I., and Pereira, B.C. (2007). Effect of light and packages on stability of spreadable processed cheese. *International Dairy Journal*, 17: 365–373.

Ammor, M.S., Argyri, A., and Nychas, G.J.E. (2009). Rapid monitoring of the spoilage of minced beef stored under conventionally and active packaging conditions using Fourier transform infrared spectroscopy in tandem with chemometrics. *Meat Science*, 81: 507–518.

Andersen, C.M., Wold, J.P., and Mortensen, G. (2006). Light-induced changes in semi-hard cheese determined by fluorescence spectroscopy and chemometrics. *International Dairy Journal*, 16: 1483–1489.

Antle, J.M. (1999). Benefits and costs of food safety regulation. *Food Policy*, 24: 605–623.

Arranz, N., Haza, A.I., García, A., Delgado, E., Rafter, J., and Morales, P. (2007). Effect of organosulfurs, isothiocyanates and vitamins C towards hydrogen peroxide-induced oxidative DNA damage (strand breaks and oxidized purines/pyrimidines) in human hepatoma cells. *Chemico-Biological Interactions*, 169: 63–71.

Artés, F., Conesa, M.A., Hernández, S., and Gil, M.I. (1999). Keeping quality of fresh-cut tomato. *Postharvest Biology and Technology*, 17: 153–162.

Arvanitoyannis, I.S. and Kourtis, L. (2001). *Food Safety and Application of HACCP to the Foods and Drinks Industry*. University Studio Press, Thessaloniki, Greece, pp. 32–58.

Arvanitoyannis, I.S., Kargaki, G., and Hadjichristodoulou, C. (2011a). Effect of three MAP compositions on the physical and microbiological properties of a low fat Greek cheese known as "Anthotyros." *Anaerobe*, 30: 1–3.

Arvanitoyannis, I.S., Kargaki, G., and Hadjichristodoulou, C. (2011b). Effect of several MAP compositions on the microbiological and sensory properties of Graviera cheese. *Anaerobe*, 30: 1–5.

Arvanitoyannis, I.S., Khah, E.M., Christakou, E.C., and Bletsos, F.A. (2005). Effect of grafting and modified atmosphere packaging on eggplant quality parameters during storage. *International Journal of Food Science and Technology*, 40: 311–322.

Audenaert, K., D'Haene, K., Messens, K., Ruyssen, T., Vandamme, P., and Huys, G. (2009). Diversity of lactic acid bacteria from modified atmosphere packaged sliced cooked meat products at sell-by date assessed by PCR-denaturing gradient gel electrophoresis. *Food Microbiology*, 27(1): 1–24.

Ayhan, Z., Eştürk, O., and Taş, E. (2008). Effect of modified atmosphere packaging on the quality and shelf life of minimally processed carrots. *Turkish Journal of Agricultural and Forestry*, 32: 57–64.

Babić, J., Cantalejo, M.J., and Arroqui, C. (2009). The effects of freeze-drying process parameters on *Broiler* chicken breast meat. *Lebensmittel Wissenschaft und Technologie—Food Science and Technology*, 42: 1325–1334.

Baert, L., Debevere, J., and Uyttendaele, M. (2009). The efficacy of preservation methods to inactivate foodborne viruses. *International Journal of Food Microbiology*, 131: 83–94.

Bai, L., Ma, C., Gong, S., and Yang, Y. (2007). Food safety assurance systems in China. *Food Control*, 18: 480–484.

Bai, J., Mielke, E.A., Chen, P.M., Spotts, R.A., Serdani, M., Hansen, J.D., and Neven, L.G. (2006). Effect of high-pressure hot-water washing treatment on fruit quality, insects, and disease in apples and pears. Part I. System description and the effect on fruit quality of 'd' Anjou' pears. *Postharvest Biology and Technology*, 40: 207–215.

Bakalivanova, T., Grigorova, S., and Kaloyanov, N. (2009). Effect of irradiation and packaging on lipid fraction of Bulgarian salami during storage. *Radiation Physics and Chemistry*, 78: 273–276.

Baker, D.A., Genigeorgis, C., Glover, J., and Razavilar, V. (1990). Growth and toxigenesis of *C. botulinum* type E in fishes packaged under modified atmospheres. *International Journal of Food Microbiology*, 10: 269–290.

Beaudry, R.M. (1999). Effect of O_2 and CO_2 partial pressure on selected phenomena affecting fruit and vegetable quality. *Postharvest Biology and Technology*, 15: 293–303.

Becker, E.M., Christensen, J., Frederiksen, C.S., and Haugaard, V.K. (2003). Front-face fluorescence spectroscopy and chemometrics in analysis of yogurt: Rapid analysis of riboflavin. *Journal of Dairy Science*, 86: 2508–2515.

Beltrán, D., Selma, M.V., Tudela, J.A., and Gil, M.I. (2005). Effect of different sanitizers on microbial and sensory quality of fresh-cut potato strips stored under modified atmosphere or vacuum packaging. *Postharvest Biology and Technology*, 37: 37–46.

Bennik, M.H.J., Peppelenbos, H.W., Nguyen-the, C., Carlin, F., Smid, E.J., and Gorris, L.G.M. (1996). Microbiology of minimally processed, modified-atmosphere packaged chicory endive. *Postharvest Biology and Microbiology*, 9: 209–221.

Bergamo, P., Fedele, E., Iannibelli, L., and Marzilla, G. (2003). Fat-soluble vitamin contents and fatty acid composition in organic and conventional Italian dairy products. *Food Chemistry*, 82: 625–631.

Beuchat, L.R. (2002). Ecological factors influencing survival and growth of human pathogens on raw fruits and vegetables. *Microbes and Infection*, 4: 413–423.

Birollo, G.A., Reinheimer, J.A., and Vinderola, C.G. (2001). Enterococci vs non-lactic acid microflora as hygiene indicators for sweetened yoghurt. *Food Microbiology*, 18: 597–604.

Bøknæs, N., Jensen, K.N., Andersen, C.M., and Martens, H. (2002a). Freshness assessment of thawed and chilled cod fillets packed in modified atmosphere using near-infrared spectroscopy. *Lebensmittel Wissenschaft und Technologie*, 35: 628–634.

Bøknæs, N., Jensen, K.N., Guldager, H.S., Østerberg, C., Nielsen, J., and Dalgaard, P. (2002b). Thawed chilled Barents sea cod fillets in modified atmosphere packaging-application of multivariate data analysis to select key parameters in good manufacturing practice. *Lebensmittel Wissenschaft und Technologie*, 35: 436–443.

Bórnez, R., Linares, M.B., and Vergara, H. (2009a). Effects of stunning with different carbon dioxide concentrations and exposure times on suckling lamb meat quality. *Meat Science*, 81: 493–498.

Bórnez, R., Linares, M.B., and Vergara, H. (2009b). Microbial quality and lipid oxidation of Manchega breed suckling lamb meat: Effect of stunning method and modified atmosphere packaging. *Meat Science*, 83: 383–389.

Boselli, E., Rodriguez-Estrada, M.T., Fedrizzi, G., and Caboni, M.F. (2009). Cholesterol photosensitised oxidation of beef meat under standard and modified atmosphere at retail conditions. *Meat Science*, 81: 224–229.
Burnett, S.L. and Beuchat, L.R. (2000). Human pathogens associated with raw produce and unpasteurized juices, and difficulties in decontamination. *Journal of Industrial Microbiology and Biotechnology*, 25: 281–287.
Buttriss, J. (2007). Nutritional properties of fermented milk products. *International Journal of Diary Technology*, 50: 21–27.
Cabeza, M.C., de la Hoz, L., Velasco, R., Cambero, M.I., and Ordóñez, J.A. (2009). Safety and quality of ready-to-eat dry fermented sausages subjected to E-beam radiation. *Meat Science*, 83: 320–327.
Cabo, M.L., Herrera, J.J.R., Sampedro, G., and Pastoriza, L. (2005). Application of nisin, CO_2 and a permeabilizing agent in the preservation of refrigerated blue whiting (*Micromesistius poutassou*). *Journal of the Sciences of Food and Agriculture*, 85: 1733–1740.
Cakir, E. and Clark, S. (2009). Swiss cheese and related products, in *The Sensory Evaluation of Dairy Products*, Eds. S. Clark, M. Costello, M.A. Drake, and F. Bodyfelt. Springer, New York, pp. 427–457.
Calderón-Montero, M., Rojas-Grau, M.A., and Martín-Belloso, O. (2008). Effect of packaging conditions on quality and shelf-life of fresh-cut pineapple (*Ananas comosus*). *Postharvest Biology and Technology*, 50: 182–189.
Campaniello, D., Bevilacqua, A., Sinigaglia, M., and Corbo, M.R. (2008). Chitosan: Antimicrobial activity and potential applications for preserving minimally processed strawberries. *Food Microbiology*, 25: 992–1000.
Cava, R., Ladero, L., González, S., Carrasco, A., and Ramírez, M.R. (2009a). Effect of pressure and holding time on colour, protein and lipid oxidation of sliced dry-cured Iberian ham and loin during refrigerated storage. *Innovative Food Science and Emerging Technologies*, 10: 76–81.
Cava, R., Tárrega, R., Ramírez, R., and Carasco, J.A. (2009b). Decolouration and lipid oxidation changes of vacuum-packed Iberian dry-cured loin treated with E-beam irradiation (5 kGy and 10 kGy) during refrigerated storage. *Innovative Food Science and Emerging Technologies*, 10: 495–499.
Chan, S.H., Gray, J.I., Gomma, E.A., Harte, B.R., Kelly, P.M., and Buckley, D.J. (1993). Cholesterol oxidation in whole milk powders as influenced by processing and packaging. *Food Chemistry*, 47: 321–328.
Chandan, R.C. (2008). Role of milk and dairy foods in nutrition and health, in *Dairy Processing and Quality Assurance*, Eds. R.C. Chandan, A. Kilara, and N.P. Shah. Wiley-Blackwell, New Delhi, India, pp. 411–428.
Charles, F., Guillaume, C., and Gontard, N. (2008). Effect of passive and active modified atmosphere packaging on quality changes of fresh endives. *Postharvest Biology and Technology*, 48: 22–29.
Charles, F., Sanchez, J., and Gontard, N. (2006). Absorption kinetics of oxygen and carbon dioxide scavengers as part of active modified atmosphere packaging. *Journal of Food Engineering*, 72: 1–7.
Chauhan, O.P., Raju, P.S., Dasgupta, D.K., and Bawa, A.S. (2006). Instrumental textural changes in banana (var. Pachbale) during ripening under active and passive modified atmosphere. *International Journal of Food Properties*, 9: 237–253.
Chavez-Franco, S.H. and Kader, A.A. (1993). Effects of CO_2 on ethylene biosynthesis in 'Bartlett' pears. *Postharvest Biology and Technology*, 3: 183–190.
Chen, J.H. and Hotchkiss, J.H. (1991). Effect of dissolved carbon dioxide on the growth of psychrotrophic organisms in cottage cheese. *Journal of Dairy Science*, 74: 2941–2945.
Chen, J.H. and Hotchkiss, J.H. (1992). Effect of temperature, sodium chloride and pH in cottage cheese in MAP. *Journal of Dairy Science*, 76: 972–977.

Chen, J.H. and Hotchkiss, J.H. (1993). Growth of *Listeria monocytogenes* and *Clostridium sporogenes* in cottage cheese in modified atmosphere packaging. *Journal of Dairy Science*, 76: 972–977.

Chonhenchob, V., Chantarasomboon, Y., and Singh, S.P. (2007). Quality changes of treated fresh-cut tropical fruits in rigid modified atmosphere packaging containers. *Packaging Technology and Science*, 20: 27–37.

Chun, H.H., Kim, J.Y., Lee, B.D., Yu, D.J., and Song, K.B. (2009). Effect of UV-C irradiation on the inactivation of inoculated pathogens and quality of chicken breasts during storage. *Food Control*, 21: 276–280.

Codex Committee on Food Hygiene (CCFH, CX/FH 04/5). (2004a). Proposed draft process by which the committee on food hygiene could undertake its work in microbiological risk assessment/risk management, Alinorm 04/27/13.

Codex Committee on Food Hygiene (CCFH, CX/FH 04/6). (2004b). Proposed draft principles and guidelines for the conduct of microbiological risk management, Alinorm 04/27/13.

Conte, A., Gammariello, D., Di Giulio, S., Attanasio, M., and Del Nobile, M.A. (2009). Active coating and modified-atmosphere packaging to extend the shelf life of Fior di Latte cheese. *Journal of Dairy Science*, 92: 887–894.

Corbo, M.R., Altieri, C., Bevilacqua, A., Campaniella, D., D'Amato, D., and Sinigaglia, M. (2005). Estimating packaging atmosphere-temperature effects on the shelf life of cod fillets. *European Food Research and Technology*, 220: 509–513.

Dalgaard, P., Madsen, H.L., Samieian, N., and Emborg, J. (2006). Biogenic amine formation and microbial spoilage in chilled garfish (*Belone belone belone*)—Effect of modified atmosphere packaging and previous frozen storage. *Journal of Applied Microbiology*, 101: 80–95.

Dalgaard, P., Mejlholm, O., and Huss, H.H. (1997). Application of an iterative approach for development of a microbial model predicting the shelf-life of packed fish. *International Journal of Food Microbiology*, 38: 169–179.

Daş, E., Gürakan, G.C., and Bayindirli, A. (2006). Effect of controlled atmosphere storage, modified atmosphere packaging and gaseous ozone treatment on the survival of *Salmonella* Enteritidis on cherry tomatoes. *Food Microbiology*, 23: 430–438.

De Jong, P. (2008). Thermal processing of milk, in *Advanced Dairy Science and Technology*, Eds. T.J. Britz and R.K. Robinson. Blackwell Publishing, Oxford, U.K., pp. 1–34.

Debevere, J. and Boskou, G. (1996). Effect of modified atmosphere packaging on the TVB/TMA-producing microflora of cod fillets. *Food Microbiology*, 31: 221–229.

Dechemi, S., Benjelloun, H., and Lebeault, J.M. (2005). Effect of modified atmosphere on the growth and extracellular enzyme activities of psychrotrophs in raw milk. *Engineering Life Sciences*, 5(4): 350–356.

Del Nobile, M.A., Conte, A., Incoronato, A.L., and Panza, O. (2009a). Modified atmosphere packaging to improve the microbial stability of ricotta. *African Journal of Microbiology Research*, 3(4): 137–142.

Del Nobile, M.A., Corbo, M.R., Speranza, B., Sinigaglia, M., Conte, A., and Caroprese, M. (2009b). Combined effect of MAP and active compounds on fresh blue fish burger. *International Journal of Food Microbiology*, 135: 281–287.

Del Nobile, M.A., Gammariello, D., Conte, A., and Attanasio, M. (2009c). A combination of chitosan, coating and modified atmosphere packaging for prolonging Fior di latte cheese shelf life. *Carbohydrate Polymers*, 78: 151–156.

Dermiki, M., Ntzimani, A., Badeka, A., Savvaidis, I.N., and Kontominas, M.G. (2008). Shelf-life extension and quality attributes of the whey cheese "Myzithra kalathaki" using modifies atmosphere packaging. *Lebensmittel Wissenchaft und Technologie*, 41: 284–294.

Devlieghere, F., Vermeulen, A., and Debevere, J. (2004). Chitosan: Antimicrobial activity, interactions with food components and applicability as a coating on fruit and vegetables. *Food Microbiology*, 21: 703–714.

Diez, A.M., Björkroth, J., Jaime, I., and Rovira, J. (2009a). Microbial, sensory and volatile changes during the anaerobic cold storage of *Morcilla de Burgos* previously inoculated with *Weissella viridescens* and *Leuconostoc mesenteroides*. *International Journal of Food Microbiology*, 131: 168–177.

Diez, A.M., Santos, E.M., Jaime, I., and Rovira, J. (2009b). Effectiveness of combined preservation methods to extend the shelf-life of *Morcilla de Burgos*. *Meat Science*, 81: 171–177.

Du, C.-J. and Sun, D.-W. (2006). Learning techniques used in computer vision for food quality evaluation: A review. *Journal of Food Engineering*, 72: 39–55.

Economou, T., Pournis, N., Ntzimani, A., and Savvaidis, I.N. (2009). Nisin–EDTA treatments and modified atmosphere packaging to increase fresh chicken meat shelf-life. *Food Chemistry*, 114: 1470–1476.

Edgar, R. and Aidoo, K.E. (2001). Microflora of blanched minimally processed fresh vegetables as components of commercial chilled ready-to-use meals. *International Journal of Food Science and Technology*, 36: 107–110.

Egan, M.B., Raats, M.M., Grubb, S.M., Eves, A., Lumbers, M.L., Dean, M.S., and Adams, M.R. (2007). A review of food safety and food hygiene training studies in the commercial sector. *Food Control*, 18: 1180–1190.

Eliot, S.C., Vuillemard, J.C., and Emind, J.P. (1998). Stability of shredded Mozzarella cheese under modified atmosphere. *Journal of Food Science*, 63(6): 1075–1080.

Elortondo, F.J.P., Ojeda, M., Albisu, M., Salmerón, J., Etayo, I., and Molina, M. (2007). Food quality certification: An approach for the development of accredited sensory evaluation methods. *Food Quality and Preference*, 18: 425–439.

Emborg, J., Laursen, B.G., and Dalgaard, P. (2005). Significant histamine formation in tuna (*Thunnus albacares*) at 2°C—Effect of vacuum- and modified atmosphere-packaging on psychrotolerant bacteria. *International Journal of Food Microbiology*, 101: 263–279.

Erturk, E. and Picha, D.H. (2007). Effect of temperature and packaging film in nutritional quality of fresh-cut sweet potatoes. *Journal of Food Quality*, 30: 450–465.

Fadela, C., Abderrahim, C., and Ahmed, B. (2009). Physico-chemical and rheological properties of yoghurt manufactured with ewe's milk and skim milk. *African Journal of Biotechnology*, 8(9): 1938–1942.

Favati, F., Galgano, F., and Pace, A.M. (2007). Shelf-life evaluation of portioned provolone cheese packaged in protective atmosphere. *Lebensmittel Wissenchaft und Technologie*, 40: 480–488.

Fonseca, S.C., Oliveira, F.A.R., and Brecht, J.K. (2002). Modelling respiration rate of fresh fruits and vegetables for modified atmosphere packages: A review. *Journal of Food Engineering*, 52: 99–119.

Franzetti, L., Martinoli, S., Piergiovanni, L., and Galli, A. (2001). Influence of active packaging on the shelf-life of minimally processed fish products in a modified atmosphere. *Packaging Technology Science*, 14: 267–274.

Gallagher, E., Kunkel, A., Gormley, T.R., and Arendt, E.K. (2003). The effect of dairy and rice powder addition on loaf and crumb characteristics, and on shelf life (intermediate and long-term) of gluten-free breads stored in a modified atmosphere. *European Food Research and Technology*, 218: 44–48.

Gammariello, D., Conte, D., Attanasio, M., and Del Nobile, M.A. (2009). Effect of modified atmospheres on microbiological and sensorial properties of Apulian fresh cheeses. *African Journal of Microbiology Research*, 3(7): 370–378.

Gil-Izquierdo, A., Conesa, M.A., Ferreres, F., and Gil, M.I. (2002). Influence of modified atmosphere packaging on quality, vitamin C, and phenolic content of artichokes (*Cynara scolymus* L.). *European Food Research and Technology*, 215: 21–27.

Giménez, B., Roncalés, P., and Beltrán, J.A. (2002). Modified atmosphere packaging of filleted rainbow trout. *Journal of the Sciences of Food and Agriculture*, 82: 1154–1159.

Gómez, P.A. and Artés, F. (2005). Improved keeping quality of minimally fresh processed celery sticks by modified atmosphere packaging. *Lebensmittel Wissenschaft und Technologie*, 38: 323–329.

Gonzales-Fandos, E., Sanz, S., and Olarte, C. (2000). Microbiological, physicochemical and sensory characteristics of Cameros cheese packages under modified atmospheres. *Food Microbiology*, 17: 407–414.

González-Aguilar, G.A., Ayala-Zavala, J.F., Ruiz-Cruz, S., Acedo-Félix, E., and Díaz-Cinco, M.E. (2004). Effect of temperature and modified atmosphere packaging on overall quality of fresh-cut bell peppers. *Lebensmittel Wissenschaft und Technologie*, 37: 817–826.

Goulas, A.E. and Kontominas, M.C. (2007a). Effect of modified atmosphere packaging and vacuum packaging on the shelf-life of refrigerated chub mackerel (*Scomber japonicus*). *European Food Research and Technology*, 224: 545–553.

Goulas, A.E. and Kontominas, M.G. (2007b). Combined effect of light salting, modified atmosphere packaging and oregano essential oil on the shelf-life of sea bream (*Sparus aurata*): Biochemical and sensory attributes. *Food Chemistry*, 100: 287–296.

Haenlein, G.F.W. (2004). Goat milk in human nutrition. *Small Ruminant Research*, 51(2): 155–163.

Haila, K., Kumpulainen, J., Häkkinen, U., and Tahvonen, R. (1992). Sugar and organic acid contents of vegetables consumed in Finland during 1988–1989. *Journal of Food Composition and Analysis*, 5: 100–107.

Halkier, B. and Holm, L. (2006). Shifting responsibilities for food safety in Europe: An introduction. *Appetite*, 47: 127–133.

Heaton, J.C. and Jones, K. (2008). Microbial contamination of fruit and vegetables and the behaviour of enteropathogens in the phyllosphere: A review. *Journal of Applied Microbiology*, 104: 613–626.

Henning, D.R., Baer, R.J., Hassan, A.N., and Dave, R. (2006). Major advances in concentrated and dry milk products, cheese, and milk fat-based spreads. *Journal of Dairy Science*, 89: 1179–1188.

Hintlian, C.B. and Hotchkiss, J.H. (1986). The safety of modified atmosphere packaging: A review. *Food Technology*, 40: 70–76.

Holm, V.K., Mortensen, G., Vishart, M., and Petersen, M.A. (2006). Impact of poly-lactic acid packaging material on semi-hard cheese. *International Dairy Journal*, 16: 913–939.

Hong, C.M., Wendorff, W.L., and Bradley, R.L. (1995). Effects of packaging and lighting on pink discoloration and lipid oxidation of annattio-colored cheeses. *Journal of Dairy Science*, 78: 1896–1902.

Hotchkiss, J.H., Chen, J.H., and Lawless, H.T. (1999). Combined effects of carbon dioxide addition and barrier films on microbial and sensory changes in pasteurized milk. *Journal of Dairy Science*, 82: 690–695.

Hotchkiss, J.H., Werner, B.G., and Lee, E.Y.C. (2006). Addition of carbon dioxide to dairy products to improve quality: A comprehensive review. *Comprehensive Reviews in Food Science and Food Safety*, 5: 158–168.

Hovda, M.B., Sivertsvik, M., Lunestad, B.T., Lorentzen, G., and Rosnes, J.T. (2007). Characterisation of the dominant bacterial population in modified atmosphere packaged farmed halibut (*Hippoglossus hippoglossus*) based on 16S rDNA-DGGE. *Food Microbiology*, 24: 362–371.

Hu, L., Lu, H., Liu, Q., Chen, X., and Jiang, X. (2005). Overexpression of *mtl*D gene in transgenic *Populus tomentosa* improves salt tolerance through accumulation of mannitol. *Tree Physiology*, 25: 1273–1281.

Huffman, L.M. and Harper, W.J. (1999). Maximizing the value of milk through separation technologies. *Journal of Dairy Science*, 82: 2238–2244.

Huss, H.H. (1997). Control of indigenous pathogenic bacteria in seafood. *Food Control*, 8(2): 91–98.

ICMSF. (2002). Microorganisms in foods 7, in *Microbiological Testing in Food Safety Management*. Kluwer Academic/Plenum Publishers, New York.

Irtwange, S.V. (2006). Application of modified atmosphere packaging and related technology in postharvest handling of fresh fruits and vegetables. *Agricultural Engineering International: The CIGR E-Journal*, 8: 1–13.

Jakobsen, M. and Risbo, J. (2009). Carbon dioxide equilibrium between product and gas phase of modified atmosphere packaging systems: Exemplified by semihard cheese. *Journal of Food Engineering*, 92: 285–290.

Jeng, H.Y.J. and Fang, T.J. (2003). Food safety control system in Taiwan-the example of food service sector. *Food control*, 14: 317–322.

Jevŝnik, M., Hlebec, V., and Raspor, P. (2008). Consumers' awareness of food safety from shopping to eating. *Food Control*, 19: 737–745.

Jin, H.H. and Lee, S.Y. (2007). Combined effect of aqueous chlorine dioxide and modified atmosphere packaging on inhibiting *Salmonella Typhimurium* and *Listeria monocytogenes* in mungbean sprouts. *Institute of Food Technologists*, 72: 441–445.

Jones, R.B., Faragher, J.D., and Winkler, S. (2006). A review of the influence of postharvest treatments on quality and glycosinolate content in broccoli (*Brassica oleracea* var. *italica*) heads. *Postharvest Biology and Technology*, 41: 1–8.

Joseph, P., Suman, S.P., Mancini, R.A., and Beach, C.M. (2009). Mass spectrometric for aldehyde adduction in carboxymyoglobin. *Meat Science*, 83: 339–344.

Juric, M., Bertelsen, G., Mortensen, G., and Petersen, M.A. (2003). Light-induced colour and aroma changes in sliced, modified atmosphere packaged semi-hard cheeses. *International Dairy Journal*, 13: 239–249.

Kader, A.A. (1980). Prevention of ripening in fruits by use of controlled atmosphere. *Institute of Food Technology*, 34: 51–54.

Kailasapathy, K. (2008). Chemical composition, physical and functional properties of milk and milk ingredients, in *Dairy Processing and Quality Assurance*, Eds. R.C. Chandan, A. Kilara, and N.P. Shah. Blackwell Publishing, New Delhi, India, pp. 75–103.

Kandeepan, G., Anjaneyulu, A.S.R., Kondaiah, N., Mendiratta, S.K., and Lakshmanan, V. (2009). Effect of age and gender on the processing characteristics of buffalo meat. *Meat Science*, 83: 10–14.

Karagül-Yüceer, Y., Wilson, J.C., and White, C.H. (2001). Formulation and processing of yogurt affect the microbial quality of carbonated yogurt. *Journal of Dairy Science*, 84: 543–550.

Karakaş, B. and Yıldız, F. (2007). Peroxidation of membrane lipids in minimally processed cucumbers packaged under modified atmospheres. *Food Chemistry*, 100: 1011–1018.

Karatapanis, A.E., Badeka, A.V., Riganakos, K.A., Savvaidis, I.N., and Kontominas, M.G. (2006). Changes in flavour volatiles of whole pasteurized milk as affected by packaging material and storage time. *International Dairy Journal*, 16: 750–761.

Keoleian, G.A., Phipps, A.W., Dritz, T., and Brachfeld, D. (2004). Life cycle environmental performance and improvement of a yogurt product delivery system. *Packaging Technology and Science*, 17: 85–103.

Keyser, M., Müller, I.A., Cilliers, F.P., Nel, W., and Gouws, P.A. (2008). Ultraviolet radiation as a non-thermal treatment for the inactivation of microorganisms in fruit juice. *Innovative Food Science and Emerging Technologies*, 9: 348–354.

Kilic, B. (2009). Current trends in traditional Turkish meat products and cuisine. *Lebensmittel Wissenschaft und Technologie—Food Science and Technology*, 42: 1581–1589.

Kim, S.J. and Ishii, G. (2007). Effect of storage temperature and duration on glycosinolate, total vitamin C and nitrate contents in rocket salad (*Eruca sativa* Mill.). *Journal of the Sciences of Food and Agriculture*, 87: 966–973.

Kim, Y.H., Keeton, J.T., Hunt, M.C., and Savelli, J.W. (2009a). Effects of L- or D-lactate enhancement on the internal cooked colour development and biochemical characteristics of beef steaks in high-oxygen modified atmosphere. *Food Chemistry*, 119: 1–24.

Kim, Y.H., Keeton, J.T., Smith, S.B., Maxim, J.E., Yang, H.S., and Savell, J.W. (2009b). Evaluation of antioxidant capacity and colour stability of calcium lactate enhancement on fresh beef under highly oxidizing conditions. *Food Chemistry*, 115: 272–278.

Kim, Y.H., Keeton, J.T., Yang, H.S., Smith, S.B., Sawyer, J.E., and Savell, J.W. (2009c). Color stability and biochemical characteristics of bovine muscles when enhanced with L- or D-potassium lactate in high-oxygen modified atmospheres. *Meat Science*, 82: 234–240.

Kim, J.G., Luo, Y., and Gross, K.C. (2004). Effect of package film on the quality of fresh-cut salad savoy. *Postharvest Biology and Technology*, 32: 99–107.

Kim, J.G., Luo, Y., Tao, Y., Saftner, R.A., and Gross, K.C. (2005). Effect of initial oxygen concentration and film oxygen transmission rate on the quality of fresh-cut romaine lettuce. *Journal of the Sciences of Food and Agriculture*, 85: 1622–1630.

Kim, J.B., Yousef, A.E., and Khadre, M. (2003). Ozone and its current and future application in the food industry. *Advances in Food and Nutrition Research*, 45: 168–218.

Klaiber, R.G., Baur, S., Wolf, G., Hammes, W.P., and Carle, R. (2005). Quality of minimally processed carrots as affected by warm water washing and chlorination. *Innovative Food Science and Emerging Technologies*, 6: 351–362.

Kozová, M., Kalaĉ, P., and Pelikániva, T. (2009a). Changes in the content of biologically active polyamines during beef storage and cooking. *Meat Science*, 81: 607–611.

Kozová, M., Kalaĉ, P., and Pelikánivá, T. (2009b). Contents of biologically active polyamines in chicken meat, liver, heart and skin after slaughter and their changes during meat storage and cooking. *Food Chemistry*, 116: 419–425.

Kristensen, D., Orlien, V., Mortensen, G., Brockhoff, P., and Skibsted, L.H. (2000). Light-induced oxidation in sliced Havarti cheese packaged in modified atmosphere. *International Dairy Journal*, 10: 95–103.

Kumar, P. and Mishra, H.N. (2004). Storage stability of mango soy fortified yoghurt powder in two different packaging materials: HDPP and ALP. *Journal of Food Engineering*, 65: 569–576.

Lakakul, R., Beaudry, M.R., and Hernandez, J.R. (1999). Modeling respiration of apple slices in modified-atmosphere packages. *Journal of Food Science*, 64: 105–110.

Lalitha, K.V., Sonaji, E.R., Manju, S., Jose, L., Gopal, T.K.S., and Ravisankar, C.N. (2005). Microbiological and biochemical changes in pearl spot (*Etroplus surutensis* Bloch) stored under modified atmospheres. *Journal of Applied Microbiology*, 99: 1222–1228.

Laurenzio, P., Malinconico, M., Pizzano, R., Manzo, C., Piciocchi, N., Sorrentino, A., and Volpe, M.G. (2006). Natural polysaccharide-based gels for dairy food preservation. *Journal of Dairy Science*, 89: 2856–2864.

Lee, S.K. and Kader, A.A. (2000). Preharvest and postharvest factors influencing vitamin C content of horticultural crops. *Postharvest Biology and Technology*, 20: 207–220.

Lee, K.S., Woo, K.L., and Lee, D.S. (1994). Modified atmosphere packaging for green chili peppers. *Packaging Technology and Science*, 7: 51–58.

Lemoine, M.L., Civello, P.M., Chaves, A.R., and Martínez, G.A. (2008). Effect of combined treatment with hot air and UV-C on senescence and quality parameters of minimally processed broccoli (*Brassica oleracea* L. var. *Italica*). *Postharvest Biology and Technology*, 48: 15–21.

Leroy, F., Vasilopoulos, C., Van Hemelryck, S., Falony, G., and De Vuyst, L. (2009). Volatile analysis of spoiled, artisan-type, modified-atmosphere-packaged cooked ham stored under different temperatures. *Food Microbiology*, 26: 94–102.

Liu, C.-L., Hsu, C.-K., and Hsu, M.-M. (2007). Improving the quality of fresh-cut pineapples with ascorbic acid/sucrose pretreatment and modified atmosphere packaging. *Packaging Technology and Science*, 20: 337–343.

Lu, S. (2009). Effects of bactericides and modified atmosphere packaging on shelf-life of Chinese shrimp (*Fenneropenaeus chinensis*). *Lebensmittel Wissenschaft und Technologie—Food Science and Technology*, 42: 286–291.

Luciano, G., Monahan, F.J., Vasta, V., Pennisi, P., Bella, M., and Priolo, A. (2009). Lipid and colour stability of meat from lambs fed fresh herbage or concentrate. *Meat Science*, 82: 193–199.

Mancini, R.A., Ramanathan, R., Suman, S.P., Konda, M.K.R., Joseph, P., Dady, G.A., Naveena, B.M., and López-López, I. (2010). Effects of lactate and modified atmospheric packaging on premature browning in cooked ground beef patties. *Meat Science*, 85: 339–346.

Maniar, A.B., Marcy, J.E., Bishop, J.R., and Duncan, S.E. (1994). Modified atmosphere packaging to maintain direct-set cottage cheese quality. *Journal of Food Science*, 59(6): 1305–1308.

Mannheim, C.H. and Soffer, T. (1996). Shelf-life extension of cottage cheese by modified atmosphere packaging. *Lebensmittel Wissenschaft und Technologie*, 29: 767–771.

Maqsood, S. and Benjakul, S. (2010). Synergistic effect of tannic acid and modified atmospheric packaging on the prevention of lipid oxidation and quality losses of refrigerated striped catfish slices. *Food Chemistry*, 121: 29–38.

Marrero, A. and Kader, A.A. (2006). Optimal temperature and modified atmosphere for keeping quality of fresh-cut pineapples. *Postharvest Biology and Technology*, 39: 163–168.

Martínez, J.A. and Artés, F. (1999). Effect of packaging treatments and vacuum-cooling on quality of qinter harvested iceberg lettuce. *Food Research International*, 32: 621–627.

Masdtromatteo, M., Lucera, A., Sinigaglia, M., and Corbo, M.R. (2009). Combined effects of thymol, carvanol and temperature on the quality of non conventional poultry patties. *Meat Science*, 83: 246–254.

Masniyom, P., Benjakul, S., and Visessanguan, W. (2002). Shelf-life extension of refrigerated seabass slices under modified atmosphere packaging. *Journal of the Sciences of Food and Agriculture*, 82: 873–880.

McConnell, R.Y., Truong, V.D., Walter, Jr. W.M., and McFeeters, R.F. (2005). Physical, chemical and microbial changes in shredded sweet potatoes. *Journal of Food Processing and Preservation*, 29: 246–267.

Mejlholm, O. and Dalgaard, P. (2002). Antimicrobial effect of essential oils on the seafood spoilage micro-organism *Photobacterium phosphoreum* in liquid media and fish products. *Letters in Applied Microbiology*, 34: 27–31.

Meng, J. and Doyle, M.P. (2002). Introduction. Microbiological food safety. *Microbes and Infection*, 4: 395–397.

Metin, S., Erkan, N., Baygar, T., and Ozden, O. (2002). Modified atmosphere packaging of fish salad. *Fisheries Science*, 68: 204–209.

Mexis, S.F., Chouliara, E., and Kontominas, M.G. (2011). Quality evaluation of grated Graviera cheese stored at 4 and 12°C using active and modified atmosphere packaging. *Packaging Technology Science*, 24: 15–29.

Mohan, C.O., Ravishankar, C.N., Gopal, T.K.S., Lalitha, K.V., and Kumar, K.A. (2010). Effect of reduced oxygen atmosphere and sodium acetate treatment on the microbial quality changes of seer fish (*Scomberomorus commerson*) steaks stored in ice. *Food Microbiology*, 27: 526–534.

Mortensen, G., Bertelsen, G., Mortensen, B.K., and Stapelfeldt, H. (2004). Light-induced changes in packaged cheeses—A review. *International Dairy Journal*, 14: 85–102.

Mortensen, G., Sørensen, J., and Stapelfeldt, H. (2003). Effect of modified atmosphere packaging and storage conditions on photooxidation of sliced Havarti cheese. *European Food Research and Technology*, 216: 57–62.

Muir, D.D. (1996). The shelf-life of dairy products: 2. Raw milk and fresh products. *Journal of the Society of Dairy Technology*, 49(2): 44–48.

Munsch-Alatossava, P., Gursoy, O., and Alatossava, T. (2010). Potential of nitrogen gas (N_2) to control psychrotrophs and mesophiles in raw milk. *Microbiological Research*, 165: 122–132.

Murcia, M.A., Jiménez-Monreal, A.M., García-Diz, L., Carmona, M., Maggi, L., and Martínez-Tomé, M. (2009). Antioxidant activity of minimally processed (in modified atmospheres), dehydrated and ready-to-eat vegetables. *Food Chemical Toxicology*, 47: 2103–2110.

Mussa, D.M., Ramaswany, H.S., and Smith, J.P. (1999). High pressure (HP) destruction kinetics of *Listeria monocytogenes* Scott A in raw milk. *Food Research International*, 31(5): 343–350.

Nam, P.H., Alejandra, B., Frédéric, H., Didier, B., Olivier, S., and André, P. (2008). A new quantitative and low-cost determination method of nitrate in vegetables, based on deconvolution of UV spectra. *Talanta*, 76: 936–940.

Nauta, M., Hill, A., Rosenquist, H., Brynestad, S., Fetsch, A., Van der Logt, P., Fazil, A., Christensen, B., Katsma, E., Borck, B., and Havelaar, A. (2009). A comparison of risk on *Campylobacter* in broiler meat. *International Journal of Food Microbiology*, 129: 107–123.

Notermans, S. and Veld, P. (1994). Microbiological challenge testing for ensuring safety of food products. *International Journal of Food Microbiology*, 24: 33–39.

Nguyen, T.B.T., Ketsa, S., and van Doorn, W.G. (2004). Effect of modified atmosphere packaging on chilling-induced peel browning in banana. *Postharvest Biology and Technology*, 31: 313–317.

Nielsen, B.R., Stapelfeldt, H., and Skibster, L.H. (1997). Differentiation between 15 whole milk powders in relation to oxidative stability during accelerated storage: Analysis of variance and canonical variable analysis. *International Dairy Journal*, 7: 589–599.

Nilsson, L., Huss, H.H., and Gram, L. (1997). Inhibition of *Listeria monocytogenes* on cold-smoked salmon by nisin and carbon dioxide atmosphere. *International Journal of Food Microbiology*, 38: 217–227.

O'Connor, C.B. (1995). *ILRI Training Manual 1*. Rural Dairy Technology. Addis Ababa, Ethiopia.

O'Mahony, F.C., O'Riordan, T.C., Papkovsakaia, N., Kerry, J.P., and Papkovsky, D.B. (2006). Non-destructive assessment of oxygen levels in industrial modified atmosphere packaged cheddar cheese. *Food Control*, 17: 286–292.

Odriozola-Serrano, I., Soliva-Fortuny, R., and Martín-Belloso, O. (2008). Effect of minimal processing on bioactive compounds and color attributes of fresh-cut tomatoes. *Lebensmittel Wissenschaft und Technologie*, 41: 217–226.

Odriozola-Serrano, I., Soliva-Fortuny, R., and Martín-Belloso, O. (2010). Changes in bioactive composition of fresh-cut strawberries stored under superatmospheric oxygen, low-oxygen or passive atmospheres. *Journal of Food Composition and Analysis*, 23: 37–43.

Ogier, J.C., Son, O., Gruss, A., Tailliez, P., and Delacroix-Buchet, A. (2002). Identification of the bacterial microflora in dairy products by temporal temperature gradient gel electrophoresis. *Applied and Environmental Microbiology*, 68(8): 3691–3701.

Olarte, C., Gonzalez-Fandos, E., and Sanz, S. (2001). A proposed methodology to determine the sensory quality of a fresh goat's cheese (Cameros cheese): Application to cheeses packaged under modified atmospheres. *Food Quality and Preference*, 12: 163–170.

Olarte, C., González-Fandos, E., Giménez, M., Sanz, S., and Portu, J. (2002). The growth of *Listeria monocytogenes* in fresh goat cheese (Cameros cheese) packaged under modified atmospheres. *Food Microbiology*, 19: 75–82.

Oyugi, E. and Buys, E.M. (2007). Microbiological quality of shredded cheddar cheese packaged in modified atmospheres. *International Journal of Dairy Technology*, 60(2): 89–95.

Ozdemir, M. and Floros, J.D. (2004). Active food packaging technologies. *Critical Reviews in Food Science and Nutrition*, 44: 185–193.

Özogul, F. and Özogul, Y. (2006). Biogenic amine content and biogenic amine quality indices of sardines (*Sardina pilchardus*) stored in modified atmosphere packaging and vacuum packaging. *Food Chemistry*, 99: 574–578.

Özogul, F., Polat, A., and Özogul, Y. (2004). The effects of modified atmosphere packaging and vacuum packaging on chemical, sensory and microbiological changes of sardines (*Sardina pilchardus*). *Food Chemistry*, 85: 49–57.

Pantazi, D., Papavergou, A., Pournis, N., Kontominas, M.G., and Savvaidis, I.N. (2008). Shelf-life of chilled fresh Mediterranean swordfish (*Xiphias gladius*) stored under various packaging conditions: Microbiological, biochemical and sensory attributes. *Food Microbiology*, 25: 136–143.

Papaioannou, G., Chouliara, I., Karatapanis, A.E., Kontominas, M.G., and Savvaidis, I.N. (2007). Shelf-life of a Greek whey cheese under modified atmosphere packaging. *International Dairy Journal*, 17: 358–364.

Paris, A., Bacci, C., Salsi, A., Bonardi, S., and Brindani, F. (2004). Microbial characterization of organic dairy products: Stracchino and ricotta cheeses. *Annali della Facoltà di Medicina Veterinaria Di Parma*, 24: 317–325.

Parra, V., Viguera, J., Sánchez, J., Peinado, J., Espárrago, F., Gutierrez, J.I., and Andrés, A.I. (2010). Modified atmosphere packaging and vacuum packaging for long period chilled storage of dry-cured Iberian ham. *Meat Science*, 84: 760–768.

Pastoriza, L., Cabo, M.L., Bernámdez, M., Sampedro, C., and Herrera, J.J.R. (2002). Combined effects of modified atmosphere packaging and lauric acid on the stability of pre-cooked fish during refrigerated storage. *European Food Research and Technology*, 215: 189–193.

Pastoriza, L., Sampedro, G., Herrera, J.J., and Cabo, M.L. (1996). Effect of modified atmosphere packaging on shelf-life of iced fresh hake slices. *Journal of the Sciences of Food and Agriculture*, 71: 541–547.

Pastoriza, L., Sampedro, G., Herrera, J.J., and Cabo, M.L. (1998). Influence of sodium chloride and modified atmosphere packaging on microbiological, chemical and sensorial properties in ice storage of slices of hake (*Merluccius merluccius*). *Food Chemistry*, 61(1/2): 23–28.

Pennacchia, C., Ercolini, D., and Villani, F. (2009). Development of a real-time PCR assay for the specific detection of *Brochothrix thermosphacta* in fresh and spoiled raw meat. *International Journal of Food Microbiology*, 134: 230–236.

Petracek, P.D., Joles, W.D., Shiraze, A., and Cameron, A.C. (2002). Modified atmosphere packaging of sweet cherry (*Prunus avium* L., ev. 'Sams') fruit: Metabolic responses to oxygen, carbon dioxide, and temperature. *Postharvest Biology and Technology*, 24: 259–270.

Pettersen, M.K., Eie, T., and Nilsson, A. (2005). Oxidative stability of cream cheese stored in thermoformed trays as affected by packaging material, drawing depth and light. *International Dairy Journal*, 15: 355–362.

Peri, C. (2006). The universe of food quality. *Food Quality and Preference*, 17: 3–8.

Pintado, M.E. and Malcata, F.X. (2000). Optimization of modified atmosphere packaging with respect to physicochemical characteristics of *Requeijão*. *Food Research International*, 33: 821–832.

Pirovani, M.E., Güemes, D.R., Piagentini, A.M., and Di Pentima, J.H. (1997). Storage quality of minimally processed cabbage packaged in plastic films. *Journal of Food Quality*, 20: 381–389.

Pluta, A., Ziarno, M., and Kruk, M. (2005). Impact of modified atmosphere packing on the quality of grated Mozzarella cheese. *Polish Journal of Food Nutrition Science*, 14/55(2): 117–122.

Poli, B.M., Messini, A., Parisi, G., Scappini, F., Vigiani, V., Giorgi, G., and Vincenzini, M. (2006). Sensory, physical, chemical and microbiological changes in European sea bass (Dicentrarchus labrax) fillets packed under modified atmosphere/air or prepared from whole fish stored in ice. *International Journal of Food Science and Technology*, 41: 444–454.

Prior, R.L., Cao, G., Martin, A., Sofic, E., McEwen, J., O'Brien, C., Lischner, N. et al. (1998). Antioxidant capacity as influenced by total phenolic and anthocyanin content, maturity, and variety of Vaccinium species. *Journal of Agricultural and Food Chemistry*, 46: 2686–2693.

Radziejewska-Kubzdela, E., Czapski, J., and Czaczyk, K. (2007). The effect of packaging conditions on the quality of minimally processed celeriac flakes. *Food Control*, 18: 1191–1197.

Ramamoorthi, L., Toshkov, S., and Brewer, M.S. (2009). Effects of carbon monoxide-modified atmosphere packaging and irradiation on *E. coli* K12 survival and raw beef quality. *Meat Science*, 83: 1–16.

Ranasinghe, L., Jayawardena, B., and Abeywickrama, K. (2005). An integrated strategy to control post-harvest decay of Embul banana by combining essential oils with modified atmosphere packaging. *International Journal of Food Science and Technology*, 40: 97–103.

Reis, F.A.R.S., Rocha, S.M., Barros, A.S., Delgadillo, I., and Coimbra, M.A. (2009). Establishment of the volatile profile of 'Bravo de Esmolfe' apple variety and identification of varietal markets. *Food Chemistry*, 113: 513–521.

Remón, S., Ferrer, A., Marquina, P., Burgos, J., and Oria, R. (2000). Use of modified atmosphere to prolong the postharvest life of Burlat cherries at two different degrees of ripeness. *Journal of the Science of Food and Agriculture*, 80: 1545–1552.

Renault, P., Houal, L., Jacquemin, G., and Chambroy, Y. (1994). Gas exchange in modified atmosphere packaging. 2: Experimental results with strawberries. *International Journal of Food Science and Technology*, 29: 379–394.

Rico, D., Martín-Diana, A.B., Barat, J.M., and Barry-Ryan, C. (2007). Extending and measuring the quality of fresh-cut fruit and vegetables: A review. *Trends in Food Science & Technology*, 18: 373–386.

Rocculi, P., Cocci, E., Romani, S., Sacchetti, G., and Rosa, M.D. (2009). Effect of 1-MCP treatment and N_2O MAP on physiological and quality changes of fresh-cut pineapple. *Postharvest Biology and Technology*, 51: 371–377.

Rocculi, P., Romani, S., and Rosa, M.D. (2005). Effect of MAP with argon and nitrous oxide on quality maintenance of minimally processed kiwifruit. *Postharvest Biology and Technology*, 35: 319–328.

Rocha, A.M.C.N., Barreiro, M.G., and Morais, A.M.M.B. (2004). Modified atmosphere package for apple 'Bravo de Esmolfe.' *Food Control*, 15: 61–64.

Rodriguez-Aguilera, R., Oliveira, J.C., Montanez, J.C., and Mahajan, P.V. (2009). Gas exchange dynamics in modified atmosphere packaging of soft cheese. *Journal of Food Engineering*, 95: 438–445.

Rolls, B.A. and Porter, J.W.G. (1973). Some effects of processing and storage on the nutritive value of milk and milk products. *Proceedings of the Nutritional Society*, 32(9): 9–15.

Rosnes, J.T., Kleiberg, G.H., Sivertsvik, M., Lunestad, B.T., and Lorentzen, G. (2006). Effect of modified atmosphere packaging and superchilled storage on the shelf-life of farmed ready-to-cook spotter wolf-fish (*Anarhichas minor*). *Packaging Technology and Science*, 19: 325–333.

Ross, T., Rasmussen, S., Fazil, A., Paoli, G., and Summer, J. (2009). Quantitative risk assessment of *Listeria monocytogenes* in ready-to-eat meats in Australia. *International Journal of Food Microbiology*, 131: 128–137.

Rudi, K., Flateland, S.L., Hanssen, J.F., Bengtsson, G., and Nissen, H. (2002). Development and evaluation of a 16S ribosomal DNA array-based approach for describing complex microbial communities in ready-to-eat vegetable salads packed in a modified atmosphere. *Applied and Environmental Microbiology*, 68: 1146–1156.

Saint-Eve, A., Lévy, C., Le Moigne, M., Ducruet, V., and Souchon, I. (2008). Quality changes in yogurt during storage in different packaging materials. *Food Chemistry*, 110: 285–293.

Sánchez-Molinero, F., García-Regueiro, J.A., and Arnau, J. (2010). Processing of dry-cured ham in a reduced-oxygen atmosphere: Effects on physicochemical and microbiological parameters and mite growth. *Meat Science*, 84: 400–408.

Sandhya, S. (2010). Modified atmosphere packaging of fresh produce: Current status and future needs. *Lebensmittel Wissenschaft und Technologie—Food Science and Technology*, 43: 381–392.

Sanguinetti, A.M., Secchi, N., Del Caro, A., Stara, G., Roggio, T., and Piga, A. (2009). Effectiveness of active and modified atmosphere packaging on shelf life extension of a cheese tart. *International Journal of Food Science and Technology*, 44: 1192–1198.

Sanz, S., Olarte, C., Echávarri, J.F., and Ayala, F. (2007). Evaluation of different varieties of cauliflower for minimal processing. *Journal of the Science of Food and Agriculture*, 87: 266–273.

Scannell, A.G.M., Hill, C., Ross, R.P., Marx, S., Hartmeier, W., and Arendt, K. (2000). Development of bioactive food packaging materials using immobilized bacteriocins Lacticin 3147 and Nisaplin. *International Journal of Food Microbiology*, 60: 241–249.

Schafer, K.S. and Kegley, S.E. (2002). Persistent toxic chemicals in the US food supply. *Journal Epidemiology and Community Health*, 56: 813–817.

Schirmer, B.C., Heiberg, R., Eie, T., Møretrø, T., Maugesten, T., Carlehøg, M., and Langsrud, S. (2009). A novel packaging method with a dissolving CO_2 headspace combined with organic acids prolongs the shelf life of fresh salmon. *International Journal of Food Microbiology*, 133: 154–160.

Sengun, I.Y. and Karapinar, M. (2005). Effectiveness of household natural sanitizers in the elimination of *Salmonella typhimurium* on rocket (*Eruca sativa* Miller) and spring onion (*Allium cepa* L.). *International Journal of Food Microbiology*, 98: 319–323.

Serrano, M., Martínez-Romero, D., Castillo, S., Guillén, F., and Valero, D. (2005). The use of natural antifungal compounds improves the beneficial effect of MAP in sweet cherry storage. *Innovative Food Science and Emerging Technologies*, 6: 115–123.

Serrano, M., Martinez-Romero, D., Guillén, F., Castillo, S., and Valero, D. (2006). Maintenance of broccoli quality and functional properties during cold storage as affected by modified atmosphere packaging. *Postharvest Biology and Technology*, 39: 61–68.

Shah, N.S. and Nath, N. (2008). Changes in qualities of minimally processed litchis: Effect of antibrowning agents, osmo-vacuum drying and moderate vacuum packaging. *Lebensmittel Wissenschaft und Technologie*, 41: 660–668.

Simón, A., González-Fandos, E., and Rodríguez, D. (2008). Effect of film and temperature on the sensory, microbiological and nutritional quality of minimally processed cauliflower. *International Journal of Food Science and Technology*, 43: 1628–1636.

Simpson, R., Acevedo, C., and Almonacid, S. (2009). Mass transfer of CO_2 in MAP systems: Advances for non-respiring foods. *Journal of Food Engineering*, 92: 233–239.

Siro, I., Devlieghere, F., Jacxsens, L., Uyttendaele, M., and Debevere, J. (2006). The microbial safety of strawberry and raspberry fruits packaged in high-oxygen and equilibrium-modified atmospheres compared to air storage. *International Journal of Food Science and Technology*, 41: 93–103.

Sivertsvik, M. (2007). The optimized modified atmosphere for packaging of pre-*rigor* filleted farmed cod (*Gadus morhua*) in 63 ml/100 ml oxygen and 37 ml/100 ml carbon dioxide. *Lebensmittel Wissenschaft und Technologie*, 40: 430–438.

Sivertsvik, M., Jeksrud, W.K., and Rosnes, J.T. (2002). A review of modified atmosphere packaging of fish and fishery products-significance of microbial growth, activities and safety. *International Journal of Food Science and Technology*, 37: 107–127.

Slifko, T.R., Smith, H.V., and Rose, J.B. (2000). Emerging parasite zoonose associated with water and food. *International Journal for Parasitology*, 30: 1379–1393.

Solano-Lopez, C.E., Ji, T., and Alvarez, V.B. (2005). Volatile compounds and chemical changes in ultrapasteurized milk packages in polyethylene terephthalate containers. *Journal of Food Science*, 70(6): 407–412.

Soldatou, N., Nerantzaki, A., Kontominas, M.G., and Savvaidis, I.N. (2009). Physicochemical and microbiological changes of "Souvlaki"—A Greek delicacy lamb meat product: Evaluation of shelf-life using microbial, colour and lipid oxidation parameters. *Food Chemistry*, 113: 36–42.

Soliva-Fortuny, R.C., Ricart-Coll, M., and Martin-Belloso, O. (2005). Sensory quality and internal atmosphere of fresh-cut Golden Delicious apples. *International Journal of Food Science and Technology*, 40: 369–375.

Somboonkaew, N. and Terry, L.A. (2010). Physiological and biochemical profiles of imported litchi fruit under modified atmosphere packaging. *Postharvest Biology and Technology*, 56: 246–253.

Sørheim, O., Westad, F., Larsen, H., and Alvseike, O. (2009). Colour of ground beef as influenced by raw materials, addition of sodium chloride and low oxygen packaging. *Meat Science*, 81: 467–473.

Stolzenbach, S., Leisner, J.J., and Byrne, D.V. (2009). Sensory shelf life determination of a processed meat product 'rullepølse' and microbial metabolites as potential indicators. *Meat Science*, 83: 285–292.

Suman, S.P., Mancini, R.A., Ramanathan, R., and Konda, M.R. (2009). Effect of lacate-enhancement, modified atmosphere packaging, and muscle source on the internal cooked colour of beef steaks. *Meat Science*, 81: 664–670.

Suparlan, Ir. and Itoh, K. (2003). Combined effects of hot water treatment (HWT) and modified atmosphere packaging (MAP) on quality of tomatoes. *Packaging Technology and Science*, 16: 171–178.

Taniwaki, M.H., Hocking, A.D., Pitt, J.I., and Fleet, G.H. (2001). Growth of fungi and mycotoxin production on cheese under modified atmospheres. *International Journal of Food Microbiology*, 68: 125–133.

Tano, K., Oulé, M.K., Doyon, G., Lencki, R.W., and Arul, J. (2007). Comparative evaluation of the effect of storage temperature fluctuation on modified atmosphere packages of selected fruit and vegetables. *Postharvest Biology and Technology*, 46: 212–221.

Tassou, C.C. and Boziaris, J.S. (2002). Survival of *Salmonella enteritidis* and changes in pH and organic acids in grated carrots inoculated or not with *Lactobacillus* sp and stored under different atmospheres at 4°C. *Journal of the Science of Food and Agriculture*, 82: 1122–1127.

Tenorio, M.D., Villanueva, M.J., and Sagardoy, M. (2004). Changes in carotenoids and chlorophylls in fresh green asparagus (*Asparagus officinalis* L) stored under modified atmosphere packaging. *Journal of the Science of Food and Agriculture*, 84: 357–365.

Thorsen, L., Budde, B.B., Koch, A.G., and Klingberg, T.D. (2009). Effect of modified atmosphere and temperature abuse on the growth from spores and cereulide production of *Bacillus weihenstephanensis* in a cooked chilled meat sausage. *International Journal of Food Microbiology*, 130: 172–178.

Tian, S.-P., Jiang, A.-L., Xu, Y., and Wang, Y.-S. (2004). Responses of physiology and quality of sweet cherry fruit to different atmospheres in storage. *Food Chemistry*, 87: 43–49.

Torrieri, E., Cavella, S., Villani, F., and Masi, P. (2006). Influence of modified atmosphere packaging on the chilled shelf-life of gutted farmed bass (*Dicentrarchus labrax*). *Journal of Food Engineering*, 77: 1078–1086.

Traill, W.B. and Koenig, A. (2009). Economic assessment of food safety standards: Costs and benefits of alternative approaches. *Food Control*, 21: 1611–1619.

Trobetas, A., Badeka, A., and Kontominas, M.G. (2008). Light-induced changes in grated graviera hard cheese packaged under modified atmospheres. *International Dairy Journal*, 18: 1133–1139.

Tudela, J.A., Espín, J.C., and Gil, M.I. (2002). Vitamin C retention in fresh-cut potatoes. *Postharvest Biology and Technology*, 26: 75–84.

Ukuku, D.O. (2004). Effect of hydrogen peroxide treatment on microbial quality and appearance of whole and fresh-cut melongs contaminated with *Salmonella* spp. *International Journal of Food Microbiology*, 95: 137–146.

Uyttendaele, M., Busschaert, P., Valero, A., Geeraerd, A.H., Vermeulen, A., Jacxsens, L., Goh, K.K., De Loy, A., Van Impe, J.F., and Devlieghere, F. (2009). Prevalence and challenge tests of *Listeria monocytogenes* in Belgian produced and retailed mayonnaise-based deli-salads, cooked meat products and smoked fish between 2005 and 2007. *International Journal of Food Microbiology*, 133: 94–104.

Vaikousi, H., Biliaderis, C.G., and Koutsoumanis, K.P. (2009). Applicability of a microbial Time Temperature Indicator (TTI) for monitoring spoilage of modified atmosphere packed minced meat. *International Journal of Food Microbiology*, 133: 272–278.

Van Asselt, A.J. and Te Giffel, M.C. (2009). Hygiene practices in liquid milk dairies, in *Milk Processing and Quality Management*, Eds. A.Y. Tamine. Blackwell Publishing, Ayr, U.K., pp. 237–253.

Van Schothorst, M., Zwietering, M.H., Ross, T., Buchanan, R.L., and Cole, M.B. (2009). Relating microbiological criteria to food safety objectives and performance objectives. *Food Control*, 20: 967–979.

Van Veen, T.W.S. (2005). International trade and food safety in developing countries. *Food Control*, 16: 491–496.

Varoquaux, P., Albagnac, G., The, C.N., and Françoise, V. (1996). Modified atmosphere packaging of fresh beansprouts. *Journal of the Science of Food and Agriculture*, 70: 224–230.

Veltman, R.H., Kho, R.M., van Schaik, A.C.R., Sanders, M.G., and Oosterhaven, J. (2000). Ascorbic acid and tissue browning in pears (*Pyrus communis* L. cvs Rocha and Conference) under controlled atmosphere conditions. *Postharvest Biology and Technology*, 19: 129–137.

Vergara, H., Bórnez, R., and Linares, M.B. (2009). CO_2 stunning procedure on Manchego light lambs: Effect on meat quality. *Meat Science*, 83: 517–522.

Viuda-Martos, M., Ruiz-Navajas, Y., Fernández-López, J., and Pérez-Álvarez, J.A. (2009). Effect of orange dietary fibre, oregano essential oil and packaging conditions on shelf-life of bologna sausages. *Food Control*, 21: 436–443.

Walstra, P., Wouters, J.T.M, and Geurts, T.J. (2006). *Dairy Science and Technology*. Taylor & Francis, New York.

Wandel, M. and Bugge, A. (1997). Environmental concern in consumer evaluation of food quality. *Food Quality and Preference*, 8: 19–26.

Wang, C.Y. and Qi, L. (1997). Modified atmosphere packaging alleviates chilling injury in cucumbers. *Postharvest Biology and Technology*, 10: 195–200.

White, C.H. (1993). Rapid methods for estimation and prediction of shelf-life of milk and dairy products. *Journal of Dairy Science*, 76: 3126–3132.

Wold, J.P., Jørgensen, K., and Lundby, F. (2002). Nondestructive measurement of light-induced oxidation in dairy products by fluorescence spectroscopy and imaging. *Journal of Dairy Science*, 85: 1693–1704.

Zagory, D. (1998). An update on modified atmosphere packaging of fresh produce. *Packaging International*, 117: 1–5.

Zagory, D. (1999). Effects of post-processing handling and packaging on microbial populations. *Postharvest Biology and Technology*, 15: 313–321.

Zakrys, P.I., O'Sullivan, M.G., Allen, P., and Kerry, J.P. (2009). Consumer acceptability and physicochemical characteristics of modified atmosphere packed beef steaks. *Meat Science*, 81: 720–725.

Zhang, H., Kong, B., Xiong, Y.L., and Sun, X. (2009). Antimicrobial activities of spice extracts against pathogenic and spoilage bacteria in modified atmosphere packaged fresh pork and vacuum packaged ham slices stored at 4°C. *Meat Science*, 81: 686–692.

Zhang, M., Xiao, G., and Salokhe, V.M. (2006). Preservation of strawberries by modified atmosphere packages with other treatments. *Packaging Technology and Science*, 19: 183–191.

Zwietering, M. (2005). Practical considerations on food safety objectives. *Food Control*, 16: 817–823.

Part III

Applications of MAP in Foods of Animal Origin

4 Fish and Seafood

Ioannis S. Arvanitoyannis and Alexandros Ch. Stratakos

CONTENTS

4.1 INTRODUCTION

Fish and shellfish are highly perishable and their deterioration is primarily due to bacterial action (Skura, 1991; Colby et al., 1993). Fish has a relatively short shelf life (12 days) under refrigerated conditions, not displaying quality hazards in hygiene when properly packaged (Oetterer, 1999). There has been a recent interest in prolonging the shelf life of fish due to the increase in demand for fresh products, which has led to a greater variety of products being packaged under modified atmosphere (MA), in which air composition is altered or "modified" (Lioulas, 1988; Manju et al., 2007). Modified atmosphere packaging (MAP) and vacuum-packed (VP) systems can further improve the shelf life, organoleptic quality, and product range of seafood. Carbon dioxide, oxygen, and nitrogen are the common gases mostly used in MA systems (Metin et al., 2002), but some noble gases such as Arg have also been studied as a substitute of nitrogen (Gimenez et al., 2002). The factors that must be considered when determining the ideal gas concentration are (i) the gas-to-fish ratio, (ii) the fish species, (iii) the packaging method, (iv) initial microbial contamination, and (v) temperature. Low temperatures, for example, inhibit the growth of *Clostridium botulinum* because the packaging method prevents contamination and maintains the atmosphere, while a specific gas mixture retards the growth of spoilage bacteria (Stammen et al., 1990). The use of the MAP presents the following advantages: lengthening of the products' shelf life, drop of economic losses, cost reductions by distributing the product over great distances because lesser shipments are required, and the supply of better quality products (Sivertsvik et al., 2002).

The aim of this work was to review the effects of MAP and VP on seafood shelf life, chemical and organoleptic properties, and the survival and growth of spoilage and pathogenic microorganisms.

4.2 FISH STORED UNDER MAP

Rainbow trout vacuum packed in polyethylene film was successfully stored in CO_2 at 180 kPa for 8 days compared with 7 days in air (1 bar) at 1°C. Storage of the VP trout, at −12°C, in CO_2 at 1.8 bar did not improve storage life over that attainable by storage of the VP trout held in air (~100 kPa) at −12°C (Partmann, 1981). Kosak and Toledo (1981) studied the combination of a chlorine solution (1000 mg/mL free chlorine) with vacuum polyethylene packaging for mullet (*Mugil cephalus*) kept at −2°C. All treatments were organoleptically acceptable up to 14 days of storage. According to Scott et al. (1984), the storage life of snapper (*Chrysophrys auratus*) was doubled by either VP or elevated CO_2 atmosphere storage at 3°C compared with storage in the air. They stated that although MA extended the storage life of snapper, commercial use of the technique was dubious because of the substantial variability in the quality of trawler-caught snapper.

Post et al. (1985) found that cod fillets packed under CO_2 and stored at 4°C had a storage life of 43 days, while at 8°C the storage life was 23 days. Temperature abuse markedly decreased the storage life of the CO_2-packed fillets at 4°C. This study and the studies of Gibson (1985) and Ogrydziak and Brown (1982) clearly demonstrated the need for strict temperature control for fish fillets packaged and MAP. Cann et al. (1985) found that Atlantic salmon (*Salmo salar*) had longer storage life at 5°C in 60% CO_2 or 40% N_2 atmosphere than when vacuum packed. According to Barnett et al. (1982), whole, eviscerated coho Salmon (*Oncorhynchus kisutch*) was stored for 3 weeks in 90% CO_2 or 10% air atmosphere at 0°C. Sockeye salmon (*Oncorhynchus nerka*) fillets displayed longer storage life at −1°C than at 1°C when packaged in CO_2 atmosphere (Powrie et al., 1987). The storage life of cultured trout (*Salmo gairdneri*) was doubled by storage in 80% CO_2 or 20% N_2 atmosphere. Trout stored in air was spoiled within 12 days, whereas trout stored in MAP (1.7°C) was still of very good quality after 14 days of storage and of rather fair quality after 20 days. The raw trout was considered marginally acceptable after 25 days at 1.7°C, but, after cooking, the trout was rated as good quality (Barnett et al., 1987). According to Statham and Bremner (1985), fillets of trevalla (*Hyperoglyphe porosa*) packed in 100% CO_2 had a storage life of 8–16 days at 4°C longer than trevalla fillets stored under aerobic conditions. A CO_2-to-fish ratio (v/v) of 4:1 was applied. Psychotrophic bacteria grew rapidly in the air-stored trevalla, reaching a population of 10^9 CFU/cm^2 within 8 days, whereas the psychotrophic population was less than 10^7 CFU/cm^2 at 8 days in the CO_2-packed fillets. Lactic acid bacteria (LAB) formed the dominant microflora in the

CO_2-packed trevalla. Cod fillets (*Gadus morhua*) were stored in bulk at 0°C in 25% CO_2 or 75% N_2 atmosphere. The MAP cod had a storage life of at least 8 extra days compared with cod stored in air at 0°C (12 days). Moreover, cod fillets that were stored under vacuum and under 40% CO_2/30% N_2/30% O_2 at 0°C, 5°C, and 10°C showed an acceptable storage life of 14, 6, and 3 days in MAP and 10, 4, and 2 days under VP at 0°C, 5°C, and 10°C, respectively (Villemure et al., 1986). Sharp et al. (1986) compared several potassium sorbate (0%, 2.5%, 5%) treatments and barrier bag permeabilities (low, intermediate, high) with CO_2 storage at 3°C of Great Lakes whitefish (*Coregonus clupeaformis*) dipped in 5% potassium sorbate and found them to have potential for longer shelf life. On the contrary, potassium sorbate treatment gave no advantage if a storage life of 5 days or less was required. The storage life end point of fillets that were not treated with potassium sorbate was between 10 and 14 days, whereas fillets treated with 2.5% and 5% potassium sorbate had a storage life in excess of 15 days at 3°C. As to the films' permeability, all packing methods enabled a shelf life of 15 days, whereas low barrier packaging promoted a shorter shelf life. Snapper fillets showed the best storage life when packed in a CO_2 atmosphere and stored at −1°C. This was better than storage in 60% N_2, 40% CO_2, or VP. The storage life of the fillets in CO_2 at −1°C was reported to be 2.25 times that in CO_2 at 3°C (Scott et al., 1986). Samples of smoked salmon of different hygienic quality were inoculated with low (6 CFU/g) and high (600 CFU/g) levels of a mixture of three strains of *Listeria monocytogenes*, after which they were VP and stored at 4°C for up to 5 weeks. *L. monocytogenes* grew well during storage in all the inoculated sample groups. The growth was, however, slightly more rapid in the fish with the better hygienic quality, thereby indicating that the richer initial bacterial flora in some fish samples might to some extent have inhibited the growth of *L. monocytogenes*. The smoked salmon was still sensorially acceptable after 4 weeks. All three strains were determined after 4 weeks in the fish with the better quality, while only two strains were recovered after the same time from the poorer quality salmon (Rorvik et al., 1991). Dalgaard et al. (1993) showed that with CO_2 values of 2%, 3%, 29%, 48%, and 97% (the remaining gas in all cases is N_2) a shelf life of 14, 13, 16, 20, and 15–16 days was obtained, respectively, for cod fillets stored at 0°C. With 97% CO_2, the shelf life was only 15–16 days, and the sensory panel described the texture of the fillets with 97% CO_2 as soft and crumbling. In addition, under vacuum, stored fillets had a shelf life of 13–14 days. Furthermore, with CO_2 values of 2%, 3%, 29%, 48%, and 97% the pH values were 6.8, 6.8, 6.8, 6.7, and 6.5, respectively, at the end of the storage period. Reddy et al. (1994) evaluated the effect of MA (75% CO_2/25% N_2, 50% CO_2/50% N_2, 25% CO_2/75% N_2) on the shelf life of tilapia (*Tilapia* spp.) fillets at 4°C. They observed that tilapia fillets packed in 75% CO_2/25% N_2 exhibited an increased shelf life of more than 25 days, with acceptable sensory characteristics. Challenge studies were carried out to evaluate the safety of value-added raw and cooked seafood nuggets inoculated with 10^3 CFU/g of *L. monocytogenes*. The nuggets were packaged in air, or 100% CO_2, with and without an Ageless SS oxygen absorbent and stored at 4°C or 12°C. Most products maintained acceptable appearance throughout storage, but nuggets stored at 12°C developed sharp, acidic odors by day 28. In nuggets stored at 4°C, pH decreased from 7.0 to 4.4–5.2, whereas at 12°C, pH decreased to 4.2–4.4 (Lyver et al., 1998). Ordóñez et al. (2000) stored hake (*Merluccius merluccius*) in atmospheres containing 20% and 40% CO_2 and in air at 2°C ± 1°C. It was reported that the shelf life of hakes increased to 4 and 11 days under 20% and 40% CO_2, respectively. Metin et al. (2002) studied the shelf life of fish salads (rainbow trout *Oncorhynchus mykiss*) in MA and compared with those of air-packaged (control) products. Samples were evaluated at 0, 7, and 14 days at 4°C, respectively. The sensory quality of MAP groups was significantly higher than that of the control group. Control packages were below the limit of acceptability at the 7th day of storage. However, MAP samples (O_2:CO_2:N_2 = 1:7:12 and CO_2:N_2 = 3:7) were not rejected until the 14th day of storage. The pH was 6.5 at the beginning of the experiment but increased over storage. Its level reached 7.2 for the control group, 6.8 for the A group (O_2:CO_2:N_2 = 1:7:12), and 7.0 for the B group (30% CO_2/70% N_2) by the 14th day of storage. Gimenez et al. (2002) investigated filleted rainbow trout (*O. mykiss*) packaged in overwrap, vacuum, and various gas mixtures, stored at 1°C ± 1°C. The gas mixtures used were 20% O_2/50% CO_2/30% N_2, 10% O_2/50% CO_2/40% N_2, 10% O_2/50% CO_2/40% Ar, 20% O_2/50% CO_2/30% Ar, 30% O_2/50% CO_2/20% N_2, and 30% O_2/50% CO_2/20% Ar. MAP samples increased the shelf life (20 days) when compared with vacuum and overwrap packaging (16 and 4 days, respectively). The sensory quality was shown to deteriorate more rapidly in vacuum and overwrap packaging than under MAP, and gas mixtures with 20% and 30% O_2 were allocated lower scores for color and fresh fish flavor intensity. Lalitha et al. (2005) studied fresh pearl spot (*Etroplus suratensis* Bloch) and found that on the basis of the attributes like skin and flesh color, odor, taste and texture, the shelf life of samples packed under air and 40% CO_2/30% O_2/30% N_2 amounted to 12–14 days. In the case of 40% CO_2/60% O_2, 50% CO_2/50% O_2, and 70% CO_2/30% O_2, the shelf life was only 19 days. A maximum shelf life of 21 days was obtained with 60% CO_2/40% O_2 concentration. Pournis et al. (2005) studied fresh open sea red mullet (*Mullus surmuletus*) packaged in four different atmospheres: M1, 10%/20%/70% (O_2/CO_2/N_2); M2, 10%/40%/50% (O_2/CO_2/N_2); M3, 10%/60%/30% (O_2/CO_2/N_2); and in air. All fish were kept at 4°C ± 0.5°C for 14 days. Sensory analyses revealed that the limit of sensorial acceptability was reached after ca. 6 days for the samples packaged in air, 8 days for the M1 and M3 samples, and after 10 days for the M2 samples. Values of pH for air and M1, M2, and M3 gas mixture packaged mullet samples were in the range of ca. 6.96–7.25 with no significant differences between packaging treatments. Torrieri et al. (2006) investigated the effect of MAP on gutted farmed bass (*Dicentrarchus labrax*) when stored at 3°C for up to 9 days. Gutted farmed bass was packed with six different atmospheres (A) 0% O_2/70% CO_2/30% N_2, (B) 20% O_2/70% CO_2/10% N_2, (C) 30% O_2/60% CO_2/10% N_2, (D) 40% O_2/60% CO_2/0% N_2, (E) 30% O_2/50% CO_2/20% N_2, and (F) 21% O_2/0% CO_2/79% N_2. The growth of aerobic mesophilic bacteria (AMB) and Enterobacteriaceae was retarded when gutted bass was packed under atmospheres (B) and (E). After 9 days of storage at 3°C, AMB and Enterobacteriaceae reached, respectively, 2.5×10^6 CFU/g and 5.2×10^4 CFU/g for the atmosphere (B) and 2.2×10^7 CFU/g and 8.0×10^5 CFU/g for the atmosphere (E). However, with the gutted bass with air as initial gas composition (F), the contamination was higher than 10^8 CFU/g for the AMB and of about 10^7 CFU/g for Enterobacteriaceae after 9 days of storage. By applying atmosphere (D), the same results of atmosphere (F) were obtained for AMB. Poli et al. (2006) studied European sea bass (*D. labrax*) fillets packed under 40% CO_2:60% N_2

(MAP) and air (AIR) or prepared from the whole ungutted fish stored in ice (round). On the whole, the shelf life of the three kinds of product could be roughly calculated as 10, 8, and 7 days after slaughtering for round, MAP, and AIR fillets, respectively. Raw round fillets displayed the best quality, while the MAP fillets had higher sensorial scores, lower pH values and microbiological counts, but higher lightness values than those of the corresponding AIR fillets. Manju et al. (2007) reported that VP in conjunction with 2% sodium acetate/potassium sorbate can be safely used to extend the shelf life of Pearl spot (*E. suratensis*) and Black Pomfret (*Parastromateus niger*) samples up to 15 and 16 days, respectively, compared to air and VP samples, which were acceptable only up to 8 and 10 days, respectively. Goulas and Kontominas (2007b) studied the effect of MAP (MAP1:70% CO_2/30% N_2 and MAP2: 50% CO_2/30% N_2/20% O_2) and VP, on the shelf life of chub mackerel (*Scomber japonicus*) fillets, stored at 2°C±0.5°C. It was reported that raw chub mackerel fillets stored in the presence of air remained acceptable up to ca. 11 days, VP and MAP2 samples up to ca. 15–16 days, whereas MAP1 samples remained acceptable up to ca. 20–21 days of storage. It is noteworthy that, flesh texture and flesh color of all packaged samples received scores above or equal to the acceptability limit up to ca. 13–14 days of storage. Ruiz-Capillas et al. (2003) studied hake slices packed with air atmosphere and atmospheres containing 60% CO_2/15% O_2/25% N_2, 40% CO_2/40% O_2/15% N_2, and 60% CO_2/40% O_2. All samples including the ones stored in air scored well up to day 13 of storage, although scores tended to be higher in the case of MA samples. Furthermore, the pH was very similar in all samples up to day 10 of storage. Thereafter, pH began to decline; in some samples this drop continued up to day 20. TVBN values of hake slices packed with air atmosphere and atmospheres containing 60% CO_2/15% O_2/25% N_2, 40% CO_2/40% O_2/15% N_2, and 60% CO_2/40% O_2 behaved in a very similar way to TMAN. Significant differences were detected at day 13 of storage. The upper acceptable limit for TVBN was reached at day 17, except for 60% CO_2/40% O_2, where the limit was reached by day 20.

The effect of MAP on fish quality and safety is summarized in Table 4.1.

4.2.1 Fish Treated with a Combination of MAP and Other Technologies

Trout fillets inoculated with 10^5 *C. botulinum* type E Beluga spores/g of fish were VP on 0.75 mm PE film and subjected to irradiation with electrons (10 MeV) at doses 0, 1, and 2 kGy. No toxin formation was reported at 0°C and the fish became toxic long after spoilage occurred at 5°C. At 10°C irradiated (1 and 2 kGy) fillets became toxic before spoilage was observed whereas nonirradiated fillets spoiled before becoming toxic. Irradiation, in this case, increased the hazard potential of the VP fillets stored at 10°C (Hussain et al., 1977). According to Powrie et al. (1987), pretreating Sockeye salmon (*O. nerka*) fillets with an acidic solution (1% citric acid, 1% ascorbic acid, 0.5% calcium chloride) and then packaging in a CO_2 atmosphere and storing at −1°C or 1°C improved fillet color but did not ameliorate or detract from the sensory properties. Treatment of Sockeye salmon (*O. nerka*) with 1% (w/v) potassium sorbate and a antioxidant dip (0.2% sodium erythorbate, 0.2% citric acid, and a 0.5% sodium chloride) prior to packaging in an MA of 60% CO_2, 35% N_2, and 5% O_2 lead to a slight improvement in quality after 18 days at 1°C compared with salmon stored

TABLE 4.1
Impact of MAP on Sensory Quality and Shelf Life of Fish

Species	Gas Composition	Sensory Quality	Shelf Life (Days)	Storage Temperature	References
Trout *S. gairdneri*	Air		12	1.7°C	Barnett
Trout *S. gairdneri*	80% CO_2/20% N_2	Very good quality		1.7°C	et al. (1987)
Trout *S. gairdneri*	80% CO_2/20% N_2	Fair quality		1.7°C	
Trout *S. gairdneri*	80% CO_2/20% N_2	Marginally acceptable		1.7°C	
Sockeye salmon *O. nerka*	CO_2		Longer storage	1°C	Powrie
Sockeye salmon *O. nerka*	CO_2		life at −1°C	−1°C	et al. (1987)
Cod fillets *G. morhua*	25% CO_2, 75% N_2		20 days	0°C ± 1°C	Villemure
Cod fillets *G. morhua*	Air		12 days	0°C ± 1°C	et al. (1986)
Cod fillets *G. morhua*	40% CO_2, 30% N_2, 30% O_2		14 days	0°C	
Cod fillets *G. morhua*	40% CO_2, 30% N_2, 30% O_2		6 days	5°C	
Cod fillets *G. morhua*	40% CO_2, 30% N_2, 30% O_2		3 days	10°C	
Cod fillets *G. morhua*	Vacuum		10 days	0°C	
Cod fillets *G. morhua*	Vacuum		4 days	5°C	
Cod fillets *G. morhua*	Vacuum		2 days	10°C	
Cod fillets	Packed under CO_2		43 days	4°C	Post
Cod fillets	Packed under CO_2		23 days	8°C	et al. (1985)
Herring fillets	40% CO_2/30% O_2/30% N_2		8 days	0°C	Cann et al. (1983)
Herring fillets	60% CO_2/40% N_2		3 days	0°C	
Herring fillets	Vacuum		13 days	0°C	
Herring fillets	Vacuum		3 days	5°C	
Hot-smoked mackerel	40% CO_2, 30% N_2, and 30% O_2		8 days	0°C	

Hot-smoked mackerel	Vacuum		13 days	0°C	
Hot-smoked mackerel	60% CO_2, 40% N_2		16 days	0°C	
Hot-smoked mackerel	Vacuum		17 days	0°C	
Cold-smoked jack mackerel (*Trachurus declivir*)	Vacuum		Storage life in excess of 30 days	0°C	Fletcher et al. (1988)
Seafood nuggets	100% CO_2	Most products maintained acceptable appearance throughout storage, but nuggets stored at 12°C developed sharp, acidic odors by day 28		4°C	Lyver et al. (1998)
Seafood nuggets	Air			12°C	
Fresh garfish spring	Air	Dark color, ammonia-like, oxidized, and nauseous flavor, dry and soft texture	15 days	0°C	Dalgaard et al. (2006)
Fresh garfish autumn	Air		17 days	0°C	
Fresh garfish spring	40% CO_2 and 60% N_2	Dark color, sour and oxidized flavor, dry texture. Samples of spring garfish also had ammonia-like and nauseous flavors and soft texture, whereas samples of autumn garfish had putrid flavor and a tough texture	20 days	0°C	
Fresh garfish autumn	40% CO_2 and 60% N_2		38 days	0°C	
Fresh garfish spring	Air	Dark color, ammonia-like, oxidized, sour and sharp flavor, dry and soft texture	9 days	5°C	
Fresh garfish autumn	Air		7 days	5°C	
Fresh garfish spring	40% CO_2 and 60% N_2	Dark color, ammonia-like, sour and sharp flavor, dry and soft texture	9 days	5°C	
Fresh garfish autumn	40% CO_2 and 60% N_2		10 days	5°C	
Frozen/thawed garfish spring	Air	Dark color, ammonia-like, oxidized, musty and putrid flavor, dry and soft texture	10 days	5°C	
Frozen/thawed garfish autumn	Air		9 days	5°C	
Frozen/thawed garfish spring	40% CO_2 and 60% N_2	Dark color, ammonia-like, sour and musty flavor, dry and tough texture (The experiment was stopped before spoilage occurred)	17 days	5°C	
Frozen/thawed garfish autumn	40% CO_2 and 60% N_2		>16 days	5°C	

(*continued*)

TABLE 4.1 (continued)
Impact of MAP on Sensory Quality and Shelf Life of Fish

Species	Gas Composition	Sensory Quality	Shelf Life (Days)	Storage Temperature	References
Fresh cod fillets	Air		1.96 days	4°C	Corbo et al. (2005)
Fresh cod fillets	Vacuum		6.00 days	4°C	
Fresh cod fillets	65% N_2, 30% CO_2, 5% O_2		5.42 days	4°C	
Fresh cod fillets	20% CO_2, 80% O_2		2.62 days	4°C	
Fresh pearl spot (*E. suratensis* Bloch)	Air		12–14 days	0°C	Lalitha et al. (2005)
Fresh pearl spot (*E. suratensis* Bloch)	40% CO_2/30% O_2/30% N_2		12–14 days	0°C	
Fresh pearl spot (*E. suratensis* Bloch)	40% CO_2/60% O_2		19 days	0°C	
Fresh pearl spot (*E. suratensis* Bloch)	50% CO_2/50% O_2		19 days	0°C	
Fresh pearl spot (*E. suratensis* Bloch)	70% CO_2/30% O_2		19 days	0°C	
Fresh pearl spot (*E. suratensis* Bloch)	60% CO_2/40% O_2		21 days	0°C	
Fish salads (rainbow trout *O. mykiss*)	Air	The sensory quality of MAP samples significantly higher than the controls	7 days	4°C	Metin et al. (2002)
Fish salads (rainbow trout *O. mykiss*)	O_2:CO_2:N_2 = 1:7:12		14 days	4°C	
Fish salads (rainbow trout *O. mykiss*)	and CO_2:N_2 = 3:7		14 days	4°C	
Farmed spotted wolf-fish (*A. minor*)	Air		8–10 days	−1.0°C ± 0.2°C	Rosnes et al. (2006)
Farmed spotted wolf-fish (*A. minor*)	Air		6–8 days	+4.0°C ± 0.2°C	
Farmed spotted wolf-fish	60% CO_2/40% N_2	Improved odor and flavor scores	15 days	−1.0°C ± 0.2°C	
Farmed spotted wolf-fish	60% CO_2/40% N_2		13 days	+4.0°C ± 0.2°C	
Filleted rainbow trout (*O. mykiss*)	Overwrap	Sensory quality deteriorated faster in vacuum and overwrap packaging than in MAP	4 days	1°C ± 1°C	Gimenez et al. (2002)
Filleted rainbow trout (*O. mykiss*)	Vacuum	Sensory quality deteriorated faster in vacuum and overwrap packaging than in MAP	16 days	1°C ± 1°C	

Filleted rainbow trout (*O. mykiss*)	20% O_2/50% CO_2/30% N_2, 10% O_2/50% CO_2/40% N_2,10% O_2/50% CO_2/40% Ar, 20% O_2/50% CO_2/30% Ar, 30% O_2/50% CO_2/20% N_2, 30% O_2/50% CO_2/20% Ar	Sensory quality deteriorated faster in vacuum and overwrap packaging than in MAP, and gas mixtures with 20% and 30% O_2 were given lower scores for color and fresh fish flavor intensity	20 days	1°C ± 1°C	
European sea bass (*D. labrax*) fillets	40% CO_2/ 60% N_2	MAP samples had better sensorial scores than the air packaged samples	8 days	2°C ± 1°C	Poli et al. (2006)
European sea bass (*D. labrax*) fillets	Air		7 days	2°C ± 1°C	
European sea bass (*D. labrax*) fillets	Whole ungutted fish stored in ice		10 days	2°C ± 1°C	
Chub mackerel (*S. japonicus*) fillets	Air		11 days	2°C ± 0.5°C	Goulas and Kontominas (2007b)
Chub mackerel (*S. japonicus*) fillets	70% CO_2/30% N_2		20–21 days	2°C ± 0.5°C	
Chub mackerel (*S. japonicus*) fillets	50% CO_2/30% N_2/20% O_2		15–16 days	2°C ± 0.5°C	
Chub mackerel (*S. japonicus*) fillets	Vacuum		15–16 days	2°C ± 0.5°C	
Cod fillets	2% CO_2/N_2	Sour	14 days	0°C	Dalgaard et al. (1993)
Cod fillets	3% CO_2/N_2	Sour	13 days	0°C	
Cod fillets	29% CO_2/N_2	Sour	16 days	0°C	
Cod fillets	48% CO_2 /N_2	Sour	20 days	0°C	
Cod fillets	97% CO_2/N_2	Sour, soft, and crumbling	15–16 days	0°C	
Cod fillets	Vacuum		13–14 days	0°C	
Red mullet (*M. surmuletus*)	Air		6 days	4°C ± 0.5°C	Pournis et al. (2005)
Red mullet (*M. surmuletus*)	10% O_2/20% CO_2/70% N_2		8 days	4°C ± 0.5°C	
Red mullet (*M. surmuletus*)	10% O_2/40% CO_2/50% N_2		10 days	4°C ± 0.5°C	
Red mullet (*M. surmuletus*)	10% O_2/60% CO_2/30% N_2		8 days	4°C ± 0.5°C	

in the MA without the potassium sorbate and antioxidant pretreatments. An addition of 1% CO to the gas mixture resulted in a slight decrease in the quality scores of antioxidant and potassium sorbate treated sockeye salmon (Fey and Regenstein, 1982). Prior to VP and storage at 4°C, sand flathead (*Platycephalus bassensis*) fillets were subjected to a variety of dipping treatments. Dipping of fillets in citrate buffer (pH 4.8) lowered the pH of the fillet surface and inhibited the growth of *S. putrefaciens* and development of sulfide-like odors. The acid treatment, however, caused bleaching of the fillets and the development of milky exudates. Glucose treatment (50 g/L) of the fillets did not lead to storage-life extension. This might be attributed to the low initial population of LAB on the fillets at the time of packaging. None of the treatments led to prolongation of the storage life of the sand flathead fillets, and storage life of the VP fillets was not longer than that of the air-stored control. *S. putrefaciens* formed the major portion of the microflora of untreated VP sand flathead fillets, while Enterobacteriaceae dominated in treated fillets (McMeekin et al., 1982). Licciardelo et al. (1984) studied cod fillets vacuum packed or packed in 60% CO_2, 40% air atmosphere in barrier bags followed by treatment with 1 kGy gamma irradiation and subsequent storage on ice. The CO_2 packed, irradiated cod fillets retained quality attributes longer than the VP product, which retained their quality longer than air-packed fillets. Statham et al. (1985) reported that fillets of morwong (*Nemadactylus macropterus*), a whitefish, had the longest storage life (13 days) when dipped in 10% polyphosphate and 1.2% potassium sorbate solutions followed by storage under 100% CO_2 barrier bags at 4°C. Individual dips followed by VP or sorbate and polyphosphate dips combined with VP were not as effective as the combined dips with the CO_2 MAP. Przybylski et al. (1989) studied commercially cultured channel catfish (*Ictalurus punctatus*) packed in PE bags with atmospheres of 100% air, 100% CO_2 or 80% CO_2, and 20% air treated with 0, 0.5, or 1.0 kGy of gamma irradiation, respectively. The packaged samples were stored at 0°C–2°C for up to 30 days. The lowest psychotrophic plate counts were obtained with 100% CO_2 atmosphere. Enhanced radiation doses decrease the microbial population in fish sampled after 20 days of storage. The atmosphere in the packages had no significant effect on the development of thiobarbituric acid (TBA) reactive substances (TBARS) in the catfish at each of the three radiation doses. Santos and Regenstein (1990) found that white hake (*Urophysis tenuis*) fillets treated with 0.5% erythorbic acid (ERA) and VP (in low-barrier film) deteriorated in quality at −7°C more rapidly than fillets stored in air without the ERA treatment. ERA hastened the rate of deterioration of white hake, and the effect was further enhanced by VP. Moreover, they found that mackerel (*Scomber scombrus*) fillets treated with 0.5% solution of ERA prior to VP, in a low-barrier film, improved storage life at −7°C. VP in conjunction with the antioxidant treatment provided better protection for the mackerel than did glazing of the antioxidant-treated fillets. According to Taylor et al. (1990)

> The treatment of fish with nisin just prior to packaging in a CO_2 atmosphere MA delayed the onset of toxin production by *C. botulinum*. Nisin-treated fillets spoiled at the same rate at 26°C and 10°C as untreated fillets. Nisin displayed no inhibitory effect on the spoilage microflora of cod, herring and smoked mackerel fillets packed under the CO_2 MA, even though *C. botulinum* toxinogenesis was considerably retarded.

Kosak and Toledo (1981) studied the application of a chlorine solution (1000 mg/mL free chlorine) in conjunction with vacuum PE packaging for mullet (*M. cephalus*) kept at −2°C. All treatments were found to be organoleptically acceptable up to 14 days of storage. According to Pastoriza et al. (1998) sodium chloride dips exhibited further inhibition of biochemical, microbiological, and sensory deterioration of MAP (50% CO_2/45% N_2/5% O_2) stored hake slices (*M. merluccius*). As a result, the total volatile base (TVB) and total viable count (TVC) values were significantly lower than those in fish stored solely under MAP conditions. Moreover, sensory properties were scored significantly higher. Consequently, the shelf life of hake slices was extended for 2 days when stored under MAP conditions (12 days shelf life) and for 8 days (18 days shelf life), if dipped in sodium chloride solutions before MAP storage. Prentice and Sainz (2000) found that washing grass carp (*Ctenopharyngodon idella*) fillets with sodium hypochlorite and brine and VP helped it reach shelf life periods of 30 and 60 days, when the product was kept under refrigeration temperatures of 8°C and 2°C, respectively. Franzetti et al. (2001) investigated the effectiveness of an innovative foam plastic tray, provided with absorbents for volatile amines and liquids, on fillets of sole (*Solea solea*), steaks of cod (*M. merluccius*), and whole cuttlefish (*Sepia filluxi*) packed under an MA (40% CO_2/60% N_2) at 3°C. The TMA for sole fillets with MAP (40% CO_2/60% N_2) combined with the innovative tray was 15 μg compared to 115 μg when stored under 40% CO_2/60% N_2. For cod steaks, the TMA for MAP (40% CO_2/60% N_2) and innovative tray was 15 and 130 μg, respectively. In the case of cuttlefish combined with the innovative tray, the TMA amounted to 7.4 and 158 μg for cuttlefish stored under 40% CO_2/60% N_2. It was also found that the innovative tray displayed strong inhibitory effect on the microbial growth, especially of H_2S-producing and Gram-negative bacteria. Chouliara et al. (2004) investigated irradiated (1 and 3 kGy) VP sea bream (*Sparus aurata*) fillets stored under refrigeration. Of the chemical indicators of spoilage, TMA values of nonirradiated, salted sea bream increased slowly to 8.87 mg N/100 g flesh, whereas for irradiated, salted samples significantly lower values were obtained, reaching a final TMA value of 6.17 and 4.52 mg N/100 g flesh at 1 and 3 kGy, respectively. TVBN values increased slowly attaining a value of 60.52 mg N/100 g for nonirradiated, salted sea bream during refrigerated storage. However, the corresponding values for irradiated fish were lower and amounted to 48.13 and 37.21 mg N/100 g muscle at 1 and 3 kGy, respectively. TBA values for irradiated, salted sea bream samples were higher than respective nonirradiated (salted) fish and augmented slowly until day 28 of storage, reaching final values of 1.01 (nonirradiated, salted), 2.15 (1 kGy), and 3.26 mg malonaldehyde/kg flesh (3 kGy), respectively. Altieri et al. (2005) evaluated the microbiological and sensory characteristics of biopreserved packed fresh plaice (*Pleuronectes platessa*), over storage. Samples treated with a *Bifidobacterium bifidum* strain and thymol as preservatives were stored at different temperatures (4°C and 12°C) and package atmospheres (in air, under vacuum, and 65% N_2, 30% CO_2, 5% O_2). Application of lower storage temperature prolonged the lag phases of TVC, thereby confirming that the calculated shelf lives were positively affected by the lower storage temperature. Furthermore, in the samples packaged in air and vacuum, the addition of thymol as antimicrobial agent enhanced the microbiological shelf life values up to 2.4 and 7 days, respectively. Rosnes et al. (2006) evaluated MAP combined with superchilling

(−1°C) as a mild preservation method for farmed spotted wolf fish (*Anarhichas minor*). Portions were packaged in air and in 60% CO_2/40% N_2 (60%:40%) atmosphere, at superchilled (−1.0°C±0.2°C) or chilled (+4.0°C±0.2°C) temperatures. MA-packaged wolf fish had improved odor and flavor scores, accompanied with a higher drip loss than fish stored in air. Sallam et al. (2007) studied the Pacific saury (*Cololabis saira*), brined (12% NaCl brine solution) or marinated (12% NaCl + 2% acetic acid; or 12% NaCl + 3% acetic acid solutions) followed by VP and storage at 4°C for 90 days. The pH of the raw Pacific saury fillets used in this study was 6.32. Brining resulted in small but significant reduction of the initial pH (6.07±0.04). The latter increased over storage. Marinating with either 2% or 3% acetic acid displayed a pronounced drop in the initial pH by about 2 units. Psychrotrophic bacterial count (PTC) in raw Pacific saury was 3.95 $\log_{10}$ CFU/g. The initial PTC was substantially reduced after the marinating process, while the brining process did not. By the end of the storage period (day 90), brined fillets exhibited considerably higher PTC of 8.31 $\log_{10}$ CFU/g, whereas the fillet samples marinated with 2% or 3% acetic acid gave much lower counts of 5.16 and 4.75 $\log_{10}$ CFU/g, respectively. According to Sallam et al. (2007) by the end of the storage period of Pacific saury (*C. saira*), a significant increase in TVBN values to relatively high levels of 34.6, 24.1, and 20.5 mg/100 g were detected for fillets marinated with 0%, 2%, and 3% acetic acid, respectively. Furthermore, by the end of the storage period, the brined fillets presented a high level of TMA (8.37 mg/100 g), whereas fillets marinated in 2% and 3% acetic acid showed significantly lower TMA values of 5.52 and 4.47 mg/100 g, respectively. Regarding TBA, significantly higher values of 2.82, 1.88, and 1.61 mg MA/kg, in comparison with the initial values, had been attained by the end of the storage for brined, 2% acetic acid-, and 3% acetic acid-marinated fillets, respectively. Cultured sea bream (*S. aurata*) fillets, stored at 4°C±0.5°C, had higher TVBN (mg N/100 g) and TMAN values stored in air (31.5±0.7 after 12 days of storage) followed by salted fillets stored in air (22.9±0.5). The TVBN values for MAP (40% CO_2/30% O_2/30% N_2) salted, MAP salted 0.4% oregano oil, and MAP salted 0.8% oregano oil amounted to 21.3±0.6, 19.5±0.4, and 19.8±0, respectively. The TMAN values (mg N/100 g), after 12 days of storage, of fillets stored in air and of salted fillets stored in air were 1.12±0.05 and 0.68±0.04, respectively. The TMAN values for MAP (40% CO_2/30% O_2/30% N_2) salted, MAP salted 0.4% oregano oil, and MAP salted 0.8% oregano oil were 0.65±0.05, 0.46±0.03, and 0.50±0.05, respectively. The values indicated the preservative effect of oregano oil. All raw sea bream fillet samples were assigned sensory scores during the first 15–16 days of storage. The salted samples remained acceptable up to ca. 20–21 days while the MAP salted samples up to ca. 27–28 days of storage. The oregano oil addition in MAP salted samples yielded a distinct but pleasant flavor, thereby contributing to a considerable slower process of fish spoilage given that the fillets treated with 0.8% (v/w) oregano oil were still sensory acceptable after 33 days of storage (Goulas and Kontominas, 2007a). Wang et al. (2008) investigated the effect of combined application of MAP and superchilled storage on the shelf life of fresh cod loins. Fresh cod loins were packed in polystyrene boxes and in MA (CO_2/N_2/O_2: 50%/45%/5%) on day 3 post catch and stored at chilled (1.5°C) and superchilled (−0.9°C) temperatures. Superchilled storage alone compared with traditional chilled storage in polystyrene boxes increased the total shelf life of cod loins from 9 to 16 or 17 days. Chilled MA packaging increased the shelf life from 9 to 14 days, and when MAP and superchilled storage were combined, a synergistic effect was observed and the shelf life was further extended to at least 21 days. Moreover, the characteristic fresh and sweet taste can be maintained longer under such conditions. However, superchilled MA packed cod loins had more a meaty texture compared with other sample groups after 7 days of storage.

Fish treated with a combination of MAP and other technologies is exhibited in Table 4.2.

4.2.2 Quality of Fish Stored under MAP

Mitsuda et al. (1980) reported that dipping fillets of farmed hamachi (*Seriola aurevittata*) in 5% sodium chloride for 1 min prior to packaging under CO_2 and storage at 3°C maintained the color and texture of the tissue during a 7-day storage period. Carbon dioxide was shown to be better than N_2 for maintenance of flesh texture. Sodium chloride displayed a better performance than potassium chloride as the dipping solution. Rockfish (*Sebastes* spp.) fillets were stored in an MA of 80% CO_2/20% O_2 for 20 days at 4°C. In vitro protein digestibility of air-stored fillets dropped while that of MAP fillets remained the same as that of fish at zero time. The computed protein efficiency ratio of the fillets stored under MA for 14 days was similar to that of fresh fillets (Morey et al., 1982). Chen et al. (1984) reported that farmed rainbow trout packaged in VP or under CO_2 showed considerably less lipid oxidation than trout stored under air at 1°C–2°C. MAP with elevated CO_2, followed by VP and air storage, was strongly inhibitory to lipid oxidation but also caused the greatest loss of carotenoid content. The cooked flavor of VP trout was better than that of trout packed under CO_2. Villemure et al. (1986) stored cod fillets under vacuum and under 40% CO_2/30% N_2/30% O_2 at 0°C, 5°C, and 10°C. In all cases, psychotrophic populations were under 10^7 CFU/g at the time when cooked flavor of the fillets reached 5.5 on a 10-point rating scale. Cooked flavor was found to deteriorate at the rate of 0.25, 0.60, and 1.00 units/day in MAP, whereas the deterioration rate under VP amounted to 0.3, 0.7, and 1.2 units/day at 0°C, 5°C, and 10°C, respectively. Belleau and Simard (1987) studied stored (6 days) fillets of Greenland halibut (*Reinhardtius hippoglossoides*) at 0°C in atmospheres containing the following N_2:CO_2 ratios: 100:0, 75:25, 50:50, 25:75, and 0:100. The optimal gas mixture was 25% CO_2–25% N_2 because texture and pH were least affected and exudate formation was minimal. The greatest exudate formation occurred with fillets stored in 100% CO_2. Cod fillets (*G. morhua*) were packed under MA, with four different gas compositions (60% CO_2/10% O_2/30% N_2, 60% CO_2/20% O_2/20% N_2, 60% CO_2/30% O_2/10% N_2, 60% CO_2/40% O_2), and stored at 6°C. Both TMA and TVB were continuously produced over 7 days of storage. From 5th day onward, the rate of TMA production differs depending on the packaging atmosphere. As a consequence of TVB production, a slight increase in pH was noticed over the first 4 days of storage. Diffusion of CO_2 in the fish muscle shows a countereffect on the pH increase by TVB production, thereby leading to stabilization of pH around 6.7 (Debevere and Boskou, 1996). Hurtado et al. (2000) evaluated the shelf life of refrigerated (2°C–3°C) VP hake (*M. capensis*) slices, which were

TABLE 4.2
Effect of MAP in Combination with Other Technologies on Fish

Species	Gas Composition	Other Technology	Sensory Assessment	Shelf Life (Days)	Volatile Basic Nitrogen/Trimethylamine Nitrogen	General Quality	Storage Temperature	References
Trout (*S. gairdneri*)	80% CO_2/20% N_2	2.3% potassium sorbate prior to MAP		~1 week			1.7°C	Barnett et al. (1987)
Sockeye salmon (*O. nerka*)	Packed under CO_2	Pretreatment of fillets with an acidic solution (1% citric acid, 1% ascorbic acid, 0.5% calcium chloride)	The combination of the two methods did not improve or detract from the sensory properties				−1°C or 1°C	Powrie et al. (1987)
Cod fillets	60% CO_2, 40% air	Treatment with 1 kGy gamma irradiation and subsequent storage on ice				Retained quality attributes longer than the vacuum-packed fillets, which retained their quality longer than air-packed fillets		Licciardelo et al. (1984)

Great Lakes whitefish (*C. clupeaformis*)	CO_2	Potassium sorbate treatment (2.5% and 5%)	>15		3°C	Sharp et al. (1986)
Great Lakes whitefish (*C. clupeaformis*)	CO_2	No treatment	10–14		3°C	
Morwong (*N. maropterus*), a whitefish	100% CO_2 barrier bags	First, dipped in 10% polyphosphate and 1.2% potassium sorbate solutions	13		4°C	Statham et al. (1985)
Commercially cultured channel catfish (*I. punctatus*)	100% air, 100% CO_2, or 80% CO_2 and 20% air	0, 0.5, or 1.0kGy of gamma irradiation		The lowest psychotrophic plate counts were obtained with 100% CO_2 atmosphere. Increased radiation doses decrease the microbial population in fish sampled after 20 days of storage	0°C–2°C for up to 30 days	Przybylski et al. (1989)

(*continued*)

TABLE 4.2 (continued)
Effect of MAP in Combination with Other Technologies on Fish

Species	Gas Composition	Other Technology	Sensory Assessment	Shelf Life (Days)	Volatile Basic Nitrogen/Trimethylamine Nitrogen	General Quality	Storage Temperature	References
White hake (*U. tenuis*) fillets	Vacuum packaged (in low-barrier film)	0.5% ERA prior to vacuum packaging				Deteriorated in quality more rapidly than fillets stored in air without the ERA treatment. ERA hastened the rate of deterioration of white hake and the effect was magnified by vacuum packaging	−7°C	Santos and Regenstein (1990)
Mackerel (*S. scombrus*) fillets	Vacuum packaged (in low-barrier film)	0.5% solution of ERA prior to vacuum packaging		Enhanced storage life		Vacuum packaging with the antioxidant treatment offered better protection for samples than did glazing of the antioxidant-treated ones	−7°C	

Trout fillets inoculated with 10^5 *C. botulinum* type E Beluga spores/g of fish	Vacuum	Irradiation with electrons (10 MeV) at doses 1 and 2 kGy	No toxin formation was observed	0°C	Hussain et al. (1977)
Trout fillets inoculated with 10^5 *C. botulinum* type E Beluga spores/g of fish	Vacuum	Irradiation with electrons (10 MeV) at doses 1 and 2 kGy	The fish became toxic long after spoilage occurred	5°C	
Trout fillets inoculated with 10^5 *C. botulinum* type E Beluga spores/g of fish	Vacuum	Irradiation with electrons (10 MeV) at doses 1 and 2 kGy	Fillets became toxic before spoilage was observed	10°C	
Trout fillets inoculated with 10^5 *C. botulinum* type E Beluga spores/g of fish	Vacuum	No irradiation treatment	Fillets spoiled before they became toxic	10°C	
Cod, herring, and smoked mackerel fillets	CO_2	Treatment with nisin prior to packaging	Delayed the onset of toxin production by *C. botulinum*. Nisin-treated fillets spoiled at the same rate at 26°C and 10°C as untreated fillets		Taylor et al. (1990)
Cod, herring, and smoked mackerel fillets	CO_2	No other treatment			

(*continued*)

TABLE 4.2 (continued)
Effect of MAP in Combination with Other Technologies on Fish

Species	Gas Composition	Other Technology	Sensory Assessment	Shelf Life (Days)	Volatile Basic Nitrogen/Trimethylamine Nitrogen	General Quality	Storage Temperature	References
Mullet (*M. cephalus*)	Vacuum	Chlorine solution (1000 mg/mL free chlorine)	All treatments were organoleptically acceptable up to 14 days of storage				−2°C	Kosak and Toledo (1981)
Salted cultured sea bream (*S. aurata*) fillets	40% CO_2/30% O_2/30% N_2	No treatment		27–28	21.3 ± 0.6 mg N/100 g/0.65 ± 0.05 mg N/100 g		4°C ± 0.5°C	Goulas and Kontominas (2007a)
Salted cultured sea bream (*S. aurata*) fillets	40% CO_2/30% O_2/30% N_2	0.4% oregano oil		28	19.5 ± 0.4 mg N/100 g/0.46 ± 0.03 mg N/100 g		4°C ± 0.5°C	
Salted cultured sea bream (*S. aurata*) fillets	40% CO_2/30% O_2/30% N_2	0.8% oregano oil		33	19.8 ± 0 mg N/100 g/0.68 ± 0.04 mg N/100 g		4°C ± 0.5°C	
Hake slices *M. merluccius* (control)	Air stored	—		10	TVB increased significantly, up to 85.23 mg 100/g after 2 weeks of storage		2°C ± 1°C	Pastoriza et al. (1998)

Hake slices *M. merluccius*	50% CO_2/45% N_2/5% O_2	—	12	TVB values significantly lower than the control after 5 days of storage	2°C ± 1°C	
Hake slices *M. merluccius*	50% CO_2/45% N_2/5% O_2	Dipped in NaCl prior to packaging	15	TVB values of control samples were threefold higher than NaCl-dipped fish stored under MAP after 14 days ice storage	2°C ± 1°C	
Hake slices *M. merluccius*	50% CO_2/45% N_2/5% O_2	Dipped in sodium chloride prior to packaging	18	Reduced TVB values significantly during storage period	2°C ± 1°C	
Black Pomfret (*P. niger*)	Air	—	8	The TVBN values of vacuum-packed samples were lower than those of the air-packed samples in both the species	In ice	Manju et al. (2007)
Black Pomfret (*P. niger*) and Pearl spot (*E. suratensis*)	Vacuum	—	10		In ice	

(*continued*)

TABLE 4.2 (continued)
Effect of MAP in Combination with Other Technologies on Fish

Species	Gas Composition	Other Technology	Sensory Assessment	Shelf Life (Days)	Volatile Basic Nitrogen/Trimethylamine Nitrogen	General Quality	Storage Temperature	References
Black Pomfret (*P. niger*)	Vacuum	2% sodium acetate		15	The TVBN contents of potassium sorbate–treated samples were slightly lower than sodium acetate–treated samples for both the species		In ice	
Pearl spot (*E. suratensis*)	Vacuum	2% potassium sorbate		16			In ice	
Salted packaged sea bream fillets (*S. aurata*)	Vacuum	Not irradiated		14–15	TVB nitrogen values increased to 60.52 mg N/100 g/TMA values increased slowly to 8.87 mg N/100 g flesh	Irradiation affected populations of bacteria, namely, *Pseudomonas* spp., H_2S-producing bacteria, *Brochothrix thermosphacta*, Enterobacteriaceae, and LAB. The	Under refrigeration	Chouliara et al. (2004)

Salted packaged sea bream fillets (*S. aurata*)	Vacuum	Irradiation (1 kGy)		27–28	TVB nitrogen values of 48.13 mg N/100 g muscle, /TMA reached a final value of 6.17 and (100 g^{-1} flesh	effect was more pronounced at the higher dose (3 kGy) applied	Under refrigeration	
Salted packaged sea bream fillets (*S. aurata*)	Vacuum	Irradiation (3 kGy)		27–28	TVB nitrogen values of 37.21 mg N/100 g muscle/TMA reached a final value of 4.52 mg N/100 g flesh		Under refrigeration	
Pacific saury (*C. saira*)	Vacuum	Brined (12% NaCl brine solution)	The overall acceptability was significantly higher in marinated versus brined fish		TVBN was 34.6 mg/100 g TMA was 8.37 mg/100 g	By the end of the storage period (day 90), brined fillets exhibited a much higher PTC of 8.31 log_{10} CFU/g	4°C	Sallam et al. (2007)
Pacific saury (*C. saira*)	Vacuum	Marinated 12% NaCl + 2% acetic acid	No significant differences were detected for the sensory		TVBN was 24.1 mg/100 g TMA 5.52 mg/100 g	Exhibited much lower counts of 5.16 log_{10} CFU/g,	4°C	

(*continued*)

TABLE 4.2 (continued)
Effect of MAP in Combination with Other Technologies on Fish

Species	Gas Composition	Other Technology	Sensory Assessment	Shelf Life (Days)	Volatile Basic Nitrogen/Trimethylamine Nitrogen	General Quality	Storage Temperature	References
Pacific saury (*C. saira*)	Vacuum	Marinated 12% NaCl + NaCl + 3% acetic acid solutions	differences in attributes between the two marinating conditions		TVBN was 20.5 mg/100 g/ TMA 4.47 mg/100 g	Exhibited much lower counts of 4.75 $\log_{10}$ CFU/g	4°C	
Sole fillets (*S. solea*)	40% CO_2/ 60% N_2	Innovative foam plastic tray, provided with absorbents for volatile amines and liquids		10	TMA 15 μg	6.1×10^8 total microbial count after 10 days of storage	3°C	Franzetti et al. (2001)
Sole fillets (*S. solea*)	40% CO_2/ 60% N_2	—		—	TMA 115 μg	5×10^8 total microbial count after 10 days of storage	3°C	
Steaks of cod (*M. merluccius*)	40% CO_2/ 60% N_2	Innovative foam plastic tray, provided with absorbents for volatile amines and liquids		10	TMA 15 μg	6×10^8 total microbial count after 10 days of storage	3°C	

Steaks of cod (*M. merluccius*)	40% CO_2/60% N_2	—		—	TMA 130 μg	4×10^8 total microbial count after 10 days of storage	3°C	
Fresh cod loins		Chilled storage (1.5°C)	The characteristic fresh and sweet taste can be maintained longer under superchilled storage under MAP	9	Super chilled MA packed cod loins had more meaty texture compared to other sample groups after 7 days of storage		1.5°C	Wang et al. (2008)
Fresh cod loins		Superchilled storage (−0.9°C)		16–17			−0.9°C	
Fresh cod loins	CO_2/N_2/O_2:50%/45%/5%)	Superchilled storage (−0.9°C)		21			−0.9°C	

found to be organoleptically acceptable up to the 43rd day of storage. Low TMA amounts and slight increase in drip were verified after 15 days of storage. According to Metin et al. (2002), the initial TVBN value of fish salad (rainbow trout *O. mykiss*) was 11.5 mg/100 g of fish. At the 7th day of storage, the TVBN value of control (air stored) samples was higher than that of the MAP groups. This value reached 25.54 mg/100 g of fish for the control group, 20.12 mg/100 g of fish for group A (O_2:CO_2:N_2 = 1:7:12), and 22.31 mg/100 g of fish for group B (CO_2:N_2 = 3:7) by the 14th day of storage. Gimenez et al. (2002) investigated filleted rainbow trout (*O. mykiss*) packaged in overwrap, vacuum, and gas mixture conditions stored at 1°C ± 1°C. The gas mixtures used were 20% O_2/50% CO_2/30% N_2, 10% O_2/50% CO_2/40% N_2,10% O_2/50% CO_2/40% Ar, 20% O_2/50% CO_2/30% Ar, 30% O_2/50% CO_2/20% N_2, and 30% O_2/50% CO_2/20% Ar. Lipid oxidation measured as TBARS was higher in MAP than in VP. Significant differences were reported between gas mixtures with 20% and 30% O_2 than those with 10% O_2. VP displayed the lowest TBARS values. Hypoxanthine levels improved considerably in vacuum and overwrap packaging. However, the production of hypoxanthine in MAP substantially slowed down, with no significant differences observed in different mixtures. MAP was very effective in reducing the TVB production. MAP practically enhanced the shelf life compared with vacuum and overwrap packaging, which was 16 and 4 days, respectively. Atlantic mackerel (*S. scombrus*) fillets stored and packed in MA at −2°C were evaluated by Hong et al. (1996). MAP storage increased the shelf life to 21 days, causing a slight increase in TVBN and TMA amounts. According to Lalitha et al. (2005) TVBN content of fresh pearl spot (*E. suratensis* Bloch) stored at 0°C ranged from 12.8 to 22.4 mg N/100 g flesh for pearl spot stored under air during 2 weeks of storage in ice. In gas mixtures of 40% CO_2/30% O_2/30% N_2 and 40% CO_2/60% O_2, the same level was reached after 19 days and for other gas mixtures (50% CO_2/50% O_2, 60% CO_2/40% O_2, 70% CO_2/30% O_2), after 29 days. The pH of fresh fish muscle was about 6.28. On the day of spoilage, the pH value reached almost 6.8 in fish packed under air and MAs. According to Dondero et al. (2004), the TVB of vacuum cold-smoked salmon (*S. salar*) increased from initial values of 22.1–25.8 mg TVBN/100 g to 31.8, 29.3, 29.8, 30.0, and 29.9 mg TVBN/100 g at 0°C, 2°C, 4°C, 6°C, and 8°C after 26, 21, 20, 10, and 7 days, respectively. Initially, K values were between 35.7% and 53.4% and, at the end of shelf life, these values were 72.6%, 69.6%, 97.8%, 63.2%, and 69.8% after 26, 21, 20, 10, and 7 days at 0°C, 2°C, 4°C, 6°C, and 8°C, respectively. Furthermore, increasing levels of Hypoxanthine were determined from initial values of 0.29–1.17 µmol/g of salmon to final values of 1.03, 0.67, 2.24, 0.85, and 0.66 µmol/g of salmon after 26, 21, 20, 10, and 7 days at 0°C, 2°C, 4°C, 6°C, and 8°C, respectively. Concentrations of TMA ranged from 2.9 to 3.5 mg TMAN/100 g in fresh cold-smoked salmon to 10.2, 7.3, 7.5, 7.4, and 7.7 mg TMAN/100 g at 0°C, 2°C, 4°C, 6°C, and 8°C after 26, 21, 20, 10, and 7 days, respectively. Sardine (*Sardina pilchardus*) was studied under MAP (60% CO_2:40% N_2) and VP, by Ozogul et al. (2004). The shelf life of sardine was determined to be 12 days in MAP, 9 days in vacuum, and 3 days in air. Bacteria grew most quickly in sardine stored in air, followed by those in vacuum, and the lowest counts were for MAP. The concentration of histamine increased, and its level reached over 20 mg/100 g for fish stored in air, 13 mg/100 g for vacuum, and 10 mg/100 g for MAP at 15 days.

The highest concentration of TMA was obtained from sardine stored in air, followed by sardine in vacuum, and the lowest in MAP. The formation of TVBN increased with storage time. When the TVC had reached 10^6 CFU/g, the TVBN content was found to be approximately 15 mg/100 g muscle for all storage conditions. MAP was combined with freeze-chilling to extend the shelf life of raw whiting, mackerel, and salmon fillets/portions. The MAP packs for mackerel and salmon (60% N_2/40% CO_2), and for whiting (30% N_2/40% CO_2/30% O_2) maintained their shape during freeze-chilling whereas packs with 100% CO_2 displayed concave sides. The chosen chilled shelf life of 5–7 days in the MAP trials was vindicated by the results as the products were near the end of their shelf life after 5 (whiting and mackerel) and 7 (salmon) days. This compares with shelf lives of 3 and 5 days, respectively, for freeze-chilled fillets in air. After 5 days of storage, whiting, mackerel, and salmon had TVBN of 24.2, 21.0, and 17.5 mg N/100 g. Moreover, after 5 days of storage for mackerel, the TMA value was 8.52 mg N/100 g (Fagan et al., 2004). Masniyom et al. (2005) determined the changes in collagen of sea bass (*Lates calcarifer*) muscle treated with and without pyrophosphate (PP) during storage in MAP (80% CO_2, 10% O_2, and 10% N_2) at 4°C. No changes in acid-soluble collagen (ASC) and pepsin-soluble collagen (PSC) of sea bass muscle with and without PP treatment were observed during storage in MAP up to 21 days. For sea bass muscle stored under an air atmosphere, ASC increased, whereas PSC decreased with a concomitant loss in the firmness. According to Pournis et al. (2005), the values of TMAN (milligrams of N/100 g of fish muscle) for fresh open sea red mullet (*M. surmuletus*) stored under air, 10%/20%/70% (O_2/CO_2/N_2), 10%/40%/50% (O_2/CO_2/N_2), 10%/60%/30% (O_2/CO_2/N_2) were 37.25 ± 0.95, 37.74 ± 0.92, 36.58 ± 0.62, 38.12 ± 0.94, respectively, after 14 days of storage (4°C ± 0.5°C). Moreover, TVBN (milligrams of N/100 g of fish muscle) values stored in these conditions were 144.1 ± 2.3, 107.2 ± 1.9, 104.3 ± 1.7, 90.4 ± 1.5, respectively, after 14 days of storage. Goulas and Kontominas (2007a,b), studied MAP chub mackerel (*S. japonicus*) fillets (MAP1: 70% CO_2/30% N_2 and MAP2: 50% CO_2/30% N_2/20% O_2) and VP, stored at 2°C ± 0.5°C. After 23 days of storage, the TVBN (mg N/100 g) values were 42.0 ± 0.3, 53.1 ± 0.5, 55.4 ± 0.6, and 67.8 ± 0.8 for MAP1, MAP2, VP, and air, respectively. After 23 days of storage, TMAN (mg N/100 g) values were 6.82 ± 0.46, 7.70 ± 0.26, 11.80 ± 0.39, and 13.88 ± 0.31 for MAP1, MAP2, VP, and air, respectively. Manju et al. (2007) investigated the quality changes of VP Black Pomfret (*P. niger*) and Pearl spot (*E. suratensis*) treated with sodium acetate and potassium sorbate, respectively, and subsequently stored in vacuum packs in ice during chill storage. The TVBN values of VP samples were lower than those of the air-packed samples in both the species. The TVBN values of treated vacuum samples (with sodium acetate and potassium sorbate) were found to be lower than those of the control packs in both species. Dalgaard et al. (1993) reported that with CO_2 values of 2%, 3%, 29%, 48%, and 97% (the remaining gas in all cases was N_2) a shelf life of cod fillets of 14, 13, 16, 20, and 15–16 days was obtained, respectively. With 97% CO_2, the shelf life was only 15–16 days and the sensory panel described the texture of the fillets with 97% CO_2 as soft and crumbling. Furthermore, under vacuum, stored fillets had a shelf life of 13–14 days. Moreover, by the end of the storage period, with CO_2 values of 2%, 3%, 29%, 48%, and 97% the pH values were 6.8, 6.8, 6.8, 6.7, and 6.5, respectively. Production of hypoxanthine was proportional to the production of TMA for both VP fillets and for fillets stored under MAP with 32% and 97% CO_2. The shelf life of cod fillets stored in these atmospheres was 14, 13, 16, 20, and 15–16 days, respectively. According to Erkan et al. (2006), TVBN (mg/100 g muscle) for fresh sardine (*S. pilchardus*) (stored at 4°C for 9 days) was 18.19 ± 0.19, 36.08 ± 1.3, and 35.5 ± 1.15 packed in air, in 5% O_2/35% CO_2/60% N_2 and 5% O_2/70% CO_2/25% N_2, respectively. TMAN (mg/100 g muscle) values for samples stored under aforementioned conditions were 7.74 ± 0.53, 9.69 ± 0.25, and 7.31 ± 0.29, respectively. The shelf life for samples stored in air and 5% O_2/35% CO_2/60% N_2 was 5 days, whereas samples stored in 5% O_2/70% CO_2/25% N_2 reached 7 days. Therefore, the increase in CO_2 percentage (from 35% to 70%) led to considerable shelf life extension by 2 days (40%). The effects of air, vacuum, and MAP (O_2/CO_2/N_2, 5%/70%/25%) on chemical properties of chub mackerel (*S. japonicus*) stored in cold storage (4°C) were studied by Erkan et al. (2007). At the beginning of storage, the TVBN value was 7.29 mg/100 g flesh for chub mackerel stored in air. The release of TVB increased up to 20.02–21.42 mg/100 g for chub mackerel stored in air and in VP at 9 days and 27.85 mg/100 g in MAP at 11 days of storage condition. Lipid oxidation, measured as TBA values, was higher in air and MAP than in VP samples. Significant differences were found between air–vacuum- and air–MAP-treated fish in terms of TBA value after 5 days of storage. This study showed that the shelf life of chub mackerel stored in cold storage is 9 days for air-packaged and VP fish and 12 days for MAP fish. Pantazi et al. (2008) evaluated the effect of air, vacuum, and MAP on the shelf life of chilled Mediterranean swordfish (*Xiphias gladius*). Fresh swordfish slices were stored in air, under vacuum, and MAP (40%/30%/30%, CO_2/N_2/O_2) under refrigeration (4°C) for a period of 16 days. Trimethylamine nitrogen (TMAN) values of swordfish samples stored in air, under VP, and MAP exceeded the limit value of 5 mg N/100 g fish muscle after days 7, 8–9, and 11 days of storage, respectively. In a similar trend, total volatile basic nitrogen (TVBN) for swordfish samples stored in air, under VP, and MAP exceeded the limit value of 25 mg N/100 g fish muscle after 7–8, 10, and 12 days of storage, respectively.

Figure 4.1 displays the impact of vacuum on TMAN values of chub mackerel (*S. japonicus*) stored at 4°C ± 1°C and Mediterranean swordfish (*X. gladius*) stored at 4°C.

The effect of MAP on the quality of fish versus storage time is given in Table 4.3.

4.2.3 Microflora of Fish Packed under MAP

According to Stier et al. (1981) King salmon (*Oncorhynchus tshawytscha*) inoculated with spores or vegetative cells of *C. botulinum* types B and E, stored in an MA of 60% CO_2/25% O_2, and 15% N_2 did not show any toxin development during storage at 4.4°C. The salmon spoiled within 12 days at 4.4°C and toxin was not detected after 57 days. Salmon inoculated with 100 *C. botulinum* type E spores/100 g and stored in 60% or 90% CO_2 at 10°C became toxic within 10 days but were not spoiled. There was no toxin detected after 7 days at 10°C. The 90% CO_2 atmosphere was more inhibitory to *C. botulinum* than the 60% CO_2 atmosphere. Toxin was detected earlier at both MAs as the level of the spore inoculum increased (Eklund, 1982). Molin et al. (1983) studied

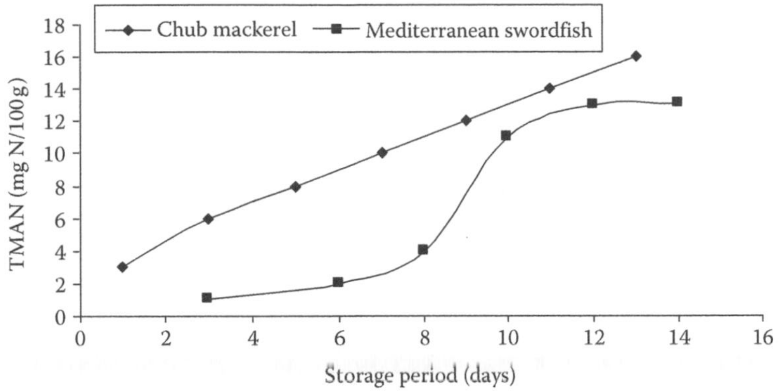

FIGURE 4.1 Effect of vacuum on TMAN of chub mackerel (*S. japonicus*) stored at 4°C ± 1°C (From Erkan, N. et al., *Eur. Food Resour. Technol.*, 222, 667, 2007) and Mediterranean swordfish (*X. gladius*) stored at 4°C. (From Pantazi, D. et al., *Food Microbiol.*, 25, 136, 2008.)

herring fillets stored in CO_2 at 2°C and found that *Lactobacillus* species dominated the microflora in CO_2 after 28 days of storage, whereas Enterobacteriaceae, Vibrionaceae, and *Lactobacillus* spp. were dominant in herring stored in N_2. According to Gray et al. (1983), perch (*Meronia americanus*), croaker (*Micropogon undulatus*), and bluefish (*Pomatomus saltatrix*) were stored in CO_2 MAP at 1.1°C and 10°C. Psychotrophic bacteria grew at a slower rate in CO_2-packed fish than on the air-stored fish on ice. The rate of growth and length of lag phase observed for the fish packed in CO_2 varied considerably with the fish species. Cann et al. (1984) found that salmon and trout inoculated with 100 *C. botulinum* type E spores/g at 10°C in 60% CO_2/ 40% N_2 or in vacuum packages spoiled before they became toxic. This is in contrast to the data of Eklund (1982). Statham and Bremner (1985) reported that psychotrophic bacteria grew rapidly in air-stored trevalla (*H. porosa*) reaching a population of 10^9 CFU/cm^2 within 8 days, whereas the psychotrophic population was less than 10^7/cm^2 at 8 days in the fillets packed in 100% CO_2. The dominant microflora in the CO_2-packed trevalla was LAB. Stenstrom (1985) investigated cod fillets packed under various gas mixtures ranging from air to 100% CO_2 in glass storage vessels with tightly sealed lids. The time taken for the microbial population on the fillets to reach 10^6 CFU/g enhanced with increasing CO_2 concentration, whereas the ratio of *Shewanella putrefaciens* dropped with increasing CO_2 concentration. Furthermore, an atmosphere of 50% CO_2 and 50% O_2 was recommended because the microbial growth exhibited a 14 day lag phase at 2°C, also because development of a dominant *Lactobacillus* microflora was favored. Moreover, the visual appearance of the cod fillets was preserved. Cod fillets were dipped in carbonic acid (pH=4.6) at 2°C (for 5–10min) before packaging in polystyrene trays and overwrapped with permeable or barrier films and vacuum packed prior to storage at 2°C. Growth of psychotrophic bacteria was most effectively controlled by barrier packaging with MA containing 25% CO_2. The obtained results indicated that the carbonic acid had similar performance in 98% CO_2 controlled atmosphere (Daniels et al., 1986). Tuna spiked with *Klebsiella oxytoca* T_2, *Morganella morganii*, or

TABLE 4.3
Effect of MAP on the Sensory and Physical Quality of Fish Stored

Species and Food Type	Gas Composition	Sensory Assessment	Shelf Life (Days)	Lipid Oxidation	Total Volatile Bases	Trimethylamine	Hypoxanthine	Storage Temperature	References
Farmed rainbow trout	Under air	Cooked flavor of VP trout was better than that of trout packed under CO_2		Both vacuum and CO_2 packages showed lesser lipid oxidation than conventionally stored trout				1°C–2°C	Chen et al. (1984)
Farmed rainbow trout	CO_2 and under vacuum			Both vacuum and CO_2 packages showed lesser lipid oxidation than conventionally stored trout					
Rockfish (*Sebastes* spp.) fillets	80% CO_2/20% O_2				In vitro protein digestibility of air-stored fillets decreased			20 days at 4°C	Morey et al. (1982)

Rockfish (*Sebastes* spp.) fillets	Air stored			In vitro protein digestibility of MAP fillets remained the same as that of the fish at zero time			20 days at 4°C	
Smoked salmon (*S. salar*)	Vacuum	26		TVB increased from initial values of 22.1–25.8 mg to 31.8 mg TVBN/100 g	Ranged from 2.9 to 3.5 mg TMAN/100 g in fresh cold-smoked salmon to 10.2 mg TMAN/100 g	Increasing levels of Hx from initial values of 0.29–1.17 μmol/g of salmon to final values to 1.03 μmol/g	0°C	Dondero et al. (2004)
Smoked salmon (*S. salar*)	Vacuum	21		TVB increased from initial values of 22.1–25.8 mg to 29.3 mg TVBN/100 g	Ranged from 2.9 to 3.5 mg TMAN/100 g in fresh cold-smoked salmon to 7.3 mg TMAN/100 g	Increasing levels of Hx from initial values of 0.29–1.17 μmol/g of salmon to final values to 0.67 μmol/g	2°C	

(*continued*)

TABLE 4.3 (continued)
Effect of MAP on the Sensory and Physical Quality of Fish Stored

Species and Food Type	Gas Composition	Sensory Assessment	Shelf Life (Days)	Lipid Oxidation	Total Volatile Bases	Trimethylamine	Hypoxanthine	Storage Temperature	References
Smoked salmon (*S. salar*)	Vacuum		20		TVB increased from initial values of 22.1–25.8 mg to 29.8 mg TVBN/100 g	Ranged from 2.9 to 3.5 mg TMAN/100 g in fresh cold-smoked salmon to 7.5 mg TMAN/100 g	Increasing levels of Hx from initial values of 0.29–1.17 µmol/g of salmon to final values to 2.24 µmol/g	4°C	
Smoked salmon (*S. salar*)	Vacuum		10		TVB increased from 22.1–25.8 mg to 30.0 mg TVBN/100 g	Ranged from 2.9 to 3.5 mg TMAN/100 g in fresh cold-smoked salmon to 7.4 mg TMAN/100 g	Levels of Hx increased from 0.29–1.17 µmol/g of salmon to 0.85 and µmol/g	6°C	
Smoked salmon (*S. salar*)	Vacuum		7		TVB increased from initial values of 22.1–25.8 mg to 29.9 mg TVBN/100 g	Ranged from 2.9 to 3.5 mg TMAN/100 g in fresh cold-smoked salmon to	Increasing levels of Hx from initial values of 0.29–1.17 µmol/g of salmon to	8°C	

					7.7 mg TMAN/100 g	final values of 0.66 µmol/g		
Cod fillets (*G. morhua*)	60% CO_2–10% O_2–30% N_2, 60% CO_2–20% O_2–20% N_2, 60% CO_2–30% O_2–10% N_2, 60% CO_2–40% O_2	4		The production of TVB was continuous in all mixtures	TMA was continuously produced during 7 days of storage in MAP. Higher levels of oxygen delay the production of TMA		6°C	Debevere and Boskou (1996)
Fresh pearl spot (*E. suratensis* Bloch)	Air	12–14		TVBN ranged from 12.8 to 22.4 mg N/100 g during 2 weeks of storage			0°C	Lalitha et al. (2005)
Fresh pearl spot (*E. suratensis* Bloch)	40% CO_2/30% O_2/30% N_2	12–14		TVBN ranged from 12.8 to 22.4 mg			0°C	
Fresh pearl spot (*E. suratensis* Bloch)	40% CO_2/60% O_2	19		N/100 g reached after 19 days			0°C	
Fresh pearl spot (*E. suratensis* Bloch)	50% CO_2/50% O_2	19		TVBN ranged from 12.8 to 22.4 mg			0°C	
Fresh pearl spot (*E. suratensis* Bloch)	60% CO_2/40% O_2	21		N/100 g after 29 days			0°C	

(*continued*)

TABLE 4.3 (continued)
Effect of MAP on the Sensory and Physical Quality of Fish Stored

Species and Food Type	Gas Composition	Sensory Assessment	Shelf Life (Days)	Lipid Oxidation	Total Volatile Bases	Trimethylamine	Hypoxanthine	Storage Temperature	References
Fresh pearl spot (*E. suratensis*)	70% CO_2/30% O_2		19					0°C	
Fish salads (rainbow trout *O. chus mykiss*)	Air		7		Initial TVBN value of fish salad was 11.5 mg/100 g of fish and reached 25.54 mg/100 g of fish			4°C	Metin et al. (2002)
Fish salads (rainbow trout *O. mykiss*)	O_2:CO_2:N_2 = 1:7:12 and CO_2:N_2 = 3:7		14		Initial TVBN value of samples was 11.5 mg/100 g of fish and reached 20.12 mg/100			4°C	
Fish salads (rainbow trout *O. mykiss*)			14		Initial TVBN value of samples was 11.5 mg/100 g of fish and reached 22.31 mg/100 g of fish			4°C	

Sardine (*Sardina pilchardus*)	Air	3		TVBN content increased up to 15 mg/100 g	The highest concentration of TMA was obtained from sardine stored in air, followed by sardine stored in vacuum and the lowest in MAP	4°C	Ozogul et al. (2004)
Sardine (*S. pilchardus*)	60% CO_2:40% N_2	12		TVBN content increased up to 19 mg/100 g for sardine		4°C	
Sardine (*S. pilchardus*)	Vacuum	9		TVBN content increased up to 17 mg/100 g		4°C	
Filleted rainbow trout (*O. mykiss*)	Overwrap	4	Highest		Hypoxanthine levels increased significantly in vacuum and overwrap packaging	1°C ± 1°C	Gimenez et al. (2002)
Filleted rainbow trout (*O. mykiss*)	Vacuum	16	Higher in MAP than in vacuum packaging			1°C ± 1°C	
Filleted rainbow trout (*O. mykiss*)	20% O_2/50% CO_2/30% N_2, 10% O_2/50% CO_2/40% N_2,10% O_2/50% CO_2/40% Ar, 20% O_2/50%	20		MAP was very effective in reducing TVB production with no differences between gas mixtures	The production of hypoxanthine in MAP was effectively slowed down, with no significant differences	1°C ± 1°C	

(continued)

TABLE 4.3 (continued)
Effect of MAP on the Sensory and Physical Quality of Fish Stored

Species and Food Type	Gas Composition	Sensory Assessment	Shelf Life (Days)	Lipid Oxidation	Total Volatile Bases	Trimethylamine	Hypoxanthine	Storage Temperature	References
	CO_2/30% Ar, 30% O_2/50% CO_2/20% N_2, 30% O_2/50% CO_2/20% Ar						being observed in different mixtures		
Mackerel	Air		3		—	—		2°C–4°C	Fagan et al. (2004)
Whiting	Air		3		—	—		2°C–4°C	
Salmon	Air		5		—	—		2°C–4°C	
Mackerel	60% N_2/40% CO_2		5		After 5 days of storage	After 5 days of storage 8.52 mg N/100 g		2°C–4°C	
Mackerel	100% CO_2		5		21.0 mg N/100 g			2°C–4°C	
Salmon	60% N_2/40% CO_2		7		After 5 days	No effect on TMA		2°C–4°C	
Salmon	100% CO_2		7		Storage 17.5 mg N/100 g			2°C–4°C	
Whiting	30% N_2/40% CO_2/30% O_2		5		After 5 days of storage	No effect on TMA		2°C–4°C	
Whiting	100% CO_2		5		24.2 mg N/100 g			2°C–4°C	

Chub mackerel (*S. japonicus*) fillets	Air	11		After 23 days of storage TVBN (mg N/100 g) value was 42.0 ± 0.3	After 23 days of storage TMAN (mg N/100 g) value was 6.82 ± 0.46	2°C ± 0.5°C	Goulas and Kontominas (2007a,b)
Chub mackerel (*S. japonicus*) fillets	70% CO_2/30% N_2	20–21		After 23 days of storage (TVBN) (mg N/100 g) value was 53.1 ± 0.5	After 23 days of storage (TMAN) (mg N/100 g) value was 7.70 ± 0.26	2°C ± 0.5°C	
Chub mackerel (*S. japonicus*) fillets	50% CO_2/30% N_2/20% O_2	15–16		After 23 days of storage TVBN (mg N/100 g) value was 55.4 ± 0.6	After 23 days of storage TMAN (mg N/100 g) value was 11.80 ± 0.39	2°C ± 0.5°C	
Chub mackerel (*S. japonicus*) fillets	Vacuum	15–16		After 23 days of storage TVBN (mg N/100 g) value was 67.8 ± 0.8	After 23 days of storage TMAN (mg N/100 g) value was 13.88 ± 0.31	2°C ± 0.5°C	
Cod fillets	2% CO_2/N_2	14	Sour		TMA was 31 mg N/100 g	0°C	Dalgaard et al. (1993)
Cod fillets	3% CO_2/N_2	13	Sour		TMA was 30 mg N/100 g	0°C	
Cod fillets	29% CO_2/N_2	16	Sour		TMA was 28 mg N/100 g	0°C	
Cod fillets	48% CO_2/N_2	20	Sour		TMA was 34 mg N/100 g	0°C	
Cod fillets	97% CO_2/N_2	15–16	Sour, soft, and crumbling		TMA 10 mg N/100 g	0°C	

(*continued*)

TABLE 4.3 (continued)
Effect of MAP on the Sensory and Physical Quality of Fish Stored

Species and Food Type	Gas Composition	Sensory Assessment	Shelf Life (Days)	Lipid Oxidation	Total Volatile Bases	Trimethylamine	Hypoxanthine	Storage Temperature	References
Red mullet (*M. surmuletus*)	Air		6		TVBN 37.25 ± 0.95 mg N/100 g after 14 days of storage	TMAN 144.1 ± 2.3 mg N/100 g after 14 days of storage		4°C ± 0.5°C	Pournis et al. (2005)
Red mullet (*M. surmuletus*)	10% O_2/20% CO_2/70% N_2		8		TVBN 37.74 ± 0.92 mg N/100 g after 14 days of storage	TMAN 107.2 ± 1.9 mg N/100 g after 14 days of storage		4°C ± 0.5°C	
Red mullet (*M. surmuletus*)	10% O_2/40% CO_2/50% N_2		10		TVBN 36.58 ± 0.62 mg N/100 g after 14 days of storage	TMAN 104.3 ± 1.7 mg N/100 g after 14 days storage		4°C ± 0.5°C	
Red mullet (*M. surmuletus*)	10% O_2/60% CO_2/30%/N_2		8		TVBN 38.12 ± 0.94 mg N/100 g after 14 days of storage	TMAN 90.4 ± 1.5 mg N/100 g, respectively, after 14 days of storage		4°C ± 0.5°C	
Sardine (*S. pilchardus*)	Air		5		TVBN (mg/100 g muscle) 18.19 ± 0.19	TMAN (mg/100 g muscle) was 7.74 ± 0.53		4°C ± 1°C	Erkan et al. (2006)

Sardine (*S. pilchardus*)	5% O_2/35% CO_2/60% N_2		5		TVBN (mg/100 g muscle) 36.08 ± 1.3	TMAN (mg/100 g muscle) was 9.69 ± 0.25		4°C ± 1°C	
Sardine (*S. pilchardus*)	5% O_2/70% CO_2/25% N_2		7		TVBN (mg/100 g muscle) 35.5 ± 1.15	TMAN (mg/100 g muscle) was 7.31 ± 0.29		4°C ± 1°C	
Chub mackerel (*S. japonicus*)	Air	For the first 4 days, sensory scores were excellent. For days 4–8, all samples were characterized as moderate	9	TBA values were higher in air and MAP than in vacuum-packaged samples. Significant differences were found between air–vacuum- and air–MAP-treated fish in TBA values after 5 days of storage	The release of TVB increased up to 20.02–21.42 mg/100 g for samples stored in air and vacuum packaged at 9 days and 27.85 mg/100 g in MAP at 11 days of storage			4°C	Erkan et al. (2007)
Chub mackerel (*S. japonicus*)	Vacuum		9					4°C	
Chub mackerel (*S. japonicus*)	O_2/O_2/N_2, 5%/70%/25%		12					4°C	
Mediterranean swordfish (*X. gladius*)	Air	Odor and taste scores for all packaging regimes showed a similar decreasing	7 (based on sensory scores)	TBA values of samples stored in air, under vacuum and MAP varied little and were between 0.6 and	TVBN for samples stored under air, VP, and MAP exceeded the limit value of 25 mg N/100 g	TMAN values of samples stored under air, VP, and MAP exceeded the limit value of 5 mg N/100 g		4°C	Pantazi et al. (2008)

(continued)

TABLE 4.3 (continued)
Effect of MAP on the Sensory and Physical Quality of Fish Stored

Species and Food Type	Gas Composition	Sensory Assessment	Shelf Life (Days)	Lipid Oxidation	Total Volatile Bases	Trimethylamine	Hypoxanthine	Storage Temperature	References
		acceptability		2.2 mg MDA/kg muscles	fish muscle after 7–8, 10, and 12 days of storage, respectively	fish muscle after days 7, 8–9, and 11 days of storage, respectively			
Mediterranean swordfish (*X. gladius*)	Under vacuum		9 (based on sensory scores)					4°C	
Mediterranean swordfish (*X. gladius*)	40%/30%/30%, CO_2/N_2/O_2		11–12 (based on sensory scores)					4°C	

Hafnia alvei T_8 followed by VP contained more histamine than spiked tuna that was not VP. Low storage temperature was the most important factor in controlling histamine production in the spiked tuna samples (Wang and Ogrysdiak, 1986). Garcia et al. (1987) and Garcia and Genigeorgis (1987) studied salmon (*O. tshawytscha*) fillets and salmon flesh homogenates inoculated with spores of *C. botulinum* types B, E, and F and showed that homogenates became toxic after 60 days at 4°C when vacuum packed but not when packed under 100% CO_2 or 70% CO_2, and 30% air. The fillets did not become toxic at 4°C, but at 8°C they became toxic within 6 days in vacuum packages and within 9 days under 100% CO_2, while odor scores indicated that the salmon was of acceptable quality. These studies showed that *C. botulinum* produces toxin more readily in the vacuum packaged salmon than in the salmon packaged under 100% CO_2 or 70% CO_2. Raised catfish (*Silurus glanis*) were studied during the storage period by Manthey et al. (1988). The storage time was considered to be 20 days. On the 27th day of storage, fish fillets showed a total anaerobic bacteria count of 10^8/cm^2 of fish skin and only 10^5/g of muscle. The TBARS amounts varied from 0.73 to 1.98 mg of malonaldehyde/kg. In addition, trimethylamine (TMA) amounts were low. According to Fugii et al. (1989)

> sardines (*Sardinops melanostictus*) were stored at 5°C in barrier bags which contained the following atmospheres after sealing: air, 20% N_2/80% CO_2 and 80% N_2/20% CO_2. Growth of anaerobic and aerobic bacteria was slowest in sardines packed under 20% N_2/80% CO_2. Initially the microflora was predominantly Vibrionaceae, *Moraxella* and *Acinetobacter*. After 10 days storage in air *Moraxella*, *Acinetobacter*, *Lactobacillus* and *Streptococcus* formed the majority of the microflora on the sardines, while microflora on sardines stored for 10 days under 80% N_2/20% CO_2 was predominantly *Lactobacillus* and *Streptococcus* sp

Microflora on sardines stored for 10 days under 20% N_2/80% CO_2 was predominantly unidentified cocci as well as *Lactobacillus* and *Streptococcus* sp. (Fugii et al., 1990). Lilly and Kauter (1990) studied 1074 samples of commercial VP freshwater and saltwater fish, produced in the United States from 27 types of fish, to mild temperature abuse at 12°C for 12 days. None of the samples was found positive for *C. botulinum* toxin. The authors claimed that either the fish in the packages sampled did not contain any *C. botulinum* spores or the spores were unable to grow out and produce toxin within the 12 days of temperature abuse. Leung et al. (1992) studied the development of *L. monocytogenes* and *Aeromonas hydrophila* in VP catfish (*I. punctatus*) fillets, stored at 4°C for 16 days. Although no increase was reported in the *L. monocytogenes* population, a rapid increase was found in the *A. hydrophila* population in products packed under air. According to Williams et al. (1995)

> fresh catfish (*Ictalurus nebulosus*) fillets treated with 0.1 and 2.0% sodium lactate solution adjusted to 5.5 pH, then vacuum packed and stored at 1.11-1°C. The shelf-life of the fillets treated with 2% sodium lactate increased from 4 to 7 days

Davies (1995) examined rainbow trout and cod and reported that in no case the growth or survival of any of the pathogens examined was higher than that in the aerobically stored control and frequently growth was reduced in MAP. The inhibition of growth by the MAs dropped markedly with increasing temperature. For both cod and trout, the higher carbon dioxide containing atmosphere (80% CO_2/20% N_2 for trout and 60% CO_2/30% N_2/10% O_2 for cod) was generally more inhibitory than the lower CO_2 containing atmosphere. Cod fillets (*G. morhua*) were packed under MA, with four different gas compositions (60% CO_2/10% O_2/30% N_2, 60% CO_2/20% O_2/20% N_2, 60% CO_2/30% O_2/10% N_2, 60% CO_2/40% O_2) and stored at 6°C. The number of the total aerobic bacteria for all gas atmospheres applied displayed a negligible increase. Storage at low temperature (6°C) in conjunction with 60% CO_2 effectively inhibited the bacterial growth. No difference can be observed between the total aerobic counts for the different gas mixtures. The number of anaerobic bacteria did not increase over the first 3 days, which can be attributed to the inhibitory effect of CO_2, at low temperature (6°C). From the fourth day onward, an exponential increase in the anaerobic count was recorded and reached the same levels as for the total. LAB was augmented by 1–2 logarithmic units over 7 days in MAP. No difference in the LAB plate counts among the gas mixtures was reported. Hydrogen sulfide-producing bacteria increased 2–3 log units in 7 days starting from the 3rd day (Debevere and Boskou, 1996). Reddy et al. (1997) investigated the shelf life and the potential for toxin production by *C. botulinum* type E in retail-type packages of fresh aquacultured salmon fillets packaged in high-barrier film bags under various atmospheres (100% air, an MA containing 75% CO_2:25% N_2, and vacuum) and stored under both refrigeration (4°C) and temperature-abuse conditions (8°C and 16°C). Toxin development preceded sensory spoilage at 16°C storage for fillets packaged in MA. Toxin development coincided with sensory spoilage or was slightly delayed for the fillets packaged in all the atmospheres at 8°C storage. At 4°C, none of the fillets packaged in either of the atmospheres developed toxin, even 20 days after spoilage as determined by sensory characteristics. The shelf life of fillets packaged in all atmospheres decreased with increase in storage temperature from 4°C to 16°C. Cai et al. (1997) studied catfish (*I. punctatus*) packaged in O_2-permeable film, in 80% CO_2 and 20% N_2 MA and stored at 4°C with a mixture of four E-type *C. botulinum* strains. The toxin was detected after 9 and 18 days in the O_2-permeable film and MA packaging, respectively. The deterioration preceded the toxin production in all packing methods. Boskou and Debevere (1997) isolated two strains of *Shewanella* spp. from cod fillets packed in MA (60% CO_2, 30% O_2, 10% N_2). One of the strains was identified as *S. putrefaciens*. The other strain could not be fully identified but was determined as a *Shewanella* spp. different from *S. putrefaciens*. The growth and trimethylamine oxide (TMAO)-reducing activity of the *Shewanella*-like strain can be inhibited when higher proportions of CO_2 together with as high as possible proportions of O_2 are introduced into the packaging atmosphere. It is suggested that introduction of a combination of 60%–70% CO_2 and 30%–40% O_2 into the packaging atmosphere will prevent TMA production by *Shewanella* spp. The quality attributes of filleted rainbow trout and Baltic herring in overwrap packages (polystyrene or wood fiber), VP, and gas packages (35% CO_2 + 32.5% Ar + 32.5% N_2, 35% CO_2 + 65% Argon, or 40% CO_2 + 60% N_2) stored at 2°C were compared. Mesophilic bacteria grew most quickly in overwrap-packed fillets and most slowly in vacuum- and gas-packed fillets. Coliform bacteria grew faster in overwrap- and vacuum-packed fish than in gas-packed fish. The sensory quality of both fillets deteriorated faster in overwrap and vacuum packages than in gas packages. The sensory quality of trout and herring in the three types of gas packages was quite similar (Randell et al., 1997). Lyhs et al. (1998) investigated the spoilage flora of vacuum-packaged, salted, cold-smoked rainbow trout fillets, with or without the addition of nitrate or nitrite, stored at 4°C and 8°C. Of 620 isolates, LAB were the major fraction (76%), predominating in all samples of spoiled product. Eighty-five isolates were found to belong to the family *Enterobacteriaceae*, with 45 of those being *Serratia plymuthica*. Eleven isolates from the nitrate treated samples stored at 8°C were identified as *Pseudomonas aeruginosa*. The types of lactic acid and other bacteria in the spoilage flora were generally reduced by the addition of nitrate or nitrite to fillets. Lyver et al. (1998) carried out studies to evaluate the safety of value-added raw and cooked seafood nuggets inoculated with 10^3 CFU/g of *L. monocytogenes*. Nuggets were packaged in both air and 100% CO_2 and stored at 4°C or 12°C. *Bacillus* spp. and LAB counts increased to 10^2 and 10^7 CFU/g, respectively, in raw seafood nuggets, whereas only *Bacillus* spp. reached 10^4 CFU/g in cooked nuggets by 28 days. With the exception of nuggets packaged in 100% CO_2, counts of *L. monocytogenes* increased to approximately 10^7 CFU/g in nuggets stored at both 4°C and 12°C after 28 days. Lyhs et al. (1999) characterized and identified a total of 405 LAB isolated from spoiled, vacuum-packaged, salted, sodium nitrite– or potassium nitrate–treated, cold-smoked rainbow trout stored at 4°C or 8°C. The *Leuconostoc* clusters and the *Lactobacillus sakei/Lactobacillus curvatus* clusters formed the two main groups. Only one isolate was identified as *Lactobacillus plantarum*, and no *Carnobacterium* strains were discovered. The relative proportion of *Leuconostoc mesenteroides* subsp. *mesenteroides* was higher in all samples stored at 4°C. Most of the *Leuconostoc citreum* were found in the samples stored at 8°C, particularly in the nitrite-treated samples. Studies were conducted to determine the effect of various levels of headspace oxygen (0%–100%, balance CO_2) or film oxygen transmission rate (OTR) on the time to toxicity in MAP fresh trout fillets challenged with *C. botulinum* type E (10^2 spores/g) and stored under moderate temperature abuse conditions (12°C). In all cases, trouts were toxic within 5 days, irrespective of the initial levels of oxygen in the package headspace. However, spoilage preceded toxigenesis. Packaging of trout fillets in low gas barrier films, with OTRs ranging from 4,000 to 10,000 cc/m²/day at 24°C and 0% relative humidity, had no effect on time required for toxicity in all MAP trout fillets. All fillets were toxic within 4–5 days and spoilage again preceded toxigenesis. It was thereby shown that the addition of headspace O_2 hardly affected the time required for toxigenesis or spoilage (Dufresne et al., 2000). Cod (*G. morhua*) fillets kept in air at 0°C for 1–8 days before MA packing and frozen stored at −20°C and −30°C for 6 weeks were evaluated by Boknaes et al. (2000). The amounts of *Photobacterium phosphoreum* were 2.3 and 5.8 UFC/g, after 1 and 8 days at 0°C, respectively. Storage at −20°C and −30°C reduced the amount of *P. phosphoreum* down to undetectable limits. Only fillets kept for 8 days at 0°C, and at −30°C afterward, displayed significant increase in *P. phosphoreum* growth, over defrosting at 2°C. Growth and toxin production by *C. botulinum* types A–E in shrimp tissue homogenates stored under vacuum at temperatures of 30°C, 15°C, 10°C, and 4°C were analyzed. All the inoculated mullet tissue homogenates held at 30°C under vacuum turned toxic within 2 days. At 15°C, toxin was detected in packs inoculated with *C. botulinum* types A and B after 5 days, in packs inoculated with types C and D after 7 days, whereas packs inoculated with type E developed toxin after 3 days. At 10°C, only type E grew and produced toxin after 8 days in inoculated (1000 type E spores) VP mullet. Toxin production by type E was delayed for a period of 25 days at 4°C in VP mullet and shrimp tissue homogenates inoculated with 1000 spores (Lalitha and Gopakumar, 2001). Lyhs et al. (2001) investigated VP "gravad" rainbow trout slices during storage at 3°C and 8°C and found that at the time of spoilage, after 27 and 20 days of storage at 3°C and 8°C, respectively, both mesophilic viable counts (MVC) and psychrotrophic viable counts (PVC) reached 10^6–10^7 CFU/g at 3°C and 10^7–10^8 CFU/g at 8°C. H_2S-producing bacteria constituted a high proportion of the PVCs, and LAB counts were lower than the other determined bacterial counts. The shelf lives of the rainbow trout slices based on microbiological and sensory analyses were 20 days and 18 days at 3°C and 8°C, respectively. Fifty-four packages (each one belonging to a different lot) of VP cold-smoked salmon (30) and trout (24) obtained at retail level after 3 weeks of storage at 2°C ± 1°C. *Psychrotrophic clostridia* ranged between 1.71 and 2.21 log CFU/g. *Salmonella*, *Escherichia coli*, and *L. monocytogenes* were not detected in any sample. *Listeriae* other than *L. monocytogenes* were isolated from three packages. Levels of *Staphylococcus aureus* lower than 4 log CFU/g were also found in three packages. Among 377 bacteria randomly isolated from aerobic 25°C plate counts, LAB predominated, with *Carnobacterium* (*Carnobacterium piscicola*) and *Lactobacillus* (eight species) being the genera most frequently found. The second and third major groups were Enterobacteriaceae and Micrococcaceae, respectively (Gonzalez-Rodrıguez et al., 2002). Tryfinopoulou et al. (2002) identified a total of 10^6 *Pseudomonas* strains isolated from *S. aurata* stored under different temperatures (at 0°C, 10°C, and 20°C) and packaging conditions (air and a MA of 40% CO_2/30% N_2/30% O_2). *Pseudomonas lundensis* was the predominant species, followed by *Pseudomonas fluorescens*, while *Pseudomonas fragi* and *Pseudomonas putida* were detected less frequently. *Pseudomonad* strains dominated under air conditions, while LAB and *Brochothrix thermosphacta* dominated under MAP. Different storage conditions appear to determine the selection of pseudomonads in gilt-head sea bream. Metin et al. (2002) found that microbiological counts of MAP fish salads (rainbow trout *O. mykiss*) generally increased more slowly than those of the air-packaged control samples. Initially, the average value of aerobes was 2.7 log CFU/g. This value increased after 7 days and reached 6.9 log CFU/g for control, 5.8 log CFU/g for MAP (O_2:CO_2:N_2 = 1:7:12, and 3.8 log CFU/g for MAP (CO_2:N_2 = 3:7). TVC of all sea bass (*L. calcarifer*) samples increased with increasing time of storage at 4°C. TVC of sea bass slices stored in air increased rapidly from an initial value of 10^4–10^8 CFU/g within 15 days and was generally higher than that of sea bass slices kept under CO_2-enriched atmosphere. The shelf life of sea bass slices packaged in 80%–100% CO_2 atmosphere could be extended to more than 20 days at 4°C, compared with 9 days to those packed in air. Moreover, higher counts of LAB were generally observed in samples kept in CO_2-enriched atmosphere compared with those stored in air. However, LAB counts in samples with 100% CO_2 were lower than those in any other samples (Masniyom et al., 2002).

Figure 4.2 displays the impact of MAP (80% CO_2/20% N_2) on the LAB population of mussels (*Mytilus galloprovincialis*) and sea bass (*L. calcarifer*); both fishery products were stored at 4°C.

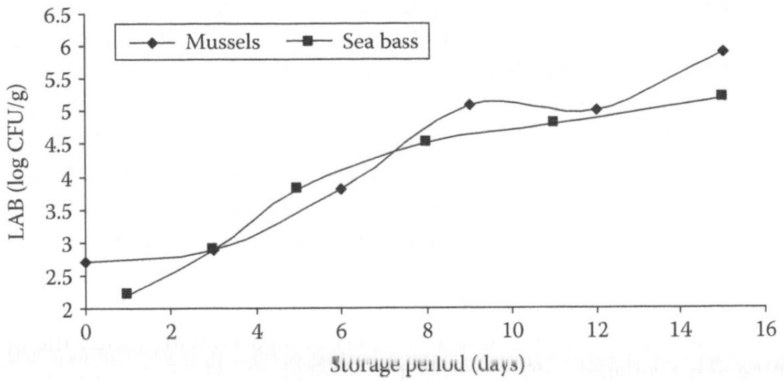

FIGURE 4.2 Effect of MAP (80% CO_2/20% N_2) on the LAB population of mussels (*M. galloprovincialis*) (From Goulas, A.E. et al., *J. Appl. Microbiol.*, 98, 752, 2005) and sea bass (*L. calcarifer*). (From Masniyom, P. et al., *J. Food Sci. Agric.*, 82, 873, 2002); both fishery products were stored at 4°C.

Franzetti et al. (2003) investigated the composition of lactic acid population in salmon, tuna, and swordfish packaged in two different MA, MAP1 (80% O_2/20% N_2) and MAP2 (40% CO_2/60% N_2), at 4°C for 6 days. The isolates were heterofermentative rods belonging to *Carnobacterium*, *Lactobacillus*, and cocci of the *Leuconostoc* genus. The microorganisms found varied with the kind of seafood and the gas composition of the MA: in MAP1, richer in oxygen than MAP2, *Carnobacterium* spp. represents the prevalent microbial group, especially in tuna and swordfish, whereas MAP2 seems to favor *Lactobacillus* spp. *Cocci*, belonging to *Leuconostoc* spp., which were dominant in salmon independently of gas composition. Dondero et al. (2004) evaluated changes in the quality of VP cold-smoked salmon (*S. salar*) during storage at different temperatures (0°C, 2°C, 4°C, 6°C, and 8°C), and total aerobic and anaerobic counts and *Lactobacillus* spp., showed significant correlation with the storage time, temperature, and sensory quality. The total aerobic count at the beginning of storage ranged from 150 to 174×10^3 CFU/g. At the end of shelf life, after 26, 21, 20, 10, and 7 days at 0°C, 2°C, 4°C, 6°C, and 8°C, respectively, levels of 185×10^4, 303×10^5, 450×10^4, $<300 \times 10^6$, and 760×10^3 CFU/g were reached. Pathogenic microorganisms (*C. botulinum*, *Salmonella*, Coliform, *S. aureus*, and *L. monocytogenes*) were not detected during the time of storage. According to Fagan et al. (2004) the TVC of whiting samples stored in air, 30% N_2/40% CO_2/30% O_2, and 100% CO_2 was 4.81 log CFU/g, 4.48 log CFU/g, and 4.34 log CFU/g, respectively. For mackerel samples stored in air, 60% N_2/40% CO_2, and 100% CO_2, the TVC was 4.88 log CFU/g, 4.18 log CFU/g, and 3.99 log CFU/g, respectively. Finally, for salmon samples stored in air, 60% N_2/40% CO_2, and 100% CO_2, the TVC was 6.23 log CFU/g, 5.04 log CFU/g, and 4.53 log CFU/g, respectively. Ozogul and Ozogul (2004) studied the effects of slaughtering methods (percussive stunning and death in ice slurry) on the quality of rainbow trout stored in ice and MAP (40% CO_2/30% N_2 and 30% O_2). No significant differences were reported in TVC of fish stored in ice and MAP, regardless of the different slaughter methods applied; fish packed in MAP displayed a reduction in bacterial counts compared with fish held in ice throughout the study. The shelf life of rainbow trout slaughtered by percussive stunning was approximately 15 or 16 days in ice and approximately 16 or 17 days in MAP, whereas the shelf life of trout slaughtered by the ice slurry method was 14 or 15 days in ice and 15 or 16 days in MAP. Chouliara et al. (2004) investigated the effect of gamma irradiation (1 and 3 kGy) on the shelf life of salted, VP sea bream (*S. aurata*) fillets stored under refrigeration. Nonirradiated, salted, VP fish served as control samples. Irradiation affected strongly populations of bacteria, namely, *Pseudomonas* spp., H_2S-producing bacteria, *B. thermosphacta*, Enterobacteriaceae, and LAB. The effect was more pronounced at the highest applied dose (3 kGy). On the basis of sensorial evaluation, a shelf life of 27–28 days was obtained for VP, salted sea bream irradiated at 1 or 3 kGy compared with a shelf life of 14–15 days for the nonirradiated, salted sample. Corbo et al. (2005) studied fresh cod fillets packaged in air, under vacuum, in an MA with low oxygen concentration (5%), and in an MA with high oxygen concentration (80%) monitored during storage at different isothermal conditions from 4°C to 12°C. At 4°C, cod fillets packaged under vacuum and in an MA with low oxygen concentration (65% N_2/30% CO_2/5% O_2) had a higher shelf life of 6 and 5.4 days, calculated from the growth of the total bacterial count, than the samples packaged in air and MAP 20% CO_2, 80% O_2 (1.96 and 2.62, respectively). During the aerobic storage of fresh pearl spot (*E. suratensis* Bloch), the aerobic counts reached the highest levels (7.3 log10 CFU/g) within 15 days, The greatest inhibitory effect on aerobic bacterial growth was observed in a mixture 60% CO_2/ 40% O_2 at 0°C. Packaging under gas mixtures 40% CO_2/60% O_2, 60% CO_2/ 40% O_2, and 70% CO_2/30% O_2 delayed and suppressed the growth of H_2S-producing bacteria during the first 3 weeks. The growth of LAB in pearl spot packed under MAs was slow, and their count remained almost constant in gas mixture 60% CO_2/40% O_2 after 3 weeks. Yeasts and molds were less numerous than bacteria in samples stored under air and MAs. Initial counts of approximately 2.8 log CFU/g increased slowly but remained low throughout the storage period (Lalitha et al., 2005). Storage trials were carried out with fresh and thawed garfish fillets in air or in MAP (40% CO_2 and 60% N_2). *P. phosphoreum* was responsible for histamine formation (>1000 ppm) in chilled fresh garfish. The use of MAP did not reduce the histamine formation. Strongly histamine-producing *P. phosphoreum* isolates formed 2080–4490 ppm at 5°C, whereas below 60 ppm were formed by other *P. phosphoreum* isolates. Frozen storage inactivated *P. phosphoreum* and consequently markedly reduced histamine formation in thawed garfish at 5°C. In conclusion it can be said *P. phosphoreum* can produce above 1000 ppm of histamine in chilled fresh garfish stored both in air and MAP. Freezing inactivates *P. phosphoreum*, extends shelf life, and significantly reduces histamine formation in thawed MAP garfish during chilled storage (Dalgaard et al., 2006). According to Emborg et al. (2005) at ~2°C both psychrotolerant *M. morganii*-like bacteria and *P. phosphoreum* were able to form toxic histamine concentrations in chilled VP tuna steaks. To improve the safety of fresh tuna it is suggested (i) that VP of lean tuna loins should be replaced with MAP with ~40% CO_2/~60% O_2 and (ii) that the shelf life of fresh MAP tuna should be equivalent to less than 14 days at 2°C. A limited shelf life would also reduce potential problems with *L. monocytogens* and *C. botulinum* type E reaching critical concentrations in chilled fresh MAP tuna. No formation of histamine was reported in naturally contaminated fresh MAP tuna with 40% CO_2/60% O_2 over 28 days of storage at 1°C. For red mullet (*M. surmuletus*), Pournis et al. (2005) found that when packaged in four different atmospheres: air, 10%/20%/70% (O_2/CO_2/N_2), 10%/40%/50% (O_2/CO_2/N_2), and 10%/60%/30% (O_2/CO_2/N_2) the TVC values were 8.43 ± 0.41, 8.01 ± 0.66, 7.66 ± 0.54, 7.55 ± 0.61 log CFU/g, after 14 days of storage(4°C ± 0.5°C). H_2S-producing bacteria and pseudomonads were part of the mullet microflora, and their growth was partly inhibited under MAP conditions. Between these two bacterial groups, H_2S-producing bacteria (including *S. putrefaciens*) were dominant toward the end of the storage period, regardless of the packaging conditions. According to Rosnes et al. (2006), there was no significant difference in APC of wolf fish (*A. minor*) packaged in air at −1°C and 4°C after 1 and 4 days, but the numbers were lower at −1°C after 6 and 8 days. Furthermore, for MA-packaged products (60% CO_2/40% N_2), no significant differences were found between −1°C and 4°C during the first 4 days. However, the numbers were lower at days 8 and 11 for products at −1°C compared to storage at 4°C. MA-packaged wolf fish had lower APC at 6, 8, 11, 13, and 15 days compared with air at −1°C and also lower counts at 6, 8, 11, and 13 days at 4°C. A shelf life of 15 days was reached at −1°C for MA-packaged fish compared with 8–10 days in air. At 4°C the shelf life was 13 days in MA and 6–8 days in air. European sea bass (*D. labrax*) fillets were packed under 40% CO_2: 60% N_2 (MAP) and air (AIR) or prepared from the whole ungutted fish stored in ice (round). Coliform count was <2.4 log CFU/g in MAP and round fillets and 2.8 log CFU/g in AIR fillets. The TVC growth was generally at lower levels in round and MAP fillets than in AIR fillets. Specifically, in round and MAP fillets, a slight increase in TVC started by the 5th day up to values just under 6 log CFU/g on the 8th day after packaging, while in AIR fillets TVC reached 8.6 log CFU/g. Among the different bacterial groups, streptococci and H_2S-producing bacteria showed the highest increase (Poli et al., 2006). Erkan et al. (2006) reported that for fresh sardine (*S. pilchardus*) stored at 4°C ± 1°C, under air, 5% O_2/35% CO_2/60% N_2, and 5% O_2/70% CO_2/25% N_2, there was an initial lag phase with no significant growth of all microorganisms during the first 3 days of storage. Logarithmic increases occurred between days 3 and 5. Total AMB counts (log CFU/g) were 3.7, 3.5, and 3.7, respectively. H_2S-producing bacteria counts (log CFU/g) were 3.7, 3.5, and 3, respectively, after 9 days of storage. The shelf life for samples stored in air and 5% O_2/35% CO_2/60% N_2 reached 5 days, while for samples stored in 5% O_2/70% CO_2/25% N_2 reached 7 days. Atlantic halibut (*Hippoglossus hippoglossus*) portions were packaged with gas mixtures of CO_2:N_2 and CO_2:O_2 (50%:50%) and with air as a reference (control). The packages were stored at 4°C. The shelf life was 10–13 days when stored in air and between 13 and 20 days for MA packages, with oxygen-enriched packages suggested as the better gas mixture, based on microbial growth and sensory scores. *S. putrefaciens* was determined sporadically and at low concentrations. The main bacterial microbiota in both MA-packaged and air-stored, farmed halibut were found to be *P. phosphoreum*, *Pseudomonas* spp., and *B. thermosphacta*. *S. putrefaciens* was not detected by molecular methods during this experiment, confirming the plate count results (Hovda et al., 2007). Erkan et al. (2007) studied the impact of air, vacuum, and MAP (O_2/CO_2/N_2, 5%/70%/25%) on microbiological properties of (*S. japonicus*) stored in cold storage (4°C).

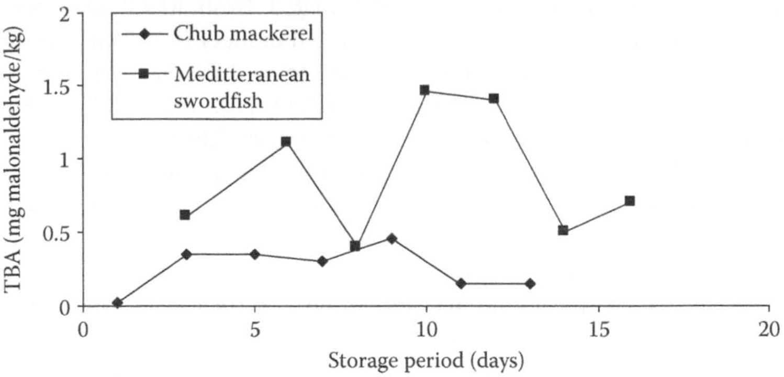

FIGURE 4.3 Effect of vacuum on TBA values of chub mackerel (*S. japonicus*) stored at 4°C ± 1°C (From Erkan, N. et al., *Eur. Food Resour. Technol.*, 222, 667, 2007) and Mediterranean swordfish (*X. gladius*) stored at 4°C. (From Pantazi, D. et al., *Food Microbiol.*, 25, 136, 2008.)

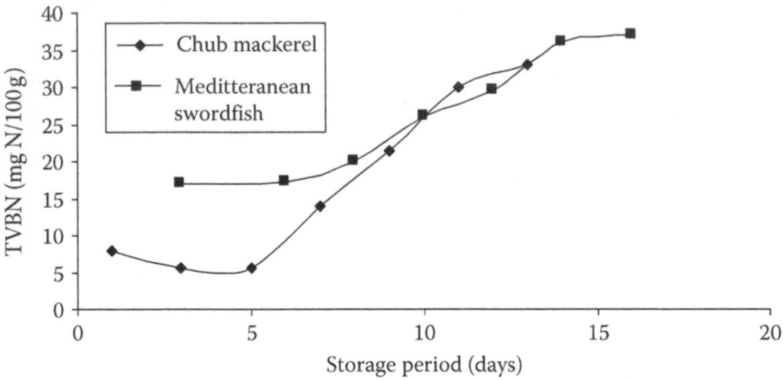

FIGURE 4.4 Effect of vacuum on TVBN of chub mackerel (*S. japonicus*) stored at 4°C ± 1°C (From Erkan, N. et al., *Eur. Food Resour. Technol.*, 222, 667, 2007) and Mediterranean swordfish (*X. gladius*) stored at 4°C. (From Pantazi, D. et al., *Food Microbiol.*, 25, 136, 2008.)

The initial mesophilic aerobic bacteria counts of all group samples (<3 log CFU/g) indicate acceptable fish quality. Mesophilic counts for air–VP and MAP samples exceeded 6 log CFU/g after 7 and 11 days of cold storage, respectively.

Figures 4.3 and 4.4 display the effect of vacuum on TBA values of chub mackerel (*S. japonicus*) stored at 4°C ± 1°C and Mediterranean swordfish (*X. gladius*) stored at 4°C.

The growth of microflora of fish packed under MAP is summarized in Table 4.4.

4.3 CRUSTACEANS AND SHELLFISH

4.3.1 Crustaceans and Shellfish Stored under MAP

Cooked crab claws (*Cancer pagurus*) packed in an atmosphere of 40% CO_2/30% N_2/30% O_2 had a storage life of 10 days at 0°C and 6 days at 5°C. The air-stored

TABLE 4.4
Effect of MAP on the Microflora and Sensory Assessment of Fish

Species	Gas Composition	Sensory Assessment	Shelf Life (Days)	Microflora	Storage Temperature	References
Cod fillets	50% CO_2 and 50% O_2			The microbial growth exhibited a 14 day lag phase, development of a dominant *Lactobacillus* microflora was favored, the visual appearance of the cod fillets was preserved	2°C	Stenstrom (1985)
Fillets of trevalla (*H. poporosa*)	100% CO_2		8–16 at 4°C longer than trevalla fillets stored under aerobic conditions	Psychotrophic population was less than 107/cm^2 at 8 days in the CO_2- packed fillets. LAB formed the dominant microflora	4°C	Statham and Bremner (1985)
Fillets of trevalla (*H. poporosa*)	Air stored			Psychotrophic bacteria grew rapidly reaching a population of 109 CFU/cm^2 within 8 days	4°C	
Salmon fillets inoculated with spores of *C. botulinum* types B, E, and F	Vacuum	Odor scores indicated that the salmon was of acceptable quality		Did not become toxic at 4°C	4°C	Garcia et al. (1987)
Salmon fillets inoculated with pores of *C. botulinum* types B, E, and F	Vacuum			Became toxic within 6 days	8°C	
Salmon fillets inoculated with pores of *C. botulinum* types B, E, and F	100% CO_2			Became toxic within 9 days	8°C	

(*continued*)

TABLE 4.4 (continued)
Effect of MAP on the Microflora and Sensory Assessment of Fish

Species	Gas Composition	Sensory Assessment	Shelf Life (Days)	Microflora	Storage Temperature	References
Salmon flesh omogenates inoculated with spores of *C. botulinum* types B, E, and F	Vacuum packed			Became toxic	60 days at 4°C	Garcia and Genigeorgis (1987)
Salmon flesh homogenates inoculated with spores of *C. botulinum* types B, E, and F	100% CO_2 or 70% CO_2 and 30% air			Did not become toxic	60 days at 4°C	
King salmon (*O. tshawytscha*) inoculated with spores or vegetative cells of *C. botulinum* types B and E	60% CO_2/25% O_2, and 15% N_2.			The salmon spoiled within 12 days at 4.4°C and toxin was not detected after 57 days	4.4°C	Stier et al. (1981)
Salmon inoculated with 100 *C. botulinum* types E spores/100 g	60% or 90% CO_2			Became toxic within 10 days but were not spoiled. There was no toxin detected after 7 days at 10°C. The 90% CO_2 atmosphere was more inhibitory to *C. botulinum* than the 60% CO_2 atmosphere	10°C	Eklund (1982)
Salmon and trout inoculated with 100 *C. botulinum* types E spores/g	% CO_2/40% N_2 or in vacuum packages			Spoiled before they became toxic	10°C	Cann et al. (1984)

1074 samples of commercial freshwater and saltwater fish	Vacuum-packaged			None of the samples were positive for *C. botulinum* toxin. The authors stated that either the fish in the packages sampled did not contain *C. botulinum* spores or the spores were unable to grow out and produce toxin	Mild temperature abuse at 12°C for 12 days	Lilly and Kauter (1990)
Aquacultured salmon fillets	100% air, an MA containing 75% CO_2:25% N_2, and vacuum		The shelf life of fillets packaged in all atmospheres decreased with increase in storage temperature from 4°C to 16°C	None of the fillets packaged in either of the atmospheres developed toxin, even 20 days after spoilage as determined by sensory characteristics	4°C	Reddy et al. (1997)
Aquacultured salmon fillets	100% air, an MA containing 75% CO_2:25% N_2, and vacuum			Toxin development coincided with sensory spoilage or was slightly delayed for the fillets packaged in all the atmospheres at 8°C storage	8°C	
Fresh aquacultured salmon fillets				Toxin development preceded sensory spoilage storage for fillets packaged in MA	16°C	
Fresh trout fillets challenged with *C. botulinum* type E (102 spore/g)	Various levels of headspace oxygen (0%–100%, balance CO_2)			In all cases, trout were toxic within 5 days, irrespective of the initial levels of oxygen in the package headspace. However, spoilage preceded toxigenesis		Dufresne et al. (2000)
Fresh trout fillets challenged with *C. botulinum* type E (10^2 spore/g)	Film OTR ranging from 4,000 to 10,000 cc/m^2/day at 24°C and 0% relative humidity			Had no effect on time to toxicity in all MAP trout fillets. Fillets were toxic within 4–5 days and spoilage again preceded toxigenesis		

(*continued*)

TABLE 4.4 (continued)
Effect of MAP on the Microflora and Sensory Assessment of Fish

Species	Gas Composition	Sensory Assessment	Shelf Life (Days)	Microflora	Storage Temperature	References
Salmon, tuna, and swordfish	MAP1 (80 O_2/20 N_2) and MAP2 (40 CO_2/60 N_2)			The microorganisms found varied with the kind of seafood and the gas composition of the MAs: in MAP1, richer of oxygen than MAP2, *Carnobacterium* spp. represents the prevalent microbial group, especially in tuna and swordfish. Cocci, belonging to *Leuconostoc* spp., were dominant in salmon independently of gas composition	4°C for 6 days	Franzetti et al. (2003)
Raw and cooked seafood nuggets inoculated with 10^3 CFU/g of *L. monocytogenes*	Air, or 100% CO_2			*Bacillus* spp. and LAB counts increased to 10^2 and 107 CFU/g, respectively, in raw seafood nuggets. Only *Bacillus* spp. reached 10^4 CFU/g in cooked nuggets by 28 days. With the exception of nuggets packaged in 100% CO_2 counts of *L. monocytogenes* increased to approximately 10^7 CFU/g in nuggets stored at both 4°C and 12°C after 28 days	4°C or 12°C	Lyver et al. (1998)
Mullet tissue homogenates	Vacuum			All samples became toxic within 2 days (*C. botulinum* types A–E)	30°C	Lalitha and Gopakumar (2001)

Mullet tissue homogenates	Vacuum		Toxin was detected in packs inoculated with *C. botulinum* types A and B after 5 days, in packs inoculated with types C and D after 7 days, whereas packs inoculated with type E developed toxin ad toxin after 3 days	15°C	
Mullet tissue homogenates	Vacuum		Only type E grew and produced toxin after 8 days	10°C	
Mullet tissue homogenates	Vacuum		Toxin production by type E was delayed for a period of 25 days	4°C	
Cold-smoked salmon (*S. salar*)	Vacuum	26	At the end of shelf 185×10^4 CFU/g were reached	0°C	Dondero et al. (2004)
Cold-smoked salmon (*S. salar*)	Vacuum	21	At the end of shelf 303×10^5 CFU/g were reached	2°C	
Cold-smoked salmon (*S. salar*)	Vacuum	20	At the end of shelf 450×10^4 CFU/g were reached	4°C	
Cold-smoked salmon (*S. salar*)	Vacuum	10	At the end of shelf $<300 \times 10^6$ CFU/g were reached	6°C	
Cold-smoked salmon (*S. salar*)	Vacuum	7	At the end of shelf 760×10^3 CFU/g were reached	8°C	
Spoiled vacuum-packaged, salted, sodium nitrite- or potassium nitrate-treated, cold-smoked rainbow trout	Vacuum		Only one isolate was identified as *L. plantarum* and no *Carnobacterium* strains were discovered. The relative proportion of *L. mesenteroides* subsp. *mesenteroides*	At 4°C or 8°C	Lyhs et al. (1999)

(continued)

TABLE 4.4 (continued)
Effect of MAP on the Microflora and Sensory Assessment of Fish

Species	Gas Composition	Sensory Assessment	Shelf Life (Days)	Microflora	Storage Temperature	References
				was higher in all samples stored at 4°C. Most of the *L. citreum* were found in the samples stored at 8°C, and particularly in the nitrite-treated samples		
Cod fillets (*G. morhua*)	60% CO_2–10% O_2–30% N_2, 60% CO_2–20% O_2–20% N_2, 60% CO_2–30% O_2–10% N_2, 60% CO_2–40% O_2		4	A poor increase was determined in the number of the total aerobic bacteria for all of the gas atmospheres applied. LAB increased by 1–2 log units during 7 days in MAP. There is no difference in the LAB plate counts among the gas mixtures. Hydrogen sulfide producing bacteria increased 2–3 log units in 7 days starting from the 3rd day	6°C	Debevere and Boskou (1996)
Smoked salmon with poor hygienic quality were inoculated with low (6 CFU/g) levels of a mixture of three strains of *L. monocytogenes*	Vacuum	The smoked salmon was still sensorially acceptable after 4 weeks		Growth was slightly faster in the fish with the better hygienic quality. All three strains were found after 4 weeks in the fish with the better quality, while only two strains were recovered after the same time from the poorer quality salmon	4°C for up to 5 weeks	Rorvik et al. (1991)

Smoked salmon with better hygienic quality were inoculated with high (600 CFU/g) levels of a mixture of three strains of *L. monocytogenes*	Vacuum			4°C for up to 5 weeks	
Fresh pearl spot (*E. suratensis* Bloch)	Air	12–14	The aerobic counts reached highest levels (7.3 log10 CFU/g) within 15 days	0°C	Lalitha et al. (2005)
Fresh pearl spot (*E. suratensis* Bloch)	60% CO_2/40% O_2	21	The growth of LAB in pearl spot packed under MAP was slow and their count remained almost constant in gas mixture 60% CO_2/40% O_2 after 3 weeks	0°C	
Fresh pearl spot (*E. suratensis* Bloch)	40% CO_2/60% O_2	19	Delayed and suppressed growth of H_2S producing bacteria during the first 3 weeks	0°C	
Fresh pearl spot (*E. suratensis* Bloch)	70% CO_2/30% O_2	19	Delayed and suppressed growth of H_2S producing bacteria during the first 3 weeks	0°C	
Fish salads (rainbow trout *O. mykiss*)	Air	7	Initially, the average value of aerobes was 2.7 log CFU/g and reached 6.9 log CFU/g	4°C	Metin et al. (2002)
Fish salads (rainbow trout *O. mykiss*)	O_2:CO_2:N_2 = 1:7:12	14	Initially, the average value of aerobes was 2.7 log CFU/g and reached 5.8 log CFU/g	4°C	
Fish salads (rainbow trout *O. mykiss*)	CO_2:N_2 = 3:7	14	Initially, the average value of aerobes was 2.7 log CFU/g and reached 3.8 log CFU/g	4°C	

(continued)

TABLE 4.4 (continued)
Effect of MAP on the Microflora and Sensory Assessment of Fish

Species	Gas Composition	Sensory Assessment	Shelf Life (Days)	Microflora	Storage Temperature	References
Fifty-four packages of vacuum-packed cold-smoked salmon (30) and trout (24)	Vacuum			Psychrotrophic clostridia ranged between 1.71 and 2.21 log CFU/g. *Salmonella*, *E. coli*, and *L. monocytogenes* were not detected. *Listeriae* other than *L. monocytogenes* were isolated from three packages. Levels of *S. aureus* lower than 4 log CFU/g were also found in three packages	2°C ± 1°C	Gonzalez-Rodr ıguez et al. (2002)
Chilled packed tuna steaks	Vacuum			Both psychrotolerant *M. morganii*-like bacteria and *P. phosphoreum* were able to form toxic histamine concentrations	~2°C	Emborg et al. (2005)
Chilled packed tuna steaks	40% CO_2/60% O_2			No formation of histamine during 28 days of storage	1°C	
Salted, cold-smoked rainbow trout fillets	Vacuum with or without the addition of nitrate or nitrite			LAB were the major fraction, predominating in all samples of spoiled product. Eighty-five isolates were found to belong to the family *Enterobacteriaceae*, with 45 of those being *S. plymuthica*. Eleven isolates from the nitrate-treated samples stored at 8°C were identified as *P. aeruginosa*. The types of lactic	4°C and 8°C	Lyhs et al. (1998)

			acid and other bacteria in the spoilage flora were generally reduced by the addition of nitrate or nitrite to fillets		
European sea bass (*D. labrax*) fillets	40% CO_2/60% N_2	8	Coliform count was <2.4 log CFU/g a slight increase in the TVC began from the 5th day up to values just under 6 log CFU/g on the 8th day after packaging	2°C ± 1°C	Poli et al. (2006)
European sea bass (*D. labrax*) fillets	Air	7	Coliform count was 2.8 log CFU/g	2°C ± 1°C	
European sea bass (*D. labrax*) fillets	Whole ungutted fish stored in ice (round)	10	Coliform count was <2.4 log CFU/g a slight increase in the TVC began from the 5th day up to values just under 6 log CFU/g on the 8th day after packaging	2°C ± 1°C	
"Gravad" rainbow trout slices	Vacuum	20	At the time of spoilage, after 27 days both MVC and PVC reached 10^6–10^7 CFU/g	3°C	Lyhs et al. (2001)
"Gravad" rainbow trout slices	Vacuum	18	At the time of spoilage, after 20 days both MVC and PVC reached 10^7–10^8 CFU/g	8°C	
Mackerel	Air	3	4.88 log CFU/g TVC	2°C–4°C	Fagan et al. (2004)
Whiting	Air	3	4.81 log CFU/g TVC	2°C–4°C	
Salmon	Air	5	6.23 log CFU/g TVC	2°C–4°C	
Mackerel	60% N_2/40% CO_2	5	4.18 log CFU/g TVC	2°C–4°C	
Mackerel	100% CO_2	5	3.99 log CFU/g TVC	2°C–4°C	
Salmon	60% N_2/40% CO_2	7	5.04 log CFU/g TVC	2°C–4°C	
Salmon	100% CO_2	7	4.53 log CFU/g TVC	2°C–4°C	

(continued)

TABLE 4.4 (continued)
Effect of MAP on the Microflora and Sensory Assessment of Fish

Species	Gas Composition	Sensory Assessment	Shelf Life (Days)	Microflora	Storage Temperature	References
Whiting	30% N_2/40% CO_2/30% O_2		5	4.48 log CFU/g TVC	2°C–4°C	
Whiting	100% CO_2		5	4.34 log CFU/g TVC	2°C–4°C	
Rainbow trout (percussive stunning)	Ice		15 or 16	There were no significant differences in TVC of fish stored in ice and MAP, regardless of the different slaughter methods used, fish packed in MAP showed a reduction in bacterial counts compared to fish held in ice throughout the study	2°C ± 2°C	Ozogul and Ozogul (2004)
	40% CO_2, 30% N_2, and 30% O_2		16 or 17		2°C ± 2°C	
Rainbow trout (death in ice slurry)	Ice		14 or 15		2°C ± 2°C	
	40% CO_2, 30% N_2, and 30% O_2		15 or 16		2°C ± 2°C	
Red mullet (*M. surmuletus*)	Air		6	TVC value 8.43 ± 0.41 log CFU/g	4°C ± 0.5°C	Pournis et al. (2005)
Red mullet (*M. surmuletus*)	10% O_2/20% CO_2/70% N_2		8	TVC value 8.01 ± 0.66 log CFU/g	4°C ± 0.5°C	
Red mullet (*M. surmuletus*)	10% O_2/40% CO_2/50% N_2		10	TVC value 7.66 ± 0.546 log CFU/g	4°C ± 0.5°C	
Red mullet (*M. surmuletus*)	10% O_2/60% CO_2/30%/N_2		8	TVC value 7.55 ± 0.61 log CFU/g	4°C ± 0.5°C	
Sardine (*S. pilchardus*)	Air		5	Total AMB counts (log CFU/g) were 3.70 and H_2S-producing bacteria counts (log CFU/g) were 3.70	4°C ± 1°C	Erkan et al. (2006)

Sardine (*S. pilchardus*)	5% O_2/35% CO_2/60% N_2		5	Total AMB counts (log CFU/g) were 3.48 and H_2S-producing bacteria counts (log CFU/g) were 3.47	4°C ± 1°C	
Sardine (*S. pilchardus*)	5% O_2/70% CO_2/25% N_2		7	Total AMB counts (log CFU/g) were 3.70 and H_2S-producing bacteria counts (log CFU/g) were 3	4°C ± 1°C	
Gilt-head sea bream (*S. aurata*)	Air			*Pseudomonad* strains dominated	0°C, 10°C, and 20°C	Tryfinopoulou et al. (2002)
Gilt-head sea bream (*S. aurata*)	40% CO_2–30% N_2–30% O_2			LAB and *B. thermosphacta* dominated	0°C, 10°C, and 20°C	
Atlantic halibut (*H. hippoglossus*)	Air	CO_2:O_2 (50%:50%) showed better sensory scores	10–13	*P. phosphoreum* and *Pseudomonas* spp. were found to dominate in the halibut. *B. thermosphacta* was found in most samples at the end of the storage period. *S. putrefaciens* was found sporadically and in low concentrations	4°C	Hovda et al. (2007)
Atlantic halibut (*H. hippoglossus*)	CO_2:N_2 and CO_2:O_2 (50%:50%)		13–20		4°C	

claws had a storage life of 5 days at 0°C and 3 days at 5°C. The MA had only a minor inhibitory effect on the growth of *S. putrefaciens* and *B. thermosphacta* compared to the air-stored controls (Cann et al., 1985). According to Bremner and Statham (1987), MAP extended the storage life of scallops (*Pecten alba*) packed in barrier bags that were evacuated and backflushed with CO_2 prior to storage at 4°C. Scallops under MAP had a storage life of 22 days compared with 10 days for those stored in air at 4°C. Shucked scallops (*Pecten maximus*) harvested in Scotland had a storage life of 7.3 days at 0°C and 4 days at 5°C when packaged in MA of 40% CO_2/30% NO_2/30% O_2 (Cann et al., 1985). Dheeragool (1989) reported that pink prawns (*Pandalus platyceros*) harvested off the coast of British Columbia had a longer storage life when CO_2 was used for the MA compared with N_2. According to Goulas et al. (2005), a shelf life of ca. 14–15 days was achieved for mussel samples based on odor and taste scores packaged under 80% CO_2/20% N_2 gas mixture. This gas composition extended had a beneficial effect on the shelf life of mussels by extending it for ca. 5–6 days, as compared with the control samples (8–9 days), and retained higher sensory scores than all other packaged samples, while the other two MAP compositions (50% CO_2/50% N_2 and 40% CO_2/30% N_2/30% O_2) used extended the shelf life of mussels by ca. 3 days (reaching a shelf life of 10–11 days). All mussel flesh samples were packaged in LDPE/PA/LDPE barrier pouches and stored at 4°C.

Table 4.5 exhibits the crustaceans and Shellfish stored under MAP.

4.3.2 Microflora of Crustaceans and Shellfish Packed under MAP

Lannelongue et al. (1982) reported that microbial growth on brown shrimp (*Penaeus aztecus*) harvested from the Gulf of Mexico was much slower as the CO_2 concentration of the MA increased. Cooked and peeled freshwater crayfish (*Procambaris clarkii*) tail meat was stored on ice in air, 100% CO_2 or 80% CO_2, and 20% air for 21 days. The 80% CO_2/20% air atmosphere led to the lowest microbial counts (aerobic psychotrophs and anaerobes) at the end of the storage period (Gerdes et al., 1989). Bremner and Statham (1987) evaluated the effect of addition of *L. plantarum* to scallops (*P. alba*) prior to VP and storage at 4°C. It was shown that lactobacilli did not suppress the growth of spoilage bacteria and as a result storage life did not increase. According to Dheeragool (1989), sulfide-producing bacteria were the primary spoilage-causing agent in N_2 packed prawns (*P. platyceros*). According to Parkin and Brown (1983), cooked whole Dungeness crab (*Cancer magister*) was stored in 80% CO_2/20% air at 1.7°C. Total aerobic psychotrophs on the MA-stored crabs remained below 10^4 CFU/g of tissue, while aerobic control samples held at 1.7°C had psychotrophic counts exceeding 10^6 CFU/g after 14 days of storage. An increase in the *L. monocytogene* population was observed in vacuum- and MA-packaged shrimp (*P. platyceros*) after 21 days of storage in ice (Harrison et al., 1991). Dorsa et al. (1993) found that *L. monocytogene* growth was inhibited at 6°C, but temperature-abuse (12°C) conditions for short periods induced a rapid growth of this microorganism in lobster (*Procambarus clarkii*). Kimura et al. (2000) studied scallop adductor muscle packed with an atmosphere of 100% O_2, 80% O_2/20% CO_2, 60% O_2/40% CO_2, and air, and stored at 5°C. The growth of bacteria in the samples

TABLE 4.5
Effect of MAP on the Shelf Life of Crustaceans and Shellfish versus Storage Time

Species	Gas Composition	Sensory Assessment	Shelf Life (Days)	Reference Material	Storage Temperature	References
Pink prawns (*P. platyceros*)	CO_2		A longer storage life was achieved when CO_2 was used compared with N_2			Dheeragool (1989)
Pink prawns (*P platyceros*)	N_2					
Scallops (*P. alba*)	Packed in barrier bags that were evacuated and backflushed with CO_2		22		4°C	Bremner and Statham (1987)
Scallops (*P. alba*)	Air		10		4°C	
Mussel samples	80% CO_2/20% N_2	Highest sensory scores for 80% CO_2/20% N_2 packaged samples	14–15	Low-density polyethylene/ polyamide/ low-density polyethylene barrier pouches	4°C	Goulas et al. (2005)
Mussel samples	Control samples (air/ water)		8–9		4°C	
Mussel samples	50% CO_2/50% N_2		10–11		4°C	
Mussel samples	40% CO_2/30% N_2/30% O_2				4°C	

was inhibited in the atmosphere containing CO_2 gas at the ratio of 20% or 40%. However, the quality of scallop adductor muscle could not be preserved due to the rapid development of rigor. In the sample packed with 100% O_2, it was noteworthy that the drop in ATP and pH, the development of rigor, the increase in octopine, and, moreover, the increase in TVC were prolonged for nearly 2 days, compared with the sample packed with air. Lalitha and Gopakumar (2001) examined the growth and toxin production by *C. botulinum* types A–E shrimp tissue homogenates stored under vacuum at temperatures of 30°C, 15°C, 10°C, and 4°C. The inoculated shrimp tissue homogenates held at 30°C under vacuum became toxic within 2 days. At 15°C, toxin was detected in packs inoculated with *C. botulinum* types A and B after 5 days and in packs inoculated with types C and D after 7 days, whereas packs inoculated with type E developed toxin after 3 days. At 10°C, only type E grew and produced toxin after 10 days, respectively, in inoculated (1000 type E spores) VP shrimp tissue homogenates. Furthermore, toxin production by type E was retarded for a period of 34 days, respectively, at 4°C in VP shrimp tissue homogenates inoculated with 1000 spores. Franzetti et al. (2003) investigated the composition of lactic acid population in shrimps packaged in two different MA, MAP1 (80% O_2/20% N_2), and MAP2 (40% CO_2/60% N_2), at 4°C for 6 days. The isolates were heterofermentative rods belonging to *Carnobacterium, Lactobacillus*, and cocci of the *Leuconostoc* genus. They reported that microorganisms varied with the kind of seafood and the gas composition of the MA: in MAP1, richer in oxygen than MAP2, *Carnobacterium* spp. represents the prevalent microbial group in shrimps whereas MAP2 seems to favor *Lactobacillus* spp. Dalgaard et al. (2003) found that for cooked and brined MAP (different mixtures of CO_2 and N_2 with less than 1% O_2) shrimps the dominant parts of spoilage associations were *Enterococcus* spp. at 15°C and 25°C, and for *Carnobacterium* spp. at 0°C, 5°C, and 8°C. Microbiological results revealed that for mussel samples packaged under 80% CO_2/ 20% N_2 and vacuum, microbial growth was delayed compared with that of air-packaged samples. The effect was more pronounced for TVC, *Pseudomonas* spp., LAB, and H_2S-producing bacteria. TVC was reduced by 0.9–1.0 log CFU/g, *Pseudomonas* spp. by 0.7–0.8 log CFU/g, LAB by 1.0–2.2 log CFU/g, and H_2S-producing bacteria by 0.7–1.2 log CFU/g. Enterobacteriaceae were not significantly affected by MAP conditions (Goulas et al., 2005).

Figure 4.5 displays the impact of MAP (80% CO_2/20% N_2) on the TVC of mussels (*M. galloprovincialis*) and sea bass (*L. calcarifer*); both fishery products were stored at 4°C.

Mejlholm et al. (2005) evaluated the growth of *L. monocytogenes* and shelf life of cooked and peeled shrimps in MAP. *B. thermosphacta* and *Carnobacterium maltaromaticum* were responsible for sensory spoilage of cooked and peeled MAP

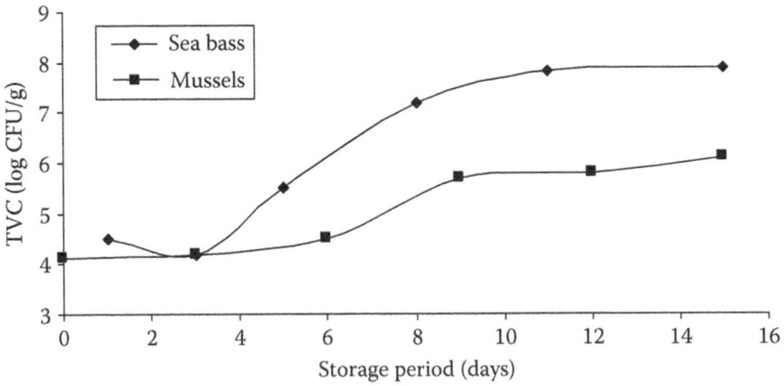

FIGURE 4.5 Effect of MAP (80% CO_2/20% N_2) on the TVC of mussels (*M. galloprovincialis*) (From Goulas, A.E. et al., *J. Appl. Microbiol.*, 98, 752, 2005) and sea bass (*L. calcarifer*). (From Masniyom, P. et al., *J. Food Sci. Agric.*, 82, 873, 2002); both fishery products were stored at 4°C.

(50% CO_2/30% N_2/20% O_2) shrimps. In challenge tests, growth of *L. monocytogenes* was observed at all of the storage temperatures studied. At 5°C and 8°C, the concentration of *L. monocytogenes* increased more than a 1000 times before the product became sensory spoiled, whereas this was not observed at 2°C. Frozen storage had only minor inhibiting effect on the growth of *L. monocytogenes* in thawed product.

Table 4.6 summarizes the microflora of crustaceans and shellfish packed under MAP.

4.3.3 Quality of Crustaceans and Shellfish Stored under MAP

Cooked whole Dungeness crab (*C. magister*) was stored in 80% CO_2/20% air at 1.7°C, and the pH of the crab meat under MA decreased during storage while that of the aerobic control increased. The highest amount of exudates was formed in the MAP crabmeat, probably due to the effect of pH on the water-holding capacity of the muscle proteins (Parkin and Brown, 1983). Matches and Layrisse (1985) studied the shelf life of shrimp (*P. platyceros*) kept under 100% CO_2 controlled atmosphere. They observed moderate discoloration, not associated with undesirable smells, differing from fish exposed to air, which was not in an acceptable state. The high CO_2 concentration delayed the appearance of black spots caused by enzymatic action. Dheeragool (1989) showed that drip loss was higher but color retention was better in the CO_2-packed prawns (*P. platyceros*) in comparison with N_2-packed prawns. Cooked freshwater crayfish (*Pasifastacus leniusculus*) packed under 80% CO_2/20% air for 21 days at 4°C was similar in sensory properties to freshly cooked crayfish, while aerobically stored crayfish was fishy in odor and flavor after 14 days at 4°C. Ammonia and TMA in the aerobic controls augmented but remained relatively static in the MAP crayfish (Wang and Brown, 1983). According to Gerdes et al. (1989) the ammonia and TMA concentrations, in cooked and peeled freshwater crayfish (*P. clarkii*) tail meat, enhanced more rapidly in air and very slowly in 100% CO_2-packed tail meat. Although the 100% CO_2-packed tail meat was of better overall quality, sensory panelists detected that this was due to carbonic acid and lactic acid production. Goulas et al. (2005) showed that for mussel flesh samples the total VBN and TMA nitrogen values remained lower than the proposed acceptability limits of 35 mg N/100g and 12 mg N/100g, respectively, after 15 days of storage. Both the VP- and air-packaged mussel samples exceeded these limits. The TBA values of all MAP (50% CO_2/50% N_2, 80% CO_2/20% N_2, 40% CO_2/30% N_2/30% O_2) and VP mussels remained lower than the proposed acceptability limit of 1 mg malondialdehyde/kg. The air-packaged samples exceeded this limit. All samples retained desirable sensory characteristics during the first 8 days of storage. Chen and Xiong (2008) examined three types of packaging systems regarding their impact on the storage stability (2°C) of precooked and peeled red claw crayfish (*Cherax quadricarinatus*) tails. MAP (80% CO_2/10% O_2/10% N_2) prevented a pH rise, purge loss, and texture toughening or softening of stored red claw crayfish when compared with PVCP or VP. Lipid oxidation was minimal in all the packaging systems tested.

The quality of crustaceans and shellfish stored under MAP is exhibited in Table 4.7.

TABLE 4.6
Effect of MAP on the Microflora of Crustaceans and Shellfish with Storage Time

Species	Initial Gas Mix	Shelf Life (Days)	Microflora	Storage Temperature	References
Scallops (*P. alba*)	Vacuum		Addition of *L. plantarum* to scallops (*P. alba*) prior to vacuum packaging showed that lactobacilli did not suppress growth of spoilage bacteria; as a result storage life was not increased	4°C	Bremner and Statham (1987)
Prawns (*P. platyceros*)	N_2 packed		Sulfide-producing bacteria were the primary spoilage causing agent		Dheeragool (1989)
Brown shrimp (*P. aztecus*)	CO_2		Microbial growth was slower as the CO_2 concentration of the MA increased		Lannelongue et al. (1982)
Cooked and peeled freshwater crayfish (*P. clarkii*) tail meat	Air		The 80% CO_2/20% air atmosphere led to the lowest microbial counts (aerobic psychotrophs and anaerobes) at the end of the storage period	Stored on ice for 21 days	Gerdes et al. (1989)
Cooked and peeled freshwater crayfish (*P. clarkii*) tail meat	80% CO_2/20% air				
Cooked whole Dungeness crab (*C. magister*)	80% CO_2/20% air		Total aerobic psychotrophs remained below 10^4 CFU/g of tissue	1.7°C	Parkin and Brown (1983)
Cooked whole Dungeness crab (*C. magister*)	Air		Aerobic control samples had psychotrophic counts exceeding 10^6CFU/g after 14 days of storage	1.7°C	
Cooked crab claws (*C. pagurus*)	40% CO_2/30% N_2/30% O_2	10	The MA had only a minor inhibitory effect on the growth of *S. putrefaciens* and *B. thermosphacta* compared with the air-stored controls	0°C	Cann et al. (1985)

Cooked crab claws (*C pagurus*)	40% CO_2/30% N_2/30% O_2	6		5°C	
Cooked crab claws (*C. pagurus*)	Air	5		0°C	
Cooked crab claws (*C. pagurus*)	Air	3		5°C	
Scallops	Vacuum-packaged		Did not form toxin in vacuum-packaged scallops	4°C and 10°C	Fletcher et al. (1988a)
Scallops	Vacuum-packaged		Only *C. botulinum* type A formed toxin in scallops but the scallops were spoiled before toxin was detected	At 27°C	
Scallop adductor muscle	100% O_2		The decrease in ATP and pH, the development of rigor, the increase in octopine, and the increase in TVC were prolonged for nearly 2 days, compared with the sample packed with air	5°C	Kimura et al. (2000)
Scallop adductor muscle	20% CO_2/80% air		The growth of bacteria in the samples was inhibited. However, the quality of scallop adductor muscle could not be preserved due to the rapid development of rigor	5°C	
Scallop adductor muscle	40% CO_2/60% air			5°C	
Mussel flesh	80% CO_2/20% N_2		Microbial growth was delayed compared with that of air-packaged samples. TVC was reduced by 0.9–1.0 log CFU/g, *Pseudomonas* spp. by 0.7–0.8 log CFU/g, LAB by 1.0–2.2 log CFU/g, and H_2S-producing bacteria by 0.7–1.2 log CFU/g. Enterobacteriaceae were not significantly affected by MAP conditions		Goulas et al. (2005)
Mussel flesh	Vacuum				
Cooked and brined shrimps	MAP (different mixtures of CO_2 and N_2 with less than 1% O_2)		The dominant parts of spoilage associations were *Enterococcus* spp. at 15°C and 25°C, and *Carnobacterium* spp. at 0°C, 5°C, and 8°C		Dalgaard et al. (2003)

(*continued*)

TABLE 4.6 (continued)
Effect of MAP on the Microflora of Crustaceans and Shellfish with Storage Time

Species	Initial Gas Mix	Shelf Life (Days)	Microflora	Storage Temperature	References
Cooked and peeled Shrimps	50% CO_2, 30% N_2 and 20% O_2		The concentration of *L. monocytogenes* increased more than a 1000-fold before the product became sensory spoiled whereas this was not observed at 2°C. *B. thermosphacta* and *C. maltaromaticum* were responsible for sensory spoilage	At 5°C and 8°C	Mejlholm et al. (2005)
Shrimps	MAP1 (80 O_2/20 N_2) and MAP2 (40 CO_2/60 N_2)		In MAP1, *Carnobacterium* spp. represents the prevalent microbial group in shrimps whereas MAP2 seems to favor *Lactobacillus* spp.	At 4°C for 6 days	Franzetti et al. (2003)
Shrimp tissue homogenates	Vacuum		Became toxic within 2 days (*C. botulinum* types A–E)	30°C	Lalitha and Gopakumar (2001)
Shrimp tissue homogenates	Vacuum		Toxin was detected in packs inoculated with *C. botulinum* types A and B after 5 days, in packs inoculated with types C and D after 7 days, whereas packs inoculated with type E developed toxin after 3 days	15°C	
Shrimp tissue homogenates	Vacuum		Only type E grew and produced toxin after 10 days	10°C	
Shrimp tissue homogenates	Vacuum		Toxin production by type E was delayed for a period of 34 days	4°C	

TABLE 4.7
Effect of MAP on the Quality and Sensory Assessment of Crustaceans and Shellfish

Species	Gas Mix	Sensory Assessment	Lipid Oxidation	Quality (in General)	Storage Temperature	References
Cooked freshwater crayfish (*P. leniusculus*)	80% CO_2/20% air	Similar in sensory properties to freshly cooked crayfish		Ammonia and trimethylamine remained relatively static in the MAP	For 21 days at 4°C	Wang and Brown (1983)
Cooked freshwater crayfish (*P. leniusculus*)	Aerobically stored	Fishy in odor and flavor after 14 days at 4°C		Ammonia and trimethylamine in the aerobic controls increased		
Cooked and peeled freshwater crayfish (*P. clarkii*) tail meat	100% CO_2	Better overall quality but sensory panelists detected due to carbonic acid and lactic acid production		Ammonia and trimethylamine concentrations increased more rapidly in air and slowest in 100% CO_2 packed tail meat	Stored on ice for 21 days	Gerdes et al. (1989)
Cooked and peeled freshwater crayfish (*Procambaris clarkii*) tail meat	Air					
Cooked whole Dungeness crab (*C. magister*)	80% CO_2/20% air			The pH of the crab meat under MA decreased during storage while that of the aerobic control increased. The highest amount of exudates was formed in the MAP crab meat, probably due to the effect of pH on the water-holding capacity of the muscle proteins	1.7°C	Parkin and Brown (1983)

(continued)

TABLE 4.7 (continued)
Effect of MAP on the Quality and Sensory Assessment of Crustaceans and Shellfish

Species	Gas Mix	Sensory Assessment	Lipid Oxidation	Quality (in General)	Storage Temperature	References
Mussels flesh samples	50%/50% CO_2/N_2, 80%/20% CO_2/N_2, and 40%/30%/30% CO_2/N_2/O_2	All samples retained desirable sensory characteristics during the first 8 days of storage		The TBA value remained lower than the proposed acceptability limit of 1 mg malondialdehyde/kg. The air-packaged samples exceeded this limit	4°C	Goulas et al. (2005)
Mussels flesh samples	Vacuum				4°C	
Pandalus borealis	100% N_2 or with atmospheric air		Pronounced lipid oxidation during frozen storage as compared with packaging in modified air		17°C	Bak et al. (1999)
Pandalus borealis	100% N_2 or with atmospheric air		The TBARS in shrimps packed in atmospheric air were significantly higher than TBARS in samples packed in modified air		−17°C	
Pandalus borealis	Air		The concentration of astaxanthin in light-exposed, atmospheric air-packed shrimps decreased to less than one-third of the initial content during the 12 months of frozen storage. Less but still a significant loss in astaxanthin content was also seen in atmospheric air-packed samples stored in darkness		−17°C	

Pandalus borealis	100% N_2		Only minor changes were seen in MAP samples exposed to light or stored in darkness		−17°C	
Precooked and peeled red claw crayfish (*C. quadricarinatus*) tails	80% CO_2/10% O_2/10% N_2		Lipid oxidation was minimal in all the packaging systems tested	Prevented pH rise, purge loss, and texture toughening or softening of stored red claw crayfish.	2°C	Gong et al. (2008)

4.3.4 Crustaceans and Shellfish Treated with Combination of MAP and Other Technologies

Goncalves et al. (2003) investigated the storage of deepwater pink shrimp (*Parapenaeus longirostris*) in MA. Two gas mixtures were tested (40% CO_2/30% O_2/30% N_2 and 45% CO_2/5% O_2/50% N_2), combined with sulfites-based pretreatment, in comparison with air storage. Generally, both atmospheres preserved the shrimp quality up to 9 days compared with 4–7 days of ice storage (only with pretreatment), although it seems that atmosphere containing 45% CO_2/5% O_2/50% N_2 was the most effective. Concerning the atmosphere treatment, only the shrimp packed in 40% CO_2/30% O_2/30% N_2 was rejected, by 70% of the panelists, at the end of storage period. Moreover, for each sampling day, the shrimp packed under 45% CO_2/5% O_2/50% N_2 had higher acceptability, except on day 7, when both batches had the same acceptability.

4.4 CEPHALOPODS

The composition of lactic acid population in cuttlefish packaged in two different MAs, MAP1 (80% O_2/20% N_2) and MAP2 (40% CO_2/60% N_2), at 4°C for 6 days was investigated. The isolates were heterofermentative rods belonging to *Carnobacterium, Lactobacillus*, and cocci of the *Leuconostoc* genus. MAP2 seems to favor *Lactobacillus* spp. Cocci, belonging to *Leuconostoc* spp., were dominant in cuttlefish independently of gas composition (Franzetti et al., 2003). Albanese et al. (2005) studied the effect of MAP with and without a moisture adsorbent, on the quality of squid (*Sepia officinalis*) during chilled storage. After 11 days of storage at 3°C, TMAN augmented from 0.4 mg/100 g in fresh samples to 24 mg/100 g in MAP with adsorbent and 35 mg/100 g in the other samples. TVBN increased from 18 mg/100 g in fresh samples to 112 mg/100 g in MAP with adsorbent and 158 mg/100 g in control samples.

Table 4.8 displays the crustaceans and shellfish treated with a combination of MAP and other technologies.

4.5 CONCLUSIONS

Fish and seafood commodities are very susceptible to microbial contamination, and it is of great importance to optimize their shelf life. Two of the most promising techniques in this direction are packaging under MA and vacuum. The former has the advantage that it provides practically fresh fish and seafood whereas the latter requires the application of freezing temperatures as well. Although the choice of the gas composition to be applied heavily depends on the fish species and fat, some of the most promising compositions were the following: N_2/CO_2/O_2: 80/10/10, 50/45/5, and 60/35/5. Furthermore, it is encouraging that no toxin formation was detected under these storage conditions.

TABLE 4.8
Effect of MAP in Combination with Other Technologies on Crustaceans and Shellfish

Species	Gas Composition	Other Technology	Sensory Assessment	Shelf Life (Days)	Microflora	Storage Temperature	References
Deepwater pink shrimp (*P. longirostris*)	40% CO_2/30% O_2/30% N_2, and 45% CO_2/5% O_2/50% N_2	Sulfites-based pretreatment	Unacceptable, by 70% of the panelists, at the end of storage. Moreover, for each sampling day, the shrimp packed under 45% CO_2/5% O_2/50% N_2 had higher acceptability, except on day 7	9 days		1.6°C ± 0.4°C	Goncalves et al. (2003)
Deepwater pink shrimp (*P. longirostris*)		Sulfites-based pretreatment		4–7 days of ice storage		1.6°C ± 0.4°C	
Cuttlefish (*S. officinalis*)	MAP1 (80% O_2/20% N_2) and MAP2 (40% CO_2/60% N_2)				The isolates were heterofermentative rods belonging to *Carnobacterium*, *Lactobacillus*, and cocci of the *Leuconostoc* genus. MAP2 seems to favor *Lactobacillus* spp., Cocci, belonging to *Leuconostoc* spp., which were dominant in cuttlefish independently of gas composition	4°C for 6 days	Franzetti et al. (2003)
Cuttlefish (*S. officinalis*)	40% CO_2/60% N_2			10 days	2.7×10^7 total microbial count after 10 days of storage	3°C	Franzetti et al. (2001)

REFERENCES

Albanese, D., Cinquanta, L., Lanorte, M.T., and Di Matteo, M. 2005. Squid (*Sepia officinalis*) stored in active packaging: Some chemical and microbiological changes. *Italian Journal of Food Science*, **17**(3): 325–332.

Altieri, C., Speranza, B., Del Nobile, M.A., and Sinigaglia, M. 2005. Suitability of bifidobacteria and thymol as biopreservatives in extending the shelf life of fresh packed plaice fillets. *Journal of Applied Microbiology*, **99**: 1294–1302.

Bak, L.S., Andersen, A.B., Andersen, E.M., and Bertelsen, G. 1999. Effect of modified atmosphere packaging on oxidative changes in frozen stored cold water shrimp (*Pandalus borealis*). *Food Chemistry*, **64**: 169–175.

Barnett, H.Z., Conrad, J.W., and Nelson, R.W. 1987. Use of laminated high and low density polyethylene flexible packaging to store trout (*Salmon gairdneri*) in a modified atmosphere. *Journal of Food Protection*, **50**: 645–651.

Barnett, H.Z., Stone, F.E., Roberts, G.C., Hunter, P.G., Nelson, R.W., and Kwork, J. 1982. A study in the use of a high concentration of CO_2 in a modified atmosphere to preserve fresh salmon. *Marine Fishers Review*, **44**(33): 7–11.

Belleau, L. and Simard, R.E. 1987. Effects d'atmospheres de dioxide de carbone et d'azote sur des fillets de poisson. *Sciences des Aliments*, **7**: 433–436.

Boknaes, N., Österberg, C., Nilsen, J., and Dalgaard, P. 2000. Influence of freshness and frozen storage temperature on quality of thawed cod fillets stored in modified atmosphere packaging. *Lebensmittel-Wissenschaft und Technologie*, **33**: 244–248.

Boskou, G. and Debevere, J. 1997. Reduction of trimethylamine oxide by *Shewanella* spp. under modified atmospheres in vitro. *Food Microbiology*, **14**: 543–553.

Bremner, H.A. and Statham, J.A. 1987. Packaging in CO_2 extends shelf life of scallops. *Food Technology of Australia*, **39**: 177–179.

Cai, P., Harrison, M.A., Huang, Y.W., and Silva, J.L. 1997. Toxin production by *Clostridium botulinum* type E in packaged channel catfish. *Journal of Food Protection*, **60**: 1358–1363.

Cann, D.C., Houston, L.G., Smith, G.L., Thompson, A.B., and Craig, A. 1984. *Studies of Salmonids and Stored under a Modified Atmosphere*. Torry Research Station, Aberdeen, U.K.

Cann, D.C., Houston, N.C., Taylor, L.Y., Stroud, G., Early, J.C., and Smith, G.L. 1985. *Studies of Shellfish Packed and Stored under a Modified Atmosphere*. Torry Research Station, Aberdeen, U.K.

Cann, D.C., Smith, G.L., and Huston, N.C. 1983. *Further Studies on Marine Fish Stored under Modified Atmosphere Packaging*. Torry Research Station, Aberdeen, U.K., 61pp.

Chen, H.C., Meyers, S.P., Hardy, R.W., and Biede, S.L. 1984. Color stability of astaxanthin pigmented rainbow trout under various packaging conditions. *Journal of Food Science*, **49**: 1337–1340.

Chen, G. and Xiong, Y.L. 2008. Shelf-stability enhancement of precooked red claw crayfish (*Cherax quadricarinatus*) tails by modified $CO_2/O_2/N_2$ gas packaging. *Lebensmittel-Wissenschaft und Technologie*, **41**: 1431–1436.

Chouliara, I., Savvaidis, I.N., Panagiotakis, N., and Kontominas, M.G. 2004. Preservation of salted, vacuum-packaged, refrigerated sea bream (*Sparus aurata*) fillets by irradiation: Microbiological, chemical and sensory attributes. *Food Microbiology*, **21**: 351–359.

Colby, J.-W., Enriquez-Ibarra, L., and Flick, G.J. Jr. 1993. Shelf life of fish and shellfish. In: *Shelf Life Studies of Foods and Beverages—Chemical, Biological, Physical, and Nutritional*, pp. 85–143, Charalambous, G., Ed., Amsterdam, the Netherlands: Elsevier.

Corbo, M.R., Altieri, C., Bevilacqua, A., Campaniello, D., Amato, D.D., and Sinigaglia, M. 2005. Estimating packaging atmosphere—Temperature effects on the shelf life of cod fillets. *European Food Resource and Technology*, **220**: 509–513.

Dalgaard, P., Gram, L., and Huss, H.H. 1993. Spoilage and shelf-life of cod fillets packed in vacuum or modified atmospheres. *International Journal of Food Microbiology*, **19**: 283–294.

Dalgaard, P., Madsen, H.L., Samieian, N., and Emborg, J. 2006. Biogenic amine formation and microbial spoilage in chilled garfish (*Belone belone belone*)—Effect of modified atmosphere packaging and previous frozen storage. *Journal of Applied Microbiology*, **101**: 80–95.

Dalgaard, P., Vancanneyt, M., Vilalta, E.N., Swings, J., Fruekilde, P., and Leisner, J.J. 2003. Identification of lactic acid bacteria from spoilage associations of cooked and brined shrimps stored under modified atmosphere between 0°C and 25°C. *Journal of Applied Microbiology*, **94**: 80–89.

Daniels, J.A., Krishnamurthi, R., and Rizvi, S.S.H. 1986. Effects of carbonic acid dips and packaging films in the shelf life of fresh fish fillets. *Journal of Food Science*, **51**: 929–931.

Davies, A.R. 1995. Fate of food-borne pathogens on modified-atmosphere packaged meat and fish. *International Biodeterioration & Biodegradation*, **36**: 407–410.

Debevere, J. and Boskou, G. 1996. Effect of modified atmosphere packaging on the TVB/TMA-producing microflora of cod fillets. *International Journal of Food Microbiology*, **31**: 221–229.

Dheeragool, P. 1989. Modified atmosphere packaging of pink prawns (*Pandalus platyceros*). M.Sc. thesis. University of British Columbia, Vancouver, British Columbia, Canada.

Dondero, M., Cisternas, F., Carvajal, L., and Simpson, S. 2004. Changes in quality of vacuum-packed cold-smoked salmon (*Salmo salar*) as a function of storage temperature. *Food Chemistry*, **87**: 543–550.

Dorsa, W.J., Marshall, D.L., Moddy, M.W., and Hackney, C.R. 1993. Low temperature growth and thermal inactivation of *Listeria monocytogenes* in precooked crawfish tail meat. *Journal of Food Protection*, **56**: 106–109.

Dufresne, I., Smith, J.P., Liu, J.N., Tarte, I., Blanchfield, B., and Austin, J.W. 2000. Effect of headspace oxygen and films of different oxygen transmission rate on toxin production by *Clostridium botulinum* type E in rainbow trout fillets stored under modified atmospheres. *Journal of Food Safety*, **20**(3): 157–175.

Eklund, M.W. 1982. Significance of *Clostridium botulinum* in fishery products preserved short of sterilization. *Food Technology*, **36**(12): 107–112, 115.

Emborg, J., Laursen, B.G., and Dalgaard, P. 2005. Significant histamine formation in tuna (*Thunnus albacares*) at 2°C—Effect of vacuum—And modified atmosphere—Packaging on psychrotolerant bacteria. *International Journal of Food Microbiology*, **101**: 263–279.

Erkan, N., Ozden, O., Alakavuk, D.U., Yildirim, S.Y., and Inugur, I. 2006. Spoilage and shelf life of sardines (*Sardina pilchardus*) packed in modified atmosphere. *European Food Resource and Technology*, **222**: 667–673.

Erkan, N., Ozden, O., and Inugur, M. 2007. The effects of modified atmosphere and VP on quality of chub mackerel. *International Journal of Food Science and Technology*, **42**: 1297–1304.

Fagan, J.D., Gormley, T.R., and Ui Mhuircheartaigh, M.M. 2004. Effect of modified atmosphere packaging with freeze-chilling on some quality parameters of raw whiting, mackerel and salmon portions. *Innovative Food Science and Emerging Technologies*, **5**: 205–214.

Fey, M.S. and Regenstein, J.M. 1982. Extending self life of fresh red hake and salmon using CO_2–O_2 modified atmosphere and potassium sorbate at 1°C. *Journal of Food Science*, **47**: 1048–1054.

Fletcher, E.C., Murrell, W.G., Statham, J.A., Stewart, G.J., and Bremer, J.A. 1988a. Packaging of scallops with sorbate: An assessment of the hazard from *Clostridium botulinum*. *Journal of Food Science*, **53**(2): 349–352, 358.

Fletcher, G.C., Summers, G., and van Veghel, P.W.C. 1998b. Levels of histamine and histamine-producing bacteria in smoked fish from New Zealand markets. *Journal of Food Protection*, **61**(8): 1064–1070.

Franzetti, L., Mrtinoli, S., Piergiovanni, L., and Gali, A. 2001. Influence of active packaging on the shelf life of minimally processed fish products in a modified atmosphere. *Packaging Technology and Science*, **14**: 267–274.

Franzetti, L., Scarpellini, M., Mora, D., and Galli, A. 2003. *Carnobacterium* spp. in seafood packaged in modified atmosphere. *Annals of Microbiology*, **53**: 189–198.

Fugii, T., Hirayama, M., Okuzumi, M., Nishino, H., and Yokohama, M. 1989. Shelf-life studies on fresh sardine packaged with carbon dioxide gas mixture. *Bulletin of the Japanese Society for the Science of Fish*, **55**: 1971–1975.

Fugii, T., Hirayama, M., Okuzumi, M., Nishino, H., and Yokoyama, M. 1990. The effect of storage in carbon dioxide–nitrogen gas mixture on the microbial flora of sardines. *Nippon Suisan Gakaishi*, **56**: 837.

Garcia, G.W. and Genigeorgis, C.A. 1987. Quantitative evaluation of *Clostridium botulinum* nonproteolytic types B, E and F growth in fresh salmon tissue homogenates stored under modified atmospheres. *Journal of Food Protection*, **50**: 390–397, 400.

Garcia, G.W., Genigeorgis, C.A., and Lindroth, S. 1987. Risk of growth production and toxin production by *Clostridium botulinum* nonproteolytic types B, E and F in salmon fillets stored under modified atmospheres at low and abused temperatures. *Journal of Food Protection*, **50**: 330–336.

Gerdes, D.L., Hoffstein, J.J, Finerty, M.W., and Grodner, R.M., 1989. The effects of elevated CO_2 atmospheres on the shelf life of freshwater crayfish (*Procambaris clarkii*) tail meat. *Lebensmittel-Wisseschaft und Technologie*, **22**: 315–318.

Gibson, D.M. 1985. Predicting the shelf life of packaged fish from conductance measurements. *Journal of Applied Bacteriology*, **58**: 465–470.

Gimenez, B., Roncales, P., and Beltran, J.A. 2002. Modified atmosphere packaging of filleted rainbow trout. *Journal of Food Science and Agriculture*, **82**: 1154–1159.

Gonçalves, A., López-Caballero, M.E., and Nunes, M.L. 2003. Quality changes of deepwater pink shrimp (*Parapenaeus longirostris*) packed in modified atmospheres. *Journal of Food Science*, **68:** 2586–2590.

Gonzalez-Rodrıguez, M.-N., Sanz, J.-J., Santos, J.-A. Otero, A., and García-Lopez, M.-L. 2002. Numbers and types of microorganisms in vacuum-packed cold-smoked freshwater fish at the retail level. *International Journal of Food Microbiology*, **77**: 161–168.

Goulas, A.E., Chouliara, I., Nessi, E., Kontominas, M.G., and Savvaidis, I.N. 2005. Microbiological, biochemical and sensory assessment of mussels (*Mytilus galloprovincialis*) stored under modified atmosphere packaging. *Journal of Applied Microbiology*, **98**: 752–760.

Goulas, A.E. and Kontominas, M.G. 2007a. Combined effect of light salting, modified atmosphere packaging and oregano essential oil on the shelf-life of sea bream (*Sparus aurata*): Biochemical and sensory attributes. *Food Chemistry*, **100**: 287–296.

Goulas, A.E. and Kontominas, M.G. 2007b. Effect of modified atmosphere packaging and VP on the shelf-life of refrigerated chub mackerel (*Scomber japonicus*): Biochemical and sensory attributes. *European Food Resource and Technology*, **224**: 545–553.

Gray, R.J.H., Hoover, D.G., and Muir, A.M. 1983. Attenuation of microbial growth on modified atmosphere-packaged fish. *Journal of Food Protection*, **46**: 610–613.

Harrison, M.A., Huang, Y.W., Chao, C.H., and Shineman, T. 1991. Fate of *Listeria monocytogenes* on packaged, refrigerated, and frozen seafood. *Journal of Food Protection*, **54**: 524–527.

Hong, L.C., Leblanc, E.L., Hawrysh, Z.J., and Hardin, R.T. 1996. Quality of Atlantic mackerel (*Scomber scombrus* L.) fillets during modified atmosphere storage. *Journal of Food Science*, **61**: 646–651.

Hovda, M.B., Sivertsvik, M., Lunestad, B.T., Lorentzen, G., and Rosnes, J.T. 2007. Characterisation of the dominant bacterial population in modified atmosphere packaged farmed halibut (*Hippoglossus hippoglossus*) based on 16S rDNA-DGGE. *Food Microbiology*, **24**: 362–371.

Hurtado, J.L., Montero, P., and Borderías, A.J. 2000. Extension of shelf life of chilled hake (*Merluccius capensis*) by high pressure. *International of Food Science and Technology*, **6**: 243–249.

Hussain, A., Ehlerrmann, D., and Diehl, J. 1977. Comparison of toxin production by *Clostridium botulinum* type E in irradiated and unirradiated vacuum-packed trout (*Salmo gairdneri*). *Archive fuer Lebensmittelhygiene*, **28**: 23–27.

Kimura, M., Narita, M., Imamura, T., Ushio, H., and Yamanaka, H. 2000. High quality control of scallop adductor muscle by different modified atmosphere packaging. *Nippon Suisan Gakkaishi*, **66**(3): 475–480.

Kosak, P.H. and Toledo, R.T. 1981. Effects of microbiological decontamination on the storage stability of fresh fish. *Journal of Food Science*, **46**: 1012–1014.

Lalitha, K.V. and Gopakumar, K. 2001. Growth and toxin production by *Clostridium botulinum* in fish (*Mugil cephalus*) and shrimp (*Penaeus indicus*) tissue homogenates stored under vacuum. *Food Microbiology*, **18**: 651–657.

Lalitha, K.V., Sonaji, E.R., Manju, S., Jose, L., Gopal, T.K.S., and Ravisankar, C.N. 2005. Microbiological and biochemical changes in pearl spot (*Etroplus suratensis* Bloch) stored under modified atmospheres. *Journal of Applied Microbiology*, **99**: 1222–1228.

Lannelongue, M., Finne, G., Hanna, M., Nickelson, R., and Vanderzant, G. 1982. Storage characteristics of brown shrimp (*Penaeus aztecus*) stored in retail-packages containing CO_2 enriched atmospheres. *Journal of Food Science*, **47**: 911–913, 923.

Leung, C., Huang, Y., and Harrison, M.A. 1992. Fate of *Listeria monocytogenes* and *Aeromonas hydrophila* on packaged channel catfish fillets stored at 4°C. *Journal of Food Protection*, **55**: 728–730.

Licciardelo, J.J., Ravesi, E.M., Tuhkuken, B.E., and Racicot, L.D. 1984. Effect of some potentially synergistic treatments in combination with 100 krad irradiation on the iced shelf life of cod fillets. *Journal of Food Science*, **49**: 1341–1346, 1375.

Lilly, T., Jr. and Kautter, D.A. 1990. Outgrowth of naturally occurring Clostridium botulinum in vacuum-packaged fresh fish. *Journal of the Association of Official Analytical Chemists*, **73**: 211–212.

Lioulas, T.S. 1988. Challenges of controlled and modified atmosphere packaging: A food company's perspective. *Food Technology*, **42**: 78–86.

Lyhs, U., Bjorkroth, J., Hyytia, E., and Korkeala, H. 1998. The spoilage flora of vacuum-packaged, sodium nitrite or potassium nitrate treated, cold-smoked rainbow trout stored at 4°C or 8°C. *International Journal of Food Microbiology*, **45**: 135–142.

Lyhs, U., Bjorkroth, J., and Korkeala, H., 1999. Characterisation of lactic acid bacteria from spoiled, vacuum-packaged, cold-smoked rainbow trout using ribotyping. *International Journal of Food Microbiology*, **52**: 77–84.

Lyhs, U., Lahtinen, J., Fredriksson-Ahoma, M., Hyytia-Trees, E., Elfing, K., and Korkeal, H. 2001. Microbiological quality and shelf-life of vacuum-packaged "gravad" rainbow trout stored at 3°C and 8°C. *International Journal of Food Microbiology*, **70**: 221–230.

Lyver, A., Smith, J.P., Tarte, I., Farber, J.M., and Nattress, F.M. 1998. Challenge studies with *Listeria monocytogenes* in a value-added seafood product stored under modified atmospheres. *Food Microbiology*, **15**: 379–389.

Manju, S., Srinivasa Gopal, T.K., Jose, L., Ravishankar, C.N., and Kumar, K.A. 2007. Nucleotide degradation of sodium acetate and potassium sorbate dip treated and vacuum packed Black Pomfret (*Parastromateus niger*) and Pearlspot (*Etroplus suratensis*) during chill storage. *Food Chemistry*, **102**: 699–706.

Manthey, M., Karnop, G., and Rehbein, H. 1988. Quality changes of European catfish (*Silurus glanis*) from warm-water aquaculture during storage on ice. *International Journal of Food Science and Technology*, **23**: 1–9.

Masniyom, P., Benjakul, S., and Visessanguan, W. 2002. Shelf-life extension of refrigerated seabass slices under modified atmosphere packaging. *Journal of Food Science and Agriculture*, **82**: 873–880.

Masniyom, P., Benjakul, S., and Visessanguan, W. 2005. Collagen changes in refrigerated sea bass muscle treated with pyrophosphate and stored in modified—Atmosphere packaging. *European Food Resource and Technology*, **220**: 322–325.

Matches, J.R. and Layrisse, M.E. 1985. Controlled atmosphere storage of spotted shrimp (*Pandalus platyceros*). *Journal of Food Protection*, **48**: 709–711.

McMeekin, T.A., Husle, L., and Bremner, H.A. 1982. Spoilage association of vacuum packed sand flathead (*Platycaphalus bassensis*) fillets. *Food Technology of Australia*, **34**: 278–282.

Mejlholm, O., Boknæs, N., and Dalgaard, P. 2005. Shelf life and safety aspects of chilled cooked and peeled shrimps (*Pandalus borealis*) in modified atmosphere packaging. *Journal of Applied Microbiology*, **99**: 66–76.

Metin, S., Erkan, N., Baygar, T., and Ozden, O. 2002. Modified atmosphere packaging of fish salad. *Fisheries Science*, **68**: 204–209.

Mitsuda, H., Nakajima, K., Mizuno, H., and Kawai, F. 1980. Use of sodium chloride solution and carbon dioxide for extending shelf life of fish fillets. *Journal of Food Science*, **45**: 661–665.

Molin, G., Stenstrom, M., and Ternstrom, A. 1983. The microbial flora of herring fillets after storage in carbon dioxide, nitrogen or air at 2°C. *Journal of Applied Bacteriology*, **55**: 49–56.

Morey, K.S., Satterlee, L.D., and Brown, W.D. 1982. Protein quality of fish in modified atmospheres as predicted by the C-PER assay. *Journal of Food Science*, **47**: 1399–1400, 1409.

Oetterer, M. 1999. Agroindútrias beneficiadoras de pescado cultivado—Unidades modulares e polivalentes para implantação, com enfoque nos pontos críticos e higiênicos e nutricionais. Tese (Livre-docente)—Escola Superior de Agricultura Luiz de Queiroz, Universidade de São Paulo, Piracicaba, Brazil, 198pp.

Ogrydziak, D.M. and Brown, W.D. 1982. Temperature effects in modified-atmosphere storage of seafood. *Food Technology*, **36**: 86–96.

Ordóñez, J.A., López-Gálvez, D.E., Fernández, M., Hierro, E., and Hoz, L.D. 2000. Microbial and physicochemical modifications of hake (*Merluccius merluccius*) steaks stored under carbon dioxide enriched atmospheres. *Journal of Food Science and Agriculture*, **80**: 1831–1840.

Ozogul, Y. and Ozogul, F. 2004. Effects of slaughtering methods on sensory, chemical and microbiological quality of rainbow trout (*Onchorynchus mykiss*) stored in ice and MAP. *European Food Resource and Technology*, **219**: 211–216.

Ozogul, F., Polat, F., and Ozogul, Y. 2004. The effects of modified atmosphere packaging and VP on chemical, sensory and microbiological changes of sardines (*Sardina pilchardus*). *Food Chemistry*, **85**: 49–57.

Pantazi, D., Papavergou, A., Pournis, N., Kontominas, M.G., and Savvaidis, I.N. 2008. Shelf-life of chilled fresh Mediterranean swordfish (*Xiphias gladius*) stored under various packaging conditions: Microbiological, biochemical and sensory attributes. *Food Microbiology*, **25**: 136–143.

Parkin, K.L. and Brown, W.D. 1983. Modified atmosphere storage of Dungeness crab (*Cancer magister*). *Journal of Food Science*, **48**: 370–374.

Partmann, W. 1981. Untersuchungen zur lagerung von ver packten regenbogenforellen in luft und kohlendioxid. *Fleischtwirtsch*, **61**: 625–629.

Pastoriza, L., Sampedro, G., Herrera, J.J., and Cabo, M.L. 1998. Influence of sodium chloride and modified atmosphere packaging on microbiological, chemical and sensorial properties in ice storage of slices of hake (*Merluccius merluccius*). *Food Chemistry*, **61**(1/2): 23–28.

Poli, B.M., Messini, A., Parisi, G., Scappini, F., Vigiani, V., Giorgi, G., and Vincenzini, M. 2006. Sensory, physical, chemical and microbiological changes in European sea bass (*Dicentrarchus labrax*) fillets packed under modified atmosphere/air or prepared from whole fish stored in ice. *International Journal of Food Science and Technology*, **41**: 444–454.

Post, L.S., Lee, D.A., Solberg, M., Furgang, D., Specchio, J., and Graham, C. 1985. Development of botuxinal toxin and sensory deterioration during storage of vacuum and modified atmosphere packaged fish fillets. *Journal of Food Science*, **50**: 990–996.

Pournis, N., Papavergou, A., Badeka, A., Kontominas, M.G., and Savvaidis, I.N. 2005. Shelf-life extension of refrigerated Mediterranean mullet (*Mullus surmuletus*) using modified atmosphere packaging. *Journal of Food Protection*, **68**(10): 2201–2207.

Powrie, W.D., Skura, B.J., and Wu, C.H. 1987. Energy conservation by storage of muscle and plant products at latent zone and modulated subfreezing temperatures. Final report, ERDAF file 0145B.01916-EP25, Agriculture, Canada.

Prentice, C. and Sainz, R. 2000. Desenvolvimento de um produto minimamente processado a base de carpa-capim (*Ctenopharyngodon idella*). In: *17th Congresso Brasileiro de Ciência e Tecnologia de Alimentos*, Fortaleza, Brazil 2000. *Resumos*, **3**(11): 116.

Przybylski, L.A., Finerty, M.W., Grodner, R.M., and Gerdes, D.L. 1989. Extension of shelf life of fresh channel catfish fillets using modified atmosphere packaging and low dose irradiation. *Journal of Food Science*, **54**: 269–273.

Randell, K., Hattula, T., and Ahvenainen, R. 1997. Effect of packaging method on the quality of rainbow trout and Baltic herring fillets. *Lebensmittel-Wisseschaft und Technologie*, **30**: 56–61.

Reddy, N.R., Schreider, C.L., Buzard, K.S., Skinner, G.E., and Armstrong, D.J. 1994. Shelf life of fresh tilapia fillets packaged in high barrier film with modified atmospheres. *Journal of Food Science*, **59**: 260–264.

Reddy, N.R., Solomon, H.M., Yep, H., Roman, M.G., and Rhodehamel, E.J. 1997b. Shelf life and toxin development by *Clostridium botulinum* during storage of modified-atmosphere-packaged fresh aquacultured salmon fillets. *Journal of Food Protection*, **60**(9): 1055–1063.

Rorvik, L.M., Yndestad, M., and Skjerve, E. 1991. Growth of *Listeria monocytogenes* in vacuum-packed, smoked salmon, during storage at 4°C. *International Journal of Food Microbiology*, **14**: 111–118.

Rosnes, J.T., Kleiberg, G.H., Sivertsvik, M., Lunestad, B.T., and Lorentzen, G. 2006. Effect of modified atmosphere packaging and superchilled storage on the shelf-life of farmed ready-to-cook spotted wolf-fish (*Anarhichas minor*). *Packaging Technology and Science*, **19**: 325–333.

Ruiz-Capillas, C., Saavedra, A., and Moral, A., 2003. Hake slices stored in retail packages under modified atmospheres with CO_2- and O_2-enriched gas mixes. *European Food Resource and Technology*, **218**: 7–12.

Sallam, Kh.I., Ahmed, A.M., Elgazzar, M.M., and Eldaly, E.A. 2007. Chemical quality and sensory attributes of marinated Pacific saury (*Cololabis saira*) during vacuum-packaged storage at 4°C. *Food Chemistry*, **102**: 1061–1070.

Santos, E.E.M. and Regenstein, J.M. 1990. Effects of VP, glazing and erythorbic acid on the shelf life frozen white hake and mackerel. *Journal of Food Science*, **55**: 64–70.

Scott, D.N., Fletcher, G.C., and Hogg, M.G. 1986. Storage of snapper fillets in modified atmospheres at −1°C. *Food Technology of Australia*, **38**: 234–238.

Scott, D.N., Fletcher, G.C., and Summers, G. 1984. Modified atmosphere and vacuum packing of snapper fillets. *Food Technology of Australia*, **36**: 330–332.

Sharp, W.F., Norback, J.P., and Stuider, D.A. 1986. Using a new measure to define shelf life of fresh whitefish. *Journal of Food Science*, **51**: 936–939, 959.

Sivertsvik, M., Jeksrud, W.K., and Rosnes, J.T. 2002. A review of modified atmosphere packaging of fish and fishery products—Significance of microbial growth, activities and safety. *International Journal of Food Science and Technology*, **37**: 107–127.

Skura, B.J. 1991. Modified atmosphere packaging of fish and fish products. In: *Modified Atmosphere Packaging of Food*, pp. 148–167, Ooraikul, B. and Stiles, M.E., Eds., Chichester, U.K.: Ellis Horwood Limited.

Stammen, K., Gerdes, D., and Caporaso, F. 1990. Modified atmosphere packaging of seafood. *Critical Reviews in Food Science and Technology*, **29**: 301–331.

Statham, J.A. and Bremner, H.A. 1985. Acceptability of trevalla (*Hyperoglyphe poporosa*) after storage in carbon dioxide. *Food Technology of Australia*, **37**: 212–215.

Statham, J.A., Bremner, H.A., and Quarmby, A.R. 1985. Storage of morwong (*Nemadactylus maropterus*) in combinations of polyphosphate, potassium sorbate and carbon dioxide at 4°C. *Journal of Food Protection*, **48**: 585–589.

Stenstrom, I. 1985. Microbial flora of cod fillets (*Gadus morhua*) stored at 2°C in different mixtures of carbon dioxide and nitrogen/oxygen. *Journal of Food Protection*, **48**: 585–589.

Stier, R.F., Bell, L., Ito, K.A., Shafer, B.D., Brown, L.A., Seeger, M.L., Allen, B.H., Porcuna, M.N., and Lerke, P.A. 1981. Effect of modified atmosphere storage on *C. botulinum* toxigenesis and the spoilage microflora of salmon fillets. *Journal of Food Science*, **46**: 1639–1642.

Taylor, L.Y., Cann, D.D., and Welch, B.J. 1990. Antibotulinal properties of nisin in fresh fish packaged in an atmosphere of carbon dioxide. *Journal of Food Protection*, **53**: 953–957.

Torrieri, E., Cavella, S., Villani, F., and Masi, P. 2006. Influence of modified atmosphere packaging on the chilled shelf life of gutted farmed bass (*Dicentrarchus labrax*). *Journal of Food Engineering*, **77**: 1078–1086.

Tryfinopoulou, P., Tsakalidou, E., and Nychas, G.-J.E. 2002. Characterization of *Pseudomonas* spp. associated with spoilage of gilt-head sea bream stored under various conditions. *Applied and Environmental Microbiology*, **68**(1): 65–72

Villemure, G., Simard, R.E., and Picard, G. 1986. Bulk storage of cod fillets and gutted cod (*Gadus morhua*) under carbon dioxide atmosphere. *Journal of Food Science*, **51**: 317–320.

Wang, M.Y. and Brown, W.D. 1983. Effects of elevated CO_2 atmosphere on storage of freshwater crayfish (*Pasifastacus leniusculus*). *Journal of Food Science*, **48**: 15–162.

Wang, M.Y. and Ogrysdiak, D.M. 1986. Residual effect of storage in an elevated carbon dioxide atmosphere in the microbial flora of rock cod (*Sebastes spp.*). *Applied and Environmental Microbiology*, **52**: 727–732.

Wang, T., Sveinsdóttir, K., Magnússon, H., and Martinsdóttir, E. 2008. Combined application of modified atmosphere packaging and superchilled storage to extend the shelf life of fresh cod (*Gadus morhua*) loins. *Journal of Food Science*, **73**(1): 11–19.

Williams, S.K., Rodrick, G.E., and West, R.L. 1995. Sodium lactate affects shelf life and consumer acceptance of fresh catfish (*Ictalurus nebulosus, marmoratus*) fillets under simulated retail conditions. *Journal of Food Science*, **60**: 636–639.

5 Fresh and Processed Meat and Meat Products

Ioannis S. Arvanitoyannis and
Alexandros Ch. Stratakos

CONTENTS

5.1 INTRODUCTION

Meat industries have shown increasing interest in the development of suitable techniques to extend shelf life and to improve consumer acceptance of products of animal origin while maintaining nutritional quality and ensuring safety (Chiavaro et al., 2008). Moreover, the need for fresh food supply to distant markets has increased the interest in procedures for extending the shelf life of products. Modified atmosphere packaging (MAP) is used in fresh meat to inhibit the microbial spoilage and to maintain the red color of meat (Brody, 1989; O'Connor-Shaw and Reyes, 2000). Color and appearance are major factors in consumer purchase decisions because they are often indicators of meat freshness (Brewer et al., 2002). The gases normally used are O_2, CO_2, and N_2, whose effects and roles in the preservation of meat quality have been extensively reported (Church, 1994). Furthermore, there is limited use of other gases such as carbon monoxide (CO) (Gill, 1996). The choice of gas mixtures used is influenced by the product sensitivity to oxygen (O_2) and carbon dioxide (CO_2), the color stabilizing requirements, and the microbiological flora capable of growing on the product (Church, 1994). Lactic acid bacteria (LAB) are the dominant microbiota in modified atmosphere packaged meats (Shaw and Harding, 1984; McMullen and Stiles, 1989). In high-O_2 MAP (>50% O_2), the O_2 maintains the muscle pigments in the desirable oxymyoglobin state, but this form of MAP gives a short shelf life.

Low-O_2 MAP (1%–10% O_2 and 50%–90% CO_2) inhibits the bacterial spoilage with a longer shelf life but deteriorates the color (O'Connor-Shaw and Reyes, 2000).

The objective of this chapter is to provide a detailed review of the effects of MAP of meat products alone or combined with other preservation methodologies (irradiation, antioxidants).

5.2 EFFECT OF MAP ON PORK

5.2.1 Quality

Huisman et al. (1994) found that rosemary (0.05% of total weight) added to cooked pork meatballs retarded the formation of warmed-over flavor (WOF) during chill storage. After cooking, the meatballs were packaged using five atmosphere packaging conditions: (i) air, (ii) 5% O_2/95% N_2, (iii) 3% O_2/97% N_2, (iv) 1% O_2/99% N_2, and (v) 100% N_2. The combination of decreased O_2 in the packages and the addition of rosemary gave significantly better sensory scores and significantly lower amount of TBARs. Boneless pork loins were assigned equally to five treatments: 100% CO_2, 50% CO_2/50% N_2, 25% CO_2/ 75% N_2, 25% CO_2/65% N_2/10% O_2, and vacuum. Loin sections were packaged in low O_2 permeability bags and then stored in darkness at 1°C for up to 22 days. Retail chops were cut from the sections and displayed in oxygen-permeable film under light at 3°C for 3 additional days. Sections stored in 25% CO_2/65% N_2/10% O_2 had more surface graying and greening, stronger off-odor compared to sections from the other four treatments. Displayed chops from sections stored in 25% CO_2/65% N_2/10% O_2 also had graying/greening at an outer layer of the chops. Off-odor of chops was most pronounced for treatments with 10% O_2 and vacuum. The lowest drip loss was determined in the samples stored in vacuum (Sørheim et al., 1996). According to Sørheim et al. (1997) pale, soft, exudative (PSE), and normal pork loins were deboned and packaged in MAP of 100% CO_2 containing 0%, 0.5%, and 1% residual oxygen. The meat was stored at 3°C, first under MAP for 21 days, followed by 5 additional days under retail display conditions with access to air. Before packaging, PSE loins were lighter and less red than those of normal meat. After MAP storage, the drip loss was twice as high from PSE meat as compared with normal meat. PSE meat was not more discolored after MAP storage than normal meat. Lastly, discoloration was observed on both PSE and normal meat with 0.5% O_2 and even more clearly with 1% O_2. The effects of carbon monoxide in modified atmosphere packaging (MAP-CO) for pork were investigated by Krause et al. (2003). Eighty pork loins were stored in four packaging environments (aerobic-overwrap, vacuum, MAP without CO, and MAP-CO). Overall, Hunter a* values were significantly greater in MAP-CO than aerobic packages. However, MAP-CO did not reduce purge loss. Results revealed significantly improved color stability and sensory scores with the use of CO. Ahn et al. (2003) evaluated the irradiation (0, 5, and 10 kGy) and MAP effects on emulsion-type cooked pork sausage stored at CO_2 (100%), N_2 (100%), or 25% CO_2/75% N_2 at 4°C. It was determined that gamma irradiation can reduce the residual nitrite in pork sausage, while residual ascorbic acid (AA) content was nearly affected by irradiation. Furthermore, it was determined that irradiation caused reduction of redness in sausage. In MAP conditions, CO_2 or CO_2/N_2 packaging was effective in reducing residual nitrite and in inhibiting loss of red color of samples compared with N_2 packaging. Martinez et al. (2005) examined the effects of different concentrations of CO_2 and the presence of low levels of CO on fresh pork sausages. Fresh sausages were packaged in different atmospheres containing $\%O_2/\%CO_2/\%N_2$: 0/20/80, 0/60/40, 40/20/40, 40/60/0, 80/20/0 and 0.3% CO/30% CO_2/rest argon. The packs were stored for 20 days at 2°C ± 1°C in the dark. It was found that increasing concentrations of CO_2 promoted oxidation of both myoglobin and lipids. Therefore, preservation of color and odor was best achieved using atmospheres containing low CO_2 concentrations (20%). Fresh pork sausages were packaged in atmospheres, using the following mixtures ($\%O_2/\%CO_2/\%N_2$): 0/20/80, 0/20/80 + O_2 scavenger, 20/20/60, 40/20/40, 60/20/20, and 80/20/0. In addition, samples were subjected to vacuum packaging or overwrap with O_2-permeable film. They were stored for 20 days at 2°C ± 1°C in the dark. It was determined that the higher the oxygen concentration, the higher were TBARs values and also their rate of increase. Samples packaged with 80% O_2 showed the highest values of TBARs. On the other hand, samples under vacuum and with O_2 scavengers had the lowest values. TBARs values of overwrapped sausages were also high during the first 8 days of storage, but their rate of increase decreased strongly after that. Discoloration of fresh sausages markedly increased throughout storage following all packaging treatments, except those including no oxygen (vacuum and 0% O_2 plus scavenger). Oxygen-free samples were given the lowest scores throughout the storage period. At day 16 of storage, only overwrap, vacuum, and 0% O_2 plus scavenger samples were considered acceptable based on color scores (Martinez et al. 2006c) (Figure 5.1).

Fresh pork sausages containing natural colorants, red yeast rice powder (*Monascus purpureus*), or a crude red beetroot (*Beta vulgaris*) juice or commercial betanin (E-162), at different concentrations, were packaged in MAP 80% O_2 and 20% CO_2 and stored in dark for 20 days. Sausages with red yeast, red beetroot juice, and betanin had lower L* and h* and higher a* and a*/b* values than control samples. Color properties of samples with red beetroot were the closest to

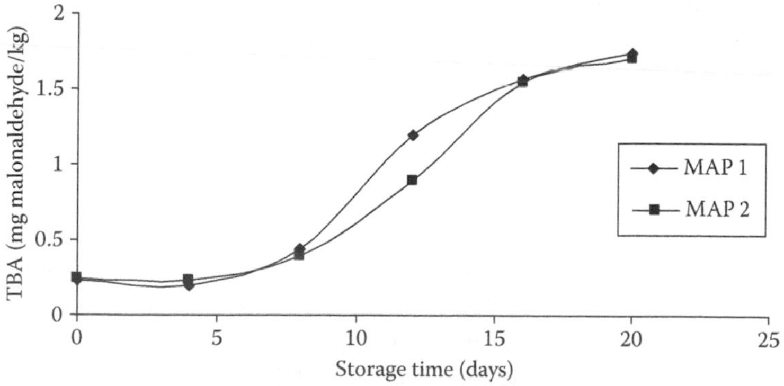

FIGURE 5.1 Effect of MAP1 (80% O_2/20% CO_2) (From Martinez, L. et al., *Meat Sci.*, 69, 493, 2005) and MAP2 (60% O_2/20% CO_2/20% N_2) (From Martinez, L. et al., *J. Sci. Food Agric.*, 86, 500, 2006a) on the TBA values of pork sausages stored at 2°C ± 1°C in the dark.

control sausages, whereas sausages with red yeast rice had significantly lower b* values (Martinez et al., 2006a). Martinez et al. (2006b) studied fresh pork sausages formulated with and without salt, and with antioxidant mixtures containing either rosemary extract, green tea powder, pu-erh tea infusion, or borage meal and their mixtures with AA, which were packaged in an atmosphere containing four parts O_2 with one part CO_2 (v/v) and stored in dark. It was found that NaCl alone in fresh pork sausages decreased the L* value and promoted lipid oxidation but protected sausages from rapid discoloration (decrease of a*). On the other hand, green tea powder, pu-erh tea infusion, and borage meal effectively inhibited lipid oxidation, dependent on the concentration used, but didn't protect sausages from color loss, even when AA was also used. Furthermore, rosemary extract in combination with AA inhibited lipid oxidation and delayed sausage discoloration. The shelf life of salted fresh pork sausages was 16 days. Wicklund et al. (2006) compared CO and high-oxygen MAP effects on the quality of enhanced pork. Pork loins were enhanced to 10.5% over initial weight to contain 0.3% salt and 0.4% phosphate (either sodium tripolyphosphate [STP] or a blend of STP and sodium hexametaphosphate). Chops were packaged in atmospheres containing 0.4% CO/30.0% CO_2/69.6% N_2 or 80% O_2/20% CO_2, aged in the dark. The acceptability of pork flavor, juiciness, and overall acceptability scores were slightly higher for the CO-MAP pork chops when compared with the high-oxygen MAP chops containing STP. The raw pork chops packaged in CO were darker and redder than the ones packaged in high-oxygen MAP. According to Zhang and Sundar (2005), freshly slaughtered pork meat was packed in MAP containing 5%, 15%, 25%, 35%, 45%, and 55% O_2 along with 20% CO_2 in each package and stored for 8 days at 4°C. In addition, nitrogen was used as a filler gas. There was an increase in TBARs values with the increase in oxygen concentration of the package, although, below 25% O_2 level the TBARs values were relatively lower and also significantly different from each other. Moreover, TVBN decreased as the O_2 concentration increased but TVBN value decreased sharply from 17.5 (25% O_2) to 9.7 (45% O_2). The color score of the fresh meat increased with the increasing concentration of oxygen in the package. At 5% and 15% oxygen levels, the color acceptability scores showed that the redness was significantly lost when compared with the normal meat. Raines et al. (2006) studied the effects of AA and Origanox™, a natural herb–derived antioxidant (OG) in different packaging systems to prevent pork lumbar vertebrae marrow discoloration. Concentrations of 1.25%, 1.875%, or 2.5% AA; 0.15% OG+0.30% AA; and 0.225% OG+0.45% AA were applied in 0.5 mL aliquots to 2.54 cm thick pork lumbar vertebrae sections. The following gas mixtures were used: (1) high-oxygen (HiOx) (80% O_2, 20% CO_2) MAP; (2) ultralow-oxygen (ULOx) (70% N_2, 30% CO_2) MAP containing an activated oxygen scavenger; and (3) polyvinyl chloride (PVC) overwrap film. Samples in HiOx MAP treated with 1.875% or 2.50% AA had superior visual color to HiOx and PVC control by day 8. Samples treated with 1.875% or 2.50% AA in ULOx MAP had the least discoloration by day 8. According to Veberg et al. (2006), minced turkey thighs and pork *semimembranosus* muscles were stored for 7 and 12 days at 4°C in high-oxygen MAP and vacuum. Pork meat was more stable against lipid oxidation, with TBARs values <0.2 mg MDA/kg, and no development of fluorescent lipid oxidation products was detected. Lund et al. (2007b) studied the effect of MAP (70% O_2/30% CO_2) and skin packaging (no oxygen) on protein oxidation and texture of porcine *longissimus dorsi* during storage for 14 days at 4°C. Tenderness was reduced in *l. dorsi* slices stored in high-oxygen atmosphere compared with ones packaged under no oxygen. Additionally, myosin was found to form intermolecular cross-links due to packaging in a high-oxygen atmosphere, and the amount of free thiols decreased indicating that protein oxidation and formation of cross-linked proteins result in less tender meat. Moreover, it was determined that the packaging atmospheres did not affect protein carbonyl content. De Santos et al. (2007) evaluated the effects of gas atmosphere in the package, refrigerated storage time, and endpoint cooking temperature on interior cooked color of injection-enhanced pork chops. Enhanced chops were packaged in 0.36% CO/20.34% CO_2 (CO-MAP), 80% O_2/20% CO_2 (HO-MAP), or PVC-overwrapped (PVC-OW, controls), stored at 4°C for 0, 12, 19, or 26 days, displayed for 2 days, and then cooked to six endpoint temperatures (54°C, 60°C, 63°C, 71°C, 77°C, and 82°C). Chops packaged in CO-MAP had the highest a* values. Furthermore, the lowest hue angles occurred in chops cooked to lower endpoint temperatures. Chops in CO-MAP had lower hue angles and higher chroma than those in HO-MAP and PVC-OW. Lastly, above 71°C, hue angle and chroma increased. Lund et al. (2008) investigated the effect of the presence of oxygen in MAP (100% N_2, atmospheric air, and 80% O_2/20% N_2) on the breaking strength of single muscle fibers isolated from pork stored at 4°C. It was determined that the breaking strength of porcine single muscle fibers did not differ significantly among the different gas compositions. However, porcine single muscle fibers isolated from day 1 (control) were found to be significantly weaker than fibers isolated from pork samples stored for 2 days in different atmospheres. Lastly, protein oxidation could not be detected.

5.2.2 Microflora

Viana et al. (2005) studied pork loin samples stored (5°C ± 0.5°C) in nylon polyethylene plastic bags using different MAPs: vacuum, 100% CO_2, 99% CO_2 and 1% CO, 100% O_2 or 100% CO followed by vacuum after 1 h of exposure. MAP with 100% O_2 showed the lowest shelf-life extension with greater growth of aerobic microorganisms, especially *Pseudomonas*. LAB prevailed in all other MAPs. Moreover, MAP with CO_2 or CO produced no additional antimicrobial effect over vacuum packaging (Viana et al., 2005). Huang et al. (2005) examined the retail shelf life of pork chops dipped in 500 ppm AA, 250 ppm citric acid, or no acid dip and stored at 1°C before simulated retail display in MAP with gas exchange or air-permeable packaging after vacuum pouch storage. The 80% N_2/20% CO_2 in MAP was exchanged with 80% O_2/20% CO_2, and chops were removed from vacuum packages and overwrapped with permeable film (VP-PVC) on the seventh day before simulated retail display at 4°C. Log numbers of psychrotrophic microorganisms were higher on VP-PVC samples than for chops in MAP on days 12 and 14. Psychrotrophic counts on AA-treated samples were decreased compared with citric acid or no dipping on pork during simulated retail display. Pork chops in MAP with gas exchange had decreased psychrotrophic counts when compared with conventional vacuum and overwrap packaging systems. Martinez et al. (2006c) investigated fresh pork sausages packaged in MAP with the following gas mixtures (%O_2/%CO_2/%N_2): 0/20/80, 0/20/80 + O_2

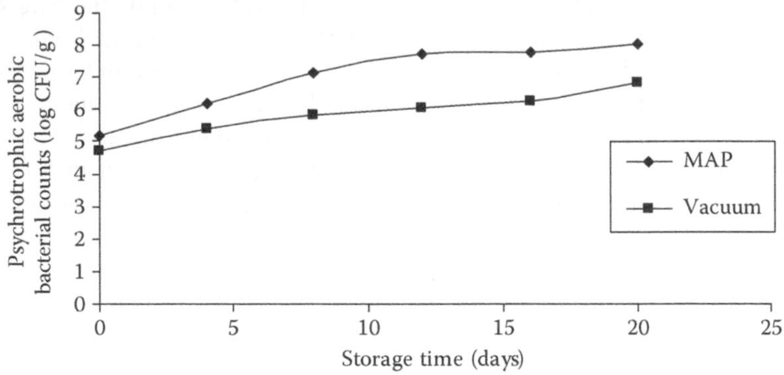

FIGURE 5.2 Effect of MAP (80% O_2/20% CO_2) (From Martinez, L. et al., *Meat Sci.*, 69, 493, 2005) and vacuum (From Martinez, L. et al., *Food Chem.*, 94, 219, 2006c) on the psychrotrophic aerobic bacterial counts of pork sausages stored at 2°C ± 1°C in the dark.

scavenger, 20/20/60, 40/20/40, 60/20/20, and 80/20/0. Samples were also packaged under vacuum and overwrapped with O_2-permeable film. They were stored for 20 days at 2°C ± 1°C in the dark. Aerobic psychrotroph counts on overwrapped sausages were found to be significantly higher than on any other sample throughout storage. Samples stored without oxygen, either under vacuum or in the presence of an O_2 scavenger, showed the lowest values. Samples stored without oxygen, either under vacuum or in the presence of an O_2 scavenger, showed the lowest values, which did not reach 10^7 CFU/g, even after 20 days of storage (Figure 5.2).

Fresh pork sausages treated with different natural antioxidants (rosemary, AA, and black pepper), packaged in 80% O_2/20% CO_2 atmosphere, and displayed at 2°C ± 1°C under different lightings (darkness, standard fluorescent, low-UV color-balanced lamp, and standard fluorescent plus a UV-filter) were investigated. The initial counts, corresponding to fresh pork sausages before packaging, were near 5 $\log_{10}$ CFU/g. From the fourth day onward, samples displayed under the standard fluorescent light showed higher counts of the total psychrotrophic aerobic bacteria than the rest of the sausages under different lighting conditions. The results revealed that microbial growth appeared to be enhanced by illumination with fluorescent light. Display under the standard fluorescent tube greatly decreased the shelf life of fresh pork sausages (6 days), whereas samples displayed under low-UV, fluorescent + UV-filter, or darkness presented a shelf life of 8–10 days (Martinez et al., 2007). Rubio et al. (2007b) studied salchichon, a dry fermented sausage, with high unsaturated fat content, packed under vacuum and MAP (20% CO_2/80% N_2). The sausages were made with pork meat and pork backfat obtained from pigs fed with three different diets (control diet-CO, high-oleic diet-HO, and high linoleic diet-HL). Salchichon type significantly affected the mesophilic aerobic bacteria, anaerobic bacteria, psychrotrophs, Enterococci, LAB, and micrococcaceae, but on the other hand, enterobacteria, pseudomonads, and yeast and mold counts were not affected. Furthermore, the lowest counts were found on CO salchichon. Lopez-Mendoza et al. (2007) inoculated raw ground pork with a strain of *Listeria monocytogenes*, and samples were distributed in lots. Half of the lots were stored aerobically, and the remaining ones were packaged using the MAP (30% CO_2+70% O_2). A different combination of nisin (N) and/or lactic acid (LA) was added to each lot (300 ppm N, 500 ppm N; 2% LA; 300 ppm N and 2% LA; 500 ppm N and 2% LA). All samples were stored at 4°C for 21 days (samples with MAP) or 7 days (samples stored aerobically). The inactivation of *L. monocytogenes* in raw ground pork stored aerobically was achieved mainly with the combination of 500 ppm N+LA (0.89 log); however, in samples with MAP, we could reduce the *L. monocytogene* population 3.45 log with the addition of LA, and the combination N+LA increases the inactivation of the other 0.5 log. Fresh pork sausages, overwrapped with oxygen-permeable films or packed under MAP (65% O_2 and 35% CO_2), immediately after production or after 1 day storage at 0°C, were stored either at 2°C in a domestic refrigerator or at 5°C in a refrigerated display cabinet beyond the commercial terms. Micrococcaceae were enabled to grow up to about 4 $\log_{10}$ CFU/g in both storage conditions. LAB increased up to about 4 $\log_{10}$ CFU/g in the refrigerator and more markedly to about 6 $\log_{10}$ CFU/g during storage at a higher temperature. Additionally, yeast and molds showed a similar trend to LAB, increasing more markedly in the display cabinet (Chiavaro et al., 2008). Zhang et al. (2009) studied the antimicrobial activity of mixed rosemary/liquorice extracts spray-applied to inoculated fresh pork packaged under MAP (80% O_2 and 20% CO_2) and stored at (4°C). The *L. monocytogenes* count on spice-treated pork decreased 2.9, 3.1, and 3.6 logs at the concentration of 2.5, 5.0, and 10.0 mg/mL, respectively, after 28 days storage, compared with the control. It was determined that the higher the concentration of extracts, the better the inhibition of *L. monocytogenes*. Furthermore, the growth of mesophilic aerobic bacteria, *Pseudomonas* spp., and total coliforms growth was also effectively inhibited by the combined extracts of rosemary and liquorice. Again, the inhibition of those microorganisms was dose dependent (Table 5.1).

5.3 EFFECT OF MAP ON BEEF

5.3.1 Quality

The use of atmospheres with low concentrations of CO (0.1%–1%), in combination with O_2 (24%), high CO_2 (50%), and N_2 (25%–25.9%), for preserving chilled beef steaks was studied by Luno et al. (2000). The atmosphere used as reference contained 70% O_2/20% CO_2/10% N_2. It was found that CO concentrations of 0.5%–0.75% were able to extend the shelf life by 5–10 days at 1°C ± 1°C, as demonstrated by delayed metmyoglobin formation (less than 40% of total myoglobin after 29 days of storage), stabilization of red color, maintenance of fresh meat odor, and significant slowing of lipid oxidation. Smiddy et al. (2002) studied beef stored under MAP (60% N_2/40% CO_2) and vacuum to determine if oxygen could be detected by the oxygen sensor in raw and cooked vacuum-packaged beef and in cooked MAP beef over various display periods and to also determine the impact that oxygen levels might have on lipid oxidation in beef. Oxygen contents detected ranged from 1.15% to 1.26% and 0.07% to 0.55% in MAP and vacuum-packed samples, respectively. Samples containing greatest levels of oxygen were most oxidized with the cooked samples being significantly more oxidized than raw samples. According to Djenane et al. (2002),

TABLE 5.1
Impact of MAP on Pork

Food Type	Gas Mix	Other Technology	Storage Temperature	Shelf Life	Sensory Assessment	Microflora	Quality	References
Fresh pork sausages	$\%O_2/\%CO_2/\%N_2$: 0/20/80		2°C ± 1°C in the dark		Preservation of color and odor of fresh pork sausages packaged in modified atmosphere was best achieved using atmospheres containing low CO_2 concentrations (20%) rather than high (60%)	Fresh sausages packaged in the atmosphere containing the highest oxygen and lowest carbon dioxide concentrations showed the highest counts of psychrotrophic aerobe counts	Increasing concentrations of CO_2 promoted oxidation of both myoglobin and lipids	Martinez et al. (2005)
Fresh pork sausages	$\%O_2/\%CO_2/\%N_2$: 0/60/40		2°C ± 1°C in the dark					
Fresh pork sausages	$\%O_2/\%CO_2/\%N_2$: 40/20/40		2°C ± 1°C in the dark					
Fresh pork sausages	$\%O_2/\%CO_2/\%N_2$: 40/60/0		2°C ± 1°C in the dark					
Fresh pork sausages	$\%O_2/\%CO_2/\%N_2$: 80/20/0		2°C ± 1°C in the dark					
Fresh pork sausages	0.3% CO/30% CO_2/rest argon		2°C ± 1°C in the dark					
Pork loin	Vacuum		5°C ± 0.5°C		Pork loins in 99% CO_2/1% CO MAP obtained the highest consumer acceptance scores after 24 h of storage. These samples and those treated with CO and then vacuum packaged received the greatest acceptance scores even after 20 days of storage		The 1% CO/99% CO_2 atmosphere was best for preserving the desirable pork loin color, and the L* and a* values remained similar to the fresh meat values using this MAP	Viana et al. (2005)
Pork loin	100% CO_2		5°C ± 0.5°C					
Pork loin	99% CO_2 and 1% CO		5°C ± 0.5°C					
Pork loin	100% O_2 or 100% CO followed by vacuum after 1 h of exposure		5°C ± 0.5°C					

Fresh pork sausages	$\%O_2/\%CO_2/\%N_2$: 0/20/80		2°C ± 1°C		Discoloration of fresh sausages markedly increased throughout storage following all packaging treatments, except those including no oxygen (vacuum and 0% O_2 plus scavenger) Oxygen-free samples were given the lowest scores	The higher the oxygen concentration, the higher were TBARs values and also their rate of increase. Fresh sausages packaged with 80% O_2 showed the highest values of TBARs. On the other hand, samples under vacuum and with O_2 scavenger had the lowest values	Martinez et al. (2006c)
Fresh pork sausages	$\%O_2/\%CO_2/\%N_2$: 0/20/80 + O_2 scavenger		2°C ± 1°C				
Fresh pork sausages	$\%O_2/\%CO_2/\%N_2$: 20/20/60		2°C ± 1°C				
Fresh pork sausages	$\%O_2/\%CO_2/\%N_2$: 40/20/40		2°C ± 1°C				
Fresh pork sausages	$\%O_2/\%CO_2/\%N_2$: 60/20/20		2°C ± 1°C				
Fresh pork sausages	$\%O_2/\%CO_2/\%N_2$: 80/20/0		2°C ± 1°C				
Cooked pork meat balls	Air	Rosemary	5°C			The combination of decreased O_2 atmosphere in the packages and addition of rosemary resulted in a significantly lower amount of TBARs	Huisman et al. (1994)
Cooked pork meat balls	5% O_2/95% N_2	Rosemary	5°C				
Cooked pork meat balls	3% O_2/97% N_2	Rosemary	5°C				
Cooked pork meat balls	1% O_2/99% N_2	Rosemary	5°C				
Cooked pork meat balls	100% N_2	Rosemary	5°C				

(*continued*)

TABLE 5.1 (continued)
Impact of MAP on Pork

Food Type	Gas Mix	Other Technology	Storage Temperature	Shelf Life	Sensory Assessment	Microflora	Quality	References
PSE and normal pork loins	100% Carbon dioxide containing 0%, 0.5%, and 1% residual oxygen		For 21 days, followed by 5 additional days under retail display conditions with access to air		Discoloration was observed on both PSE and normal meat with 0.5% O_2 and even more clearly with 1% O_2		The drip loss after MAP storage was twice as high from PSE meat as from normal meat. PSE meat was not more discolored after MAP storage than normal meat	Sørheim et al. (1997)
Raw ground pork	30% CO_2/70% O_2	Lactic acid	4°C			Reduction of *L. monocytogene* population by 3.45 log		Lopez-Mendoza et al. (2007)
Raw ground pork	30% CO_2/70% O_2	Lactic acid and nisin	4°C			Reduction of *L. monocytogene* population by 3.95 log		

Freshly slaughtered pork meat	5%, 15%, 25%, 35%, 45%, and 55% O_2 along with 20% CO_2. Nitrogen was used as a filler gas	4°C	The color score of the fresh meat increased with the increasing concentration of oxygen in the package	There was lower microbial count at high oxygen concentration. The maximum growth was observed at the O_2 concentrations of 15% and 25%	There was an increase in TBARs values with the increase in oxygen concentration of the package. TVBN decreased as the O_2 concentration increased. The TVBN value decreased sharply from 17.5 at 25% O_2 to 9.7 at 45% O_2	Zhang and Sundar (2005)
Fresh pork sausages	65% O_2 and 35% CO_2	2°C in a domestic refrigerator		Micrococcaceae were enabled to grow up to almost 4 $\log_{10}$ CFU/g for both storage conditions. LAB were initially 1 $\log_{10}$ CFU/g and increased up to about 4 $\log_{10}$ CFU/g in the refrigerator and		Chiavaro et al. (2008)
Fresh pork sausages	65% O_2 and 35% CO_2	5°C in a refrigerated display cabinet beyond the commercial terms				

(*continued*)

TABLE 5.1 (continued)
Impact of MAP on Pork

Food Type	Gas Mix	Other Technology	Storage Temperature	Shelf Life	Sensory Assessment	Microflora	Quality	References
						more markedly to about 6 log_{10} CFU/g during storage at a higher temperature. Yeast and molds showed similar LAB trends, increasing more markedly in the display cabinet		
Porcine *l. dorsi*	100% N_2		4°C				The breaking strength of porcine single muscle fibers did not differ significantly among the different gas compositions	Lund et al. (2008)
Porcine *l. dorsi*	Air		4°C					
Porcine *l. dorsi*	80% O_2/20% N_2		4°C					

Emulsion-type cooked pork sausage	CO_2 (100%), N_2 (100%), or 25% CO_2/75% N_2	0, 5, and 10 kGy	4°C	CO_2 or CO_2/N_2 packaging was effective for reducing residual nitrite and for inhibiting loss of red color of sausage compared with N_2 packaging	Ahn et al. (2003)
Fresh pork sausages	80% O_2 and 20% CO_2		2°C ± 1°C	Sausages with red yeast, red beetroot juice, and betanin had lower L* and h* and higher a* and a*/b* values than control samples	Martinez et al. (2006a,b)
Fresh pork sausages	80% O_2 and 20% CO_2	Red yeast rice powder (*M. purpureus*)	2°C ± 1°C		
Fresh pork sausages	80% O_2 and 20% CO_2	Crude red beetroot (*B. vulgaris*) juice	2°C ± 1°C	Color properties of sausages with red beetroot were the closest to control sausages	
Fresh pork sausages	80% O_2 and 20% CO_2	Commercial betanin (E-162)	2°C ± 1°C		

(*continued*)

TABLE 5.1 (continued)
Impact of MAP on Pork

Food Type	Gas Mix	Other Technology	Storage Temperature	Shelf Life	Sensory Assessment	Microflora	Quality	References
Porcine *l. dorsi*	70% O_2/30% CO_2		4°C				Tenderness was reduced in samples stored in high-oxygen atmosphere compared with packaging without oxygen. Myosin was found to form intermolecular cross-links due to packaging in a high-oxygen atmosphere and the amount of free thiols decreased. Protein carbonyl content was not affected by the packaging atmospheres	Lund et al. (2007a,b)

Fresh pork	80% O_2 and 20% CO_2	Mixed rosemary/liquorice extracts (2.5 mg/mL)	4°C	*L. monocytogene* count decreased 2.9 logs. *Pseudomonas* spp. count decreased 1.6 logs. Total coliform count decreased 0.6 logs	Zhang et al. (2009)
Fresh pork	80% O_2 and 20% CO_2	Mixed rosemary/liquorice extracts (5.0 mg/mL)	4°C	*L. monocytogene* count decreased 3.1 logs. *Pseudomonas* spp. count decreased 2.1 logs. Total coliform count decreased 0.8 logs	
Fresh pork	80% O_2 and 20% CO_2	Mixed rosemary/liquorice extracts (10.0 mg/mL)	4°C	*L. monocytogene* count decreased 3.6 logs. *Pseudomonas* spp. count decreased 2.6 logs. Total coliform count decreased 1.2 logs	

fresh beefsteaks were sprayed on the surface with vitamin C (500 ppm), taurine (50 mM), rosemary (1000 ppm), and vitamin E (100 ppm), the three latter in combination with 500 ppm of vitamin C, packaged in MAP (70% O_2+20% CO_2+10% N_2) and stored at 1°C ± 1°C for 29 days. Both combinations of vitamin C with either rosemary extract or taurine extended the shelf life of fresh beef steaks by about 10 days. Myoglobin oxidation and lipid oxidation were mostly delayed by rosemary in combination with vitamin C. The combination of vitamins E and C was significantly less effective than any other in delaying oxidation, but its effect was more intense than that of vitamin C alone. Treatment with rosemary combined with vitamin C resulted in the most intense red color at the end of storage. Furthermore, treatment with taurine and vitamin C had a lower protective effect of redness than rosemary, while the combination of vitamin E and vitamin C was less effective in maintaining red color (Figure 5.3).

Sanchez-Escalante et al. (2003) studied the combined effect of MAP (70% O_2/20% CO_2/10% N_2) with the addition of lycopene-rich tomato pulp (LRTP), oregano, and AA and their mixtures on the stability of beef patties stored at 2°C. The addition of LRTP showed antioxidant activity when it was mixed with AA and oregano. AA showed a poor antioxidant activity, since it maintained TBA values under 2.0 mg MA/kg only for 8 days. Lipid oxidation increased rapidly with increasing time in all samples, but the concentration of TBARs in antioxidant-containing samples was significantly lower than that in the control. The effect of oregano was concentration dependent, showing a higher inhibition at 500 ppm. However, the presence of 200 ppm oregano and AA exerted the same effect than 500 ppm oregano. The effect of two packaging cycles (double flush, 28 s; triple flush, 55 s) on the color stability of beef in retail trays with oxygen scavengers within a mother pack (50% CO_2/ 50% N_2) was determined by Isdell et al. (2003). Steaks from six muscles, *M. longissimus dorsi*, *M. psoas major*, *M. semimembranosus*, *M. gluteus medius*, *M. semitendinosus*, and *M. biceps femoris*, were examined. It was determined that both packaging cycles were equally acceptable in terms of visual and odor acceptability. Based on

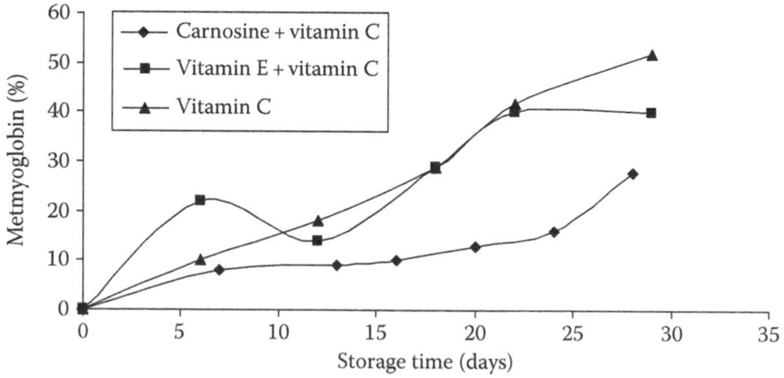

FIGURE 5.3 Metmyoglobin formation (%) of beef steaks treated with vitamin C (500 ppm), vitamin E (100 ppm) + vitamin C (500 ppm) (From Djenane, D. et al., *Food Chem.*, 76, 407, 2002), and with carnosine (50 mM) + vitamin C (500 ppm) (From Djenane, N. et al., *Food Chem.*, 85, 453, 2004) stored under MAP (70% O_2/20% CO_2/10% N_2) at 1°C ± 1°C.

reflectance measurements (R_{630}–R_{580}), *M. longissimus dorsi* and *M. semitendinosus* steaks had a display life of 4 days after storage for up to 6 weeks. *M. psoas major* and *M. semimembranosus* steaks had a display life as long as fresh controls after 4 weeks, whereas *M. gluteus medius* and *M. biceps femoris* steaks had a shorter display life than fresh steaks. Fresh beefsteaks, either sprayed on the surface with a solution of rosemary and vitamin C or not sprayed, were packaged in MAP (%O_2 + 20% CO_2 + 10% N_2) and displayed at 1°C ± 1°C without illumination or illuminated by a standard fluorescent lamp, a low-UV, color-balanced lamp, or a fluorescent lamp with a UV filter. Steaks not exposed to UV radiation had a* values that were significantly higher than those of steaks illuminated with a standard supermarket fluorescent lamp. Treatment with rosemary and vitamin C combined with the absence of UV radiation resulted in an intense red color at the end of display. TBARs formation was most intense in beef steaks displayed under conventional light. TBARs was strongly inhibited by treatment with rosemary together with AA, showing significant differences throughout the whole display time with untreated beef steaks, both displayed in the presence of UV radiation. UV-free lighting resulted in a significant inhibition of lipid oxidation; however, this effect was evident only after 10 days of display. Furthermore, the simultaneous action of the antioxidant mixture and the absence of UV radiation gave rise to a more pronounced inhibitory effect of TBARs formation from day 5 of display onward (Djenane et al., 2003) (Figure 5.4).

Sørheim et al. (2004) studied ground beef stored for 4 days in 60% CO_2/39.6% N_2/0.4% CO and vacuum or in 100% CO_2, 50% CO_2/50% N_2, 20% CO_2/80% N_2, 100% N_2, and vacuum. Cooking loss of ground beef patties was found to be higher of all CO_2 treatments than non-CO_2 treatments. Storage of raw ground beef in CO_2 caused a concentration-dependent decrease in raw meat pH of up to 0.12 units in

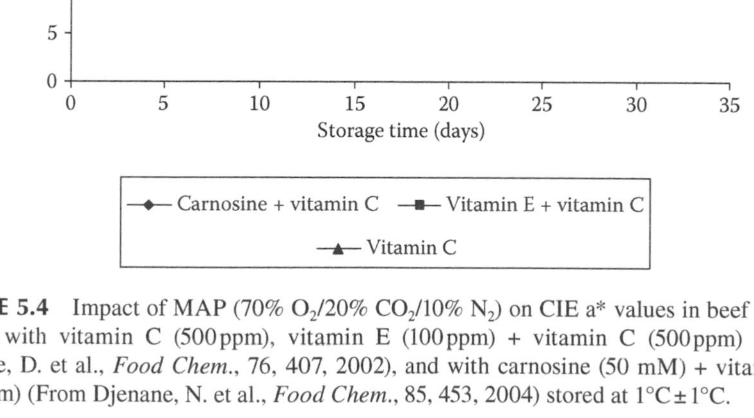

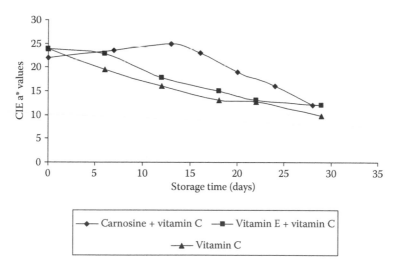

FIGURE 5.4 Impact of MAP (70% O_2/20% CO_2/10% N_2) on CIE a* values in beef steaks treated with vitamin C (500 ppm), vitamin E (100 ppm) + vitamin C (500 ppm) (From Djenane, D. et al., *Food Chem.*, 76, 407, 2002), and with carnosine (50 mM) + vitamin C (500 ppm) (From Djenane, N. et al., *Food Chem.*, 85, 453, 2004) stored at 1°C ± 1°C.

100% CO_2. Lastly, the hardness of cooked ground beef was not affected by CO_2 exposure. Seyfert et al. (2004) studied beef round muscles (injection-enhanced to 6%), packaged in high-oxygen (HiOx) or ultralow-oxygen (ULOx) modified atmospheres, stored 7 days and displayed 2 days (HiOx) or stored 16 days and displayed 1 day (ULOx) at 0°C, and cooked to 71.1°C. It was determined that the raw internal color for steaks in HiOx was lighter, redder, more yellow, and saturated, and had more oxymyoglobin and less deoxymyoglobin than steaks in ULOx. Moreover, the cooked internal color of steaks from HiOx appeared prematurely brown and was darker, less red, yellow, and saturated, and had more denatured myoglobin than steaks from ULOx. Hunt et al. (2004) compared the color of ground beef stored in 0.4% CO, 30% CO_2, and 69.6% N_2 (but removed from the modified atmosphere before display) with the product displayed immediately after packaging in PVC film, which contained only air. Storage of ground beef for up to 35 days in 0.4% CO resulted in typical initial bloomed color. Storage of ground beef in 0.4% CO decreased color stability in comparison to product exposed to only oxygen, whereas color life increased for tenderloin and inside *semimembranosus* muscles. The impact of carnosine (β-*alanine-L-histidine*) (50 mM), carnitine (50 mM) and L-ascorbic acid (500 ppm) solutions on the shelf life of fresh beef steaks packaged in modified atmosphere (70% O_2/20% CO_2/10% N_2) at 1°C ± 1°C was studied by Djenane et al. (2004). The results showed that the combination of carnosine with AA was the most effective in delaying metmyoglobin formation. Carnosine gave a significant inhibition of TBARs formation. A small inhibition of TBARs production was observed in samples treated with carnitine with respect to the control. On the other hand, the combination of carnitine and AA led to an absence of inhibition of TBARs formation. Beef steaks treated with a combination of carnosine and AA had lower off-odor and discoloration scores compared with other samples throughout the storage period, followed by carnosine and AA alone. Mancini et al. (2005) found that packaging atmospheres containing high levels of oxygen promote beef bone marrow discoloration, particularly in lumbar vertebrae. Compared with 80% O_2, removal of oxygen from packages will limit marrow discoloration during both storage of vertebrae (2 weeks at 4°C) and display of ribs (1 week at 1°C). Combining 0.4% CO with oxygen exclusion could promote a bright red lumbar vertebrae color for 6 weeks after packaging. The study of John et al. (2005) compared the effect of different packaging systems on beef steaks. Samples were either vacuum packaged or placed in the following MAPs: (1) 0.4% CO, 30.3% CO_2, and 69.3% N_2, or (2) 80% O_2 and 20% CO_2, and stored at 2°C. For steaks in a high-oxygen atmosphere, mean TBA values increased from 1.02 at day 7 to 4.40 after 21 days storage. However, steaks packaged in vacuum or CO did not show this large variation in TBA values during storage. Furthermore, premature browning and rancidity associated with steaks packaged in 80% O_2 was prevented by packaging in 0.4% CO or vacuum. Fresh minced beef muscle (*M. longissimus dorsi*) was supplemented with tea catechins (TC) at levels of 0, 200, 400, 600, 800, and 1000 mg/kg minced muscle. Treated samples were held in a refrigerated (4°C) display cabinet under aerobic or MAP (80:20, O_2:CO_2) conditions for 7 days. The addition of catechins significantly reduced lipid oxidation in fresh minced beef patties under both aerobic and MAP packaging conditions. The addition of catechins at a level of 200 mg/kg meat improved the color stability in beef patties and prolonged the color shelf life by 2 days. Catechins at levels 600, 800, and 1000 mg/kg did not show color stabilizing effect in beef patties (Tang et al., 2006). The quality aspects of sliced dry-cured beef “Cecina de Leon” preserved in vacuum and the following gas mixtures (20%/80% CO_2/N_2 and 80%/20% CO_2/N_2) were studied by Rubio et al. (2006). The samples were stored for 210 days at 6°C. Sensory parameters (color, odor, taste, hardness, juiciness, and acceptability) decreased during storage, and samples stored in 80%/20% CO_2/N_2 presented lower scores than samples stored in vacuum and 20%/80% CO_2/N_2 at the end of storage. No changes were observed in lightness (L*), redness (a*), and yellowness (b*) in vacuum- and gas-packaged samples throughout storage. Stetzer et al. (2007) determined how CO versus traditional high-oxygen MAP treatments affects beef quality. Strip loins were enhanced with a commercial phosphate–salt solution. Steaks were packaged in atmospheres containing either 0.4% CO/30% CO_2/69.6% N_2 (HiOx) or 80% O_2/20% CO_2 (HiOx), aged in the dark for 12 and 26 days, and placed in a lighted retail display case. After 14 days of storage, the evaluation revealed that CO treatment had some effect on raw color and surface sheen and on interior cooked color compared with the high-oxygen atmosphere. Rubio et al. (2007a) studied the effect of vacuum and MAP (20% CO_2/ 80% N_2 and 80% CO_2/20% N_2) on a dry-cured beef product called “Cecina de leon.” Packaged product was stored at 6°C for 210 days. No differences in pH and a_w values were found between the three systems of packaging at the end of storage. No differences in lightness (L*) and yellowness (b*) values were found between packaging systems at any storage time. Redness (a*) showed a pronounced initial fading within the first 15 days in the vacuum samples and within the second month in the gas-packaged samples. In the study, sodium erythorbate and AA were compared as a means to stabilize surface color of bone-in beef steaks in high-oxygen modified atmosphere (80% O_2 and 20% CO_2). Color parameters were determined after packaging (display at 1°C). Both the lumbar vertebrae and *longissimus lumborum* of bone-in strip loins were topically treated with either AA or sodium erythorbate (0%, 0.05%, 0.1%, 0.5%, 1.0%, or 1.5%, wt/wt basis). The results showed that sodium erythorbate was as effective as AA for inhibiting vertebrae discoloration. Either of the reducing agents at 0.5%, 1.0%, or 1.5% improved vertebrae redness when compared with 0%, 0.05%, and 0.1) (Mancini et al., 2007). The effect of rosemary extract and ascorbate/citrate (1:1) in combination with MAP (100% N_2, 80% O_2/20% N_2) on protein and lipid oxidation in minced beef patties during storage in the dark for up to 6 days at 4°C was investigated by Lund et al. (2007a). Both lipid and protein oxidation during storage was found to be increased due to the high levels of oxygen. The antioxidants used were not able to decrease the development of protein carbonyls in any of the packaging condition. Furthermore, the rosemary extract had no antioxidative effect on protein oxidation in either of the packaging conditions, while a slight increase in protein carbonyl content in beef with rosemary extract added and stored in 80% O_2/20% N_2 was found compared with beef without the addition of antioxidant. Ascorbate/citrate had a prooxidative effect on proteins in both atmospheres. In high-oxygen atmospheres, both antioxidants protected the fresh red meat color with ascorbate/citrate being more efficient than the rosemary extract, whereas no effect of antioxidant on meat color was found in beef patties stored in 100% N_2. The effects of aerobic packaging, vacuum, or MAP (65% N_2/35% CO_2) on the quality of Turkish pastirma were studied by Gok et al. (2008). With aerobic packaging, TBARs increased throughout storage up to day 120. TBARs in both vacuum and MAP increased initially from, respectively, 0.88 and 0.86 mg malonaldehyde/kg on day 0 to 1.66 and 1.54 mg malonaldehyde/kg on day 60, remaining relatively constant up to day 120. Over the storage time, the hexanal content increased, with aerobic packaging showing the greatest increase and MAP the lowest increase. MAP preserved color better than vacuum or aerobic packaging, which was reflected in color scores. Lund et al. (2008) studied the effect of high-oxygen atmosphere (80% O_2/20% N_2) packaging on mechanical properties of single muscle fibers from bovine *l. dorsi*. Storage of bovine *l. dorsi* for 48 h in the presence of oxygen significantly increased the breaking strength of single muscle fibers when compared with storage in a 100% N_2 atmosphere. The effect on cholesterol and lipid susceptibility to oxidation was investigated by Ferioli et al. (2008) in commercial minced beef held under MAP (80% O_2/20% CO_2). 7a-Hydroxycholesterol, 7b-hydroxycholesterol, and 7-ketocholesterol were the more abundant cholesterol oxidation products (COPs) identified. COPs significantly increased in raw beef during storage: after 1, 8, and 15 days since packaging COPs After 8 and 15 days at 4°C, COPs were detected at levels two and five times higher with respect to muscle stored for 1 day after packaging. Kozova et al. (2009) studied polyamine content changes in beef loin (*l. lumborum*) cuts. Samples were stored aerobically, vacuum packaged, and packaged in a MAP (70% N_2 and 30% CO_2, v/v). Only spermine was detected, and putrescine and spermidine contents were below the detection limits. In vacuum-packaged loins stored up to 21 days, the initial spermine level decreased. Specifically, on day 21, spermine content decreased to about 85% of the initial level. However, in samples stored under MAP, the decrease in spermine content was reduced to about 80% of the initial value at the end of storage (day 21). For air-packaged samples, spermine levels remained almost stable until the end of storage (day 9). Bovine steaks (*M. longissimus lumborum* and *M. psoas major*) were injection enhanced with L- or D-lactate and packaged under MAP with a high oxygen (80% O_2/ CO_2), stored 9 days at 2°C and then displayed for 5 days at 1°C. The results showed that enhancement with L-lactate resulted in less color deterioration and higher a* and chroma values than nonenhanced controls. Furthermore, it was shown that L-lactate enhancement significantly increases NADH concentration and total reducing activity of *M. longissimus lumborum* and *M. psoas*. It was concluded that L-lactate enhancement can be successfully used to improve muscles with lower color stability packaged under MAP with a high oxygen concentration (Kim et al., 2010). On the other hand, the study of Suman et al. (2009) examined the effect of lactate enhancement, on the internal cooked color of beef steaks (*l. lumborum* and psoas major muscles). The steaks were packaged under vacuum, MAP (80% O_2/20% CO_2), or CO-MAP (0.4% CO/19.6% CO_2/80% N_2) and stored at 1°C. It was determined that lactate-enhancement at 2.5% level resulted in darker interiors of cooked steaks. Moreover, interior cooked redness decreased during storage for steaks in vacuum and MAP, but it was stable for steaks in CO. This study provides evidence that a combination of lactate enhancement and CO MAP can minimize premature browning in whole-muscle beef steaks. Furthermore, the effects of L- or D-lactate on internal cooked color development of beef steaks packaged in high-oxygen (80% O_2/20% CO_2) MAP were investigated by Kim et al. (2009). Cooked steaks (70°C) enhanced with 2.5% L-lactate/phosphate maintained higher a*/b* ratios, lower hue values, higher total reducing activity and NADH concentration, and lower percent myoglobin denaturation than the controls and D-lactate injected samples. On the other hand, enhancement with 2.5% D-lactate did not affect cooked color, total reducing activity, NADH, or the percent myoglobin denaturation.

5.3.2 Microflora

Franco-Abuín et al. (1997) investigated the behavior of *L. monocytogenes* and *L. innocua* in raw minced meat packaged under MAP. The gas mixtures tested were 100% CO_2; 65% CO_2, 25% O_2, 10% N_2, and 20% CO_2, 80% O_2. Samples containing minced meat were inoculated or un-inoculated with *L. monocytogenes* and *L. innocua* and were stored at 4°C for 18 days. The results showed that the 100% CO_2 atmosphere was the most effective for the inhibition of growth of both species. None of the gas mixtures were found to be bactericidal. The levels of *L. innocua* recovered from all the modified atmospheres tested were always lower than those of *L. monocytogenes*. According to Gill and Badoni (2003), a 40 kg lot of manufacturing beef used for the production of ground beef products was divided into two batches. One batch was pasteurized by immersion in water at 85°C for 60 s; the other batch was not pasteurized. Both batches were then ground. The ground meat was packed in overwrapped trays, which were master packaged under a modified atmosphere of 70% O_2/ 30% CO_2. The master packs were stored at 2°C for up to 12 days. At the time of pack preparation and at 2 day intervals, a master pack containing pasteurized and another pack containing unpasteurized meat were opened and retail packs from each master pack were displayed at 4°C for 3 days. Samples for microbiological analysis were obtained at the times of opening master packs and at the end of display. After either a period of storage or a period of storage and display, the numbers of bacteria recovered from pasteurized meat were less than the numbers recovered from unpasteurized meat. The color of pasteurized meat was perceived as being paler than that of unpasteurized meat, but discoloration was similar or less, and retail appearance was similar or better for pasteurized than unpasteurized meat at all times. The odors of displayed, pasteurized meat were generally less intense and more acceptable than those of unpasteurized meat. Djenane et al. (2003) investigated the effect of treatment with a rosemary extract in combination with vitamin C, together with display under UV-free lighting, on the display life of beef steaks packaged under MAP (70% O_2/20% CO_2/10% N_2). Microbial growth was enhanced in steaks displayed under the conventional light, showing an increasing difference with that of steaks displayed under all other lighting conditions. Only small mostly nonsignificant differences were evident between those displayed in the dark and under UV-free lighting. Steaks displayed under conventional light and treated with rosemary and AA led to a significant decrease in microbial counts from day 15; differences were mostly not significant in steaks displayed in the dark or under UV-free lighting. In another study, beef patties were treated with AA, oregano extract, LRTP, and their mixtures and packaged in MAP (70% O_2/20% CO_2/10% N_2) at 2°C. All samples showed a significant increase in total psychrotrophic counts along storage. Differences among treatments were not significant, with the exception of samples treated with the oregano extract alone (500 ppm). The remaining samples containing oregano presented a similar behavior to that shown by samples containing LRTP and AA. The addition of LRTP, alone or mixed, did not significantly affect the psychrotrophic bacterial counts (Sanchez-Escalante et al., 2003). Xu et al. (2006) examined the influence of natural preservative and MAP on the shelf life of sliced beef ham stored at 9°C under fluorescence lighting conditions with vacuum-packaged products as controls. The results show that the number of Lactobacilli in the vacuum-packaged products grew in a high rate. Furthermore, it was found that the growth of LAB is not retarded in MAP (60%CO_2+40%N_2). The natural preservative (150 mg/kg nisin+0.3% tea polyphenols) had significant bacteriostatic activities against Lactobacilli, and the effects of bacteriostatic were better by dipping after slicing than that by incorporation before curing. The influence of storage period and packaging method on the microbiological quality of sliced dry-cured beef Cecina de Leon was studied by Rubio et al. (2006). Enterobacteria, enterococci, and pseudomonads counts were significantly inhibited under 20%/80% CO_2/N_2 and 80%/20% CO_2/N_2 compared with vacuum packaging. Mesophilic aerobic counts on Cecina de Leon slices packaged under vacuum increased at 90 days of storage. However, up to 60 days, no changes were observed in these counts in samples under 20%/80% CO_2/N_2 and a decrease was found when slices were packed under 80%/20% CO_2/N_2. A similar trend was observed for anaerobic and psychrotrophic counts. Rubio et al. (2007a) investigated the shelf life of commercial Cecina de Leon, a dry-cured beef product, packaged in vacuum and in CO_2/N_2 atmospheres (20%/80% CO_2/N_2 and 80%/20% CO_2/N_2). All samples were stored at 6°C for 210 days. Mesophilic aerobic, anaerobic, and psychrotrophic numbers on Cecina de Leon portions packaged under vacuum and in gas mixtures remained nearly constant (4–6 log CFU/g) during storage. Pseudomonad counts were significantly inhibited when the Cecina de Leon was packaged in gas mixtures. On the other hand, pseudomonad counts on Cecina de Leon packaged under vacuum showed no statistical differences at the end of storage compared to those obtained at the beginning. The effects of aerobic packaging, vacuum packaging, or MAP (65% N_2/35% CO_2) on a Turkish dry-cured beef product (pastirma) were investigated by Gok et al. (2008). Generally, pastirmas packaged under MAP had the lowest Enterobacteriacea counts followed by those packaged in vacuum and under air. Moreover, MAP gave lower yeast and mold counts than the other treatments. Lastly, MAP yielded lower *Pseudomonas* counts and LAB than either aerobic packaging or vacuum. The objective of the study of Ramamoorthi et al. (2009) was to evaluate the combined effects of irradiation and carbon monoxide in modified atmosphere packaging (CO-MAP) (0.4% CO/20% CO_2/79.6% N_2) on fresh beef loins stored at 4°C for 28 days. In both packaging conditions, irradiation at 0.5 and 1.0 kGy reduced the initial total plate count from about 5.3 to about 2 $\log_{10}$ CFU. However, 1.5 and 2.0 kGy reduced the total plate counts of samples below the detection limit. Furthermore, irradiation of CO-packaged samples at 0.5 and 1.0 kGy reduced total coliform counts to 3.1 and 1.6 $\log_{10}$ CFU/g, respectively. These counts did not change throughout storage. Finally, irradiation at 1.5 or 2.0 kGy gave total coliform counts below the detection limits during the whole storage period for air and CO-MAP samples (Table 5.2).

TABLE 5.2
Impact of MAP on Beef

Food Type	Gas Mix	Other Technology	Storage Temperature	Shelf Life	Sensory Assessment	Microflora	Quality	References
Chilled beef steaks	CO (0.1%–1%), in combination with O_2 (24%), high CO_2 (50%), and N_2 (25%–25.9%)		1°C±1°C		Maintenance of fresh meat odor (no variation of sensory score after 24 days)	Bacterial counts showed that all atmospheres greatly reduced total aerobic population numbers, including *B. thermosphacta*. LAB, were not affected	Significant slowing of lipid oxidation	Luno et al. (2000)
Raw minced beef	100% CO_2		4°C			The 100% CO_2 atmosphere was the most effective for the inhibition of growth of both species. None of the gas mixtures were bactericidal		Franco-Abuín et al. (1997)
Raw minced beef	65% CO_2, 25% O_2, 10% N_2		4°C					
Raw minced beef	20% CO_2, 80% O_2		4°C					
Sliced beef ham	60% CO_2/40% N_2	Natural preservative	9°C			The natural preservative has significant	a* values of products are not affected by natural preservative. MAP or	Xu et al. (2006)

(continued)

TABLE 5.2 (continued)
Impact of MAP on Beef

Food Type	Gas Mix	Other Technology	Storage Temperature	Shelf Life	Sensory Assessment	Microflora	Quality	References
		(150 mg/kg nisin + 0.3% tea polyphenols)				bacteriostatic activities against Lactobacilli	natural preservatives had little effect on lipid oxidation	
Fresh beef steaks	70% O_2 + 20% CO_2 + 10% N_2	Rosemary (1000 ppm) with vitamin C (500 ppm)	1°C ± 1°C			Lower counts during the whole period of storage	Most effective in delaying myoglobin oxidation and lipid oxidation; most intense red color at the end of storage, showing the highest a* values	Djenane et al. (2002)
Fresh beef steaks	70% O_2 + 20% CO_2 + 10% N_2	Vitamins E (100 ppm) and C (500 ppm)	1°C ± 1°C				Significantly less effective in delaying meat oxidation, though its effect was more intense than that of vitamin C alone	
Beef patties	70% O_2/20% CO_2/10% N_2	Treated with AA, oregano extract, LRTP, and their mixtures	2°C	Oregano extended the shelf life of beef patties by about 8 days. LRTP, even in	Odor of almost all samples, except the control and those with AA and LRTP alone, remained	All samples showed a significant increase in total psychrotrophic counts along storage. Differences among treatments were	The addition of LRTP showed antioxidant activity when it was mixed with AA and oregano. The formation of surface metmyoglobin was	Sanchez-Escalante et al. (2003)

				combination with oregano, gave rise to an extension of about 4 days	acceptable for 16 days	not significant; with the exception of samples treated with the oregano extract alone (500 ppm), samples containing oregano presented a similar behavior to that shown by samples containing LRTP and AA	significantly delayed by the addition of oregano alone and its mixtures with AA and/or LRTP. AA showed a poor antioxidant activity. The presence of 200 ppm oregano and AA exerted the same effect than 500 ppm oregano	
Bovine *l. dorsi*	80% O_2/20% N_2		4°C				Significantly increased the breaking strength of single muscle fibers when compared with storage in a 100% N_2 atmosphere	Lund et al. (2008)
Bovine *l. dorsi*	100% N_2		4°C					
Beef steaks	0.4% CO/30% CO_2/69.6% N_2		4°C				CO treatment had some effect on raw color and surface sheen and on interior cooked color compared with the high-oxygen atmosphere	Stetzer et al. (2007)
Beef steaks	80% O_2/20% CO_2		4°C					

(*continued*)

TABLE 5.2 (continued)
Impact of MAP on Beef

Food Type	Gas Mix	Other Technology	Storage Temperature	Shelf Life	Sensory Assessment	Microflora	Quality	References
Cecina de Leon (a dry-cured beef product)	20%/80% CO_2/N_2		6°C		The sensory properties of CL stored in 20%/80% CO_2/N_2 were slightly less acceptable than that of the samples packed under vacuum and under 80%/20% CO_2/N_2 at 210 days of storage	Pseudomonad counts were significantly inhibited	The pH and a_w values remained constant during storage. No differences in lightness (L*) and yellowness (b*) values were found between packaging systems at any storage time. Redness (a*) showed a pronounced initial fading within the first 15 days in the VP samples, and within the second month in the gas-packaged samples	Rubio et al. (2007a,b)
Cecina de Leon (a dry-cured beef product)	80%/20% CO_2/N_2		6°C			Pseudomonad counts were significantly inhibited		
Cecina de Leon (a dry-cured beef product)	Vacuum		6°C			Pseudomonad counts showed no differences at the end of storage compared to those obtained at the beginning		
Fresh beef steaks	70% O_2/20% CO_2/10% N_2	Carnosine (β-*alanine-L-histidine*) (50 mM)	1°C ± 1°C			Psychrotrophic aerobe counts in all samples gradually increased along storage (results not shown). They reached values	Significant inhibition of TBARs formation	Djenane et al. (2004)

						around 7 log_{10} CFU/ cm^2 in all samples after 28 days of storage		
Fresh beef steaks	70% O_2/20% CO_2/10% N_2	Carnitine (50 mM)	1°C ± 1°C		The combination of carnosine and AA were given lower scores than any other steaks throughout the whole storage period, both for off odor and discoloration, followed by carnosine and AA alone, which also differed significantly from the control	Psychrotrophic aerobe counts in all samples gradually increased along storage. They reached values around 7 log_{10} CFU/cm^2 in all samples after 28 days of storage	Small inhibition of TBARs production with carnitine with respect to the control	
Fresh beef steaks	70% O_2/20% CO_2/10% N_2	Carnosine (β-*alanine*-L-*histidine*) (50 mM) and L-ascorbic acid (500 ppm) solutions	1°C ± 1°C				Most effective in delaying metmyoglobin formation	
Fresh beef steaks	70% O_2/20% CO_2/10% N_2	Carnitine (50 mM) and L-ascorbic acid (500 ppm) solutions	1°C ± 1°C				Absence of inhibition of TBARs formation	
Beef steaks	Vacuum		2°C				TBA values increased	John et al. (2005)

(*continued*)

TABLE 5.2 (continued)
Impact of MAP on Beef

Food Type	Gas Mix	Other Technology	Storage Temperature	Shelf Life	Sensory Assessment	Microflora	Quality	References
Beef steaks	0.4% CO, 30.3% CO_2, and 69.3% N_2		2°C				from 1.02 at day 7 to 4.40 after 21 days storage. Steaks packaged in vacuum or CO did not exhibit this large variation in TBA values during storage	
Bone-in beef steaks	80% O_2 and 20% CO_2		2°C					
Bone-in beef steaks	80% O_2 and 20% CO_2	AA (0%, 0.05%, 0.1%, 0.5%, 1.0%, or 1.5%, wt/wt basis)	1°C				Sodium erythorbate was as effective as AA for inhibiting vertebrae discoloration. Either reducing agent at 0.5%, 1.0%, or 1.5% improved vertebrae redness compared with 0%, 0.05%, and 0.1%	Mancini et al. (2007)
Bone-in beef steaks	80% O_2 and 20% CO_2	Sodium erythorbate (0%, 0.05%, 0.1%, 0.5%, 1.0%, or 1.5%, wt/wt basis)	1°C					
Fresh minced beef muscle (*M. longissimus dorsi*)	Air	Supplementation with TC at levels of 0, 200, 400, 600, 800, and 1000 mg/kg	4°C				TC at levels of 200 mg/kg delayed the formation of MetMb. TC at levels 600, 800, and 1000 mg/kg did	Tang et al. (2006)

Fresh minced beef muscle (*M. longissimus dorsi*)	80:20, O_2:CO_2	Supplementation with TC at levels of 0, 200, 400, 600, 800, and 1000 mg/kg	4°C				not show color stabilizing effect in beef patties	
Turkish pastirma	Air		4°C		MAP preserved color better than VP or AP, which was reflected in color scores. Also, with increased storage time, taste scores decreased and became lowest on day 120. Texture, appearance, and acceptability scores showed similar decreasing trends with storage time	Samples packaged under MAP had the lowest Enterobacteriacea counts followed by those packaged in VP and AP. MAP had lower yeast and mold counts than AP or VP. MAP yielded lower *Pseudomonas* counts than the other packaging methods. MAP resulted in lower LAB than in any other treatment	TBARs increased throughout storage from 0.90 mg malonaldehyde/kg on day 0–2.80 mg malonaldehyde/kg on day 120.	Gok et al. (2008)
Turkish pastirma	Vacuum		4°C				TBARs increased initially from, respectively, 0.88 malonaldehyde/kg on day 0–1.66 malonaldehyde/kg on day 60, remaining relatively constant up to day 120	
Turkish pastirma	65% N_2/35% CO_2		4°C				TBARs increased initially from, respectively, 0.86 mg malonaldehyde/kg on day 0–1.54 mg malonaldehyde/kg on day 60, remaining relatively constant up to day 120	

(continued)

TABLE 5.2 (continued)
Impact of MAP on Beef

Food Type	Gas Mix	Other Technology	Storage Temperature	Shelf Life	Sensory Assessment	Microflora	Quality	References
Minced beef	80% O_2/20% CO_2		4°C				7a-Hydroxycholesterol, 7b-hydroxycholesterol, and 7-ketocholesterol were the more abundant COPs. Cooking did not affect cholesterol oxidation in freshly packaged minced beef but led to a rise in COPs	Ferioli et al. (2008)
Sliced dry-cured beef Cecina de Leon	20%/80% CO_2/N_2		6°C		Color, odor, taste, hardness, juiciness, and acceptability decreased during storage, and samples stored in 80%/20% CO_2/N_2 presented lower scores than samples stored in the	Microbial counts at 60 days of the gas-packaged samples were lower than the vacuum-packed ones; they were never higher than the spoilage limit (7 log ufc/g)	Slight increase in pH was observed throughout storage. No changes were observed in lightness (L*), redness (a*), and yellowness (b*) in vacuum- and gas-packaged samples during storage	Rubio et al. (2006)
Sliced dry-cured beef Cecina de Leon	80%/20% CO_2/N_2		6°C					
Sliced dry-cured beef Cecina de Leon	Vacuum		6°C					

					other two systems (vacuum and 20%/80% CO_2/N_2) at the end of storage		
Minced beef patties	100% N_2	Rosemary extract and ascorbate/ citrate (1:1)	4°C			No effect of antioxidant on meat color was found in beef patties stored in 100% nitrogen. The antioxidants used were not able to decrease the development of protein carbonyls	Lund et al. (2007a,b)
Minced beef patties	80% O_2/20% N_2	Rosemary extract and ascorbate/ citrate (1:1)	4°C			Increase in lipid and protein oxidation during storage. The antioxidants used were not able to decrease the development of protein carbonyls. Both antioxidants protected the fresh red meat color, with ascorbate/citrate being more efficient than the rosemary extract	

(continued)

TABLE 5.2 (continued)
Impact of MAP on Beef

Food Type	Gas Mix	Other Technology	Storage Temperature	Shelf Life	Sensory Assessment	Microflora	Quality	References
Fresh beef loins	Air	0, 0.5, 1.0, 1.5, or 2.0 kGy	4°C		The a* value of MAP samples was 40% higher than that of air-packaged samples	Irradiation at 0.5 and 1.0 kGy reduced initial TPC counts from about 5.3 to about 2 $\log_{10}$ CFU. 1.5 and 2.0 kGy, irradiation resulted in TPC below the detection level. Irradiation at 1.5 or 2.0 kGy resulted in total coliform counts below the detection limits throughout storage		Ramamoorthi et al. (2009)

5.4 EFFECT OF MAP ON LAMB MEAT

Soldatou et al. (2009) studied fresh Souvlaki-type lamb meat packaged under air, vacuum, and MAP (30% CO_2/70% N_2 and 70% CO_2/30% N_2) and stored at 4°C for 13 days. Souvlaki color parameters were not negatively affected by either vacuum or MAP conditions. MAP (70% CO_2/30% N_2) and vacuum were the most effective treatments for the inhibition of total viable counts, *Pseudomonas* spp., yeasts and *Brochothrix thermosphacta*. LAB and Enterobacteriaceae found in the microbial flora of Souvlaki, however, increased during storage under all treatments. The data clearly showed that the use of vacuum and MAP (70% CO_2/30% N_2) extended the shelf-life of samples by approximately 4–5 days, compared with samples stored under air.

5.5 CONCLUSIONS

One of the most important applications of MAP falls in the field of meat and meat products. This particular application is closely linked to undesirable coloration that may negatively affect the consumer's behavior toward purchasing this meat. Implementation of MAP was also effective in restricting the growth of microorganisms such as *Pseudomonas* spp., yeasts, and *B. thermosphacta*. Combination of MAP with other preservation technologies such as irradiation and antioxidants (lycopene, vitamin A and C, oregano) considerably reduced the oxidation of meat/meat lipid, thereby suggesting that application of the so-called hurdle technology is beneficial to the packaged meat/meat product.

REFERENCES

Ahn, H.-J., Jo, C., Lee, J.W., Kim, J.H., Kim, K.H., and Byun, M.W. 2003. Irradiation and modified atmosphere packaging effects on residual nitrite, ascorbic acid, nitrosomyoglobin, and color in sausage. *Journal of Agricultural and Food Chemistry*, 51: 1249–1253.

Brewer, M.S., Jensen, J., Prestat, C., Zhu, L.G., and McKeith, F.K. 2002. Visual acceptability and consumer purchase intent of pumped pork loin roasts. *Journal of Muscle Foods*, 13(1): 53–68.

Brody, A.L. 1989. *Controlled Modified Atmosphere Vacuum Packaging of Food*. Trumbull, CT: Food & Nutrition Press Inc.

Chiavaro, E., Zanardi, E., Bottari, E., and Ianieri, A. 2008. Efficacy of different storage practices in maintaining the physicochemical and microbiological properties of fresh pork sausage. *Journal of Muscle Foods*, 19: 157–174.

Church, N. 1994. Developments in modified-atmosphere packaging and related technologies: A review. *Trends in Food Science & Technology*, 5: 345–352.

De Santos, F., Rojas, M., Lockhorn, G., and Brewer, M.S. 2007. Effect of carbon monoxide in modified atmosphere packaging, storage time and endpoint cooking temperature on the internal color of enhanced pork. *Meat Science*, 77: 520–528.

Djenane, D., Martinez, L., Sanchez-Escalante, A., Beltran, J.A., and Roncales, P. 2004. Antioxidant effect of carnosine and carnitine in fresh beef steaks stored under modified atmosphere. *Food Chemistry*, 85: 453–459.

Djenane, D., Sanchez-Escalante, A., Beltran, J.A., and Roncales, P. 2002. Ability of α-tocopherol, taurine and rosemary, in combination with vitamin C, to increase the oxidative stability of beef steaks packaged in modified atmosphere. *Food Chemistry*, 76: 407–415.

Djenane, D., Sanchez-Escalante, A., Beltran, J.A., and Roncales, P. 2003. Extension of the shelf life of beef steaks packaged in a modified atmosphere by treatment with rosemary and displayed under UV-free lighting. *Meat Science*, 64: 417–426.

Ferioli, F., Caboni, M.F., and Dutta, P.C. 2008. Evaluation of cholesterol and lipid oxidation in raw and cooked minced beef stored under oxygen-enriched atmosphere. *Meat Science*, 80(3): 681–685.

Franco-Abuín, C.M., Rozas-Barrero, J., Romero-Rodríguez, M.A., Cepeda-Sáez, A., and Fente-Sampayo, C. 1997. Effect of modified atmosphere packaging on the growth and survival of *Listeria* in raw minced beef. *Food Science and Technology International*, 3(4): 285–290.

Gill, C.O. 1996. Extending the storage life of raw chilled meats. *Meat Science*, 43(Supplement): 99–109.

Gill, C.O. and Badoni, M. 2003. Effects of storage under a modified atmosphere on the microbiological and organoleptic qualities of ground beef prepared from pasteurized manufacturing beef. *International Journal of Food Science and Technology*, 38: 233–240.

Gok, V., Obuz, E., and Akkaya, L. 2008. Effects of packaging method and storage time on the chemical, microbiological, and sensory properties of Turkish pastirma—A dry cured beef product. *Meat Science*, 80(2): 335–344.

Huang, N.-Y., Ho, C.-P., and McMillin, K.W. 2005. Retail shelf-life of pork dipped in organic acid before modified atmosphere or vacuum packaging. *Journal of Food Science*, 70(8): 382–387.

Huisman, M., Madsen, H.L., Skibsted, L.H., and Bertelsen, G. 1994. The combined effect of rosemary (*Rosmarinus officinalis* L.) and modified atmosphere packaging as protection against warmed over flavour in cooked minced pork meat. *Zeitschrift fuȑ LebensmittelUntersuchung und-Forschung A*, 198(1): 57–59.

Hunt, M.C., Mancini, R.A., Hachmeister, K.A., Kropf, D.H., Merriman, M., Delduca, G., and Milliken, G. 2004. Carbon monoxide in modified atmosphere packaging affects color, shelf life, and microorganisms of beef steaks and round beef. *Journal of Food Science*, 69(1): 45–52.

Isdell, E., Allen, P., Doherty, A., and Butler, F. 2003. Effect of packaging cycle on the colour stability of six beef muscles stored in a modified atmosphere mother pack system with oxygen scavengers. *International Journal of Food Science and Technology*, 38: 623–632.

John, L., Cornforth, D., Carpenter, C.E., Sorheim, O., Pettee, B.C., and Whittier, D.R. 2005. Color and thiobarbituric acid values of cooked top sirloin steaks packaged in modified atmospheres of 80% oxygen, or 0.4% carbon monoxide, or vacuum. *Meat Science*, 69: 441–449.

Kim, Y.H., Keeton, J.T., Hunt, M.C., and Savell, J.W. 2010. Effects of L- or D-lactate enhancement on the internal cooked colour development and biochemical characteristics of beef steaks in high oxygen modified atmosphere. *Food Chemistry,* 119(3): 918–922.

Kim, Y.H., Keeton, J.T., Yang, H.S., Smith, S.B., Sawyer, J.E., and Savell, J.W. 2009. Color stability and biochemical characteristics of bovine muscles when enhanced with L- or D-potassium lactate in high-oxygen modified atmospheres. *Meat Science*, 82: 234–240.

Kozova, M., Kalac, P., and Pelikanova, T. 2009. Changes in the content of biologically active polyamines during beef loin storage and cooking. *Meat Science*, 81: 607–611.

Krause, T.R., Sebranek, J.G., Rust, R.E., and Honeyman, M.S. 2003. Use of carbon monoxide packaging for improving the shelf life of pork. *Journal of Food Science*, 68(8): 2596–2603.

Lopez-Mendoza, M.C., Ruiz, P., and Mata, C.M. 2007. Combined effects of nisin, lactic acid and modified atmosphere packaging on the survival of *Listeria monocytogenes* in raw ground pork. *International Journal of Food Science and Technology*, 42: 562–566.

Lund, M.N., Christensen, M., Fregil, L., Hviid, M.S., and Skibsted, L.H. 2008. Effect of high-oxygen atmosphere packaging on mechanical properties of single muscle fibres from bovine and porcine *Longissimus dorsi*. *European Food Research and Technology*, 227(5): 1323–1328.

Lund, M.N., Hviid, M.S., and Skibsted, L.H. 2007a. The combined effect of antioxidants and modified atmosphere packaging on protein and lipid oxidation in beef patties during chill storage. *Meat Science*, 76: 226–233.

Lund, M.N., Lametsch, R., Hviid, M.S., Jensen, O.N., and Skibsted, L.H. 2007b. High-oxygen packaging atmosphere influences protein oxidation and tenderness of porcine *Longissimus dorsi* during chill storage. *Meat Science*, 77: 295–303.

Luno, M., Roncalés, P., Djenane, D., and Beltrán, J.A. 2000. Beef shelf life in low O_2 and high CO_2 atmospheres containing different low CO concentrations. *Meat Science*, 55(4): 413–419.

Mancini, R.A., Hunt, M.C., Hachmeister, K.A., Kropf, D.H., and Johnson, D.E. 2005. Exclusion of oxygen from modified atmosphere packages limits beef rib and lumbar vertebrae marrow discoloration during display and storage. *Meat Science*, 69: 493–500.

Mancini, R.A., Hunt, M.C., Seyfert, M., Kropf, D.H., Hachmeister, K.A., Herald, T.J., and Johnson, D.E. 2007. Comparison of ascorbic acid and sodium erythorbate: Effects on the 24h display colour of beef lumbar vertebrae and *Longissimus lumborum* packaged in high-oxygen modified atmospheres. *Meat Science*, 75: 39–43.

Martinez, L., Cilla, I., Beltran, J.A., and Roncales, P. 2006a. Comparative effect of red yeast rice (*Monascus purpureus*), red beet root (*Beta vulgaris*) and betanin (E-162) on colour and consumer acceptability of fresh pork sausages packaged in a modified atmosphere. *Journal of the Science of Food and Agriculture*, 86: 500–508.

Martinez, L., Cilla, I., Beltran, J.A., and Roncales, P. 2006b. Antioxidant effect of rosemary, borage, green tea, pu-erh tea and ascorbic acid on fresh pork sausages packaged in a modified atmosphere: Influence of the presence of sodium chloride. *Journal of the Science of Food and Agriculture*, 86: 1298–1307.

Martinez, L., Cilla, I., Beltran, J.A., and Roncales, P. 2007. Effect of illumination on the display life of fresh pork sausages packaged in modified atmosphere: Influence of the addition of rosemary, ascorbic acid and black pepper. *Meat Science*, 75: 443–450.

Martinez, L., Djenane, D., Cilla, I., Beltran, J.A., and Roncales, P. 2005. Effect of different concentrations of carbon dioxide and low concentration of carbon monoxide on the shelf-life of fresh pork sausages packaged in modified atmosphere. *Meat Science*, 71: 563–570.

Martinez, L., Djenane, D., Cilla, I., Beltran, J.A., and Roncales, P. 2006c. Effect of varying oxygen concentrations on the shelf-life of fresh pork sausages packaged in modified atmosphere. *Food Chemistry*, 94: 219–225.

McMullen, L.M. and Stiles, M.E. 1989. Storage life of selected meat sandwiches at 4°C in modified gas atmospheres. *Journal of Food Protection*, 52: 792–798.

O'Connor-Shaw, R.E. and Reyes, V.G. 2000. Use of modified atmosphere packaging. In: *Encyclopedia of Food Microbiology* (eds. R.K. Robinson, C.A. Batt, and P.D. Patel), pp. 410–416. London, U.K.: Academic Press.

Raines, C.R., Dikeman, M.E., Grobbel, J.P., and Yancey, E.J. 2006. Effects of ascorbic acid and Origanoxe in different packaging systems to prevent pork lumbar vertebrae discoloration. *Meat Science*, 74: 267–271.

Ramamoorthi, L., Toshkov, S., and Brewer, M.S. 2009. Effects of carbon monoxide-modified atmosphere packaging and irradiation on *E. coli* K12 survival and raw beef quality. *Meat Science*, 83(3): 358–365.

Rubio, B., Martinez, B., Gonzalez-Fernandez, C., Garcia-Cachan, M.D., Rovira, J., and Jaime, I. 2006. Influence of storage period and packaging method on sliced dry cured beef "Cecina de Leon": Effects on microbiological, physicochemical and sensory quality. *Meat Science*, 74: 710–717.

Rubio, B., Martinez, B., Gonzalez-Fernandez, C., Garcia-Cachan, M.A., Rovira, J., and Jaime, I. 2007a. Effect of modified atmosphere packaging on the microbiological and sensory quality on a dry cured beef product: "Cecina de leon." *Meat Science*, 75: 515–522.

Rubio, B., Martinez, B., Sanchez, M.J., Garcνa-Cachan, M.D., Rovira, J., and Jaime, I. 2007b. Study of the shelf life of a dry fermented sausage "salchichon" made from raw material enriched in monounsaturated and polyunsaturated fatty acids and stored under modified atmospheres. *Meat Science*, 76: 128–137.

Sanchez-Escalante, A., Torrescano, G., Djenane, D., Beltran, J.A., and Roncales, P. 2003. Combined effect of modified atmosphere packaging and addition of lycopene rich tomato pulp, oregano and ascorbic acid and their mixtures on the stability of beef patties. *Food Science and Technology International*, 9(2): 77–84.

Seyfert, M., Hunt, M.C., Mancini, R.A., Kropf, D.H., and Stroda, S.L. 2004. Internal premature browning in cooked steaks from enhanced beef round muscles packaged in high-oxygen and ultra-low oxygen modified atmospheres. *Journal of Food Science*, 69(2): 142–146.

Shaw, B.G. and Harding, C.D. 1984. A numerical taxonomic study of lactic acid bacteria from vacuum-packed beef, pork, lamb, and bacon. *Journal of Applied Bacteriology*, 56: 25–40.

Smiddy, M., Fitzgerald, M., Kerry, J.P., Papkovsky, D.B., O'Sullivan, C.K., and Guilbault, G.G. 2002. Use of oxygen sensors to non-destructively measure the oxygen content in modified atmosphere and vacuum packed beef: Impact of oxygen content on lipid oxidation. *Meat Science*, 61: 285–290.

Soldatou, N., Nerantzaki, A., Kontominas, M.G., and Savvaidis, I.N. 2009. Physicochemical and microbiological changes of "Souvlaki"—A Greek delicacy lamb meat product: Evaluation of shelf-life using microbial, colour and lipid oxidation parameters. *Food Chemistry*, 113: 36–42.

Sørheim, O., Erlandsen, T., Nissen, H., Lea, P., and Høyem, T. 1997. Effects of modified atmosphere storage on colour and microbiological shelf life of normal and pale, soft and exudative pork. *Meat Science*, 47(1–2): 147–155.

Sørheim, O., Kropf, D.H., Hunt, M.C., Karwoski, M.T., and Warren, K.E. 1996. Effects of modified gas atmosphere packaging on pork loin colour, display life and drip loss. *Meat Science*, 43(2): 203–212.

Sørheim, O., Ofstad, R., and Lea, P. 2004. Effects of carbon dioxide on yield, texture and microstructure of cooked ground beef. *Meat Science*, 67: 231–236.

Stetzer, A.J., Wicklund, R.A., Paulson, D.D., Tucker, E.M., Macfarlane, B.J., and Brewer, M.S. 2007. Effect of carbon monoxide and high oxygen modified atmosphere packaging (MAP) on quality characteristics of beef strip steaks. *Journal of Muscle Foods*, 18: 56–66.

Suman, S.P., Mancini, R.A., Ramanathan, R., and Konda, M.R. 2009. Effect of lactate enhancement, modified atmosphere packaging, and muscle source on the internal cooked colour of beef steaks. *Meat Science*, 81: 664–670.

Tang, S.Z., Ou, S.Y., Huang, X.S., Li, W., Kerry, J.P., and Buckley, D.J. 2006. Effects of added tea catechins on colour stability and lipid oxidation in minced beef patties held under aerobic and modified atmospheric packaging conditions. *Journal of Food Engineering*, 77: 248–253.

Veberg, A., Sorheim, O., Moan, J., Iani, V., Juzenas, P., Nilsen, A.N., and Wold, J.P. 2006. Measurement of lipid oxidation and porphyrins in high oxygen modified atmosphere and vacuum-packed minced turkey and pork meat by fluorescence spectra and images. *Meat Science*, 73: 511–520.

Viana, E.S., Gomide, L.A.M., and Vanetti, M.C.D. 2005. Effect of modified atmospheres on microbiological, color and sensory properties of refrigerated pork. *Meat Science*, 71: 696–705.

Wicklund, R.A., Paulson, D.D., Tucker, E.M., Stetzer, A.J., DeSantos, F., Rojas, M., MacFarlane, B.J., and Brewer, M.S. 2006. Effect of carbon monoxide and high oxygen modified atmosphere packaging and phosphate enhanced, case-ready pork chops. *Meat Science*, 74: 704–709.

Xu, B., Liu, Z., Ren, F., and Sun, Y. 2006. Effects of natural preservative and modified atmosphere packaging on shelf-life of sliced beef ham. *Nongye Gongcheng Xuebao/Transactions of the Chinese Society of Agricultural Engineering*, 22(3): 143–147.

Zhang, H., Kong, B., Xiong, Y.L., and Sun, X. 2009. Antimicrobial activities of spice extracts against pathogenic and spoilage bacteria in modified atmosphere packaged fresh pork and vacuum packaged ham slices stored at 4°C. *Meat Science*, 81: 686–692.

Zhang, M. and Sundar, S. 2005. Effect of oxygen concentration on the shelf-life of fresh pork packed in a modified atmosphere. *Packaging Technology and Science*, 18: 217–222.

6 Poultry

Ioannis S. Arvanitoyannis and
Alexandros Ch. Stratakos

CONTENTS

6.1 INTRODUCTION

Consumption of chicken has increased in many countries due to its beneficial health effects and nutritional value (Chouliara et al., 2007). Poultry meat is very susceptible to spoilage, and its stability and microbiological safety are based on a combination of various factors in order not to be contaminated by microorganisms (Chun et al., 2009). The spoilage of fresh poultry products, apart from being an economic burden to the producer, may also cause health problems, since poultry meat is likely to contain pathogenic microorganisms (Geornaras et al., 1998). The main bacterial pathogens occurring in chicken are *Escherichia coli*, *Salmonella* spp., *Listeria monocytogenes*, and *Campylobacter jejuni*, present in the intestinal microflora of the chicken (Anang et al., 2007). Application of MAP has the potential to suppress microbial growth and extend the shelf life of foods (Zardetto, 2005). According to Brody (1996), "nearly a third of all fresh poultry in North America is master-packaged in bulk under modified atmospheres for distribution to retail grocery and hotel, restaurant and institutional outlets." The success of MAP depends on several factors such as gas mixture composition, initial microbial meat quality, temperature control, packaging properties, and the efficiency of equipment used (Taylor, 1996).

The objective of the present chapter is to provide a review of the data available on the effect of MAP, on chemical and microbiological quality of poultry meat, either on its own or in conjunction with other preservation methodologies.

6.2 EFFECT OF MAP ON THE QUALITY OF TURKEY

According to Bagorogoza et al. (2001) "fresh skinless turkey breasts packaged under air or nitrogen were either irradiated (2.4–2.9 kGy) or not and stored at 2°C for 3 days. Irradiation treatment had strong impact on color, odor, flavor, and the levels of α-tocopherol as well. Irradiated samples had a more intense pink color and acrid/irradiation odor. Alpha-tocopherol levels were reduced because of irradiation treatment. Moreover, irradiated cooked turkey had less turkey flavor, but stronger metallic flavor than nonirradiated samples. After the third day of storage, cooked samples originally packaged in air displayed a low intensity stale flavor that was absent in nitrogen-packaged samples." Mechanically deboned turkey meat (MDTM) was stored in different packaging materials (film produced with natural antioxidant (α-tocopherol) or synthetic antioxidant) at −20°C for 12 months in vacuum (VP), MA, or air. One-half of the samples were thawed at 4°C for 24 h after 1 month of storage and then refrozen. Results revealed that VP and MAP samples had lower thiobarbituric acid reactive substances (TBARS) values and hexanal content than air-packaged samples. Hexanal content and TBARS values increased with storage time, and the highest levels were recorded after 6 months of storage. The largest increase was obtained in the presence of oxygen. MDTM stored in packages where a natural antioxidant (α-tocopherol) was used in production of one of the PE layers displayed, in almost every instance, the lowest TBARS values and hexanal content when stored in VP or MAP. Furthermore, TBARS values and hexanal content showed no dependence of the temperature profile over storage (Pettersen et al., 2004). According to Ntzimani et al. (2008), low levels of BAs were recorded throughout the entire storage period, with the exception of histamine, tyramine, and tryptamine for smoked turkey fillets over storage under air, VP, skin, and two modified atmospheres MAP_1 (30% CO_2/70% N_2) and MAP_2 (50% CO_2/50% N_2), at 4°C ± 0.5°C, for a period of 30 days. Values for these three BAs were the highest for air-packaged samples (32.9, 25.0, and 4.1 mg/kg, respectively) and the lowest for skin-packaged samples (11.9, 4.3, and 2.8 mg/kg, respectively) after 30 days storage. Although BAs are not considered to be responsible for spoilage off-odors in smoked turkey fillets, the obtained results revealed that their concentrations, especially those of histamine, tyramine, and tryptamine, correlated well with the development of off-odors. Tryptamine, histamine, and tyramine could be used as chemical indicators of turkey meat spoilage. Furthermore, odor analysis revealed that the shelf life of all turkey samples, irrespectively of MAP conditions, amounted to approximately 27–30 days with slightly better sensory characteristics obtained for skin-packaged samples. On the contrary, the shelf life of samples packaged in air was 22–23 days. Fraqueza et al. (2008) investigated the shelf life of turkey meat of different color categories (such as pale, soft, exudative (PSE)-like, intermediate, and dark) packaged in air and MA (50% CO_2 and 50% N_2) (0°C ± 1°C). Over storage, light color meat exhibited no increase in TVB-N but on the other hand intermediate and dark meat showed a significant increase from day 5 to day 12 of storage. This trend was much more clear in the case of dark meat. Sliced turkey meat packaged with 50% CO_2 and 50% N_2 did not reveal any significant differences of TVB-N either between different color categories or during storage time. The results obtained clearly demonstrated that the shelf life of sliced turkey meat under MAP was 20 days for intermediate and light color meat and 13 days for dark meat.

6.3 EFFECT OF MAP ON THE MICROFLORA OF TURKEY

Juneja et al. (1996) studied *Clostridium perfringens* occurring in samples of sterile ground turkey to assess its growth under MAP. Samples were packaged under various atmospheres ($CO_2/O_2/N_2$: 75/5/20, 75/10/15, 75/20/5, 25/20/55, 50/20/30) and stored at 4°C, 15°C, and 28°C. Cyclic and static temperature abuse of the product was also investigated. The study revealed that the growth of *C. perfringens* was the slowest under 25%–50% CO_2/20% O_2/balance N_2 at 15°C and 28°C. At 4°C no growth was reported for up to 28 days. Temperature abuse (storage at 28°C) of refrigerated products for 8 h did not allow the *C. perfringens* growth. It was shown that the application of 25%–50% CO_2/20% O_2/balance N_2 may extend the shelf life of turkey, but it cannot be relied upon to eliminate the risk of *C. perfringens* food poisoning if the proper refrigeration conditions are not applied. Ground turkey was gamma irradiated at 5°C to 0, 1.5, and 2.5 kGy and inoculated (~100 CFU/g) after irradiation with a cocktail of *L. monocytogenes* ATCC 7644, 15313, 49594, and 43256. The meat was then packaged in air-permeable pouches or under atmospheres containing 30% or 53% CO_2, 19% O_2, and 51% or 24% N_2 and stored at 7°C for up to 28 days. A dose of 2.5 kGy prolonged the time for the total plate count (TPC) to reach 10^7 CFU/g from 4 to 19 days compared with that for nonirradiated turkey in air-permeable pouches. Moreover, *L. monocytogenes* did not grow more rapidly over storage at 7°C on irradiated than on nonirradiated raw ground turkey, and there was a concentration-dependent inhibition of its growth by CO_2 (Thayer and Boyd, 2000). Lawlor et al. (2000) investigated the ability of nonproteolytic *Clostridium botulinum* type B spores to grow and produce toxin in cooked, uncured turkey packaged under MAP at refrigeration and mild to moderate abuse temperatures. Cook-in-bag turkey breast were packaged in O_2-impermeable bags under two MAs (100% N_2 and 30% CO_2:70% N_2) and stored at 4°C, 10°C, and 15°C. At 10°C and 15°C, toxin was detected by day 14 and day 7, without being affected by packaging atmosphere composition. At 4°C, toxin was detected by day 14 in samples packaged under 100% N_2 and by day 28 in samples packaged under 30% CO_2:70% N_2. It is noteworthy that in any storage temperature, toxin detection preceded or coincided with development of spoilage sensory characteristics, thereby revealing the potential for consumption of toxic products when evidence of spoilage is not observed. According to Mano et al. (2000) "*Aeromonas hydrophila* was inoculated onto turkey meat slices. Inoculated and control samples were packaged in MA (100% N_2, 20/80 and 40/60 CO_2/O_2) or in air in plastic bags (1°C and 7°C). Packaging in MAP resulted in a strong inhibition of bacterial growth at 18°C, particularly in samples stored under CO_2/O_2 atmospheres. *A. hydrophila* grew on turkey meat stored in 100% N_2 at 1°C and 7°C. Likewise, growth of this microorganism was detected on turkey stored in 20/80 CO_2/O_2 at 7°C. No growth was observed in turkey samples stored under 40/60 CO_2/O_2 at both temperatures." On the other hand, *Brocthothrix thermosphacta* grew in all investigated atmospheres and temperatures, except for turkey meat at 7°C in 40%/60% CO_2/O_2. Pexara et al. (2002) examined the shelf life of cured, cooked, sliced turkey fillets stored under vacuum (UV) and MAP (80% CO_2/20% N_2; 60% CO_2/20% O_2/20% N_2; 0.4% CO/80% CO_2/rest N_2; 1% CO/80% CO_2/rest N_2; 0.5% CO/24% O_2/50% CO_2/ rest N_2 and 100% N_2) at 4°C and 10°C. It was shown that the microbial population was nearly only lactic acid bacteria (LAB). Moreover, the study revealed that the average shelf life was 2 and 1 weeks at 4°C and 10°C, respectively. No differences between VP and 80% CO_2/20% N_2 were found after 25 days at 4°C. As revealed by the other five tested MAP stored samples, insignificant differences in growth of LAB were reported except for the mixture (60% O_2/20% CO_2/20% N_2) that displayed the most rapid growth. The other applied mixtures had all the same effects as VP on the growth of the spoilage flora at 4°C. An inhibitory effect of the MA was found for 0.5% CO/50% CO_2/24% O_2/N_2. Gas mixture composition did not affect the growth at 10°C in comparison with the growth at 4°C. However, the lowest total count and LAB count at 10°C were observed in 100% N_2. The study of Dhananjayan et al. (2006) assessed how deep MAP bactericidal effects penetrate into a ground meat patty. Patties made from freshly ground turkey breasts were subjected to two MAP treatments of high CO_2 (97%) or high O_2 (80% O_2, 20% CO_2). Total plate and LAB counts were determined for the top, middle, and bottom depth of the patties. Overall counts were lower in a high-CO_2 atmosphere when compared with the high-O_2 MA. Samples stored under a high-CO_2 atmosphere exhibited slower bacterial growth in the top layer compared with the middle and bottom layers. TPC did not differ in layers for patties packaged in a high-O_2 atmosphere. Moreover, LAB counts augmented in the high-O_2 MA but no similar trend was recorded in CO_2-packaged patties. Therefore, high-CO_2 MAP retarded the growth of total bacteria as well as LAB. Moreover, there was slower growth in the top meat layer exposed to CO_2 compared with interior layers. Rajkumar et al. (2007) investigated the effect of MA (80% O_2/20% CO_2) packaging, VP, and aerobic packaging on the total viable count (TVC) and anaerobic count, pH, drip loss, and odor score of fresh and stored turkey meat (4°C ± 1°C). TVC, anaerobic counts, drip loss, and odor score were the lowest in samples packaged under MAP. It was found that turkey meat packaged under MAP was kept safely up to 14 days of storage at 4°C ± 1°C based on the desirable TVC, anaerobic counts, and drip loss. However, turkey meat packaged under UV could be safely kept up to 21 days of storage. In another study Ntzimani et al. (2008) examined the formation of biogenic amines in smoked turkey fillets over storage under air, VP, skin, and two modified atmospheres (MAP_1: 30% CO_2/70% N_2 and MAP_2: 50% CO_2/50% N_2), at 4°C ± 0.5°C, for a period of 30 days. *Pseudomonas* spp. and Enterobacteriaceae did not increase in skin-packaged fillets and MAP_2, which remained under the detection limit until day 30 of storage. *Pseudomonas* spp. and Enterobacteriaceae for the rest of the packaging treatments remained below 5 log CFU/g throughout storage. On the other hand, LAB were dominant throughout the storage period, regardless of the packaging conditions (gas composition). Finally, mesophiles reached 7 log CFU/g after ca. 19–20 days for the air- and skin-packed samples, 22–23 days for the MAP_2 and VP samples, and 25–26 days for the MAP_1 packed samples.

The effect of MAP on LAB of precooked chicken fillets and smoked turkey breasts is displayed in Figure 6.1.

Table 6.1 shows the effect of MAP on physical and sensory properties of turkey meat.

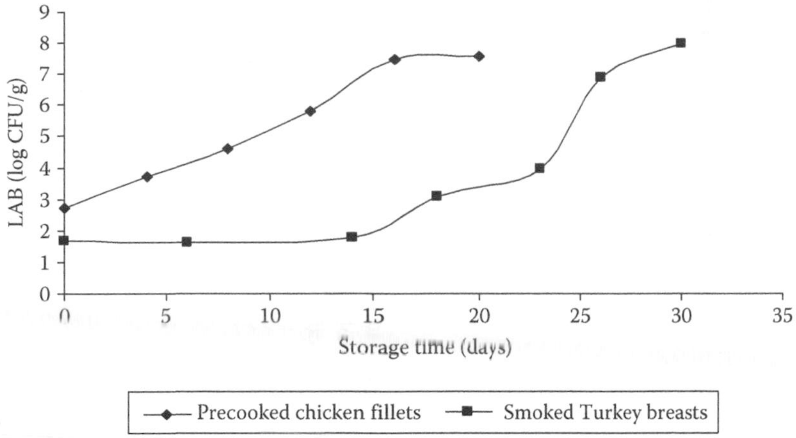

FIGURE 6.1 Effect of MAP (30% CO_2/70% N_2) on the LAB population on precooked chicken fillets (From Patsias, A. et al., *Food Microbiol.*, 23, 423, 2006) and smoked turkey breasts (From Ntzimani, A.G. et al., *Food Microbiol.*, 25, 509, 2008) stored at 4°C.

6.4 EFFECT OF MAP ON THE QUALITY OF CHICKEN

Patsias et al. (2006) investigated the effect of MAP including air on the shelf life and quality of precooked chicken meat product stored at 4°C. The gas mixtures used were M1: 30%/70% (CO_2/N_2), M2: 60%/40% (CO_2/N_2), and M3: 90%/10% (CO_2/N_2). Values of pH for air and M1, M2, and M3 gas mixture packaged chicken samples varied within the range 6.25–6.42. TBA values of air and MAP samples decreased up to day 8 of refrigerated storage, but after that TBA values augmented significantly for air-packaged chicken samples while those obtained for M1, M2, and M3 samples remained relatively constant. TBA values in all samples were lower or equal to 3.0 mg malonaldehyde/kg over the entire storage period of 20 days. Balamatsia et al. (2007) examined the formation of volatile amines (TVB-N, TMA-N) in chicken breast fillets packaged under air (A), VP, and two MAPs (M1: 30%/65%/5%: CO_2/N_2/O_2 and M2 65%/30%/5%: CO_2/N_2/O_2) at 4°C. Although the TMA-N content was initially low, after day 3 of storage, TMA-N values of the control and VP samples were significantly higher than those of MAP samples. Final TMA-N values for all chicken samples packaged under MAPs exhibited significantly lower values than those of samples packaged in air and UV. In a similar trend, the TVB-N values of chicken also increased from an initial value of ca. 20.5 mg N/100 g to end values of 54.5, 45.8, 43.1, and 29.6 mg N/100 g for air, VP, M1, and M2 packaged samples, respectively. Keokamnerd et al. (2008) tested four commercial rosemary oleoresin preparations that were added to ground chicken thigh meat and then packaged in 80% O_2/20% CO_2 MA trays. The rosemary preparations differed either in oil and water solubility, or dispersion properties, or both. The obtained results revealed that the addition of rosemary to ground chicken had an overall positive effect on raw meat appearance during storage and cooked meat flavor. Oxidation was retarded in meat with added rosemary as shown by the lower TBA values, hexanal concentrations, and sensory scores. Color (redness) was more stable in meat with added rosemary compared with

TABLE 6.1
Impact of MAP on Physical and Sensory Properties of Turkey Meat

Food Type	Gas Mix	Storage Temperature	Shelf Life	Sensory Assessment	Microflora	Quality	References
Ground turkey	$CO_2/O_2/N_2$: 75/5/20	4°C, 15°C, and 28°C			The growth of *C. perfringens* was slowest under 25%–50% CO_2/20% O_2/balance N_2 at 15°C and 28°C. There was no growth at 4°C for up to 28 days		Juneja et al. (1996)
Ground turkey	$CO_2/O_2/N_2$: 75/10/15	4°C, 15°C, and 28°C					
Ground turkey	$CO_2/O_2/N_2$: 75/20/5	4°C, 15°C, and 28°C					
Ground turkey	$CO_2/O_2/N_2$: 25/20/55	4°C, 15°C, and 28°C					
Ground turkey	$CO_2/O_2/N_2$: 50/20/30	4°C, 15°C, and 28°C					
Turkey meat	Aerobic packaging	4°C ± 1°C	—	Based on the odor score, turkey packaged under UV was better	TVC and anaerobic counts were lowest in samples packaged under modified atmosphere	Drip loss was the lowest in samples packaged under modified atmosphere	Rajkumar et al. (2007)
Turkey meat	80% O_2/20% CO_2	4°C ± 1°C	14 days				
Turkey meat	VP	4°C ± 1°C	21 days				
Cooked, uncured turkey	100% N_2	15°C			Botulinal toxin was detected by day 7, independent of packaging atmosphere		Lawlor et al. (2000)
Cooked, uncured turkey	30% CO_2:70% N_2	15°C					
Cooked, uncured turkey	100% N_2	10°C			Botulinal toxin was detected by day 14, independent of packaging atmosphere		
Cooked, uncured turkey	30% CO_2:70% N_2	10°C					
Cooked, uncured turkey	100% N_2	4°C			Botulinal toxin was detected by day 14		

Cooked, uncured turkey	30% CO_2:70% N_2	4°C			Botulinal toxin was detected by day 28		
Smoked turkey fillets	Air	4°C ± 0.5°C	22–23 days based on sensory scores	Similar odor scores for samples stored under all packaging treatments, except for aerobic conditions, show that the type of MAP had no significant influence on odor of all samples. Similar results were obtained for taste attributes	*Pseudomonas* spp. and Enterobacteriaceae remained below 5 log CFU/g throughout storage. Mesophiles reached 7 log CFU/g after ca. 19–20 days	Values for histamine, tyramine, and tryptamine were the highest for air-packaged samples (32.9, 25.0, and 4.1 mg/kg, respectively) and the lowest for skin-packaged samples (11.9, 4.3, and 2.8 mg/kg, respectively) after 30 days of storage	Ntzimani et al. (2008)
Smoked turkey fillets	VP	4°C ± 0.5°C	27–30 days based on sensory scores		*Pseudomonas* spp. and Enterobacteriaceae remained below 5 log CFU/g throughout storage. Mesophiles reached 7 log CFU/g 22–23 days		
Smoked turkey fillets	Skin packaged	4°C ± 0.5°C	27–30 days based on sensory scores		*Pseudomonas* spp. and Enterobacteriaceae did not increase during storage. Mesophiles reached 7 log CFU/g after ca. 19–20 days		

(*continued*)

TABLE 6.1 (continued)
Impact of MAP on Physical and Sensory Properties of Turkey Meat

Food Type	Gas Mix	Storage Temperature	Shelf Life	Sensory Assessment	Microflora	Quality	References
Smoked turkey fillets	MAP_1 (30% CO_2/70% N_2)	4°C ± 0.5°C	27–30 days based on sensory scores		*Pseudomonas* spp. and Enterobacteriaceae did not increase during storage. Mesophiles reached 7 log CFU/g 25–26 days		
Smoked turkey fillets	MAP_2 (50% CO_2/50% N_2)	4°C ± 0.5°C	27–30 days based on sensory scores		*Pseudomonas* spp. and Enterobacteriaceae remained below 5 log CFU/g throughout storage. Mesophiles reached 7 log CFU/g 22–23 days		
Turkey meat	100% N_2	1°C and 7°C			Growth of *A. hydrophila*		Mano et al. (2000)
Turkey meat	20%/80% CO_2/O_2	1°C and 7°C			Growth of *A. hydrophila*		
Turkey meat	40%/60% CO_2/O_2	1°C and 7°C			No growth of *A. hydrophila* No growth of *B. thermosphacta* at 7°C		
Turkey meat (PSE)	50% CO_2/50% N_2	0°C ± 1°C	20 days			During storage time, light color meat did not have an increase in TVB-N but intermediate and dark meat showed a significant increase from day 5 to day 12 of storage, particularly dark meat	Fraqueza et al. (2008)
Turkey meat (PSE)	50% CO_2/50% N_2	0°C ± 1°C	20 days				
Turkey meat (PSE)	50% CO_2/50% N_2	0°C ± 1°C	13 days				

Cured, cooked, sliced turkey fillets	80% CO_2/20% N_2; 60% CO_2/20% O_2/20% N_2; 0.4% CO/80% CO_2/rest N_2; 1% CO/80% CO_2/rest N_2; 0.5% CO/24% O_2/50% CO_2/rest N_2 and 100% N_2	4°C	2 weeks		Gaseous mixtures did not affect the growth at 10°C in comparison with the growth at 4°C. However, the lowest total count and LAB count at 10°C were observed in 100% N_2. No differences between VP and 80% CO_2/20% N_2 were noted after 25 days at 4°C	The a* values of VP packed sausages and sausages packed under modified atmosphere of 80% CO_2/20% N_2 were insignificantly affected by the storage temperature. Sausages packed in high O_2 concentration were less red than the other sausages and discoloration was significantly greater at 10°C than at 4°C. The bright red color of the sausages was more stable in 100% N_2	Pexara et al. (2002)
Cured, cooked, sliced turkey fillets	80% CO_2/20% N_2; 60% CO_2/20% O_2/20% N_2; 0.4% CO/80% CO_2/rest N_2; 1% CO/80% CO_2/rest N_2; 0.5% CO/24% O_2/50% CO_2/rest N_2 and 100% N_2	10°C	1 week				
Cured, cooked, sliced turkey fillets	VP	4°C	2 weeks				
Cured, cooked, sliced turkey fillets	VP	10°C					

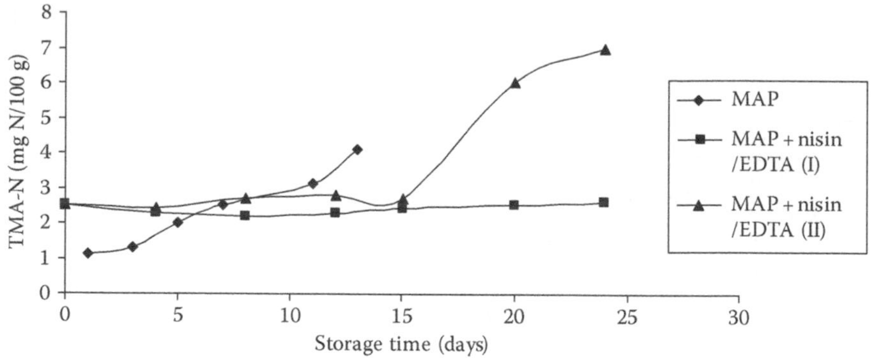

FIGURE 6.2 Effect of MAP (65% CO_2/30% N_2/5% O_2) (From Balamatsia, C.C. et al., *Food Chem.*, 104, 1622, 2007) and MAP (65% CO_2/30% N_2/5% O_2) combined with NIS/EDTA (1500 IU g/50 mM) treatment (From Economou, T. et al., *Food Chem.*, 114, 1470, 2009) on the TMA-N values of chicken meat stored at 4°C.

meat without rosemary, as reflected in redness values, hue angles, and visual scores. It was also shown that of the four rosemary preparations tested, the oil-soluble, most concentrated preparation was the most effective in maintaining meat quality compared with the other preparations tested.

The impact of several MAP compositions in the presence or absence of nisin (NIS) on the TMA-N values of chicken meat is exhibited in Figure 6.2.

6.5 EFFECT OF MAP ON THE MICROFLORA OF CHICKEN

Fresh chicken breast meats inoculated with *Yersinia enterocolitica* and *A. hydrophila* were packaged in glass jars either containing different compositions of MAs (100% CO_2 and 80% CO_2/20% N_2) or in VP or containing air and stored at 3°C ± 1°C and 8°C ± 1°C. The results displayed that the growth of *Y. enterocolitica* and *A. hydrophila* were considerably retarded following MAP storage. However, the pathogens were capable of growth in MAP and VP storage at both temperatures. No differences were recorded concerning the total aerobic bacterial counts, between the values for chicken breast meats stored in different atmospheres. Packaged meat in CO_2 was found to have the greatest inhibitory impact on the growth of psychrotrophic aerobic bacteria during the first 3 days.

MAP and VP storage allowed LAB to augment at higher rate than those samples stored under air. The effect of MAP storage was most pronounced at 3°C ± 1°C (Ozbas et al., 1996). The effect of CO_2 (100%), N_2 (100%), CO_2/O_2 (20%:80%) or VP at 3°C and 10°C on the microbial flora in skinless poultry breast fillets or thigh meat was studied by Kakouri and Nychas (1994). LAB and *B. thermosphacta* were the predominant organisms in samples stored in VP, CO_2, and N_2. Furthermore, Pseudomonads grew only in CO_2/O_2 packaging systems. Jimenez et al. (1997) analyzed chicken breasts with skin packaged in air, VP, or under MAP of 30% CO_2/70% N_2 and 70% CO_2/30% N_2. Pseudomonad growth was reduced in both types of MAP but it grew well in air and VP. On the other hand, the growth of Lactobacilli,

Enterobacteriaceae, and *B. thermosphacta* was not inhibited under MAP. The atmosphere of 70% CO_2/30% N_2 prolonged the shelf life up to 21 days compared to 5 days for samples under air. Jimenez et al. (1999) treated samples of chicken breasts with skin with a 1% acetic acid solution or untreated and packaged in a 70% CO_2/30% N_2 MA. The application of MAP, or MAP combined with low pH (acetic acid), was sufficient to suppress the growth of pseudomonads and extend the lag phase of lactobacilli up to day 7. A similar growth trend was observed in TVC and in enterobacteria only when acetic acid solution and MAP were used together. Hugas et al. (1998) examined the effect of the bacteriocinogenic *Lactobacillus sakei* CTC494 in the production of sakacin K by *Lb. sakei CTC494* against Listeria in chicken breasts, packaged in oxygen-permeable film, under VP and MAP (20% CO_2:80% O_2) stored at 7°C. Listeria inhibition in raw chicken breasts could not be reached with the sole application of VP or MAP. However, Listeria's growth was inhibited to different extents in all packaging systems by the inoculation of *Lb. sakei* CTC494 or sakacin K. The highest inhibition was recorded in the VP samples of poultry breasts. Cosby et al. (1999) investigated the impact of treatment with disodium ethylenediaminetetraacetate (EDTA) and NIS in conjunction with storage under MAP or VP on the shelf life of poultry. Chicken drummettes were soaked with various combinations of EDTA and NIS for 30 min at 15°C and stored at 4°C. Samples treated with EDTA–NIS stored UV had significantly lower total aerobic plate (TAP) counts (TAPC) than untreated ones stored under air. As regards the shelf life, the EDTA–NIS treatment increased it by a minimum of 4 days when packaged under aerobic conditions and a maximum of 9 days when VP. The effect of NIS and EDTA treatments on the shelf life of fresh chicken meat stored under MAP (65% CO_2/30% N_2/5% O_2) at 4°C was evaluated by Economou et al. (2009). NIS, in combination with EDTA, was able to reduce *B. thermosphacta* and LAB populations by approximately 1–2 logs. Moreover, the application of MAP combined with NIS–EDTA treatments resulted in organoleptic extension of samples. Cegielska-Radziejewska and Pikul (2001) determined the effect of MAP (100% N_2 or 70% N_2, and 30% CO_2), compared with air packaging, on the quality and shelf life of medium comminuted poultry meat sausage stored at temperatures of 1°C ± 1°C and 7°C ± 1°C. At both examined temperatures, the sliced sausage packaged with MA containing N_2 and CO_2 resulted in a lower count of psychrophilic aerobic bacteria and substantially lower rate of unfavorable changes of taste and odor, in comparison with sausage samples packaged with N_2 only or with air atmosphere. On the other hand, storage at 7°C ± 1°C resulted in twice as large reduction of the shelf life of sliced sausage packaged with the mixture of N_2 and CO_2. The replacement of 30% nitrogen with CO_2 in packaging, in comparison with 100% N_2 atmosphere, led to the extension of product shelf life at 1°C ± 1°C temperature by 1 week. The formation of biogenic amines was studied by Rokka et al. (2004) in broiler chicken cuts stored under MAP (80% CO_2/20% N_2). Tyramine formation rate was considerably affected by the storage temperature. The levels of tyramine increased just after 5 days of storage provided the storage temperature was higher than 6.1°C. Putrescine concentration increased after 7 days whereas the cadaverine levels increased after 9 days. No formation of putrescine and cadaverine was detected below 6.1°C. Tyramine was formed over the storage at low temperatures, although the formation rate was slower than that above 6.1°C. The results obtained showed that the formation of tyramine seemed to be highly related to the increase in the aerobic mesophilic viable count. According to Al-Haddad et al. (2005) "chilled breasts of chicken were inoculated with *Salmonella infantis* or *Pseudomonas aeruginosa* and then subjected to one of the following treatments: (i) exposure to gaseous ozone (>2000 ppm for up to 30 min), (ii) storage under 70% CO_2:30% N_2, and (iii) exposure to gaseous ozone (>2000 ppm for 15 min) followed by storage under 70% CO_2:30% N_2. Gaseous ozone reduced the counts of salmonellae by 97% and pseudomonads by 95%, whereas indigenous coliforms remained unaffected. MAP reduced the cell count of *S. infantis* by 72% following initial exposure and then stabilized, coliforms grew, but *Ps. aeruginosa* behaved like *S. infantis*, which showed an initial reduction of 58% followed by stability. Gaseous ozone treatment followed by gas packaging allowed the survival of *S. infantis*, *Ps. aeruginosa*, and coliforms over 9 days at 7°C." Patsias et al. (2006) investigated the effect of MAP on shelf-life extension of precooked chicken meat product stored at 4°C. The gas mixtures applied were of the following composition: M1: 30%/70% (CO_2/N_2), M2: 60%/40% (CO_2/N_2), and M3: 90%/10% (CO_2/N_2). Aerobically packaged samples were used as controls. TVC of precooked chicken product reached 7 log CFU/g, after days 12 and 16 of storage (air and M1 samples), respectively. The M2 and M3 gas mixture packaged samples did not reach this value throughout the 20 days storage. LAB, and to a lesser degree *B. thermosphacta*, were part of the natural microflora of precooked chicken samples stored in air and under MAP, reaching 7.0–8.1 log CFU/g at the end of storage. Moreover, Enterobacteriaceae counts were below 2 log CFU/g in all chicken samples regardless of the packaging conditions throughout storage. Fresh chicken meat (breast fillet) was packaged under four different atmospheres: air (A), VP, and two modified atmospheres (MAPs), specifically M1, 30%/65%/5% ($CO_2/N_2/O_2$) and M2, 65%/30%/5% ($CO_2/N_2/O_2$). All packaged samples were stored at 4°C ± 0.5°C for 15 days. VP and M1 and M2 gas mixtures were the most effective in delaying the development of aerobic spoilage microbial flora. Pseudomonas initial population was ca. 4.2 log CFU/g while the count of 7 log CFU/g was exceeded between days 7–11 of storage (control samples), whereas VP, M1, and M2 chicken samples did not reach this population throughout the 15 days of storage. Initial LAB count was ca. 3.9 log CFU/g while a count of 7 log CFU/g was exceeded on day 5 for air and on day 12 for VP samples. Equally, the M1 and M2 gas mixture-packaged samples did not achieve this amount throughout the 15 day storage period. Enterobacteriaceae and *B. thermosphacta* were also the dominant bacterial groups in chicken spoilage, producing lower counts for VP and MAP samples in comparison to samples packaged under air throughout storage. Moreover, yeasts were less numerous than bacteria in chicken samples stored in air, UV, and MAPs (Balamatsia et al., 2007).

The effect of MAP on chicken meat physical, microbiological, and sensory properties is given in Table 6.2.

6.6 CONCLUSIONS

It was shown that by applying the appropriate MAP composition of gases ($CO_2/N_2/O_2$: 65/30/5, CO_2/N_2: 90/10) it is possible to extend the poultry shelf life up to 20 days. At the aforementioned gas compositions, the microorganism growth (LAB,

TABLE 6.2
Impact of MAP on Chicken Meat Physical, Microbiological, and Sensory Properties

Food Type	Gas Mix	Other Technology	Storage Temperature	Shelf Life	Sensory Assessment	Microflora	Quality	References
Ground chicken thigh meat	80% O_2/20% CO_2	Rosemary preparations (which differed in oil and water solubility, dispersion properties, or both)			The addition of rosemary to ground chicken had an overall positive effect on raw meat appearance during storage and cooked meat flavor	Rosemary addition had no effect on bacterial growth	Oxidation was slowed in meat with added rosemary as indicated by lower TBA values, lower hexanal concentrations	Keokamnerd et al. (2008)
Fresh chicken breast meats	Air		3°C ± 1°C and 8°C ± 1°C			No differences for total aerobic bacterial counts, between the different atmospheres. Packaged meat in CO_2 had the greatest inhibitory effect on psychrotrophic aerobic bacteria. LAB of samples stored under MAP conditions and in VP increased rapidly compared to samples stored in air		Ozbas et al. (1996)
Fresh chicken breast meats	VP		3°C ± 1°C and 8°C ± 1°C					
Fresh chicken breast meats	100% CO_2 and 80% CO_2/20% N_2		3°C ± 1°C and 8°C ± 1°C					

(continued)

TABLE 6.2 (continued)
Impact of MAP on Chicken Meat Physical, Microbiological, and Sensory Properties

Food Type	Gas Mix	Other Technology	Storage Temperature	Shelf Life	Sensory Assessment	Microflora	Quality	References
Chicken breasts with skin	Air		4°C	5 days		Pseudomonads grew well		Jimenez et al. (1997)
Chicken breasts with skin	30% CO_2/70% N_2		4°C			Growth of Lactobacilli, Enterobacteriaceae, and *B. thermosphacta* not affected. Pseudomonads grew well in air and under VP, but growth was suppressed in both types MAP		
Chicken breasts with skin	70% CO_2/30% N_2		4°C	21 days		Growth of Lactobacilli, Enterobacteriaceae, and *B. thermosphacta* was not affected. Pseudomonads grew well in air or under VP, but growth was suppressed in both types MAP		
Skinless poultry breast fillets or thigh meat	CO_2 (100%)		3°C and 10°C			LAB and *B. thermosphacta* were the predominant organisms in samples stored in VP packs, CO_2, and N_2. Pseudomonads grew only in CO_2/O_2 packaging systems		Kakouri and Nychas (1994)

Skinless poultry breast fillets or thigh meat	N_2 (100%)		3°C and 10°C			
Skinless poultry breast fillets or thigh meat	CO_2/O_2 (20%: 80%)		3°C and 10°C			
Skinless poultry breast fillets or thigh meat	VP		3°C and 10°C			
Chicken drummettes	Air		4°C		Significantly lower TAPC than untreated controls stored under aerobic conditions	Cosby et al. (1999)
Chicken drummettes	VP	Treatment with a disodium ethylenediaminetetra and NIS for 30 min at 15°C	4°C	9 days		

(*continued*)

TABLE 6.2 (continued)
Impact of MAP on Chicken Meat Physical, Microbiological, and Sensory Properties

Food Type	Gas Mix	Other Technology	Storage Temperature	Shelf Life	Sensory Assessment	Microflora	Quality	References
Chicken breasts with skin	70% CO_2/30% N_2	Treatment with 1% acetic acid solution	4°C		Acetic acid decontamination suppressed the off odor production in samples stored 21 days according to sensory evaluations	Suppression of growth of pseudomonads and extend the lag phase of lactobacilli up to day 7. Also growth delay was observed in TVCs and in enterobacteria		Jimenez et al. (1999)
Chicken breast fillets	Air		4°C±0.5°C	6–7 days		VP and MAP gas mixtures were the most effective for delaying the development of aerobic spoilage microbial flora. Enterobacteriaceae and *B. thermosphacta* were also dominant bacterial groups in chicken spoilage producing lower counts for VP and MAP samples in comparison to air-packaged samples	TMA-N values of the control and VP-packaged samples were significantly higher than those of MAP chicken and increased at a higher rate	Balamatsia et al. (2007)
Chicken breast fillets	VP		4°C±0.5°C	9–10 days				
Chicken breast fillets	M1, 30%/65%/5% $CO_2/N_2/O_2$		4°C±0.5°C	9–10 days				

Broiler chicken cuts	80% CO_2/20% N_2	3°C–22°C				Increase in tyramine after 5 days of storage above 6.1°C. Increase in putrescine after 7 days. Increase in cadaverine after 9 days. Tyramine formation (slow) over storage at low temperatures	Rokka et al. (2004)
Precooked chicken meat	Air	4°C	16 days	Air-packaged chicken samples received higher overall acceptability scores than M1, M2, and M3 gas-packaged samples up to day 12, whereas after this time period this trend was reversed continuing throughout storage	LAB and to a lesser degree *B. thermosphacta*, constituted part of the natural microflora of samples stored in air and under MAP, reaching 7.0–8.1 log CFU/g at the end of storage. Moreover, Enterobacteriaceae counts were below 2 log CFU/g in all samples irrespective of the packaging conditions	TBA values in all cases remained lower or equal to 3.0 mg malonaldehyde/kg during the entire storage period of 20 days	Patsias et al. (2006)
Precooked chicken meat	M1: 30%/70% (CO_2/N_2)	4°C	20 days				
Precooked chicken meat	M2: 60%/40% (CO_2/N_2)	4°C	Never reached the limit of overall acceptability				
Precooked chicken meat	M3: 90%/10% (CO_2/N_2)	4°C					

(continued)

TABLE 6.2 (continued)
Impact of MAP on Chicken Meat Physical, Microbiological, and Sensory Properties

Food Type	Gas Mix	Other Technology	Storage Temperature	Shelf Life	Sensory Assessment	Microflora	Quality	References
Chicken meat	65% CO_2/30% N_2/5% O_2	—	4°C	—				Economou et al. (2009)
Chicken meat	65% CO_2/30% N_2/5% O_2	500 IU/g NIS	4°C	1–2 days extension				
Chicken meat	65% CO_2/30% N_2/5% O_2	1500 IU/g NIS	4°C	3–4 days extension		Affected populations of mesophilic bacteria, Pseudomonas sp., *B. thermosphacta* LAB, and Enterobacteriaceae		
Chicken meat	65% CO_2/30% N_2/5% O_2	500 IU/g NIS–10 mM EDTA	4°C	3–4 days extension				
Chicken meat	65% CO_2/30% N_2/5% O_2	1500 IU/g NIS–10 mM EDTA	4°C	7–8 days extension			Significant effect on the formation of volatile amines, TMA-N, and TVB-N	
Chicken meat	65% CO_2/30% N_2/5% O_2	500 IU/g NIS–50 mM EDTA	4°C	13–14 days extension	Maintained acceptable odor attributes even up to 24 days			
Chicken meat	65% CO_2/30% N_2/5% O_2	1500 IU/g NIS–50 mM EDTA	4°C	9–10 days extension	Maintaining acceptable odor attributes even up to 20 days			

Enterobacteriace, *B. Thermosphacta*) remained within acceptable limits. Therefore, both MAP and VP can be effectively used to prolong the shelf life of poultry.

REFERENCES

Al-Haddad, K.S.H., Al-Qassemi, R.A.S., and Robinson, R.K. 2005. The use of gaseous ozone and gas packaging to control populations of *Salmonella infantis* and *Pseudomonas aeruginosa* on the skin of chicken portions. *Food Control*, 16: 405–410.

Anang, D.M., Rusul, G., Bakar, J., and Ling, F.H. 2007. Effects of lactic acid and lauricidin on the survival of *Listeria monocytogenes*, *Salmonella enteritidis* and *Escherichia coli* O157:H7 in chicken breast stored at 4°C. *Food Control*, 18(8): 961–969.

Bagorogoza, K., Bowers, J., and Okot-Kotber, M. 2001. The effect of irradiation and modified atmosphere packaging on the quality of intact chill-stored turkey breast. *Journal of Food Science*, 66(2): 367–372.

Balamatsia, C.C., Patsias, A., Kontominas, M.G., and Savvaidis, I.N. 2007. Possible role of volatile amines as quality-indicating metabolites in modified atmosphere-packaged chicken fillets: Correlation with microbiological and sensory attributes. *Food Chemistry*, 104: 1622–1628.

Brody, A.L. 1996. *Envasado de alimentos en atmós feras controladas, modifiadas y a vacio*. Editorial Acribia, S.A., Zaragoza Espana, Spain.

Cegielska-Radziejewska, R. and Pikul, J. 2001. Effects of gas atmosphere, storage temperature and time on the quality and shelf-life of sliced poultry sausage. *Archiv fur Geflugelkunde*, 65(6): 274–280.

Chouliara, E., Karatapanis, A., Savvaidis, I.N., and Kontominas, M.S. 2007. Combined effect of oregano essential oil and modified atmosphere packaging on shelf-life extension of fresh chicken breast meat, stored at 4°C. *Food Microbiology*, 24(6): 607–617.

Chun, H.H., Kim, J.Y., Lee, B.D., Yu, D.J., and Song, K.B. 2009. Effect of UV-C irradiation on the inactivation of inoculated pathogens and quality of chicken breasts during storage. *LWT—Food Science and Technology*, 42: 1325–1334.

Cosby, D.E., Harrison, M.A., Toledo, R.T., and Craven, S.E. 1999. Vacuum or modified atmosphere packaging and EDTA-nisin treatment to increase poultry product shelf life. *Journal of Applied Poultry Research*, 8(2): 185–190.

Dhananjayan, R., Han, I.Y., Acton, J.C., and Dawson, P.L. 2006. Growth depth effects of bacteria in ground turkey meat patties subjected to high carbon dioxide or high oxygen atmospheres. *Poultry Science*, 85(10): 1821–1828.

Economou, T., Pournis, N., Ntzimani, A., and Savvaidis, I.N. 2009. Nisin–EDTA treatments and modified atmosphere packaging to increase fresh chicken meat shelf-life. *Food Chemistry*, 114: 1470–1476.

Fraqueza, M.J., Ferreira, M.C., and Barreto, A.S. 2008. Spoilage of light (PSE-like) and dark turkey meat under aerobic or modified atmosphere package: Microbial indicators and their relationship with total volatile basic nitrogen. *British Poultry Science*, 49(1): 12–20.

Geornaras, I., de Jesus, A., van Zyl, E., and von Holy, A. 1998. Bacterial populations associated with the dirty area of the South African poultry abattoir. *Journal of Food Protection*, 61: 700–703.

Hugas, M., Pages, F., Garriga, M., and Monfort, J.M. 1998. Application of the bacteriocinogenic *Lactobacillus sakei* CTC494 to prevent growth of Listeria in fresh and cooked meat products packed with different atmospheres. *Food Microbiology*, 15: 639–650.

Jimenez, S.M., Salsi, M.S., Tiburzi, M.C., Rafaghelli, R.C., and Pirovani, M.E. 1999. Combined use of acetic acid treatment and modified atmosphere packaging for extending the shelf-life of chilled chicken breast portions. *Journal of Applied Microbiology*, 87: 339–344.

Jimenez, S.M., Salsi, M.S., Tiburzi, M.C., Rafaghelli, R.C., Tessi, M.A., and Coutaz, V.R. 1997. Spoilage microflora in fresh chicken breast stored at 4°C: Influence of packaging methods. *Journal of Applied Microbiology*, 83(5): 613–618.

Juneja, V.K., Marmer, B.S., and Call, J.E. 1996. Influence of modified atmosphere packaging on growth of *Clostridium perfringens* in cooked turkey. *Journal of Food Safety*, 16: 141–150.

Kakouri, A. and Nychas, G.J.E. 1994. Storage of poultry meat under modified atmospheres or vacuum packs: Possible role of microbial metabolites as indicator of spoilage. *Journal of Applied Bacteriology*, 76(2): 163–172.

Keokamnerd, T., Acton, J.C., Han, I.Y., and Dawson, P.L. 2008. Effect of commercial rosemary oleoresin preparations on ground chicken thigh meat quality packaged in a high-oxygen atmosphere. *Poultry Science*, 87(1): 170–179.

Lawlor, K.A., Pierson, M.D., Hackney, C.R., Claus, J.R., and Marcy, J.E. 2000. Nonproteolytic *Clostridium botulinum* toxigenesis in cooked turkey stored under modified atmospheres. *Journal of Food Protection*, 63(11): 1511–1516.

Mano, S.B., Ordonez, J.A., and Garcia de Fernando, G.D. 2000. Growth/survival of natural flora and *Aeromonas hydrophila* on refrigerated uncooked pork and turkey packaged in modified atmospheres. *Food Microbiology*, 17: 657–669.

Ntzimani, A.G., Paleologos, E.K., Savvaidis, I.N., and Kontominas, M.G. 2008. Formation of biogenic amines and relation to microbial flora and sensory changes in smoked turkey breast fillets stored under various packaging conditions at 4°C. *Food Microbiology*, 25: 509–517.

Ozbas, Z.Y., Vural, H., and Aytac, S.A. 1996. Effects of modified atmosphere and vacuum packaging on the growth of spoilage and inoculated pathogenic bacteria on fresh poultry. *Zeitschrift fur Lebensmittel—Untersuchung und-Forschung A*, 203(4): 326–332.

Patsias, A., Chouliara, I., Badeka, A., Savvaidis, I.N., and Kontominas, M.G. 2006. Shelf-life of a chilled precooked chicken product stored in air and under modified atmospheres: Microbiological, chemical, sensory attributes. *Food Microbiology*, 23: 423–429.

Pettersen, M.K., Mielnik, M.B., Eie, T., Skrede, G., and Nilsson, A. 2004. Lipid oxidation in frozen, mechanically deboned turkey meat as affected by packaging parameters and storage conditions. *Poultry Science*, 83(7): 1240–1248.

Pexara, E.S., Metaxopoulos, J., and Drosinos, E.H. 2002. Evaluation of shelf life of cured, cooked, sliced turkey fillets and cooked pork sausages 'piroski'-stored under vacuum and modified atmospheres at +4 and +10°C. *Meat Science*, 62: 33–43.

Rajkumar, R., Dushyanthan, K., Asha Rajini, R., and Sureshkumar, S. 2007. Effect of modified atmosphere packaging on microbial and physical qualities of turkey meat. *American Journal of Food Technology*, 2(3): 183–189.

Rokka, M., Eerola, S., Smolander, M., Alakomi, H.L., and Ahvenainen, R. 2004. Monitoring of the quality of modified atmosphere packaged broiler chicken cuts stored in different temperature conditions B. Biogenic amines as quality-indicating metabolites. *Food Control*, 15: 601–607.

Taylor, A.S. 1996. Modified atmosphere packing of meat. In: *Meat Quality and Meat Packaging*, Part II. EC/CE/AMST, Utrecht, the Netherlands, pp. 301–311.

Thayer, D.W. and Boyd, G. 2000. Reduction of normal flora by irradiation and its effect on the ability of *Listeria monocytogenes* to multiply on ground turkey stored at 7°C when packaged under a modified atmosphere. *Journal of Food Protection*, 63: 1702–1706.

Zardetto, S. 2005. Effect of modified atmosphere packaging at abuse temperature on the growth of *Penicillium aurantiogriseum* isolated from fresh filled pasta. *Food Microbiology*, 22: 367–371.

7 Milk and Dairy Products

Ioannis S. Arvanitoyannis and Georgios Tziatzios

CONTENTS

7.1 INTRODUCTION

Milk is the source of a wide range of proteins required for nutrition purposes for the most promising new food products nowadays. Isolated milk proteins are natural, trusted food ingredients with excellent functionality (Huffman and Harper, 1999). Milk fat is an excellent dietary source of retinol, TH, and β-carotene, which display their antioxidant activity in biological tissues as well as in foods. β-Carotene and retinol scavenge both singlet oxygen and lipoperoxides, thereby considerably limiting the oxidation of fatty acid (Bergamo et al., 2003). The importance of goat milk and its products, such as yogurt, cheese, and powder, in human nutrition can be classified along the following three axes: (1) feeding more starving and malnourished people in the developing world than from cow milk; (2) treating people affected with cow milk allergies and gastrointestinal disorders; and (3) fulfilling the gastronomic needs of connoisseur consumers, an expanding market share in many developed countries (Haenlein, 2004).

Fermented milks were first prepared more than 2000 years ago. Allowing milk to ferment naturally results in an acidic product that does not spoil. Fermented milks are wholesome and readily digestible; examples of such products are yogurt, *kefir*, *koumiss*, and acidophilus milk (O'Connor, 1995). The nutritional value of fermented milk products such as yogurt is assumed to be similar to the milk they are made from (Alm, 1982). The nutritional value of these products is related to the milk utilized. The aforementioned dairy products are rich in protein, vitamins, and minerals, and there may be eventual presence of other ingredients (milk powder, sugar, fruit puree, and fruit extracts), whereas the microorganisms used can considerably affect the texture and organoleptic characteristics (Buttriss, 2007; Gambelli et al., 1999). Yogurt is considered to be more nutritious than many other fermented milk products because it contains a high level of milk solids in addition to nutrients developed over the fermentation process (Fadela et al., 2009). Lactic acid bacteria (LAB) in fermented milk products have been suggested as a possible supplementary source of vitamin in human nutrition (Alm, 1982).

Many factors can and do affect the shelf life of dairy products. Some of these factors are (1) raw milk quality, (2) high temperature short term (HTST) pasteurizer operation, (3) cleanliness and sanitation of pasteurization lines, (4) extent of sanitation of pasteurized milk storage tanks, (5) sanitation and "protective" design of filling machines, (6) the state of cleanliness of the overall plant environment, and (7) adequate temperature control. Therefore, killing psychrotrophic bacteria that can contaminate milk in conjunction with temperature (high or low) can become a crucial factor (White, 1993). The shelf life of whole pasteurized milk in Greece is 5 days (Zygoura et al., 2003), whereas the shelf life of yogurt is very short, that is, 1 day at ambient temperature (25°C–35°C) and about 4–5 days at 7°C. Improvements in the shelf life of yogurt can be brought about by its dehydration and conversion into a shelf stable powder. Yogurt can be dried by freeze, spray, microwave, or convective drying methods, which do account for the viability and activity of the yogurt bacteria (Kumar and Mishra, 2004). The shelf life of fresh Cameros cheese under refrigeration is only some days (Olarte et al., 2002). On the other hand, according to Mannheim and Soffer (1996), the shelf life of cottage cheese, in the absence of chemical preservatives and storage under proper refrigeration (3°C–4°C), is 14–21 days. The shelf life of soft cheeses amounts to approximately 4 weeks when stored at 4°C (Rodriguez-Aguilera et al., 2009). The shelf life of Mozzarella in brine is very short, approximately from 5 to 7 days. Grated Graviera cheese packaged in transparent containers has a relative short shelf life of approximately 2–3 weeks under refrigeration greatly depending on its salt content (Trobetas et al., 2008).

Both yield and quality of milk products, including cheeses, ice cream and yogurt mixes, cultured products, and related products, can be affected by the condition of raw milk. Significant changes do occur as a result of microbial growth in raw milk over transport and holding (Hotchkiss et al., 2006). Milk and dairy products are highly nutritious media, in which microorganisms can grow and cause spoilage. The contents and types of microorganisms in milk and dairy products depend heavily on the microbial quality of the raw materials, the conditions under which the products are produced, and the temperature and duration of storage (Van Asselt and Te Giffel, 2009). Two main groups of bacteria are involved in cheese manufacture and ripening. The first group consists of starter bacteria (mainly *Lactococcus*) added to milk during cheese manufacturing. Approximately 10^9 CFU of starter bacteria per gram is present in the final product. The second group consists of adventitious microorganisms (contaminants, secondary microflora) from the environment, which contaminate the milk or cheese curd during manufacture and ripening. This group includes numerous species of LAB (*Lactobacillus*, *Pediococcus*, *Enterococcus*, and *Leuconostoc*) and surface cheese bacteria (*Micrococcus* and *Staphylococcus*). The presence of enterococci in dairy products has long been considered an index of insufficient sanitary conditions over milk production and processing. In fermented dairy products, the potential presence of *Enterococcus faecalis, Enterococcus feacium*, and *Enterococcus bovis* mainly depends on each individual factory and product because one has to take into account the locally existing conditions (Birolo et al., 2000).

The numerous hydrolytic enzymes expressed by this secondary microflora presumably affect proteolysis and lipolysis during cheese ripening and thus may contribute to cheese maturation (Ogier et al., 2002). The most detrimental result from the release of lipolytic and proteolytic enzymes is the collapse of the protein structure of the micelles and/or hydrolysis of lipids (Hotchkiss et al., 2006). Proteolysis is the most complicated of the primary events during cheese ripening, especially in internally ripened cheeses by bacteria. It results in the formation of peptides, free amino acids, and free ammonia. Peptides contribute to both flavor and texture, whereas the free amino acids and ammonia contribute to flavor and sapidness and control the pH as well. However, erroneous or excessive proteolysis may lead to the formation of peptides and free amino acids that can potentially cause off-flavors (Seisa et al., 2004). Milk and dairy products are very susceptible to oxidation. Dairy products are prone to light oxidation because of the presence of riboflavin (vitamin B_2). This strong photosensitizer is capable of absorbing both visible and UV light and transferring this energy into highly reactive forms of oxygen (Mestdagh et al., 2005).

It is believed that cheese making was discovered accidentally and initially developed in Iraq circa 7000–6000 BC. The migration of populations due to famines, conflicts, and invasions fielded its further spreading. Some representative examples of these migrations are the development of Swiss cheeses by the Helvetian tribe in Switzerland and the introduction of cheese making into England by the Romans. Various cheese varieties were developed per region because of the different prevailing agricultural conditions in each country. Currently, there are more than 2000 recognized varieties of cheese (O'Connor, 1995).

The proteins of cheese are slightly inferior in quality to those of whole milk owing to the loss of the whey proteins with their higher content of the sulfur amino acids. Other losses in the partition between curd and whey include part of the calcium and significant proportions of the original nicotinic acid, vitamin B, biotin, vitamin BIZ, folic acid, and vitamin C, and most of the lactose. Ripening decomposes the residual vitamin C although—depending on the culture used—several of the B vitamins may be synthesized, mainly in the outer cheese layers (Rolls and Porge, 1973).

Cheese, in view of its high susceptibility to spoilage, must be well preserved in terms of its freshness and quality by picking the right kind of package. Each cheese type has its own requirements for packaging. However, there are some standard demands applicable in all cheese types:

- Protection and preservation of the product from the effects of oxygen to avoid the growth of molds and oxidation of fats. Special attention must be paid to the tightness of the packaging seams. The latter is important than the packaging material itself.
- Cheese must be well protected from drying by the package.
- Package should be able to withstand mechanical stresses.
- Packaging material should permeate carbon dioxide properly (Ahtiainen, 2009).

Processed cheese is not a preserved food but a "semipreserved food" with a limited shelf life. The maximal shelf-life guarantee to be given for premium grade processed cheese should not exceed 3–4 months, when the product is packaged in plastic foils. However, products stored in metal cans or tubes may have longer shelf lives. Processed cheese products usually retain their good quality for up to 6–12 months at room temperature. It is noteworthy that there are strong differences with regard to the shelf life of the following products stored at room temperature: 8 weeks for slices, 20 weeks for small portions, more than 1 year for products packed in tubes or cans (Schär and Bosset, 2002). In cheese, flavor and taste are, mainly, generated by the starters during the ripening stage. Proteolysis and lipolysis are the first steps of the elaboration of a large number of taste and odor compounds directly involved in the sensory quality of cheeses (Molimard et al., 1997). The sensory quality of ripened cheeses is determined by the technological parameters and the initial characteristics of the raw material. Diet modifies the chemical composition of milk and, thus, affects the sensory quality of milk products. In particular, milk produced from cows fed on pasture has different fatty acid composition and volatile compounds than milk produced from cows fed on hay or grain. Hard and semihard cheeses made from raw milk ripen more rapidly and have a more intense flavor and a different qualitative flavor than other cheeses made from pasteurized or microfiltered milk (Buchin et al., 1998).

Apart from traditional glass bottles and coated paperboard cartons, all-plastic containers have been used for pasteurized milk packaging. Problems with all-plastic containers used in the studies cited here include light transmission and oxygen permeability. It should be noted, however, that oxidative reactions (auto-oxidation) have been reported to take place in milk packaged even in coated paperboard cartons, which were found to be more or less impermeable to oxygen. More recently, polyethylene terephthalate (PET) and coextruded high-density polyethylene (HDPE) bottles have been used for fresh milk packaging (Karatapanis et al., 2006). PET is a versatile plastic material normally used in the bottling of soft beverages, and it has been introduced as a packaging material for fluid milk. PET offers several advantages as a food-packaging material, including transparency, light weight, resistance, and recyclability. Moreover, pigmented PET enhances its versatility by protecting the food from light, which, in turn, helps to protect food flavor against light-induced lipid oxidation. One potentially negative attribute of PET packaging, however, is that its oxygen permeability may be a factor in the development of oxidized off-flavors in UP milk over time (Solano-Lopez et al., 2005). More recently, PET and coextruded HDPE bottles have been used for fresh milk packaging. On the other hand, pigmented HDPE bottles, both monolayer and multilayer, at a higher thickness than current PET, are finding their way into the fresh milk packaging market. Both provide excellent convenience through easy opening and reclosing, thus minimizing recontamination (Karatapanis et al., 2006).

It is well known that packaging materials such as polyethylene (PE) and polystyrene (PS) are gas permeable and allow the diffusion of oxygen into yogurt during storage. PS, as a package material, is used in the manufacture of plastic cups for dairy products, such as ice cream or yogurt cups (Talwalkar and Kailasapathy, 2004; Zabaniotou and Kassidi, 2003). The survival of *Lactobacillus acidophilus* improved over a 35-day period in yogurts that were packaged in glass bottles as compared to when the yogurt was packaged in plastic cups. The oxygen content in yogurts stored in plastic cups was enhanced due to the permeation of oxygen, whereas the yogurts contained in the glass bottles retained a low oxygen environment. This led to the suggestion that prevention of oxygen toxicity in probiotic bacteria is feasible provided that yogurts are packed in glass containers. Although effective, glass jars are rather inconvenient and impractical because of their high cost and handling hazards. On the other hand, PE and PS do not have sufficient oxygen barrier properties and are thus inappropriate to prevent oxygen ingress into yogurt during storage. Yogurts when packaged in PS-based packaging reinforced with an added gas-barrier layer (Nupak™) were found to demonstrate no increase in their dissolved oxygen levels. In comparison, the dissolved oxygen in yogurts packaged with conventional PS tubs was seen to rise steadily over the shelf life (Talwalkar and Kailasapathy, 2004). The two main packaging methods used by cheese manufacturers are vacuum packaging (VP) and gas-flushed (GP) packaging. GF and heat sealing (HS) are used to package cheese shreds and cubed cheeses to minimize aggregation and loss of individual identity of cheese particle (Agarwal et al., 2005). It is anticipated that packaging of dairy products in transparent packaging materials may lead to off-flavor formation, discoloration, nutrient loss, and formation of toxic compounds (Mortensen et al., 2002). Dairy products are frequently exposed to light during retail storage and display. This exposure can substantially affect the quality of these products, especially when packed in transparent films or containers (Wold et al., 2002). Oxygen permeability of the plastic film was closely associated with surface oxidation in Cheddar cheese. Oxygen was the primary factor in initiating lipid oxidation triggered by lighting. A gold lamp or yellow shield has been put forward as replacement for the current white fluorescent light tubes because they caused fewer light-induced flavors in milk product. Aluminum-laminated film and film containing a UV-screening material retained a more desirable cheese flavor than did cheeses wrapped in PE film (Hong et al., 1995). Efficient cheese packaging is dependent on a number of important parameters, such as the use of starter cultures in cheese production; the type of cheese, that is, stabilized (cream, Feta), active (semi-soft, hard), or ripened (Brie); and the initial contamination and storage conditions (Gammariello et al., 2009). Currently, the packaging of Mozzarella is made from rigid or flexible films of multilayer material, packages fabricated from PE/paper laminated films or, more recently, tetra pack-type packages (Laurienzo et al., 2006). To maintain soft cheese craft and farm appearance, they are usually wrapped in paper and occasionally inserted in wooden-foil boxes, as popularized by French cheeses such as Camembert. Moreover, they are commonly wrapped in a three-layer film consisting of a wax layer, in contact with the cheese surface, a paper layer, and an outer layer of varnish. This combination leads to a material with low O_2 and water vapor permeabilities (Rodriguez-Aguilera et al., 2009).

Due to environmental concerns, the application of aluminum and metalized foils has gradually dropped, and the packaging industry focuses on downgauging the materials. This has led to enhanced use of transparent packaging materials, often with decreased barriers against light and oxygen, thereby increasing the risk of light-induced oxidation (Andersen et al., 2006).

7.2 RAW MILK

The addition of pure CO_2 to milk made possible a rapid drop in the closely linked pH that reached rapidly the critical value of 6. The initial pH after the addition of a mixture of 50% CO_2 and 50% N_2 ranged from 6.6 to 6.7, which was close to the normal range for milk (6.7–6.8). The pH of milk treated 50% CO_2 and 50% N_2 remained close to the initial value until day 6 of storage at 7°C and then dropped slightly to reach a final value of 6.2 at day 9 (Dechemi et al., 2005). The bacterial microflora was held at its original level in the milk stored at 6.0°C with the higher nitrogen rate, N_1, for 11 days. At 7.0°C, the control, C, exhibited a 5-log-unit enhancement within 10 days, irrespective of the subgroup of bacteria examined. At this temperature, N_1-treated milk displayed a 2-log-unit increase, and N_2 showed a 3.5-log-unit rise in bacterial counts over the same 10-day period. At 12.0°C, the growth of the total bacteria or psychrotrophs was nearly unchanged for 2 days under the higher flow rate, N_1. At day 3, the total counts were still below 10^5 CFU/mL, measuring at 3.6×10^4 CFU/mL under N_1 conditions, whereas the control had already reached its peak at 6×10^8 CFU/mL. N_2 also exerted an inhibitory effect on the mesophilic bacteria, to some extent (Munsch-Alatossava et al., 2007).

7.3 MILK AND CREAM POWDER

Milk powder manufacturing is a simple process now carried out on a large scale. It involves the gentle removal of water at the lowest possible cost under stringent hygiene conditions while retaining all the desirable natural properties of the milk: color, flavor, solubility, and nutritional value. The milk powder contains lactose (38%), protein (26%), fat (26%), and ash (6%) in the same proportions as fluid milk: The milk powder is produced in three forms, full cream (26% fat), partially skimmed (8%–24% fat), and skimmed (1.5% fat) milk powders for animal food in which fat was not more than 1.5% (El Khier et al., 2009). A steady decrease in the Hunter L^*-value was recorded with storage time after an initial, small increase. The Hunter a^*-value increased during storage while the Hunter b^*-value, after an initial decrease, was enhanced from day 8 of storage to day 43. The same qualitative changes of the Hunter L^*- and b^*-values were observed for the reconstituted milk as in the powder during storage. However, the a^*-value of the reconstituted milk decreased over storage (Nielsen et al., 1997). Lipid oxidation including the generation of cholesterol oxidation products was the highest in samples processed by high NO_X. Oxygen absorbers effectively controlled cholesterol oxidation during the 6 month storage period even in those samples from the high NO_X drying system. The oxidation of cholesterol in whole milk powders can be minimized by the use of oxygen-impermeable packaging systems containing oxygen absorbers or by using drying processes that generate low levels of NO_X (Chan et al., 1993). The oxidation of cream and milk powder proceeded very slowly when stored in darkness at 30°C. Oxygen concentration of 0.3 mL/L in the headspace was not low enough to protect the sample from sensory changes during 25 weeks of storage. Even the samples with low oxygen concentration were not significantly different according to sensory analysis after either 25 or 45 weeks of storage because the oxygen available for oxidation was sufficient to cause a change in sensory perception compared to the milk powder (Anderson and Lingnert, 1998).

7.4 FRESH CHEESE

7.4.1 Cameros Cheese

Cameros cheese is a fresh fat cheese made from pasteurized goat milk. It takes its name from the Cameros geographical area in the province of La Rioja (Spain). Based on microbial determinations, it can be concluded that the application of packaging systems under 100% CO_2 atmosphere for Cameros cheese inhibits the development of microorganisms involved in its spoilage and considerably extended the shelf life of this product. Mixtures of CO_2/N_2 delay the microbial growth and psychrotroph growth was lower when the CO_2 concentration increased (3.81 ± 0.28 log CFU/g in 20% CO_2, 3.00 ± 0.19 log CFU/g in 40% CO_2 2.43 ± 0.24 log CFU/g in 50% CO_2, 0.22 ± 0.15 log CFU/g in 100% CO_2) and cheeses packaged under MAs only reached mesophile populations above 7 log CFU/g on day 28 of storage depending on the conditions, with the exception of cheeses packaged under 100% CO_2, which resulted in value of 5.42 ± 0.53 log CFU/g (Gonzalez-Fandos et al., 2000; Olarte et al., 2002). After 14 days of storage, the sensory characteristics of the control cheeses were unacceptable for all the parameters investigated. However, the overall score for cheeses stored in 40% and 50% CO_2 were not altered substantially retaining a reasonable acceptability until the end of the storage period. The 100% CO_2 atmosphere had a very negative effect on the sensory quality specially in taste. With regard to Cameros

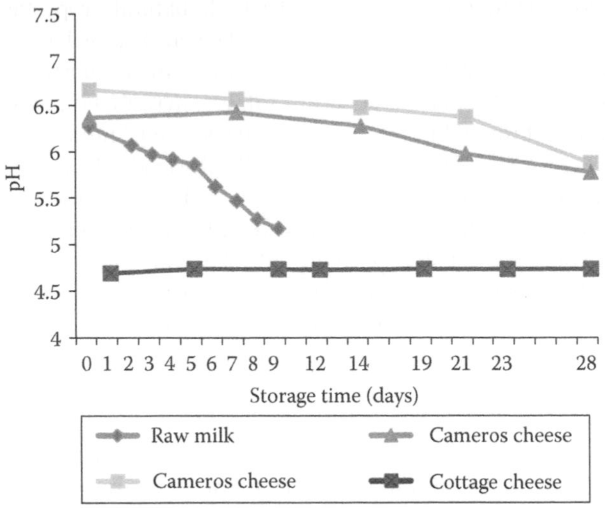

FIGURE 7.1 Effect of MAP (50% CO_2/50% N_2) on pH of raw milk (♦) (From Dechemi, S. et al., *Eng. Life Sci.*, 5(4), 350, 2005) and on cameros cheese (■) (From Gonzalez-Fandos, E. et al., *Food Microbiol.*, 17, 407, 2000), effect of MAP (40% CO_2/60% N_2) (▲) on cameros cheese (From Olarte, C. et al., *Food Microbiol.*, 19, 75, 2002), and effect of MAP (100% CO_2) on cottage cheese (x). (From Maniar, B. et al., *J. Food Sci.*, 59(3), 1305, 1994.)

cheese, packaging in 50% CO_2/50% N_2 and 40% CO_2/60% N_2 was the most effective for maintaining good sensory characteristics especially in taste and odor. However, texture and appearance were negatively affected under these conditions (Olarte et al., 2001) (Figure 7.1).

7.4.2 Cottage Cheese

Cottage cheese is a fresh, unripened product produced from pasteurized skimmed cows' milk. The standard shelf life of cottage cheese is estimated to be 10–21 days. The growth of contaminating yeasts and molds induces undesirable changes in flavor, odor, texture, and appearance. MAP has been applied to successfully prolong the shelf life of cottage cheese and improve the shelf life of sliced and shredded cheeses (Grove, 1998). Sensory evaluation indicated that cottage cheese flushed with air displayed low flavor scores after day 19. Cottage cheese tasted acidic and appeared gel-like and an increase in nonidentified off-flavors was reported. Cottage cheese flushed with MAs had no definite flavor or appearance defects after 28 days. Cottage cheese flushed with 100% CO_2 had highest scores, considering all attributes, with a mean score at 28 days of 4.1, followed by cottage cheese flushed with 75% CO_2/25% N_2, (x=3.6) and air (x=2.9), respectively. An optimum headspace of about 25% (v/v) flushed with pure CO_2 extended the shelf life of cottage cheese at 8°C by about 150% without altering the sensory properties or causing any other negative effect (Maniar et al., 1994; Mannheim and Soffer, 1996). No difference in yeast and mold counts over a period of 6 days between CO_2 flushed jars and controls was reported.

There was a significant difference in the odor of cheeses after 6 days. Cheese from control jars had a very bad, not sour but putrid odor possibly due to bacteria growth such as *Pseudomonades*, while the cottage cheese in CO_2-flushed jars maintained its good natural odor (Mannheim and Soffer, 1996). Total counts of bacteria in control samples increased by 3 log cycles after 4 days, while in the CO_2-flushed jars the bacteria remained in the lag phase for about 12 days followed by an increase of about 1.5 log cycles until 21 days. In the conventionally packaged cottage cheese, *Listeria monocytogenes* increased from 10^4 to 10^7 CFU/g after lag phases of 28 and 7 days at 4°C and 7°C, respectively. In contrast, *L. monocytogenes* failed to grow in cottage cheese packaged with CO_2 and stored at 4°C up to 63 days and increased from 10^4 to 10^5 CFU/g in products packaged with CO_2 at 7°C. Initial counts of LAB were 5.50×10^6 CFU/g. An increase to 1.32×10^7 CFU/g for cottage cheese packaged under air over the 28 days was observed, while counts remained almost unchanged for cottage cheese treated with MAs (Chen and Hotchkiss, 1993; Maniar et al., 1994; Mannheim and Soffer, 1996).

7.4.3 "Fior di latte"

The combination of chitosan, active coating, and MAP increased the shelf life of the packaged "Fior di latte" to 5 days. As regards the coliform growth, the obtained data suggest that chitosan, coating, and MAP alone did not seem to affect, to a great extent, the cell growth cycle, whereas the use of active coating effectively inhibited the growth of these microorganisms. The presence of chitosan also enhanced the antimicrobial efficacy of the active coating. MAP itself hardly affected the color of the packaged cheese. On the contrary, *Pseudomonas* was affected by coating and MAP as well as by active coating (Del Nobile et al., 2009a).

7.4.4 Requeijão Cremoso

Mesophilic (maximum of 4.0×10^4 CFU/g), psychrotrophic (maximum of 4.0×10^4 CFU/g), and spore-forming aerobic mesophilic (maximum of 3.2×10^3 CFU/g) microorganisms and molds and yeasts (maximum of 1.3×10^2 CFU/g) were found to be either not present or present in very low numbers, independent of the package type and storage condition (under light and in the dark) (Vercelino Alves et al., 2007). The minimization of release of every free fatty acid (FFA) at 4°C as storage time elapses requires increases in the content of CO_2 in the flushing gas. It was therefore shown that plain CO_2 completely inhibits lipolysis at all storage temperatures. Higher storage temperature or longer storage time displayed increases in the rate and extent of lactose metabolism, respectively. However, the composition of the initial flushing gas did not play an important role in lactose and lactic acid contents, as it did in the case of inner pH (Pintado and Malcata, 2000).

The degree of overall sensory quality loss depended both on the package type and whether the product was exposed to light or stored in the dark. The loss of overall quality of Requeijão Cremoso is significantly greater when the product is exposed to light over storage. The occurring changes affect the lipid phase and cause the product to be rejected by sensory analysis (Vercelino Alves et al., 2007).

7.4.5 Giuncata and Primosale Cheese

A slight decrease in pH values was recorded for all tested samples. The decrease by the end of storage is typical of fresh cheese produced without a starter culture. The average moisture dropped during storage for all cheese samples. There were no significant differences between cheese packaged in air and cheese packaged under vacuum (UV). Similar results were also recorded for the MAP samples. All samples displayed an increasing trend: the initial microbial count was about 6.00 log CFU/g becoming dominant toward the end of the storage period, regardless of packaging conditions (about 8.50 log CFU/g). The MAP conditions did not affect the growth of typical dairy microorganisms. As far as *Pseudomonas* spp. counts are concerned, the vacuum and the control samples displayed a gradual growth of spoilage microorganisms. Among the MAP samples 50:50 CO_2:N_2 and 90:10 CO_2:N_2 compositions were the most effective in reducing *Pseudomonas* spp. counts. No molds were detected on samples over the entire storage period. With regard to yeasts, they were detected in similar amounts in the entire MAP and VP samples with a microbial load lower than the control sample (Gammariello et al., 2009).

7.4.6 Cheddar Cheese

Cheddar cheese is a popular cheese originating from England in the town of Cheddar in the sixteenth century. It has color variations from pale to deep yellow/orange while the flavor can be mild and creamy for the mild-Cheddar to strong and biting for the mature Cheddar. It is also described as having a slightly nutty/walnut flavor. The differences in texture and flavor of Cheddar are closely linked to the length of the ripening period. Mild Cheddar can be sold for 3–4 months and its texture is close and firm yet pliable and breaks down smoothly when small portions are kneaded between the fingers. Mature Cheddar is usually 12–14 months old endowed with an intense flavor and the texture of the cheese at this stage is harder. Cheddar cheese is a rennet coagulated cheese. The production of Cheddar cheese involves the mixture of milk, rennet, microorganisms, and salt (Oyugi and Buys, 2007).

In packaged cheese samples, the levels of aerobic microbes were maintained at approximately 10^8–10^9 CFU/g. The low levels of O_2 and the high levels of CO_2 maintain bacteria in the lag phase of their growth. When the protective atmosphere is compromised, the microbial numbers seem to increase to ~10^{12} CFU/g (O'Mahony et al., 2006). The LAB counts in the shredded Cheddar cheese in the 80% CO_2/17% N_2/3% O_2 and in the in the 73% CO_2/27% N_2 atmosphere were log 7.2–7.6 CFU/g initially and during week 16, the LAB counts were log 6.8–7.0 CFU/g. The average mold population in the shredded Cheddar cheese in the 80% CO_2/17% N_2/3% O_2 atmosphere was 0.3 log CFU/g and in the cheese in the 73% CO_2/27% N_2 atmosphere the average value was 0.2 log CFU/g (Oyugi and Buys, 2007). The ability to conduct 100% quality control ensures that all samples with enhanced levels of oxygen could be detected. Residual oxygen was monitored over 4 months of product shelf life at +4°C along with measurement of microbial growth (O'Mahony et al., 2006).

7.5 PASTA FILATA

7.5.1 Mozzarella Cheese

Mozzarella cheese was originally manufactured from high-fat buffalo milk in the Battipaglia region of Italy. Nowadays, Mozzarella cheese is soft, white, and unripened and made from cow milk, buffalo milk, and even milk powder (El Owni and Osman, 2009). Mozzarella cheese is a member of the pasta filata (stretched curd) cheese family, which involves the principle of skillfully stretching the curd in hot water to get the smooth texture in cheese (Conte et al., 2007; El Owni and Osman, 2009).

The initial pH drop of the investigated cheese samples was related to degradation of lactose to lactic acid. Upon exhaustion of lactose supply, the pH stabilized. Following 6 weeks of storage, the rise in pH might be due to acid metabolism of certain microorganisms or its neutralization by protein degradation products. In the final stage of storage, as a result of higher (i.e., room) temperature, an intensive protein proteolysis might have been taking place that caused a rise in pH and total acidity at the same time. The highest increase in the acidity (a fall of pH and a rise in the titratable acidity) was recorded in cheese samples packed under carbon dioxide, while the lowest increase was reported in cheese samples packed in the reduced pressure air atmosphere or the atmosphere of nitrogen (Pluta et al., 2005). LAB counts varied little with all counts within a 1 log range, between 6.73 log CFU/g (air, week 1) and 7.70 log/g (50/50 CO_2/N_2, week 4). The means of each treatment were similar: 7.09 for 1 atm, 7.34 for 100% CO_2, and the others varied between 7.41 and 7.49 log CFU/g. The initial count (4.36 log CFU/g) reached 7 log CFU/g after 3 weeks and then stabilized around 7.2 log CFU/g. The means of MAP were also similar between 6.68 and 6.82 log CFU/g in 100% CO_2 and vacuum, respectively. Mesophilic bacteria counts remained relatively stable and varied between 6.84 log CFU/g (air, week 1) and 7.82 log CFU/g (100% N_2, week 2) Mesophilic counts of cheeses packaged under air dropped during the first week, thereby confirming the inhibition of LAB by oxygen (Eliot et al., 1998).

7.6 WHEY CHEESE

7.6.1 "Anthotyros"

The use of both MAP conditions extended the shelf life of fresh "Anthotyros" cheese stored at 4°C by ca. 10 days (30% CO_2/70% N_2) or 20 days (70% CO_2/30% N_2), and by ca. 2 days (30% CO_2/70% N_2) and 4 days (70% CO_2/30% N_2) at 12°C, with cheese maintaining satisfactory sensory characteristics. Of the two MAs, the 70/30 (CO_2/N_2) gas mixture was the most effective for the inhibition of total volatile counts (TVC), perhaps due to the inhibitory effect of the higher concentration of CO_2 on microbial growth. Inhibition of TVCs was greater at 4°C than at 12°C, probably due to the enhanced solubility of CO_2 at lower temperatures. Counts of LAB followed the same pattern as TVCs, with low counts for gas mixtures [(30% CO_2/70% N_2) or (70% CO_2/30% N_2)] at both temperatures (Papaioannou et al., 2007).

7.6.2 "Myzithra Kalathaki"

Myzithra cheese was packaged in three different atmospheres: 20% CO_2/80% N_2, 40% CO_2/60% N_2, and 60% CO_2/40% N_2. Of the three MAP samples, 40% CO_2/60% N_2 and 60% CO_2/40% N_2 were substantially more effective in reducing Enterobacteriaceae counts, which at the end of the storage period ranged between 3.6 and 3.8 log CFU/g. Yeasts and molds for all MA whey cheese samples remained below the detection limit of the method (1 log CFU/g) until day 13 of storage. As regards LAB of the MAP samples, those packaged under 40% CO_2 gave the lowest counts as compared with all other treatments, reaching the value of 7 log CFU/g on day 42 of storage (Dermiki et al., 2008) (Figures 7.2 through 7.5).

7.6.3 Ricotta and Stracchino Cheese

The values of total coliforms, in handicraft Ricotta cheese samples packaged in MAP, ranged from a minimum of 1.5×10^2 CFU/g to a maximum of 3.6×10^7 CFU/g at the end of shelf life. Ricotta cheese packaged in normal atmosphere showed similar coliform values, even if the mean value was a little higher. Total coliforms were isolated from 80.0% of handicraft Stracchino cheese samples packaged under normal atmosphere, with values ranging between 3.6×10^2 and 4.9×10^5 CFU/g at

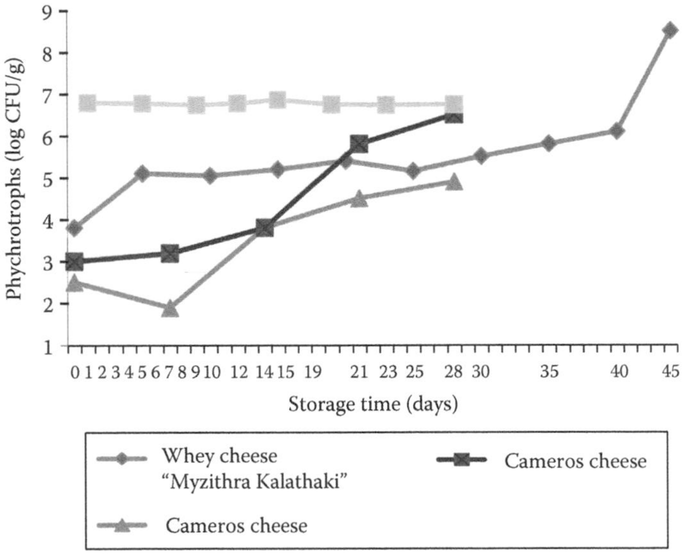

FIGURE 7.2 Effect of MAP (40% CO_2/60% N_2) on phychrotrophs on whey cheese "Myzithra Kalathaki" (♦) (From Dermiki, M. et al., *LWT Food Sci. Technol.*, 41, 284, 2008) and on cameros cheese (▲) (From Olarte, C. et al., *Food Microbiol.*, 19, 75, 2002), effect of MAP (50% CO_2/50% N_2) on cameros cheese (■) (From Gonzalez-Fandos, E. et al., *Food Microbiol.*, 17, 407, 2000), and effect of MAP (100% CO_2) on cottage cheese (x). (From Maniar, B. et al., *J. Food Sci.*, 59(3), 1305, 1994.)

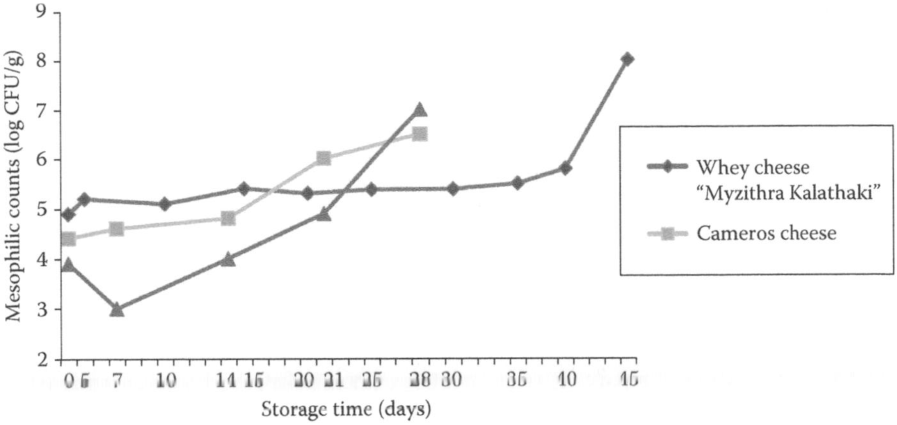

FIGURE 7.3 Effect of MAP (40% CO_2/60% N_2) on mesophilic counts on whey cheese "Myzithra Kalathaki" (◆) (From Dermiki, M. et al., *LWT Food Sci. Technol.*, 41, 284, 2008) and on cameros cheese (▲) (From Olarte, C. et al., *Food Microbiol.*, 19, 75, 2002), effect of MAP (50% CO_2/50% N_2) on cameros cheese (■). (From Gonzalez-Fandos, E. et al., *Food Microbiol.*, 17, 407, 2000.)

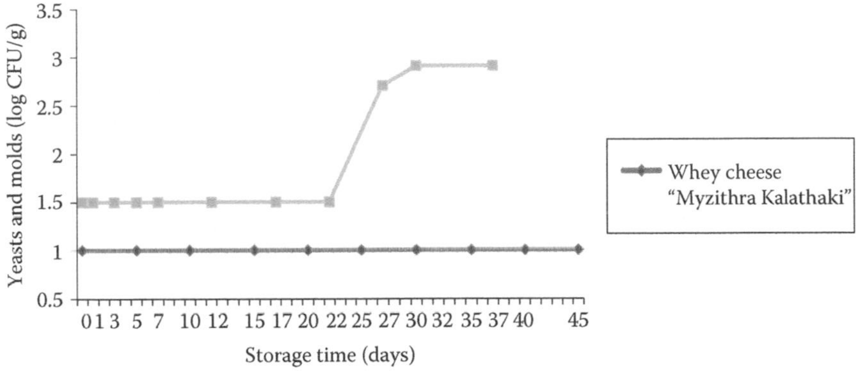

FIGURE 7.4 Effect of MAP (40% CO_2/60% N_2) on yeasts and molds on whey cheese "Myzithra Kalathaki" (◆) (From Dermiki, M. et al., *LWT Food Sci. Technol.*, 41, 284, 2008) and effect of MAP (70% CO_2/30% N_2) on whey cheese "Anthotyros" (■). (From Papaioannou, G. et al., *Int. Dairy J.*, 17, 358, 2007.)

time zero and between 2.3×10^3 and 4.0×10^6 CFU/g at the end of the shelf life. The products packaged in MAP displayed recoveries comprising between 2.5×10^2 and 4.3×10^5 CFU/g at the production time and between 1.5×10^2 and 4.3×10^6 CFU/g at the end of the shelf life. Finally, only one industrial stracchino cheese sample (7.1%) was found to have been contaminated by total coliforms (10^2 CFU/g) (Paris et al., 2004).

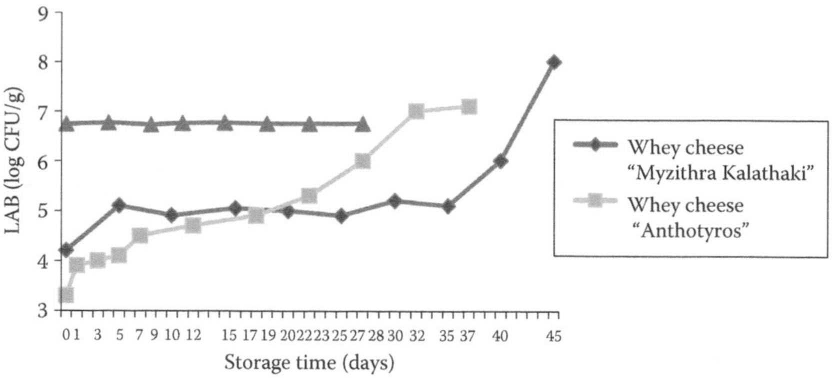

FIGURE 7.5 Effect of MAP (40% CO_2/60% N_2) on LAB on whey cheese "Myzithra Kalathaki" (◆) (From Dermiki, M. et al., *LWT Food Sci. Technol.*, 41, 284, 2008) and effect of MAP (70% CO_2/30% N_2) on whey cheese "Anthotyros" (■) (From Papaioannou, G. et al., *Int. Dairy J.*, 17, 358, 2007) effect of MAP (100% CO_2) on cottage cheese (▲). (From Maniar, B. et al., *J. Food Sci.*, 59(3), 1305, 1994.)

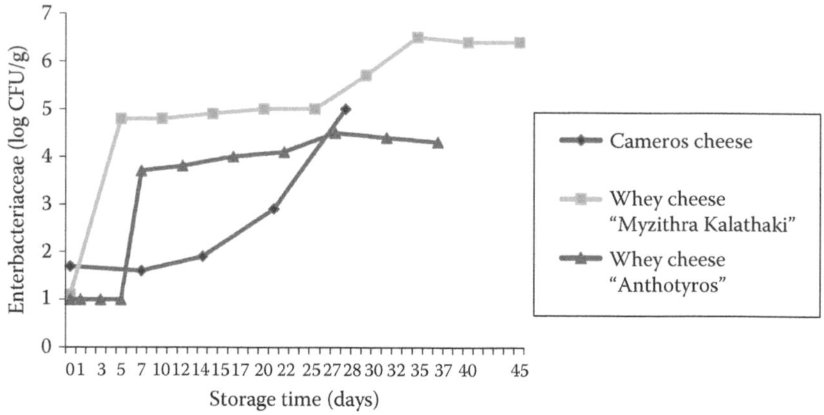

FIGURE 7.6 Effect of MAP (50% CO_2/50% N_2) on enterobacteriaceae on whey cheese cameros cheese (◆) (From Gonzalez-Fandos, E. et al., *Food Microbiol.*, 17, 407, 2000), effect of MAP (40% CO_2/60% N_2) on LAB on whey cheese "Myzithra Kalathaki" (■) (From Dermiki, M. et al., *LWT-Food Sci. Technol.*, 41, 284, 2008), and effect of MAP (70% CO_2/30% N_2) on whey cheese "Anthotyros" (▲). (From Papaioannou, G. et al., *Int. Dairy J.*, 17, 358, 2007.)

The shelf-life values for Ricotta cheese were 1.14 ± 0.34, 0.55 ± 0.5, 0.36 ± 0.6, and 3.37 ± 0.7, respectively, for 50:50 (CO_2:N_2), 70:30 (CO_2:N_2), 95:5 (CO_2:N_2), and samples. No differences were recorded by comparing L^* (lightness) values and a^* values (redness). On the contrary, substantial differences between samples stored under ordinary atmosphere and MAP were recorded in the yellow index (b^*) (Del Nobile et al., 2009b) (Figure 7.6).

7.7 SEMIHARD CHEESE

7.7.1 Sliced Samso Cheese

Light exposure resulted in significantly decreased yellowness and increased redness of sliced Samso cheese packaged in 0% (100% N_2), 20% (80% N_2), or 100% CO_2 (0% N_2), during storage. Cheese stored in 100% CO_2 and exposed to light had significantly lower L^* values (lightness) than cheeses stored under the other storage and packaging combinations. The sensory panel described cheeses stored in the dark in 0% and 20% CO_2 as most buttermilk-like (odor and taste) and buttery (odor). However, cheese stored in 100% CO_2 and exposed to light was described by the sensory panel as rancid (taste and odor) and of a dry/crumbly texture (Juric et al., 2003).

7.7.2 Provolone Cheese

The mixture of 30% CO_2 and 70% N_2 prolonged Provolone cheese shelf life by 50% in comparison with VP, bringing it up to 280 days. The mesophilic aerobic bacteria count and LAB were similar in samples packaged with 100% CO_2 or VP and significantly lower than those determined in all the other samples. The lowest amount of LAB was found in the samples conditioned with 30% CO_2 that also had the highest pH compared with all other cheeses. As far as yeasts and molds are concerned, cheeses stored under 20% or 30% CO_2 exhibited a significantly higher yeast content in comparison with the samples packaged with the other tested systems. The mold count was approximately equal to that for control (VP) samples, thereby showing that a CO_2 saturated atmosphere is not more effective than VP in limiting their growth (Favati et al., 2007).

7.7.3 Sliced Havarti Cheese

Light exposure conditions and residual oxygen contents had significant effect on the color of cheeses. An initial drop followed by a rise in a^* values during storage was recorded for the 0.6% and 0.1% residual oxygen. Cheeses packed with oxygen absorbers (0.01% residual oxygen) had lower b^* values than did cheeses packaged in 0.6% residual oxygen. The changes on two one-dimensional color axes, a^* and b^*, were not counterbalanced, and overall changes in hue during storage were recorded. However, changes were not related to the combination of light source and residual oxygen content (Mortensen et al., 2002). The results of the sensory evaluation vs. time are described as sensory attributes. For all attributes apart from sour odor, there was a significant treatment effect, that is, effect of light exposure. For all attributes except for sweet and sour odor and sour taste, a significant storage time effect was recorded. The peroxide value (PV) was recorded as an index of lipid oxidation. The PV were found to be rather variable within a^* range between 0.5 and 1.25 mEq O_2/kg lipid for both treatments during the storage period. Application of statistical analysis revealed no significant difference between cheeses stored under light or in the dark. The reported variation found for the PV of the samples could be possibly attributed to the natural variations occurring among the cheeses (Kristensen et al., 2000) (Figures 7.7 through 7.9).

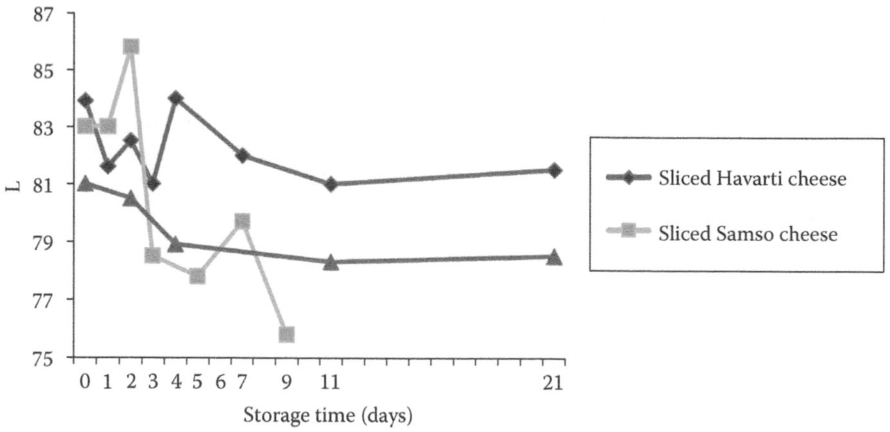

FIGURE 7.7 Effect of MAP (25% CO_2/75% N_2) on L (lightness) on sliced Havarti cheese (♦) (From Kristensen, D. et al., *Int. Dairy J.*, 10, 95, 2000), effect of MAP (100% CO_2) on sliced Samso cheese (■) (From Juric, M. et al., *Int. Dairy J.*, 13, 239, 2003), and effect of MAP (100% CO_2) on grated Graviera cheese (▲). (From Trobetas, A. et al., *Int. Dairy J.*, 18, 1133, 2008.)

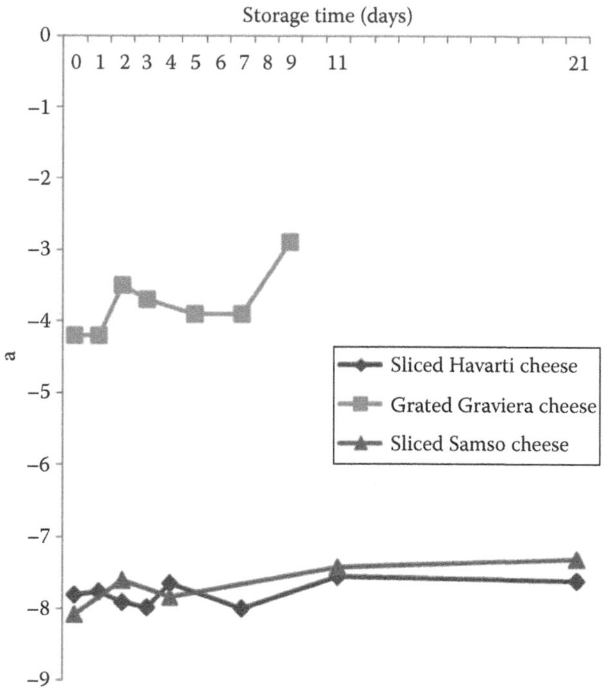

FIGURE 7.8 Effect of MAP (25% CO_2/75% N_2) on a (redness) on sliced Havarti cheese (♦) (From Kristensen, D. et al., *Int. Dairy J.*, 10, 95, 2000), effect of MAP (100% CO_2) on grated Graviera cheese (■) (From Trobetas, A. et al., *Int. Dairy J.*, 18, 1133, 2008), and effect of MAP (100% CO_2) on sliced Samso cheese (▲). (From Juric, M. et al., *Int. Dairy J.*, 13, 239, 2003.)

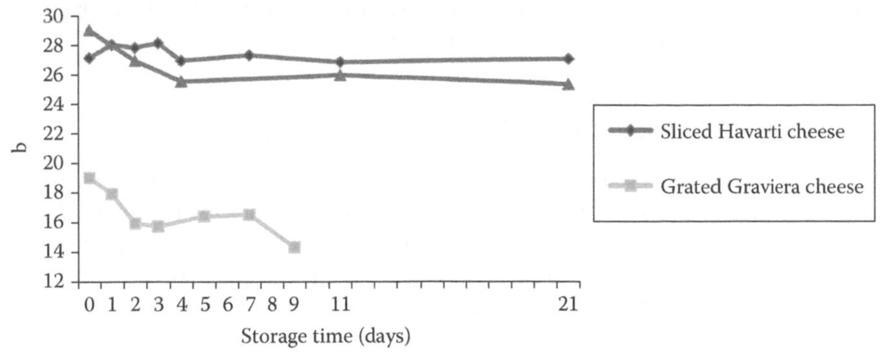

FIGURE 7.9 Effect of MAP (25% CO_2/75% N_2) on b (yellowness) on sliced Havarti cheese (♦) (From Kristensen, D. et al., *Int. Dairy J.*, 10, 95, 2000), effect of MAP (100% CO_2) on grated Graviera cheese (■) (From Trobetas, A. et al., *Int. Dairy J.*, 18, 1133, 2008), and effect of MAP (100% CO_2) on sliced Samso cheese (▲). (From Juric, M. et al., *Int. Dairy J.*, 13, 239, 2003.)

7.8 HARD CHEESE

7.8.1 Grated Graviera

No statistically significant differences in shelf life were recorded for samples stored under 100% CO_2 either in the dark or under light (3 weeks in the dark vs. 2.8 weeks under light). For samples stored under 100% N_2 and 50% N_2/50% CO_2, differences were at least 9 weeks vs. 4 weeks, respectively. Thus, the shelf life of grated Graviera cheese may be maximized by using MAP (either 100% N_2 or 50% N_2/50% CO_2) in combination with a packaging material impermeable to light. For samples packaged under 100% CO_2, the sensory analysis gave the lowest score with regard to taste. The latter was described as "bitter" after week 5 of storage (Trobetas et al., 2008).

7.8.2 Parmigiano Reggiano Cheese

Packed portioned Parmigiano Reggiano cheese was monitored during 3 months of storage at 4°C. Packaging conditions were UV and in MAs 50:50 and 30:70 CO_2/N_2 ratios. All samples underwent proteolysis together with changes in the textural and sensory characteristics. The UV-packed sample revealed the occurrence of an oil dropping up phenomenon, which induced significant changes in product characteristics such as an increase in cohesion, sourness, and yellowness. The MA-packed samples displayed different textural behavior for 30:70 CO_2/N_2 ratio samples evolving toward a more cohesive and friable structure than 50:50 CO_2/N_2 ratios. A similar observation was recorded for the flavor profile after 90 days of storage, revealing a softer taste for the MAP cheese than the unpacked cheese. Parmigiano Reggiano cheese hardness exhibited discordant results according to the testing method (compression, shear, or shear–compression), regardless of whether it involved the fracture of cheese structure or not (Romani et al., 2002).

7.9 WHITE CHEESES, CHEESE, AND CHEESECAKES

The white cheeses analyzed were deliberately contaminated with *L. monocytogenes*. The average contamination level of white cheeses prior to packing reached 6.9×10^2 CFU/g. In control samples, the population of *L. monocytogenes* was reduced by 2.1×10^2 CFU/g after 7 days. After 14 and 21 days of storage, a further drop was recorded in the population of that pathogen, that is, to 5×10 and 1.5×10 CFU/g, respectively. Over the storage period, a decrease in the cell number of *L. monocytogenes* was also observed in the VP samples and those packed with oxygen absorber, to a level close to that recorded for the control sample. The number of *Enterococcus* spp. *streptcoocci* in the white cheeses analyzed was low (3.5×10 CFU/g on average) and was hardly altered over the storage of white cheeses in different packaging. Similarly, the number of coli group bacteria and that of *Escherichia coli* bacteria remained low and unchangeable during storage (Panfil-Kuncewicz et al., 2006). All fungi displayed growth in 20% or 40% CO_2 with 1% or 5% O_2, whereas growth was strongly reduced in the presence of 20%–80% CO_2. Mycotoxin production considerably declined but not totally annihilated in 20% or 40% CO_2 and 1% or 5% O_2, respectively (Taniwaki et al., 2001). Cheesecakes were packaged under different N_2/CO_2 ratios (70/30 and 20/80) (MAP batches). MAP prolonged the shelf life of samples packed under 30% and 80% CO_2 up to 14 and 34 days, respectively, whereas tarts stored in air spoiled after 7 days. It was shown that application of MAP led to a significant increase in hardness after 14 days of storage (Sanguinetti et al., 2009).

7.10 SOFT CHEESE

Soft surface-mold ripened farmhouse cheeses made from pasteurized milk and having a high moisture content are common craft products in various countries, including Ireland. Since these cheeses have a short shelf life, in most cases they do require controlled refrigerated conditions over distribution and sale. The quality and safety are impaired during storage due to fat oxidations, microbiological contamination, and loss of moisture and aroma (Rodriguez-Aguilera et al., 2009). Spoilage was recognized as welling of packets or as visible growth on the surface of the cheese. In total, 18 yeast species were isolated from the soft cheese. *Torulaspora delbrueckii, Candida parapsilosis, Pichia fermentans, Pichia norvegensis,* and *Pichia membranaefaciens* were most commonly isolated in spoiled soft cheese from Dairy A. *Yarrowia lipolytica, Debaryomyces hansenii, Pichia guilliermondii, Cryptococcus sp., and Rhodotorula sp.* were most common on cheese from Dairy B. Yeast occurrence on soft cheese from Dairy A was most likely caused by recontamination from the production and packing area. Yeast occurrence on soft cheese from Dairy B was associated with the yeasts found on the material used for decoration (Westall and Filtenborg, 1998) (Figure 7.10).

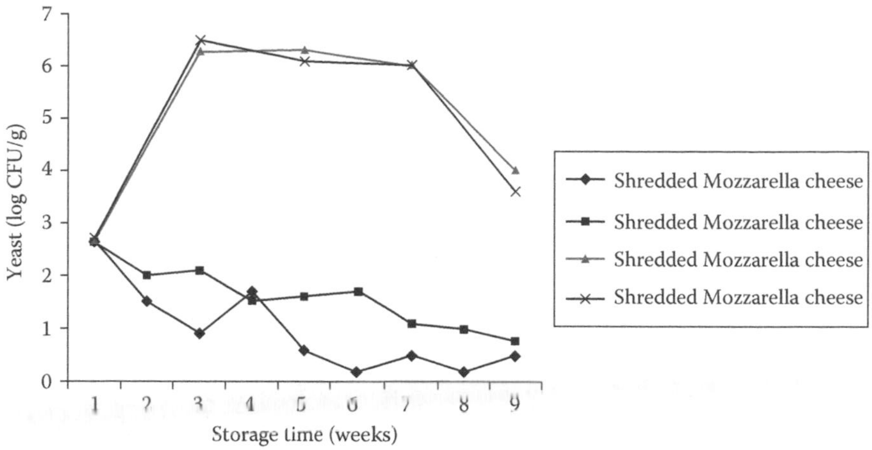

FIGURE 7.10 Effect of MAP (50%CO_2/50%N_2) on yeast on shredded Mozzarella cheese (♦) (From Eliot, C.S. et al., *J. Food Sci.*, 63(6), 1075, 1998), and (Pluta et al., 2005), effect of MAP (10% CO_2/90%N_2) on yeast on shredded Mozzarella cheese (▲) (From Eliot, C.S. et al., *J. Food Sci.*, 63(6), 1075, 1998), (■) and effect of MAP (70%CO_2/30%N_2) on shredded Mozzarella cheese (x). (Pluta et al., 2005.)

7.11 YOGURT

Yogurt is prepared with acidification of milk and fortified with the addition of dried skim milk, in the presence of LAB (Muir, 1996). Refrigerated milk properly acidified with carbon dioxide could be satisfactorily used for yogurt manufacture without any observable change in sensory properties. Gassing with carbon dioxide enhanced the shelf life of yogurt. Carbonization of the finished yogurt took place by a process involving postproduction homogenization at a pressure of 175 kg/cm^2. Combination of postincubation heat-treatment and gas flushing with carbon dioxide improved the keeping quality of yogurt for more than 4 weeks at 15°C. It has been noted that incorporation of carbon dioxide at lower levels had no effect on the bacterial population of yogurt (Sarkar, 2006).

7.12 CONCLUSIONS

The majority of dairy products are susceptible to oxidation and alteration, and their shelf life is rather limited unless supplementary heat treatment or modified atmosphere is applied. Although the MAP conditions may vary from one dairy product to another depending mainly on their fat content, some typical gas compositions are CO_2/N_2:100/0, 90/10, 80/20, 70/30, 50/50, 40/60, 20/80, and 0/100. In case oxygen is the third component, its content will be in the range 3%–5%. The presence of 100% CO_2 is not suggested for the fear of packaging collapse due to CO_2 absorbance by the dairy product. Another option is packaging UV, which can also considerably prolong the shelf life of dairy products (Table 7.1).

TABLE 7.1
Effect of MA and Packaging Material on Dairy Products Shelf-Life Prolongation

Species	Gas Composition: Product Ratio	Sensory Assessment	Shelf Life (Days)	Shelf-Life Extension (Days)	Reference Material	Storage Temperature	References
Cameros cheese	20% CO_2/80% N_2		7 days	14 days	The plastic films used were provided by Dixie (Dixie, Bern, Switzerland) with a CO_2 permeability of less than 13 $cm^3/m^2/24$ h at 1 atm and O_2 permeability of 5 $cm^3/m^2/24$ h at 1 atm	4°C	Olarte et al. (2002)
Cameros cheese	40% CO_2/60% N_2	Reasonable acceptability	7 days	—			
Cameros cheese	100% CO_2	Negative effect in sensory quality	7 days	—			
Cameros cheese	50% CO_2/50% N_2	Good sensory characteristics	7 days	21 days	The plastic films used were provided by Dixie (Dixie, Bern, Switzerland) with a CO_2 permeability of less than 13 $cm^3/m^2/24$ h at 1 atm and O_2 permeability of 5 $cm^3/m^2/24$ h at 1 atm	3°C–4°C	Gonzalez-Fandos et al. (2000)
Cheddar cheese	73% CO_2/27% N_2	—	120 days 210 days with oxygen scavenging films	360 days with and without oxygen scavenging films	Two laminate packaging films and oxygen scavenging films A laminate film (control film), which consisted of Bx nylon/linear low-density polyethylene (LLDPE)/low-density polyethylene (LDPE)/ LLDPE	5°C ± 1°C	Oyugi and Buys (2007)

	80% CO_2/17% N_2/3% O_2		120 days 210 days with oxygen scavenging films with an oxygen scavenger Ciba® SHELPLUS™ O_2 (Ciba Specialty Chemicals, Sweden) incorporated into its multilayer structure at 3% of its total weight	210 days 360 days with oxygen scavenging films with an oxygen scavenger Ciba SHELPLUS O_2 (Ciba Specialty Chemicals, Sweden) incorporated into its multilayer structure at 3% of its total weight	It had an oxygen transmission rate (OTR) < 20 mL/m²/24 h/atm at 22°C and 75% relative humidity (RH) Another laminate film with an oxygen scavenger Ciba SHELPLUS O_2 incorporated into its multilayer structure at 3% of its total weight. It consisted of Ex nylon/linear LDPE/LDPE with master batch containing O_2/LLDPE It had an OTR < 20 mL/m²/24 h/atm at 22°C and 75% RH		
Cheddar cheese	70% CO_2/30% N_2	Slight changes in calibration properties after long-term exposure to food	—	Residual oxygen (<0.75%) was monitored over 120 days of product shelf life	Samples were packaged using triplex type laminate (Amcor Flexibles, Europe) consisting of Polyester (PET) on the outside/Print-adhesive/PET middle layer/Hot Melt adhesive (for reseal)/Spec.PE (for peel). There is a coating of either PVDC (saran polyvinylidene chloride) or	4°C	O'Mahony et al. (2006)

(continued)

TABLE 7.1 (continued)
Effect of MA and Packaging Material on Dairy Products Shelf-Life Prolongation

Species	Gas Composition: Product Ratio	Sensory Assessment	Shelf Life (Days)	Shelf-Life Extension (Days)	Reference Material	Storage Temperature	References
					EVOH (Ethylene vinyl alcohol copolymer) on the inside of the outer layer as a gas barrier, oxygen barrier <2 cm^3/m^2/24 h/1 bar/O_2 0% HR and N_2 100% HR		
Cheddar cheese	60% N_2/40% CO_2	—	—	24 days	PS/EVOH/PE cellulose-based packaging paper was assessed for the preparation of bioactive inserts, designed for concomitant use with MAP. Plastic film (70:30, PE:PA) of the type used to produce vacuum pouches was also assessed for its ability to adsorb bacteriocin	4°C	Scannell et al. (2000)
Cheese	20% CO_2/5% O_2	—	~14 days	~15 days	Barrier plastic bags (polypropylene:ethylene vinyl alcohol:polypropylene)	25°C	Taniwaki et al. (2001)

	20% CO_2/1% O_2	—			were placed in PE plastic bags to permit air exchange while limiting a_w changes		
	40% CO_2/5% O_2	—					
	40% CO_2/1% O_2	—					
Cheese tart	70% N_2/30% CO_2 O_2 from 0% to maximum 0.40% 20% N_2/80% CO_2 O_2 from 0 to maximum 0.40%	Sensory analysis evidenced a significant decrease in almost all considered parameters, especially for the experimental batch that withstood prolonged storage The overall acceptability was always above the threshold for the entire storage period	After 7 days	Up to 14 days Up to 34 days	The tarts were packaged inside multistrate (EVOH/PS/PE) gas barrier trays, (two pieces of cake for each tray) (Aerpack B5-30, Coopbox Italia, Reggio Emilia, Italy) and wrapped with a multistrate (EVOH/OPET/PE) gas and water barrier film with a thickness of 5[illegible] μm (EOM 360B, Sealed Air, the United States). The gas transmission rates for the tray were as follows: O_2, 1.07 $cm^3/m^2/24h$/bar at 23°C; CO_2, 5.35 $cm^3/m^2/24h$/bar at 23°C; water vapor, 63 $g/m^2/24h$/bar at 38°C. The gas transmission rates for the film were as follows: O_2, 4 $cm^3/m^2/24h$/bar at 23°C; CO_2, 5.35 $cm^3/m^2/24h$/bar at 23°C;	20°C	Sanguinetti et al. (2009)

(continued)

TABLE 7.1 (continued)
Effect of MA and Packaging Material on Dairy Products Shelf-Life Prolongation

Species	Gas Composition: Product Ratio	Sensory Assessment	Shelf Life (Days)	Shelf-Life Extension (Days)	Reference Material	Storage Temperature	References
					water vapor, 9 g/m²/24 h/bar at 38°C		
Cottage cheese	Pure CO_2	Without altering sensory properties or causing any other negative effect	14–17 days under proper refrigeration of 3°C–4°C	Extended shelf-life of cottage cheese by about 150%	330 mL glass jars with white cap twist-off lids; 250 mL PS cups with heat sealable aluminum covers; high-barrier trays made from a laminate of polypropylene/ethyl-vinylalcohol/polypropylene with a laminate cover film that consisted of aluminum/polypropylene	8°C	Mannheim and Soffer (1996)
Cottage cheese	35% CO_2	—	21–28 days	The direct addition of CO_2 to enhance shelf life was feasible by 100%–200% without altering the sensory properties	PS tubs overwrapped and without overwrap high-barrier heat shrink film. The headspace of packages was sampled (100 µL) using a gastight syringe	4°C–7°C	Chen and Hotchkiss (1993)

Cream cheese	—	Cream cheese stored in trays made of A-PET/PE offered the best protection against oxidation with respect to sensory flavor (acidulous and sunlight flavor notes)	—	—	Transparent and black Multipet with polymer combination of amorphous polyethylene terephthalate + polyethylene (A-PET/PE). The thickness of the sheet before thermoforming was 750 and 550 mm, respectively Another polymer combination was white, PS + EVOH copolymer + PE with a thickness of 750 mm and the third polymer combination was white PP/PE with a thickness of 600 mm. The lid film used in the experiment was 52 mm thick transparent PET/AlOx/PE film	4°C	Pettersen et al. (2005)
Crottin de Chavignol cheese	20% CO_2, 80% N_2	Significant difference in MAP for some textural characteristics as hardness, gumminess, and chewiness	180 days	—	—	−25°C	Esmer et al. (2009)
Giuncata cheese	50% CO_2/50% N_2	Had the same behavior	—	—	All portions (200 g) of Giuncata were packaged in commercially available bags with a thickness of 95 μm	8°C	Gammariello et al. (2009)

(*continued*)

TABLE 7.1 (continued)
Effect of MA and Packaging Material on Dairy Products Shelf-Life Prolongation

Species	Gas Composition: Product Ratio	Sensory Assessment	Shelf Life (Days)	Shelf-Life Extension (Days)	Reference Material	Storage Temperature	References
					These were obtained by laminating a nylon layer and polyolefin layer and have an OTR of 50 mL/m^2/24·h at 1 atm, measured a 23°C and 75% RH		
Fior di latte cheese	30% CO_2/5% O_2, and 65% N_2	MAP did not affect the color of the packaged cheese	—	MAP increased the shelf life of the packaged "Fior di latte" to 1 day	Bags that were obtained by laminating a nylon layer and a polyolefin layer and have an OTR of 50 mL/m^2/24 h at 1 atm, measured at 23°C and 75% RH and a water vapor transmission rate (WVTR) of 1.64 g/m^2/24 h at 1 atm, measured at 23°C and 85% RH	4°C	Del Nobile et al. (2009a)
Grater graviera (hard cheese)	100% CO_2 100% N_2	Light-protected samples retained their desirable sensory attributes longer than their light-exposed counterparts, while taste was a more sensitive attribute than odor in cheese evaluation	~17 days (in the dark) 14 days (under light)	21 days (in the dark) ~18–19 days (under light) 270 days (in the dark) 120 days (under light)	LDPE/polyamide (PA)/LDPE barrier pouches 80 g per pouch, 75 μm thickness having an OTR of 52.2 mL/m^2/day/atm and a water vapor transmission rate of 1.29 g/m^2/day	4°C	Trobetas et al. (2008)

	50% CO_2/50% N_2			270 days (in the dark) 120 days(under light)			
Kashkaval	90% CO_2/10% N_2	The hardness of samples packed UV was higher than that of samples packed under atmospheric or MA, when packed in the same material	—	—	Chosen packaging materials: Cryovac foil, thickness 70 μm, usually used for packing of cut hard cheeses, and two domestic foils: the combined foil PP lacquered with PVDC lacquer/poly-ethylene, PP (FVDC)PE	—	Lazić and Curakovic (1997)
Milk powder	70% N_2/30% CO_2	Powder and reconstituted milk turned darker and yellower with storage time. Fluorescent oxidation products were formed at a constant rate from carbonyl compounds and protein amino groups Milk powders were best distinguished by their color and by their content of β-lactoglobulin and α-lactalbumin	Fresher powders (corresponding to accelerated storage times up to 25 days) and older powders (accelerated storage times from 25 to 43 days)		The milk powders were placed in PVC boxes without lids and stored in one of the two identical incubators at 50°C over a saturated aqueous solution of magnesium chloride (a_w 0.31)	50°C	Nielsen et al. (1998)

(continued)

TABLE 7.1 (continued)
Effect of MA and Packaging Material on Dairy Products Shelf-Life Prolongation

Species	Gas Composition: Product Ratio	Sensory Assessment	Shelf Life (Days)	Shelf-Life Extension (Days)	Reference Material	Storage Temperature	References
Milk powder	Low and high levels for NO_x (O_2 concentration in PE pouches and the glass vials without and with oxygen was 20.1%, 20.2%, and 0.03%)	—	Oxygen absorbers controlled cholesterol oxidation over the 6 month storage period	—	Pouched oxygen absorbers Polyethylene pouches O (PE) (Ziploic freezer bags 5 in. × 7 in. 2 mil 50 µm thickness and crimp sealed glass vials (with and without oxygen absorbers) (100 mL) were sealed by aluminum crimp on caps with teflon septa	20°C ± 1°C and 40°C ± 1°C	Chan et al. (1993)
Mozzarella cheese	50% N_2/50% CO_2 70% CO_2/30% N_2	The sensory evaluation of grated Mozzarella cheese after 4 weeks of cold storage revealed no statistically significant difference between the packing variants employed	13 days	—	The packaging material consisted of laminated PE/PA bags	10°C ± 0.5°C	Pluta et al. (2005)
Parmigiano Reggiano cheese	50% N_2/50% CO_2 (MA1)	The MAP samples showed different textural behavior	Commercial shelf life of this product was 3 months	—	Into high-barrier plastic pouches (nylon/ PE, 30 mm nylon and 120 mm of PE)	4°C	Romani et al. (2002)

	30% CO_2/70% (MA2)	(MA2 sample evolved toward a more cohesive and friable structure than MA1). Similar evolution was recorded for the flavor profile, after 90 days of storage			The permeability data of the plastic film were the following: 0.42 O_2, 1.18 CO_2 (ASTM D 1434), 0.09 H_2O (ASTM E 94)		
Primosale cheese	10% CO_2/90% N_2	Not good sensorial quality in all mixtures	—	—	All portions (200 g) of Primosale were packaged in commercially available bags with a thickness of 95 µm. These were obtained by laminating a nylon layer and polyolefin in layer and have an OTR of 50 mL/m²/24 h at 1 atm, measured at 23°C and 75% RH	8°C	Gammariello et al. (2009)
Provo-lone cheese	10% CO_2/90% N_2	—	190 days	100 days	Pouches made of PA and PE (20 mm PA/80 mm PE) and having an oxygen permeability of 50 cm³/m²/24 h/bar³ at 23°C	4°C–8°C	Favati et al. (2007)
	20% CO_2/80% N_2			118 days			
	30% CO_2/70% N_2			280 days			
	100% CO_2			175 days			
Raw milk	The N_2 gas was 99.999% pure	Under N_2, at 6.0°C and 7.0°C, the pH values were acceptable. The studies performed at 12.0°C showed excessive bacterial	At 6.0°C 11 days At 12°C 4 days	—	Each of the two flasks was connected to a flow meter via 0.2 µm sterile filters	6.0°C, 7.0°C, and 12.0°C ± 0.1°C	Munsch-Alatossava et al. (2010)

(continued)

TABLE 7.1 (continued)
Effect of MA and Packaging Material on Dairy Products Shelf-Life Prolongation

Species	Gas Composition: Product Ratio	Sensory Assessment	Shelf Life (Days)	Shelf-Life Extension (Days)	Reference Material	Storage Temperature	References
		growth after 4 days and the resulting acidification of the milk would exclude it from further use					
Raw milk	100% N_2	Nitrogen gas has no effect on the milk quality	—	—	The flasks were treated immediately at 7°C by adjusting the flow rates of the mixture to 100 mL/min during 25 min, so that the residual oxygen content of the milk decreased by two levels	7°C ± 0.2°C	Dechemi et al. (2005)
	50% CO_2 and 50% N_2	50% CO_2 and 50% N_2 had strong inhibitory effects on the enzyme activities		2 days for 50% CO_2 and 50% N_2 and 1 day for 100% CO_2 for all microbiological groups			
	100% CO_2	CO_2 could be accompanied by a decrease in pH and a significant acidification					
Requeijão (Portuguese whey cheese)	100% N_2	Good at 4°C	—	All types of package can increase the shelf life of whey cheese	The headspace of sterile bags was flushed with the appropriate gas at 550 mbar, and automatic sealing took place after 1.5 s	4°C, 8°C, and 12°C	Pintado and Malcata (2000)
	50% CO_2 and 50% N_2	Good at 4°C	—				
	100% CO_2	Good at 4°C					

Requeijão Cremoso (processed cheese)	—	Detrimental effect of light on all the sensory attributes while no significant differences were observed between the samples of Requeijão contained in the different package types tested in the course of 60 days of storage in the absence of light	8 days loss the overall quality in PP cup, 10 days in Coex squeeze tube, 15 days in SG 4 days in PE squeeze	The incorporation of an oxygen barrier into the Coex squeeze tube enhances the shelf life of the product under light for more than 1 week	Glass, 275 mL glass, 208 mL, cup thermoformed from PP, 322 mL (PP cup/250 g), Coex squeeze tube, 269 mL, (Coex tube/186 g) PE squeeze tube, 269 mL (PE tube/186 g)	10°C	Vercelino Alves et al. (2007)
Ricotta	50% CO_2/50% N_2 CO_2/N_2 50%/50%		14 days at 2°C Low storage temperatures can obviously affect microbial growth of fresh whey cheese. Shelf-life values of 33, 13.5, and 5.5 days for commercial Ricotta cheese packaged in PE bags and kept at 6°C, 17°C, and 25°C, respectively	1.14 ± 0.34 for ordinary atmosphere 0.55 ± 0.5 for 50% CO_2 /50% N_2 0.36 ± 0.6 for 70% CO_2 /30% N_2 and 3.37 ± 0.7 for 95% CO_2 /5% N_2. The longest shelf life was obtained with MAP containing 95% of CO_2	Packaged in nylon-based high-barrier multilayer plastic bags (250 mm × 350 mm and thickness 200 μm), having an OTR of 30 $cm^3/m^2/24\,h$ at 1 atm (measured at 23°C and 75% RH), and a WVTR of 1 $g/m^2/24\,h$ (measured at 23°C and 85% RH)	4°C	Del Nobile et al. (2009b)

(continued)

TABLE 7.1 (continued)
Effect of MA and Packaging Material on Dairy Products Shelf-Life Prolongation

Species	Gas Composition: Product Ratio	Sensory Assessment	Shelf Life (Days)	Shelf-Life Extension (Days)	Reference Material	Storage Temperature	References
Samso (semihard cheese sliced)	20% CO_2/80% N_2	Most buttermilk-like (odor and taste) buttery	—	—	A thermoformed transparent laminate of PA and PE (80 mm PA/120 mm PE) with an oxygen permeability of 13 cm^3/m^2/24 h at 23°C, 50% RH, and 1 atm	5°C (dark-stored)	Juric et al. (2003)
	100% N_2	(odor)		—			
	100% CO_2	Rancid and having a dry/crumbly texture				5°C (light-stored)	
Semihard cheese	50% CO_2/50% N_2	An elevated partial pressure of carbon dioxide in the gas phase might occur (in this case 57% [V/V]) with potential unwanted effects on the cheese quality, such as changes in texture and flavor	30 days	—	High-barrier pouches measuring 13.5 cm × 20.5 cm (12 μm PETP/9 μm ALU/75 μm) with very low gas permeability specified as an OTR of <0.1 cm^3/m^2/24 h/atm, corresponding to a CO_2 transmission rate of <0.025 cm^3/m^2/24 h/atm and a maximum loss of CO_2 over the packaging material	5°C	Jakobsen and Risbo (2009)
Sliced Havarti cheese	(0.6% or 0.01%)	Both storage and packaging parameters may reduce the effect of light-induced oxidation	—	—	Transparent material	3.6°C ± 1.0°C or 9.2°C ± 0.7°C	Andersen et al. (2006)

Sliced Havarti cheese	25% CO_2/75% N_2/0.4% O_2	All eight sensory attributes (sweet odor, buttery odor, rancid odor, sour odor, sweet taste, buttery taste, rancid taste, sour taste) were highly associated with each other, as shown by the fact that 95.6% of the total variation was explained by the first component	21 days	—	Conventional packaging materials consisting of a thermoformed transparent dome made of polyester, a thermoformed burgundy colored PS base, and with a polyester barrier layer. The OTR of the dome was determined to 0.034 cm^3 per package (24 h, 23°C, 0%/50% RH) according to standard method (ASTM F 1307). The OTR of the barrier layer was 60 cm^3/m^2 (24 h, 23°C, 5%/95% RH) according to the manufacturer	5°C	Kristensen et al. (2000)
St.Kilian cheese (soft cheese)	19% O_2 and 10% CO_2	The temperature and O_2 concentration have a significant effect on the gas exchange rate, whereas the effect of CO_2 concentration was negligible, thereby confirming the effect of refrigeration conditions on problems under MAP	Approximately 28 days under refrigerated conditions	—	Batches	4°C, 12°C, and 20°C	Rodriguez-Aguilera et al. (2009)

(continued)

TABLE 7.1 (continued)
Effect of MA and Packaging Material on Dairy Products Shelf-Life Prolongation

Species	Gas Composition: Product Ratio	Sensory Assessment	Shelf Life (Days)	Shelf-Life Extension (Days)	Reference Material	Storage Temperature	References
Stratiatella cheese	50% CO_2/50% N_2 0% O_2	Prolongation of sensorial acceptability limit	—	—	Traditional tubs	8°C	Gammariello et al. (2009)
	75% CO_2/25% N_2 0% O_2						
	30% CO_2/65% N_2 5% O_2		—	—			
Whey cheese "Anthotyros"	30% CO_2/70% N_2	Cheese maintaining good sensory characteristics	Less than 7 days	10 days at 4°C 2 days at 12°C	PS boxes containing ice (for the cheese's transportation) and PE barrier pouches were heat-sealed using a vacuum-sealer connected to the gas mixer	4°C and 12°C	Papaioannou et al. (2007)
	70% CO_2/30% N_2		Less than 7 days	20 days at 4°C 4 days at 12°C			
Whey Cheese "Myzithra Kalathaki"	20% CO_2/80% N_2	Good sensory characteristics until day 20 of storage	10–12 days	8–10 days	Plastic pouches composed of co-extruded LDPA/PA/LDPE 75 mm in thickness with an OTR of 52.2 mL/(m^2 day atm) and a WVTR of 2.4 g/(m^2 day), measured using the Oxtran 2/20 and Permatran 3/31 permeability testers	4°C ± 0.5°C	Dermiki et al. (2008)
	40% CO_2/60% N_2	Good sensory characteristics for 30 days of storage		14–16 days			
	60% CO_2/40% N_2			18–20 days			

REFERENCES

Ahtiainen, H. 2009. Paperboard cup with a window new packaging solutions, MSc thesis, Faculty of Technology, Lappeenranta University of Technology, Lappeenranta, Finland.

Agarwal, S., Costello, M., and Clark, S. 2005. Gas-flushed packaging contributes to calcium lactate crystals in Cheddar cheese. *Journal of Dairy Science* 88: 3773–3783.

Alm, L. 1982. Effect of fermentation on lactose, glucose and galactose content in milk and suitability of fermented milk products for lactose intolerant individuals. *Journal of Dairy Science* 65: 346–352.

Andersen, Ch.M., Wold, J.P., and Mortensen, G. 2006. Light-induced changes in semi-hard cheese determined by fluorescence spectroscopy and chemometrics. *International Dairy Journal* 16: 1483–1489.

Anderson, K. and Lingnert, H. 1998. Influence of oxygen and copper concentration on lipid oxidation in rapeseed oil. *Journal of the American Oil Chemist's Society* 75 (8): 1041–1046.

Birolo, L., Tutino, M.L., Fontanella, B., Gerday, C., Mainolfi, K., Pascarella, S., Sannia, G., Vinci, F., and Marino, G. 2000. Aspartate aminotransferase from the antarctic bacterium Pseudoalteromonas haloplanktis TAC125: Cloning, expression, properties, and molecular modelling. *European Journal of Biochemistry* 267: 2790–2802.

Bergamo, P., Fedele, E., Iannibelli, L., and Marzillo, G. 2003. Fat-soluble vitamin contents and fatty acid composition in organic and conventional Italian dairy products. *Food Chemistry* 82: 625–631.

Buchin, S., Delague, V., Duboz, G., Berdague, L.J., Beuvier, E., Pochet, S., and Grappin, R. 1998. Influence of pasteurization and fat composition of milk on the volatile compounds and flavor characteristics of a semi-hard cheese. *Journal of Dairy Science* 81: 3097–3108.

Buttriss, J. 2007. Nutritional properties of fermented milk products. *International Journal of Dairy Technology* 50 (1): 21–27.

Chan, S.H., Gray, J.I., Gomaa, E.A., Harte, B.R., Kelly, P.H., and Buckley, D.J. 1993. Cholesterol oxidation in whole milk powders as influenced by processing and packaging. *Food Chemistry* 47: 321–328.

Chen, J.H. and Hotchkiss, J.H. 1993. Growth of *Listeria monocytogenes* and *Clostridium sporogenes* in cottage cheese in modified atmosphere packaging. *Journal of Dairy Science* 76 (4): 972–977.

Conte, A., Scrocco, C., Sinigaglia, M., and Del Nobile, M.A. 2007. Innovative active packaging systems to prolong the shelf life of Mozzarella cheese. *Journal of Dairy Science* 90: 2126–2131.

Dechemi, S., Benjelloun, H., and Lebeault, M.J. 2005. Effect of modified atmospheres on the growth and extracellular enzymes activities on psychrotrophs in raw milk. *Engineering Life of Science* 5 (4): 350–356.

Del Nobile, A.M., Conte, A., Incoronato, A.L., and Panza, O. 2009b. Modified atmosphere packaging to improve the microbial stability of Ricotta. *African Journal of Microbiology Research* 3 (4): 137–142.

Del Nobile, A.M., Gammariello, D., Conte, A., and Attanasio, M. 2009a. A combination of chitosan, coating and modified atmosphere packaging for prolonging Fior di latte cheese shelf life. *Carbohydrate Polymers* 78: 151–156.

Dermiki, M., Ntzimani, A., Badeka, A., Savvaidis, N.I., and Kontomina, M.G. 2008. Shelf-life extension and quality attributes of the whey cheese "Myzithra Kalathaki" using modified atmosphere packaging. *LWT—Food Science and Technology* 41: 284–294.

Eliot, C.S., Vuillemard, C.J., and Emond, P.J. 1998. Stability of shredded Mozzarella cheese under modified atmospheres. *Journal of Food Science* 63 (6): 1075–1080.

El Khier, S., Khalid, M., and Yagoub Abu El Gasim, A. 2009. Quality assessment of milk powders packed in Sudan. *Pakistan Journal of Nutrition* 8 (4): 388–391.

El Owni, O.A.O. and Osman, E.S. 2009. Evaluation of chemical composition and yield of Mozzarella cheese using two different methods of processing. *Pakistan Journal of Nutrition* 8 (5): 684–687.

Esmer, O., Balkir, P., and Seckin, A.K. 2009. Changes in chemical, textural and sensory characteristics of Crottin de Chavignol cheese manufactured from frozen curd and packaged under modified atmosphere. *Milchwissenschaft* 64 (2): 184–187.

Fadela, C., Abderrahim, C., and Bensoltane, A. 2009. Physico-chemical and rheological properties of yoghurt manufactured with ewe's milk and skim milk. *African Journal of Biotechnology* 8 (9): 1938–1942.

Favati, F., Galgano, F., and Pace, M.A. 2007 Shelf-life evaluation of portioned provolone cheese packaged in protective atmosphere. *LWT—Food Science and Technology* 40: 480–488.

Gambelli, L., Manzi, P., Panfili, G., Vivanti, V., and Pizzoferrato, L. 1999. Constituents of nutritional relevance in fermented milk products commercialised in Italy. *Food Chemistry* 66 (3): 353–358.

Gammariello, D., Conte, A., Attanasio, M., and Del Nobile, A.M. 2009. Effect of modified atmospheres on microbiological and sensorial properties of Apulian fresh cheeses. *African Journal of Microbiology Research* 3 (7): 370–378.

Gonzalez-Fandos, E., Sanz, S., and Olarte, C. 2000. Microbiological, physicochemical and sensory characteristics of Cameros cheese packaged under modified atmospheres. *Food Microbiology* 17: 407–414.

Grove, T.M. 1998. Use of antimycotics, modified atmospheres, and packaging to affect mold spoilage in dairy products, PhD dissertation submitted to the Faculty of the Food Science and Technology, Virginia Polytechnic Institute and State University, Blacksburg, VA.

Haenlein G.F.W. 2004. Goat milk in human nutrition. *Small Ruminant Research* 51: 155–163.

Hong, M.C., Wendorff, L.W., and Bradley, L.R. 1995. Effects of packaging and lighting on pink discoloration and lipid oxidation of annatto-colored cheeses. *Journal of Dairy Science* 78: 1896–1902.

Hotchkiss, J.H., Werner, B.G., and Lee, E.Y.C. 2006. Addition of carbon dioxide to dairy products to improve quality: A comprehensive review. *Comprehensive Reviews in Food Science and Food Safety* 5 (4): 158–168.

Huffman, M.L. and Harper, J.W. 1999. Symposium: Marketing dairy value through technology maximizing the value of milk through separation technologies. *Journal of Dairy Science* 82: 2238–2244.

Jakobsen, M. and Risbo, J. 2009. Carbon dioxide equilibrium between product and gas phase of modified atmosphere packaging systems: Exemplified by semi-hard cheese. *Journal of Food Engineering* 92: 285–290.

Juric, M., Bertelsen, G., Mortensen, G., and Petersen, A.M. 2003. Light-induced colour and aroma changes in sliced, modified atmosphere packaged semi-hard cheeses. *International Dairy Journal* 13: 239–249.

Karatapanis, E.A., Badeka, V.A., Riganakos, A.K., Savvaidis, N.I., and Kontominas, G.M. 2006. Changes in flavour volatiles of whole pasteurized milk as affected by packaging material and storage time. *International Dairy Journal* 16: 750–761.

Kristensen, D., Orlien, V., Mortensen, G., Brockhoff, P., and Skibsted, H.L. 2000. Light-induced oxidation in sliced Havarti cheese packaged in modified atmosphere. *International Dairy Journal* 10: 95–103.

Kumar, P. and Mishra, H.N. 2004. Storage stability of mango soy fortified yoghurt powder in two different packaging materials: HDPP and ALP. *Journal of Food Engineering* 65: 569–576.

Laurienzo, P., Malinconico, M., Pizzano, R., Manzo, C., Piciocchi, N., Sorrentino, A., and Volpe, G.M. 2006. Natural polysaccharide-based gels for dairy food preservation. *Journal of Dairy Science* 89: 2856–2864.

Lazić, V. and Curaković, M. 1997. Influence of packaging on the rheological characteristics of Kashkaval. *Acta Alimentaria* 26 (2): 153–161.

Maniar, B., Marcy, E.J., Bishop, J.R., and Duncan, S.E. 1994. Modified atmosphere packaging to maintain direct-set cottage cheese quality. *Journal of Food Science* 59 (3): 1305–1308, 1327.

Mannheim, C.H. and Soffer, T. 1996. Shelf-life extension of cottage cheese by modified atmosphere packaging. *LWT—Food Science and Technology* 29: 767–771.

Mestdagh, F., de Meulenaer, B., de Clippeleer, J., Devlieghere, F., and Huyghebaert, A. 2005. Protective influence of several packaging materials on light oxidation of milk. *Journal of Dairy Science* 88: 499–510.

Molimard, P., Lesschaeve, I., Issanchou, S., Brousse, M., and Spinnler, H.E. 1997. Effect of the association of surface flora on the sensory properties of mould-ripened cheese. *Lait* 77: 181–187.

Mortensen, G., Sorensen, J., and Stapelfeldt, H. 2002. Effect of modified atmosphere packaging and storage conditions on photooxidation of sliced Havarti cheese. *European Food Research and Technology* 216 (1): 57–62.

Muir, D.D. 1996. The shelf-life of dairy products: 2. Raw milk and fresh products. *Journal of the Society of Dairy Technology* 49 (2): 44–48.

Munsch-Alatossava, P. and Alatossava, T. 2007. Antibiotic resistance of raw milk-associated psychrotrophic bacteria. *Microbiology Research*, 162: 115–123.

Munsch-Alatossava, P., Gursoy, O., and Alatossava, T. 2010. Potential of nitrogen gas (N_2) to control psychrotrophs and mesophiles in raw milk. *Microbiological Research* 165 (2): 122–132.

Nielsen, B.J., Borgen, A., Nielsen, G.C., and Scheel, C. 1998. Strategies for controlling seed-borne diseases in cereals and possibilities for reducing fungicide seed treatments. In *The Brighton Conference—Pest and Diseases,* pp. 893–900. http://www.agrologica.dk/publikationer/phd.appendiks5.htm (accessed on May 13, 2011).

Nielsen, R.B., Stapelfeldt, H., and Skibsted, L.H. 1997. Differentiation between 15 whole milk powders in relation to oxidative stability during accelerated storage: Analysis of variance and canonical variable analysis. *International Dairy Journal* 7: 589–599.

O'Connor, C.B. 1995. *Rural Dairy Technology*. International Livestock Research Institute, Addis Ababa, Ethiopia.

Ogier, J.-C., Olivier, S., Gruss, A., Tailliez, P., and Delacroix-Buchet, A. 2002. Identification of the bacterial microflora in dairy products by temporal temperature gradient gel electrophoresis. *Applied and Environmental Microbiology* 68 (8): 3691–3701.

Olarte, C., Gonzalez- Fandos, E., Gimenez, M., Sanz, S., and Portu, J. 2002. The growth of *Listeria monocytogenes* in fresh goat cheese (Cameros cheese) packaged under modified atmospheres. *Food Microbiology* 19: 75–82.

Olarte, C., Gonzalez- Fandos, E., and Sanz, S. 2001. A proposed methodology to determine the sensory quality of a fresh goat's cheese (Cameros cheese): Application to cheeses packaged under modified atmospheres. *Food Quality and Preference* 12: 163–170.

O'Mahony, C.F., O'Riordan, T.C., Papkovskaia, N., Kerry, P.J., and Papkovsky, B.D. 2006. Non-destructive assessment of oxygen levels in industrial modified atmosphere packaged cheddar cheese. *Food Control* 17: 286–292.

Oyugi, E. and Buys, M.E. 2007. Microbiological quality of shredded cheddar cheese packaged in modified atmospheres. *International Journal of Dairy Technology* 60 (2): 89–95.

Panfil-Kuncewicz, H., Staniewski, B., Szpendowski, J., and Nowak, H. 2006. Application of active packaging to improve the shelf life of white cheeses. *Polish Journal of Food and Nutrition Sciences* 15: 165–168.

Papaioannou, G., Chouliara, I., Karatapanis, E.A., Kontominas, G.M., and Savvaidis, N.I. 2007. Shelf-life of a Greek whey cheese under modified atmosphere packaging. *International Dairy Journal* 17: 358–364.

Paris, A., Bacci, C., Salsi, A., Bonardi, S., and Brindani, F. 2004. Microbial characterization of organic dairy products: *Stracchino and Ricotta* cheeses. *Ann Fac Medic Vet di Parma* XXIV: 317–325.

Pettersen, M.K., Eie, T., and Nilsson, A. 2005. Oxidative stability of cream cheese stored in thermoformed trays as affected by packaging material, drawing depth and light. *International Dairy Journal* 15: 355–362.

Pintado, E.M. and Malcata, X.F. 2000. Optimization of modified atmosphere packaging with respect to physicochemical characteristics of Requeijão. *Food Research International* 33: 821–832.

Pluta, A., Ziarno, M., and Kruk, M. 2005. Impact of modified atmosphere packing on the quality of grated Mozzarella cheese. *Polish Journal of Food and Nutrition Sciences* 14/55 (2): 117–122.

Rodriguez-Aguilera, R., Oliveira, C.J., Montanez, C.J., and Mahajan, V.P. 2009. Gas exchange dynamics in modified atmosphere packaging of soft cheese. *Journal of Food Engineering* 95 (3): 438–445.

Rolls, B.A. and Porge, G.W.J. 1973. Some effects of processing and storage on the nutritive value of milk and milk products. *Proceedings of the Nutrition Society* 32: 9–15.

Romani, S., Sacchetti, G., Pittia, P., Pinnavaia, G.G., and Dalla Rosa, M. 2002. Physical, chemical, textural and sensorial changes of portioned *Parmigiano reggiano* cheese packed under different conditions. *Food Science and Technology International* 8 (4): 203–211.

Sanguinetti, A.-M., Secchi, N., Del Caro, A., Stara, G., Roggio, T., and Piga, A. 2009. Effectiveness of active and modified atmosphere packaging on shelf life extension of a cheese tart. *International Journal of Food Science and Technology* 44: 1192–1198.

Sarkar, S. 2006. Shelf-life extension of cultured milk products. *Nutrition & Food Science* 36 (1): 24–31.

Scannell, A.G.M., Hill, C., Ross, P.R., Marx, S., Hartmeier, W., and Arendt, K.E. 2000. Development of bioactive food packaging materials using immobilised bacteriocins Lacticin 3147 and Nisaplin. *International Journal of Food Microbiology* 60: 241–249.

Schär, W. and Bosset, O.J. 2002. Chemical and physico-chemical changes in processed cheese and ready-made fondue during storage: A review. *Lebensmittel-Wissenschaft und-Technologie* 35: 15–20.

Seisa, D., Osthoff, G., Hugo, C., Hugo, A., Bothma, C., and van der Merwe, J. 2004. The effect of low-dose gamma irradiation and temperature on the microbiological and chemical changes during ripening of cheddar cheese. *Radiation Physics and Chemistry* 69: 419–431.

Solano-Lopez, E.C., Taehyun, J.I., and Alvarez, B.V. 2005. Volatile compounds and chemical changes in ultrapasteurized milk packaged in polyethylene terephthalate containers. *Journal of Food Science* 70 (6): 407–412.

Talwalkar, A. and Kailasapathy, K. 2004. The role of oxygen in the viability of probiotic bacteria with reference to *L. acidophilus* and *Bifidobacterium spp. Current Issues Intestinal Microbiology* 5: 1–8.

Taniwaki, M.H., Hocking, A.D., Pitt, J.I., and Fleet, G.H. 2001. Growth of fungi and mycotoxin production of cheese under modified atmospheres. *International Journal of Food Microbiology* 68: 125–133.

Trobetas, A., Badeka, A., and Kontominas, G.M. 2008. Light-induced changes in grated Graviera hard cheese packaged under modified atmospheres. *International Dairy Journal* 18: 1133–1139.

Van Asselt, A.J. and Te Giffel, M.C. 2009. Hygiene practices in liquid milk dairies. In: *Milk Processing and Quality Management*, A.Y. Tamime (ed.), Wiley Blackwell Publishing Ltd., Oxford, U.K.

Vercelino Alves, R.M., Van Dender, G.F.A., Jaime, B.M., Moreno Izildinha, S.B.M., and Pereira, C. 2007. Beatriz effect of light and packages on stability of spreadable processed cheese. *International Dairy Journal* 17: 365–373.

Westall, S. and Filtenborg, O. 1998. Spoilage yeasts of decorated soft cheese packed in modified atmosphere. *Food Microbiology* 15: 243–249.

White, C.H. 1993. Rapid methods of shelf-life of for estimation and prediction milk and dairy products. *Journal of Dairy Science* 76: 3126–3132.

Wold, P.J., Jorgensen, K., and Lundby, F. 2002. Nondestructive measurement of light-induced oxidation in dairy products by fluorescence spectroscopy and imaging. *Journal of Dairy Science* 85: 1693–1704.

Zabaniotou, A. and Kassidi, E. 2003. Life cycle assessment applied to egg packaging made from polystyrene and recycled paper. *Journal of Cleaner Production* 11: 549–559.

Zygoura, P., Moyssiadi, T., Badeka, A., Kondyli, E., Savvaidis, I., and Kontominas, M.G. 2003. Shelf life of whole pasteurized milk in Greece: Effect of packaging material. *Food Chemistry* 87 (1): 1–9.

Part IV

Applications of MAP in Foods of Plant Origin

8 Cereals

Ioannis S. Arvanitoyannis and
Konstantinos Kotsanopoulos

CONTENTS

8.1 INTRODUCTION

The use of packaging in the food supply chain is very important and is an essential part of food processing. It can be used for various purposes such as the assurance of protection of food from external infections (microorganisms), surrounding conditions (like atmosphere gases, water activity, etc.), and the appropriate labeling of foods (La Storia et al., 2008).

Dried cereals comprise a large variety of foods such as ready-to-eat grain foods, breakfast cereals, crackers, wafers, cookies, granola bars, dried snacks, all types of dry bakery mixes, and grain ingredients such as flours, rice, oats, corn meal, corn grits, and dried pasta (Cook and Johnson, 2009).

Following harvesting, storage and processing do take place in food processing plants. Subsequently, foods are transported to stores, ending up in supermarkets. During this process, they are constantly exposed to risks (such as infections of various microorganisms and pests) that degrade their quality and potentially threaten consumers' health. It should be noted that 5%–10% of products in developed and 35% in developing countries are destroyed annually because of pest attacks. The cost of damages from pests, fungi, and toxins in the United States can amount to $1,000,000,000 (Campbell et al., 2004).

8.1.1 Modified Atmosphere Packaging

The replacement of the air in a container with a different gas or a combination of gases is called modified atmosphere packaging (MAP). The most commonly used gases are carbon dioxide (CO_2) and nitrogen (N_2). MAP is simultaneously applied with other agents to control microbial growth and release toxic substances originated from the metabolism of microorganisms. CO_2 has a major role in MAP technique because it prevents the growth of both bacteria and fungi (Zardetto, 2005).

Lifetime prolongation of products packaged under MAP and the adequate preservation of their quality for a long period of time have contributed to extensive use of this technology. The number of MAP products sold worldwide increased from 1.6 billion packs in 1990 to 2.6 billion in 1997 (Smith and Day, 2003). Nowadays, an extensive variety of edible cereals is produced to meet human nutritional requirements. These products include baked goods, refrigerated dough, pasta, dried cereal products, snack foods, etc. A variety of changes (physical, chemical, and microbiological) occurring in these foods considerably affect sensory properties such as flavor, taste, and appearance and render them unacceptable for human consumption. Microorganisms can be found everywhere and potentially cause damage to nutrients and health problems to consumers (Cook and Johnson, 2009).

MAP is often used for processing products such as cereals of all kinds and dried fruits. Various changes in the proportion of gases contained in a package (removal of oxygen and access of carbon dioxide and nitrogen) can be implemented instead of smoking and can effectively help to suppress harmful insects and microorganisms (Riudavets et al., 2009). Two of the most common harmful organisms usually occurring in food plants are *Plodia interpunctella* and *Ephestia kuehniella*. These species affect a vast variety of products like grain-based food, dried fruits, and their by-products (Riudavets et al., 2009).

According to Iconomou et al. (2006),

> Controlled or modified atmosphere storage is the storage of grains with the composition of the intergranular air controlled in some manner. Changing the relative concentrations of O_2, N_2 and CO_2, the quality of the grain may be preserved for a longer time than storage under ambient air conditions

The rates of the main gases in environment range at the following levels: 78% nitrogen, 0.035% carbon dioxide, and 21% oxygen. Any change in this ratio inside food packages (usually by reducing the rate of oxygen and increasing the proportion of the other two essential gases) leads to restriction of the growth of harmful organisms because the majority of the latter are aerobic (Iconomou et al., 2006).

8.1.2 Cereal Grains

It is generally accepted that a relatively limited number of pests, bacteria, and fungi are liable for the largest percentage of cereal infection. Among the most common organisms included are fungi such as *Aspergillus*, which affects grain-based foods; *Claviceps*, affecting corn and grain; and *Fusarium*, which usually affects corn. These microorganisms are responsible for the deterioration of many bakery products as they can grow at relatively low levels of water activity. Generally, many types of microorganisms can contribute to the infection of foods. It is important to note that different microorganisms occurring in cereals depend heavily on the initial microbiological load of raw materials, processing procedures, general conditions of hygiene over production, transportation, and retailing conditions (Cook and Johnson, 2009).

8.2 MAP APPLICATION ON CEREAL PRODUCTS

Many types of cereal products are used both for human and animal consumption and can be effectively preserved for many months. From harvesting until consumption, cereals can be infected by a variety of agents such as insects, microorganisms, and ambient hazards that induce various physical, chemical, and sensory changes, which result in extensive produce quality deterioration (Iconomou et al., 2006).

The stringent restrictions imposed on the use of chemical methods have led cereal producers to the adoption of MAP for preserving a wide variety of foods. Successful maintenance of cereal products using MAP is mainly affected by the permeability of the grain bins to gases. However, further investigation is required to enhance the salability of cereal produces. Because of the intensive research already carried out on MAP technology and its applications, it can finally respond to modern consuming patterns. Active packaging technology can also be used in conjunction with MAP to enhance organoleptic features and shelf life of food products. Furthermore, "time–temperature indicators" can be used to indicate the time until the expiry of the product's shelf life and have been proven to be significant tools in the protection of consumer health (Jayas and Jeyamkondan, 2002) (Table 8.1).

MAP and controlled atmosphere (CA) packaging have been widely applied with relatively acceptable results to protect bulk grains under storage against various pests. Nevertheless, these technologies are not often the first option in the milling and processing industry, due to their exorbitant cost of application in comparison with methyl bromide. Moreover, these procedures not only have to be monitored but also sealing is essential for achieving adequate and safe management in view of possible contamination problems. Still, technologies like MAP, vacuum sealing, or treatments with low pressure are applicable to small-scale or ad hoc implementations (Campbell et al., 2004).

In recent years, the investigation of more applicable ways for creating and applying MAP has proven essential in minimizing potential losses from insects, pests, and microorganisms. Peanuts were among the first products in the United States where MAP was applied to protect them and rapidly adopted by cereal industries (Bell et al., 1993).

It is standard practice to reckon active packaging as a MAP field, because it modifies the gas composition inside the package to prolong the products' lifetime, improving at the same time their organoleptic properties and maintaining their safety. As regards cereals, active packages could be applied to enhance taste and increase purchasing power (La Storia et al., 2008).

Many types of cereals, like cold cereals and milled grain, are not adequately maintained in single packs or sealed bags even when oxygen absorbers are used to reduce the oxygen concentration and are therefore unsuitable for canning (Hagan,

TABLE 8.1
Effect of Several Gas Compositions on Shelf Life and Micro- and Macroflora and Fauna of Grains

Food Type	Gas Mix	Storage Temperature	Shelf Life	Micro- and Macroflora and Fauna	Quality	References
Grain	60% CO_2	4°C–7°C	Exposed for 14 days	All adult *Cryptolestes ferrugineus* and *Sitophilus granarius* were killed		Bell et al. (1993)
Rice	CO_2	30°C	42 days		Increase in yellowness [CO_2] had no significant effect	Gras et al. (1989)
Corn	2% O_2		360 days	All adult Lepidoptera were killed; *Aspergillus glaucus*, *A. famarii*, and *Penicillium* sp. were not present in seeds stored under CA; molds observed	The flour acidity was kept stable	Iconomou et al. (2006)
Wheat	8% O_2		180 days	*Sitophilus oryzea* population was killed after the first 6 days under MAP		
Fresh filled pasta	30% CO_2/70% N_2	15°C	4–42 days	*P. aurantiogriseum*, molds	Growth rate decreased and lag phase increased compared with growth in air	Zardetto (2005)
	50% CO_2/50% N_2		11 days			
	100% N_2		2–3 days			
	100% CO_2		61 days (16 days before the treatment)		No growth of organisms observed	
Milled rice	4 bar pressure CO_2 100%	25°C	1–2 days	*S. zeamais*	99% mortality of all stages of the insects	Noomhorm et al. (2009)
	6 bar pressure CO_2 100%		0–38 days			

2003). Active packages must be able to accommodate substances like antioxidants, moisture absorbers, carbon dioxide, and ethanol releasers, to enhance both shelf life and nutritive value. Oxidative reactions are responsible for extensive damages in food quality and, as a consequence, for significant economic losses. Additives like butylated hydroxyanisole (BHA) and butylated hydroxytoluene (BHT) are extensively applied in polymer processing to effectively restrain these harmful changes. Their usage is very effective for the preservation of dry breakfast cereals and several types of crackers (Duncan and Webster, 2009) (Figure 8.1).

Japan was the first country where an oxygen absorber was incorporated into a packet to preserve food products. This change in package atmosphere occurs through chemical reactions, and the method is included in MAP technique. Oxygen absorbers are much more effective than simple replacement of atmosphere gases, because they minimize the problem of product staling, caused by gas losses due to poorly sealed packaging (Kotsianis et al., 2002). Nevertheless, even after MAP packaging, there is always a small percentage of residual oxygen inside the package and can potentially negatively affect the cereal products. Oxygen absorbers can be used with or without MAP to remove this portion, eliminating the damaging factors and reducing the necessary packaging time. Generally, MAP is used in the first place in commercial procedures to reduce most of O_2, and then a scavenger is applied to remove any remaining O_2 (Vermeiren et al., 1999).

During cereal storage, moisture can be easily released. This moisture is the ideal environment for the growth of different species of fungi that produce various

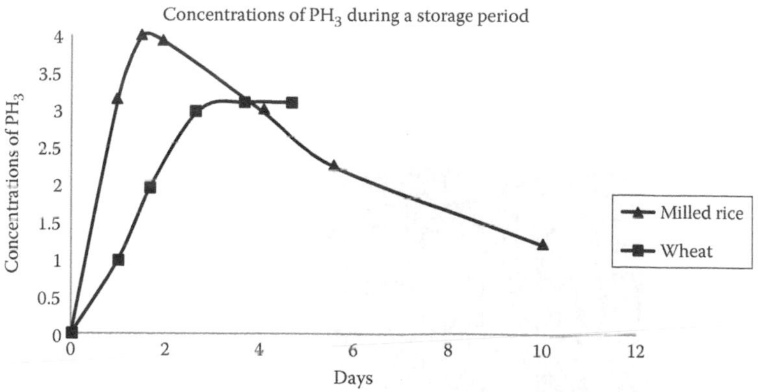

FIGURE 8.1 Observed concentrations of PH_3 in fumigations of milled rice in bags at 27°C and 14% moisture under PE sheeting and modeled concentrations of PH_3 released (From Bengston, M. et al., Efficacy of phosphine fumigations on bagged milled rice under polyethylene sheeting in Indonesia, in *Proceedings of an International Conference on Controlled Atmosphere and Fumigation in Stored Products*, E.J. Donahaye, S. Navarro, and A. Varnava (Eds.), Princto Ltd., Nicosia, Cyprus, pp. 225–233, 1996) in comparison with PH_3 concentrations at 30°C in sealed bin with wheat. (From Reed, C., Influence of grain temperature of efficacy of fumigation in leaky bins, in *CAF International Conference on Controlled Atmosphere and Fumigation in Stored Products*, E.J. Donahaye, S. Navarro, and A. Varnava (Eds.), Princto Ltd., Nicosia, Cyprus, pp. 235–242, 1997.)

mycotoxins. Aflatoxins (type of mycotoxins) have been detected in maize, rice, and other cereals; but they have been rarely found in sorghum. On the other hand, deoxyvarenol, a toxin often found in sorghum, is considered to be responsible for causing various toxic effects to rodents (Waniska et al., 2001). The primary toxins occurring in different kinds of cereal products are aflatoxins, fumonisins, and deoxynivalenol, the main metabolic products of the microorganisms *Aspergillus flavus* and *A. parasiticus, Fusarium verticillioides, and F. graminearum*, respectively (Cook and Johnson, 2009). Studies carried out in Pakistan revealed that MAP and VP are effective techniques in protecting cereal products against potential hazards and can considerably lengthen the storage period (Saleemullah et al., 2006) (Figures 8.2 and 8.3).

Various tests were carried out applying a 100 ton sealed box to explore the effectiveness of two distinct MAP. The gas applied in the first MAP was CO_2 whilst in the second derivatives of the propane combustion were applied. The results pointed out that the two gases are effective in protecting products from insects and microorganisms. The atmosphere conditions had an increased effectiveness when MAP was generated by the addition of gas in the headspace rather than in any other place of the package. Propane combustion products were not only 10 times cheaper than CO_2 but much more effective as well (Bell et al., 1993).

It has been reported that most cereal products tend to absorb CO_2 when exposed to enhanced concentrations of this gas. Because of this absorption, the pressure in the package headspace tends to decrease. Manufacturers should always use properly made and adequately resistant packages to prevent products from being damaged by careless handling. The CO_2, previously absorbed by produces, escapes into the atmosphere after the opening of the package without causing any damage to foods (Hagan, 2003).

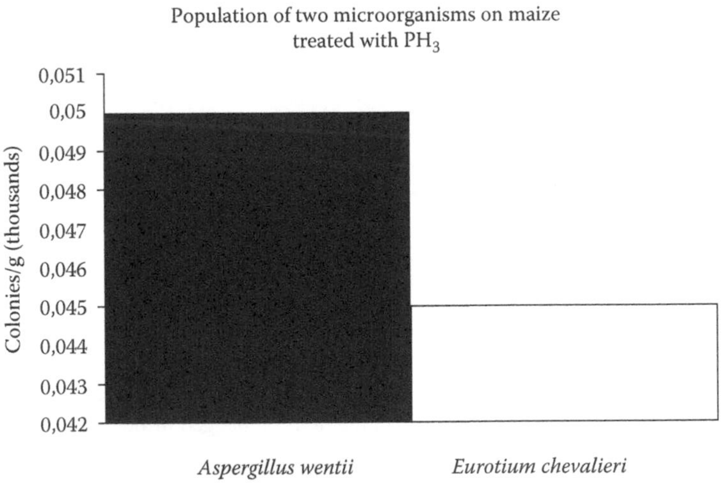

FIGURE 8.2 Population of *Aspergillus wentii* and *Eurotium chevalieri* on maize treated with PH_3. (From Dharmaputra, O.S., Effect of controlled atmospheres and fumigants on storage fungi—A review of research activities at seameo biotrop, in *CAF International Conference on Controlled Atmosphere and Fumigation in Stored Products*, E.J. Donahaye, S. Navarro, and A. Varnava (Eds.), Princto Ltd., Nicosia, Cyprus, pp. 199–265, 1997.)

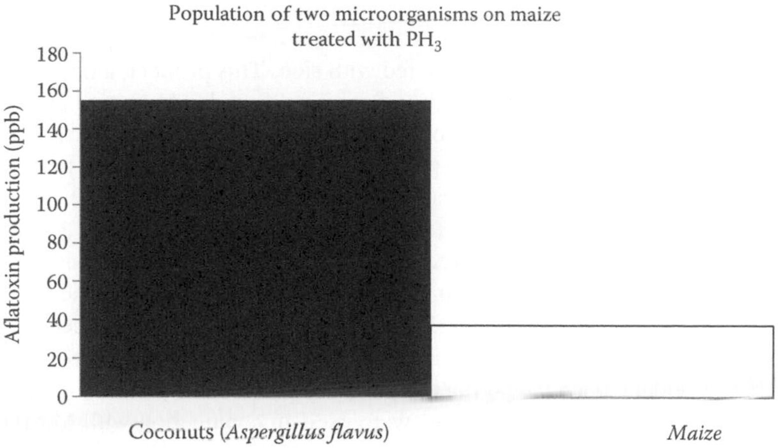

FIGURE 8.3 Aflatoxin B1 production after treatment with concentration of PH_3: 1.5 mg/L for 5 days on 10% coconut extract medium in comparison with aflatoxin B1 production on maize treated with CO_2. (From Dharmaputra, O.S., Effect of controlled atmospheres and fumigants on storage fungi—A review of research activities at seameo biotrop, in *CAF International Conference on Controlled Atmosphere and Fumigation in Stored Products*, E.J. Donahaye, S. Navarro, and A. Varnava (Eds.), Princto Ltd., Nicosia, Cyprus, pp. 199–265, 1997.)

Iconomou et al. (2006) investigated the qualitative characteristics of different cereal products stored under MAP or CA (2% O_2 for 1 year and 8% O_2 for 6 months). It was shown that atmospheres with a high percentage of N_2 stabilized the pH of raw materials over the whole experimental period and significantly increased the sprouting capacity of produces. In summary, wheat and maize grains were effectively preserved and products were efficiently protected against insects and microorganisms over a satisfactory period of time (Iconomou et al., 2006).

Kim et al. (2003) reported that O_3 application can effectively destroy *Bacillus* spp. and *Micrococcus* spp. units (bacteria often occurring on cereals), protecting several cereal products from these harmful microorganisms. The efficacy of O_3 application greatly depends on the available free surface of the product. The greater the free surface of the product, the more prolonged the exposure and larger the quantity of active gas that are required to effectively inactivate and kill bacteria and fungi, respectively. Therefore, this treatment should be applied, if possible, on produces before milling to minimize the functional costs.

MAP normally requires a drop in O_2 percentage inside the package. In the case of cereal packaging, a reduction in the rate of O_2 from 20% to 2%–3% is efficient to kill all kinds of harmful pests. Various parasites found in several cereal products require only 1 month to multiply and give rise to the next generation. Because of this short period between two generations and the rapid and facile growth and spread of insects and other small animals that infest cereals, it is necessary for MAP to be immediately implemented even on the slightest suspicion of infection. The drop in oxygen concentration, down to 1%, can be reached by simultaneous sealing of the package. Under these conditions, no molds can continue expanding (Iconomou et al., 2006).

8.2.1 Rice

Billions of people nowadays are mainly fed with rice. This product, along with corn and wheat, constitutes the source of 49% of calories provided to human population. To be more specific, rice provides about 23% of calories, thereby corresponding to one-fourth of the calories provided to people worldwide, whereas lesser percentages are occupied by the other two categories of grains (Subudhi et al., 2006).

A widely common variety of rice, the brown rice, was the subject of the study of Ott (1988). This product was preserved with various MAP (i.e., CO_2) and VP in an attempt to examine the most significant changes of two of its organoleptic characteristics: taste and smell. The maximum shelf life was recorded for rice placed in sealed plastic packages and stored at low temperature (3°C). The reduction in organoleptic spoilage was evident when the product was stored under MAP (like CO_2) and VP.

Several pressures (1, 4, 6, and 8 bar) were used in conjunction with MAP (CO_2) to preserve milled rice against *Sitophilus zeamais* (Coleoptera: Curculionidae). This insect is responsible for extensive damages to stored grains. *S. zeamais* commonly occurs in maize and rice, and its larvae damage produces by developing within an individual grain, eating it away from the inside out until maturity, and then reproduce releasing more harmful larvae. Application of high pressures (4, 6, 8 bar) minimized the required time of exposure needed to kill 99% of all insects, from approximately 6.5 days to 29, 9.0, and 4.8 h correspondingly. Infant insects were found to be more resistant than adults, but to the highly increased pressure of 6 and 8 bar, the percentage of mortality showed no difference between infant and adult stages. By preserving the product under MAP in combination with high-pressure application, qualitative characteristics were better maintained, thereby enhancing the freshness of the product (Noomhorm et al., 2009).

Bacillus cereus is a pathogenic bacterium usually occurring in foods and is responsible for several foodborne diseases. Symptoms of eating contaminated foods may vary from nausea and vomiting to diarrhea. These symptoms are attributed to two types of enterotoxins produced by the bacteria and do take place a few hours after eating contaminated food. Cereals such as rice may contain such microorganisms even after thermal processing. Resistant spores of the bacterium survive cooking and bacteria appear after prolonged storage of food at ambient temperature (Sutherland et al., 1996). According to Parra et al. (2010), the application of four different MAPs (60/40 = 60% N_2 + 40% CO_2; 70/30 = 70% N_2 + 30% CO_2; 80/20 = 80% N_2 + 20% CO_2; argon = 70% argon + 30% CO_2) at low temperature (4°C ± 1°C) effectively protected food products against this bacterium.

The Maillard reaction was first discovered by Camille Maillard. In this reaction, an amino group in an amino acid forms condensation products with aldehydes. Rice yellowing is considered to take place due to this reaction. Gras et al. (1989) measured the yellowing values of two rice cultivars originated from Australia. The obtained values were modeled; and the effects of water activity, MAP (CO_2), oxygen, and temperature were recorded. Temperature and moisture had a great effect on yellowing, whereas CO_2 seemed to have no effect at all. Following the theoretical calculations, after 42 days (at 30°C), an important enhancement in yellowness is anticipated to occur.

8.2.2 Wheat

Wheat is a cereal grain belonging to Poaceae, a family in the Class of Liliopsida (the monocots) of the flowering plants. *Triticum aestivum* L. em. Thell, known as bread wheat, is a plant of great importance (Varshney et al., 2006).

In an experiment by Cofie-Agblor et al. (1995), glass vials were used to study the ability of wheat to absorb CO_2 at several temperatures of 0°C, 10°C, 20°C, and 30°C and different moisture contents. Increase in temperature resulted in reduction of CO_2 absorption, whereas the increase in moisture content from 12% to 18% at 20°C rather caused absorption enhancement. Moreover, the sorption equilibrium took place within 2.5 days at 30°C.

Cereals and the products produced from their processing are often affected by the fungus *Penicillium aurantiogriseum*. This fungus has been observed in various products such as those consisting mainly of wheat and pasta. The optimal growth temperature of *P. aurantiogriseum* has been detected as 23°C, but it is capable of withstanding a wide range of temperatures (−2°C to 30°C). It has been observed that an increase in CO_2 in the growth environment of this microorganism contributes to the enhancement of its growth. Moreover, it has previously been reported that this microorganism continues growing after MAP application, in CO_2 concentrations of up to 30%. This gas increases the lag phase of the fungus but decreases the growth rate (Zardetto, 2005).

Cook and Johnson (2009) studied the application of MAP on fresh pasta. Gas-impermeable cans were used for packaging of these products. After storing the pasta at 30°C, the growth of *Clostridium botulinum* was observed in many cans. This fact complicates the mechanisms of harmful microorganism destruction.

Rodríguez et al. (2000) investigated the possibility of storage of sliced wheat flour bread under MAP, in various a_w, humidities, and pH levels. The effectiveness of four combinations of gases (100% N_2, 20% CO_2/80% N_2, 50% CO_2/50% N_2, and air control) was examined to determine whether they could be used for the preservation of the product under study. In bread, no preservation additives were used, MAP (CO_2:N_2/50:50) was shown to be very effective, enhancing the shelf life of the product by 117% (at 22°C–25°C) or 158% (at 15°C–20°C) depending on storage temperature. Application of preservatives led to shelf-life increases by 116% (at 22°C–25°C) or 167% (at 15°C–20°C), in 100% N_2.

Inclusion of ethanol (1°C–2%) inside the packages and replacement of packaging air from CO_2 can prolong the shelf life of sliced bread for more than 3 weeks (room temperature). Furthermore, VP or MAP using N_2 and CO_2 reduce bread oxidation and protect products from harmful microorganisms by decreasing the oxygen content. The main disadvantage of these technologies is the high cost of packaging machinery and gases (high purity) (Cook and Johnson, 2009).

Gastón et al. (2009) modeled the temperature and humidity variations caused by seasonal fluctuations. This model can effectively predict and itemize these changes and has already been successfully applied to wheat kept in silobags. Grain respiration and calculation of CO_2 and O_2 fluctuation rates are included in the mathematical determinations of this model, and the increase and reduction of CO_2 and O_2 variations showed very small differences in comparison with field measures. To be more specific, as concerns wheat, 100 days after placement of the product into silobags, deflection of CO_2 and O_2 concentrations were calculated to be 1.8% and 0.6%, respectively. However, wet wheat presented more significant deviations (5% or above).

According to Bell et al. (1993), the Central Science Laboratory of Slough (United Kingdom) constructed a self-cooled propane combustion unit to purify wheat farm facilities. After using it, O_2 concentration fell to 1% protecting wheat products effectively. The unit is capable of modifying 10–11 m^3 gas/h, but additional operations, like sealing and covering the gas admission points with sheets of polyethylene (PE), are required.

8.2.3 Maize or Corn

Corn or maize is a cereal belonging to the family Poaceae. It originated from the American continent where it was cultivated 5500 years ago by Incas, Maya, and Aztecs. While corn is a main source of food in numerous countries, its nutritional value is low in comparison with other cereals. Corn grains constitute the raw material that industries commonly employ for the production of alcohol after adequate treatment of the produces. Moreover, corn-processing residues comprise a great amount of biomass. The United States has the world's largest production (285 million tons/ year), followed by China, Brazil, and Mexico (Hongwei, 2006).

Application of a CO_2-modified atmosphere in sweet corn packages was shown to effectively preserve products from harmful fungi (Rodov et al., 2000). The product was stored at 2°C, while CO_2 concentration inside the package augmented due to corn respiration. Transfer of products at a temperature above refrigeration leads to enhanced respiration of the produces. This problem can be overcome with removing the liner for a short period. In this way, an appropriate MA is maintained and the quality of the product is preserved. After storing the product for 2 weeks at 2°C and for 4 additional days at 20°C, only a low microbial growth was recorded and quality and organoleptic characteristics were effectively preserved.

Samapundo et al. (2007b) investigated how MAs can possibly affect the life cycle of fungi *F. verticilliodes* and *F. proliferatum* as well as fumosin B1 excretion. After determination of the original CO_2 content and its interplay with a_w on life-cycle and fumosin B1 modulation, it was reported that only 10% of the initial CO_2 was sufficient to stop toxin excretion by *F. verticilliodes. F. proliferatum* seemed to be less susceptible to small atmosphere changes and suspended toxin production processes at much higher concentrations of CO_2. Nevertheless, it is obvious that MAs using CO_2 can be easily applied to protect maize adequately both from fungi and mycotoxins over storage.

Samapundo et al. (2007a) demonstrated that oxygen should better be removed completely over maize storage under MA conditions. Only in this way products can be effectively protected from fungi and mycotoxins. Moreover, in the same study, the effect of VP on fungi life cycle was studied as well. According to the obtained results, in MAP there was no difference observed in colonies with a decrease in O_2 concentration down to 2%. On the contrary, VP had a significant impact on fungi growth rates and managed to effectively protect products by neutralizing the harmful microorganisms.

8.2.4 Oat

Oat is a cereal grain belonging to the family Gramineae (Poaceae). Its provenance has not been fully identified (Murphy and Hoffman 1992). Several scientists of different backgrounds have given different names to this cereal product depending on the principles or standards applied in each taxonomic categorization (Rines et al., 2006).

Vermeiren et al. (1999) clearly and beyond any doubt demonstrated that an active packaging consisting of high-density polyethylene (HDPE) rich in BHT (a concentration of up to 0.32%) can prolong the expiry date of cereals, in comparison with packaging in a less BHT-impregnated material. Forty-two days after the start of the experiment, the film was clear of BHT, and 19% of its initial concentration had been transferred to the grains, thereby protecting effectively the product from oxidative reactions (Vermeiren et al. 1999).

Bagdan (2000) aimed at examining the usage and potential application of antioxidants and MAP, on prolongation of the expiry date of a breakfast cereal product, rich in omega-3 fatty acids. Two different types of MAPs were applied to effectively store the product: a N_2-enriched and an O_2-free (using an oxygen absorber) atmosphere. Specimens were kept at 21°C and 35°C preserving the products efficiently for approximately a year.

Decorticated oats were examined after storage under different conditions (various O_2 concentrations, temperature, and lighting) to evaluate the organoleptic properties of the product. Samples were stored for 3 and 10 months at 38°C and 23°C, respectively. The differences recorded in the oxygen quantity of the headspace of the package were in satisfactory agreement with the organoleptic changes of the product (Larsen et al., 2005).

Larsen et al. (2003) investigated the effects of application of MAP (high N_2 concentration with simultaneous decrease in O_2 percentage) and other factors like temperature and lightness on the sensory properties of decorticated oat products, during 3 months. After careful determination of O_2 concentration in the headspace of the container, scientists managed to select the most suitable containers anticipated to maintain very low levels of O_2 in their interior, during an acceptable period. The gas in the headspace of the packages was subjected to different variations of light (at 23°C), and the obtained results revealed that the food absorbed the entire amount of oxygen remaining inside the package. Similar results were obtained after carrying out the experiment using raw oat.

Cereal products like wheat, barley, oats, and canola were exposed to high concentrations of CO_2 to measure the absorption of this gas over 6 days. During the experiment, 250 g of grains were put in 500 mL bottles and subjected to two different CAs (49% and 70% CO_2). Gas absorption was reduced steadily with rising temperature. At 20°C, granola's rate of absorption displayed a sharp increase. At almost the same ratio of water activity (a_w), oat absorbed more CO_2 than barley.

Increase in a_w caused a drop in absorption especially for wheat grains. At 25°C, hulless barley and barley grains showed almost the same absorption but, under the same conditions, hulless oats and oats exhibited entirely different absorption behavior (Cofie-Agblor et al., 1998).

8.3 CONCLUSIONS

The main gas used in CA of grains is usually CO_2 although propane combustion gases were also used effectively. Gas absorption by cereal grains was reduced steadily with increasing temperature. Apart from CA, MAP was another promising alternative especially in terms of the following gas composition: 2% O_2 (or less), and MA of N_2-CO_2: 100/0, 80/20, 70/30, and 60/40 and argon/CO_2: 70/30.

As regards the MA preservation of processed cereals (pasta) or cereal-based products (bread), some suggested representative gas compositions were as follows: N_2/CO_2: 100/0, 80/20, 50/50.

Although the application of the aforementioned gas composition proved to be beneficial to the shelf-life extension of these produces, a further combination of MA with active packaging or preservative(s) led to even more promising results (longer shelf-life extension up to 167%) depending on the preservation temperature. Obviously, the shelf-life prolongation was positively affected by lower temperatures (~15°C) and in the presence of preservatives. In other words, application of hurdle technology led to the best and most promising results.

REFERENCES

Bagdan, G.C. (2000). *Shelf-Life Extension Studies on an Omega-3 Enriched Breakfast Cereal.* MSc thesis, Department of Food Science and Agricultural Chemistry, McGill University, Ottawa, Ontario, Canada, pp. 1–77.

Bell, C.H., Chakrabarti, B., Llewellin, B., and Wontner-Smith, T.J. (1993). A comparison of carbon dioxide and burner gas as replacement atmospheres for control of grain pests in a welded steel silo. *Postharvest Biology and Technology*, 2: 241–253.

Bengston, M., Sidik, M., Halid, H., and Alip, A. (1996). Efficacy of phosphine fumigations on bagged milled rice under polyethylene sheeting in Indonesia. In *Proceedings of an International Conference on Controlled Atmosphere and Fumigation in Stored Products*, Eds.: E.J. Donahaye, S. Navarro, and A. Varnava. Printco Ltd., Nicosia, Cyprus, pp. 225–233.

Campbell, J.F., Arthur, F.H., and Mullen, M.A. (2004). Insect management in food processing facilities. *Advances in Food and Nutrition Research*, 48: 239–295.

Cofie-Agblor, R., Muir, W.E., Jayas, D.S., and White, N.D.G. (1998). Carbon dioxide sorption by grains and canola at two CO_2 concentrations. *Journal of Stored Products Research*, 34(2/3): 159–170.

Cofie-Agblor, R., Muir, W.E., Sinicio, R., Cenkowski, S., and Jayas, D.S. (1995). Characteristics of carbon dioxide sorption by stored wheat. *Journal of Stored Products Research*, 31(4): 317–324.

Cook, F.K. and Johnson, B.L. (2009). Microbiological spoilage of cereal products. In *Compendium of the Microbiological Spoilage of Foods and Beverages. Food Microbiology and Food Safety*, Eds.: W.H. Sperber and M.P. Doyle. Springer-Verlag, New York, pp. 223–244.

Dharmaputra, O.S. (1997). Effect of controlled atmospheres and fumigants on storage fungi—A review of research activities at seameo biotrop. In *CAF International Conference on Controlled Atmosphere and Fumigation in Stored Products*, Eds.: E.J. Donahaye, S. Navarro, and A. Varnava. Princto Ltd., Nicosia, Cyprus, pp. 199–265.

Duncan, S.E. and Webster, J.B. (2009). Sensory impacts of food-packaging interactions. *Advances in Food and Nutrition Research*, 56: 17–64.

Gastón, A., Abalone, R., Bartosik, R.E., and Rodríguez, J.C. (2009). Mathematical modeling of heat and moisture transfer of wheat stored in plastic bags (silobags). *Biosystems Engineering*, 104: 72–85.

Gras, P.W., Banks, H.J., Bason, M.L., and Arriola, L.P. (1989). A quantitative study of the influences of temperature, water activity and storage atmosphere on the yellowing of milled rice. *Journal of Cereal Science*, 9: 77–89.

Hagan, A.T. (2003). *Prudent Food Storage: Questions and Answers*. Version 4.0, Gainesville, FL (accessed on May 18, 2011).

Hongwei, C. (2006). Maize. In: *Cereal and Millets*, Ed.: K. Chittaranjan. Springer, Philadelphia, PA, pp. 135–153.

Iconomou, D., Athanasopoulos, P., Arapoglou, D., Varzakas, T., and Christopoulou, N. (2006). Cereal quality characteristics as affected by controlled atmospheric storage conditions. *American Journal of Food Technology*, 1(2): 149–157.

Jayas, D.S. and Jeyamkondan, S. (2002). Modified atmosphere storage of grains meats fruits and vegetables. *Biosystems Engineering*, 82(3): 235–251.

Kim, J.G., Yousef, A., and Khadre, M.A. (2003). Ozone and it's current and future application in the food industry. *Advances in Food and Nutrition Research*, 45: 168–218.

Kotsianis, I.S., Giannou, V., and Tzia, C. (2002). Production and packaging of bakery products using MAP technology. *Trends in Food Science and Technology*, 13: 319–324.

Larsen, H., Lea, P., and Rodbotten, M. (2005). Sensory changes in extruded oat stored under different packaging, light and temperature conditions. *Food Quality and Preferences*, 16: 573–584.

Larsen, H., Magnus, E.M., and Wicklund, T. (2003). Effect of oxygen transmission rate of the packages, light, and storage temperature on the oxidative stability of extruded oat packaged in nitrogen atmosphere. *Journal of Food Science*, 68: 1100–1108.

La Storia, A.D., Mauriello, P.G., and Musso, S.S. (2008). Development and application of antimicrobial food packaging. In *Tesi si Dottorato di Ricerca in Scienze e Tecnologie delle Produzioni Agro-Alimentari XX ciclo*, Naples, Italy, pp. 1–113.

Murphy, J.P. and Hoffman, L.A. (1992). The origin, history, and production of oat. In *Oat Science and Technology*, Eds.: H.G. Marshall and M.E. Agronomy Monograph 33. ASA and CSSA, Madison, WI, pp. 1–28.

Noomhorm, A., Sirisoontaralak, P., Uraichuen, J., and Ahmad, I. (2009). Effects of pressurized carbon dioxide on controlling *Sitophilus zeamais* (*Coleoptera*: *Curculionidae*) and the quality of milled rice. *Journal of Stored Products Research*, 45: 201–205.

Ott, D.B. (1988). The effect of packaging on vitamin stability in cereal grain products—A review. *Journal of Food Composition and Analysis*, 1: 189–201.

Parra, V., Viguera, J., Sánchez, J., Peinado, J., Espárrago, F., Gutierrez, J.I., and Andrés, A.I. (2010). Modified atmosphere packaging and vacuum packaging for long period chilled storage of dry-cured Iberian ham. *Meat Science*, 84: 760–768.

Reed, C. (1997). Influence of grain temperature of efficacy of fumigation in leaky bins. In *CAF International Conference on Controlled Atmosphere and Fumigation in Stored Products*, Eds.: E.J. Donahaye, S. Navarro, and A. Varnava. Princto Ltd., Nicosia, Cyprus, pp. 235–242.

Rines, H.W., Molnar, S.J., Tinker, N.A., and Phillips, R.L. (2006). Oat. In *Cereal and Millets*, Ed.: K. Chittaranjan. Springer, Philadelphia, PA, pp. 211–242.

Riudavets, J., Castañe, C., Alomar, O., Pons, M.J., and Gabarra, R. (2009). Modified atmosphere packaging (MAP) as an alternative measure for controlling ten pests that attack processed food products. *Journal of Stored Products Research*, 45: 91–96.

Rodov, V., Copel, A., Aharoni, N., Aharoni, Y., Wiseblum, A., Horev, B., and Vinokur, Y. (2000). Nested modified-atmosphere packages maintain quality of trimmed sweet corn during cold storage and the shelf life period. *Postharvest Biology and Technology*, 18: 259–266.

Rodríguez, M., Medina, L.M., and Jordano, R. (2000). Effect of modified atmosphere packaging on the shelf life of sliced wheat flour bread. *Nahrung*, 44: 247–252.

Saleemullah, A.I., Iqtidar, A.K., and Hamidullah, S. (2006). Aflatoxin contents of stored and artificially inoculated cereals and nuts. *Food Chemistry*, 98: 699–703.

Samapundo, S., Meulenaer, B.D., Atukwase, A., Debevere, J., and Devlieghere, F. (2007a). The influence of modified atmospheres and their interaction with water activity on the radial growth and fumonisin B(1) production of *Fusarium verticillioides* and *F. proliferatum* on corn. Part II: The effect of initial headspace oxygen concentration. *International Journal of Food Microbiology*, 113: 339–345.

Samapundo, S., Meulenaer, B.D., Atukwase, A., Debevere, J., and Devlieghere, F. (2007b). The influence of modified atmospheres and their interaction with water activity on the radial growth and fumonisin B_1 production of *Fusarium verticillioides* and *F. proliferatum* on corn. Part I: The effect of initial headspace carbon dioxide concentration. *International Journal of Food Microbiology*, 114: 160–167.

Smith, A.G. and Day, B.P.F. (2003). *Effect of Modified-Atmosphere Packaging on Food Quality*. Elsevier Science Ltd., Amsterdam, the Netherlands, pp. 1157–1163.

Subudhi, P.K., Sasaki, T., and Khush, G.S. (2006). Rice. In *Cereal and Millets*, Ed.: K. Chittaranjan. Springer, Philadelphia, PA, pp. 1–78.

Sutherland, J.P., Aherne, A., and Beaumont, A.L. (1996). Preparation and validation of a growth model for *Bacillus cereus*: The effects of temperature, pH, sodium chloride and carbon dioxide. *International Journal of Food Microbiology*, 30: 359–372.

Varshney, R.K., Balyan, H.S., and Langridge, P. (2006). Wheat. In *Cereal and Millets*, Ed.: K. Chittaranjan. Springer, Philadelphia, PA, pp. 79–134.

Vermeiren, L., Devlieghere, F., van Beest, M., de Kruijf, N., and Debevere, J. (1999). Developments in the active packaging of foods. *Trends in Food Science and Technology*, 10: 77–86.

Waniska, R.D., Venkatesha, R.T., Chandrashekar, A., Krishnaveni, S., Bejosano, F.P., Jeoung, J., Jayaraj, J., Muthukrishnan, S., and Liang, G.H. (2001). Antifungal proteins and other mechanisms in the control of sorghum stalk rot and grain mold. *Journal of Agriculture and Food Chemistry*, 49: 4732–4742.

Zardetto, S. (2005). Effect of modified atmosphere packaging at abuse temperature on the growth of *Penicillium aurantiogriseum* isolated from fresh filled pasta. *Food Microbiology*, 22: 367–371.

9 Minimally Processed Vegetables

Ioannis S. Arvanitoyannis and Achilleas Bouletis

CONTENTS

9.1 INTRODUCTION

Vegetables play an important role in the human diet, because they provide the human body with vitamins (vitamin A, niacin, thiamin, riboflavin, and ascorbic acid) and inorganic salts (calcium, phosphorus, iron, potassium, sodium), and are of low calorie value (Bletsos, 2002; Arvanitoyannis et al., 2005). The major components of vegetables are water, carbohydrates, proteins, and lipids. Water represents 70%–95% of fresh weight, protein varies from 1% to 8%, and lipids from 0.1% to 1% (Ooraikul and Stiles, 1990).

In view of the inherent difficulty in preserving vegetables for a long period, several preservation techniques were introduced, but the first of practical use in the food industry was the canning process invented by Nicolas Appert. A variety of materials have been used for food packaging such as glass, paper and board, and plastics. Polymeric films and containers have substantial advantages over metal, glass, and paper in terms of versatility of shape, size, and structural properties and their light weightedness, toughness, low material cost, and generally less energy demand for manufacturing and transportation (Eskin and Robinson, 2001). Being initiated by the oil boom of the 1960s and 1970s and the concept of a low-cost plastic alternative to the metal can, there has since been an extensive commercialization of various attractive alternative packaging formats with an emphasis always placed on functionality and variety (Maskell, 1991).

Consumer's demand for additive-free fruits and vegetable and of high overall quality and safety has seen a pronounced rise in the recent years (Artes et al., 2007). This was due to improvements occurring in socioeconomic standards, education, and the role of media in uncovering a high number of dietary scandals. These changes motivated the food industry to adopt sophisticated minimal processing and preservation techniques to preserve both the quality and the safety of the final product.

The International Fresh-cut Produce Association (IFPA) defines fresh-cut products as fruit or vegetables that have been trimmed and/or peeled and/or cut into 100% usable product that is bagged or prepackaged to offer consumers high nutrition, convenience, and flavor while still maintaining its freshness (Lamikanra, 2002). Processing of vegetables induces a rapid physiological deterioration, biochemical changes, and microbial degradation of the product even when only slight processing operations can be used (O'Beirne and Francis, 2003), which may result in degradation of the color, texture, and flavor (Kabir, 1994; Varoquaux and Wiley, 1994; Rico et al., 2007).

Shelf life may be defined as the period of time from harvest to manufacture to consumption that a food product remains safe and wholesome under recommended production and storage conditions (Irtwange, 2006). The shelf life of vegetables after harvest is strongly influenced by intrinsic factors such as respiration rate, ethylene production and sensitivity, transpiration, and compositional changes.

9.1.1 Physiological Factors Affecting Shelf Life of Minimally Processed Vegetables

The metabolic process in the cells of vegetables comprising the catabolism of complex organic compounds such as sugars, organic acids, amino acids, and fatty acids with the ensuing production of energy through oxidation–reduction enzymic reactions is known as respiration (Ooraikul and Stiles, 1990). The rate of deterioration of harvested commodities is strongly related to the respiration rate (Fallik and Aharoni, 2004; Irtwange, 2006).

Respiration is further affected by the stage of maturity of the commodity. Vegetables comprise an extensive variety of plant organs (roots, tubers, seeds, bulbs, fruits, sprouts, stems, and leaves) that have dissimilar metabolic activities and consequently various respiration rates. Even different cultivars of the same product can occasionally display different respiration rates (Gran and Beaudry, 1992; Song et al., 1992; Fonseca et al., 2002).

Ethylene, a plant hormone, plays a large role in shelf life and can cause a marked increase in respiration rates and enhance ripening and senescence (Nguyen-the and Carlin 1994; Lin and Zhao, 2007). In some commodities, accelerated aging and the triggering of ripening can take place following exposure to ethylene concentrations as low as 0.1 mL/L (Lee et al., 1995; FDA/CFSAN, 2001). Flesh firmness of the fruits was shown to decrease with ripening (Manolopoulou and Papadopoulou, 1998; Fallik et al., 2001; Johnston et al., 2001; Rocha et al., 2004), because of the rise in ethylene production (Arvanitoyannis et al., 2005). Batu and Thompson (1998) report that ethylene triggers the ripening of tomatoes and is linked with an abrupt change in the physiology of tomato fruits at the beginning of ripening. Moreover, endogenous ethylene accelerates senescence in leaves of some, but not all species (Able et al., 2003, 2005), and that senescence is primarily expressed by yellowing, which is generally associated with losses in quality and marketability (Koukounaras et al., 2007).

Transpiration is the evaporation of water from plant tissues. Water loss is a very important cause of produce deterioration, with severe consequences. Elazar (2004) states that water loss is, first, a loss of marketable weight and then, adversely affects appearance (wilting and shriveling) (Irtwange, 2006). Metabolic processes including respiration and transpiration are particularly temperature dependent (Lin and Zhao, 2007; Tano et al., 2007). Moisture losses of 3%–6% make the product unacceptable for sale or consumption (Villanueva et al., 2005), and for mushrooms, the acceptable weight loss is about 2% (Villaescusa and Gil, 2003).

Maturation and ripening of plants induces many changes in pigments. Some may continue after, or start only at, harvest. These changes, though they start or continue after harvest, which may be either desirable or undesirable, can occur as loss of chlorophyll, development of carotenoids, and development of anthocyanins and other phenolic compounds (Fallik and Aharoni, 2004; Irtwange, 2006). The loss of green color in florets has been attributed to chlorophyll degradation and is closely linked to respiration rate, ethylene production, and lipid peroxidation (King and Morris, 1994; Zhuang et al., 1995; Serrano et al., 2006). One of the main causes of quality losses in minimally fresh processed fruits is enzymatic browning. Altered phenol metabolism is considered to be involved in leaf browning of lettuce (Saltveit, 2000). The first step in phenol metabolism is the conversion of the amino acid L-phenylalanine to trans-cinnamic acid by means of the enzyme phenylalanine ammonia lyase (PAL) (Degl'Innocenti et al., 2007). The next step is the oxidation of phenolic compounds catalyzed by the polyphenol oxidase enzyme (PPO); the resulting colorless quinones are later on polymerized leading to melanins. These substances display brown, reddish, or black coloration (Artes et al., 2007). As a result, some researchers have suggested that the activity of PAL may be a marker for shelf life in fresh-cut products (Lopez-Galvez et al., 1996; Degl'Innocenti et al., 2005, 2007). Enzymatic browning is one of the most important compositional changes in vegetables because it has a great impact on visual quality, thereby having adverse effect on the marketability of the product.

9.1.2 Chemical Hazards

Residues of five or more persistent toxic chemicals in a single food item were not unusual, with the most commonly found POPs being p,p'-Dichlorodiphenyltrichloroethane (DDT) and its metabolites (found in 21% of samples tested in 1998 and 22% in 1999), and dieldrin (found in 10% of samples tested in 1998 and 12% in 1999) (Schafer and Kegley, 2002). The U.S.D.A (U.S. Department of Agriculture) reported that for a period of 10 years (1993–2003), approximately 65% of the fresh fruit and vegetable samples contained detectable pesticide residues in the washed, edible tissues. The percentage of commodities with detectable residues varied substantially ranging from <1% (onions) up to 97% (nectarines) (Punzi et al., 2005). Analysis of six seasonal vegetables by Kumari et al. (2002) revealed that the tested samples were 100% contaminated with low but still measurable amounts of insecticide residues. In Spain, Gonzalez-Rodriguez et al. (2008) found that pesticide residues were determined above the maxima residue limits (MRL) in 15 out of the 75 analyzed samples, with a total of 18 violations of the MRL. The highest concentrations of fungicides were detected in lettuce, and the highest concentrations of insecticides were reported in Swiss chard. Proper implementation of procedures like washing, blanching, and peeling of vegetables lowered the pesticide levels considerably, from 50% to 100% in most cases (Chavarri et al., 2005). Heavy metals, such as cadmium, copper, lead, chromium, and mercury, stand for important environmental pollutants and can be accumulated in high concentrations in vegetables (Islam et al., 2007) as a result of contaminated soil and the fertilizing techniques applied (Alloway et al., 1990).

9.1.3 Microbiological Parameters Affecting Shelf Life

The concept of minimal processing of fresh-cut products excludes the application of any microbial preventive technique that could affect their quality. Minimally processed vegetables are microbial carriers, and some of them can be a serious hazard to public health. The presence of high microbial populations leads to shortened shelf life, and the lower the initial bacterial counts on produce, the better the produce quality and the longer its shelf life (Zagory, 1999).

Fresh vegetables normally have an elaborate spoilage microflora, attributed to intense contact with various types of microorganisms during growth and postharvest handling. Therefore, the numbers of microorganisms determined on vegetables are highly variable. Initial mesophilic counts of all of asparagus samples dropped within the range 10^4–10^5 CFU/g, which agrees with those found by Zagory (1999) (Villanueva et al., 2005). Eighty to ninety percentage of bacteria are Gram-negative rods, predominantly *Pseudomonas, Enterobacter,* or *Erwinia* species (Francis et al., 1999). Jacxsens et al. (2003) reported that the type of spoilage and quality deterioration in vivo depends on the type of vegetable. Lactic acid bacteria (LAB) have been detected in mixed salads and could predominate when held at abuse (30°C) temperatures (Francis et al., 1999). Especially, the outgrowth of LAB can be accompanied with production of organic acids such as lactic acid (LA) and acetic acid. Carlin et al. (1989) and Kakiomenou et al. (1996) isolated *Leuconostoc mesenteroides*, as the main spoiler of grated carrots. *Leuconostoc* spp. are heterofermentative and produces, next to LA, also ethanol and CO_2. Furthermore, high amounts of yeasts (>10^5 CFU/g) (*Candida* spp.) can release an off-flavor of fresh-cut produce due to the production of CO_2, ethanol, organic acids, and volatile esters (Jacxsens et al., 2003). Both *Aspergillus flavus* and *Aspergillus parasiticus* are abundant in nature and, under favorable environmental conditions, can grow and produce aflatoxin on various substrates (Ellis et al., 1993).

9.1.4 Pathogenic Organisms

Aeromonas spp. occurs in water, soil, feces, and on vegetation (McMahon and Wilson, 2001). *Aeromonas* was isolated from a wide range of fresh produce including sprouted seeds, asparagus, broccoli, cauliflower, carrot, celery, cherry tomatoes, courgette, cucumber, lettuce, mushroom, pepper, turnip, and watercress (Merino et al. 1995). Thirty-four percent of organic vegetables were detected to be contaminated with *Aeromonas* (McMahon and Wilson, 2001) Compared with 26% of conventionally cultivated vegetables (Neyts et al., 2001; Heaton and Jones,). *Aeromonas hydrophila* is responsible for a broad spectrum of infections in humans, and *Aeromonas* spp. have been linked epidemiologically with travelers' diarrhea (Francis et al., 1999).

Campylobacter jejuni is the most common cause of gastrointestinal illness worldwide, affecting more than 2 million people in the United States and 50,000 throughout England and Wales yearly (Evans et al. 2003), but the majority of cases are usually sporadic (Heaton and Jones, 2008). *Campylobacter* has been isolated from various produce items sampled from farmers' markets in Canada and from mushrooms sampled from retail markets in the United States. Although consumption of contaminated food of animal origin, particularly poultry, is largely responsible for infection, *Campylobacter* enteritis has also been linked to lettuce or salads (Harris et al., 2003) and outbreaks have been related to sweet potatoes, cucumber, melon, and strawberries. Kumar et al. (2001) isolated *C. jejuni* from spinach, fenugreek, lettuce, radish, parsley, green onions, potatoes, and mushrooms (Heaton and Jones, 2008).

Enterotoxigenic *Escherichia coli* is a common cause of travelers' diarrhea, an illness sometimes experienced when visiting developing countries. Raw vegetables are thought to be a common cause of travelers' diarrhea (Harris et al., 2003). Dairy cattle have been identified as a potential reservoir for *E. coli O157:H7.* It has been shown that there is a high probability for cross-contamination of meats, and other types of foods with *E. coli 0157:H7* during processing, handling, and marketing are substantial. Contamination of raw salad vegetables with *E. coli 0157:H7* would most likely occur during the assembling of ready-to-eat (RTE) meals that also include beef or other potential carriers of the organism. The probability of *E. coli 0157:H7* being on raw vegetables originating from agronomic systems applying irrigation with contaminated water must not be disregarded either (Abdul-Raouf et al., 1993).

Listeria spp. are abundant in the environment and can be isolated from soil, water, vegetation, the feces of livestock, and vegetation irrigated with contaminated water. The potential of environmental *Listeria* to contaminate fresh produce, resulting in enteric infection, has long been recognized, and Harvey and Gilmour (1993) suggested that the main cause for this was processing. Beuchat (1998) summarized a number of surveys referring to the presence of *Listeria monocytogenes* on cucumber, peppers, potato, radish, leafy vegetables, beansprout, broccoli, tomato, and cabbage at point-of-sale (POS) (Heaton and Jones, 2008). Concerns about potential pathogen contamination in produce were focused on *L. monocytogenes* due to its ability to grow at refrigeration temperatures. Beuchat and Brakett (1990) investigated that populations of *L. monocytogenes* dropped upon contact with raw carrots. Kakiomenou et al. (1998) claimed that *L. monocytogenes* inoculated in carrot and lettuce salad survived under refrigeration and MA, but did not grow.

Salmonellae have been isolated from many types of raw fruits and vegetables. Outbreaks of salmonellosis have been associated to various vegetables, including tomatoes and bean sprouts, thereby making it imperative to apply hygienic practices when handling them (Beuchat/WHO, 1998). The results reported by Das et al. (2006) demonstrate that *Salmonella enteritidis* can survive and grow during the storage of tomatoes depending on the location site of the pathogen on fruit, suspension cell density, and storage temperature. Drosinos et al. (2000) showed that as far as the survival of *S. enteritidis* in such low pH products (tomatoes and cheese) is concerned, this could be possibly attributed to its adaptation by enhancing its capacity for pH homeostasis.

Spores of *Clostridium* species, including *Clostridium botulinum* and *Clostridium perfringens*, as well as spores of enterotoxigenic *Bacillus cereus*, are commonly found in soil, so their occasional presence in fruits and vegetables should not be unexpected (Beuchat and Ryu, 1997). However, there is a hazard of spore forming bacteria to public health only when produce is handled in a manner that enables germination of spores and growth of vegetative cells. Of particular concern are vegetables packaged under MA (Harris et al., 2003). The high rate of respiration of salad vegetables can lead to an anaerobic environment in film-wrapped packages, thereby encouraging the growth of *C. botulinum* and botulinal toxin production. Botulism has been associated with coleslaw prepared from packaged, shredded cabbage and chopped garlic in oil (Beuchat/WHO, 1998). *C. perfringens* was linked with one outbreak epidemiologically associated with the consumption of salad (Harris et al., 2003).

The foodborne transmission of giardiasis was suggested in the 1920s, and anecdotal evidence from other outbreaks has frequently implicated food handlers and contaminated fruit and vegetables. The first foodborne outbreak of giardiasis in the United States was reported in 1979. The outbreaks, affecting 217 individuals, between 1979 and 1990, are linked with contamination by food handlers, and include foods such as fruit salad, raw vegetables, lettuce, onions, and tomatoes. Suspected outbreaks of foodborne cryptosporidiosis have been recorded for travelers visiting Mexico, the United Kingdom, and Australia; and the suspect foods included salads (Slifko et al., 2000).

Virus particles of *norovirus* and hepatitis A virus (*HAV*) are excreted by an infected host or released in vomit. Therefore, there is a high probability of these being present in sewage and fecally contaminated water. During the period 1992–1999, viruses accounted for a similar number (ca. 20%) of produce-related infections as *Salmonella* (Seymour and Appleton, 2001). It is worth noting that enteric viruses have a low infective dose and remain active even after exposure to low pH (<3) (Seymour and Appleton, 2001) and temperature abuses. Irrigation with sewage-contaminated water was associated with *HAV* outbreaks related to consumption of lettuce (Seymour and Appleton, 2001) and spring onions (Heaton and Jones, 2008). An enhanced incidence of domestic *HAV* without any evident source of infection in Sweden and a small outbreak in late spring 2001 revealed that consumption of imported rocket salad was closely associated with the disease. Imported salad has been considered responsible for *HAV* outbreaks in other countries and may prove to be a severe problem in case the foods come from countries where this disease is endemic due to low immunity of local population (Nygard et al., 2001).

9.1.5 Intervention Methods

The two most widely used sanitizers for decontaminating fresh produce are liquid chlorine and hypochlorite. Chlorine compounds are usually applied at levels of 50–200 ppm free chlorine and with contact times of less than 5 min (Francis and O'Beirne, 2002; Watada and Qui, 1999). Although washing with chlorinated water has been traditionally employed in vegetable decontamination, several reports have refuted its effectiveness (Adams et al., 1989; Beuchat, 1999; Li et al., 2001; Rico et al., 2007).

The main advantages of chlorine dioxide (ClO_2) over HOCl comprise reduced reactivity with organic matter and higher activity at neutral pH. One of the typical problems related to chlorine dioxide is its stability. ClO_2 is advantageous over HOCl in view of its fewer organohalogens although its oxidizing power is reported as 2.5 times that of chlorine. A maximum of 3 ppm is allowable for contact with whole produce. Any treatment of produce with chlorine dioxide must be always accompanied with water rinsing or blanching, cooking, or canning (FDA/CFSAN, 2001). Barakat et al. (2007) reported an approximately a 4.5 log CFU/g reduction per strawberry of all examined bacteria (*E. coli O157:H7*, *L. monocytogenes*, and *Salmonella enterica*) with treatment with 5 mg/L ClO_2 for 10 min.

Organic acids (e.g., LA, citric acid, acetic acid, tartaric acid) are strong antimicrobial agents against psychrophilic and mesophilic microorganisms in fresh-cut fruit and vegetables (Bari et al., 2005; Rico et al., 2007). In a publication by Akbas and Imez (2007) on iceberg lettuce, the maximal reduction for *E. coli* (about 2 $\log_{10}$ CFU/g) was obtained for samples dipped in lactic or citric acids. The maximal reduction for *L. monocytogenes* (about 1.5 $\log_{10}$ CFU/g) was reached for samples dipped in LA. Chemicals containing SH-groups including sulfites are often used to avoid browning in vegetables such as potatoes. However, the application of these compounds in fresh-cut commodities can cause bronchial asthma (Beltran et al., 2005a).

Juven and Pierson (1996) reviewed research reports on the application of antimicrobial H_2O_2 in the food industry. H_2O_2 has bactericidal and inhibitory activity, thanks to its oxidizing properties and its capacity to generate other cytotoxic oxidizing species. The sporicidal activity of H_2O_2 in conjunction with rapid breakdown makes it a desirable sterilant for use on some food contact surfaces, and packaging materials in aseptic filling operations (FDA/CFSAN, 2001).

Ozone is well known for its strong antimicrobial activity with high reactivity, penetrability, and spontaneous decomposition to a nontoxic product (Grass et al., 2003; Kim et al., 1999). It was shown that treatment with ozone had a beneficial effect in prolonging the storage life of fresh noncut commodities such as broccoli, cucumber, pears, raspberries, and strawberries by reducing microbial populations and by oxidation of ethylene (Beuchat et al., 1998; Kim et al., 1999; Rico et al., 2007). Apart from the effectiveness on microorganisms, excessive use of ozone may considerably affect the surface color of some fruits and vegetables such as peaches, carrots, and broccoli florets (Das et al., 2006).

Electrolyzed water (EW), also known as electrolyzed oxidizing water, is conventionally generated by electrolysis of aqueous sodium chloride (Kim et al., 2000). Acidic EW (pH 2.1–4.5) has a strong bactericidal effect against pathogens and spoilage microorganisms. It has been shown to be more effective than chlorine due to a high oxidation reduction potential (ORP) and led to higher efficiency in decreasing viable aerobes than ozone on whole lettuce, but this occurred at the expense of produce quality when applied on fresh-cut vegetables (Rico et al., 2007). Ongeng et al. (2006) reported that washing the vegetables for 1 min in electrolyzed oxidizing water led to 1.9, 1.2, and 1.3 log reductions of psychrotrophs, LAB, and *Enterobacteriacae*, respectively, whereas washing for 5 min led to 3.3, 2.6, and 1.9 log reductions.

In nature, there are a large number of different types of antimicrobial compounds very crucial for the natural defense of all kinds of living organisms (Rauha et al., 2000). Vinegar and lemon juice containing acetic and citric acids naturally could be regarded as alternative disinfectants to eliminate or at least to decrease pathogens without provoking any health risk to consumers. Treatment of rocket leaves with fresh lemon juice and vinegar led to a significant reduction of *Salmonella typhimurium*, whereas the maximal reduction to undetectable level was attained with 15 min treatment (Sengun and Karapinar, 2005). Research by Devlieghere et al. (2004) revealed that chitosan had a strong bactericidal effect especially against Gram (−) bacteria when applied as a coating on lettuce and strawberries. Both cinnamon oil and clove oil added at 2% in potato dextrose agar (PDA) completely inhibited the growth of seven mycotoxigenic molds (*A. flavus*, *A. parasiticus*) up to 21 days and the growth of yeasts. Cinnamon oil and clove oil could inhibit several bacteria including *Lactobacillus* sp., *Salmonella* sp., *Pseudomonas striafaciens*, and *C. botulinum* (Matan et al., 2006).

9.1.6 Physical Treatments for Minimally Processed Vegetables

Blanching not involving any chemical treatment can decrease initial mesophilic counts of leafy salads by more than 3 log CFU/g and *Enterobacteriaceae* counts by less than 1 log CFU/g. However, blanching itself introduces obnoxious alterations in the product by the loss of nutrients through thermal degradation, diffusion, and leaching, enhances power consumption, and produces effluents (Rico et al., 2007). Suparlan and Itoh (2003) reported that hot water treatment (HWT) could be applied as disinfectant for tomatoes prior to storage under modified atmosphere packaging (MAP) to reduce microbial growth, cracking, and decay due to excessive water vapor inside the package.

Ionizing irradiation is a nonthermal technology that effectively suppresses foodborne pathogens in fresh vegetables. Although in some cases the low doses of irradiation can inactivate some radiation-sensitive pathogens, to achieve a 5 log reduction, higher doses of radiation must be applied. At those doses, irradiation may trigger undesirable changes in quality, such as softening, browning, and loss of vitamin C (Fan et al., 2003).

Ultraviolet light (UV-C) acts as an antimicrobial agent directly due to DNA damage and indirectly to induction of resistance mechanisms in different fruit and vegetables against pathogens. Exposure to UV-C initiates the synthesis of health-promoting compounds such as anthocyanins and stilbenoids. Two more advantages are the low cost and the requirement of simple equipment (Rico et al., 2007).

Seymour et al. (2002) recorded that reductions in *S. typhimurium* attached to iceberg lettuce obtained by cleaning with water, chlorinated water, ultrasound with water, and ultrasound with chlorinated water were 0.7, 1.7, 1.5, and 2.7 logs, respectively. However, the fresh produce industry is unlikely to decide on applying ultrasound technology because of the high cost involved in capital investment and process optimization.

External leaves of whole lettuce were determined to have counts approximately 1 log cycle higher than inner leaf layers (Adams et al., 1989). Results by Toivonen and Stan (2004) revealed that washing has a catalytic effect on quality indices of green pepper slices. This effect is unlikely to be affected by the suppression of stress-related compounds released during the cutting operation. The higher the number of washes after cutting, the greater the retention of firmness of the pepper slices during storage.

9.1.7 Use of Modified Atmosphere Packaging in Minimally Processed Vegetables

MAP is the alteration of the gaseous environment due to respiration (passive MAP [PMAP]) or addition and removal of gases from food packages (active MAP [AMAP]) to manipulate the levels of O_2 and CO_2 (Das et al., 2006). MAP was first introduced in the mid-to-late 1940s because of its capability of decreasing the O_2 levels adequately to retard the ripening of apple fruit. The main drawback of MAP application in the early studies was the lack of accurate control of O_2 levels in the package. However since then, there has been a great improvement in terms of types and properties of polymers to provide a wider range of physical properties such as gas permeability, tensile strength, flexibility, printability, and clarity. Therefore, effective MA packaging systems were developed for a number of commodities (Mir and Beaudry, 2002).

In the literature, the terms modified atmosphere (MA) and controlled atmosphere (CA) are used interchangeably. They both differ in the degree of control exerted over the atmosphere composition. In MA storage, the gas composition changes initially and changes dynamically depending on the respiration rate of the food product and permeability of film or storage structure surrounding the food product. In CA storage, the gas atmosphere is continuously controlled throughout the storage period (Jayas and Jeyamkondan, 2002).

The MAP influences the physiology of vegetables, leading to changes in life span, as well as physicochemical, sensory, and microbiological characteristics. To design a suitable MAP package, the factors to be considered are the commodity type, temperature, optimal O_2 and CO_2 partial pressures, respiration rate, product weight, atmosphere outside the package, and the permeability of packaging film to gases (Arvanitoyannis et al., 2005).

In all preservation techniques, there are many critical parameters that affect storage life and quality of the product. Storage temperature, relative humidity, gas composition, the physiology of products, and the package material are factors that must be thoroughly examined, carefully evaluated, and selected for the most beneficial results of MAP.

Postharvest tissue metabolism is greatly affected by storage temperature and has a strong effect on the senescence of pak choy leaves (Koukounaras et al., 2007). Chilling injury is the result of physiological response to low temperature and the ensuing symptoms influencing product acceptability. For maximal shelf life, a temperature range between 13°C and 20°C was the most appropriate for tomatoes and when fruits stored at temperatures greater than 20°C had a short shelf life and could be subject to decay (Batu and Thompson, 1998). Chauhan et al. (2006) found that banana samples with passive as well as active modes of MAP by means of specific gas mixture flushing and partial vacuum packaging (VP) displayed prolonged periods of shelf life at 14°C. Ayala-Zavala et al. (2004) reported that strawberry fruit quality was maintained longer at 0°C than at 5°C or 10°C over a 13 day postharvest period, but the strawberry fruit stored at lower temperatures had lower antioxidant capacity and total phenolics and anthocyanin concentrations (Shin et al., 2007). It was shown that storage temperature substantially affected the microbial growth.

Drosinos et al. (2000) reported a declining trend of *S. enteritidis* in MAP diced tomatoes stored at 4°C whereas it survived storage at 10°C and deduced that *S. enteritidis* cannot grow under MA storage at refrigeration temperatures but it may consist of a risk when a temperature abuse is produced in a commercial chain (Das et al., 2006). The effect of temperature is extremely important in package design, and continuous and perforated films greatly differ in their response to temperature changes. The O_2 and CO_2 permeability of continuous films increases with temperature, while the diffusion of gases through perforations is not susceptible to temperature changes. For instance, O_2 permeation through LDPE can enhance 200% from 0°C to 15°C, but the exchange of O_2 through perforations will increase only 11% across this temperature range (Mir and Beaudry, 2002).

Oxygen, carbon dioxide, and nitrogen are most often used in MAP/CAS. Other gases such as nitrous and nitric oxides, sulfur dioxide, ethylene, chlorine, as well as ozone and propylene oxide have been suggested and investigated experimentally. The aforementioned gases are not used commercially due to safety concerns and regulatory and cost considerations. These gases are combined in three ways for use in MA: inert blanketing using N_2, semireactive blanketing using CO_2/N_2 or $O_2/CO_2/N_2$, or fully reactive blanketing using CO_2 or CO_2/O_2 (FDA/CFSAN, 2001).

The effects of MAP are based on the often observed slowing of plant respiration in low O_2 environments. As the concentration of O_2 inside the package falls below 10%, respiration starts to slow down. This respiration suppression continues until O_2 reaches approximately 2%–4% for most produces. If oxygen drops to O_2 contents lower than 2%–4% (depending on product and temperature), fermentative metabolism replaces normal aerobic metabolism and off-flavors, off-odors, and undesirable volatiles are produced (Zagory, 1995). Therefore, the recommended percentage of O_2 in an MA for fruits and vegetables for both safety and quality issues lies in the range of 1% and 5%. However, it is accepted that the oxygen level will eventually reach levels below 1% in MAP produce (FDA/CFSAN, 2001).

Allende et al. (2002) reported that treatment of super atmospheric O_2 is an effective means for inhibiting both microbial growth and enzymatic discoloration and preventing anaerobic fermentation reactions. Allende et al. (2004b) revealed that addition of super atmospheric O_2 to the packages alleviated tissue injury in addition to reducing microbial growth. MAP was shown to be beneficial in maintaining the quality of fresh-cut baby spinach. The combined treatment of high O_2 levels and 10–20 kPa CO_2 may provide adequate suppression of microbial growth and prolonged shelf life. However, high O_2 levels could generate reactive oxygen species that damage the vital cellular macromolecules, thereby inhibiting the microbial growth when oxidative stresses has the upper hand over the cellular antioxidant protection systems. Concentrations higher than 25 kPa are regarded as explosive, and special precautions and care have to be taken on the work floor (Escalona et al., 2006).

Of the three major gases used in MAP or CA, CO_2 is the most important because of its direct antimicrobial activity due to induced changes in cell membrane function comprising effects on nutrient uptake and absorption, inhibition of enzymes or drop in the rate of enzyme reactions, penetration of bacterial membranes resulting in intracellular pH changes, and modifications in the physicochemical properties of proteins (Das et al., 2006). Reports on other vegetables under MAP with high CO_2 concentrations revealed chlorophyll loss retardation. It is reported that high CO_2 and low O_2 concentrations decrease considerably the breakdown of chlorophyll to phaeophytin (Gomez and Artes, 2005). Nitrogen has three applications in MAP: (i) displacement of O_2 to delay oxidation, (ii) retardation of the growth of aerobic spoilage organisms, and (iii) as a filler to keep package conformity (FDA/CFSAN, 2001).

Plant tissues tend to lose moisture when the RH is below 99%–99.5%. Generally, water loss leads to visible wilting or wrinkling of the surface of most commodities when it exceeds 4%–6% of the total fresh weight. The majority of MAP films are rather impermeable to water. The RH is close to saturation in the majority of continuous or perforated film packages (Mir and Beaudry, 2002).

Alternative approaches to provide high oxygen transmission rates (OTR), especially in applications where there is limited package surface area for gas exchange, have included films with holes or pores. Another application is the use of microporous and microperforated films, which allow much more rapid gas exchange than would normally be possible through plastic films. However, it should be pointed out that films with pores or small holes have several physical limitations with regard to physicochemical properties. CO_2 diffuses through plastic films two to six times more rapidly than O_2. As a result, CO_2 exits a package more rapidly than O_2 enters. This leads to equilibrium atmospheres of low O_2 and relatively low CO_2 (Zagory, 1998).

Edible coatings have long been used to retain quality and prolong the shelf life of some fresh fruits and vegetables, such as citrus fruits, apples, and cucumbers. Fruits or vegetables are usually coated by dipping in or spraying with a variety of edible materials, thereby forming a semipermeable membrane on the surface for suppressing respiration and controlling moisture loss. A variety of edible materials, including lipids, polysaccharides, and proteins, either alone or in combinations, have been applied to produce edible coatings (Lin and Zhao, 2007). Yun et al. (2006) reported that microporous earthenware can be a component of packaging material and when the earthenware sheet was combined with a plastic box and used in packaging for strawberries and enoki mushrooms at 5°C, its unique permeability properties developed an MA that was beneficial for maintaining the quality of the produce.

MAP can be used in conjunction with oxygen scavengers to eliminate oxygen contained in the packaging headspace and in the product or permeating through the packaging material during storage. Oxygen scavengers can both slow down or limit deterioration because of product component oxidation and/or growth of microorganisms or survival of insects. The presence of O_2 scavengers can also have a favorable effect on O_2-sensitive respiring products such as fresh or minimally processed fruits and vegetables (Charles et al., 2006).

Control of RH within the package by means of moisture absorbers has a strong impact on the quality of the mushrooms since sorbitol promoted deterioration and enhancing levels of silica gel augmented the weight loss. Inclusion of hygroscopic compounds in the packaging does not improve quality parameters sufficiently to justify their acceptance by the consumers, as reported in the case of *Agaricus* (Villaescusa and Gil, 2003).

Essential oils are well-known inhibitors of microorganisms. Cinnamon oil and clove oil are both natural preservative and flavoring substances not harmful when consumed in food products. Matan et al. (2006) showed that a high concentration of CO_2 (40%) and low concentration of O_2 (<0.05%) with the volatile gas phase of cinnamon oil and clove oil may be appropriate as an active packaging system for retarding the microorganism growth on intermediate-moisture foods.

Different treatments have been assessed in reducing browning in fresh-cut products, such as the application of antioxidant compounds (e.g., sulfites), calcium salts to maintain membrane integrity, chemical inhibitors of polyphenoloxidase (PPO) and/or peroxidases, or the use of MAP to exclude oxygen. Ascorbic acid is a highly effective inhibitor of tissue enzymatic browning because it reduces quinones to phenolic compounds, thereby preventing the synthesis of the brown color pigments (Degl'Innocenti et al., 2007).

Removing ethylene from the storage atmosphere is desirable for obtaining better firmness retention and quality. In many cases, the cost of removal is higher than the obtained benefits. Ethylene can be removed either with chemical oxidation with potassium permanganate or with catalytic oxidation (Jayas and Jeyamkondan, 2002).

Chitosan, a deacetylated form of chitin, can also be used as an antimicrobial film to cover fresh fruits and vegetables. Chitosan activates several defense processes in the host tissue, acts as a water-binding agent, and inhibits various enzymes (Devlieghere et al., 2004). Alginate and calcium–alginate films were used as coating materials for mushrooms (Kim et al., 2006).

9.2 LETTUCE

Kim et al. (2005) sliced, washed, dried, and packaged romaine lettuce leaves in films with OTRs of 8.0 and 16.6 pmol/s m^2 Pa, and with initial headspace O_2 of 0, 1, 2.5, 10, and 21 kPa and stored at 5°C. With 8.0 OTR-packaged lettuce pieces, ≤1 kPa initial headspace O_2 treatments triggered an essentially anaerobic environment within the packages and enhanced acetaldehyde and ethanol accumulation and off-odor development. Augmenting O_2 concentration above 1 kPa in 8.0 OTR packages decreased fermentative volatile production.

Fan et al. (2003) investigated the impact of warm water treatment (dipping in either 5°C or 47°C water for 2 min) prior to irradiation (0, 0.5, 1, or 2 kGy) of iceberg lettuce and packaging under PMAP (films with OTR: 4000 mL/h m^2) at 3°C. Samples irradiated at 0.5 and 1 kGy displayed similar firmness and vitamin C (4–8 μg/g) and antioxidant contents (1600–1800 nmol/g FRAP at 5°C) as the controls after 14 and 21 days of storage except for 1 kGy samples dipped at 47°C, which exhibited lesser antioxidant contents than controls at 14 days of storage.

Ozonated water treatment combined with AMAP storage (4 kPa O_2 and 12 kPa CO_2) did not induce any changes in individual phenolic compounds. Water-washed samples stored in MAP revealed a drop in vitamin C content to reach a 75% reduction

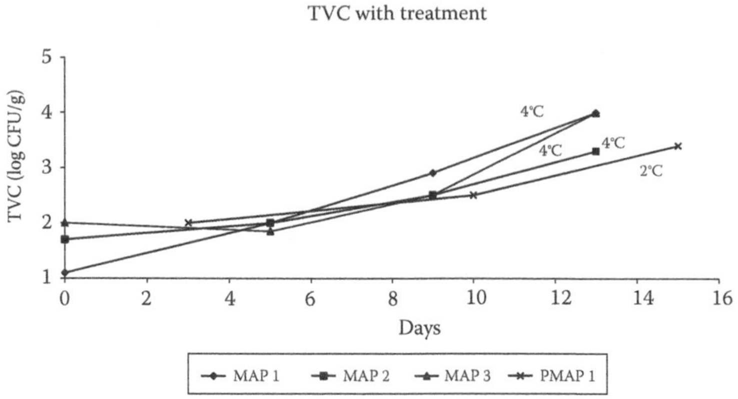

FIGURE 9.1 Changes in TVC of iceberg lettuce under MAP and various treatments before packaging vs. storage time (MAP 1 [4% O_2/12% CO_2] and rinsing with chlorine [80 mg/L], MAP 2 [4% O_2/12% CO_2] and washing with ozonated water of 10 mg/L activated by ultraviolet C [UV-C], MAP 3 [4% O_2/12% CO_2] and washing with ozonated water 20 mg/L (From Beltran, D. et al., *J. Agric. Food Chem.,* 53, 5654, 2005a), PMAP 1 [created by 44 μm polyolephine film] and γ-irradiation with 0.16–0.22 kGy 2 days after packaging. (From Hagenmaier, R. and Baker, R., *J. Agric. Food Chem.,* 45, 2864, 1997.))

by the end of the storage. Under AMAP, the external appearance was preserved for all the washing treatments. The most effective treatments with bactericidal effects were ozone 20 and ozone 10 activated with UV-C (1.8 and 2.5 log CFU reductions of mesophiles in AMAP packages at 4°C), which were as effective as chlorine (Beltran et al., 2005a). In Figure 9.1, it is clear that chlorine treatment gave the best results in lowering the initial microbial load, but due to its short-term action, the final results by the end of the storage are rather poor. Both γ-irradiation and ozonated water keep a standard antibacterial action during storage, but the latter has a slight advantage at the 13th day.

Evaluation of the microbial quality of "Lollo Rosso" lettuce over the entire production chain revealed that washing considerably diminished the microbial counts (3 log CFU/g for psychrotrophic and LAB and almost 2 log CFU/g for coliforms) and packaging under MAP at 5°C with initial gas atmosphere of 3 kPa O_2 and 5 kPa CO_2 led to a shelf life of 6 days (mesophiles and psychrotrophs increased from 5 to 8 log CFU/g after 7 days). After 7 days, strong off-odors were detected probably due to high final gas concentrations (Allende et al., 2004a).

The shelf life of chlorinated, γ-irradiated, and stored under PMAP iceberg lettuce was shown to increase with enhanced ethanol content for all samples, irrespective of whether they were irradiated. Ethanol tended to be higher for irradiated samples (800, 1500, 1900 ppm for 0, 0.2, and 0.5 kGy irradiation, respectively, after 13 day storage at 2°C). However, the increase in ethylene content due to irradiation was much smaller than that initiated by storage at higher temperature. Moreover, fewer microorganisms were detected in irradiated samples than in the control (Hagenmaier and Baker, 1997).

Chlorine, irradiation, and ozonated water of 20 mg/L concentration are the most effective coliform treatments in reducing the initial load and growth rate. Activation

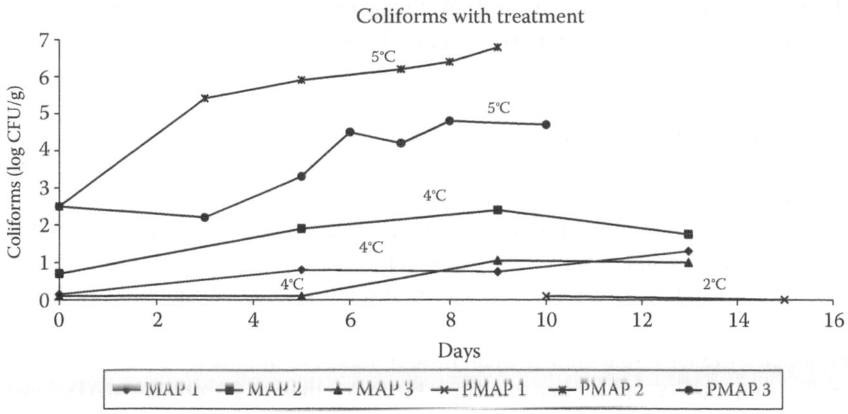

FIGURE 9.2 The effect of MAP and various treatments on Coliforms for different varieties of lettuce vs. storage time (MAP 1 [4% O_2/12% CO_2] and rinsing with chlorine [80 mg/L], MAP 2 [4% O_2/12% CO_2] and washing with ozonated water of 10 mg/L activated by UV-C, MAP 3 [4% O_2/12% CO_2] and washing with ozonated water 20 mg/L (From Beltran, D. et al., *J. Agric. Food Chem.*, 53, 5654, 2005a), PMAP 1 [created by 44 μm polyolephine film] and γ- irradiation with 0.16–0.22 kGy 2 days after packaging for iceberg lettuce (From Hagenmaier, R. and Baker, R., *J. Agric. Food Chem.*, 45, 2864, 1997), PMAP 2 [with a steady state of 2%–10% O_2/5%–12% CO_2] and UV-C treatment of 8.14 kJ/m^2 corresponding dose for "Red Oak Leaf" lettuce (From Allende, A. and Artes, F., *Lebensm. Wiss. Technol.*, 36, 779, 2003a), PMAP 3 [with bioriented polypropylene used as a packaging film] and UV-C treatment of 8.14 kJ/m^2 corresponding dose for "Lollo Rosso" lettuce. (From Allende, A. and Artes, F., *Food Res. Int.*, 36, 739, 2003b.))

of ozonated water with UV-C has poor antimicrobial results compared with higher ozone concentrations. UV-C treated lettuces were shown to be more susceptible to coliform colonization with results varying from one cultivar to another (Figure 9.2).

The effects of cutting method and packaging film (bioriented PP [BOPP] and a polyolefin [PO]) on the sensory quality of butterhead lettuce stored at 5°C were investigated by Martinez et al. (2008). Lettuce packaged in PD-961 and cut manually displayed less browning than the rest. Lettuce cut with a knife and packaged in BOPP exhibited the highest number of the dark and necrotic spots and was exposed to higher CO_2 concentrations. Manually cut lettuce exhibited lower tissue damage than the ones cut with knife, thereby displaying lesser deterioration rates.

Challenge studies to assess the effect of temperature abuse during storage period were conducted by Chua et al. (2008). They found that MAP (with initial atmosphere of 1 kPa O_2) enabled all six *rpoS* genes of enterohemorrhagic *E. coli* isolates inoculated on romaine lettuce to induce gastric acid resistance over the 8 day storage period if the temperature was higher than 15°C. However, no acid resistance was induced for enterohemorrhagic *E. coli* isolates inoculated on MAP-stored lettuce stored at temperatures lower than 10°C.

The effect of temperature, antimicrobial dips (100 ppm chlorine or 1% citric acid solution for 5 min), and gas atmosphere (gas flush with 100% N_2 or PMAP) on *Listeria innocua* and *L. monocytogenes* inoculated on iceberg lettuce was studied by Francis and O'Beirne (1997). Both N_2 flushing and use of antimicrobial dips at 8°C augmented the survival and growth of *Listeria* populations. The *Listeria* counts dropped by almost 1 log cycle in N_2 packs, lowered by 1 log for packs with MAP, and decreased by <1.5 log in air.

Temperature and plant variety were of great importance in *L. innocua* population with iceberg lettuce to have decreased the bacterial numbers and reduce storage temperature to minimize the microbial growth (Figure 9.3).

Martinez and Artes (1999) investigated winter harvested iceberg lettuce after it was vacuum cooled and stored under passive (perforated PP, PP 25, 30, and 40 μm) and AMAP (5% O_2 and 0% CO_2). The optimal treatments for decreasing pink rib disorder were both active and PMAP (index < 1.71). Heads packed in nonperforated PP bags maintained a higher visual quality than those unpacked. Among MAP treatments, the overall visual quality was higher in 40 mm PP MAP than in 25 mm PP MAP.

The effect of UV-C (0.4, 0.81, 2.44, 4.07, and 8.14 kJ/m^2) treatment and storage at 5°C under MAP (BOPP film used) conditions on quality of lettuce was investigated by Allende and Artes (2003a). All the applied UV-C radiation decreased psychrotrophic growth about 0.5–2 log unit CFU/g and extended the shelf life of the product by 2 days or even longer, depending on the applied UV-C radiation doses. It is noteworthy that no substantial difference was reported between UV-C-treated and untreated lettuce based on the sensory properties during storage.

According to Figure 9.4, the UV-C treatment is much more effective than gaseous chlorine treatment in terms of reducing psychrotroph numbers. This conclusion is

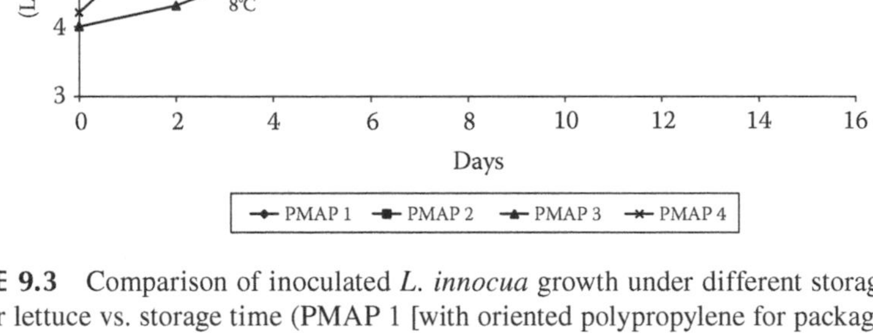

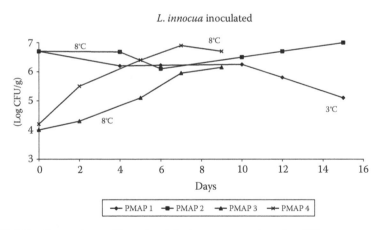

FIGURE 9.3 Comparison of inoculated *L. innocua* growth under different storage conditions for lettuce vs. storage time (PMAP 1 [with oriented polypropylene for packaging film] and inoculation with *L. monocytogenes* and *L. innocua* strains, PMAP 2 [with oriented polypropylene for packaging film] and inoculation with *L. monocytogenes* and *L. innocua* strains for iceberg lettuce (From Francis, G.A. and O'Beirne, D., *Int. J. Food Sci. Technol.*, 32, 141, 1997), PMAP 3 [with oriented polypropylene as a packaging film] and inoculation with *E. Coli* and *L. innocua* for razor blade cut iceberg lettuce, PMAP 4 [with oriented polypropylene as a packaging film] and inoculation with *E. Coli* and *L. innocua* for razor blade–cut butterhead lettuce. (From Gleeson, E. and O'Beirne, D., *Food Control*, 16, 677, 2005.))

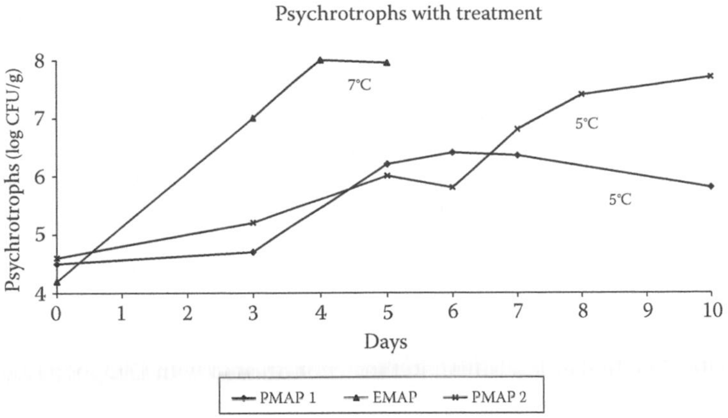

FIGURE 9.4 Psychrotroph counts for treated varieties of lettuce under MAP vs. storage time (PMAP 1 [with a steady state of 2%–10% O_2/5%–12% CO_2] and UV-C treatment of 8.14 kJ/m² corresponding dose for "Red Oak Leaf" lettuce (From Allende, A. and Artes, F., *Lebensm. Wiss. Technol.*, 36, 779, 2003a), EMAP (2%–4% O_2/9% CO_2), with ClO_2 gas treatment after an immersion in an L-cysteine solution for iceberg lettuce (From Gomez-Lopez, V.M. et al., *Int. J. Food Microbiol.*, 121, 74, 2008), PMAP 2 [with bioriented polypropylene used as a packaging film] and UV-C treatment of 8.14 kJ/m² corresponding dose for "Lollo Rosso" lettuce. (From Allende, A. and Artes, F., *Food Res. Int.*, 36, 739, 2003b.))

further corroborated by the lower storage temperature of UV-C treated samples, ideal for psychrotroph growth.

The impact of various O_2 levels from 0 to 100 kPa in conjunction with 0, 10, and 20 kPa CO_2 on the respiration metabolism of greenhouse grown fresh-cut butter lettuce stored at 1°C, 5°C, and 9°C was studied by Escalona et al. (2006). Gas composition with high CO_2 levels (20 kPa) probably triggered a metabolic disorder thereby augmenting the respiration rate of lettuce. The O_2 consumption rate at all treatments revealed a significant rise when increasing the temperature to 9°C and was two to three times higher than at 1°C.

The effect of packaging atmosphere (PMAP with OPP film used with or without N_2 flushing) and storage temperature (3°C and 8°C) on total ascorbic acid (TAA) of iceberg lettuce was studied by Barry Ryan and O'Beirne (1999). They reported that higher levels of TAA were maintained in samples prepared by tearing the lettuce into strips (19 mg/100 g at day 10). PMAP enhanced TAA retention over that in unsealed bags and retention levels were increased further by N_2 flushing (10% higher).

Ares et al. (2008) studied the effect of both passive (PP film) and active (initial gas mixture of 5% O_2 and 2.5% CO_2) MAP on sensory shelf life of butterhead lettuce leaves, stored at 5°C and 10°C. The results showed an increase in sensory deterioration rate proportional to storage temperature increase. At 10°C, AMAP was as effective as PMAP in preserving quality. On the contrary, at 5°C, lettuce in AMAP showed lower deterioration rate and higher sensory shelf life than those in PMAP.

The MAs were passively developed in packages stored at 4°C (monooriented PP [MOPP] film, PE trays overwrapped with a multilayer polyolefin [PO] or plasticized PVC film) after dipping in an ascorbic acid (0.3%) and citric acid (0.3%) solution by Pirovani et al. (1998). Neither the 0.3% ascorbic plus 0.3% citric acid dipping nor the application of MAP affected the microorganism populations considerably. The visual sensory quality of the product was maintained much better in OPP bags compared to the rest of polymeric materials.

Iceberg lettuce treated with gaseous chlorine oxide and cysteine (0.5% solution) and stored under MAP at 7°C (initial atmosphere of 2%–4% O_2 and 9% CO_2) was investigated by Gomez-Lopez et al. (2008). Despite the recorded initial reduction in microorganisms due to ClO_2, APC and psychrotroph counts, reached in the samples treated with ClO_2 higher levels than in those non-treated with ClO_2 before the third day. Immersion in the L-cysteine solution was obligatory prior to ClO_2 treatment to avoid the formation of brown pigments.

Allende and Artes (2003b) evaluated the combination of UV-C radiation and PMAP (bioriented polypropylene [BOPP] film used) for preservation of Lollo Rosso lettuce stored at 5°C. The best scores for browning and overall visual quality were linked to the application of the two highest UV-C doses (4.06 and 8.14 kJ/m^2). All UV-C doses reduced the growth of psychrotrophs (>1 log CFU/g), coliform, and yeast (>0.4 kJ/m^2). However, significant differences were reported only by using the highest dose.

Gaseous chlorine oxide is the most effective treatment for yeasts with 1 log CFU difference from the other treatments. The two cultivars of lettuce displayed the same behavior regarding yeast growth with UV-C treatment with only a certain difference recorded by the end of storage (Figure 9.5).

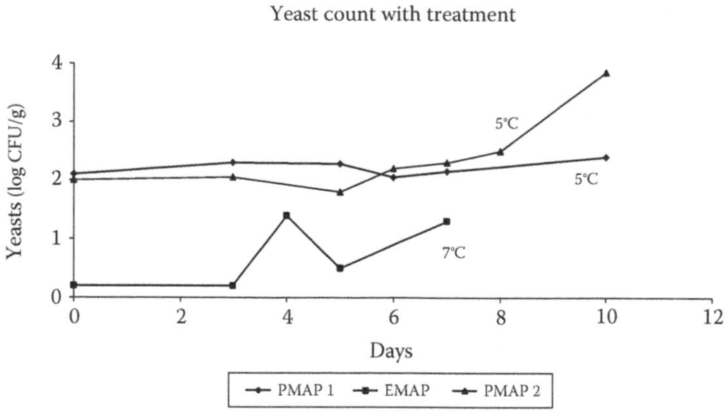

FIGURE 9.5 Yeast count for treated varieties of lettuce under passive or active MAP vs. storage time (PMAP 1 [with a steady state of 2%–10% O_2/5%–12% CO_2) and UV-C treatment of 8.14 kJ/m^2 corresponding dose for "Red Oak Leaf" lettuce (From Allende, A. and Artes, F., *Lebensm. Wiss. Technol.*, 36, 779, 2003a), EMAP [2%–4% O_2/9% CO_2), with ClO_2 gas treatment after an immersion in an L-cysteine solution for iceberg lettuce (From Gomez-Lopez, V.M. et al., *Int. J. Food Microbiol.*, 121, 74, 2008), PMAP 2 [with bioriented polypropylene used as a packaging film] and UV-C treatment of 8.14 kJ/m^2 corresponding dose for "Lollo Rosso" lettuce. (From Allende, A. and Artes, F., *Food Res. Int.*, 36, 739, 2003b.))

Experiments were carried out by Bidawid et al. (2001) to study the effect of various MA (CO_2:N_2 at 30:70, 50:50, 70:30, and 100% CO_2) on the survival rate of *HAV* on romaine lettuce stored both at 4°C and at room temperature. Data obtained indicated that MAP hardly affected *HAV* survival when present on the surface of produce incubated at 4°C. A slight increase in virus survival on lettuce was recorded when high CO_2 levels were used.

Francis and O'Beirne (2001) reported on the survival and growth of inoculated *L. monocytogenes* and *E. coli O157:H7* on shredded iceberg lettuce during storage at 4°C and 8°C and under passively MAs (oriented PP film used). Populations of *L. monocytogenes* and *E. coli O157:H7* did not change considerably on lettuce stored at 4°C. *E. coli O157:H7* generally survived and grew more rapidly than *L. monocytogenes* on lettuce (2.5 and 1.5 log CFU/g for storage at 8°C, respectively).

In Figure 9.6, there is a comparison of mesophile growth when other microorganisms have already been inoculated. Since temperature is an important factor for microbial growth (higher temperatures result in greater mesophile populations), different combinations of inoculated bacteria lead to different total viable count (TVC) numbers, probably because their byproducts or their nature do not allow them to grow. Low oxygen atmospheres showed more encouraging results in decreasing mesophile populations.

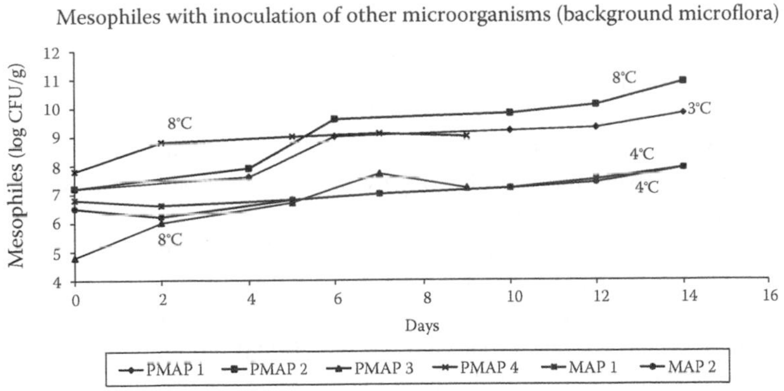

FIGURE 9.6 TVC for different varieties of lettuce with inoculation of other microorganisms under MAP vs. storage time (PMAP 1 [with oriented polypropylene for packaging film] and inoculation with *L. monocytogenes* and *L. innocua* strains, PMAP 2 [with oriented polypropylene for packaging film] and inoculation with *L. monocytogenes* and *L. innocua* strains for iceberg lettuce (From Francis, G.A. and O'Beirne, D., *Int. J. Food Sci. Technol.*, 32, 141, 1997), PMAP 3 [with oriented polypropylene as a packaging film] and inoculation with *E. coli* and *L. innocua* for razor blade cut iceberg lettuce, PMAP 4 [with oriented polypropylene as a packaging film] and inoculation with *E. coli* and *L. innocua* for razor blade cut butterhead lettuce (From Gleeson, E. and O'Beirne, D., *Food Control*, 16, 677, 2005), MAP 1 [2.1% O_2/4.9% CO_2] with inoculation with *S. enteritidis*, MAP 2 [2.1% O_2/4.9% CO_2] with inoculation with *L. monocytogenes*. (From Kakiomenou, K. et al., *World J. Microbiol. Biotechnol.*, 14, 383, 1998.))

Jacxsens et al. (2003) investigated the effect of MA (3% O_2 and 2%–5% CO_2) in conjunction with two types of films (low and high permeability) to assess the quality of mixed lettuce (mixture of endive, curled endive, radicchio lettuce, *lollo rosso* and *lollo bionta* lettuces). On day 6, mixed lettuce stored under equilibrium modified atmosphere (EMA) (with a high gas permeability film) was shown to be unacceptable by the trained panel. The mixed lettuce, stored in the BOPP, was unacceptable for consumption on day 4. The odor and the taste were among the first parameters to deteriorate.

It has been shown that the slicing method to be applied over storage (8°C) is of great importance on subsequent growth and survival of inoculated *L. innocua* and *E. coli* on sliced iceberg and butterhead lettuce. *L. innocua* grew faster and *E. coli* survived better on vegetables sliced with blades that induced the most damage to cut surfaces. The slicing method also influenced the growth of background microflora. For example, razor-sliced vegetables were found to have lower counts than other treatments (1.4 log cycles lower than manually torn iceberg lettuce on day 9) (Gleeson and O'Beirne, 2005).

Gomez-Lopez et al. (2005) investigated the effect of intense light pulses (ILP) decontamination on the shelf life of minimally processed lettuce stored at 7°C in equilibrium MAP (EMAP) (films with OTR: 2290 mL/kg h). They found that control samples reached the microbial acceptability limit before or at day 3, while treated ones did after the third day. Samples became sensorially unacceptable at day 3, due to their low overall visual quality.

Fresh lettuce was inoculated with *S. enteritidis* and *L. monocytogenes* and was stored under MAP with initial head-spaces of 4.9% CO_2/2.1% O_2/93% N_2 and 5% CO_2/5.2% O_2/89.8% N_2. LAB were the predominant organisms in all samples. The pH dropped significantly over the storage of vegetables. *S. enteritidis* was reduced considerably with storage (Kakiomenou et al., 1998).

Active MA conditions with relatively low oxygen levels and low storage temperatures can effectively maintain LAB growth to an acceptable level regardless of inoculated bacteria (Figure 9.7).

The evaluation of temperature dependence on the shelf life of mixed lettuce (20% endive, 20% curled endive, 20% radicchio lettuce, 20% *lollo rosso*, and 20% *lollo bionta* lettuces) under MAP (3% O_2, 5% CO_2, 92% N_2), as affected by microbial proliferation, was carried out by Jacxsens et al. (2002a). Spoilage was due to contamination of psychrotrophs while LAB and yeasts were not that important. The end of shelf life of mixed lettuce, based on sensory properties, was reached after 9, 7, 5, and 3 days when stored at 2°C, 4°C, 7°C, and 10°C, respectively.

Jacxsens et al. (2002b) also studied the impact of temperature fluctuations (TF) (temperatures from 5°C to 20°C) in a simulated cold distribution chain (representative of commercial conditions) on the quality of EMA (3% O_2 and 5% CO_2) packaged processed mixed lettuce. Inoculated *L. monocytogenes* survived on lettuce and *A. caviae* managed to grow (0.58 log CFU/g growth rate). Samples were judged unacceptable on day 5 (after 1 day of storage in the consumers' refrigerator) based on unacceptable color formation.

A synopsis of lettuce species, sensory assessment, metabolic changes, and reached shelf-life prolongation when stored under MAP, is given in Table 9.1.

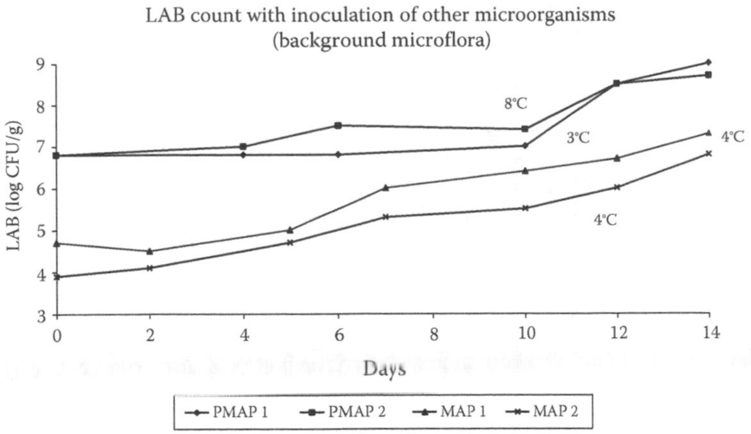

FIGURE 9.7 LAB count for lettuce with inoculation of other microorganisms under MAP vs. storage time (PMAP 1 [with oriented polypropylene for packaging film] and inoculation with *L. monocytogenes* and *L. innocua* strains, PMAP 2 [with oriented polypropylene for packaging film] and inoculation with *L. monocytogenes* and *L. innocua* strains (From Francis, G.A. and O'Beirne, D., *Int. J. Food Sci. Technol.*, 32, 141, 1997), MAP 1 [2.1% O_2/4.9% CO_2] with inoculation with *S. enteritidis*, MAP 2 [2.1% O_2/4.9% CO_2] with inoculation with *L. monocytogenes*. (From Kakiomenou, K. et al., *World J. Microbiol. Biotechnol.*, 14, 383, 1998.))

9.3 CHICORY ENDIVE

Airtight containers with continuous flow of gas mixture (1.5% O_2 and 20% CO_2/0% O_2 and 20% CO_2) were used to control growth of inoculated *L. monocytogenes* on chicory endives stored at 8°C. The growth of the psychrotrophic pathogen *L. monocytogenes*, inoculated on the product, was not restricted. Both growth rate and final population density were slightly higher under 20% CO_2 and 0% O_2 than when 1.5% O_2 was present with the CO_2 (Bennik et al., 1996).

From Figure 9.8, it can be concluded that irradiation reduced initial microbial levels and growth rates under all gas compositions applied. Anoxic conditions and high CO_2 concentrations were advantageous to *L. monocytogenes* growth while lower amounts of CO_2 and higher O_2 levels kept microbial population at starting (low) level. Low temperatures do not favor rapid growth of *Listeria* on chicory endive.

The effect of hurdle technology that is irradiation (0.3 or 0.6 kGy) in conjunction with passive (laminated foil/plastic barrier bag) or AMAP (5% O_2 and 5% CO_2/10% O_2 and 10% CO_2) on survival of inoculated *L. monocytogenes* and sensory analysis of endive stored at 4°C was investigated by Niemira et al. (2005). A dose of 0.6 kGy resulted in reductions of 3.09, 2.41, and 2.53 $\log_{10}$ CFU/g of *L. monocytogenes* and 2.34, 2.20, and 2.26 $\log_{10}$ CFU/g of total microflora in control, MAP A (5% O_2/5% CO_2), and MAP B (10% O_2/10% CO_2) samples, respectively.

Carlin et al. (1996) investigated the storage of minimally processed fresh broad-leaved endive at 3°C and 10°C in MAs containing air, 10% O_2/10% CO_2, 10%

TABLE 9.1
Synopsis of Lettuce Species, Sensory Assessment, Metabolic Changes and Reached Shelf-Life Prolongation When Stored under MAP

Species and Food Type	Initial Gas Mix	Packaging Material	Treatment before Packaging	Storage Temperature (°C) and Storage Period (Days)	Color	Microflora	Texture–Weight Loss	Sensory Analysis	Shelf-Life (Days) Extension	References
Romaine lettuce (*Lactuca Sativa L.*)	The samples were flushed with 0, 1, 2.5, 10, or 21 kPa O_2	(1) PP with OTR: 8 pmol/s m^2 Pa (2) PP with OTR: 16.6 pmol/s m^2 Pa	The samples were sliced, cut, washed in 100 mg/mL chlorine solution (NaOCl)	5°C—14 days	Barely detectable discoloration developed on 8.0 OTR-packages with <21 kPaO_2 while clear discoloration occurred on samples with 21 kPa initial O_2 More severe discoloration on samples packaged with 16.6 OTR films		CO_2 injury, as indicated by brown staining mainly on outer and immature leaf surfaces, occurred on 8.0 OTR-packaged samples by day 10	Ethanol accumulation in 16.6 OTR packages was less than half of that in 8.0 OTR flushed with ≤10 kPa O_2. Electrolyte leakage was higher in packages treated with 8.0 than in 16.6 OTR packages. By day 10, off-odor was higher in 8.0 OTR samples with ≤10 kPa O_2 than in any other samples	At the end of storage, overall quality was higher in 8.0 OTR-packaged samples flushed with ≥2.5 kPa O_2 Overall quality of 16.6 OTR-packaged samples was low due to severe discoloration regardless of initial O_2	Kim et al. (2005)

Iceberg lettuce *cv. Sharpshooter*	Irradiated samples had a faster initial decrease in O_2 levels and were relatively stable thereafter. Irradiated samples had lower O_2 levels, regardless of water treatment temperature. Fast gas changes due to stimulation of respiration by irradiation	(1) Film bags (E-300, Cryovac) with an OTR of 4000 cm^3/h/m^2	Lettuce was dipped in either 5°C or 47°C water for 2 min, packaged in MA film bags and exposed to 0, 0.5, 1, or 2 kGy γ-radiation	3°C—21 days	Lettuce lost greenness and became darker over storage. Lettuce samples dipped at 47°C had better quality than those dipped at 5°C. Surface browning decreased with higher radiation doses		Firmness decreased with increasing radiation doses at the first day of storage in lettuce dipped at 5°C. Lettuce treated at 47°C had lower firmness than at 5°C. Lettuce irradiated at 2 kGy had higher cellular leakage	At 7 days of storage, vitamin C increased with higher radiation dose in lettuce at 47°C whereas antioxidant content decreased. Warm water dipping decreased the lettuce's ability to synthesize phenolics and antioxidants. Reduction in visual quality was slower in irradiated samples	After 21 days of storage, lettuce dipped at 5°C and irradiated at 2 kGy and lettuce treated at 47°C and irradiated at 0.5 or 1 kGy had better quality than the rest of lettuce samples	Fan et al. (2003)
Iceberg lettuce (*Lactuca Sativa L.*)	(1) 4% O_2/12% CO_2 AMAP and at the end of storage was 0.5%–2% O_2/18%–22% CO_2	PE terephthalate (PET)-PP (PP) multilayer film with OTR: 4.2 10^{-13} mol/s/m^2/Pa	Lettuce was washed at 4°C with three ozonated water dips (10, 20, and 10 activated by UV-C light	4°C—13 days	Promotion of browning not observed for any washing solutions. Highest degree of browning	Ozone and MAP and chlorine and MAP slowed microbial growth in more than ozone, chlorine, and MAP alone	Neither chlorine nor ozone affected the texture of fresh-cut lettuce over storage	The visual quality of lettuce was excellent for all treatments. After 9 days of storage, lettuce washed with	Ozonated water is alternative sanitizer to chlorine for fresh-cut lettuce due to good maintenance of sensorial quality	Beltran et al. (2005a)

(*continued*)

TABLE 9.1 (continued)
Synopsis of Lettuce Species, Sensory Assessment, Metabolic Changes and Reached Shelf-Life Prolongation When Stored under MAP

Species and Food Type	Initial Gas Mix	Packaging Material	Treatment before Packaging	Storage Temperature (°C) and Storage Period (Days)	Color	Microflora	Texture–Weight Loss	Sensory Analysis	Shelf-Life (Days) Extension	References
			mg/L min total ozone dose), and compared with chlorine rinses		reported in air-stored samples	Coliforms were reduced 4.9 log units after 9 days by chlorine, ozone 20, and MAP		water or chlorine and stored in air showed a decrease in visual quality. The phenolic content of lettuce was reduced by MAP. The content of vitamin C decreased over storage, especially under MAP	and browning control with no effect on antioxidant constituents	
Red pigmented lettuce (*Lactuca sativa*, "Lollo Rosso")	3% O_2/5% CO_2 AMAP	Bags of 35 μm PP film	Lettuce was washed with chlorinated water, shredded, rinsed, and centrifuged	5°C for 7 days		Mesophilic and psychrotrophic bacterial counts increased from 5 to 8 log CFU/g after 7 days. LAB increased		After 7 days of storage, severe off-odors were detected. Sensory quality of fresh processed	When fresh processed "*Lollo Rosso*" lettuce is processed and stored under the current conditions, the	Allende et al. (2004a)

			prior to packaging			from 3 to 6 log CFU/g after 7 days		"*Lollo Rosso*" lettuce decreased during shelf life	self-life of the product is not longer than 6 days	
Iceberg lettuce (*Lactuca Sativa L. var Raleigh-Patriot*)		44 µm thick PO laminate with O_2 and CO_2 permeabilities of 3,800 and 13,000 mL/m²/day/atm	Lettuce was washed with chlorinated water (0.8–2.0 ppm of free chlorine) and γ-irradiated with 0.1–0.5 kGy 2 days after packaging	2°C ± 2°C—10 days		The higher the dosage, the lower the microbial population, although increasing the dose above 0.2 kGy had decreasing benefit	High irradiation doses affected the texture of lettuce, reducing the shearing force	Ethanol content increased with storage time for all samples in sealed bags, whether or not irradiated	It appears feasible to combine chlorination with irradiation at 0.15–0.5 kGy to produce fresh-cut, chopped lettuce with reduced microbial population	Hagenmaier and Baker (1997)
Butterhead lettuce (*Lactuca sativa* L., cv. Wang)	PMAP: (1) 14% O_2/5% CO_2 (2) 16% O_2/1.2% CO_2	(1) BOPP with OTR: 2,000–3,000 and CDTR: 6,000–7,000 (2) PO PD-961 with OTR: 6,000–8,000 and CDTR: 19,000–22,000 mL/m² day atm	Treated with chlorinated water (200 ppm total chlorine) for 10 min. The leaves were cut with a sharp knife or manually	5°C ± 0.5°C for 17 days	Less browning associated with the use of PD-961 film over BOPP. Similar conclusion for the hand-cut alternative over the use of a knife		Weight loss increased with storage. Significant decrease in maximal force and compression area with storage	Dark stains developed on 8th day for lettuce stored in BOPP film and cut with a knife, on 10th day for lettuce stored both in BOPP and PD-961 films when cut manually, and on 17th day for lettuce in PD-961 film and cut with the hand		Martinez et al. (2008)

(continued)

TABLE 9.1 (continued)
Synopsis of Lettuce Species, Sensory Assessment, Metabolic Changes and Reached Shelf-Life Prolongation When Stored under MAP

Species and Food Type	Initial Gas Mix	Packaging Material	Treatment before Packaging	Storage Temperature (°C) and Storage Period (Days)	Color	Microflora	Texture–Weight Loss	Sensory Analysis	Shelf-Life (Days) Extension	References
Romaine lettuce (*Lactuca Sativa L.*)	AMAP with evacuation until O_2 is 1% and 4% CO_2		The leaves were washed in 100 mg/mL total chlorine solution (NaOCl) and inoculated with five strains of hemorrhagic *E. Coli* (*O157:H7, O26:H11, O55:H7, O91:H21, O111:H12*)	5°C, 10°C, 15°C, 20°C for 7 days		MAP enabled all six *rpoS*-defective isolates to induce acid resistance over the 8 day storage period if t ≥ 15°C. No acid resistance induced for MAP-stored lettuce at t ≤ 10°C or under aerobic conditions				Chua et al. (2008)
Iceberg lettuce (*Lactuca Sativa L. var Alladin.*)	(1) PMAP (2) AMAP with 4% O_2 and 0% CO_2 with gas flush with N_2	OPP	Shredded lettuce was washed in a chlorine (100 ppm. for 5 min) or citric acid (1%, 5 min dip)	3°C and 8°C—14 days		Antimicrobial dips gave better survival of *L. innocua*. At 3°C, flushing with N_2 extended the microbial			Increase in *L. innocua* in samples treated with antimicrobial dip maybe due to reduction in	Francis and O'Beirne (1997)

			solution. Then, the leaves wee inoculated with *L. innocua* and *L. monocytogenes* strains			survival, while counts in MAP packs were not different from the samples packaged in air			naturally occurring microflora thereby giving competitive advantage to *L. innocua*	
Winter harvested iceberg lettuce (*Lactuca sativa L. cv. Coolguard*)	(1) AMAP with N_2 flush and 5% O_2 and 0% CO_2 (2) PMAP	(1) Perforated PP (22 µm) (2) PP (25, 30, 40 µm)	All the treatments vacuum cooled	2°C for 14 days and 12°C for 2.5 days (shelf-life period)	Vacuum-cooling had a favorable effect on reducing pink rib but only during the shelf-life period		Lowest weight losses in MAP (less than 0.1%), while head under perforated PP showed weight losses less than 0.93%	Heart-leaf injury increased over shelf-life testing in non vacuum-cooled lettuces. Russet spotting severity was higher in naked lettuce, perforated PP and 40 µm PP MAP	Heads packed in unperforated PP bags maintained a higher visual quality by the end of experiment, than those unpacked. The under MAP quality was in 40 µm PP	Martinez and Artes (1999)
"Red Oak Leaf" lettuce (*Lactuca sativa L.*)	(1) 2%–10% O_2/5%–12% CO_2	BOPP with OTR: 1,800 mL/m² day atm	UV-C treatment with (1) 0.41, (2) 0.81, (3) 2.44, (4) 4.07, and (5) 8.14 corresponding doses (kJ/m²)	5°C for 9 days	Analysis browning and color of the product showed no difference between product UV-C treated and untreated control	UV-C radiation reduced psychrotrophic growth about 0.5–2 log unit CFU/g and yeast growth with higher reductions with higher dosages UV-C did not affect LAB		No difference between UV-C-treated and untreated lettuce for sensory properties during storage		Allende and Artes (2003a)

(*continued*)

TABLE 9.1 (continued)
Synopsis of Lettuce Species, Sensory Assessment, Metabolic Changes and Reached Shelf-Life Prolongation When Stored under MAP

Species and Food Type	Initial Gas Mix	Packaging Material	Treatment before Packaging	Storage Temperature (°C) and Storage Period (Days)	Color	Microflora	Texture–Weight Loss	Sensory Analysis	Shelf-Life (Days) Extension	References
Butterhead lettuces (*Lactuca sativa* L.) cv. Zendria	Active CA with 0%–100% O_2 and 0%, 10%, 20% CO_2			1.5°C and 9°C for 10 days	High CO_2 levels of 20 kPa cause physiological stress and increase more the respiration rate than the lettuce stored in air atmosphere (20 kPa O_2 and 0 kPa CO_2)			The visual quality of lettuce maintained for longer time at 5°C and even more at 1°C. Lettuce exposed to low O_2 levels and moderate to high CO_2 levels had higher respiration rate	80 kPa O_2 must be used in (MAP) to avoid fermentation of fresh-cut butterhead lettuce in combination with 10–20 kPa CO_2 for respiration rate reduction	Escalona et al. (2006)
Spanish iceberg lettuce (cultivar *Salodin*)	(1) PMAP. (2) Nitrogen flush	35 μm thick OPP with OTR: 1,200 mL/m² day atm and CDTR: 4,000 mL/m² day atm	Samples were shredded into 6 mm wide pieces, either manually or by machine. Shredded lettuce samples	3°C and 8°C for 10 days				Higher levels of TAA were maintained in samples prepared with manually tearing. MAP increased TAA	Manually prepared lettuce samples had better appearance scores. Flushing with nitrogen improved acceptability	Barry-Ryan and O'Beirne (1999)

			dipped for 5 min in 100 ppm chlorine solution					level. The latter were reduced at higher storage temperature		
Butterhead lettuce (*Lactuca sativa L.*, cv Wang)	(1) PMAP (2) AMAP with 5% O_2 and 2.5% CO_2	PP (PP) (40 μm thickness)	Chlorinated water (200 ppm total chlorine) for 10 min	5°C and 10°C for 49 and 21 days, respectively	Lettuce leaves stored at 10°C showed higher wilting appearance and browning on the midribs than those stored at 5°C		For all studied storage temperatures, weight loss reached values higher than 20%	Off-odor only appeared after 21 days of storage at 10°C and after 42 and 49 days of storage at 5°C. AMAP slowed down sensory deterioration rate at 5°C when compared with PMAP	If lettuce leaves were stored at 5°C, the use of active MA increased the shelf life of lettuce leaves by 20% with respect to passive MA	Ares et al. (2008)
Iceberg lettuce	PMAP	MOPP film and with a multilayer PO (RD-106) or PVC film	Rinsed with tap water (0.2 mg/L total chlorine) for 4 min. Half of shredded lettuce dipped in a 0.3% ascorbic acid plus 0.3% citric acid solution	4°C for 8 days	The main attribute of the end of shelf life was enzymatic browning of sliced surfaces with pVC-PE and RD-106-PE trays	Neither 0.3% ascorbic plus 0.3% citric acid dip treatment nor MAP, affected the microorganism growth		Samples in OPP bags were better in general appearance, wilting and browning than the rest packaging investments	The shelf life of shredded lettuce in OPP bags exceeded 8 days	Pirovani et al. (1997)

(continued)

TABLE 9.1 (continued)
Synopsis of Lettuce Species, Sensory Assessment, Metabolic Changes and Reached Shelf-Life Prolongation When Stored under MAP

Species and Food Type	Initial Gas Mix	Packaging Material	Treatment before Packaging	Storage Temperature (°C) and Storage Period (Days)	Color	Microflora	Texture–Weight Loss	Sensory Analysis	Shelf-Life (Days) Extension	References
Iceberg lettuce (*Lactuca sativa* var. capitata L.)			Immersion for 1 min in a solution of 0.5% HCl-L-cysteine monohydrate and treatment with ClO_2		Immersion in the L-cysteine solution before ClO_2 treatment avoided the development of brown pigments	Treatment with ClO_2 gas (after immersion in L-cysteine solution) reduced mesophiles and psychrotroph counts. L-cysteine minimized the decontamination efficacy of ClO_2		ClO_2 treatment (after immersion in L-cysteine solution) did not cause any alteration of the sensory attributes	The shelf life of MAP lettuce from the sensorial point of view was limited to 4 days in both, untreated and treated samples	Gomez-Lopez et al. (2008)
"Lollo Rosso" lettuce (*Lactuca sativa*.)		BOPP with OTR: 1,800 mL/m² day atm	UV-C treatment with (1) 0.41, (2) 0.81, (3) 2.44, (4) 4.07, and (5) 8.14 corresponding doses (kJ/m²)	5°C—10 days	UV-C treatment decreased browning in lettuce samples	Psychrotrophic growth was reduced only with the highest UV-C doses. Coliforms were inhibited with application of highest UV-C dose. UV-C		The highest applied UV-C radiation doses (2.44, 4.07, and 8.14 kJ/m²) improved the sensory quality of product. Lettuce tissue became shinier		Allende and Artes (2003)

						stimulated LAB growth. The higher the dose, the stronger the yeast reduction	as the highest UV-C treatment was applied		
Romaine lettuce	(1) PMAP. (2) 30% CO_2 (3) 50% CO_2 (4) 70% CO_2 (5) 100% CO_2	Barrier plastic bags of low O_2 permeability (0.46–0.93 mL/100 mL day atm)	Inoculation of lettuce with *HAV* carried out with spreading 10 μL of virus-containing solution ($1.7 \cdot 10^5$ plaque forming unit)	4°C for 12 days	Browning was observed in pieces of lettuce incubated under 100% CO_2	The lowest *HAV* survival rate of 47.5% was reported on lettuce stored in a petri dish, whereas the highest survival rate was under 70% CO_2		Lettuce overall quality was much better under MAP conditions than when incubated in air	Bidawid et al. (2001)
Irish iceberg lettuce	PMAP: 3%–4% O_2 and 10%–12% CO_2	35 μm OPP	Inoculation with *L. monocytogenes* and two nontoxigenic *E. coli O157:H7* strains	4°C and 8°C for 12 days		Populations of *L. monocytogenes* and *E. coli O157:H7* increased over storage on lettuce held at 8°C but hardly changed on lettuce stored at 4°C			Francis and O'Beirne (2001)
Mixed lettuce (20% endive, 20% curled endive, 20% radicchio lettuce, 20% lollo rosso,	AMAP: 3% O_2 and 2%–5% CO_2	(1) BOPP film (30 μm), PVC coated with an OTR of 15 mL O_2/m^2 24 h atm		7°C for 13 days		Accumulation of CO_2 in BOPP packages and generated anoxic conditions prevented rapid	Lettuce, packaged under EMA, showed no metabolite production over storage while ethanol		Jacxsens et al. (2003)

(*continued*)

TABLE 9.1 (continued)
Synopsis of Lettuce Species, Sensory Assessment, Metabolic Changes and Reached Shelf-Life Prolongation When Stored under MAP

Species and Food Type	Initial Gas Mix	Packaging Material	Treatment before Packaging	Storage Temperature (°C) and Storage Period (Days)	Color	Microflora	Texture–Weight Loss	Sensory Analysis	Shelf-Life (Days) Extension	References
and 20% lollo bionta lettuces—red and green cultivar)		(2) High permeable packaging film with OTR: 2270 mL O_2/m^2 24 h atm (EMA)				outgrowth of psychrotrophic bacteria. Yeasts exceeded 10^5 CFU/g on day 6		was detected in BOPP packages. At day 6, lettuce stored under EMA was unacceptable whereas BOPP packages were non-edible on day 4		
Irish butterhead lettuce	Nitrogen flush before sealing	OPP 35 μm	Lettuce was (1) Hand torn into slices (2) Cut with blunt knife (3) Cut with razor blade Inoculation with *E. Coli* and *L. Innoqua*	8°C for 9 days		Counts of *E. Coli* and *Listeria Innoqua* on razor blade sliced lettuce were lower than counts on hand torn and blunt knife sliced lettuce			Counts of total mesophilic counts on razor-sliced butterhead lettuce were significantly lower than those of other slicing treatments	Gleeson and O'Beirne (2005)

Iceberg lettuce		OPP 35 μm	Lettuce was (1) Hand torn into slices (2) Cut with razor blade Inoculation with *E. Coli* and *L. Innoqua*	8°C for 9 days		Counts of *E. Coli* were lower on hand torn lettuce than on razor-sliced iceberg lettuce. Populations of *L. innocua* increased 1.5–2 log cycles over storage independently from the slicing treatment			Counts of total mesophilic counts on hand torn iceberg lettuce were 1.4 log cycles higher than counts on razor-sliced iceberg lettuce	
Iceberg lettuce (*Lactuca sativa var. capitata L.*)	PMAP	Film with OTR: 3529 mL O_2/m^2 24 h atm	Treatment with ILP	7°C for 9 days		Psychrotrophic count kept lower than that for the controls over the 5 days, whilst yeasts count of lettuce were higher		Overall visual quality reached unacceptable scores at day 3	ILP did not prolong the shelf life of MP iceberg lettuce but gained 1 extra storage day	Gomez-Lopez et al. (2005)
Lettuce	AMAP: (1) 2.1% O_2/4.9% CO_2 (2) 5.2% O_2/5% CO_2	PE bags	Inoculation with *S. enteritidis* and *L. monocytogenes* strains	4°C ± 0.2°C for 14 days		LAB and TVC were lower in MAP than in air. *S. enteritidis* and *L. monocytogenes* decreased during the storage in both the packaging systems			Packaging in MAP is not necessarily an extra hurdle for growth of *S. enteritidis* and *L. Monocytogenes*	Kakiomenou et al. (1998)

(continued)

TABLE 9.1 (continued)
Synopsis of Lettuce Species, Sensory Assessment, Metabolic Changes and Reached Shelf-Life Prolongation When Stored under MAP

Species and Food Type	Initial Gas Mix	Packaging Material	Treatment before Packaging	Storage Temperature (°C) and Storage Period (Days)	Color	Microflora	Texture–Weight Loss	Sensory Analysis	Shelf-Life (Days) Extension	References
Mixed lettuce (20% endive, 20% curled endive, 20% radicchio lettuce, 20% lollo rosso, and 20% lollo bionta lettuces—red and green variety)	AMAP: 3% O_2/5% CO_2	Bags with OTR: $1.04 \cdot 10^{-11}$ mol O_2/m² s Pa	Inoculation with *L. monocytogenes* strains and *A. caviae*	2°C, 4°C, 7°C, and 10°C for 11 days		Psychrotrophic counts were the limiting group of micro-organisms. *L. monocytogenes* survived at 2°C while growth was also possible at other temperatures		4°C is the optimal storage temperature for organoleptic attributes	The end of the shelf life of mixed lettuce, based on sensory properties, was reached after 9, 7, 5, and 3 days, respectively, stored at 2°C, 4°C, 7°C, and 10°C	Jacxsens et al. (2002a)
Mixed lettuce (20% endive, 20% curled endive, 20% radicchio lettuce, 20% lollo rosso, and 20% lollo biontalettuces	AMAP: 3% O_2/5% CO_2	Bags with OTR: 2026 mL O_2/m² 24 h atm	Inoculation with *L. monocytogenes* strains and *A. caviae*	T < 12°C, t = 4°C for 24 h, t = 5°C for 2 h, t = 10°C for 24 h, t = 5°C for 2 h, t = 10°C for 8 h, t = 7°C for 48 h,		Yeasts exceeded the limit at sampling 4(t = 10°C for 8 h). Psychrotrophic counts exceeded the limit at the last phase(t = 7°C). For LAB		Lettuce was rejected on day 5 (after 1 day of storage in the consumers' refrigerator, moment of sampling 6) by 50% of the		Jacxsens et al. (2002b)

red and green variety)		t = 20°C for 2 h and t = 7°C for the rest 7 days	no critical limit was reached. Slight increase in numbers of *A. caviae*	panel based on undesired color modifications	
Irish iceberg lettuce	Inoculation with four strains of *L. monocytogenes* with or without genes for glutamase decarboxylase resistance mechanism	4°C, 8°C, and 15°C	The wild strain survived better than the double mutant Δ*gad*AB strain		Francis et al. (2007)

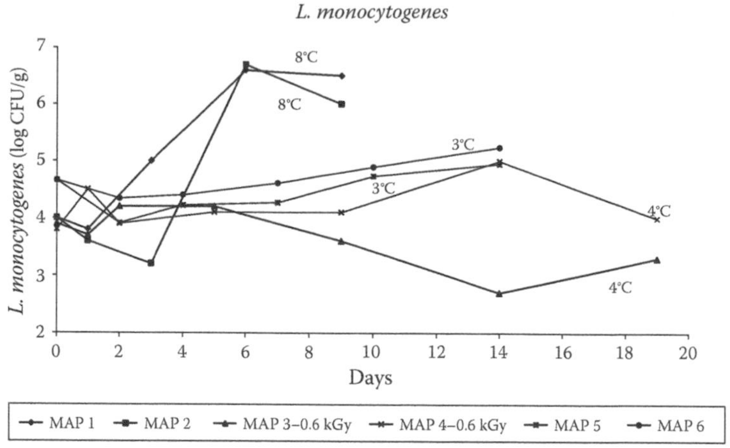

FIGURE 9.8 Growth rate of inoculated strains of *L. monocytogenes* under MAP on chicory endive (MAP 1 [20% CO_2/80% N_2], MAP 2 [20% CO_2/78.5% N_2/1.5% O_2] (From Bennik, M.H.J. et al., *Postharvest Biol. Technol.*, 9, 209, 1996), MAP 3 [5% CO_2/90% N_2/5% O_2], MAP 4 [10% CO_2/80% N_2/10% O_2] (From Niemira, B.A. et al., *Radiat. Phys. Chem.*, 72, 41, 2005), MAP 5 [30% CO_2/60% N_2/10% O_2], MAP 6 [50% CO_2/40% N_2/10% O_2]. (From Carlin, F. et al., *Int. J. Food Microbiol.*, 32, 159, 1996.))

O_2/30% CO_2, and 10% O_2/50% CO_2. Both at 3°C and at 10°C, the lowest extent of spoilage of endive leaves was recorded in the MAs including 10% O_2/10% CO_2. The rise in numbers of *L. monocytogenes* at 3°C was low (between 0.3 and 1.5 log CFU/g), whereas it was slightly higher during storage in air.

Endives stored in LDPE under AMAP in the presence of oxygen scavengers at 20°C exhibited the lowest total chlorophyll amount at days 3 and 7, thus confirming the data from sensory and image analysis on the positive oxygen scavenger effect with regard to greening and browning delay. Moreover, in AMAP, the basal parts of endives remained white/yellow and browning was avoided (Charles et al., 2008).

The effect of storage under an EMA (initial atmosphere of 3% O_2 and 2%–5% CO_2) at 7°C on the quality of shredded chicory endives was investigated by Jacxsens et al. (2003). Gram-negative *Pseudomonas* spp. dominated the spoilage flora of shredded chicory endives. The growth of the LAB on the shredded chicory endives was slow. Taste and odor displayed deterioration signs at day 13, when the microbial contamination was high and LA was detected while color was modified by day 6. It was shown that storage of chicory endive under low CO_2 and O_2 conditions was more effective than at high CO_2 atmospheres, irrespective of storage temperature for LAB risk assessment especially for the first days of storage (Figure 9.9).

In another publication the same authors (Jacxsens et al., 2001) studied the application of high oxygen atmosphere (HOA) (70%, 80%, and 95% O_2) as another means for extending the shelf life of shredded chicory endive. The overall shelf life was at least doubled when HOAs were applied compared to the low EMA packages. This

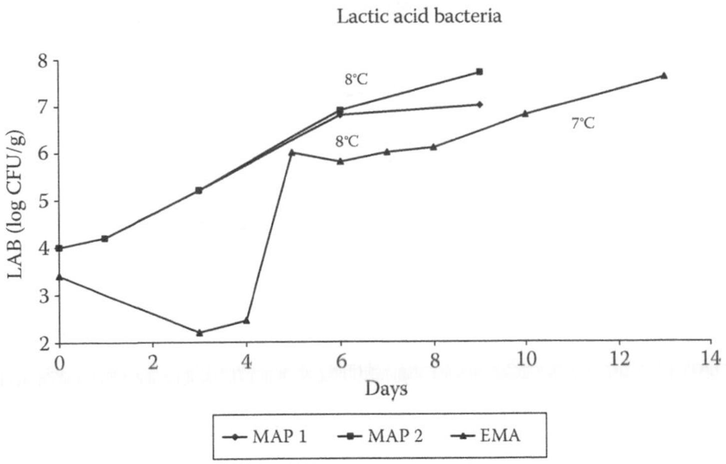

FIGURE 9.9 Changes in LAB population during storage under MAP on chicory endive vs. storage time (MAP 1 [20% CO_2/80% N_2], MAP 2 [20% CO_2/78.5% N_2/1.5% O_2] (From Bennik, M.H.J. et al., *Postharvest Biol. Technol.*, 9, 209, 1996), EMA [with initial atmosphere of 2%–5% CO_2/3% O_2]. (From Jacxsens, L. et al., *Int. J. Food Microbiol.*, 83, 263, 2003.))

was done due to the fact that the HOAs almost doubled (from 4 to 7 days) the time required by yeasts to reach their critical limit compared to the EMA conditions.

Storage at 4°C had a more marked rise in yeast population in comparison with EMA with the same atmospheric conditions at 7°C. Therefore, temperature enhancement had a negative impact on yeast growth. High-oxygen packages displayed less yeast growth than EMA ones under the same storage conditions, thus proving that O_2 enriched atmospheres are much more effective for chicory endive preservation (Figure 9.10).

The microbiological, sensorial, and chemical data for chicory endives are summarized in Table 9.2.

9.4 TOMATO

The impact of various MAP conditions (4% CO_2 + 1% or 20% O_2, 8% CO_2 + 1% or 20% O_2, or 12% CO_2 + 1% or 20% O_2) and packaging films (film A and film B [87.4 and 60 mL/h m^2 atm, respectively]) during cold storage (5°C–10°C) on the quality of fresh-cut tomato slices was investigated by Hong and Gross (2001). Tomatoes under 4% CO_2 + 1% O_2 had the highest soluble solids content (SSC) (4.26%), while the product stored under 8% CO_2 + 20% O_2 had a considerably higher pH (4.77). Slices in containers with 4% CO_2 + 20% O_2 displayed the highest visible fungal growth (87%). It was shown that MAP imparted good quality tomato slices with a shelf life of 2 weeks or more at 5°C.

Different packaging films (PE-20μ and 50μ, PVC-10μ, PP-25μ) were employed by Batu and Thompson (1998) to stimulate the ideal conditions for preserving pink tomatoes stored at 13°C. All unwrapped tomatoes were overripe and soft after 30 days. When packaging was sealed, especially with PE_{50} and PP films, the red

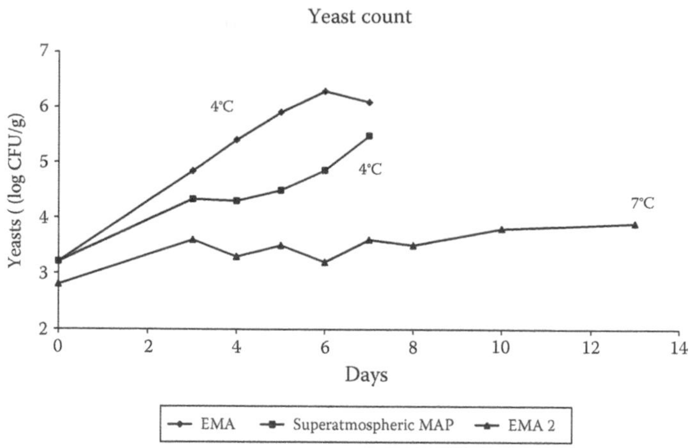

FIGURE 9.10 The effect of passive and active MAP on yeast count for chicory endive vs. storage time (EMA 1 [with initial atmosphere of 5% CO_2/3% O_2], superatmospheric MAP [95% O_2] (From Jacxsens, L. et al., *Int. J. Food Microbiol.*, 73, 331, 2001), EMA 2 [with initial atmosphere of 2%–5% CO_2/3% O_2]. (From Jacxsens, L. et al., *Int. J. Food Microbiol.*, 83, 263, 2003.))

color development of tomatoes was considerably delayed up to 30 days of storage. Furthermore, these tomatoes remained very firm even after 60 days of storage (1.5 and 1.6 N/mm, respectively).

Another form of active packaging consisted of granular-activated carbon (GAC), either on its own or impregnated with palladium as a catalyst inside tomato packages (OPP film used) under PMAP. The attributes with most significant differences were color and firmness, for which the GAC-Pd had the highest scores (1.9 and 7.3 skin and mesocarp a^* values and 3.05 N/mm, respectively). The decay recorded after 28 days was significantly lower (39%) for GAC-Pd than for tomatoes with GAC (65%) (Bailen et al., 2006).

Both passive (no CO_2) and active (high CO_2 pressure) MA packages were applied at 0°C and 5°C for tomato (slices and wedges) storage by Aguayo et al. (2004). MAP considerably diminished total plate counts (TPCs) of slices at 5°C (>5 log CFU/g for all MAP treatments), although only AMAP reduced TPCs in wedges after 14 days at 5°C. Temperature proved to be more effective than MAP treatments in drastically diminishing the microbial growth. Both MAP treatments led to better appearance and overall quality than control samples in air.

Fresh-cut tomato slices were stored at 0°C and 5°C under active (12–14 kPa O_2 + 0 kPa CO_2) MAP (Composite [Vascolan] or BOPP), with or without an ethylene absorber (EA). Visual quality scored higher for packages containing EA. Texture was excellent at 0°C while it dropped slightly at 5°C. Microbial count was very low for samples kept at 0°C (<3.6 log CFU/g). The maturity index (SSC/titratable acidity [TA]) exhibited minimal variation for composite-EAP at 0°C (14.9) and for composite at 5°C (14.5) (Gil et al., 2002).

TABLE 9.2
Microbiological, Sensorial, and Chemical Data for Chicory Endive

Species and Food Type	Initial Gas Mix	Packaging Material	Treatment before Packaging	Storage Temperature (°C) and Storage Period (Days)	Color	Microflora	Texture–Weight Loss	Sensory Analysis	Shelf-Life (Days) Extension	References
Chicory endive *Cichorium endivia L.*	(1) Air (2)1.5% O_2/20% CO_2 (3) 0% O_2/20% CO_2	Airtight containers with continuous flow of gas mixture (flow rate 200 mL/min)	Inoculation with *L. monocytogenes* by spraying 10 mL of a cell suspension (±10^5 CFU/mL) on 1 kg of produce	8°C for 9 days		No significant differences in total aerobic count *Pseudomonas* counts lower under 0% O_2 *Enterobacteria* more abundant under 0% O_2 *L. monocytogenes* growth rate was higher in low O_2 and high CO_2 conditions			The spoilage rate under air proceeded faster (2 days), whereas under MAP the same reached in 4 days	Bennik et al. (1996)
Chicory endive *Cichorium endivia L.*	(1) Air (2) 5% O_2/5% CO_2 (3) 10% O_2/10% CO_2	Laminated foil/plastic barrier bag	Sanitation with 300 ppm sodium hypochlorite sol	4°C for 19 days	Irradiated leafs in air retained color better than	Irradiation reduced initial levels of *L. monocytogenes* and total microflora	No significant differences detected from irradiation		14 days	Niemira et al. (2005)

(continued)

TABLE 9.2 (continued)
Microbiological, Sensorial, and Chemical Data for Chicory Endive

Species and Food Type	Initial Gas Mix	Packaging Material	Treatment before Packaging	Storage Temperature (°C) and Storage Period (Days)	Color	Microflora	Texture–Weight Loss	Sensory Analysis	Shelf-Life (Days) Extension	References
			Inoculation with *L. monocytogenes* strains. Irradiation with 0.3 and 0.6 kGy		non-irradiated	under all gas compositions. MAP under both atmospheres did not allow their regrowth	and in any atmosphere			
Chicory endive *Cichorium endivia L.*	(1) Air (2) 10% O_2/10% CO_2 (3) 10% O_2/30% CO_2 (4) 10% O_2/50% CO_2	Plastic boxes with perforated polymeric film as a cover placed in airtight containers with continuous gas flow	Inoculation with *L. monocytogenes* strains	3°C—10–14 days, 10°C—7 days		At 10°C *L. monocytogenes* grew better with increase in CO_2. The growth of aerobic microflora dropped with storage in CO_2 enriched MAP		The visual quality was improved by storage in the second gas composition The other samples showed extensive spoilage after storage		Carlin et al. (1995)
Endive *Cichorium intybus* L.	PMAP: 3 kPa O_2/5 kPa CO_2 with 50% reduction of	LDPE film of 50 μm thickness and macroperforated OPP	Oxygen scavenger sachets were inserted in the LDPE bags	20°C—7 days	Color changes of endives were delayed only with AMAP			Browning was delayed by PMAP Samples under	AMAP prolonged the acceptable	Charles et al. (2008)

	transient period for AMAP (with oxygen scavengers)				(Oxygen scavengers)			AMAP maintained acceptable appearance	appearance of endives from 3 days to 6 days	
Chicory endive *Cichorium endivia L.*	AMAP: 3% O_2 and 2%–5% CO_2	Packaging film with OTR: 3704 mL O_2/m^2 24 h atm		7°C for 13 days	Shredded endive is very sensitive for enzymatic discoloration and the color was the limiting sensory property at day 6	Shredded chicory endives contained only after day 8 more than 10^8 CFU/g psychrotrophic bacteria. The growth of the LAB was slow. Yeasts never exceeded 10^5 CFU/g limit		Taste and odor were unacceptable at day 13, when the microbial contamination was high		Jacxsens et al. (2003)
Chicory endive *Cichorium endivia L.*	AMAP with (1) 3% O_2 and 5% CO_2. (2) HOA (70%, 80%, and 95% O_2)	Packaging film with OTR: 1446 mL O_2/m^2 24 h atm		4°C—7 days	Enzymatic discoloration made endive unacceptable in 3 days under MAP, while no unacceptable scores were obtained for the HOA	Yeasts under low O_2 concentrations exceeded the limit at day 4 while high O_2 packages exceeded the limit at day 7		No significant difference between both methods of packaging was found for the shredded chicory endives based on the organoleptic properties	The overall obtained shelf life is at least doubled by using HOAs compared to the MAP packages	Jacxsens et al. (2001)

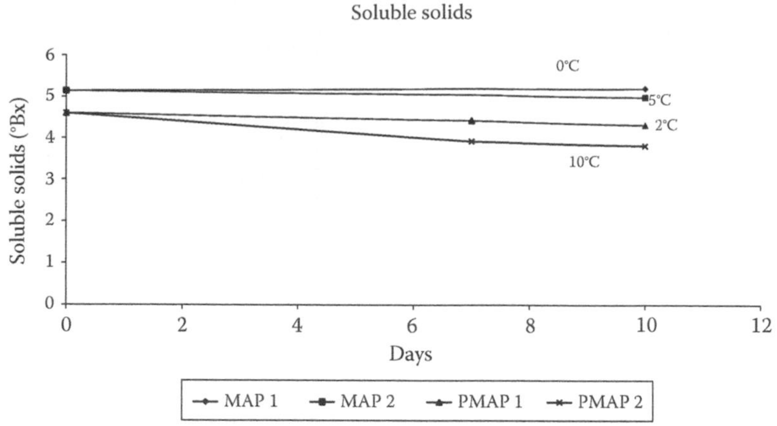

FIGURE 9.11 Alterations in soluble solids of tomatoes under different storage conditions and treatments (MAP 1 with EAP [12%–14% O_2] and addition of EA containing $KMnO_4$, MAP 2 with EAP [12%–14% O_2] and addition of EA containing $KMnO_4$ (From Gil, M.I. et al., *Postharvest Biol. Technol.*, 25, 199, 2002), PMAP 1 [with perforated polypropylene for package film] and immersion into chlorinated water [0.7 M] with $CaCl_2$ [0.09 M], PMAP 2 [with perforated polypropylene for package film] and immersion into chlorinated water [0.7 M] with $CaCl_2$ [0.09 M]. (From Artes et al., *Postharvest Biol. Technol.*, 17, 153–162, 1999.))

Figure 9.11 reveals that the higher the storage temperature, the greater is the reduction in SSC. SSC is affected by $CaCl_2$ and the applied storage conditions of PMAP, thereby undermining the quality of tomatoes.

Artes et al. (1999) investigated the effects of calcium chloride washings (0.7 mM chlorinated water with or without 0.09 M $CaCl_2$) and passive or AMAP (7.5% O_2) at 2°C and 10°C on preservation of quality of fresh-cut tomato. It was found that dipping tomato slices in calcium chloride at 2°C was beneficial in maintaining the quality of the tomato slices (4.3°Bx SSC, 4.0 pH, 0.36 g citric acid/100 mL TA for 10 days at 2°C). Passive or AMAP retarded the maturity development, particularly in tomato slices kept under AMAP.

Storage temperature is the sole differing factor that can adequately justify the significant difference in values of acidity among the analyzed samples. Higher storage temperatures induce physicochemical changes in tomatoes that augment the TA, whereas a lower temperature is in favor of the establishment of a more stable environment for the preservation of products (Figure 9.12.).

Six different tomato cultivars were packaged under MAP (5% O_2 + 5% CO_2) and kept in the fridge (4°C). The highest content of lycopene was reported for Bodar tomatoes (80.5 mg/kg fw), whereas the concentration in the other cultivars varied between 20.0 and 43.1 mg/kg fw. The initial colors of fresh-cut tomatoes were maintained reasonably well for 3 weeks under cold storage. On the other hand, vitamin C concentration displayed no substantial variations throughout storage time (80–200 mg/kg fw for the studied cultivars after 21 days [Odriozola-Serrano et al., 2008]).

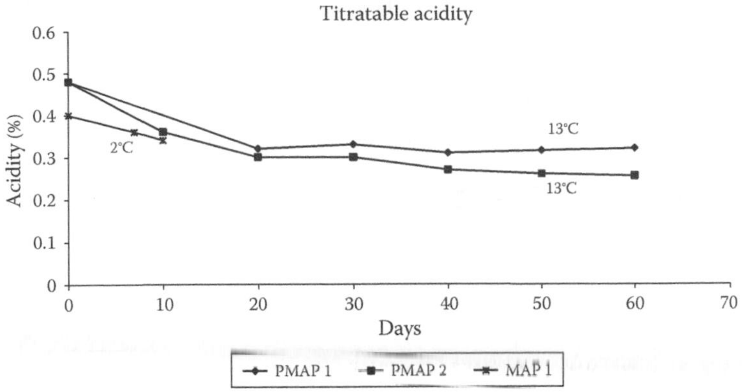

FIGURE 9.12 Changes in titratable acidity of tomatoes under MAP vs. storage time (PMAP 1 [equilibrium atmospheric conditions of 6%–7% O_2/4% CO_2) with the use of polyethylene 50 µm as a packaging film, PMAP 2 [equilibrium atmospheric conditions of 12%–13% O_2/5%–6% CO_2] with the use of polypropylene 25 µm as a packaging film (From Batu, A. and Thomson, A.K., *Trans. J. Agric. For.*, 22, 365, 1996), MAP [7.5% O_2]. (From Artes et al., *Postharvest Biol. Technol.*, 17, 153–162, 1999.))

The impact of PMAP (LDPE film), CA (5% CO_2), and gaseous ozone treatment on the survival of inoculated (low and high inoculum levels) *S. enteritidis* on cherry tomatoes was investigated by Das et al. (2006). MAP storage of tomatoes with low inoculum dose (3.0 $\log_{10}$ CFU/tomato) provided cell death in a shorter time (4 days) compared with ambient air (6 days). Ozone gas treatment of 10 mg/L at different time intervals of 5 and 15 min was found to be effective, respectively, on low and high (7 $\log_{10}$ CFU/tomato) inoculum levels of *S. enteritidis* attached for 1 h.

Suparlan and Itoh (2003) investigated the tomatoes' shelf-life extension by immersing them in hot water (42.5°C) for 30 min, packaging under PMAP (LDPE film) for 14 days at 10°C, and then at 22°C for 3 days without packaging. Combination of HWT and MAP decreased the weight loss (<0.7%), decay and mold growth (2.57 log CFU/g), inhibited the ripening process, maintained firmness, and prevented decay and cracking of tomatoes stored at 10°C.

Figure 9.13 shows that the loss of firmness is greater in untreated samples stored under PMAP than in those with GAC and GAC-Pd treatments. Moreover, the higher storage temperature may have played its role in softening of untreated tomatoes.

Dipping matures green cherry tomato in hot water (39°C for 90 min) and then storing in plastic films of various O_2, but similar CO_2 permeabilities at 15°C led to 26% reduced color development area in heat-treated fruit in the M_3 film (OTR: $3.56 \cdot 10^{-17}$ mol/s m^2 Pa) (Sayed Ali et al., 2004). HWT diminished the color by 7% and the low permeability film (12.2 kPa O_2) by 6% compared to the control. The highest h° and lowest C* values were recorded in heat-treated fruit stored under the lowest O_2 atmosphere (low permeability film).

Mature green tomatoes stored in MA containers with steady-state atmospheres of 5% O_2 and 5% CO_2 at 13°C were subjected to 10°C TF over 35 days to simulate storage and transport conditions (Tano et al., 2007). The firmness of unpackaged

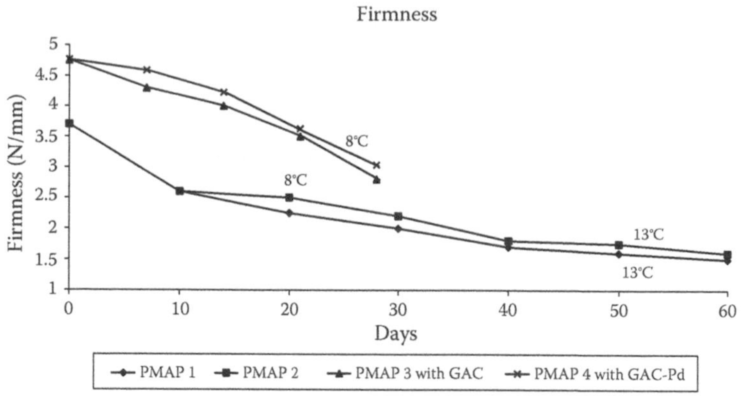

FIGURE 9.13 Changes in tomato firmness under MAP vs. storage time (PMAP 1 [equilibrium atmospheric conditions of 6%–7% O_2/4% CO_2] with the use of polyethylene 50 μm as a packaging film, PMAP 2 [equilibrium atmospheric conditions of 12%–13% O_2/5%–6% CO_2] with the use of polypropylene 25 μm as a packaging film (From Batu, A. and Thomson, A.K., *Trans. J. Agric. For.*, 22, 365, 1996), PMAP 3 [with steady-state atmosphere of 7% O_2/7% CO_2] with addition of granular activated carbon [GAC], PMAP 4 [with steady-state atmosphere of 7% O_2/7% CO_2] and with addition of GAC impregnated with palladium as a catalyst [GAC-Pd]. (From Bailen, G. et al., *J. Agric. Food Chem.*, 54, 2229, 2006.))

tomatoes was lower than those of MA at constant temperature and fluctuating temperatures (0.8, 3.9, and 2.7 N/mm, respectively). MAP storage, both under steady and fluctuating temperatures, minimized weight loss compared to air (0.5%, 0.7%, and 3.4%, respectively).

In regard to the packaging film applied, films with high gas permeability have greater weight loss than less permeable films. High storage temperatures enhance product respiration and consequently initiate loss of mass and fluids. Tomatoes with GAC or GAC-Pd displayed considerably lower weight losses than untreated or hot water–treated ones (Figure 9.14).

All the aforementioned data for tomatoes stored under MAP are summarized in Table 9.3.

9.5 CARROT

The efficiency of gamma irradiation (doses from 0.15 to 0.9 kGy) in conjunction with MAP (60% O_2 + 30% CO_2) on grated carrots inoculated with *E. coli* (6 log CFU/g) and stored at 4°C ± 1°C was investigated by Lacroix and Lafortune (2004). A complete inhibition of *E. coli* was reached over the whole storage period when samples were treated at 0.6 kGy under MAP or when treated at 0.9 kGy under air. In samples treated under air a 1–2 log CFU/g bacteria were detected between day 5 and day 15.

A comparison of different conducted treatments shown in Figure 9.15 revealed that gaseous chlorine dioxide decreased the initial microbial load more than any other reported treatment and was by far more effective than chlorine or citric acid.

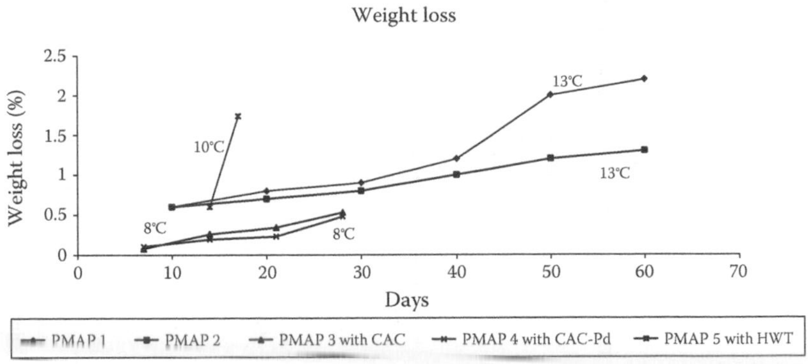

FIGURE 9.14 Weight loss of treated and untreated tomatoes under passive MAP vs. storage time (PMAP 1 [equilibrium atmospheric conditions of 6%–7% O_2/4% CO_2] with the use of polyethylene 50 μm as a packaging film, PMAP 2 [equilibrium atmospheric conditions of 12%–13% O_2/5%–6% CO_2] with the use of polypropylene 25 μm as a packaging film (From Batu, A. and Thomson, A.K., *Trans. J. Agric. For.*, 22, 365, 1998), PMAP 3 [with steady-state atmosphere of 7% O_2/7% CO_2] with addition of GAC, PMAP 4 [with steady-state atmosphere of 7% O_2/7% CO_2] and with addition of GAC impregnated with palladium as a catalyst [GAC-Pd] (From Bailen, G. et al., *J. Agric. Food Chem.*, 54, 2229, 2006), PMAP 5 [with steady-state atmosphere of 5% O_2/8% CO_2] with immersion of tomatoes in hot water (42.4°C) for 30 min prior to their packaging [HWT]. (From Suparlan, M. and Itoh, K., *Packag. Technol. Sci.*, 16, 171, 2003.))

High oxygen levels can restrict mesophile growth, thereby extending the shelf life of carrots. PMAP with a highly permeable film displayed nonpromising results compared to AMAP packages with low or high oxygen levels.

The effect of passive (PP film) and AMAP at low (5% O_2, 10% CO_2) and high oxygen concentrations (80% O_2, 10% CO_2) at 4°C on RTE peeled and sliced carrots after having been dipped into citric acid (0.1% w/v for 15 min) was studied by Ayhan et al. (2008). Taking into account the oxygen level in the headspace and the obtained sensory results, the shelf life of minimally processed carrots was 7 days for high oxygen and PMAP application. Nevertheless, it was reduced to just 2 days for MA with low oxygen.

The impact of application of edible coating (Nature Seal) before packaging (OPP or microperforated PA_{60} film) to provide MA conditions for storage of carrot discs at 4°C and 8°C was studied by Cliff-Byrnes and O'Beirne (2007). At both 4°C and 8°C, carrots packaged in PA_{60} film at 4°C and 8°C had higher L^* values compared with carrots packaged in OPP film. For PA_{60} film, headspace levels of the total volatiles were reduced by 47.6% at 4°C and by 48.9% at 8°C. The Nature Seal treatment was reported to have substantial beneficial effects on the visual quality of carrots particularly at 4°C.

Tassou and Boziaris (2002) investigated the fate of inoculated *S. enteritidis* on grated carrots stored in air, MA (2.1% O_2, 4.9% CO_2), and vacuum at 4°C in the presence and absence of *Lactobacillus* sp. LAB multiplied in all cases and a pH drop was recorded in all samples inoculated with *Lactobacillus* sp. (from 5.4–5.8 to

TABLE 9.3
Mentioned Data for Tomatoes Stored under MAP

Species and Food Type	Initial Gas Mix	Packaging Material	Treatment before Packaging	Storage Temperature (°C) and Storage Period (Days)	Color	Microflora	Texture–Weight Loss	Sensory Analysis	Shelf-Life (Days) Extension	References
Tomato slices (*Lycopersicon esculentum Mill.*)	(1) 1% O_2/4% CO_2 (2) 20% O_2/4% CO_2 (3) 1% O_2/8% CO_2 (4) 20% O_2/8% CO_2 (5) 1% O_2/12% CO_2 (6) 20% O_2/12% CO_2 (7) Air	Film A with 87.4 mL/h/m²/atm OTR at 5°C and 119.3 at 10°C Film B with 60 mL/h/m²/atm OTR at 5°C and 77.8 at 10°C		5°C–10°C—19 days	The Hunter a* value was the highest in slices in containers with composition of 4% CO_2 1% O_2	After 15 days at 5°C, no visible fungal growth was observed on slices in containers with an initial atmospheric composition of 1% O_2/12% CO_2	The slices in containers with the initial atmospheric composition of 12% CO_2/1% O_2 had the highest firmness	Ethylene concentration was higher with film B than with film A. Tomato slices with film B had less chilling injury compared to slices with film A	MAP can result in a shelf life of fresh-cut tomato slices of 2 weeks or more at 5°C	Hong and Gross (2001)
Tomato	PMAP: (1) 3% O_2/11%–13% CO_2	(1) PE-20μ (2) PE-50μ (3) PVC-10μ (4) PP-25μ	1–2 min dipping in 100 ppm of thiabendazole	13°C—60 days	Fruits sealed in plastic films changed color more		The fruits in plastic film softened more slowly	The weight loss was related to film	30 or 40 days with the MAP products	Batu and Thomson (1996)

	(2) 6%–7% O_2/4% CO_2 (3) 4% O_2/11%–12% CO_2 (4) 12%–13% O_2/5%–6% CO_2				slowly especially those in PE_{50} and in PP		than those stored unwrapped	permeability. Films with low gas permeability had less weight loss	remained edible over 60 days	
Tomato (*Lycopersicon esculentum* Mill. Cv. "Beef")	PMAP: (1) 4 kPa O_2/10 kPa CO_2 (control) (2) 7 kPa O_2/7 kPa CO_2 (with activated carbon)	20 μm thickness nonperforated OPP	Addition of GAC alone or impregnated with palladium as a catalyst (GAC-Pd) (EAs)	8°C—28 days	Changes in color were greater in control tomatoes, while tomatoes with activated carbon had their ripening process delayed		The loss of firmness was higher in control fruits than in those with GAC or GAC-Pd, especially during the first 14 days of storage	Ethylene was lower in MAP packages with GAC and especially in those with GAC-Pd. Tomatoes with GAC or GAC-Pd exhibited significantly lower weight losses and had better sensory attributes	21 days for control and 28 days for GAC-Pd	Bailen et al. (2006)

(continued)

TABLE 9.3 (continued)
Mentioned Data for Tomatoes Stored under MAP

Species and Food Type	Initial Gas Mix	Packaging Material	Treatment before Packaging	Storage Temperature (°C) and Storage Period (Days)	Color	Microflora	Texture–Weight Loss	Sensory Analysis	Shelf-Life (Days) Extension	References
Tomato slices. (*Lycopersicum esculentum Mill.*)	(1) PMAP (2) 3 kPa O_2 (3) 3 kPa O_2/4 kPa CO_2	OPP film of 35 µm thickness	1 min dipping into sodium hypochlorite solution (1.4 mg/L). The samples were sliced or divided into wedges	0°C and 5°C—14 days	A higher increase in °h of samples stored at 0°C than in those at 5°C indicating that at 5°C a slight ripening occurred	Temperature was more effective than MAP treatments in lowering microbial growth. Fungi detected only in control samples at 5°C. Yeast counts higher in control than in MAP samples		Under PMAP the C_2H_4 level was higher at 5°C than at 0°C. The effect of temperature was inhibited by AMAP. The reduction in sensorial scores was higher at 5°C than at 0°C. MAP minimized the weight losses when the temperature was higher	Only the combined use of 0°C plus passive or AMAP resulted in a shelf life of 14 days in fresh-cut tomato	Aguayo et al. (2004)

Tomato (*Lycopersicum esculentum Mill.*) cv "Durinta"	AMAP with 12–14 kPa O_2/0 kPa CO_2 was used in both packages with or without EA containing $KMnO_4$ on celite (EA)	(1) Vascolan with 80 μm thickness (2) BiOPP	1 min dipping into sodium hypochlorite solution (1.3 mM). Then the samples were sliced	0°C and 5°C—7 and 10 days	Storage at higher temperatures increased juice red color showing lower hue values. Low temperature inhibited the decline of hue values, but showed chilling injury	Microbial count was very low for samples kept at 0°C. Compared with those at 0°C in air, in samples at 5°C yeasts and molds counts increased more than 3 log	Neither storage temperature nor film promoted softening	No accumulation of C_2H_4 was observed in packages with EA throughout storage. The maturity index (SSC/TA) was affected by the type of film. Visual quality scored higher for packages with EA than for packages without EA at 5°C	For keeping quality of tomato slices up to 10 days the best storage conditions were obtained at 0°C of the film used	Gil et al. (2002)
Tomato (*Lycopersicum esculentum Mill.*) cv "Durinta"	(1) PMAP (2) 7.5% O_2	(1) Vascolan	Whole fruits were dipped into chlorinated water (1.3 mM) for 1 min. Then the samples were sliced and immersed into 2 L of	2°C and 10°C—7 and 10 days	2°C maintained the color close to that in the beginning of the experiment		A significant reduction in firmness was recorded for samples kept at 10°C compared to those stored at 2°C	Storing tomato slices at 10°C increased the rate of C_2H_4 production to be fivefold higher than that of whole fruit. Compared to 10°C,	The best treatments for preserving quality were MAP (similar performance for active and passive)	Artes et al. (1999)

(*continued*)

TABLE 9.3 (continued)
Mentioned Data for Tomatoes Stored under MAP

Species and Food Type	Initial Gas Mix	Packaging Material	Treatment before Packaging	Storage Temperature (°C) and Storage Period (Days)	Color	Microflora	Texture–Weight Loss	Sensory Analysis	Shelf-Life (Days) Extension	References
			chlorinated water (0.7 mM) with and without $CaCl_2$ (0.09 M) for 1 min at 4°C					tomato slices kept at 2°C had better visual quality, aroma and texture	followed by calcium dips	
Tomato (*Lycopersicum esculentum Mill.*) cv Rambo, Durinta, Bodar, Pitenza, Cencara, and Bola	(1) 5 kPa O_2/5 kPa CO_2	(1) ILPRA with O_2 and CO_2 permeabilities 110 and 500 cm^3/m^2 day bar at 23°C and 0% RH	Tomatoes were sliced	4°C—21 days	Neither CIE a* nor CIE b* shifted significantly with time in the tested conditions			Lycopene content varied widely among the studied cultivars. Vitamin C concentration showed no substantial variations through the storage time. Phenolics increased after 14 days of storage at	Fresh-cut tomatoes maintained the main antioxidant compounds and color parameters for 21 days at 4°C ± 1°C	Odriozola-Serrano et al. (2008)

								4°C, irrespective of the studied cultivar	
Cherry tomatoes	(1) PMAP: 6% O_2/4% CO_2 (2) CA with 5% CO_2	LDPE	Tomatoes were spot inoculated on the surface and in the stem scars with *S. enteritidis*. The second group (stem scars) had two inoculum doses, high (7 log_{10} CFU/ tomato) or low (3 log_{10}CFU). The samples were treated with ozone	7°C for 20 days and 22°C for 10 days		The death rate of *S. enteritidis* on the surfaces of tomatoes that were stored in modified atmosphere was faster than that of stored in CA at 7°C. LAB and TVC grew regardless of the atmosphere and the storage temperature	At the last 3–4 days of both MAP and CA storage periods, a light softening and darkening was observed at 22°C		Das et al. (2006)
Tomato (*Lycopersicum esculentum Mill.*) cv "Maru"	(1) PMAP: 5% O_2/8% CO_2	LDPE	HWT (immersed in hot water (42.5°C) for 30 min)	14 days at 10°C and then at 22°C for 3 days without packaging	Color development of both treated and untreated tomatoes	HWT prior to storage in MAP slightly reduced mold	The firmness of HWT tomatoes packed with plastic film was greater	Ethylene concentration inside the packages increased with storage	Suparlan and Itoh (2003)

(*continued*)

TABLE 9.3 (continued)
Mentioned Data for Tomatoes Stored under MAP

Species and Food Type	Initial Gas Mix	Packaging Material	Treatment before Packaging	Storage Temperature (°C) and Storage Period (Days)	Color	Microflora	Texture–Weight Loss	Sensory Analysis	Shelf-Life (Days) Extension	References
					packaged with plastic film was suppressed during the 2 week period of storage	growth. Both treated and untreated tomatoes packaged with plastic film resulted in higher microbial count than unpackaged tomatoes	than untreated ones. HWT before MAP prevents tomato decaying and cracking	time for both treated and untreated tomatoes. The use of MAP reduced the weight loss of tomatoes during storage for 14 days at 10°C. Combination of HWT and MAP had no significant effect on SSC		
Tomato (*Lycopersicum esculentum Mill.*) cv "Coco"	PMAP: (1) 18.2 kPa O_2/3.4–3.7 kPa CO_2 (M_1)	The O_2 permanences of M_1, M_2, and M_3 films were 7.78 ×	Immersion in 100 mg/L NaOCl for 2 min and HWT	15 days at 15°C	The lowest color development was observed in					Sayed Ali et al. (2004)

	(2) 15.2 kPa O_2/3.4–3.7 kPa CO_2 (M_2) (3) 12.2 kPa O_2/3.4–3.7 kPa CO_2 (M_3)	10^{-17}, 5.49 × 10^{-17}, 3.56 × 10^{-17} and the CO_2 permanences were 9.12 × 10^{-17}, 8.4 × 10^{-17}, and 8.9 × 10^{-17}, mol/s/m²/Pa	(immersed in hot water (39°C) for 90 min)		the fruit subjected to HWT with subsequent sealing with M_3 film				
Mature-green tomatoes (*Lycopersicon esculentum* cv. Trust)	PMAP: 5%O_2/5% CO_2	26 L plastic containers with OTR: 8.96. 10^{-12} mol/s Pa and CDTR: 63.02. 10^{-12} mol/s Pa at 13°C		(1) 13°C. (2) 13°C for 10 days and then transferred to 23°C for 2 days. After the 2 days at 23°C, they were removed to 13°C and the sequence was repeated on day 20 and 30	Ripening of tomatoes was retarded when stored under MA conditions at constant temperature		Firmness of unpackaged tomatoes was lower than that of MA at constant temperature	The increase in ethanol level was higher under MA at constant storage temperature, but the increase was more marked under MA with TF	Tano et al. (2007)

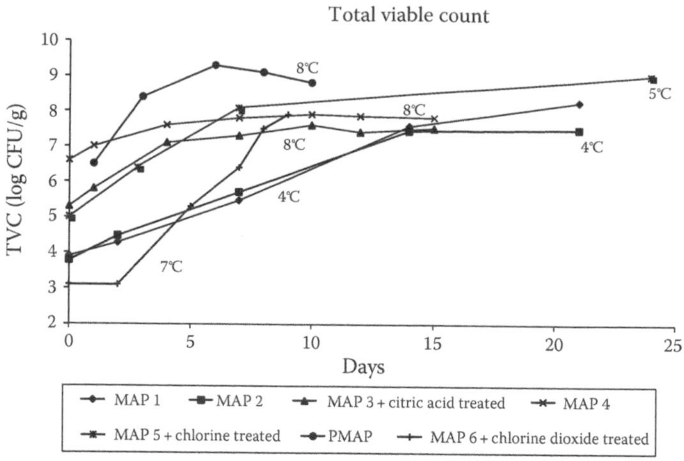

FIGURE 9.15 TVC for carrots under MAP and various treatments vs. storage time (MAP 1 [5% O_2/10% CO_2/85% N_2], MAP 2 [80% O_2/10% CO_2/10% N_2] (From Ayhan, Z. et al., *Turk. J. Agric. For.*, 32, 57, 2008), MAP 3 [50% O_2/30% CO_2/20% N_2], MAP 4 [90% O_2/10% CO_2] (From Amanatidou et al., *J. Food Sci.*, 65(1), 61, 2000), MAP 5 [3% O_2/97% N_2] (From Beuchat, L.R. and Brakett, R.E., *Appl. Environ. Microbiol.*, 56(6), 1734, 1990), PMAP [P-Plus with equilibrium atmospheric conditions of 6% O_2/15% CO_2] (From Barry-Ryan, C. et al., *J. Food Sci.*, 65(4), 726, 2000), MAP 6 [4.5% O_2/8.9% CO_2/86.6% N_2] for grated carrots. (From Gomez-Lopez, V.M. et al., *Int. J. Food Microbiol.* 116, 221, 2007a.)).

4.1–4.3). It was noteworthy that the population of *Salmonella* survived at the level of the initial inoculum.

Amanatidou et al. (2000) studied the impact of various gas compositions, mostly high O_2 and CO_2 MA (90% O_2/10% CO_2, 80% O_2/20% CO_2, 50% O_2/30% CO_2, 70% O_2/30% CO_2, 1% O_2/10% CO_2), combined with citric acid (0.1% or 0.5%), H_2O_2, chlorine, $CaCl_2$, and an alginate edible coating on the preservation of carrots. Citric acid alone or incorporated in the coating allowed color retention, and combination with $CaCl_2$ significantly reduced the initial total flora for at least 1 or 2 log CFU/g. The shelf life was prolonged by 3 days with 0.1% citric acid disinfection and coating before storage under MA (50% O_2/30% CO_2).

Shredded orange and purple carrots were stored by Alasalvar et al. (2005) under MAP (95% O_2/5% CO_2, 5% O_2/5% CO_2) at 5°C ± 2°C. The content of anthocyanin, occurring solely in purple carrots, dropped substantially in the 95% O_2 + 5% CO_2 treatment (from 5.1 to 4.6 mg/100 g). In both orange and purple carrots, the highest loss of total carotenoids was recorded in the 95% O_2/5% CO_2 treatment (6.5 and 2.7 reductions, respectively). The longest shelf life of purple carrots was 10 days under 5% CO_2 + 5% O_2.

Beuchat and Brackett (1990) investigated the effect of chlorine in conjunction with MAP (3% O_2) on shredded carrots inoculated with *L. monocytogenes* strains and stored at 5°C and 15°C. Chlorine treatment was found to decrease initial yeasts and molds and the initial mesophile population by at least 90%. Neither chlorine treatment nor MA conditions really influenced the growth of mesophilic aerobes at 5°C. Populations of *L. monocytogenes* were practically unaffected by cutting treatment, chlorine treatment, or the initial composition of gaseous atmospheres.

Barry-Ryan et al. (2000) employed packaging films of various components (OPP, Pebax with hydrophilic coating, polyether block amide, P-Plus 1, P-Plus 2) and storage temperatures (3°C and 8°C) to find a range of EMA for storage of shredded carrots. Packs with EMAs produced with P-Plus 1 and polyether films had total aerobic loads of 0.5 log CFU/g lower from day 5 onward. The highest firmness was reported for carrots stored in the P-Plus films (3.3 and 3.55 kN for P-Plus 2 at 3°C and 8°C, respectively). The film P-Plus 1 was identified to be the most suitable for carrot storage.

The shelf-life extension of grated carrots stored at 7°C was investigated in the presence of gaseous chlorine dioxide treatment and MAP (4.5% O_2 + 8.9% CO_2) (Gomez-Lopez et al., 2007a). The psychrotrophic counts attained unacceptable levels after 4 days on untreated samples but after 8 days on the treated ones. A lag phase of minimal 2 days was recorded for mesophilic aerobic bacteria, psychrotrophs, and LAB.

Atmospheres of high oxygen levels have a detrimental effect on bacterial development and reproduction rate, thus maintaining their numbers within acceptable limits. Treatment with citric acid decreased the initial LAB numbers under 1 log CFU and has a clear difference from the other packages studied. Co-existence with other inoculated microorganisms has more marked effect on LAB growth with *Salmonella* (Figure 9.16).

The inoculation of sliced carrots (with blunt or sharp machine blade and a razor blade) with *E. coli* and *L. innocua* and storage under PMAP (8°C) was studied. The

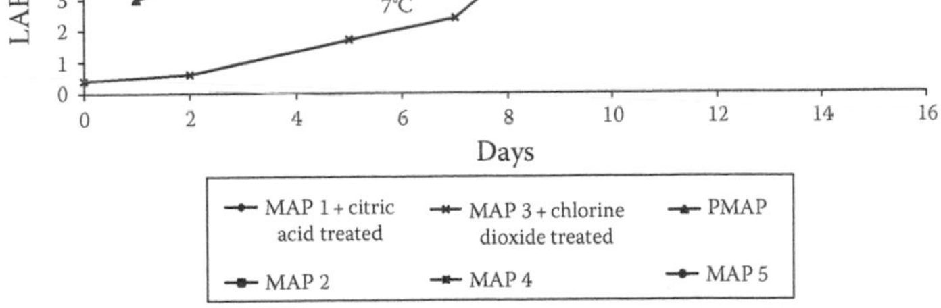

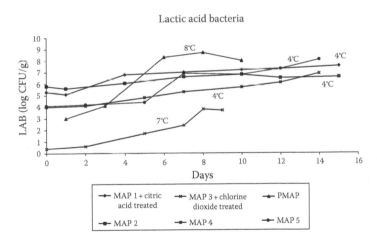

FIGURE 9.16 LAB numbers for carrots under MAP packages vs. storage time (MAP 1 [50% O_2/30% CO_2/20% N_2], MAP 2 [90% O_2/10% CO_2] (From Amanatidou et al., *J. Food Sci.*, 65(1), 61, 2000), PMAP [P-Plus with equilibrium atmospheric conditions of 6% O_2/15% CO_2] (From Barry-Ryan, C. et al., *J. Food Sci.*, 65(4), 726, 2000), MAP 3 [4.5% O_2/8.9% CO_2/86.6% N_2] for grated carrots (From Gomez-Lopez, V.M. et al., *Int. J. Food Microbiol.* 116, 221, 2007a), MAP 4 [2.1% O_2/4.9% CO_2/93% N_2 and inoculation with *Salmonella* strains], MAP 5 [2.1% O_2/4.9% CO_2/93% N_2 and inoculation with *L. Monocytogenes* strains]. (From Kakiomenou, K. et al., *World J. Microbiol. Biotechnol.*, 14, 383, 1998.))

populations of *E. coli* on sliced carrots decreased by 2–3 log cycles over the storage period. However, no significant difference between total counts on razor-sliced carrots and carrots sliced with sharp or blunt machine blade (9.4–10 log CFU/g) were reported by the end of storage (Gleeson and O'Beirne, 2005).

Pilon et al. (2006) investigated both the vitamin C content and microorganism growth on carrots stored under air, vacuum, and MAP (2% O_2 + 10% CO_2) at 1°C ± 1°C for 21 days (LDPE-BOPP film used). Vitamin C was retained until 21 days of storage for carrots in all treatments carried out (13.54–14.24 mg/100 g). The β-carotene content in the carrot packed under MA displayed the lowest values (2.851 mg/100 g). Psychrotrophs grew slightly over the storage period of the samples packed under air, vacuum, and MAP (3.18, 2.45, and 2.84 log CFU/g, respectively).

Gaseous chlorine dioxide exhibits a clear advantage over common chlorine treatment for lowering the psychrotrophic flora of carrots with differences reaching up to 2 log CFU/g. Application of very low temperatures had rather inhibiting effect even on psychrotrophs growth, but the potential of causing chilling injury to the product must be taken into account (Figure 9.17).

The survival study of inoculated *S. enteritidis* and *L. monocytogenes* on shredded carrots packed under MA (2.1% O_2/4.9% CO_2, 5.2% O_2/5% CO_2) at 4°C revealed that the numbers of *S. enteritidis* decreased during storage in both packaging systems. LAB proved to be the predominant organisms in all samples. It is noteworthy that the pH dropped significantly during the storage (Kakiomenou et al., 1998).

Summarized data are shown in Table 9.4.

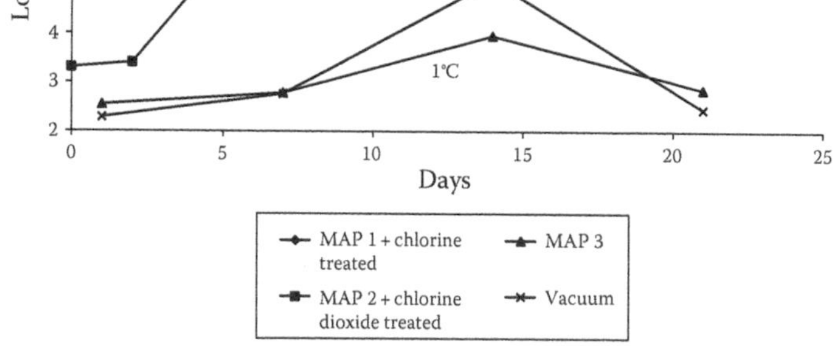

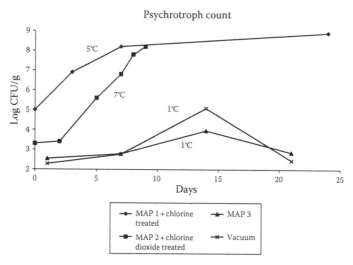

FIGURE 9.17 Psychrotroph bacteria under various package conditions and treatments for carrots vs. storage time (MAP 1 [3% O_2/97% N_2] (From Beuchat, L.R. and Brakett, R.E., *Appl. Environ. Microbiol.*, 56(6), 1734, 1990), MAP 2 [4.5% O_2/8.9% CO_2/86.6% N_2] for grated carrots (From Gomez-Lopez, V.M. et al., *Int. J. Food Microbiol.*, 116, 221, 2007a), MAP 3 [2% O_2/10% CO_2/88% N_2] and vacuum. (From Pilon, L. et al., *Ciência Tecnologia Alimentas Campinas,* 26(1), 150, 2006.))

TABLE 9.4
Synopsis of Carrot Species, Sensory Assessment, Metabolic Changes, and Reached Shelf-Life Prolongation When Stored under MAP

Species and Food Type	Initial Gas Mix	Packaging Material	Treatment before Packaging	Storage Temperature (°C) and Storage Period (Days)	Color	Microflora	Texture–Weight Loss	Sensory Analysis	Shelf-Life (Days) Extension	References
Grated Carrots (*Daucus carota*)	(1) Air (2) 60% O_2/30% CO_2	0.5 mm metallized polyester/2 mm EVA copolymer sterile bag	Gamma irradiation at a dose of 0.15, 0.3, 0.6, 0.9 kGy. Inoculation with *E. coli* (10^6 CFU/g)	4°C ± 1°C—50 days		The application of 0.6 kGy under MAP assured a complete inhibition of *E. coli* in grated carrots			The alternative combinations of MAP and irradiation treatment can be used to maintain the quality of fresh minimally processed carrots	Lacroix and Lafortune (2004)
Carrots (*Daucus carota*) cv. *Nantes*	(1) Air (2) 5% O_2/10% CO_2 (3) 80% O_2/10% CO_2	CPP–OPP with O_2 and CO_2 permeabilities 1296 and 3877 cm^3/m^2 day	Dipping into citric acid solution (0.1% w/v) for 15 min, and then sliced. The sliced carrots were dipped into citric acid solution (0.1% w/v) for 10 min	4°C—21 days	Good retention of orange color with no significant surface drying, due to effect of citric acid	There was no yeast or mold growth during the 21 days of storage in any of the applications	The texture values declined at both passive and AMAP, after 14th day, indicating significant softening	On day 14, the acceptability scores significantly decreased in all treatments applied	7 days for high oxygen and PMAP application	Ayhan et al. (2008)

(*continued*)

TABLE 9.4 (continued)
Synopsis of Carrot Species, Sensory Assessment, Metabolic Changes, and Reached Shelf-Life Prolongation When Stored under MAP

Species and Food Type	Initial Gas Mix	Packaging Material	Treatment before Packaging	Storage Temperature (°C) and Storage Period (Days)	Color	Microflora	Texture–Weight Loss	Sensory Analysis	Shelf-Life (Days) Extension	References
Carrots *(Daucus carota) cv. Nairobi*	PMAP: (1) 1% O_2/12% CO_2 at 4°C (1) 1% O_2/16% CO_2 at 8°C (2) 7% O_2/12% CO_2 at 4°C (2) 4% O_2/15% CO_2 at 8°C	(1) Monoaxial OPP microperforated 35μm (2) PA-60	Antimicrobial treatment with chlorine solution at 4°C. Carrots were then dipped in an edible coating, Nature Seal solution for 30 s	4°C or 8°C—6 days	L* values increased with storage temperature. Carrots packaged in PA-60 film had higher L*values than carrots packaged in OPP film. Carrots treated with Nature Seal had lower L* values compared with untreated			PA-60 generated atmosphere had beneficial effects on flavor and acceptability but adverse effects on visual quality (whitening). OPP had overall poor results. Edible coating improved visual quality		Cliffe-Byrnes and O'Beirne (2007)

Grated carrots	(1) PMAP. (2) 2.1% O_2/4.9% CO_2	(1) PE bags with O_2 and CO_2 permeability 1000 and 5450 mL/m^2 day bar	The carrots were immersed in alcohol flamed, peeled, again immersed in alcohol and flamed and grated with a sterile grater. The carrots were inoculated with *S. enteritidis* and *Lactobacillus* sp.	4°C—12 days		*S. enteritidis* survived under all treatments. Prevention of its growth is due to its inability to compete successfully with LAB		Samples inoculated with *Lactobacillus* sp (with *Salmonella* or not) showed a dramatic pH drop		Tassou and Boziaris (2002)
Sliced carrots cv. "*Amsterdamse bak*"	Control atmosphere (1) 90% O_2/10% CO_2 (2) 80% O_2/20% CO_2 (3) 50% O_2/30% CO_2 (4) 70%O_2/30% CO_2 (5) 1% O_2/10% CO_2	Airtight containers with continuous flush of combination of gases	Distilled water, NaOCl solution (200 mg/L), chlorine or 5% (v/v) H_2O_2, 0.1 and 0.5% (w/v) citric acid, coating (S170 + 2% $CaCl_2$), combination of the last two treatments	8°C—15 days	Citric acid alone or incorporated in the coating allowed color retention and inhibited white discoloration. Treated carrots (dipped in 0.1% citric acid) kept their characteristics for at least 12 days under 50% O_2/30% CO_2 and 1% O_2/10% CO_2	Chlorine or H_2O_2 reduced the level of Enterobacteria. Combinations of 0.1% or 0.5% citric acid and 2% $CaCl_2$ reduced initial total flora for at least 1 or 2 log CFU, respectively	Chlorine treatment hardly affected the firmness of carrots. The coating retained characteristics for 8 days. Firmness was satisfactory for samples under 1% or 50% O_2	Chlorine treatment did not affect initial pH. Under all CA, the phenolic content of treated samples was lower than that of untreated samples. The lowest ethylene was at 90% O_2/10% CO_2	Shelf life was extended from 12 to 15 days, but only when products were disinfected with 0.1% citric acid and coated prior to storage under MAs	Amanatidou et al. (2000)

(continued)

TABLE 9.4 (continued)
Synopsis of Carrot Species, Sensory Assessment, Metabolic Changes, and Reached Shelf-Life Prolongation When Stored under MAP

Species and Food Type	Initial Gas Mix	Packaging Material	Treatment before Packaging	Storage Temperature (°C) and Storage Period (Days)	Color	Microflora	Texture–Weight Loss	Sensory Analysis	Shelf-Life (Days) Extension	References
Orange and purple carrots	(1) Air (2) 5% O_2/5% CO_2 (3) 95% O_2/5% CO_2	PE bags	Disinfection for 5 min (100 ppm free chlorine solution, and then the carrots were shredded	5°C ± 2°C for 13 days				In both orange and purple carrots, the loss of total carotenoids was greater in the 95% O_2 treatment. 5% O_2 + 5% CO_2 reduced the accumulation of total phenols	Shredded purple carrot can be stored under 5% O_2 /5% CO_2 treatment for up to 10 days	Alasalvar et al. (2005)
Carrots (*Daucus carota*)	(1) 3% O_2/97% CO_2	L bags (Cryovac) with OTR of 3000 mL/m² 24 h	Shredding, chlorine treatment (into water with 200 µg/mL) and inoculation with two strains of *L. monocytogenes*	5°C and 15°C—14 days		Populations of *L. monocytogenes* decreased upon contact with raw carrots. Chlorine reduced the initial			Cutting, chlorine treatment, and MAP had no effect on growth of *L. monocytogenes* or naturally occurring microflora	Beuchat and Brakett (1990)

						population of yeasts, molds, and mesophilic organisms by at least 90%				
Carrots (*Daucus carota* L., cv. *Nantaise des Sables*)	PMAP: (1) >1% O_2/30% CO_2 (2) >1% O_2/4% CO_2 (3) >1% O_2/16% CO_2. (4) 10% O_2/10% CO_2 (5) 18% O_2/3% CO_2	(1) OPP (2) Pebax film (3) Polyether block amide (4) P-Plus 1 (5) P-Plus 2	Chlorine treatment (into water with 100 µg/mL free chlorine) for 5 min and shredding	3°C and 8°C		High CO_2 levels combined with low temperatures inhibited microbial growth. Only LAB increased over storage, coinciding with high levels of CO_2	Firmness values for shredded carrots increased in all packs up to day [illegible], whereas after day [illegible] firmness decreased	Highest appearance scores were given to product packed in P plus-1. Highest aroma scores were given to product packed in the P-Plus films	The deterioration of carrots occurred more rapidly with depletion of O_2 than by the rise in CO_2 package	Barry-Ryan et al. (2000)
Grated carrots (*Daucus Carota L.*)	(1) 4.5% O_2/8.9% CO_2	Film with O_2 permeability of 3529 mL O_2/kg h	ClO_2 gas treatment	7°C for 9 days		The reduction in psychrotrophs, LAB, and mesophilic aerobic bacteria counts was 1.88, 1.71, and 2.6 log CFU/g, respectively. The yeast counts became the same with the untreated ones in 5 days		Significant decrease of pH in the treated samples. Treatment did not impair the sensory attributes	ClO_2 gas treatment prolonged shelf life of MP carrots for one day	Gómez-López et al. (2007a)

(continued)

TABLE 9.4 (continued)
Synopsis of Carrot Species, Sensory Assessment, Metabolic Changes, and Reached Shelf-Life Prolongation When Stored under MAP

Species and Food Type	Initial Gas Mix	Packaging Material	Treatment before Packaging	Storage Temperature (°C) and Storage Period (Days)	Color	Microflora	Texture–Weight Loss	Sensory Analysis	Shelf-Life (Days) Extension	References
Irish carrots (cultivar Nairobi)		OPP 35 μm	Carrots were cut with (1) Blunt machine blade (2) Sharp machine blade (3) Razor blade	8°C for 9 days		Counts of *E. Coli* on razor-sliced carrots were around 1 log cycle lower than counts on the blunt/sharp machine blade sliced carrots			Counts of total mesophilic counts were significantly higher on carrots sliced with a machine blade	Gleeson and O'Beirne (2005)
Nantes carrots (*Daucus carota* L.)	(1) Vacuum (2) AMAP: 2% O_2/10% CO_2	BOPP/LDPE (biaxially orientated PP/ low-density PE) plastic bags	Immersion in cold water (7°C) with 100 mg/L of free chlorine at pH 7.0 for 15 min	1°C ± 1°C for 21 days		Carrots were negative for total or fecal coliforms, anaerobic mesophiles, and *Salmonella*. The count of psychrotrophs in carrots was 10^2–10^3 CFU/g		Vitamin C was maintained for all treatments. β-carotene in the carrot packed under MAP showed the lowest values		Pilon et al. (2006)

Carrots (*Daucus carota L.*)	AMAP: (1) 2.1% O_2/4.9% CO_2 (2) 5.2% O_2/5% CO_2	PE bags	Inoculation with *S. enteritidis* and *L. monocytogenes* strains	4°C ± 0.2°C for 14 days	LAB and TVC were lower in MAP than in air. *S. enteritidis* and *L. monocytogenes* dropped over storage in both packaging systems	Packaging in MAP therefore does not signify an additional hurdle for growth of *S. enteritidis* and *L. monocyto*genes compared with conventional packaging	Kakiomenou et al. (1998)

9.6 BROCCOLI

Rangkadilok et al. (2002) investigated the impact of various packaging treatments (CA with 1.5% O_2/6% CO_2, PMAP with microperforated LDPE stored at 20°C, PMAP with LDPE stored at 4°C), on broccoli storage time (10 and 25 days). No changes in glucoraphanin in MAP with no holes (5.0 μmol/g D/W) at 4°C and two microholes (5.3 μmol/g D/W) at 20°C (10 days) were reported. In the MAP at 20°C, most of the green color was maintained for 7 days while in the control, 30%–40% turned yellow by day 3.

The synergistic action of sorbitol (water absorbent) and $KMnO_4$ (ethylene absorbent) in PMAP (PD-961 film) for broccoli heads stored at 0°C–1°C was reported by De Ell et al. (2006). Broccoli heads in MAP with sorbitol had better appearance, firmness, and odor ratings after 29 days. Broccoli heads in all MAP treatments were ethanol-free and/or acetaldehyde-free for 3 weeks. Moreover, the presence of sorbitol increased the weight loss of broccoli (>1.3%) during MAP.

In Figure 9.18, it is clear that all the applied materials managed to create optimal conditions for preserving the weight of the product for the selected time periods. Macroperforations on PP (PMAP 1) did not work satisfactorily and allowed moisture loss with a total loss of 13.33% at the end of storage.

Barth and Zhuang (1996) investigated the effect of PMAP (Xironet a 105 μm film), ventpackaging (film with uniform perforations), and automatic misting (AM) on broccoli florets stored at 5°C. Moisture was kept under MAP while moisture loss was significant in VP (76%) and VP/AM and AM samples (84%). Application of VP led to lower ascorbic acid (58%) losses. The usage of MAP maintained the reduced ascorbic acid (RAA) content. MAP followed by AM is the most effective postharvest storage treatment.

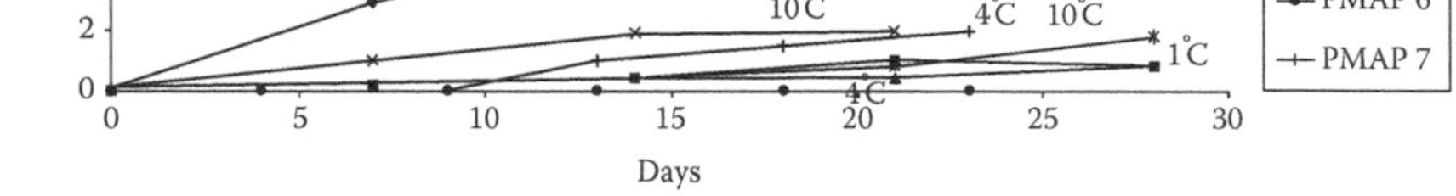

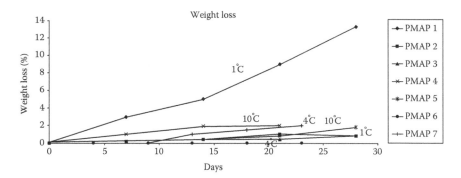

FIGURE 9.18 Changes in weight of fresh broccoli stored under modified atmosphere conditions (PMAP 1 was created with the use of macroperforated polypropylene [OTR: 1600 mL/m² day atm], PMAP 2 was created with the use of microperforated polypropylene [OTR: 2500 mL/m² day atm] and for PMAP 3, an non perforated polypropylene was used [OTR: 1600 mL/m² day atm] (From Serrano, M. et al., *Postharvest Biol. Technol.*, 39, 61, 2006), LDPE [OTR: 10960 mL/m² day atm] with EAs inside the package was the packaging film that created PMAP 4 and an LDPE [OTR: 4290 mL/m² day atm] created PMAP 5 (From Jacobsson, A. et al., *Eur. Food Res. Technol.*, 218, 157, 2004c), for PMAP 6 and 7, polyethylene bags with no and two microholes were used by Jia et al. (2008)).

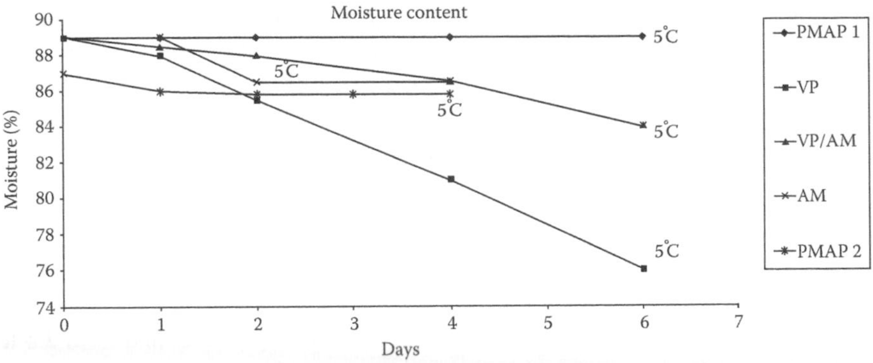

FIGURE 9.19 The effect of modified atmosphere storage on moisture content of broccoli (PD-941 film was used in PMAP 1 while in ventpackaging a "xironet" macroperforated film was used, VP/AM are ventpackaged samples stored in the misted sections of the display case and AM are the samples stored in the misted section of the display case without any atmosphere modification (From Barth, M.M. and Zhuang, H., *Postharvest Biol. Technol.*, 9, 141, 1996), PMAP 2 was created with the use of PD-941 polymeric film. (From Zhuang, H. et al., *J. Agric. Food Chem.*, 43(10), 2585, 1995.))

The comparison in Figure 9.19 clearly indicates that MA in conjunction with the use of composite films can lower considerably the moisture loss and prolong shelf life of the product. The use of misted display cases especially in products with high respiration rates like broccoli retains moisture.

Three types of film (macro-, micro-, and nonperforated PP) were applied for PMAP of broccoli at 1°C for 28 days. No significant changes in Hue angle (color) were reported for broccoli in nonperforated films by Serrano et al. (2006). The TAA remained the same and only a slight decrease was observed for macroperforated packages (from 40.7 to 30.5 mg/g). No significant changes in chlorophyll content (3.6 mg/g) were observed for nonperforated film. Broccoli packaged with microperforated and nonperforated films displayed prolonged storability up to 28 days.

Jacobsson et al. (2004a) investigated the effect of the aroma compounds on broccoli stored under different passive MA (OPP, PVC, and LDPE with ethylene absorbing sachet at 10°C for 7 days [1] or 4°C for 3 days and 10°C for the other 4 days [2]). LDPE and PVC maintained the concentration of dimethylsulfide, dimethyldisulfide, and dimethyltrisulfide (2.4, 10.3, 0.05 and 4.8, 15.5, and 0.07 ppb, respectively, for condition [2]). OPP resulted in most of the off-odors (8.1, 26.6, 0.16 ppb). TF augmented considerably the content of aroma compounds.

Jacobsson et al. (2004b) used OPP, LDPE with ethylene absorbing sachet and PVC to produce passive MA for broccoli storage at 10°C for 7 days (I) or 4°C for 3 days and 10°C for the other 4 days (II). The use of PVC under storage conditions (II) resulted in broccoli with similar taste and smell as fresh broccoli. However, at 10°C, the material was not able to maintain the appearance. Broccoli storage in LDPE was found to have the highest intensities of freshness, greenness, compactness, and evenness.

The effect of PMAP formed by 5 films (OPP, PVC, two LDPE films one of which contained an EA sachet) at 4°C and 10°C for 28 days was evaluated by Jacobsson et al. (2004c). LDPE (II without EA) maintained chlorophyll content better than the other packaging materials (25% reduction at condition II). Broccoli stored in PVC at 4°C (58)* differed from samples in OPP and LDPE II (111 and 110, respectively) and displayed a more rapid drop in required cutting energy. LDPE II (without EA) resulted in the overall longest shelf life.

Jia et al. (2008) investigated the storage of broccoli florets packaged in PE bags with no holes (M_0), 2 (M_1), and 4 (M_2) microholes at 4°C and 20°C. The best retention of color and visual quality was recorded in M_0 samples, followed by M_1 and M_2. The weight loss for those packaged in PE bags dropped to below 4% by the end of the storage at 4°C or 20°C. Application of M_0, M_1, and M_2 prolonged the shelf life from 10 (control) to 28.5, 19.1, and 15.2 days, respectively, at 4°C, and from 2.5 to 7.2, 5.6, and 4.8 days, respectively, at 20°C.

The storage of broccoli buds under PMAP (PD-941 polymer film) or AM treatment for 96 h at 5°C was studied by Zhuang et al. (1995). No significant loss of protein was observed in MAP-treated samples while in AM-treated samples, the level was 85 mg/g of dw by 96 h, about 20% less compared to the initial level. Total fatty acid (TFA) levels were reduced down to 35.71 mg in AM samples and to 42.40 mg in MAP from an initial 50.76 mg/g of dw in the first 24 h.

Artes et al. (2001) investigated the storage of broccoli florets at 1°C for 7 days and at 20°C for 2.5 days under MAP (PVC, LDPE 11, 15, and 20 μm were used) in an attempt to simulate the retail sale period. After 7 days at 1°C, the florets wrapped in PVC lost about 0.7% of their fresh weight while the losses in LDPE films were >0.01% for the same period. The lowest physiological disorders appeared in florets under LDPE 15 and 20 μm. On the whole after shelf life, among the LDPE films studied, florets under LDPE 15 showed the best results.

Broccoli stored under PMAP (with a steady-state atmosphere of 3% O_2 and 8% CO_2 at 3°C) were exposed to TF (ΔT: 10°C for 30 days), to simulate storage and transport conditions. Unpackaged broccoli revealed a decrease by 30% of its initial weight after only 30 days of storage while broccoli under MA or TF-MA lost almost 3% and 5%, respectively. No visible infection was reported in packages of broccoli at constant temperature, whereas under TF-MA, the loss because of bacterial blotches amounted to approximately 6.2% (Tano et al. 2007).

Schreiner et al. (2007) investigated the effect of PMAP (biaxial OPP with 2 [PMAP-1] and 8 [PMAP-2] micro holes) on postharvest glucosinolate dynamics in mixed broccoli and cauliflower florets stored at 8°C for 7 days. In broccoli florets stored at both MAs, the total indole glucosinolate concentration did not show any changes while aliphatic glucosinolates dropped from 2.42 to 1.65 for PMAP-1 and 1.74 for PMAP-2.

A brief summary of the aforementioned data is given in Table 9.5.

9.7 KOHLRABI

Escalona et al. (2007a) applied PMAP (with OPP and amide-PE films used after washing in an NaOCl water solution) to preserve kohlrabi sticks at 0°C for 14 days.

* The number 58, 110 and 111 represent the cutting energy needed. The cutting energy was calculated as the cutting force, F, divided by the diameter, D, of the broccoli stem.

TABLE 9.5
Brief Summary of the *Broccoli* Species When Stored under MAP

Species and Food Type	Initial Gas Mix	Packaging Material	Treatment before Packaging	Storage Temperature (°C) and Storage Period (Days)	Color	Microflora	Texture–Weight Loss	Sensory Analysis	Shelf-Life (Days) Extension	References
Broccoli heads (“*Marathon*” cv.)	(1) CA with 1.5% O_2/6% CO_2 (2) PMAP	(1) LDPE without holes and stored at 4°C (2) LDPE with microholes and stored at 20°C (both films were used for PMAP)		4°C and 20°C for 25 and 10 days	In CA storage treatment, the broccoli maintained their green color and freshness. Broccoli kept green color for 10 days at 4°C and for 7 days at 20°C			No significant changes in glucoraphanin concentration in broccoli heads stored for up to 10 days in MAP with no holes at 4°C and two microholes at 20°C		Rangkadilok et al. (2002)
Broccoli heads (*Brassica oleracea L., Italica* “*Marathon*” cv.)	PMAP: 1.3%–1.9% O_2/8.9%–10.4% CO_2	PD-961 with OTR 6,000–8,000 and CDTR 19,000–22,000 mL/m^2 24 h	(1) $KMnO_4$ 20 g (2) 2.5 g Sorbitol + $KMnO_4$ 20 g (3) 5 g Sorbitol + $KMnO_4$ 20 g	0°C–1°C for 29 days			Broccoli heads in MAP with sorbitol had higher firmness compared to the controls	Broccoli in MAP containing 20 g of sorbitol had the least amount of off-odor. Weight loss of the broccoli increased with	The addition of sorbitol with $KMnO_4$ improved broccoli quality retention	DeEll et al. (2006)

(*continued*)

TABLE 9.5 (continued)
Brief Summary of the *Broccoli* Species When Stored under MAP

Species and Food Type	Initial Gas Mix	Packaging Material	Treatment before Packaging	Storage Temperature (°C) and Storage Period (Days)	Color	Microflora	Texture–Weight Loss	Sensory Analysis	Shelf-Life (Days) Extension	References
			(4) 10 g Sorbitol + $KMnO_4$ 20 g (5) 20 g Sorbitol + $KMnO_4$ 20 g					sorbitol. MAP packages were volatile-free (ethanol and acetaldehyde) for the first 22 days		
Broccoli florets (*"Iron Duke"* cv.)	PMAP: (1) 7.5% O_2/11.2% CO_2 (2) VentPackaging (3) AM	"Xironet"		5°C—6 days	No differences in hue angle values were observed between VP and AM samples, but these values were lower than that of MAP florets		Moisture content was maintained in MAP samples	MAP resulted in retention of vitamins and other quality attributes in broccoli florets over storage. VP/AM-treated samples showed lesser POD activity, but the lowest activity occurred in MAP samples		Barth and Zhuang (1996)

Broccoli heads (*B. oleracea* L. var. *Italica Marathon* cv.)	PMAP: (1) 20 kPa O_2/0.08 kPa CO_2 (2) 14 kPa O_2/2–2.5 kPa CO_2 (3) 5 kPa O_2/6 kPa CO_2	(1) Macroperforated (Ma-P) (OTR—1,600, CDTR—3,600) (2) Microperforated (Mi-P) (OTR—2,500, CDTR—25,000) (3) Nonperforated (No-P) (OTR—1,600, CDTR—3,600) (mL/m² day atm)		1°C—28 days	A slight decrease in hue angle values was observed for Ma-P packaged broccoli, while no significant changes were found for broccoli heads in No-P films		Broccoli in Ma-P had an increase in firmness after 14 days of storage. For heads packaged either in Mi-P or No-P, slight decrease in texture was shown	For all MAP broccolis, total antioxidant activity was maintained over the experiment. Ascorbic acid retention was higher in broccoli packaged in Mi-P or No-P. No changes in chlorophyll content were observed for No-P broccoli	Broccoli packaged with Mi-P and No-P films prolonged storability up to 28 days with high quality attributes this period being only 5 days in control broccoli	Serrano et al. (2006)
Broccoli (*B. oleracea* var. *Italica Marathon* cv.)	PMAP: (1) 14% O_2/10.5% CO_2 (2) 6% O_2/7% CO_2 (3) 17.9% $O_2$4% CO_2	(1) OPP (2) LDPE with ethylene adsorbing sachet. (3) PVC		(1) 10°C for 7 days (2) 4°C for 3 days and 4 days at 10°C			No significant differences in weight loss, between the two storage conditions for broccoli packaged in materials 1–3	Storage in OPP resulted in off-odors, while storage in LDPE and PVC maintained the concentration of volatile sulfur compounds over storage		Jacobsson et al. (2004a)

(continued)

TABLE 9.5 (continued)
Brief Summary of the *Broccoli* Species When Stored under MAP

Species and Food Type	Initial Gas Mix	Packaging Material	Treatment before Packaging	Storage Temperature (°C) and Storage Period (Days)	Color	Microflora	Texture–Weight Loss	Sensory Analysis	Shelf-Life (Days) Extension	References
Broccoli (*B. oleracea* var. *Italica Marathon* cv.)	PMAP: (1) 14%–12.9% O_2/10.5%–11.1% CO_2 (2) 6%–4.2% O_2/6.8%–7.2% CO_2 (3) 17.9%–16% O_2/3.8%–4.7% CO_2 For the first and second storage temperature	(1) OPP (2) LDPE with ethylene adsorbing sachet (3) PVC		(1) 10°C for 7 days (2) 4°C for 3 days and 4 days at 10°C			Broccoli packaged in materials 1 and 2 was rated less crisp and had less chewing resistance than broccoli packaged in material 3	Broccoli stored in materials 2 and 3, and the non-packaged samples, had similar properties to the fresh sample with respect to smell and flavor. The second storage temperature had a better effect on appearance than the first	Broccoli packaged in LDPE with an EA was perceived to be the sample most similar to fresh broccoli	Jacobsson et al. (2004b)
Broccoli (*B. oleracea* var. *Italica* "Monterey"	PMAP: (1) 18.8%–15.3% O_2/2.5%–6.7% CO_2	(1) OPP (2) LDPE with ethylene adsorbing sachet		4°C and 10°C for 28 days	The longest shelf life, when 30% of the head had		Packaging materials 1 and 3 showed less weight loss		The lower temperature resulted in longer shelf life. The	Jacobsson et al. (2004c)

	(2) 13.2%–12.6% O_2/1.9%–2.5% CO_2 (3) 12.4%–9.5% O_2/2.9%–4% CO_2 (4) 20.6%–18.8% O_2/0.3%–1.3% CO_2 For the first and second storage temperature	(OTR: 10,960 and CDTR: >46,000) (3) LDPE (OTR: 4,290 and CDTR: 18,600 mL/m^2 day atm) (4) PVC			turned yellow, was obtained with OPP		for the packaged broccoli. LDPE, was the best material in respect of texture		shelf life was prolonged by 8–14 days with MAP LDPE performed best in maintaining the quality and shelf life of the broccoli	
Broccoli florets (*B. oleracea* var. *Italica* “Youxiu” cv.)	PMAP	(1) PE with no holes (2) PE with two holes (3) PE with four holes	Washing with a solution of 50 ppm NaOCl for 1 min	4°C and 20°C for 23 and 5 days, respectively	Best retention of color and visual quality were found in PE (1) samples, followed by PE (2) and PE (3)		The weight loss was lowest in PE (1) samples, followed by PE (2) and PE (3)	The loss of glucosinolates was lower under the three MAP treatments	The shelf lives of unwrapped broccoli were 10 days and 2.5 days at 4°C and 20°C. Cooling at 4°C with MA under PE (1) extended the shelf life to 28.5 days	Jia et al. (2008)
Broccoli heads (*B. oleracea* var. *Italica* “Iron Duke” cv.)	PMAP: (1) 8.7% O_2/4% CO_2	Cryovac PD-941		5°C for 96 h				MAP maintained chlorophyll in broccoli buds, reduced losses in		Zhuang et al. (1995)

(*continued*)

TABLE 9.5 (continued)
Brief Summary of the *Broccoli* Species When Stored under MAP

Species and Food Type	Initial Gas Mix	Packaging Material	Treatment before Packaging	Storage Temperature (°C) and Storage Period (Days)	Color	Microflora	Texture–Weight Loss	Sensory Analysis	Shelf-Life (Days) Extension	References
					—			polysaturated fatty acid and lipoxygenase activity		
Broccoli Florets (*B. oleracea* var. *Italica* "Shogun" cv.)	PMAP: (1) 18.3%–15.5% O_2/0.6%–1.9% CO_2 (2) 18%–15.8% O_2/0.8%–2.5% CO_2 (3) 18.7%–15.3% O_2/0.6%–3.1% CO_2 (4) 18.3%–15.1% O_2/0.9%–3% CO_2 For the first and second storage temperature	(1) PVC (2) LDPE 11 µm (3) LDPE 15 µm (4) LDPE 20 µm		1°C for 7 days and 20°C for 2.5 days to simulate a retail sale period	No significant changes in color were detected over any treatment		Over shelf life, weight losses were about 7–15 times lower in florets wrapped in LDPE films than in those wrapped in PVC	The lowest physiological disorders like yellowing and browning appeared in florets under LDPE 15 and 20	Florets under LDPE 15 showed the best results, regarding shelf life	Artes et al. (2001)

Broccoli (*B. oleracea* L. cv. Acadi)	PMAP: 3%O_2/8% CO_2	26L plastic containers with OTR: $21.76 \cdot 10^{-12}$ mol/s pa and CDTR: $61.52 \cdot 10^{-12}$ mol/s pa at 3°C	(1) 3°C (2) 3°C for 8 days and transfer to 13°C for 2 days. After 2 days at 13°C, they moved to 3°C, and the sequence was repeated at day 20 of storage	Highest retention of green color and chlorophyll under constant temperatures	Unpackaged broccoli lost about a third of its initial weight after only 30 days of storage Significant difference between weight loss under MA at constant temperature and temperature fluctuation	The effect of MA on the increase in total concentration of acetaldehyde and ethanol was very small compared to air storage. An increase of 2.6 fold was recorded under MA with TF. Under TF, loss due to bacterial blotches was estimated at 6.2% and lower than the loss at ambient air		Tano et al. (2007)
Broccoli florets cv. *Milady*	PMAP: (1) 1% O_2/21% CO_2 (2) 8% O_2/14% CO_2	(1) BOPP with two microholes (2) BOPP with eight microholes	8°C for 7 days	No color loss in any applied MA was detected over the entire storage period	The fresh weight loss at the end of both MAP was low and varied between 0.8% and 1.8%	The predominant glucosinolates were glucoraphanin and glucobrassicin	An MA of 1% O_2 + 21% CO_2 is recommended for storage of mixed-packaged broccoli florets for up to 7 days at 8°C	Schreiner et al. (2007)

Control sticks displayed weight losses two times higher (0.52%) than sticks under MAP (0.12% and 0.21% for OPP and amide-PE, respectively). SSC and TA values ranged from 6.0 to 6.6°Bx and 0.06 to 0.08 g citric acid/100 mL. Kohlrabi had higher sugar content by the end of the storage, especially in OPP samples (2.36, 2.39, and 0.89 g/100 mL for fructose, glucose, and sucrose, respectively).

Storage of kohlrabi stems under MAP (with OPP and amide PE and washing in an NaOCl water solution) for 60 days at 0°C in conjunction with a retail period was studied by Escalona et al. (2007b). The firmness was found to significantly decrease in MAP stems compared with control (88.3, 91.2, and 117.1 for amide-PE, OPP, and control samples, respectively). By the end of storage, MAP maintained the appealing appearance of the stems.

Application of PMAP (OPP 20 μm, OPP 40 μm, and amide-PE films) to kohlrabi stems and their storage at 0°C for 14 days and at 10°C for 3 days was investigated by Escalona et al. (2007c). After the retail sale period, weight losses were higher for Amide-PE, reaching up to 0.52 g/100 g. At harvest, values of SST, pH, and TA were 9.31°Bx, 6.5%, and 0.1%, respectively. After cold storage, SST and TA displayed a slight decrease. The sucrose content dropped significantly down to 0.4 g/100 mL in all treatments.

Significant information about trials on kohlrabi is given in Table 9.6.

9.8 COLESLAW MIX (80% SHREDDED CABBAGE, 20% SHREDDED CARROTS)

Coleslaw mix stored was inoculated with *L. monocytogenes* and two strains of *E. coli* and stored under PMAP (OPP film) at 4°C and 8°C for 12 days. At 4°C, populations of *E. coli O157:H7* dropped by 1–1.5 log cycles, but viable cells were still detected by the end of the storage period. *L. monocytogenes* was substantially reduced at both storage atmospheres (1.5 and 2.0 log CFU/g for storage at 8°C and 4°C, respectively) (Francis and O'Beirne, 2001).

From Figure 9.20, it was concluded that passive modified atmosphere conditions generated by the use of PA-160 microperforated film combined with low-temperature storage and chlorine treatment (immersion with agitation in a chlorine solution of 100 ppm chlorine) resulted in considerably lower values for the initial and final psychrophilic numbers.

Cliffe-Byrnes et al. (2003) investigated the effect of packaging treatments (PMAP with OPP and four microperforated OPP films, PA-120, PA-160, PA-190, and PA-210) on the quality of dry coleslaw stored at 4°C and 8°C for 9 days. OPP had higher acceptability of color scores at both temperatures. The firmness of coleslaw under OPP (less than half the initial value at seventh day) and PA-120 was lower than for the other PA films. An increase in storage temperature from 4°C to 8°C led to a reduction in shelf life for all film types.

Dry coleslaw mix was inoculated with *L. monocytogenes* and *L. innocua* and stored under PMAP (OPP and four microperforated OPP films, PA-120, PA-160, PA-190, and PA-210) at 3°C and 8°C. At 8°C, listerial counts were generally maintained at inoculation levels over the storage period with the exception of OPP in which *L. innocua* decreased to 2.8 log CFU/g. A similar trend was recorded for

TABLE 9.6
Significant Information about Trials on Kohlrabi When Stored under MAP

Species and Food Type	Initial Gas Mix	Packaging Material	Treatment before Packaging	Storage Temperature (°C) and Storage Period (Days)	Color	Microflora	Texture–Weight Loss	Sensory Analysis	Shelf-Life (Days) Extension	References
Kohlrabi sticks (*B. oleracea* L. *gongyloides* group)	PMAP: (1) 9% O_2/7% CO_2 (2) 7% O_2/9% CO_2	(1) OPP (2) Amide-PE	Washing in a water solution of 50 mg/L NaOCl for 1 min	0°C—14 days	Chroma parameters dropped after 7 days in all treatments		Control sticks showed weight losses that were twofold higher than sticks under MAP	The appearance decreased with storage time. MAP had no effect on this parameter. Sticks stored under MAP showed a better texture and taste than control	Sticks under an equilibrium atmosphere reached by amide-PE kept an acceptable sensorial quality for 14 days	Escalona et al. (2007a)

(continued)

TABLE 9.6 (continued)
Significant Information about Trials on Kohlrabi When Stored under MAP

Species and Food Type	Initial Gas Mix	Packaging Material	Treatment before Packaging	Storage Temperature (°C) and Storage Period (Days)	Color	Microflora	Texture–Weight Loss	Sensory Analysis	Shelf-Life (Days) Extension	References
Kohlrabi stems (*B. oleracea* L. *gongyloides* group)	PMAP: (1) 4.5%–5.5% O_2/11%–12% CO_2 (2) 2.5% O_2/35% CO_2	(1) OPP (2) Amide-PE	Washing in a water solution of 50 mg/L NaOCl for 1 min	0°C for 30 and 60 days and at 12°C for 3 days	MAP stems had a slight yellowing, showing lower hue value on the skin after 60 + 3 days	After 60 days at 0°C, control stems developed bacterial soft rot caused by *Erwinia* sp. and *Pseudomonas* sp.	Firmness decreased in MAP stems compared to control	After 30 + 3 days, the appearance decreased in all treatments. Stems packed in amide-PE reached the highest scores. After 60 + 3 days, MAP maintained good appearance	PMAP with amide–PE extended the shelf life of kohlrabi stems to 60 days	Escalona et al. (2007b)

Kohlrabi stems (*B. oleracea* L. *gongyloides* group)	PMAP: (1) 5% O_2/10%–11% CO_2 (2) 6% O_2/10% CO_2 (3) 5% O_2/14%–15% CO_2	(1) OPP 20 μm (2) OPP 40 μm (3) Amide-PE copolymer	Stems washed in a solution of 100 mg/L NaOCl for 1 min at 5°C	At 0°C for 14 days and at 10°C for 3 days	Weight losses were higher for amide-PE, in comparison with OPP film	After the retail sale period, sugar content decreased in control stems compared to those stored under MAP. Values of SST and TA increased over the retail sale period due to enhanced water loss at 10°C. Both appearance and turgidity of leaves were worse in the control than in MAP	Atmospheres with 5% O_2/10%–15% CO_2 kept a fresh quality of the kohlrabi stems for 2 weeks at 0°C	Escalona et al. (2007c)

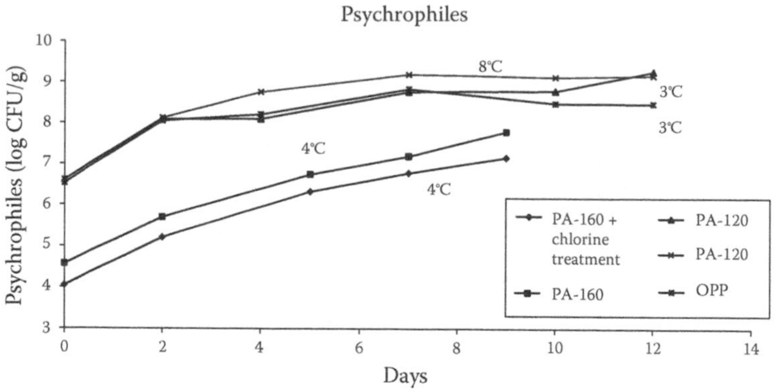

FIGURE 9.20 The effect of storage conditions created by different packaging films on psychrophilic counts for dry coleslaw mix (80% shredded cabbage, 20% shredded carrots) vs. storage time (OPP-Oriented Polypropylene, PA-120 microperforated OPP (From Bourke, P. and O'Beirne, D., *Int. J. Food Sci. Technol.*, 39, 509, 2004), PA-160 microperforated OPP. (From Cliffe-Byrnes, V. and O'Beirne, D., *Food Control*, 16, 707, 2005.))

L. monocytogenes strains. Total aerobic mesophiles reached 8 log CFU/g at day 7, whereas TPCs in OPP films were reduced down to 7.51 log CFU/g at day 12 (Bourke and O'Beirne, 2004).

Washing coleslaw mix with a chlorine solution and storing it at lower temperatures had beneficial results in reducing considerably the initial microbial load of coleslaw and lowering the growth rates of mesophilic counts. Packages using PA-160 microperforated OPP film gave better results in reducing the TVC (Figure 9.21).

The effect of chlorine treatment (washing in a 100 ppm chlorine solution for 5 min) in conjunction with application of PMAP (microperforated OPP films PA-160

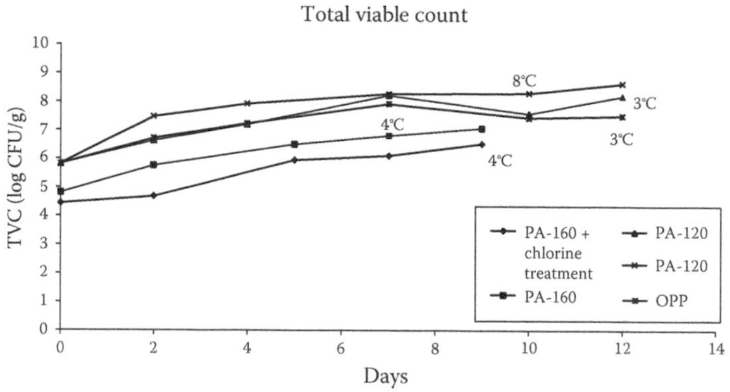

FIGURE 9.21 The effect of packaging film on TVC for dry coleslaw mix (80% shredded cabbage, 20% shredded carrots) (OPP-Oriented Polypropylene, PA-120 microperforated OPP (From Bourke, P. and O'Beirne, D., *Int. J. Food Sci. Technol.*, 39, 509, 2004), PA-160 microperforated OPP. (From Cliffe-Byrnes, V. and O'Beirne, D., *Food Control*, 16, 707, 2005.))

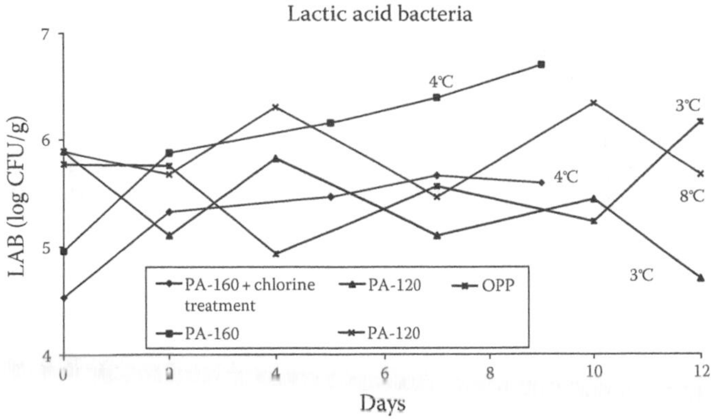

FIGURE 9.22 The number of LAB for dry coleslaw mix (80% shredded cabbage, 20% shredded carrots) under passive modified atmosphere packaging produced by different packaging films vs. storage time (OPP-Oriented Polypropylene, PA-120 microperforated OPP (From Bourke, P. and O'Beirne, D., *Int. J. Food Sci. Technol.*, 39, 509, 2004), PA-160 microperforated OPP (From Cliffe-Byrnes, V. and O'Beirne, D., *Food Control*, 16, 707, 2005.))

and PA-210 used) on coleslaw mix stored at 4°C and 8°C for 9 days was investigated by Cliffe-Byrnes and O'Beirne (2005). At 4°C, chlorine significantly decreased mesophilic bacteria by about 1 log (6.53 and 6.4 log CFU/g for PA-160 and 210, respectively). At 4°C, yeast and mold counts dropped by chlorine in both film types (>0.5 log CFU/g reduction). The most effective combination was chlorine washing in conjunction with storage in PA-160 film at 4°C.

Figure 9.22 displays that an initial wash with chlorine solution (100 ppm) significantly reduced the initial microbial load of LAB in coleslaw mix. PA-120 film and low storage temperatures restricted the growth of LAB. PA-160 gave the worst results, maybe due to the atmosphere modification that affected the other epiphytic populations, with lesser competition for the bacteria growth.

Francis et al. (2007) investigated the impact of glutamate decarboxylase (GAD) acid resistance system to survival and growth of *L. monocytogenes* LO28 in PMAP (OPP 35 μm) coleslaw stored at 4°C, 8°C, and 15°C for 12 days. Coleslaw was inoculated with a wild strain of *L. monocytogenes* and four strains with mutant genes *(ΔgadA, ΔgadB, ΔgadC, ΔgadAB)* that had negligible GAD activity. At 8°C, *ΔgadAB* populations (1.9 log CFU/g) were lower than the wild-type or other mutant strains, with numbers falling by 3 log cycles. The wild-type strain survived better than *ΔgadAB*.

The collected data from researches on coleslaw mix are summarized in Table 9.7.

9.9 BAMBOO SHOOTS

The effect of MAP (LDPE film with initial atmosphere of 2% O_2 and 5% CO_2) on the browning and shelf life of bamboo shoots stored at 10°C was evaluated by Shen et al (2006). It was shown that MAP strongly limited the formation of malondialdehyde

TABLE 9.7
Collected Data from Researches on Coleslaw Mix When Stored under MAP

Species and Food Type	Initial Gas Mix	Packaging Material	Treatment before Packaging	Storage Temperature (°C) and Storage Period (Days)	Color	Microflora	Texture–Weight loss	Sensory Analysis	Shelf-Life (Days) Extension	References
Dry coleslaw mix (80% shredded cabbage, 20% shredded carrots)	PMAP: 0%–2% O_2 and 25%–27% CO_2	35 μm OPP	Inoculation with *L. monocytogenes* and two nontoxigenic *E. coli O157:H7* strains	4°C and 8°C for 12 days		*L. monocytogenes* counts in packs of coleslaw mix decreased by 1.5 log cycles during storage at 8°C, and by approximately 2.0 log cycles when samples were held at 4°C. *E. coli O157:H7* counts dropped for 4°C and 8°C				Francis and O'Beirne (2001)
Dry coleslaw mix (Dutch cabbage cultivar	PMAP	(1) OPP (2) PA-120 (3) PA-160 (4) PA-190 (5) PA-210		4°C and 8°C for 9 days	Scores for acceptability of color dropped over storage of coleslaw packed in microperforated films, or in air		At 4°C the firmness of coleslaw remained stable For all films, whereas at 8°C, firmness of coleslaw in	The decrease rate in acceptability of appearance was greater at 8°C. PA films scored	Packaging coleslaw within OPP had negative effects on quality, withloss of firmness,	Cliffe-Byrnes et al. (2003)

Marathon and carrot cultivar *Nairobi*)					The product stored in OPP had higher acceptability scores at both temperatures		microperforated packs increased slightly. Firmness of coleslaw packed in OPP and PA-120 was lower than in PA films from day 5	higher than OPP and open packs from day 7 onward at 4°C. At 8°C, appearance scores for open packs were much lower than PA films over the entire storage time. Weight loss was higher in air and statistically greater at 8°C	high cell permeability and exudate, poor acceptability of aroma and reduced pH but good retention of color	
Dry coleslaw mix (Dutch cabbage cultivar *Marathon* and carrot cultivar *Nairobi*)	PMAP	(1) OPP (2) PA-120 (3) PA-160 (4) PA-190 (5) PA-210	Inoculation with *L. monocytogenes* and *L. Innoqua*	3°C and 8°C for 12 days		At 3°C, *L. Innoqua* counts were lower in coleslaw packaged in PA-120 than all other PA films. *L. monocytogenes* counts within PA-160 reduced compared to OPP. At 8°C listerial counts within OPP were more reduced than PA-120, PA-210				Bourke and O'Beirne (2003)

(*continued*)

TABLE 9.7 (continued)
Collected Data from Researches on Coleslaw Mix When Stored under MAP

Species and Food Type	Initial Gas Mix	Packaging Material	Treatment before Packaging	Storage Temperature (°C) and Storage Period (Days)	Color	Microflora	Texture–Weight loss	Sensory Analysis	Shelf-Life (Days) Extension	References
						Pseudomonas counts within OPP were significantly lower at both temperatures				
Dry coleslaw mix (Dutch cabbage cultivar *Marathon* and carrot cultivar *Nairobi*)		(1) PA-160 (2) PA-210	Washing for 5 min in a chlorine solution at 4°C (100 ppm chlorine)	4°C and 8°C for 9 days	At 4°C, chlorine-washed samples had higher scores for color within both films, and these effects persisted throughout storage	At 4°C, washing with chlorine (CW) reduced mesophilic counts by 1 log and psychrophilic bacteria. Reductions in initial counts for *Pseudomonas* species, molds and yeasts, at both 4°C and 8°C for both films. CW reduced loads of LAB and coliforms for both temperatures		Chlorine-washed samples scored significantly better, particularly at 4°C. Chlorine-washed coleslaw also scored better for aroma in both film types	At 8°C, shelf life was reduced to between 4 and 6 days for all treatments. Of the treatments examined, the most effective was chlorine treatment combined with PA-160 packaging at 4°C	Cliffe-Byrnes and O'Beirne (2005)

Dry coleslaw mix (80% shredded cabbage, 20% shredded carrots)	OPP bags	Inoculation with four strains of *L. monocytogenes* with or without genes for glutamase decarboxylase resistance mechanism	4°C, 8°C, and 15°C for 12 days		The wild strain survived better than the double mutant Δ*gad*AB strain		The *gadB* and *gadC* genes played the greatest role in packaged coleslaw	Francis et al. (2007)

and the activity of peroxidase and phenylalanin ammonialyase, thereby restricting browning and lignification. The polyphenol contents of MAP treatment dropped to their lowest value at the fourth day (1.1 mg/g fw) and rose and leveled off thereafter (1.2 mg/g fw). The lignin contents of control (1.33%) were 36% higher than MAP (0.97%).

9.10 GINSENG

Hu et al. (2005) stored fresh ginseng under PMAP (PVC and LDPE films used) at 0°C, 10°C, and 20°C. The highest firmness values were recorded at 0°C, especially in the PVC packages (122.2 N). Over all treatments, loss of ginseng firmness had the highest value at 20°C (102.3 and 109.3 N for PVC and LDPE, respectively). The microbial populations augmented from 5.5×10^6 CFU/g at storage of 100 days to 8.5, 10.2, and 14×10^6 CFU/g at 0°C, 10°C, and 20°C, respectively, after 150 days in PVC packages. The best quality of fresh ginseng was obtained by combination of low temperature and PVC.

American ginseng roots were treated with an antimicrobial agent (dipping in 0.5% DF-100 solution for 5 min), and stored at 2°C under various CA (2% O_2 and 2%, 5%, and 8% CO_2) or passive MA (PD-941 [high], PD-961 [medium], and PD-900 [low permeability] PO films used) conditions. No noticeable changes were recorded in appearance, ginsenosides, and saponin content (5.98 at harvest and 6.19, 6.91 for CA at 5% CO_2 and PD-961, respectively) after 3 months at CA (5% CO_2) storage and MA packages (Jeon and Lee, 1999).

Macura et al. (2001) investigated the survival of inoculated *C. botulinum* in MA (Winpac medium transmission laminate film used) packaged ginseng roots stored at 2°C, 10°C, and 21°C. No toxin was determined at 21°C, neither for control nor for inoculated roots after 6 weeks of storage. At this time, all the roots were classified as unacceptable in terms of appearance and the study was terminated. At 10°C, the number of anaerobes increased considerably at week 14, thereby leading to toxin production before the spoilage of product.

9.11 MUNGBEAN

Chlorine dioxide (dipping in 100 ppm ClO_2 solution for 5 min) combined with MAP (passive MA with a nylon–PE film used, vacuum and active MA with 100% CO_2 and 100% N_2) was applied to mungbean sprouts during storage at 5°C ± 2°C for 7 days (Jin and Lee, 2007). ClO_2 treatment proved to be very effective in decreasing both inoculated *S. typhimurium* and *L. monocytogenes* populations (3.0 and 1.5 $\log_{10}$ CFU/g reduction, respectively). Samples treated with ClO_2 and packaged UV or CO_2 gas exhibited the highest visual quality.

9.12 ARTICHOKE

Gil-Izquierdo et al. (2002) used five films (PVC, LDPE, and three microperforated polypropylene films PP_1, PP_2, and PP_3) to create proper MA conditions for artichokes stored at 5°C for 8 days. The minimum loss of vitamin C (1.8 and 3.5 mg/100 g for the edible portion) was reported for the artichokes maintained with lower CO_2 (control and PVC) levels, whereas in the other MAP with higher CO_2 levels (PP_1 and PP_2), rather large losses were recorded (5.5 and 6.2 mg/100 g, respectively).

9.13 CUCUMBER

Cucumbers stored under passive MA conditions (microperforated and intact LDPE films used) at 5°C for 18 days were studied by Wang and Qi (1997). After 18 days of storage, the nonwrapped samples lost 9.2% of their initial weight while samples from perforated bags lost 0.9% and those from sealed bags lost only 0.2%. Application of chilling stress induced strong increases in putrescine levels for all treatments, whereas the sealed fruit displayed the highest levels of putrescine (340 nmol/g).

Karakas and Yildiz (2007) investigated the effect of storage conditions (storage under CA with normal and superatmospheric [70% O_2] conditions) and physical tissue damage (bruising by dropping a weight) on membrane peroxidation in minimally processed cucumber tissue stored at 4°C or 20°C. High levels of lipid hydroperoxides (4.6 μmol H_2O_2 equivalents/g) were reported for bruised samples at 20°C. A considerable drop in lipid peroxidation was recorded under MAP with higher levels of oxygen.

9.14 SWEET POTATO

Erturk and Picha (2007) sliced sweet potatoes and stored them under MA conditions using a low (PD-900), medium (PD-961), and high (PD-941 PO) permeability film at 2°C or 8°C for 14 days. Alcohol insoluble solid content (mainly starch) dropped significantly during 14 days of storage, but at a lower rate at 2°C than 8°C. Ethanol content of slices in PD-941 and PD-961 remained at initial levels at 2°C of storage for 14 days, whereas the sucrose and total sugar content augmented considerably at 8°C over the 14 day storage period.

Shredded sweet potatoes from two major commercial cultivars (Beauregard and Hernandez) were packed in low (PD-900) and medium (PD-961) O_2 permeability bags at 4°C and flushed with gas composed of 5% O_2 and 4% CO_2 (McConnell et al. (2005). The ethanol increased up to 20 and 24 mL/L in sweet potatoes in PD-900 and PD-961 bags. No color changes were recorded for both cultivars regardless of treatment. The firmness decreased less than 5% in the PD-900 and 8% in the PD-961.

9.15 PUMPKIN

Cut pieces of pumpkin were dip treated (in citric acid [0.2%] and potassium metabisulfite [0.1%] for 3 min [treat. 1], a solution of soluble starch [0.2%] extra-pure plus calcium chloride [0.1%] for 3 min [treat. 2] and mannose [0.1%] solution, and vacuumized [25n vacuum] for 5 min [treat 3]) and stored in LDPE or PP bags, sealed with and without vacuum and stored at 5°C ± 2°C, 13°C ± 2°C, and 23°C ± 2°C.

According to Habibunnisa et al. (2001), "At 5 ± 2°C pumpkin treated with treat.1 and stored in LDPE remained in good condition for 25 days with a minimal weight loss of 0.06%, minimal loss in nutrient composition and was microbiologically safe."

9.16 CAULIFLOWER

Two different films (nonperforated PVC and oriented PP) were used to induce atmospheric modification for cauliflowers stored at 4°C and 8°C for 20 days. Simon et al. (2008) found that in PVC, mesophiles, *Pseudomonas*, and *Enterobacteriaceae* counts were lower than in the other films, at both temperatures (3.95, 4.0, and <1 log CFU/g at 4°C and 4.9, 5.1 and 3.3 at 8°C, respectively). However, it is noteworthy that at 4°C or 8°C, the mesophiles were only 0.8 and 0.79 log units lower in PP than with control.

Several cauliflower cultivars (*Abruzzi, Dulis, Casper, Serrano, Caprio, Nautilus, Beluga*, and *Arbon*) were stored under various MAP (microperforated PVC, P-Plus 120, 160, and 240 films) at 4°C for 25 days. In the case of *Dulis, Casper, Nautilus, Arbon*, and *Beluga* varieties, no significant differences were recorded in the microbial load depending on the film used and were lower than the legal maximum. No changes in weight (<0.7%) were recorded for the three treatments where P-Plus films were used (Sanz et al., 2007).

9.17 ONION

Microbial proliferation and sensory quality aspects of sliced onions were tested at different temperatures (−2°C, 4°C, and 10°C) and atmospheric conditions (LDPE with [AMAP] or without [PMAP] 40% CO_2 + 1% O_2). The onions were significantly less firm at 4°C and 10°C than at −2°C. The microbial shelf lives of the tested onions exposed or not to 40% CO_2 + 1% O_2, or at −2°C, 4°C, and 10°C, were 12.5, 9.5, 7, 12, 9, and 6 days, respectively; and their sensorial shelf lives were 12, 8, 5, 10.5, 7, and 5 days, respectively (Liu and Li, 2006).

Various packaging treatments (PMAP with LDPE and PP, AMAP with LDPE and an ethylene scavenger, and LDPE with initial gas concentration of 9.5 kPa CO_2 + 18.2 kPa O_2 and moderate VP) were applied to bunched onions stored at 10°C for 28 days to determine the optimal packing method (Hong and Kim, 2004). Samples in the moderate VP package retained better visual quality (almost constant Hunter L*) and displayed less discoloration and decay. The shelf life of MP bunched onions could be prolonged to below 14 days in MA, over 14 days in PE and PP, about 21 days in PE + ES, and at least 21 days in MVP.

9.18 CABBAGE

Cut salted Chinese cabbage with air, 100% CO_2, or 25% CO_2/75% N_2 packaging was irradiated with 0.5, 1, and 2 kGy, respectively; and the microbiological and physicochemical qualities were investigated during storage at 4°C for 3 weeks. Coliforms also decreased by irradiation (2.5–3.7 log CFU/g reduction) and the combined treatment of irradiation and MA (1.3–1.7 log CFU/g) was effective as well. Generally, gamma irradiation at 1 kGy or above was found to reduce the phenolic contents in cabbage (Ahn et al., 2005).

Gomez-Lopez et al. (2007b) immersed shredded cabbage in neutral electrolyzed oxidizing water containing 40 mg/L of free chlorine up to 5 min, and then stored under EMAP (film with 4600 mL O_2/m^2 24 h atm was used at 7°C) at 4°C and 7°C. Total aerobic plate counts, psychrotrophs, and yeasts attained the stationary phase after 6 days of storage at 4°C (7.7 log CFU/g) and showed no further increase after 13 days at 7°C (7.2 log CFU/g). A shelf-life prolongation of at least 5 days and 3 days in samples stored at 4°C and 7°C, respectively, was achieved.

The impact of PMAP (OPP [treat. 1], PE trays overwrapped with multilayer PO [treat. 2], or with a plasticized PVC [treat. 3]) and storage at 3°C on quality of cabbage was investigated by Pirovani et al. (1997). The mesophilic aerobic populations augmented from $20 \cdot 10^3$ CFU/g to $22 \cdot 10^4$ CFU/g for OPP, to $51 \cdot 10^5$ CFU/g for RD106-PE, and to $85 \cdot 10^5$ log CFU/g for PVC-PE trays after 8 days. A quality level between excellent and good for cabbage can be maintained for 6–7 days using PVC-PE or RD106-PE and for 9–10 days for OPP bags.

9.19 FENNEL

Escalona et al. (2004) investigated the effects of PMAP (OPP bags [treat. 1] and plastic baskets with OPP film [treat. 2]) to inhibit browning of the butt-end cut zone of fennel bulbs stored during 14 days at 0°C followed with complementary air storage during 3 days at 15°C. After cold storage, control and MAP bulbs exhibited weight losses lower than 0.1%. A slight decrease in SSC was recorded by the end of storage. MAP treatments were shown to preserve the visual appearance by the end of shelf life much better than control.

Fennel was washed on the butt-end cut with ascorbic (1%) and citric (5%) acids, packed in PP baskets sealed with PP film to generate an MA and stored for 14 days at 0°C followed by 4 days in air at 15°C. It was found that films and antioxidants hardly influenced the fennel firmness. In fact, the rate of bulb softening was proportional to the increase in temperature and weight loss (49 and 56 N for control and OPP samples, respectively, at the end of storage). Antioxidant solutions were shown to have no effect on delaying browning (Artes et al., 2002).

Escalona et al. (2005) investigated diced fennel washed in chlorinated water (100 mg/L) and stored under PMAP (OPP bags [treat. 1] and plastic baskets with OPP film [treat. 2]) at 0°C for 14 days. The gas composition in MAP treatments had no effect on the final product color and was not effective in delaying browning. In all treatments, by the end of storage, a slight rise in pH and a slight drop in SSC and TA values were recorded (6.6, 3.0, and 0.37 and 6.6, 2.8, and 0.37 for PP bags and baskets), compared with values at harvest (6.3, 4.1°Bx and 0.54 g of oxalic acid/L).

9.20 EGGPLANT

The effect of grafting *S. Sisymbriifolium* [gr. 1], *S. Torvum* [gr. 2], methylbromide [gr. 3], and Perlka [gr. 4] in conjunction with storage under MAP (30% CO_2 in HDPE bags) at 10°C on quality parameters of eggplant was investigated by

Arvanitoyannis et al. (2005). The interaction of treatments, storage duration, and MAP revealed no significant differences. In the air group (6.54, 6.59, 6.56, and 6.66 for gr. 1, 2, 3, and 4), pH is higher than MAP (6.64, 6.56, 6.53, and 6.52 for gr. 1, 2, 3, and 4) in all cases except for *S. Sisymbriifolium.*

9.21 SNOW PEA

Pariasca et al. (2000) examined the effects of precooling, application of PMAP (polymethyl pentene 25 μm [PMP-1] and 35 μm [PMP-2], LDPE and OPP were used) and CA (2.5, 5, and 10 kPa O_2 with 5 kPa CO_2, 0, 5, and 10 kPa CO_2 with 5 kPa O_2 were the compositions examined), and storage on snow pea pods at 5°C. Precooled pods displayed higher ascorbic acid contents than the nonprecooled ones, whereas PMP (3.5 and 3.6 mg/100 g for PMP-1 and 2) and LDPE (3.6) had higher contents than LDPE (3.0) and PP-bagged pods (1.8).

9.22 POTATO

Potatoes were peeled with (i) an abrasive peeler, (ii) a hand-peeler, and (iii) a lye solution, were treated with several antibrowning agents (dipping in 0.5% L-cysteine, 0.5% L-cysteine plus 2% citric acid mixture, 5% ascorbic acid, and 0.1% potassium metabisulfite and stored under AMAP with 9% CO_2/3% O_2), and were stored under active (9% CO_2/3% O_2, 9% CO_2 and 100% N_2 compositions) MAP at 2°C. Nitrogen flushing with a highly permeable multilayered PO packaging (PD-961EZ, PD-941, and RD-106 films) was the best, economical, and simple choice, leading to the same results (7.5% reduction in L^* value) with the other compositions (Gunes and Lee, 1997).

Beltran et al. (2005b) investigated the impact of sanitizers (sodium sulfite [SS], sodium hypochlorite [SH], Tsunami [T], ozone [O], and the combination of ozone–Tsunami [OT]) on the sensory and microbial quality of fresh-cut potatoes stored under PMAP (LDPE film was used) and VP at 4°C. No browning was detected in potatoes (UV) washed with SS, O, and OT. OT and SH retarded considerably the microbial growth, with mesophiles 0.75–1.27 logs lower than in water washed samples.

The effect of γ-irradiation dose (0–1.5 kGy), citric acid (0%–1.0%), and potassium metasulfite (KMS) dipping solution concentration (0%–1.0%) on potato cubes stored under MAP (PP film was used) at 4°C ± 1°C for 4 weeks was evaluated by Baskaran et al. (2007). At optimal irradiation dose (1.0 kGy), the highest level of L^*-value (59.8) and the lowest level of b^*-value (1.25) were obtained for the maximal concentration of citric acid (1.0%) and KMS concentration (1.0%).

9.23 CELERY

Gomez and Artes (2005) used LDPE and OPP to generate MA conditions for storage of celery sticks at 4°C for 15 days. Total sugar content dropped for air (1.10 g/100 mL) but not for MAP (1.26 and 1.29 g/100 mL for LDPE and OPP, respectively), thus revealing a lower respiration rate, which would considerably slow down the loss of sugars. Mesophiles (1.95 and 2.87 log CFU/g for LDPE and OPP), psychrotrophs (3.19 and 1.62 log CFU/g for LDPE and OPP, respectively), and molds and yeasts (1.95 and 2.41 log CFU/g for LDPE and OPP) were substantially lower for MAP than for control.

Radziejewska-Kubzdela et al. (2007) applied several CO_2 contents (0%, 5%, 10%, 20%, 30%, 50%) plus 2% O_2 (oriented PA/PE laminated film) in an effort to preserve celeriac flakes at 4°C and 15°C (only for 0%, 5%, 10% CO_2 packs). Storage of celeriac flakes packaged under MA at 15°C led to quality deterioration. Packaging of celeriac flakes under MA containing 5% or 10% CO_2 had a strong advantageous effect on their color. After 12 days at 4°C, celeriac flakes under MA containing 5% or 10% CO_2 were described with better quality.

9.24 PEPPER

The impact of VP (Saran film used) and PMAP (PD-961 PO films) at 5°C and 10°C on quality and shelf life of bell peppers was thoroughly investigated by Gonzalez-Aguilar et al. (2004). Application of MAP maintained the overall quality of fresh-cut peppers at 5°C and minimized the texture loss. The latter was more noticeable at 5°C (15.5 N) than at 10°C (13 N). When the temperature rose from 5°C to 10°C, the microorganisms augmented by more than 3 log CFU/g. Therefore, a drop in temperature was found to be a more critical factor than MAP in terms of decreasing microbial counts.

Lack of oxygen due to VP restricts the coliform growth and their number is maintained at acceptable levels by the end of storage. This was verified by the low coliform numbers determined by low oxygen AMAP. Moreover, low storage temperatures helped in peppers preservation by maintaining coliforms at a low level. The use of LDPE as a packaging film was unsuccessful since it led to the poorest results of all the storage conditions examined (Figure 9.23).

Koide and Shi (2007) employed a biodegradable film based on PLA and LDPE (perforated or not) to generate MA conditions for storage of green peppers at 10°C for 7 days. No development of yellow color on the green pepper surface was recorded and the chlorophyll content remained almost constant throughout the entire storage period. The weight loss values for green peppers by the end of storage were 2.46% ± 0.6%, 0.38% ± 0.1%, and 1.59% ± 0.4% for PLA, LDPE, and perforated LDPE, respectively.

Green chili peppers were stored under PMAP (LDPE, PVC, and cast PP [CPP] films were used) at 10°C for 14 days (Lee et al., 1994). Negligible or very slight weight loss was reported for peppers in all packages (0% for LDPE and CPP and 1% for PVC). Similar to ascorbic acid decomposition, chlorophyll content was significantly decreased for the initial 7 days of storage and only negligibly for the following 7 days of storage (15.1, 15.8, and 15.4 mg/100 g for LDPE, CPP, and PVC, respectively).

From Figure 9.24 it was concluded that green chili peppers exhibited lower values of ascorbic acid when stored either with LDPE or with CPP at 10°C. The temperature storage seems to play apparently a minor role but the decrease in ascorbic acid values was mainly derived from the nature of the product itself.

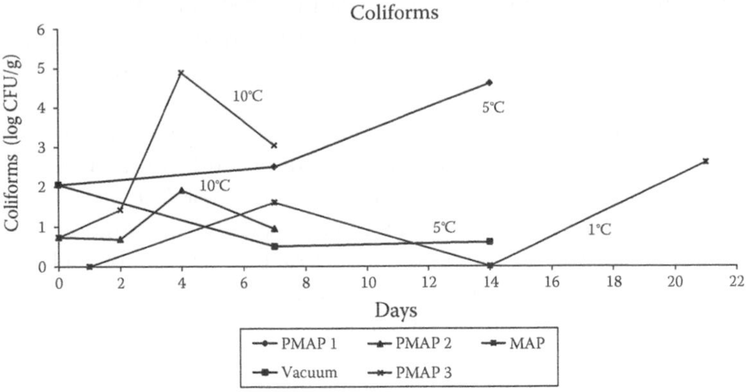

FIGURE 9.23 Changes in coliform count for green peppers under various packaging conditions (PMAP 1 [with PD-961 with OTR: 6000-8000 mL O_2/m^2 24 h atm used as a packaging film], Vacuum [with Saran used as a packaging film] after immersion in a Clorox solution [200 mL/L] (From Gonzalez-Aguilar, G.A. et al., *Lebensm. Wiss. Technol.*, 37, 817, 2004), PMAP 2 [with PLA—polylactic acid–based biodegradable film, used as a packaging film], PMAP 3 [with low density polyethylene used as a packaging film] (From Koide, S.and Shi, J., *Food Control*, 18, 1121, 2007), MAP 1 [2% O_2/10% CO_2] after immersion in a chlorine solution of 100mg/L. (From Pilon, L. et al., *Ciência Tecnologia Alimentas Campinas,* 26(1), 150, 2006.))

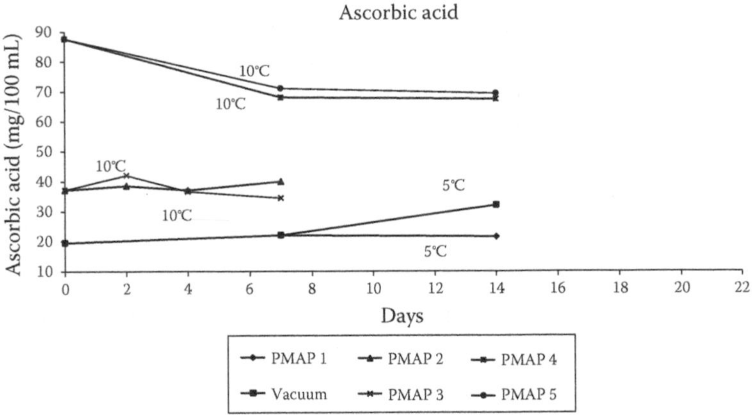

FIGURE 9.24 Ascorbic acid content of varieties of peppers under various packaging conditions (PMAP 1 [with PD-961 with OTR: 6000–000 mL O_2/m^2 24 h atm used as a packaging film], vacum [with Saran used as a packaging film] after immersion in a Clorox solution [200 mL/L] (From Gonzalez-Aguilar, G.A. et al., *Lebensm. Wiss. Technol.*, 37, 817, 2004.), PMAP 2 [with PLA—polylactic acid–based biodegradable film, used as a packaging film], PMAP 3 [with low density polyethylene used as a packaging film] (From Koide, S. and Shi, J., *Food Control*, 18, 1121, 2007.), PMAP 4 [with equilibrium atmosphere of 5.2% O_2/3.1% CO_2 and low density polyethylene used as a packaging film] for green chili peppers, PMAP 5 [with equilibrium atmosphere of 1.7% O_2/4.9% CO_2 and cast polypropylene used as a packaging film] for green chili peppers. (From Lee, K.S. et al., *Packag. Technol. Sci.*, 7, 51, 1994.))

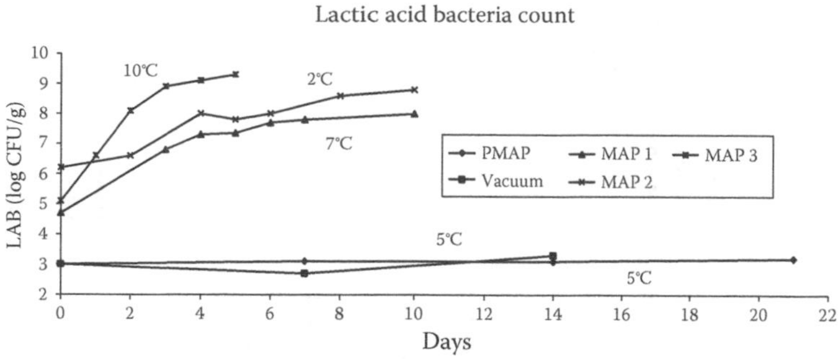

FIGURE 9.25 Changes in LAB count under different storage conditions for various pepper varieties vs. storage time (PMAP [with PD-961 with OTR: 6000–8000 mL O_2/m^2 24 h atm used as a packaging film], vacum [with Saran used as a packaging film] after immersion in a Clorox solution [200 mL/L] (From Gonzalez-Aguilar, G.A. et al., *Lebensm. Wiss. Technol.*, 37, 817, 2004), MAP 1 [3% O_2/2%–5% CO_2] for mixture of shredded green, red, and yellow bell peppers (From Jacxsens, L. et al., *Int. J. Food Microbiol.*, 83, 263, 2003), MAP 2 [3% O_2/5% CO_2] for mixture of shredded green, red, and yellow bell peppers inoculated with *L. monocytogenes* and *Aer. caviae* strains, MAP 3 [3% O_2/5% CO_2] for mixture of shredded green, red, and yellow bell peppers inoculated with *L. monocytogenes* and *Aer. caviae* strains. (From Jacxsens, L. et al., *Postharvest Biol. Technol.*, 26, 59, 2002a.))

In another experiment, pressed cardboard trays overwrapped with VF-71 PE and LDPE bags were employed toward MA storage of green chili peppers at 8°C or 24°C. After 4 and 6 weeks of storage at 8°C, the disease ratings were 1.1 and 3.9 for overwrapped fruit, and 0.6 and 2.2 for bagged fruit. The best option for prolonging the shelf life of fresh green chili peppers was employment of LDPE bags stored at 8°C (Wall and Berghage, 1996).

LAB under vacuum conditions have a bacteriostatic effect and in conjunction with PMAP (a high permeability film) at low temperatures are the ideal conditions for peppers conservation. Shredding, and temperature rising allow LAB to grow, thereby augmenting the risk of food spoilage. Moreover, the coexistence of inoculated bacteria may play a significant role by inhibiting the growth of antagonistic flora (Figure 9.25).

Sliced bell peppers were washed once, twice, or three sequential times in fresh distilled water and stored under MAP (PE film) at 7°C for 10 days (Toivonen and Stan, 2004). Soluble solids and total phenolics were found to be lower in washed slices, with phenolic levels being the lowest in slices washed three times (4.85°Bx and 0.7 μg/g, respectively). According to these results, it was clear that washing has a strong effect on physicochemical measures of quality in green pepper slices.

A synoptical presentation of MAP application to peppers is depicted in Table 9.8.

9.25 SPINACH

Spinach inoculated with *E. coli O157:H7* was packaged in four different environments (air, vacuum, 100% N_2 gas, and 100% CO_2 gas packaging) after treatment with water, 100 ppm chlorine dioxide, or 100 ppm sodium hypochlorite for 5 min and

TABLE 9.8
Synoptical Presentation of MAP Application to Peppers

Species and Food Type	Initial Gas Mix	Packaging Material	Treatment before Packaging	Storage Temperature (°C) and Storage Period (Days)	Color	Microflora	Texture–Weight Loss	Sensory Analysis	Shelf-Life (Days) Extension	References
Green bell peppers (*Capsicum annum*) California type, cv. "Meteor"	(1) 100/0 (2) 80/20 (3) 60/0 (4) 50/20 (5) 20/20 (6) 0/0 (7) 0.5/0 (8) 1/0 (9) 3/0 (10) 9/0 (11) 0/20 expressed in O_2/CO_2		The peppers were cut into dices	2°C, 7°C, 14°C for 4 days			At 14°C weight loss was always lower under CA than in air. Weight losses were 1.5–3 fold higher at 7°C and 14°C when compared to losses at 2°C	At 7°C, values of CO_2 were about 1.5–2 fold those at 2°C. The best visual appearance was under high O_2 combined with 20 kPa CO_2. Samples exposed to 0, 0.5, and 1 kPa O_2 at all temperatures as well as the 0/20/80 treatment developed off-odors		Conesa et al. (2007)
Green chili peppers (*Capsicum annuum*) New		(1) VF-71 PE (2) LDPE		8°C and 24°C	At 8°C fruits remained green for 4 weeks of storage		Peppers maintained their weight best in PE bags at 8°C		Postharvest quality maintained up to 4 weeks when fruit were packaged in	Wall and Berghage (1996)

Mexican Type									LDPE bags at 8°C	
Mixture of green, red, and yellow shredded bell peppers (*C. annuum L.*)	AMAP: 3% O_2 and 2%–5% CO_2	Packaging film with OTR: 2897 mL O_2/m^2 24 h atm		7°C for 13 days		Psychrotrophic bacteria reached 10^8 CFU/g in 3–4 days. For LAB, the limit of 10^7 CFU/g was exceeded after 3 days of storage. Yeasts exceeded 10^5 CFU/g on day 7	At day 6, a great water loss was reported, and consequently a loss of tcrispness (texture)	The concentration of fructose and sucrose was constant while glucose was consumed from day 5. Acetic acid detectable from day 5. The mixed bell peppers rejected at day 6 (acid odor and taste)		Jacxsens et al. (2003)
Bell pepper fruits (*C. annum* cv. wonder)	(1) PMAP for PD-961 EZ (2) Vacum was applied to Saran film	(1) Saran with OTR: 1.2 mL O_2/m^2 24 h atm (2) PD-961EZ with OTR: 6000–8000 mL O_2/m^2 24 h atm	Immersion in a Clorox solution (200 μL/L)	5°C and 10°C for 14 days		Fresh-cut peppers stored at 5°C presented the lowest microbial growth	Water loss was more evident in vacuum stored peppers at 10°C. The use of MAP decreased the texture loss of fresh-cut peppers, being more noticeable at 5°C	The highest accumulation of C_2H_4 was observed in the product packaged at 10°C. Peppers under vacuum and stored at 10°C exhibited the highest content of ethanol and acetaldehyde. Storage at 5°C maintained most of the existing ascorbic acid	Overall quality decreased continuously in treatments stored at 10°C followed by those strips packed under vacuum at 5°C	Gonzalez-Aguilar et al. (2004)

(continued)

TABLE 9.8 (continued)
Synoptical Presentation of MAP Application to Peppers

Species and Food Type	Initial Gas Mix	Packaging Material	Treatment before Packaging	Storage Temperature (°C) and Storage Period (Days)	Color	Microflora	Texture–Weight Loss	Sensory Analysis	Shelf-Life (Days) Extension	References
Green peppers (*C. annuum L.*)	PMAP: (1) 11.6% O_2 and 5.6% CO_2 (2) 14.8% O_2 and 2.3% CO_2	(1) PLA-based biodegradable film (2) LDPE		10°C for 7 days	The (a*/b*) value, hue angle, and chroma did not change remarkably with storage time in all treatments	The level of aerobic bacteria counts in PLA, LDPE film package showed very little change. The coliform bacteria in the PLA and LDPE package augmented by less than 1 log CFU/g.	PLA film packaging and other packaging treatments can be used to preserve the hardness of green pepper over storage		Biodegradable PLA film for MA package is an alternative for LDPE film	Koide and Shi (2007)
Green chili peppers (*C. annum*) Nogkwang variety	PMAP: (1) 5.2% O_2 and 3.1% CO_2 (2) 1.7% O_2 and 4.9% CO_2 (3) 18.4% O_2 and 0.9% CO_2	(1) LDPE (2) CPP (3) PVC		10°C for 14 days			Packaged or wrapped peppers showed no unacceptable appearance due to weight loss within 14 days	Packaged peppers showed a higher content of ascorbic acid than the control with CPP packages having the best score	The LDPE and CPP packages provided better quality retention compared with unpackaged controls	Lee et al. (1994)

Green bell peppers (*C. annuum L.*)	PMAP	PE bags with OTR: 4000 mL O_2/m^2 24 h atm	(1) No wash (2) One wash in fresh distilled water (3) Two sequential washes in fresh distilled water (4) Three sequential washes in fresh distilled water	7°C for 10 days			Firmness retention improved more for washed slices, than unwashed slices. The firmness retention improved incrementally with the number of washes	Solute leakage was reduced with the number of washes SS and total phenolics were lower in washed slices, with phenolic levels being the lowest in slices that had been washed three times. Anaerobic metabolite levels were highest in the unwashed slices	The benefits of washing are associated with removal of SS, phenolics, and acetaldehyde from the cut surfaces and had a beneficial effect on reducing physical changes in peppers	Toivonen and Stan (2004)
Magali green peppers (*C. annuum* L.)	(1) Vacum (2) AMAP: 2% O_2/10% CO_2	BOPP/LDPE (biaxially oriented PP/low-density PE) plastic bags	Immersion in cold water (7°C) with 100 mg/L of free chlorine at pH 7.0 for 15 min	1°C ± 1°C for 21 days		Peppers contained total coliforms, psychrotrophics, anaerobic mesophiles during the storage period of all treatments. Psychrotroph counts were 10^3–10^6 CFU/g		The pH average was significantly affected during storage. Vitamin C levels oscillated with a slight decrease during the period of storage. β-carotene was reduced after the first week of storage for all treatments.		Pilon et al. (2006)

(*continued*)

TABLE 9.8 (continued)
Synoptical Presentation of MAP Application to Peppers

Species and Food Type	Initial Gas Mix	Packaging Material	Treatment before Packaging	Storage Temperature (°C) and Storage Period (Days)	Color	Microflora	Texture–Weight Loss	Sensory Analysis	Shelf-Life (Days) Extension	References
Green, red, and yellow bell peppers (*C. annuum* L.)	AMAP: 3% O_2/5% CO_2	Bags with OTR:1.57·10^{-11} mol O_2/m^2 s Pa	Inoculation with *L. monocytogenes* strains and *A. caviae*	2°C, 4°C, 7°C, and 10°C for 10 days		Yeasts and LAB limited the shelf life for peppers. Decrease of *L. monocytogenes* and survival of *A. caviae* on mixed bell peppers		The bell peppers became unacceptable after 7, 9, 6, and 2 days of storage at, respectively, 2°C, 4°C, 7°C, and 10°C	4°C is the optimal storage temperature for sensory attributes	Jacxsens et al. (2002a)
Green, red, and yellow bell peppers (*C. annuum* L.)	AMAP: 3% O_2/5% CO_2	Bags with OTR: 2838 mL O_2/m^2 24 h atm	Inoculation with *L. monocytogenes* strains and *A. caviae*	T < 12°C, t = 4°C for 24 h, t = 5°C for 2 h, t = 10°C for 24 h, t = 5°C for 2 h, t=10°C for 8 h, t = 7°C for 48 h, t = 20°C for 2 h and t = 7°C for the rest of the experiment		LAB and yeasts grew very fast on peppers. TPC and yeasts exceeded their critical limits at sampling 4 (t = 10°C for 8 h). *L. monocytogenes* and *A. caviae* declined after 6 days of storage		The mixed bell peppers were rejected due to their sensorial quality (texture loss) at the moment of sampling 4 (t = 10°C for 8 h)		Jacxsens et al. (2002b)

stored at 7°C ± 2°C. Treatment with ClO_2 and NaOCl reduced significantly the levels of *E. coli O157:H7* by 2.6 and 1.1 log_{10} CFU/g, respectively. After 7 days of storage significant differences (about 3–4 log) of *E. coli O157:H7* populations were reported between samples packed in air and other packaging methods following treatment with chemical sanitizers (Lee and Baek, 2008).

Gil et al. (1999) investigated the effect of MA (12% O_2 + 7% CO_2 on day 3 and 6% O_2 + 14% CO_2 on day 7) on the antioxidant of fresh-cut spinach stored at 10°C for 7 days. After storage, the total flavonoid content remained rather stable in both air and MAP, and no degradation was reported. A pronounced increase in dehydroascorbic acid (DHAA) was recorded under MAP and had similar vitamin C content to the initial samples. On the other hand, a drop in free radical scavenging activity was reported particularly for samples under MAP.

Fresh-cut spinach was treated with citric acid and ascorbic acid solutions (from 0% up to 1%) and packaged in OPP or LDPE bags (PMAP) and stored at 4°C (Piagentini et al., 2003). The pH of fresh-cut spinach packaged in OPP and LDPE bags augmented over the refrigerated storage for any chemical treatment applied. The pH of spinach packaged in OPP bags treated with a solution of citric acid = ascorbic acid = 0.50% increased from 4.89 (t = 0) to 6.65 and 7.17 after 7 and 14 days, respectively.

Allende et al. (2004b) investigated the effect of super atmospheric O_2 (80 and 100 kPa O_2 gas flush with two PE films) and PMAP (PE_1 with high permeability and PE_2 barrier films) on quality of minimally processed baby spinach stored at 5°C. The tissue electrolyte leakage (0.29 for PMAP-PE_1, 0.39 for PMAP-PE_2, and 0.22–0.26 for superatmospheric treatments) revealed that high O_2 greatly affected the lower tissue electrolyte leakage and higher product quality scores.

9.26 ASPARAGUS

Siomos et al. (2000) studied the preservation of asparagus under MAP. White asparagus spears were over-wrapped with a 16 mm stretch film (in 5 L glass jars with ethylene free air passing through) and kept at 2.5°C, 5°C, 10°C, 15°C, 20°C, and 25°C under darkness or light for 6 days. Acetaldehyde ranged from 10 to 20 μL/L with the highest concentration being monitored at the highest temperature. The drop in SSC of the packaged spears was significantly lower than the fall of SSC of spears stored in air, at temperatures higher than 10°C.

An et al. (2006) investigated the changes in lignifying, antioxidant enzyme activities and cell wall compositions of fresh-cut green asparagus in 1 mg/L aqueous ozone pre-treated, and subsequent MAP (LDPE film used) over storage at 3°C for 25 days. Cellulose and hemicellulose contents in the control samples enhanced by approximately 359% and 283%, respectively, whereas the enhancement in the ozone treated MAP samples was 72% and 54%, respectively.

Treated asparagus (dipped in 20 ppm 6-benzylaminopurine [6-BA] for 10 min) was stored under active (LDPE 25 μm film with 10% O_2/5% CO_2 initial atmosphere) and PMAP (LDPE 15 μm film) at 2°C for 25 days. The drop in ascorbic acid of spears without 6-BA treatments (14 and 17 mg/100 g for PMAP and AMAP, respectively) was higher than the reduction of treated spears (24 and 26 mg/100 g, respectively).

Overall appearance scored higher for treated packages, particularly for those under AMAP (An et al., 2006).

Zhang et al. (2008) studied the impact of treatment with compressed argon and xenon (Ar/Xe 2/9 v:v) on asparagus spears stored at 4°C compared to MAP (AMAP with 5% O_2/5% CO_2 initial atmosphere). Ar–Xe and MAP treatments decreased the tract opening rate of asparagus spears after the sixth day of storage and preserved the chlorophyll better than cold storage (24 and 25 mg/100 g for Ar–Xe and MAP, respectively).

The microbiological quality and sensorial characteristics of white asparagus washed with chlorine or water and packaged under MAP (perforated PVC and P-Plus 160 films used) at 4°C for 15 days were reported by Simon et al. (2004). Asparagus in P-Plus displayed a slight decrease in texture. Application of PVC packages led to a mean acidity of 0.088% of citric acid, while in P-Plus, the respective value was 0.077%. Water reduced the initial coliform counts by 0.5 log cycles, while chlorine decreased them by 1 log.

Green asparagus spears were stored under refrigeration at 2°C (first treatment), MAP (OPP P-Plus film) at 2°C (second), and MAP at 10°C after 5 days at 2°C (third) until they were not regarded suitable for consumption. Villanueva et al. (2005) reported a decrease in vitamin C in all treatments (40.3%, 58.3%, and 49.1% vitamin C retention for the first, second, and third treatment, respectively). The highest levels of aerobic microorganisms were detected in samples stored under MAP at 10°C (6 log CFU/g).

The most important data concerning the storage of asparagus under MAP are summarized in Table 9.9.

9.27 MUSHROOM

Tao et al. (2006) investigated the impact of vacuum cooling prior to packaging under MAP (5% ± 1% O_2/3% ± 1% CO_2 initial atmosphere with LDPE film) or hypobaric conditions (20–30 kPa total pressure) at 4°C ± 1°C. The degree of browning under both conditions enhanced slowly reaching approximately the same value by the end of storage (1.4–1.45). Soluble solids varied considerably during storage reaching 6.5% and 6.75% after 15 days for MAP and HC, respectively.

Oyster mushrooms were washed with water, 0.5% citric acid, 0.5% calcium chloride, and 0.5% citric acid with 0.5% calcium chloride and stored under PMAP (PP, 0.075, 0.05, and 0.0375 mm LDPE and LLDPE films) at 8°C for 8 days (Jayathunge and Illeperuma, 2005). Employing the same LLDPE film and washing treatment and adding 1, 3, and 5 g of magnesium oxide in the packages (acting as CO_2 scavengers) treatments with 3 and 5 g exhibited the highest weight loss of 2.2%, whereas samples with 1 g displayed the minimum (1.9%) after 12 days.

The effect of various treatments on weight loss was evaluated in Figure 9.26. Mushrooms stored under hypobaric conditions underwent serious weight loss that reduced shelf life to only 2 days. Vacuum packages having closer contact of the film with the produce might have had a restricted effective area for water vapor transmission (WVT).

TABLE 9.9
Most Important Data Concerning the Storage of Asparagus under MAP

Species and Food Type	Initial Gas Mix	Packaging Material	Treatment before Packaging	Storage Temperature (°C) and Storage Period (Days)	Color	Microflora	Texture–Weight Loss	Sensory Analysis	Shelf Life (Days)-Life Extension	References
Asparagus spears (*Asparagus officinalis L.*)	For all temperatures, the gas composition fluctuated between 1% O_2/4.6%–7% CO_2	Stretch film with OTR: 583 and CDTR: 1750 mL/m² h atm		2.5°C, 5°C, 10°C, 15°C, 20°C, and 25°C under continuous light or darkness for 6 days	Spears stored at 20°C and 25°C developed browning, mainly at their base		Weight loss of the spears stored in MAP increased exponentially with temperature	The amount of ethanol accumulated in spear tissues was linearly related to the storage temperature. Decrease of soluble solids after 6 days of storage at temperatures ≥15°C	Spear quality was best after 6 days of storage in packages at 2.5°C and 5°C	Siomos et al. (2000)

(continued)

TABLE 9.9 (continued)
Most Important Data Concerning the Storage of Asparagus under MAP

Species and Food Type	Initial Gas Mix	Packaging Material	Treatment before Packaging	Storage Temperature (°C) and Storage Period (Days)	Color	Microflora	Texture–Weight Loss	Sensory Analysis	Shelf Life (Days)-Life Extension	References
White asparagus (*A. officinalis L. Cipres var.*)	PMAP: 15.5% O_2/5.5% CO_2	OPP P-Plus 160	Washing with 100 ppm of sodium hypochlorite	4°C for 15 days		Washing with water and with chlorine reduced initial mesophile, *Enterobacteriaceae*, coliform and psychrotroph counts *Enterobacteriaceae* counts in P-Plus were lower than those in control	Weight losses were lower in asparagus packaged in P-Plus film. Asparagus in P-Plus film showed slight decrease in texture	Ascorbic acid and sugars dropped throughout the storage period		Simon et al. (2004)
Asparagus (*A. officinalis L.*)	Active MA: (1)5% O_2/5% CO_2	(1) PVC (2) Compressed air treated and Ar–Xe treated products	(1) Compressed air treatment at room temperature at 1.1 MPa for 24 h	4°C for 18 days	MAP and Ar–Xe treatments preserved chlorophyll better than control		The weight loss of the MAP was the lowest among the four treatments	MAP and Ar–Xe treatment reduced the bract opening rate of asparagus spears. Ar–Xe	Treating asparagus spears with a 1.1 MPa mixture of Ar and Xe 2:9 (v:v)	Zhang et al. (2008)

		packed in unsealed PVC	(2) Treatment with mixed Ar and Xe at 2:9 (v:v) at room temperature at 1.1 MPa for 24 h			treatment did not reduce the loss of vitamin C	extended their shelf life	
Asparagus (*A. officinalis L.*)	Passive MA on LDPE 1 Active MA on LDPE 2%–10% O_2/5% CO_2. Treatment with 6-BA increased O_2 and decreased CO_2 compared to untreated samples	(1) LDPE 15µm (2) LDPE 25µm with AMAP	Dipping in 20 ppm 6-BA for 10 min	2°C for 24 days	Asparagus with initial 10 kPa O_2 and 5 kPa CO_2 plus 6-BA treatments had the greatest lightness (L^*), greenness (a^*), hue angle, and the least ΔE^*	6-BA on MAP had beneficial effect on ascorbic acid retention. Overall appearance scored higher for packages (LDPE) with 6-BA than without		An et al. (2006)

(*continued*)

TABLE 9.9 (continued)
Most Important Data Concerning the Storage of Asparagus under MAP

Species and Food Type	Initial Gas Mix	Packaging Material	Treatment before Packaging	Storage Temperature (°C) and Storage Period (Days)	Color	Microflora	Texture–Weight Loss	Sensory Analysis	Shelf Life (Days)-Life Extension	References
Asparagus (*A. officinalis L.*)	Passive MA	LDPE	The fresh-cut asparagus were dipped with aqueous ozone (1 mg/L)	3°C for 25 days			The increasing of the cell wall compositions (lignin, cellulose, hemicellulose) under the aqueous ozone treatment or (and) MAP dropped	The enzyme activities in fresh-cut asparagus including PAL, superoxide dismutase, ascorbate peroxidase, glutathione reductase were inhibited by aqueous ozone treatment and subsequent MAP		An et al. (2007)

Asparagus (*A. officinalis L.*)	Passive MA: (1)14% ± 1% O_2/8% ± 1% CO_2 at 2°C (2) 5% ± 1% O_2/14% ± 1% CO_2 at 10°C	OPP P-Plus with OTR: 14,000 mL/m^2 atm 24 h	(1) 2°C for 25 days (2) 2°C for 5 days and then at 10°C	Highest levels of aerobic microorganisms in samples stored under MAP at 10°C. Yeasts and molds grew in smaller numbers in the MAP samples	Weight losses detected in MAP storage were low (2% ± 0.1%). In control samples the increase in shear force was greater than MAP	The refrigerated samples showed longitudinal striation, dryness, bract opening, faster than the stalks stored under MAP. Greater retention of vitamin C in MAP samples in both temperatures compared to control	9–12 days for the refrigerated samples, 14 days for the stalks stored under MAP at 10°C and 26 days for this same technique stored at 2°C	Villanueva et al. (2005)
Asparagus (*A. officinalis L.*)	Passive MA: (1) 15% O_2/5% CO_2 at 2°C (2) 5% O_2/13% CO_2 at 10°C		(1) 2°C for 26–33 days (2) 2°C for 5 days and then at 10°C for 20 days			For MAP spears, the total oxygenated carotenoid loss was significantly smaller. Greater retention of chlorophylls and less increase of pheophytins were detected in the samples packed under MA conditions	MAP at 2°C was more effective in prolonging the shelf life up to 4 weeks	Tenorio et al. (2004)

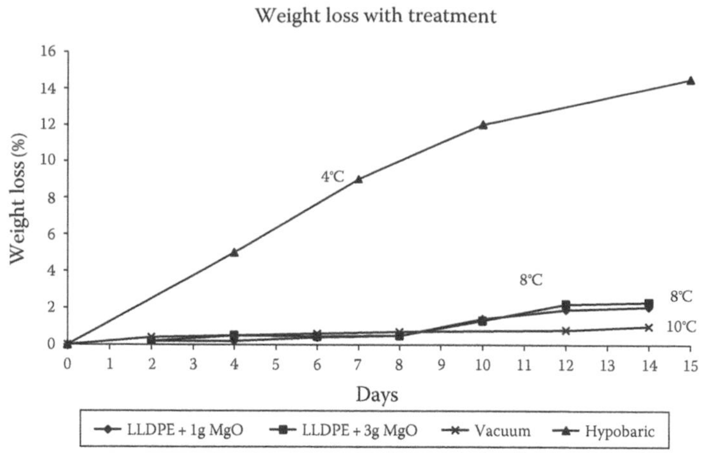

FIGURE 9.26 Weight loss in treated mushrooms stored under MAP (PMAP 1 and 2 were created with the use of linear LDPE and 1 or 3 g of MgO as an CO_2 absorber for *Pleurotus* mushrooms (From Jayathunge, L. and Illeperuma, C., *J. Food Sci.*, 70(9), 573, 2005.), for full vacuum condition, a polyolefin film was used to preserve fresh enoki mushrooms (From Kang, J.S. et al., *J. Sci. Food Agric.*, 81, 109, 2000.), hypobaric conditions were created in a room with 20–30 kPa total pressure after vacuum cooling. (From Tao, F. et al., *J. Food Eng.*, 77, 545, 2006.))

Masson et al. (2002) inoculated *Pseudomonas fluorescens* and *Candida sake* into homogenized mushrooms and stored under various gaseous atmospheres (CO_2/O_2 [25%/1%]-MAP 1 and CO_2/O_2 [50%/1%]-MAP 2) at 5°C and 10°C for 18 days. The obtained results revealed that different storage temperatures had no significant effect on the growth rate of microorganisms. Furthermore, the presence of CO_2 decreased their maximal growth rate compared to air.

Sorbitol and sodium chloride (5, 10, or 15 g) were used by Roy et al. (1996) to modify the in-package relative humidity (IPRH) of mushrooms (water irrigated and $CaCl_2$) stored in MAP (PE film) at 12°C. More rapid moisture absorption by NaCl led to almost double the weight loss as compared to packages with the same amounts of sorbitol. IPRH of 87%–90%, reached within 9 days storage, in packages with 10 and 15 g sorbitol, was considered optimum and corresponded to a surface moisture content of 90.5%–91%.

Barron et al. (2002) used a hydrophilic (wheat gluten-based material and synthetic polymer and a polyether polyamide) and a microporous film to create the optimal MA conditions for mushrooms stored at 10°C and 20°C under high RH (>92%). At 10°C, both hydrophilic films retarded the opening of the cap, contrary to microporous film. Changes in quality of the mushrooms stored at 20°C were too rapid to allow for a reliable assessment.

The effect of combined vacuum cooled (VC) and AMAP (5% ± 1% O_2/3% ± 1% CO_2 initial atmosphere with LDPE film) at 4°C ± 1°C was investigated by Tao et al. (2007). The superoxide anion generation of the mushrooms stored under MAP with VC was the lowest (>3.25/min/mg protein). Over the 2 weeks of storage, the

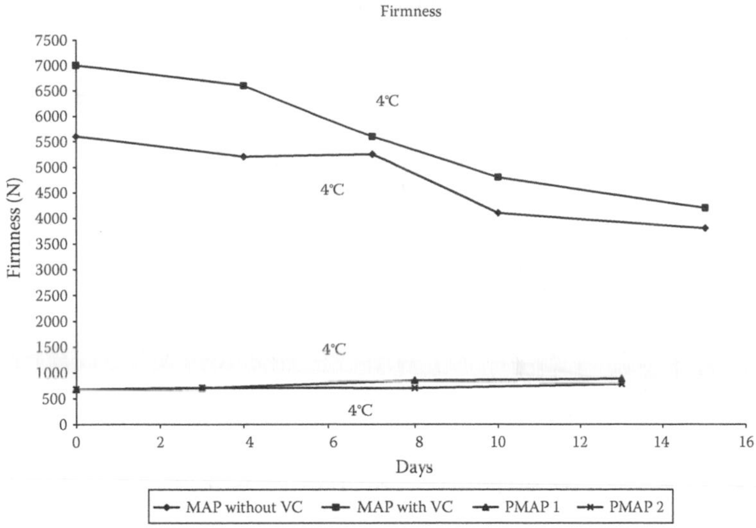

FIGURE 9.27 The effect of MAP and various treatments on firmness of mushrooms vs. storage time (MAP with initial atmosphere of 5% ± 1% O_2/3% ± 1% CO_2 and LDPE film used after vacuum cooling treatment at 5°C (From Tao, K. et al., *Postharvest Biol. Technol.*, 46, 212, 2007), PMAP 1 and 2 with microperforated oriented polypropylene used as packaging film [with 45,000 and 2,400 mL/m^2 day atm O_2 permeabilities, respectively). (From Simon, A. et al., *Int. J. Food Sci. Technol.*, 40, 94, 2005.))

mushroom stored under cooling mushroom or MAP with or without VC treatment dropped in firmness by an average of 60.44%, 58.18%, 43.19%, and 40.52%, respectively.

In Figure 9.27, the mushrooms stored under MAP with VC maintained higher firmness over the storage possibly because of the effect on enzymatic demethylation of pectins. A rise in shear force was shown in PMAP samples with more obvious results in PMAP 1 and can be attributed to higher dehydration levels.

Simon et al. (2005) applied MA conditions passively (PVC, OPP_1, and OPP_2 films with high and low O_2 permeability, respectively) for the storage of sliced mushrooms at 4°C ± 1°C for 13 days. Weight losses varied within a range of 0.5%–2% for PP_1 and PP_2 films. After 7 days of storage, the mesophile counts were 1.3–1.7 log units lower in PP_1 and PP_2 than in PVC. However, anaerobic spores (2 log CFU/g) were detected in PP_2 samples. MA containing 15% CO_2 and <0.1% O_2 restricted considerably mushroom development and toughening and reduced microbial growth.

$CaCl_2O_2$ (0.4 and 0.8 g/L) treatment and passive (LDPE with two or four perforations or PVC) or active atmosphere modification (10 mg/100 mL O_2/10 mg/100 mL CO_2) was tested for the preservation of mushrooms stored at 5°C ± 1°C for 10 days. Application of $CaCl_2O_2$ was not beneficial to color. On the contrary, there was a disadvantage with the use of higher concentration (mushrooms browner with 0.8 mg/L). The MAP generated by the perforated (4) LDPE improved the color of the under treatment sample (Kuyper et al., 1993).

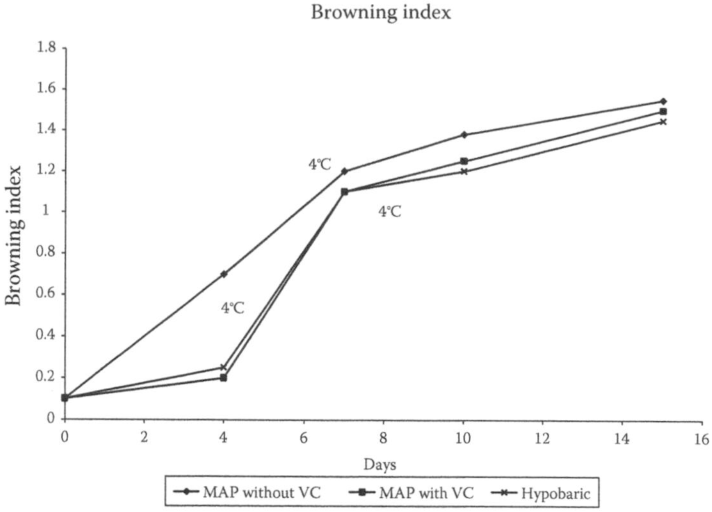

FIGURE 9.28 Changes in browning index in mushrooms stored under various conditions (MAP with initial atmosphere of 5% ± 1% O_2/3% ± 1% CO_2 and LDPE film used after vacuum cooling treatment at 5°C (From Tao, K. et al., *Postharvest Biol. Technol.*, 46, 212, 2007), hypobaric conditions were created in a room with 20–30 kPa total pressure after vacuum cooling. (From Tao, F. et al., *J. Food Eng.*, 77, 545, 2006.))

Ares et al. (2006) used a macroperforated (control) PP, an LDPE, and a PP film for MA packaging of shiitake mushrooms at 5°C for 16 days. Mushrooms stored in PP and PE developed off-odor. The latter could be due to the rapid decrease of O_2 concentration down to 1.3% within the first days of storage. Moreover, the off-odor of mushrooms stored in PP was higher than those stored in PE. The gills were significantly browner and uniform in mushrooms stored in PE and PP packages than control.

Both hypobaric conditions and MAP after VC process had a great impact on the extent of browning in mushrooms. MAP manages to retard enzymatic browning effectively, especially in the first days 6 days of storage and remains advantageous over the other treatments throughout the storage period (Figure 9.28).

Various storage temperatures (0°C, 4°C, and 7°C), MA conditions (PVC, two microperforated polypropylenes [MPP_1 and MPP_2] and an LDPE film for storage at 4°C), and different moisture absorbers (10, 15, and 20 g of sorbitol and 3, 5, 7, and 15 g of silica gel in MPP_2 film) for extending the shelf life of mushrooms were used by Villaescusa and Gil (2003). MPP_2 without silica gel was shown to be the optimal treatment, generating optimal conditions (15 kPa O_2/4 kPa CO_2), maintaining visual quality (7.0), and having 1.1 mycelium growth and 1.1% weight loss.

Antmann et al. (2008) stored shiitake mushrooms under passive (two macroperforated PE films; PEA and PEB) and active (15% O_2 and 25% O_2 in an LDPE film) atmosphere for 18 days at 5°C. Throughout storage time, mushrooms stored in films A were the firmest (1.4 N). Considering the development of the evaluated sensory attributes, mushrooms in film A displayed the lowest deterioration rate, followed by those in film B, and finally those packaged in AMAP.

Kang et al. (2000) investigated the shelf life of enoki mushrooms packaged under various conditions (full and half vacuum with RD-106 PO for packaging film, half vacuum [HV] with CPP and POs RD-106 and PD-941 for films and RD-106 packages stored at 5°C, 10°C, and 15°C) and stored at 10°C for 14 days. Mushrooms in full-vacuum (FV) packages underwent color deterioration thereby resulting in a very low score on overall sensory quality. It was shown that half-vacuumed RD-106 package is a suitable film for packaging mushrooms.

A combination of passive (LDPE and PP films) and active (5% O_2/2.5% CO_2) atmosphere conditions were applied for storage of shiitake mushrooms at 5°C for 20 days. Higher microbial counts were determined in PE packages (10^5 and $>10^4$ for mesophiles, $2{\cdot}10^3$ and $1.75{\cdot}10^3$ for LAB, $2{\cdot}10^3$ and $4{\cdot}10^3$ for yeasts and molds for PM and AM, respectively). Samples stored in PP packages ($2{\cdot}10^4$ and $>10^4$ for mesophiles, $>10^3$ and $>10^3$ for LAB, $2{\cdot}10^3$ and $3{\cdot}10^3$ CFU/g for yeasts and molds for PM and AM, respectively) displayed lower counts and a slight tendency to drop (Parentelli et al., 2007).

It is evident from Figure 9.29 that microbial growth was slower for samples stored under AMAP and especially in PE film than PMAP. High temperature seems to favor the microbial growth and limits considerably the shelf life.

Cliffe-Byrnes and O'Beirne (2008) applied antimicrobial (ClO_2 and H_2O_2 were tested in various concentrations and washing duration) in conjunction with anti-browning (sodium d-isoascorbate and H_2O_2 at various concentrations) treatments (see Table 9.10) on sliced mushrooms stored under PMAP (PA-190 OPP) at 4°C and 8°C for 7 days. Mushrooms ClO_2 treated up to 50 mg/L for 60 s exhibited lesser browning index values. "The most effective treatment proved to be 3% H_2O_2 for up to 60 s followed by a spray application of 4% sodium d-isoascorbate monohydrate or 1% H_2O_2 (reduced BI and microbial load)."

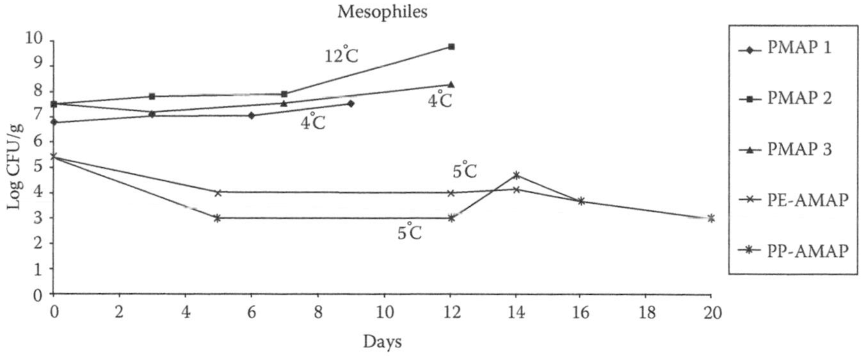

FIGURE 9.29 The effect of MAP on mesophiles for different varieties of mushrooms vs. storage time (PMAP 1 was created in polyolefin pouches with OTR: 13800 mL/m² day atm (From Roy, S. et al., *J. Food Sci.*, 61(2), 391, 1996), PMAP 2 and 3 with microperforated oriented polypropylene used as packaging film [with 45,000 and 2,400 mL/m² day atm O_2 permeabilities, respectively] (From Simon, A. et al., *Int. J. Food Sci. Technol.*, 40, 94, 2005), for PE-AMAP and PP-AMAP an initial gas mixture of 5% O_2 and 2.5% CO_2 was used with a polyethylene and polypropylene used as a packaging film, respectively. (From Parentelli, C. et al., *J. Sci. Food Agric.*, 87, 164, 2007.))

TABLE 9.10
Synopsis of Articles Concerning the Application of MAP on Mushrooms

Species and Food Type	Initial Gas Mix	Packaging Material	Treatment before Packaging	Storage Temperature (°C) and Storage Period (Days)	Color	Microflora	Texture–Weight Loss	Sensory Analysis	Shelf-Life (Days) Extension	References
White mushrooms (*Agaricus bisporus*)	(1) Hypovaric room: mushrooms were stored hypobarically in air at 20–30 kPa total pressure (2) 5% ± 1% O_2/3% ± 1% CO_2 for MAP	LDPE.	Mushrooms were vacuum cooled to 5°C	4°C ± 1°C for 14 days	The degree of browning in cooling room increased rapidly mushrooms stored under hypobaric room and MAP increased slowly		The weight loss of the mushrooms under MAP was the lowest among the three storage conditions	The membrane permeability of mushrooms stored in cooling room was the greatest (16.67%), while the membrane permeability of MAP was only 15.0% at the end of storage		Tao et al. (2006)
Oyster mushrooms (*Pleurotus* spp.)		(1) PP (2) LDPE (3) LLDPE	(1) Water (2) 0.5% citric acid (3) 0.5% calcium chloride (4) 0.5% citric acid with 0.5% calcium chloride	8°C for 8 days	Mushrooms packaged in LLDPE and washed with 0.5% calcium chloride and 0.5% citric acid showed the least off-color development			Mushrooms packaged in LLDPE washed with 0.5% calcium chloride and 0.5% calcium chloride with 0.5% citric acid showed the least off-odor development	Washing of mushroom with citric acid and $CaCl_2$ and packaging in LLDPE with 3 g of MgO as a CO_2 scavenger was successful in extending postharvest life from 6 days in to 12 days	Jayathunge and Illeperuma (2005)

Oyster mushrooms (*Pleurotus* spp.)		LLDPE with CO_2 scavengers. Magnesium oxide 1, 3, or 5 g was used as carbon dioxide scavengers	0.5% citric acid with 0.5% calcium chloride	8°C for 14 days	Mushrooms packaged with 3 and 5 g of magnesium oxide showed the highest lightness values over storage		Mushrooms treated with 1 g of magnesium oxide showed the lowest weight loss	Concentration of ethanol increased during storage and was significantly higher in the control samples than those with magnesium oxide		
Homogenized mushrooms	(1) 1% O_2/25% CO_2 (2) 1% O_2/50% CO_2	Plastic bag (TS-II) with OTR: 8 and CDTR: 15 mL/m² day atm	Mushrooms were homogenized with stomacher. Then, strains of *P. fluorescens* and *C. sake* were inoculated into them	5°C and 10°C for 18 days		Different storage temperatures had no effect on growth rate of microorganisms. Presence of CO_2 lowered their maximal growth rate compared to air		The final pH sample having 50% CO_2 was lower than other samples		Masson et al. (2002)
Mushrooms (*A. bisporus*)		PO pouches with OTR: 13,800 mL/m² day atm. Different amounts (5, 10, 15 g) of either anhydrous D-sorbitol or NaCl crystals were used as humidity absorbers	Mushrooms irrigated with 0.3% $CaCl_2$	12°C for 9 days	Irrigation with $CaCl_2$ in the presence of 15 g of sorbitol in the package had a beneficial effect on the mushrooms color		More rapid moisture absorption by NaCl resulted in almost double the weight loss compared to mushrooms packed with the same amounts of sorbitol	In-package RH was lower throughout storage in packages containing 5 and 10 g NaCl and after 3 days storage in packages containing 15 g NaCl than normal grown with the same amounts of sorbitol	RH of 87%–90%, reached within 9 days storage, in packages with 10 and 15 g sorbitol was considered optimum for mushrooms	Roy et al. (1996)

(continued)

TABLE 9.10 (continued)
Synopsis of Articles Concerning the Application of MAP on Mushrooms

Species and Food Type	Initial Gas Mix	Packaging Material	Treatment before Packaging	Storage Temperature (°C) and Storage Period (Days)	Color	Microflora	Texture–Weight Loss	Sensory Analysis	Shelf-Life (Days) Extension	References
Mushrooms (*A. bisporus L.*)	PMAP: (1) 2%–0.5% O_2/2.5%–10% CO_2 (2) 1% O_2/13%–16% CO_2 (3) 11%–10% O_2/11%–12% CO_2. For the two storage temperatures	(1) Biopolymer film (wheat-gluten-based material) with OTR: 4,160 mL/m^2 day atm and CDTR: 65,380 mL/m^2 day atm (2) Hydrophilic polyether PA film with OTR: 6,500 mL/m^2 day atm and CDTR: 56,000 mL/m^2 day atm (3) Polymeric microporous film		10°C and 20°C for 140 h	Color of mushroom measured with the L* CIE parameters was not significantly different regardless of the film used at 10°C			Both hydrophilic films delayed the opening of cap in contrast to microporous film		Barron et al. (2002)

Mushrooms (*A. bisporus*.)	5% ± 1% O_2/3% ± 1% CO_2	LDPE	Mushrooms were VC to 5°C before storage	4°C ± 1°C—15 days	The degree of browning under MA with VC treatment increased slowly		The mushrooms stored under MAP with VC maintained higher firmness over the storage	The superoxide anion generation of the mushrooms stored under MAP with VC was the lowest. Mushrooms under MAP with VC maintained higher superoxide dismutase and peroxidase activity over the storage	Both VC treatment and MAP storages extended the shelf life of mushrooms	Tao et al. (2007)
Mushrooms (*A. bisporus* L.)	PMAP: (1)13% O_2/2.5% CO_2 (2) 20% O_2/2.5% CO_2 (3) >0.1% O_2/15% CO_2	(1) PVC with OTR: 25,000 (2) OPP with OTR 45,000 mL/m^2 day atm (3) OPP with OTR 2,400 mL/m^2 day atm		4°C ± 1°C for 13 days	Film type, and MA generated, had no effect on the L* parameter	In third film after 12 days, mesophiles, psychrotrophs, and *Pseudomonas* were 2.0, 1.6, and 1.4 log units lower than control	The third film generated conditions inside the package, not allowing rise in toughening	No slice deformation and no blotches were observed in mushrooms packed in the third film		Simon et al. (2005)
Mushrooms	(1) PMAP (2) 10 mg/100 mL O_2/10 mg/100 mL CO_2	(1) LDPE with 4, 2, and no perforations (2) PVC	Treatment with 0.4–0.8 g/L $CaCl_2O_2$	5°C ± 1°C for 10 days	L-value of the combination (LDPE, four perforations, MAP, no $CaCl_2O_2$) was higher than other treatments	*Pseudomonas* and TPC were lower with $CaCl_2O_2$ treatment. Anaerobic conditions (PVC) led to lower microbial populations				Kuyper et al. (1993)

(*continued*)

TABLE 9.10 (continued)
Synopsis of Articles Concerning the Application of MAP on Mushrooms

Species and Food Type	Initial Gas Mix	Packaging Material	Treatment before Packaging	Storage Temperature (°C) and Storage Period (Days)	Color	Microflora	Texture–Weight Loss	Sensory Analysis	Shelf-Life (Days) Extension	References
Shiitake mushrooms (*Lentinula edodes*)	PMAP: (1) 1.3% O_2/9% CO_2 (2) 1.3% O_2/13.6% CO_2	(1) LDPE (2) PP (3) Perforated PP		5°C—16 days			Mushrooms stored in PP or PE had a weight loss of 5.6% after 16 days. Mushrooms stored in macroperforated PP packages were firmer than the others	Mushrooms in PP and PE developed off-odor. Off-odor of mushrooms in PP was higher than those stored in PE. The gills were browner and less uniform in mushrooms stored in PE and PP	Mushrooms stored in macroperforated bags had a higher shelf life (12 days) than those stored in sealed PE or PP bags (5 days)	Ares et al. (2006)
Mushrooms (*Pleurotus ostreatus*)	PMAP: (1) 1.5% O_2/8% CO_2 (2) 2% O_2/12% CO_2 (3) 12% O_2/7% CO_2 15% O_2/4% CO_2 was the MA of the packages that had moisture absorbers	(1) PVC (2) LDPE (3) Micro-perofarated PP (4) Macroper-forated PP. (MAP took place at 4°C for 7 days) (5) Macroper-forated PP was used with the moisture absorbers	Different amounts of silica gel and sorbitol were used as humidity absorbers in a Macro-PP package	0°C, 4°C, and 7°C for 7 days. 4°C was the temperature of packages with MA	*Pleurotus* stored at 0°C maintained the initial intense color		Texture losses occurred after 11 days storage from an initial firm texture to soft texture at 4°C and 7°C. Weight loss increased with temperature. MAP with silica gel increased weight loss	Mushrooms at 0°C had the best visual quality. Mushrooms in PVC films led to better visual quality. Aroma decreased with storage time under all MA conditions. Severe off-odors detected for samples stored in PVC and LDPE packages	Shelf life of *Pleurotus* mushrooms is feasible with low temperature and proper humidity. MAP (15% O_2/5% CO_2) was beneficial to quality for 7 days at 4°C	Villaescusa and Gil (2003)

Shiitake mushrooms (*L. edodes*)	Active MA: (1) 15% O_2 (2) 25% O_2 PMAP: (1) 20% O_2/0.5% CO_2 (2) 18% O_2/2% CO_2	(1) Macroperforated PE 1 (2) Macroperforated PE 2 (3) LDPE for active MA		5°C—18 days			Mushrooms packaged in film 2 and passive MA revealed a weight loss less than 5%. No differences in weight loss of mushrooms packaged in the two active MAs. Mushrooms stored in film 1 were the firmest. Mushrooms packed in film 2 were firmer than those packaged in active MA	Off-odors were greater in samples stored in packages filled with 25% O_2. The active MA led to a lesser reduction of mushroom deterioration rate than PMAP in film 2		Antmann et al. (2008)
Enoki mushrooms (*Flammulina velutipes*)	(2) >1% O_2/37% CO_2 (3, 4) 1.7%–3.4% O_2/2.6%–5.8% CO_2	(1) PVC (2) Cast PP (3) PO 1 (RD-106) (4) PO 2 (PD-941)		10°C—14 days	CPP and RD-106 packages preserved sensorial color quality for mushrooms stored at 10°C for 8 days				RD-106 of medium gas permeability appears to be a promising film for packaging enoki mushrooms	Kang et al. (2000)

(continued)

TABLE 9.10 (continued)
Synopsis of Articles Concerning the Application of MAP on Mushrooms

Species and Food Type	Initial Gas Mix	Packaging Material	Treatment before Packaging	Storage Temperature (°C) and Storage Period (Days)	Color	Microflora	Texture–Weight Loss	Sensory Analysis	Shelf-Life (Days) Extension	References
Enoki mushrooms (*F. velutipes*)	Passive MA for half vacuum: 1. 1%–2.4% O_2/3.5%–5.6% CO_2	PO 1	The packages were sealed under FV and HV	10°C—14 days	Mushrooms in FV MA packages suffered from color deterioration		Packages sealed with normal air resulted in higher weight loss compared to HV or FV ones	The FV packages gave a slightly worse sensory quality than the HV ones	The HV package with a reduced free volume was the best option for quality retention	
Enoki mushrooms (*F. velutipes*)	The packages had similar O_2 and CO_2 (<2%) concentrations	PO 1		5°C, 10°C, and 15°C for 14 days	Higher temperature caused darkening		Higher temperature accelerated the weight loss	Stripe elongation more profound in high temperature	Higher temperature shortened shelf life	
Shiitake mushrooms (*L. edodes*)	(A) PMAP: (1) 2% O_2/9% CO_2 (2) 2% O_2/13.6% CO_2 These two concentrations are under equilibrium (B) AMAP: 5% O_2/2.5% CO_2	(1) LDPE (2) PP		5°C for 20 days	Mushrooms in control bags had higher color values than all the packages in MAs	Aerobic mesophilic bacteria counts remained at low counts in all the storage conditions. Higher counts of LAB were found in PE packages (AM and PM), whereas samples stored in PP (AM and PM) showed lower counts	Mushrooms stored in MAP displayed a weight loss lower than 6.5% by the end storage. Mushrooms in control bags showed higher firmness and module than those stored in MA	Mushrooms stored in macroperforated bags did not develop off-odors	Both active and passive MA resulted higher mushroom deterioration rate than that of mushrooms maintained at atmospheric air	Parentelli et al. (2007)

Sliced mushrooms (*A. bisporus* L.)	No significant effect was found in the in-pack concentrations At 4°C 15% O_2/5% CO_2 At 8°C 7%–9% O_2/8%–11% CO_2	Microperforated PA-190	(1) Washing with ClO_2 for 30, 60, or 120 s at concentrations: 10, 25, 50, 75, and 100 mg/L (2) H_2O_2 (3%) for 30, 60, 120 s (3) 30 or 60 s with various concentrations (10–100 mg/L) of ClO_2 followed by H_2O_2 (3%) for 30 or 60 s (4) H_2O_2 (3%) for 30 or 60 s and treatment of sliced mushrooms with sodium d-isoascorbate (2% or 4%) for 30 or 60 s (5) Spray application of sodium disoascorbate (0.5%, 2%, or 4%) or H_2O_2 (1% or 3%)	4°C and 8°C—7 days	L^* values for mushrooms treated with up to 50 mg/L ClO_2 were higher than water washed control slices. Mushrooms washed for 30 or 60 s with H_2O_2 had higher L^* values than other treatments with H_2O_2. Mushrooms washed with 4% sodium isoascorbate had higher L^* values and lower browning indices	ClO_2 use reduced microbial counts and *Pseudomonas* bacteria. The reduction increased with concentration and washing duration. H_2O_2 achieved greater than 1-log reduction for *pseudomonads*, and both mesophilic and psychrophilic populations		No differences in appearance between the treatments with ClO_2. Mushrooms treated with H_2O_2 for 30 or 60 s scored similarly to controls for overall appearance throughout storage at 4°C	The most effective treatment in extending shelf life was 3% H_2O_2 for up to 60 s to whole mushrooms with a spray application of 4% sodium d-isoascorbate or 1% H_2O_2 to the slices followed by storage under MAs at 4°C	Cliffe-Byrnes and O'Beirne (2008)

(*continued*)

TABLE 9.10 (continued)
Synopsis of Articles Concerning the Application of MAP on Mushrooms

Species and Food Type	Initial Gas Mix	Packaging Material	Treatment before Packaging	Storage Temperature (°C) and Storage Period (Days)	Color	Microflora	Texture–Weight Loss	Sensory Analysis	Shelf-Life (Days) Extension	References
Whole and sliced mushrooms (*A. bisporus* L.)	PMAP: coating treatment stimulated respiration of the mushrooms and changed the gas composition except for packages with PD-961	(1) PVC (2) PO 1 (PD-941) (3) PO 2 (PD-961)	Sliced and whole mushrooms were spray-coated with solution of chitosan and $CaCl_2$ (2 g/100 mL deionized water)	12°C for 6 days	Coated mushrooms had lower L^* value than the control Chitosan gave higher ΔE values than other treatments for sliced mushrooms		In both whole and sliced mushrooms, PD-961 packages displayed the lowest weight losses	PD-961 and coating with chitosan maintained a low maturity index value for both whole and sliced mushrooms	PD-961 provided a successful MAP that extended shelf life up to 6 days	Kim et al. (2006)
Mushrooms (*A. bisporus*)	(1) Passive MAP (2) HOA (HOA) (70%, 80%, and 95% O_2)	BOPP with OTR: 914 mL O_2/m^2 24 h atm		4°C for 7 days	The mushroom slices under HOA were rejected after 6 days because of enzymatic discoloration, opposed to 3 days packed in P-MAP	Aerobic psychrotrophic counts higher in passive than in HOA at day 7 Microorganisms were responsible for the spoilage of the mushrooms slices		During the 7 tested days, the mushrooms slices stored under the HOA kept their fresh odor while on day 6 an unacceptable odor was detected for the P-MAP packaged mushroom slices	The overall achievable shelf life is de minimis doubled by using high O_2 atmospheres compared to the P-MAP packages	Jacxsens et al. (2001)

Mushrooms (*A. Bisporus* cv. U3 Sylvan 381)	PMAP: 5%O_2/9.5% CO_2	4 L plastic containers with OTR 5.58. 10^{-12} mol/s pa and CDTR: 13.55. 10^{-12} mol/s pa at 4°C	(1) 4°C (2) 4°C and 14°C for 2 days alternatively, and the sequence was repeated three times over the 12 day storage	Color of mushrooms better preserved under constant temperatures (higher L^* values and lower a^* values)		The increase in ethanol level was 1.9-fold higher in MA at a constant storage temperature, but it was very strong (24.4-fold) under MA with TF. Less infections recorded for packaged mushrooms at constant temperature	Tano et al. (2007)

Kim et al. (2006) investigated the shelf life of mushrooms, whole or sliced, spray-coated with solution of chitosan and $CaCl_2$ (2 g per 100 mL) and packaged with a PVC wrap or two POs (PD-941 and PD-961) at 12°C for 6 days. According to Kim et al. (2006), "coating did not significantly affect the weight loss in sliced mushrooms packed in PD-941 film. In both whole and sliced mushrooms, PD-961 packages displayed the lowest weight losses (>4%) and maintained a low maturity index (>2)."

Jacxsens et al. (2001) compared HOA (70%, 80%, and 95% O_2 in a barrier film) and PMAP (BOPP film used) for preserving sliced mushrooms at 4°C for 7 days. A difference in aerobic psychrotrophic count was detected ($6.46 \pm 0.06\ \log_{10}$ CFU/g on HOA packaged while 7.34 ± 0.87 log CFU/g on EMA). Apparently, G^- microorganisms were responsible for the spoilage of the mushrooms slices (no outgrowth of yeasts and LAB either).

The combined use of H_2O_2 with sodium isoascorbate forms an effective barrier that reduces both the initial microbial load and the growth rate of psychrotrophs. Although both ClO_2 and H_2O_2 treatments did not have the expected results, at day 7, they had the same counts with the untreated samples (Figure 9.30).

Mushrooms were stored in plastic containers (under MA conditions) and exposed to TF (4°C and 14°C for 2 days alternatively) during the 12 day storage period (Tano et al., 2007). MA storage at 4°C had the highest average reflectance value (lightness, L*) of 76.1, which dropped from an initial value of 84.3. The firmness differed (from 4.9 down to 2.4 N/mm for constant and TF) between the tested conditions. More than 25% of the mushroom cap area had symptoms of blotch diseases for mushrooms subjected to TF.

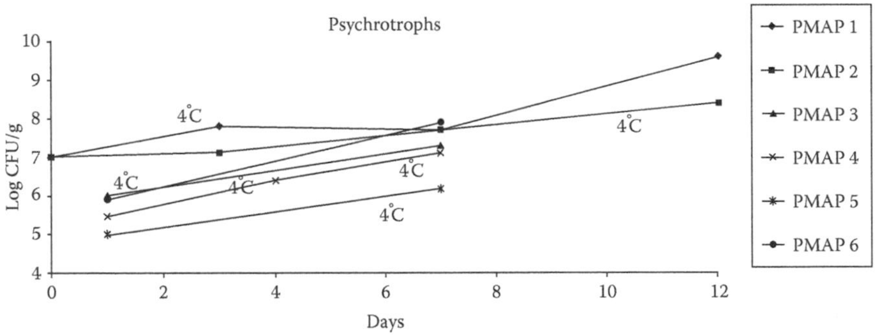

FIGURE 9.30 Changes in psychrotroph populations for mushroom samples pretreated and stored under MAP (PMAP 1 and 2 with microperforated oriented polypropylene used as packaging film [with 45,000 and 2,400 mL/m² day atm O_2 permeabilities, respectively) (From Simon, A. et al., *Int. J. Food Sci. Technol.*, 40, 94, 2005), washing with ClO_2 [50mg/mL] for 60 s and stored under PMAP 3 [PA-160 oriented polypropylene film used], washing with H_2O_2 3% for 60 s and then stored under PMAP 4 [PA-160 used], washing [60 s] into a H_2O_2 [3%] and sodium isoascorbate [30 s] [4%] solution and stored under PMAP 5 [PA-160 used] and PMAP 6 was created with the use of PA-160 film after washing with H_2O_2 [3%] for 60 s and spraying the products with sodium isoascorbate [4%]. (From Cliffe-Byrnes, V. and O'Beirne, D., *Postharvest Biol. Technol.*, 48, 283, 2008.))

A synopsis of articles concerning the application of MAP on mushrooms is given in Table 9.10.

9.28 CONCLUSIONS

Vegetables were among the first foods that MAP (Passive or Active) was applied to. The obtained results depended on two factors: MA gas composition and packaging material OTR. The shelf-life prolongation of vegetables was determined based on the microbiological and sensory analysis. Apart from nitrogen, the presence of carbon dioxide (bacteriostatic effect) and oxygen (<10%) proved to be beneficial to vegetable shelf-life extension. Moreover, the packaging materials with the best performance were mostly laminates made from BOPP, LDPE, and PET. In most cases, the combination of MA with additives (i.e., citric acid) or active compounds (i.e., calcium chloride as humidity absorber) proved to be beneficial to shelf-life prolongation.

ABBREVIATIONS

6-BA	6-benzylaminopourine
AM	automatic misting
AMAP	active modified atmosphere packaging
BOPP	bioriented polypropylene
CA	controlled atmosphere
CFU	colony forming units
CPP	cast polypropylene
CW	chlorine washing
DDT	dichlorodiphenyl trichloroethane
DHAA	dehydroascorbic acid
DW	dry weight
EA	ethylene absorber
EMA	equilibrium modified atmosphere
EW	electrolyzed water
FDA	Food and Drug Administration
FV	full vacuum
GAC	granular activated carbon
GAD	glutamate decarboxylase
HAV	*hepatitis A* virus
HC	hypobaric conditions
HDPE	high-density polyethylene
HOA	high oxygen atmosphere
HV	half vacuum
HWT	hot-water treatment
IFPA	international Fresh-cut Produce Association
ILP	intense light pulse

IPRH	in package relative humidity
KMS	potassium metasulfite
LAB	lactic acid bacteria.
LDPE	low-density polyethylene
LLDPE	linear low-density poly ethylene
MA	modified atmosphere
MAP/ CAS	modified atmosphere packaging/controlled atmosphere storage
MOPP	mono-oriented polypropylene
MRL	maxima residue limits
MVP	moderate vacuum packaging
OPP	oriented polypropylene
ORP	oxidation reduction potential
OTR	oxygen transmission rate
PA	polyamide
PAL	phenylalanine ammonia lyase
PDA	potato dextrose agar
PE	polyethylene
PLA	polylactic acid
PMAP	passive modified atmosphere packaging
PMP	polymethyl pentene
PO	polyolefin
POD	peroxidase
POS	point of sale
PP	polypropylene
PPM	parts per million
PPO	polyphenol oxidase enzyme
PVC	polyvinyl chloride
RAA	reduced ascorbic acid
RH	relative humidity
SS	soluble solids
SSC	soluble solids content
TA	titratable acidity
TAA	total ascorbic acid
TF	temperature fluctuation
TFA	total fatty acid
TPC	total plate counts
USDA	United States Department of Agriculture
UV	under vacuum
UV-C	ultraviolet
VC	vacuum cooling
VP	vacuum packaging

WHO World Health Organization
WVT water vapor transmission

REFERENCES

Abdul-Raouf, U.M., Beuchat, L.R., and Ammar, M.S. (1993). Survival and growth of *Escherichia coli O157:H7* on salad vegetables. *Applied and Environmental Microbiology*, 59: 1999–2006.

Able, A.J., Wong, L.S., Prasad, A., and O'Hare, T.J. (2003). The effects of methylcyclopropene on the of minimally processed leafy Asian vegetables. *Postharvest Biology and Technology*, 27: 157–161.

Able, A.J., Wong, L.S., Prasad, A., and O'Hare, T.J. (2005). The physiology of senescence in detached pak choy leaves (*Brassica rapa* var. *chinensis*) during storage at different temperatures. *Postharvest Biology and Technology*, 35: 271–278.

Adams, M.R., Hartley, A.D., and Cox, L.J. (1989). Factors affecting the efficacy of washing procedures used in the production of prepared salads. *Food Microbiology*, 6: 69–77.

Aguayo, E., Escalona, V., and Artes, F. (2004). Quality of fresh-cut tomato as affected by type of cut, packaging, temperature and storage time. *European Food Research and Technology*, 219: 492–499.

Ahn, H.J., Kim, J.H., Kim, J.K., Kim, D.H., Yook, H.S., and Byun, M.W. (2005). Combined effects of irradiation and modified atmosphere packaging on minimally processed Chinese cabbage (*Brassica rapa L.*). *Food Chemistry*, 89: 589–597.

Akbas, M.Y. and Olmez, H., (2007). Inactivation of *Escherichia coli* and *Listeria monocytogenes* on iceberg lettuce by dip wash treatments with organic acids. *Letters in Applied Microbiology*, 44: 619–624.

Alasalvar, C., Al-Farsi, M., Quantick, P.C., Shahidi, F., and Wiktorowicz, R. (2005). Effect of chill storage and modified atmosphere packaging (MAP) on antioxidant activity, anthocyanins, carotenoids, phenolics and sensory quality of ready-to-eat shredded orange and purple carrots. *Food Chemistry*, 89: 69–76.

Allende, A., Aguayo, E., and Artes, F. (2004a). Microbial and sensory quality of commercial fresh processed red lettuce throughout the production chain and shelf life. *International Journal of Food Microbiology*, 91: 109–117.

Allende, A. and Artes, F. (2003a). Combined ultraviolet-C and modified atmosphere packaging treatments for reducing microbial growth of fresh processed lettuce. *Lebensmittel-Wissenschaff und Technologie*, 36: 779–786.

Allende, A. and Artes, F. (2003b). UV-C radiation as a novel technique for keeping quality of fresh processed "*Lollo Rosso*" lettuce. *Food Research International*, 36: 739–746.

Allende, A., Jacxsens, L., Devlieghere, F., Debevere, J., and Artés, F. (2002). Effect of super atmospheric oxygen packaging on sensorial quality, spoilage, and *Listeria monocytogenes* and *Aeromonas caviae* growth in fresh processed mixed salads. *Journal of Food Protection*, 65: 1565–1573.

Allende, A., Luo, Y., McEvoy, J.L., Artés, F., and Wang, C.Y. (2004b). Microbial and quality changes in minimally processed baby spinach leaves stored under super atmospheric oxygen and modified atmosphere conditions. *Postharvest Biology and Technology*, 33: 51–59.

Alloway, B.J., Jackson, A.P., and Morgan, H. (1990). The accumulation of cadmium by vegetables grown on soils contaminated from a variety of sources. *The Science of the Total Environment*, 91: 223–236.

Amanatidou, A., Slump, R.A., Gorris, L.G.M., and Smid, E.J. (2000). High oxygen and high carbon dioxide modified atmospheres for shelf-life extension of minimally processed carrots. *Journal of Food Science*, 65(1): 61–66.

An, J., Zhang, M., and Lu, Q. (2007). Changes in some quality indexes in fresh-cut green asparagus pretreated with aqueous ozone and subsequent modified atmosphere packaging. *Journal of Food Engineering*, 78: 340–344.

An, J., Zhang, M., Lu, Q., and Zhang, Z. (2006). Effect of a prestorage treatment with 6-benzylaminopurine and modified atmosphere packaging storage on the respiration and quality of green asparagus spears. *Journal of Food Engineering*, 77: 951–957.

Antmann, G., Ares, G., Lema, P., and Lareo, C. (2008). Influence of modified atmosphere packaging on sensory quality of *shiitake* mushrooms. *Postharvest Biology and Technology*, 49: 164–170.

Ares, G., Lareo, C., and Lema, P. (2008). Sensory shelf life of butterhead lettuce leaves in active and passive modified atmosphere packages. *International Journal of Food Science and Technology*, 43: 1671–1677.

Ares, G., Parentelli, C., Gambaro, A., Lareo, C., and Lema, P. (2006). Sensory shelf life of *shiitake* mushrooms stored under passive modified atmosphere. *Postharvest Biology and Technology*, 41: 191–197.

Artes, F., Conesa, M.A., Hernandez, S., and Gil, M.I. (1999). Keeping quality of fresh-cut tomato. *Postharvest Biology and Technology*, 17: 153–162.

Artes, F., Escalona, V.H., and Artes-Hdez, F. (2002). Modified atmosphere packaging of fennel. *Journal of Food Science*, 67(4): 1550–1554.

Artes, F., Gomes, P.A., and Artes-Hernandez, F. (2007). Physical, physiological and microbial deterioration of minimally fresh processed fruits and vegetables. *Food Science and Technology International*, 13: 177.

Artes, F., Vallejo, F., and Martinez, J.A. (2001). Quality of broccoli as influenced by film wrapping during shipment. *European Food Research and Technology*, 213: 480–483.

Arvanitoyannis, I.S., Khah, E.M., Christakou, E.C., and Bletsos, F.A. (2005). Effect of grafting and modified atmosphere packaging on eggplant quality parameters during storage. *International Journal of Food Science and Technology*, 40: 311–322.

Ayala-Zavala, J.F., Wang, S.Y., Wang, C.Y., and Gonzalez-Aguilar, G.A. (2004). Effect of storage temperatures on antioxidant capacity and aroma compounds in strawberry fruit. *Lebensmittel-Wissenschaft und Technologie-Food Science and Technology*, 37: 687–695.

Ayhan, Z., Esturk, O., and Tas, E. (2008). Effect of modified atmosphere packaging on the quality and shelf life of minimally processed carrots. *Turkish Journal of Agriculture and Forestry*, 32: 57–64.

Bailen, G., Gullen, F., Castillo, S., Serrano, M., Valero, D., and Martinez-Romero, D. (2006). Use of activated carbon inside modified atmosphere packages to maintain tomato fruit quality during cold storage. *Journal of Agricultural and Food Chemistry*, 54: 2229–2235.

Barakat, S.M., Bhagat, M.A.R., and Linton, R.H. (2007). Inactivation kinetics of inoculated *Escherichia coli O157:H7*, *Listeria monocytogenes* and *Salmonella enterica* on strawberries by chlorine dioxide gas. *Food Microbiology*, 24: 736–744.

Bari, M.L., Ukuku, D.O., Kawasaki, T., Inatsu, Y., Isshiki, K., and Kawamoto, S. (2005). Combined efficacy of nisin and pediocin with sodium lactate, citric acid, phytic acid, and potassium sorbate and EDTA in reducing the *Listeria monocytogenes* population of inoculated fresh-cut produce. *Journal of Food Protection*, 68: 1381–1387.

Barron, C., Varoquaux, P., Guilvert, S., Gontard, N., and Gouble, B. (2002). Modified atmosphere packaging of cultivated mushroom (*Agaricus bisporus l.*) with hydrophilic films. *Journal of Food Science*, 67(1): 251–255.

Barry-Ryan, C. and O'Beirne, D. (1999). Ascorbic acid retention in shredded iceberg lettuce as affected by minimal processing. *Journal of Food Science*, 64: 3.

Barry-Ryan, C., Pacussi, J.M., and O'Beirne, D. (2000). Quality of shredded carrots as affected by packaging film and storage temperature. *Journal of Food Science*, 65(4): 726–730.

Barth, M.M. and Zhuang, H. (1996). Packaging design affects antioxidant vitamin retention and quality of *Broccoli florets* during postharvest storage. *Postharvest Biology and Technology*, 9: 141–150.

Baskaran, R., Usha Devi, A., Nayak, C.A., Kudachikar, V.B., Prakash, M.N.K., Prakash, M., Ramana, K.V.R., and Rastogi, N.K. (2007). Effect of low-dose γ-irradiation on the shelf life and quality characteristics of minimally processed potato cubes under modified atmosphere packaging. *Radiation Physics and Chemistry*, 76: 1042–1049.

Batu, A. and Thompson, A.K. (1998). Effects of modified atmosphere packaging on post harvest qualities of pink tomatoes. *Transactions of Journal of Agriculture and Forestry*, 22: 365–372.

Beltran, D., Selma, M.V., Marian, A., and Gil, M.I. (2005a). Ozonated water extends the shelf life of fresh-cut lettuce. *Journal of Agricultural and Food Chemistry*, 53: 5654–5663.

Beltran, D., Selma, M.V., Tudela, J.A., and Gil, M.I. (2005b). Effect of different sanitizers on microbial and sensory quality of fresh-cut potato strips stored under modified atmosphere or vacuum packaging. *Postharvest Biology and Technology*, 37: 37–46.

Bennik, M.H.J., Peppelenbos, H.W., Nguyen-the, C., Carlin, F., Smid, E.J., and Gorris, L.G.M. (1996). Microbiology of minimally processed, modified-atmosphere packaged chicory endive. *Postharvest Biology and Technology*, 9: 209–221.

Beuchat, L.R. (1998). Surface decontamination of fruits and vegetables eaten raw: A review. Food Safety Unit, World Health Organization, WHO/FSF/98.2.

Beuchat, L.R. (1999). Survival of enterohemorrhagic *Escherichia coli O157:H7* in bovine feces applied to lettuce and the effectiveness of chlorinated water as a disinfectant. *Journal of Food Protection*, 62(8): 845–849.

Beuchat, L.R. and Brackett, R.E. (1990). Inhibitory effects of raw carrots on *Listeria monocytogenes. Applied and Environmental Microbiology*, 56(6): 1734–1742.

Beuchat, L.R., Nail, B.V., Adler, B.B., and Clavero, M.R.S. (1998). Efficacy of spray application of chlorine in killing pathogenic bacteria on raw apples, tomatoes, and lettuce. *Journal of Food Protection*, 61: 1305–1311.

Beuchat, L.R. and Ryu, J.H. (1997). Produce handling and processing practices. *Emerging Infectious Diseases*, 3(4): 459–465.

Bidawid, S., Farber, J.M., and Sattar, S.A. (2001). Survival of *Hepatitis A* virus on modified atmosphere-packaged (MAP) lettuce. *Food Microbiology*, 18: 95–102.

Bletsos, F.A. (2002). Evaluation of new and commercial eggplant hybrids and cultivars in relation to their frozen product. *Acta Horticulturae*, 579: 89–93.

Bourke, P. and O'Beirne, D. (2004). Effects of packaging type, gas atmosphere and storage temperature on survival and growth of *Listeria spp.* in shredded dry coleslaw and its components. *International Journal of Food Science and Technology*, 39: 509–523.

Carlin, F., Nguyen-the, C., Cudennec, P., and Reich, M., (1989). Microbiological spoilage of fresh ready-to use carrots. *Science des Aliments*, 9: 371–386.

Carlin, F., Nguyen-the, C., Da Silva, A.A., and Cochet, C. (1996). Effects of carbon dioxide on the fate of *Listeria monocytogenes*, of aerobic bacteria and on the development of spoilage in minimally processed fresh endive. *International Journal of Food Microbiology*, 32: 159–172.

Charles, F., Guillaume, C., and Gontard, N. (2008). Effect of passive and active modified atmosphere packaging on quality changes of fresh endives. *Postharvest Biology and Technology*, 48: 22–29.

Charles, F., Sanchez, J., and Gontard, N. (2006). Absorption kinetics of oxygen and carbon dioxide scavengers as part of active modified atmosphere packaging. *Journal of Food Engineering*, 72: 1–7.

Chauhan, O.P., Raju, P.S., Dasgupta, D.K., and Bawa, A.S. (2006). Instrumental textural changes in banana (Var. *Pachbale*) during ripening under active and passive modified atmosphere. *International Journal of Food Properties*, 9: 237–253.

Chavarri, M.J., Herrera, A., and Arino, A. (2005). The decrease in pesticides in fruit and vegetables during commercial processing. *International Journal of Food Science and Technology*, 40: 205–211.

Chua, D., Goh, K., Saftner, R.A., and Bhagwat, A.A. (2008). Fresh-cut lettuce in modified atmosphere packages stored at improper temperatures supports enterohemorrhagic *E. coli* isolates to survive gastric acid challenge. *Journal of Food Science*, 73(3): 148–153.

Cliffe-Byrnes, V., Mc Laughlin, C.P., and O'Beirne, D. (2003). The effects of packaging film and storage temperature on the quality of a dry coleslaw mix packaged in a modified atmosphere. *International Journal of Food Science and Technology*, 38: 187–199.

Cliffe-Byrnes, V. and O'Beirne, D. (2005). Effects of chlorine treatment and packaging on the quality and shelf-life of modified atmosphere (MA) packaged coleslaw mix. *Food Control*, 16: 707–716.

Cliffe-Byrnes, V. and O'Beirne, D. (2007). The effects of modified atmospheres, edible coating and storage temperatures on the sensory quality of carrot discs. *International Journal of Food Science and Technology*, 42: 1338–1349.

Cliffe-Byrnes, V. and O'Beirne, D. (2008). Effects of washing treatment on microbial and sensory quality of modified atmosphere (MA) packaged fresh sliced mushroom (*Agaricus bisporus*). *Postharvest Biology and Technology*, 48: 283–294.

Das, E., Gurakan, G.C., and Bayındırlı, A. (2006). Effect of controlled atmosphere storage, modified atmosphere packaging and gaseous ozone treatment on the survival of *Salmonella Enteritidis* on cherry tomatoes. *Food Microbiology*, 23: 430–438.

De Ell, J.R., Toivonen, P.M.A., Cornut, F., Roger, C., and Vigneault, C. (2006). Addition of sorbitol with $KMnO_4$ improves broccoli quality retention in modified atmosphere packages. *Journal of Food Quality*, 29: 65–75.

Degl'Innocenti, E., Guidi, L., Pardossi, A., and Tognoni, F. (2005). Biochemical study of leaf browning in minimally processed leaves of lettuce (*Lactuca sativa L.* var. *acephala*). *Journal of Agricultural and Food Chemistry*, 53: 9980–9984.

Degl'Innocenti, E., Pardossi, A., Tognoni, F., and Guidi, L. (2007). Physiological basis of sensitivity to enzymatic browning in 'lettuce', 'escarole' and 'rocket salad' when stored as fresh-cut products. *Food Chemistry*, 104: 209–215.

Devlieghere, F., Vermeulen, A., and Debevere, J. (2004). Chitosan antimicrobial activity, interactions with food components and applicability as a coating on fruit and vegetables. *Food Microbiology*, 21: 703–714.

Drosinos, E.H., Tassou, C., Kakiomenou, K., and Nychas, G.J.E. (2000). Microbiological, physico-chemical and organoleptic attributes of a country tomato salad and fate of *Salmonella enteritidis* during storage under aerobic or modified atmosphere packaging conditions at 4°C and at 10°C. *Food Control*, 11: 131–135.

Elazar, R. (2004). Postharvest physiology, pathology and handling of fresh commodities. Lecture Notes. Department of Market Research. Ministry of Agriculture and Rural Development, Israel. In http://hdl.handle.net/1813/10565

Ellis, W.O., Smith, J.P., Simpson, B.K., Khanizadeh, S., and Oldham, J.H. (1993). Control of growth and aflatoxin production of *Aspergillus flavus* under modified atmosphere packaging (MAP). *Food Microbiology*, 10: 9–21.

Erturk, E. and Picha, D.H. (2007). Effect of temperature and packaging film on nutritional quality of fresh-cut sweet potatoes. *Journal of Food Quality*, 30: 450–465.

Escalona, V.H., Aguayo, E., and Artes, F. (2005). Overall quality throughout shelf life of minimally fresh processed fennel. *Journal of Food Science*, 70(1): 13–17.

Escalona, V.H., Aguayo, E., and Artes, F. (2007a). Quality changes of fresh-cut kohlrabi sticks under modified atmosphere packaging. *Journal of Food Science*, 72(5): 303–307.

Escalona, V.H., Aguayo, E., and Artes, F. (2007b). Modified atmosphere packaging improved quality of kohlrabi stems. *Lebensmittel-Wissenschaft und Technologie*, 40: 397–403.

Escalona, V.H., Aguayo, E., and Artes, F. (2007c). Extending the shelf life of kohlrabi stems by modified atmosphere packaging. *Journal of Food Science*, 72(5): 308–313.

Escalona, V.H., Aguayo, E., Gomez, P., and Artes, F. (2004). Modified atmosphere packaging inhibits browning in fennel. *Lebensmittel-Wissenschaft und Technologie*, 37: 115–121.

Escalona, V.H., Verlinden, B.E., Geysen, S., and Nicolaı, B.M. (2006). Changes in respiration of fresh-cut butterhead lettuce under controlled atmospheres using low and superatmospheric oxygen conditions with different carbon dioxide levels. *Postharvest Biology and Technology*, 39: 48–55.

Eskin, M. and Robinson, D.S. (2001). *Food Shelf Life Stability: Chemical, Biochemical, and Microbiological Changes*. Boca Raton, FL: CRC Press.

Evans, M.R., Ribeiro, C.D., and Salmon, R.L. (2003). Hazards of healthy living: Bottled water and salad vegetables as risk factors for *Campylobacter* infection. *Emerging Infectious Diseases*, 9: 1219–1225.

Fallik, E. and Aharoni, Y. (2004). Postharvest physiology, pathology and handling of fresh produce. Lecture Notes. International Research and Development course on Postharvest Biology and Technology. The Volcani Center, Israel, p. 30.

Fallik, E., Alkali-Tuvia, S., and Horev, B. (2001). Characterization of "*Galia*" melon aroma by GC and mass spectrometric sensor measurements after prolonged storage. *Postharvest Biology and Technology*, 22: 85–91.

Fan, X., Toivonen, P.M.A., Rajkowski, K.T., and Sokorai, K.J.B. (2003). Warm water treatment in combination with modified atmosphere packaging reduces undesirable effects of irradiation on the quality of fresh-cut iceberg lettuce. *Journal of Agricultural and Food Chemistry*, 51: 1231–1236.

FDA/CFSAN. (2001). Analysis and evaluation of preventive control measures for the control and reduction/elimination of microbial hazards on fresh and fresh-cut produce. http://www.fda.gov/ Food/ ScienceResearch/ ResearchAreas/ Safe Practices for Food Processes/ ucm091368.htm

Fonseca, S.C., Oliveira, F.A.R., and Brecht, J.K. (2002). Modelling respiration rate of fresh fruits and vegetables for modified atmosphere packages: A review. *Journal of Food Engineering*, 52: 99–119.

Francis, G.A. and O'Beirne, D. (1997). Effects of gas atmosphere, antimicrobial dip and temperature on the fate of *Listeria innocua* and *Listeria monocytogenes* on minimally processed lettuce. *International Journal of Food Science and Technology*, 32: 141–151.

Francis, G.A. and O'Beirne, D. (2001). Effects of vegetable type, package atmosphere and storage temperature on growth and survival of *Escherichia coli O157:H7* and *Listeria monocytogenes. Journal of Industrial Microbiology and Biotechnology*, 27: 111–116.

Francis, G.A. and O'Beirne, D. (2002). Effects of vegetable type and antimicrobial dipping on survival and growth of *Listeria innocua* and *E. coli. International Journal of Food Science and Technology*, 37(6): 711–718.

Francis, G.A., Scollard, J., Meally, A., Bolton, D.J., Gahan, C.G.M., Cotter, P.D., Hill, C., and O'Beirne, D. (2007). The glutamate decarboxylase acid resistance mechanism affects survival of *Listeria monocytogenes LO28* in modified atmosphere-packaged foods. *Journal of Applied Microbiology*, 103: 2316–2324.

Francis, G.A., Thomas, C., and O'Beirne, D. (1999). The microbiological safety of minimally processed vegetables. *International Journal of Food Science and Technology*, 34: 1–22.

Gil, M.I., Conesa, M.A., and Artes, F. (2002). Quality changes in fresh cut tomato as affected by modified atmosphere packaging. *Postharvest Biology and Technology*, 25: 199–207.

Gil, M.I., Ferreres, F., and Tomas-Barberan, F.A. (1999). Effect of postharvest storage and processing on the antioxidant constituents (flavonoids and vitamin C) of fresh-cut spinach. *Journal of Agricultural and Food Chemistry*, 47: 2213–2217.

Gil-Izquierdo, A., Conesa, M.A., Ferreres, F., and Gil, M.A. (2002). Influence of modified atmosphere packaging on quality, vitamin C and phenolic content of artichokes (*Cynara scolymus L.). European Food Research and Technology*, 215: 21–27.

Gleeson, E. and O'Beirne, D. (2005). Effects of process severity on survival and growth of *Escherichia coli* and *Listeria innocua* on minimally processed vegetables. *Food Control*, 16: 677–685.

Gomez, P.A. and Artes, F. (2005). Improved keeping quality of minimally fresh processed celery sticks by modified atmosphere packaging. *Lebensmittel-Wissenschaft und Technologie*, 38: 323–329.

Gomez-Lopez, V.M., Devlieghere, F., Bonduelle, V., and Debevere, J. (2005). Intense light pulses decontamination of minimally processed vegetables and their shelf-life. *International Journal of Food Microbiology*, 103: 79–89.

Gomez-Lopez, V.M., Devlieghere, F., Ragaert, P., and Debevere, J. (2007a). Shelf-life extension of minimally processed carrots by gaseous chlorine dioxide. *International Journal of Food Microbiology*, 116: 221–227.

Gomez-Lopez, V.M., Ragaert, P., Jeyachchandran, V., Debevere, J., and Devlieghere, F. (2008). Shelf-life of minimally processed lettuce and cabbage treated with gaseous chlorine dioxide and cysteine. *International Journal of Food Microbiology*, 121: 74–83.

Gomez-Lopez, V.M., Ragaert, P., Ryckeboer, J., Jeyachchandran, V., Debevere, J., and Devlieghere, F. (2007b). Shelf-life of minimally processed cabbage treated with neutral electrolysed oxidising water and stored under equilibrium modified atmosphere. *International Journal of Food Microbiology*, 117: 91–98.

Gonzalez-Aguilar, G.A., Ayala-Zavala, J.F., Ruiz-Cruz, S., Acedo-Felix, E., and Diaz-Cinco, M.E. (2004). Effect of temperature and modified atmosphere packaging on overall quality of fresh-cut bell peppers. *Lebensmittel-Wissenschaft und Technologie*, 37: 817–826.

Gonzalez-Rodrıguez, R.M., Rial-Otero, R., Cancho-Grande, B., and Simal-Gandara, J. (2008). Occurrence of fungicide and insecticide residues in trade samples of leafy vegetables. *Food Chemistry*, 107: 1342–1347.

Gran, C.D. and Beaudry, R.M. (1992). Determination of the low oxygen limit for several commercial apple cultivars by respiratory quotient breakpoint. *Postharvest Biology and Technology*, 3: 259–267.

Grass, M.L., Vidal, D., Betoret, N., Chiralt, A., and Fito, P. (2003). Calcium fortification of vegetables by vacuum impregnation interactions with cellular matrix. *Journal of Food Engineering*, 56: 279–284.

Gunes, G. and Lee, C.Y. (1997). Color of minimally processed potatoes as affected by modified atmosphere packaging and antibrowning agents. *Journal of Food Science*, 62(3): 572–575.

Habibunnisa, T., Baskaran, R., Prasad, R., and Shivaiah, K.M. (2001). Storage behaviour of minimally processed pumpkin (*Cucurbita maxima*) under modified atmosphere packaging conditions. *European Food Research and Technology*, 212: 165–169.

Hagenmaier, R.D. and Baker, R.A. (1997). Low-dose irradiation of cut iceberg lettuce in modified atmosphere packaging. *Journal of Agricultural and Food Chemistry*, 45: 2864–2868.

Harris, L.J., Farber, J.N., Beuchat, L.R., Parish, M.E., Suslow, T.V., Garrett, E.H., and Busta, F.F. (2003). Outbreaks associated with fresh produce—Incidence, growth, and survival of pathogens in fresh and fresh-cut produce. *Comprehensive Reviews Food Science and Food Safety*, 2(Supplement s1): 78–141.

Harvey, J. and Gilmour, A. (1993). Occurrence and characteristics of *Listeria* in food produced in Northern Ireland. *International Journal of Food Microbiology*, 19: 193–205.

Heaton, J.C. and Jones, K. (2008). Microbial contamination of fruit and vegetables and the behaviour of enteropathogens in the phyllosphere: A review. *Journal of Applied Microbiology*, 104: 613–626.

Hong, J.H. and Gross, K.C. (2001). Maintaining quality of fresh-cut tomato slices through modified atmosphere packaging and low temperature storage. *Journal of Food Science*, 66(7): 960–965.

Hong, S.I. and Kim, D. (2004). The effect of packaging treatment on the storage quality of minimally processed bunched onions. *International Journal of Food Science and Technology*, 39: 1033–1041.

Hu, W., Xu, P., and Uchino, T. (2005). Extending storage life of fresh ginseng by modified atmosphere packaging. *Journal of the Science of Food and Agriculture*, 85: 2475–2481.

Irtwange, S.V. (2006). Application of modified atmosphere packaging and related technology in postharvest handling of fresh fruits and vegetables. *Agricultural Engineering International: The CIGR Ejournal*. Invited Overview 4 (VIII) In http://hdl.handle.net/1813/10565

Islam, E., Xiao-e, Y., Zhen-li, H., and Qaisar, M. (2007). Assessing potential dietary toxicity of heavy metals in selected vegetables and food crops. *Journal of Zhejiang University Sciences B*, 8(1): 1–13.

Jacobsson, A., Nielsen, T., and Sjoholm, I. (2004a). Influence of temperature, modified atmosphere packaging, and heat treatment on aroma compounds in broccoli. *Journal of Agricultural and Food Chemistry*, 52: 1607–1614.

Jacobsson, A., Nielsen, T., Sjoholm, I., and Werdin, K. (2004b). Influence of packaging material and storage condition on the sensory quality of broccoli. *Food Quality and Preference*, 15: 301–310.

Jacobsson, A., Nielsen, T., and Sjoholm, I. (2004c). Effects of type of packaging material on shelf-life of fresh broccoli by means of changes in weight color and texture. *European Food Research and Technology*, 218: 157–163.

Jacxsens, L., Devlieghere, F., and Debevere, J. (2002a). Temperature dependence of shelf-life as affected by microbial proliferation and sensory quality of equilibrium modified atmosphere packaged fresh produce. *Postharvest Biology and Technology*, 26: 59–73.

Jacxsens, L., Devlieghere, F., and Debevere, J. (2002b). Predictive modelling for packaging design equilibrium modified atmosphere packages of fresh-cut vegetables subjected to a simulated distribution chain. *International Journal of Food Microbiology*, 73: 331–341.

Jacxsens, L., Devlieghere, F., Ragaert, P., Vanneste, E., and Debevere, J. (2003). Relation between microbiological quality, metabolite production and sensory quality of equilibrium modified atmosphere packaged fresh cut produce. *International Journal of Food Microbiology*, 83: 263–280.

Jacxsens, L., Delvieghere, F., Van Der Steen, C., and Debevere, J. (2001). Effect of high oxygen modified atmosphere packaging on microbial growth and sensorial qualities of fresh-cut produce. *International Journal of Food Microbiology*, 71: 197–210.

Jayas, D.S. and Jeyamkondan, S. (2002). Modified atmosphere storage of grains, meats, fruits and vegetables. *Biosystems Engineering*, 82(3): 235–251.

Jayathunge, L. and Illeperuma, C. (2005). Extension of postharvest life of oyster mushroom by modified atmosphere packaging technique. *Journal of Food Science*, 70(9): 573–578.

Jeon, B.S. and Lee, C.Y. (1999). Shelf-life extension of American fresh ginseng by controlled atmosphere storage and modified atmosphere packaging. *Journal of Food Science*, 64(2): 318–321.

Jia, C.G., Wei, C.J., Wei, J., Yan, G.F., Wang, B.L., and Wang, Q.M. (2008). Effect of modified atmosphere packaging on visual quality and glucosinolates of broccoli florets. *Food Chemistry*, doi:10.1016/j.foodchem.2008.09.009

Jin, H.H. and Lee, S.Y. (2007). Combined effect of aqueous chlorine dioxide and modified atmosphere packaging on inhibiting *Salmonella typhimurium* and *Listeria monocytogenes* in mungbean sprouts. *Journal of Food Science*, 72(9): 441–445.

Johnston, J.W., Hewett, E.W., Hertog, M.L.A.T.M., and Harker, F.R. (2001). Temperature induces differential softening responses in apple cultivars. *Postharvest Biology and Technology*, 23: 185–196.

Juven, B. and Pierson, M.D. (1996). Antibacterial effects of hydrogen peroxide and methods for its detection and quantitation. *Journal of Food Protection*, 59(11): 1233–1241.

Kabir, H. (1994). Fresh-cut vegetables. In *Modified Atmosphere Food Packaging*. A.L. Brods. and V.A. Herndon (Eds.), Naperville, IL: Institute of Packaging Professionals. pp. 155–160.

Kakiomenou, K., Tassou, C., and Nychas, G. (1996). Microbiological, physiochemical and organoleptic changes of shredded carrots stored under modified storage. *International Journal of Food Science and Technology*, 31: 359–366.

Kakiomenou, K., Tassou, C., and Nychas, G.J. (1998). Survival of *Salmonella enteritidis* and *Listeria monocytogenes* on salad vegetables. *World Journal of Microbiology and Biotechnology*, 14: 383–387.

Kang, J.S., Park, W.P., and Lee, D.S. (2000). Quality of *enoki* mushrooms as affected by packaging conditions. *Journal of the Science of Food and Agriculture*, 81: 109–114.

Karakas, B. and Yildiz, F. (2007). Peroxidation of membrane lipids in minimally processed cucumbers packaged under modified atmospheres. *Food Chemistry*, 100: 1011–1018.

Kim, C., Hung, Y.C., and Brackett, R.E. (2000). Roles of oxidation reduction potential in electrolyzed oxidizing and chemically modified water for the inactivation of food-related pathogens. *Journal of Food Protection*, 63: 19–24.

Kim, K.M., Ko, J.A., Lee, J.S., Park, H.J., and Hanna, M.A. (2006). Effect of modified atmosphere packaging on the shelf-life of coated, whole and sliced mushrooms. *Lebensmittel-Wissenschaff und Technologie*, 39: 364–371.

Kim, J.G., Luo, Y., Tao, Y., Saftner, R.A., and Gross, K.C. (2005). Effect of initial oxygen concentration and film oxygen transmission rate on the quality of fresh-cut romaine lettuce. *Journal of the Science of Food and Agriculture*, 85: 1622–1630.

Kim, J.G., Yousef, A.E., and Chism, G.W. (1999). Application of ozone for enhancing the microbiological safety and quality of foods—A review. *Journal of Food Protection*, 62(9): 1071–1087.

King, G.A. and Morris, S.C. (1994). Physiological changes of broccoli during early postharvest senescence and through the preharvest–postharvest continuum. *Journal of American Society of Horticultural Sciences*, 119: 270–275.

Koide, S. and Shi, J. (2007). Microbial and quality evaluation of green peppers stored in biodegradable film packaging. *Food Control*, 18: 1121–1125.

Koukounaras, A., Siomos, A.S., and Sfakiotakis, E. (2007). Postharvest CO_2 and ethylene production and quality of rocket (*Eruca sativa Mill.*) leaves as affected by leaf age and storage temperature. *Postharvest Biology and Technology*, 46: 167–173.

Kumar, A., Agarwal, R.K., Bhilegaonkar, K.N., Shome, B.R., and Bachhil, V.N. (2001). Occurrence of *Campylobacter jejuni* in vegetables. *International Journal of Food Microbiology*, 67: 153–155.

Kumari, B., Madan, V.K., Kumar, R., and Kathral, T.S. (2002). Monitoring of seasonal vegetables for pesticide residues. *Environmental Monitoring and Assessment*, 74: 263–270.

Kuyper, L., Weinert, L.A.G., and McGill, A.E.J. (1993). The effect of modified atmosphere packaging and addition of calcium hypochlorite on the atmosphere composition, color and microbial quality of mushrooms. *Lebensmittel-Wissenschaft und Technologie*, 26(1): 14–20.

Lacroix, M. and Lafortune, R. (2004). Combined effects of gamma irradiation and modified atmosphere packaging on bacterial resistance in grated carrots (*Daucus carota*). *Radiation Physics and Chemistry*, 71: 77–80.

Lamikanra, O. (2002). Preface. In *Fresh-Cut Fruits and Vegetables: Science, Technology and Market*. O. Lamikanra (Ed.), Boca Raton, FL: CRC Press.

Lee, L., Arul, J., Lencki, R., and Castaigne, F. (1995). A review on modified atmosphere packaging and preservation of fresh fruits and vegetables: Physiological basis and practical aspects—Part 1. *Packaging Technology and Science*, 8: 315–331.

Lee, S.Y. and Baek, S.Y. (2008). Effect of chemical sanitizer combined with modified atmosphere packaging on inhibiting *Escherichia coli O157:H7* in commercial spinach. *Food Microbiology*, 25: 582–587.

Lee, K.S., Woo, K.L., and Lee, D.S. (1994). Modified atmosphere packaging for green chili peppers. *Packaging Technology and Science*, 7: 51–58.

Li, Y., Brackett, R.E., Chen, J., and Beuchat, L.R. (2001). Survival and growth of *Escherichia coli O157:H7* inoculated onto cut lettuce before or after heating in chlorinated water, followed by storage at 5°C or 15°C. *Journal of Food Protection*, 64(3): 305–309.

Lin, D. and Zhao, Y. (2007). Innovations in the development and application of edible coatings for fresh and minimally processed fruits and vegetables. *Comprehensive Reviews in Food Science and Food Safety*, 6(3): 60–75.

Liu, F. and Li, Y. (2006). Storage characteristics and relationships between microbial growth parameters and shelf life of MAP sliced onions. *Postharvest Biology and Technology*, 40: 262–268.

Lopez-Galvez, G., Saltveit, M., and Cantwell, M. (1996). Wound-induced phenylalanine ammonia lyase activity: Factors affecting its induction and correlation with the quality of minimally processed lettuces. *Postharvest Biology and Technology*, 9: 223–233.

Macura, D., McCannel, A.M., and Li, M.Z.C. (2001). Survival of *Clostridium botulinum* in modified atmosphere packaged fresh whole north american ginseng roots. *Food Research International*, 34: 123–125.

Manolopoulou, H. and Papadopoulou, P. (1998). A study of respiratory and physico-chemical changes of four kiwi fruit cultivars during cool-storage. *Food Chemistry*, 63: 529–534.

Martinez, I., Ares, G., and Lema, P. (2008). Influence of cut and packaging film on sensory quality of fresh-cut butterhead lettuce (*Lactuca sativa L., cv. Wang*). *Journal of Food Quality*, 31: 48–66.

Martinez, J.A. and Artes, F. (1999). Effect of packaging treatments and vacuum-cooling on quality of winter harvested iceberg lettuce. *Food Research International*, 32: 621–627.

Maskell, A.J. (1991). Long-life ambient food packaging: A history—From the tin can to plastics and beyond. *Packaging Technology and Science*, 4: 21–28.

Masson, Y., Ainsworth, P., Fuller, D., Bozkurt, H., and Ibanoglu, S. (2002). Growth of *Pseudomonas fluorescens* and *Candida sake* in homogenized mushrooms under modified atmosphere. *Journal of Food Engineering*, 54: 125–131.

Matan, N., Rimkeeree, H., Mawson, A.J., Chompreeda, P., Haruthaithanasan, V., and Parker, M. (2006). Antimicrobial activity of cinnamon and clove oils under modified atmosphere conditions. *International Journal of Food Microbiology*, 107: 180–185.

McConnell, R.Y., Truong, V.D., Walter Jr., W.M., and McFeeters, R.F. (2005). Physical, chemical and microbial changes in shredded sweet potatoes. *Journal of Food Processing and Preservation*, 29: 246–267.

McMahon, M.A.S. and Wilson, I.G. (2001). The occurrence of enteric pathogens and *Aeromonas* species in organic vegetables. *International Journal of Food Microbiology*, 70: 155–162.

Merino, S., Rubires, X., Knochel, S., and Tomas, J.M. (1995). Emerging pathogens: *Aeromonas spp. International Journal of Food Microbiology*, 28: 157–168.

Mir, N. and Beaudry, R.M. (2002). *Modified Atmosphere Packaging*. Michigan State University East Lansing. ftp://ftp.esat.kuleuven.ac.be/pub/sista/barrero/Tesis_JJ/parameterfruits

Neyts, K., Huys, G., Uyttendaele, M., Swings, J., and Debevere, J. (2001). Incidence and identification of mesophilic *Aeromonas spp.* from retail foods. *Letters in Applied Microbiology*, 31: 359–363.

Nguyen-the, C. and Carlin, F. (1994). The microbiology of minimally processed fresh fruits and vegetables. *Critical Reviews in Food Science and Nutrition*, 34(4): 371–401.

Niemira, B.A., Fan, X., and Sokorai, K.J.B. (2005). Irradiation and modified atmosphere packaging of endive influences survival and regrowth of *Listeria monocytogenes* and product sensory qualities. *Radiation Physics and Chemistry*, 72: 41–48.

Nygård, K., Andersson, Y., Lindkvist, P., Ancker, C., Asteberg, I., Dannetun, E., Eitrem, R. et al. (2001). Imported rocket salad partly responsible for increased incidence of *hepatitis A. Eurosurveillance*, 6(10): 151–153.

O'Beirne, D. and Francis, G.A. (2003). Reducing the pathogen risk in MAP-prepared produce. In *Novel Food Packaging Techniques*. R. Ahvenainen (Ed.), Cambridge, U.K.: Woodhead Publishing Limited. pp. 231–286.

Odriozola-Serrano, I., Soliva-Fortuny, R., and Martin-Belloso, O. (2008). Effect of minimal processing on bioactive compounds and color attributes of fresh-cut tomatoes. *Lebensmittel-Wissenschaff und Technologie*, 41: 217–226.

Ongeng, D., Devlieghere, F., Debevere, J., Coosemans, J., and Ryckeboer, J. (2006). The efficacy of electrolysed oxidising water for inactivating spoilage microorganisms in process water and on minimally processed vegetables. *International Journal of Food Microbiology*, 109: 187–197.

Ooraikul, B. and Stiles, M.E. (1990). Chemical composition and structure of fruits and vegetables. *Modified Atmosphere Packaging of Food*. E. Horwood (Ed.), London, U.K.: Chapman & Hall. pp. 167–176.

Parentelli, C., Ares, G., Corona, M., Lareo, C., Gambaro, A., Soubes, M., and Lema, P. (2007). Sensory and microbiological quality of *shiitake* mushrooms in modified-atmosphere packages. *Journal of the Science of Food and Agriculture*, 87: 1645–1652.

Pariasca, J.A.T., Miyazaki, T., Hisaka, H., Nakagawa, H., and Sato, T. (2000). Effect of modified atmosphere packaging (MAP) and controlled atmosphere (CA) storage on the quality of snow pea pods (*Pisum sativum L. var. saccharatum*). *Postharvest Biology and Technology*, 21: 213–223.

Piagentini, A.M., Guemes, D.R., and Pirovani, M.E. (2003). Mesophilic aerobic population of fresh-cut spinach as affected by chemical treatment and type of packaging film. *Journal of Food Science*, 68(2): 602–606.

Pilon, L., Oetterer, M., Gallo, C.R., Spoto, M.H.F., and Jan-Mar, T. (2006). Shelf life of minimally processed carrot and green pepper. *Ciência Tecnologia Alimentas Campinas*, 26(1): 150–158.

Pirovani, M.E., Guemes, D.R., Piagentini, A.M., and Di Pentima, J.H. (1997). Storage quality of minimally processed cabbage packaged in plastic films. *Journal of Food Quality*, 20: 381–389.

Pirovani, M.E., Piagentini, A.M., Guemes, D.R., and Di Pentima, J.H. (1998). Quality of minimally processed lettuce as influenced by packaging and chemical treatment. *Joumal of Food Quality*, 22: 475–484.

Punzi, J.S., Lamont, M., Haynes, D., and Epstein, R.L. (2005). Pesticide residues on fresh and processed fruit and vegetables, grains, meats, milk, and drinking water. *Outlooks on Pest Management*, doi: 10.1564/16jun12

Radziejewska-Kubzdela, E., Czapski, J., and Czaczyk, K. (2007). The effect of packaging conditions on the quality of minimally processed celeriac flakes. *Food Control*, 18: 1191–1197.

Rangkadilok, N., Tomkins, B., Nicolas, M.E., Premier, R.R., Bennett, R.N., Eagling, D.R., and Taylor, P.W.J. (2002). The effect of post-harvest and packaging treatments on glucoraphanin concentration in broccoli (*Brassica oleracea var. italica*). *Journal of Agricultural and Food Chemistry*, 50: 7386–7391.

Rauha, J.P., Remes, S., Heinonen, M., Hopia, A., Kahkfnen, M., Kujala, T., Pihlaja, K., Vuorela, H., and Vuorela, P. (2000). Antimicrobial effect of Finnish plant extracts containing flavonoids and other phenolic compounds. *International Journal of Food Microbiology*, 56(1): 3–12.

Rico, D., Martın-Diana, A.B., Barat, J.M. and Barry-Ryan, C. (2007). Extending and measuring the quality of fresh-cut fruit and vegetables: A review. *Trends in Food Science and Technology*, 18: 373–386.

Rocha, A.M.C.N., Barreiro, M.G. and Morais, A.M.M.B. (2004). Modified atmosphere package for apple "*Bravo de Esmolfe*". *Food Control*, 15: 61–64.

Roy, S., Anantheswaran, R.C. and Beelman, R.B. (1996). Modified atmosphere and modified humidity packaging of fresh mushrooms. *Journal of Food Science*, 61(2): 391–397.

Saltveit, M.E. (2000). Wound induced changes in phenolic metabolism and tissue browning are altered by heat shock. *Postharvest Biology and Technology*, 21: 61–69.

Sanz, S., Olarte, C., Echavarri, J.F., and Ayala, F. (2007). Evaluation of different varieties of cauliflower for minimal processing. *Journal of the Science of Food and Agriculture*, 87: 266–273.

Sayed Ali, Md., Nakano, K., and Maezawa, S. (2004). Combined effect of heat treatment and modified atmosphere packaging on the color development of cherry tomato. *Postharvest Biology and Technology*, 34: 113–116.

Schafer, K.S. and Kegley, S.E. (2002). Persistent toxic chemicals in the US food supply. *Journal of Epidemiology and Community Health*, 56: 813–817.

Schreiner, M., Peters, P., and Krumbein, A. (2007). Changes of glucosinolates in mixed fresh-cut broccoli and cauliflower florets in modified atmosphere packaging. *Journal of Food Science*, 72(8): 585–589.

Sengun, I.Y. and Karapinar, M. (2005). Effectiveness of household natural sanitizers in the elimination of *Salmonella typhimurium* on rocket (*Eruca sativa Miller*) and spring onion (*Allium cepa L.*). *International Journal of Food Microbiology*, 98: 319–323.

Serrano, M., Martinez-Romero, D., Guillen, F., Castillo, S., and Valero, D. (2006). Maintenance of broccoli quality and functional properties during cold storage as affected by modified atmosphere packaging. *Postharvest Biology and Technology*, 39: 61–68.

Seymour, I.J. and Appleton, H. (2001). A review, foodborne viruses and fresh produce. *Journal of Applied Microbiology*, 91: 759–773.

Seymour, I.J., Burfoot, D., Smith, R.L., Cox, L.A., and Lockwood, A. (2002). Ultrasound decontamination of minimally processed fruits and vegetables. *International Journal of Food Science and Technology*, 37: 547–557.

Shen, Q., Kong, F., and Wang, Q. (2006). Effect of modified atmosphere packaging on the browning and lignification of bamboo shoots. *Journal of Food Engineering*, 77: 348–354.

Shin, Y., Liu, R.H., Nock, J.F., Holliday, D., and Watkins, C.B. (2007). Temperature and relative humidity effects on quality, total ascorbic acid, phenolics and flavonoid concentrations, and antioxidant activity of strawberry. *Postharvest Biology and Technology*, 45: 349–357.

Simon, A., Gonzales-Fandos, E., and Rodriguez, D. (2008). Effect of film and temperature on the sensory, microbiological and nutritional quality of minimally processed cauliflower. *International Journal of Food Science and Technology*, 43: 1628–1636.

Simon, A., Gonzalez-Fandos, E., and Tobar, V. (2004). Influence of washing and packaging on the sensory and microbiological quality of fresh peeled white asparagus. *Journal of Food Science*, 69(1): 6–12.

Simon, A., Gonzalez-Fandos, E., and Tobar, V. (2005). The sensory and microbiological quality of fresh sliced mushroom (*Agaricus bisporus L.*) packaged in modified atmospheres. *International Journal of Food Science and Technology*, 40: 943–952.

Siomos, A.S., Sfakiotakis, E.M., and Dogras, C.C. (2000). Modified atmosphere packaging of white asparagus-spears: Composition, color and textural quality responses to temperature and light. *Scientia Horticulturae*, 84: 1–13.

Slifko, T.R., Smith, H.V., and Rose, J.B. (2000). Emerging parasite zoonoses associated with water and food. *International Journal for Parasitology*, 30: 1379–1393.

Song, Y., Kim, H.K., and Yam, K.L. (1992). Respiration rate of blueberry in modified atmosphere at various temperatures. *Journal of the American Society for Horticultural Science*, 117: 925–929.

Suparlan, M. and Itoh, K. (2003). Combined effects of hot water treatment (HWT) and modified atmosphere packaging (MAP) on quality of tomatoes. *Packaging Technology and Science*, 16: 171–178.

Tano, K., Oule, M.K., Doyon, G., Lencki, R.W., and Arul, J. (2007). Comparative evaluation of the effect of storage temperature fluctuation on modified atmosphere packages of selected fruit and vegetables. *Postharvest Biology and Technology*, 46: 212–221.

Tao, F., Zhang, M., Hangqing, Y., and Jincai, S. (2006). Effects of different storage conditions on chemical and physical properties of white mushrooms after vacuum cooling. *Journal of Food Engineering*, 77: 545–549.

Tao, F., Zhang, M., and Yu, H. (2007). Effect of vacuum cooling on physiological changes in the antioxidant system of mushroom under different storage conditions. *Journal of Food Engineering*, 79: 1302–1309.

Tassou, C.C. and Boziaris, J.S. (2002). Survival of *Salmonella enteritidis* and changes in pH and organic acids in grated carrots inoculated or not with *Lactobacillus spp.* and stored under different atmospheres at 4°C. *Journal of the Science of Food and Agriculture*, 82: 1122–1127.

Tenorio, M.D., Villanueva, M.J., and Sagardoy, M. (2004). Changes in carotenoids and chlorophylls in fresh green asparagus (*Asparagus officinalis L.*) stored under modified atmosphere packaging. *Journal of the Science of Food and Agriculture*, 84: 357–365.

Toivonen, P.M.A. and Stan, S. (2004). The effect of washing on physicochemical changes in packaged, sliced green peppers. *International Journal of Food Science and Technology*, 39: 43–51.

Varoquaux, P. and Wiley, R. (1994). Biological and biochemical changes in minimally processed refrigerated fruits and vegetables. In *Minimally Processed Refrigerated Fruits and Vegetables*. R.C. Wiley (Ed.), New York: Chapman & Hall. pp. 226–268.

Villaescusa, R. and Gil, M.I. (2003). Quality improvement of *Pleurotus* mushrooms by modified atmosphere packaging and moisture absorbers. *Postharvest Biology and Technology*, 28: 169–179.

Villanueva, M.J., Tenorio, M.D., Sagardoy, M., Redondo, A., and Saco, M.D. (2005). Physical, chemical, histological and microbiological changes in fresh green asparagus (*Asparagus officinalis L.*) stored inmodified atmosphere packaging. *Food Chemistry*, 91: 609–619.

Wall, M.M. and Berghage, R.D. (1996). Prolonging the shelf-life of fresh green chile peppers through modified atmosphere packaging and low temperature storage. *Journal of Food Quality*, 19: 467–477.

Wang, C.Y. and Qi, L. (1997). Modified atmosphere packaging alleviates chilling injury in cucumbers. *Postharvest Biology and Technology*, 10: 195–200.

Watada, A.E. and Qui, L. (1999). Quality of fresh-cut produce. *Postharvest Biology and Technology*, 15: 201–205.

Yun, J.H., An, D.S., Lee, K.E., Jun, B.S., and Lee, D.S. (2006). Modified atmosphere packaging of fresh produce using microporous earthenware material. *Packaging Technology and Science*, 19: 269–278.

Zagory, D. (1998). An update on modified atmosphere packaging of fresh produce. Packaging International 117 In http://www.nsf.org/business/nsf_davis_fresh/articles_map.pdf

Zagory, D. (1999). Effects of post-processing handling and packaging on microbial populations. *Postharvest Biology and Technology*, 15: 313–321.

Zhang, M., Zhan, Z.G., Wang, S.J., and Tang, J.M. (2008). Extending the shelf-life of asparagus spears with a compressed mix of argon and xenon gases. *Lebensmittel-Wissenschaft und Technologie*, 41: 686–691.

Zhuang, H., Hildebrand, D.F., and Barth, M.M. (1995). Senescence of broccoli buds is related to changes in lipid peroxidation. *Journal of Agriculture and Food Chemistry*, 43(10): 2585–2591.

10 Fruits

Ioannis S. Arvanitoyannis, Maria Savva, and Nikoletta K. Dionisopoulou

CONTENTS

10.1 APPLICATION OF MAP TO FRUITS

10.1.1 STRAWBERRIES

Renault et al. (1994) conducted an experiment with *Fragaria ananassa Dush* cultivar strawberries in an attempt to test a new gas transport model through microperforated PP films and fruit respiration in MAP (O_2/CO_2 gas composition). Experiments were conducted initially with packs filled with either 100% NO_2 or 100% O_2. Sensory parameters such as color and firmness and pH were also investigated. It was found that MAP prolonged the shelf life of the fruit from 2 (control shelf life) up to 7 days.

Silva et al. (1999) designed a MAP procedure, to keep the desired levels of O_2 and CO_2 inside the packages containing *Fragaria ananassa* strawberries from "Oso Grande" cultivar stored at 7°C and 19°C. The packages at 7°C were filled with 10% O_2 and 11.8% CO_2, whereas the packages at 19°C were filled with 10% O_2 and 11.3% CO_2. PE bags with ethyl vinyl acetate (EVA) film 38.1 μm were used as packaging materials. Investigation of the gas composition of the samples stored at the two different temperatures revealed that the concentration of O_2 was suppressed, whereas the CO_2 content increased at both temperatures. The shelf life of strawberries at MAP reached 6 days at 7°C and only 4 days at 19°C.

Tano et al. (2008) examined the effect of temperature fluctuations (4°C–14°C) on gas composition in MAP of fresh strawberry (*Fragaria* × *ananassa Duch.* cultivar) and on its quality during 11 days of storage. The strawberries were first kept under precooled conditions at 4°C for 10 h and then stored in 4 L plastic containers filled with 6% O_2 and 15% CO_2. The percentage range of ethanol and acetaldehyde in the fruits in conjunction with measurements of loss of product, color, pH, firmness, titratable acidity (TA), and infection percentage packed in MAP revealed that the shelf life of strawberries under MAP was prolonged up to 3 days.

In 2001, Van der Steen et al. conducted an experiment using *Fragaria elvira* strawberries, packed in containers with a permeable monolayer ethylene absorbing film (thickness of 30 μm). The containers were exposed to four different systems of MAP conditions: (i) the conventional packaging in a macroperforated high-barrier film (AIR), (ii) the low O_2 MA (3–5 kPa O_2, 5–10 kPa CO_2, and balance N_2), (iii) the high O_2 modified atmosphere (HOA,) (70 kPa O_2-balance N_2): HOA in a high-barrier film, and (iv) HOA in an MA film with an adjusted film permeability. It was only after 5 days and at about 3 kPa O_2 and 5 kPa CO_2 that the high O_2 atmosphere in the MA film reached steady state. To avoid an accumulation of ethylene inside the high-barrier package, an ethylene-absorbing monolayer was used. The shelf life of strawberries subjected to the AIR treatment was shortened by the growth of molds rather than by sensory unacceptance. On the other hand, the sensory properties considerably limited the shelf life of the fruit under MAP. Although the high O_2 atmospheres retarded the growth of molds, the gradual depletion of O_2 and CO_2 accumulation in the barrier film led to a decline in sensory quality. It is noteworthy that with the MA film, the inhibitory effect on mold growth persisted because of the initial high O_2 levels but the sensory quality declined following CO_2 accumulation. Tasting stopped when the growth of molds was reported. The strawberries packaged in low O_2 MA and HOA were found to be sensorially unacceptable for taste at day 5. Fruits treated with HOA displayed lower firmness and off-flavors compared with low O_2 MA and HOA MAP similarly to the results obtained when the CO_2 content was enhanced in the packages. After day 6 all the strawberries were evaluated as being too ripe.

The microbial safety of strawberry fruits (*Fragaria ananassa Camarosa* cultivar) inoculated with *Escherichia coli*, *Salmonella* spp., and *Listeria monocytogenes* and produced under two systems, high-oxygen atmospheres (HOA) and equilibrium modified atmospheres (EMA), in the presence of an ethylene-absorbing film was investigated by Siro et al. (2003). A commonly used macroperforated bioriented-polypropylene (BOPP) (30 μm) packaging film was used. Three acid-resistant *E. coli* strains (PT 30, PP5, and PT 16 from poultry), two *L. monocytogene* strain isolates, taken from their collection (LM1A and LM234 from poultry and vegetables, respectively), and two *Salmonella* spp. strains (*Salmonella enteritidis* LMG 5570 and *Salmonella typhimurium* LMG 5569) were obtained. Both O_2 in the strawberry fruit packaged in EMA and CO_2 displayed little variation during storage (1.0%–6.4%, and 5%–8%, respectively, O_2). In the HOA packages, the oxygen concentration rapidly decreased from 95% to 10.9% in 7 days in strawberries.

Hertog et al. (1998) tried to preserve the quality of "Elsanta" strawberries (*Fragaria ananassa Duchesne* cultivar) by being unaffected by *Botrytis* infection. The spoilage, in terms of number of strawberries affected, was clearly described by a simple sigmoidal growth curve. The rate of quality decay was shown to depend on temperature and the metabolic rate of the strawberries and the parameters for the gas exchange model were estimated as a function of O_2, CO_2 gases, and temperature.

The experimental data on spoilage, gathered from different batches of strawberries at various atmospheric conditions and temperatures, could be described with the model to a level of 83%. The number of strawberries visibly affected by *Botrytis* provided a visual quality indicator. The effect of gas conditions on spoilage was analyzed by means of the concept that fruit softening is the primary event enabling botrytis infection. Although the pathogen–host interaction is most probably more complex, the applied approach still resulted in a useful working model and the shelf life of MAP tested strawberry reached 6 days.

The impact of (MAP), in combination with ozone treatment and an edible film coating, on the preservation of Fengxiang variety strawberries was investigated by Zhang et al. (2005). A combination of 2.5% O_2 with 15% CO_2 was shown to be the optimal gas composition for the MAP of strawberries. The results revealed that two treatments, No.1: polyvinyl alcohol (PVA) 134 (2%), monostearate acylglycerol (0.5%), phytic acid (0.05%), sorbitol (0.05%), sodium alginate (0.1%), absolute alcohol (8%), No.2: PVA 134 (1%), soluble starch (1%), glucose (1%), sucrose (1%), sodium alginate (0.1%), and sorbitol (0.05%), gave the best effects on strawberry quality and shelf life. The latter was extended to 8–10 days compared with 4 days that was the control's sample shelf life. In another experiment, the strawberries were treated with ozonated water. The strawberries were dried with a fan and coated with edible films followed by drying. A PVC bag-type packaging material with 48 μm thickness was used for the MAP tests. It was found that the shelf life of the strawberries after 1 and 2 treatments can be extended to 8–10 days, longer than those of single MAP-treated strawberries and control samples.

In an attempt to obtain a more extended postharvest life of *Chandler* and *Douglas* strawberry cultivars, Picon et al. (1992) tested PE two different types of packaging films: (a) perforated cellophane sheets (CS) placed on top of the baskets and fixed with elastic bands and (b) LDPE bags heat-sealed after having introduced one or more baskets per bag. Strawberries were subjected to the following treatments: (i) for CS, baskets were covered with perforated CS, (ii) for CS + PC (precooling), baskets were covered with perforated CS and precooled with forced air flow at 2.5°C and 5 m/s, (iii) LDPE bag, one basket was introduced per PE bag, (iv) for PB + CO_2, six baskets were introduced per PE bag, CO_2 was flushed until a 15% concentration was established, and (v) for PB + EA, one basket was introduced per LDPE together with two ethylene-absorber sachets. Then, all baskets were stored at 2°C with the control untreated strawberry baskets. It was concluded that LDPE treatments generally resulted in the lowest weight losses and conductivity and bruising score and the highest firmness value. The use of the ethylene absorber imparted no particular benefits to LDPEB strawberries. The MAP shelf life reached approximately 8–10 days while the control shelf life was 3 days (Table 10.1).

Nielsen and Leufven (2007) carried out an MAP experiment on strawberries from *Honeyone and Korona* cultivars. The strawberries were harvested and immediately cooled to 2°C, stored in perforated PP bags at 5°C for 10 days, and filled with 11%–14% O_2 and 9%–12% CO_2. Antimist coated oriented polypropylene (COPP) film (thickness of 35 μm) was used for the package. They managed to reduce effectively the packaging material weight loss by preventing dehydration and improved the fruits' appearance as well. However, storage under MAP affected the aroma

TABLE 10.1
Effect of MAP (Gas Composition and Packaging Material) on Shelf Life of Various Strawberry Cultivars

Fruit	MAP Composition	Packaging Material	Shelf-Life Parameters Studied	Usage of Other Parameter	Shelf Life	References
Fragaria × Ananassa	100% NO_2 or 100% CO_2	Microperforated polypropylene pack	Gas composition, color, pH, firmness		Control: 2 days MAP: 7 days	Renault et al. (1994)
"Oso Grande" cv.	7°C: 10% O_2, 11.3% CO_2 19°C: 10% O_2, 11,8% CO_2	Polyethylene bag with EVA film, 38.1 μm	Gas composition at 7°C and 19°C, firmness, sensory quality		Control: 2 days MAP: 4 days at 7°C and 2 days at 19°C	Silva et al. (1999)
Fragaria × Ananassa, "Dosh" cv.	6% O_2 15% CO_2	4 L plastic container	Gas composition, weight loss, color, pH, TA, microbial infection	Acetaldehyde, ethanol	Control: 2 days MAP: 3 days	Tano et al. (2008)
Fragaria × elvira	5% O_2 10% CO_2	Permeable monolayer ethylene film, 30 μm	Gas composition, sensory quality, firmness	Acetaldehyde, ethanol	Control: 2 days MAP: 5 days	Van der Steen et al. (2001)
Fragaria × Ananassa, "Camarosa" cv.	EMA: 3% O_2, 5% CO_2, balance N_2 HOA: 95% O_2, 5% N_2	Macroperforated BOPP film, 30 μm	Gas composition, microbial infection, quality assessment	Three *E. coli* strains (PT30, PP5, PT16), two *L. monocytogenes* (LM1A, LM234)	Control: 3 days MAP: 5 days	Siro et al. (2003)

Fragaria × Ananassa, "Duchesne" cv.	5% O_2 15% CO_2		Gas composition, microbial infection, Botrytis infection	Botrytis (mycille)	Control: 3 days MAP: 6 days	Hertog et al. (1998)
Fengxiang variety	2.5% O_2 15% CO_2	PVC film, 48 µm	Gas composition, soluble sugar, acidity, TA, anthocyanin	Two ozone treatments	Control: 2 days MAP: 6–7 days	Zhang et al. (2005)
Fragaria × Ananassa, "Chandler & Douglas" cv.	15% CO_2	Cellophane bag, polyethylene film, 30 µm	Gas composition, weight loss, color, pH, TA, firmness, electrical conductivity, anthocyanin	Precooling, ethylene absorber	Control: 3 days MAP: 8–10 days	Picon et al. (1992)
Fragaria × Ananassa "Honeoye & Korona" cv.	11%–14% O_2 9%–12% CO_2	Antimist COPP film, 38 µm	Weight, soluble sugar, pH, aroma, acidity, TA		Control: 3 days MAP: 7 days	Nielsen and Leufven (2007)

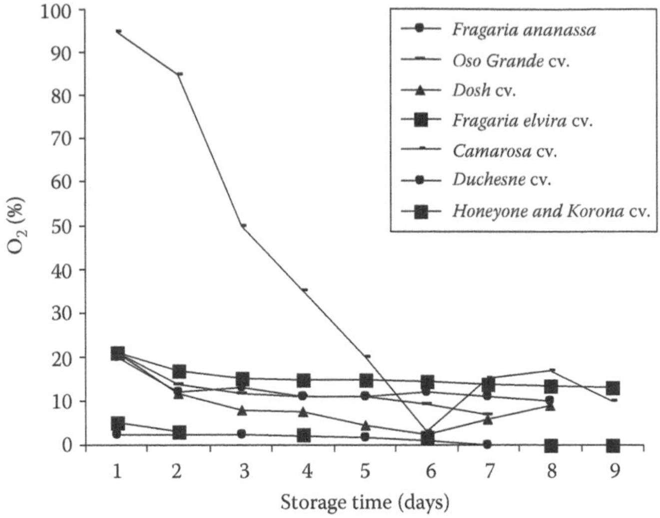

FIGURE 10.1 Oxygen composition of strawberry cultivars stored at 5°C for 10 days. (From Renault, P. et al., *Int. J. Food Sci. Technol.*, 29, 379, 1994; Silva, F.M. et al., *Postharvest Biol. Technol.*, 17, 9, 1999.)

development in *Korona* strawberries. Ethyl acetate, an indicator of fermentative metabolism, was released in large amounts in packaged *Korona*. Although this finding suggested that the altered gas composition initiated a process resulting in the accumulation of ethyl acetate in the tissue of *Korona* strawberry, the aroma production in Honeoye was not affected likewise.

The change in the oxygen concentration in strawberry cultivars packaged under MA conditions and stored at 5°C for 10 days is displayed in Figure 10.1. It is evident that the oxygen content dropped with time because of respiration activity. Figure 10.2 displays the change in carbon dioxide concentration in strawberry cultivars packaged in MA conditions and stored at 5°C for 10 days and change in firmness in strawberries under MAP and stored at 5°C for 12 days. It is obvious that the conditions of packaging greatly altered the flesh of the product. Figure 10.3 refers to the percentage of weight loss of strawberries versus storage time and firmness of apples. It is clear that in the beginning of the experiment fruits did not lose a high amount of their weight but after day 4 the percentage of loss increased substantially. The greatest loss was recorded for Camarosa cultivar at day 11.

10.1.2 Cherries

Spotts et al. (1998) investigated the effectiveness of preharvest iprodione and postharvest *Cryptococcus infirmo-miniatus* (CIM) treatments alone and in combination for the control of decay of sweet cherry fruit (*Prunus avium*). Although a single preharvest application of iprodione at 1.13 kg a.i./ha reduced brown rot in stored sweet cherry fruit in both years, better control was obtained in the cherry fruit

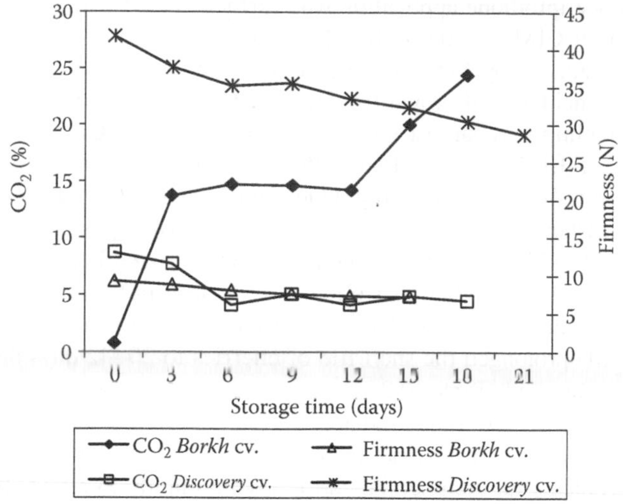

FIGURE 10.2 Change in CO_2 composition of strawberries stored at 5°C for 10 days (From Tano, K. et al., *Eur. J. Sci. Res.*, 21(2), 364, 2008) and firmness of strawberries. (From Van der Steen, C. et al., *Postharvest Biol. Technol.*, 26, 49, 2001.)

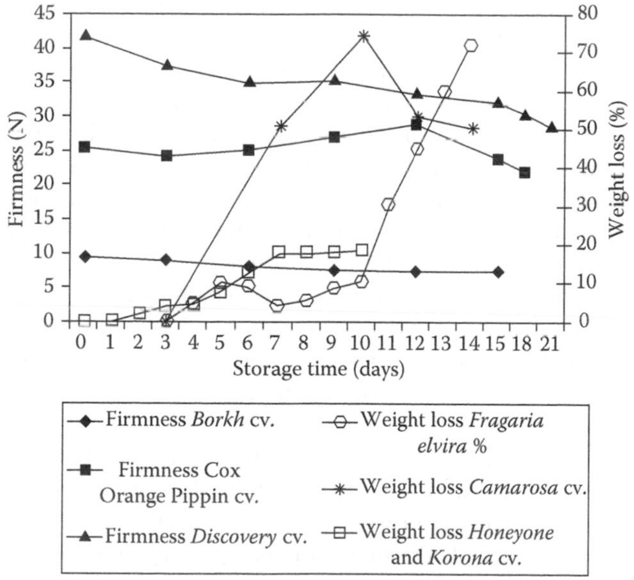

FIGURE 10.3 Weight loss of strawberries stored at 5°C (From Hertog, M.L.A.T.M. et al., *Postharvest Biol. Technol.*, 15, 1, 1998) and firmness of apple cultivars stored at 15°C versus storage time. (From Soliva-Fortuny, R.C. et al., *Int. J. Food Sci. Technol.*, 40, 369, 2003.)

when a preharvest iprodione application was further treated with a postharvest dip in a suspension of CIM containing 0.5–1.5 × 10^8 colony-forming units (CFU)/mL. Brown rot was reduced by MAP alone and combined as a result of a CIM–MAP synergism. The incidence of brown rot decreased from 41.5% in the control down to 0.4% by combining preharvest iprodione and postharvest CIM treatments with MAP. PE bags of 3.78 L were used and filled with 50% CO_2 and 50% O_2. Several other postharvest diseases, including blue mold (*Penicillium expansum* Link), Alternaria rot (*Alternaria alternata* (Fr.:Fr.) Keissl.), Rhizopus rot (*Rhizopus stolonifer* (Erhenb.:Fr.) Vuill.), and gray mold (*Botrytis cinerea* Pers.:Fr.), responsible for cherries' decay were studied as well. The CO_2 level in the MAP was found to be 11.5% after 42 days of storage. These two treatments, (preharvest iprodione and postharvest CIM) prolonged the shelf life of cherries to 20–42 days depending on the MAP system applied (Table 10.2).

Serrano et al. (2004) examined the application of natural antifungal compounds on MAP used in sweet cherries (*Prunus avium, Starking* cultivar). Sweet cherry displays severe problems for commercialization owing mainly to relatively fast decay and rapid loss of sensory quality, both for the fruit and the stem. The developed package was based on the inclusion of eugenol, thymol, menthol, or eucalyptol (pure essential oils) to trays sealed with nonoriented perforated PP film (20 μm thickness) to generate an MAP with concentrations of 10% O_2 and 20% CO_2. All cherries were stored over 16 days at 1°C and 90% RH. Steady-state atmosphere was reached only after 9 days of cold storage with 2%–3% of CO_2 and 11%–12% of O_2. When cherry quality parameters were determined, those treated with eugenol, thymol, or menthol exhibited benefits in terms of reduced weight loss, delayed color changes, and maintenance of fruit firmness. However, the weight losses were affected by the addition of natural antifungals. The addition of eugenol, thymol, or menthol led to lower reduction in chroma index, as compared to control cherries. The firmness of cherries at harvest was 1.48 ± 0.05 N/mm and over cold storage, a drop was observed but for those treated with eugenol, whose firmness remained unchanged. The final values of cherry firmness (N/mm) from high to low were eugenol (1.40), thymol (1.24), menthol (1.12), control (0.93), and eucalyptol (0.71). The microbial analysis revealed that all essential oils reduced molds and yeasts, and total aerobic mesophilic colonies by 4 and 2 log CFU compared with control, respectively.

After 16 days of cold storage, in all packages containing essential oils the microbial populations underwent a drastic reduction, which was by far more effective for mold and yeast counts (below 1.5 log CFU/g). In contrast, increases in microbial populations were recorded for control cherries, especially for mold and yeast counts, which were 4.9 log CFU/g. Serrano et al. (2004) suggested as alternative for synthetic fungicides, the active packaging addition of essential oils for improving MAP effectiveness. It was shown that the addition of eugenol, thymol, or menthol augmented the beneficial effect of MAP in terms of delaying weight loss, softening, color changes, and stem deterioration, thereby prolonging the cherry quality for longer storage periods. However, packages with eucalyptol performed worse than MAP control. All essential oils were shown to be effective in reducing the microorganism growth (damages in membrane integrity), the effect being higher for molds and yeasts than for mesophilic aerobics (Lambert et al., 2001).

TABLE 10.2
Effect of MAP (Gas Composition and Packaging Material) on Various Cherry Cultivars

Fruit	MAP Composition	Packaging Material	Shelf-Life Parameters Studied	Usage of Other Parameter	Shelf Life	References
Prunus avium	10% O_2 20% CO_2	Nonperforated oriented polypropylene film, 20 µm	Weight loss, color, firmness, sweetness, microbial infection	Antifungal treatments (eugenol, thymol, menthol, eucalyptol)	Control: 6 days MAP: 16 days	Serrano et al. (2004)
Prunus avium	14.4% O_2 7.1% CO_2	PE film, 20 µm	Brown rot (*M. fructicola*)	Preharvest fungicide, postharvest yeast	Control: 14 days MAP: 20 days at 2.8°C and 40 days at −0.5°C	Spotts et al. (2001)
Prunus avium "Lapis" cv.	18% O_2 4% CO_2	PE film, 40 µm	Acidity, pH, color, firmness, ethanol and ethylene productivity, vitamin C		Control: 14 days MAP: 20 days	Tian et al. (2003)
Navalinda sweet cherry	3% O_2 12% CO_2	Macroperforated and microperforated PP film, 35 µm	Acidity, color, firmness, rotting	Precooling method, hydrocooling	Control: 10 days MAP: 16 days	Alique et al. (2003)
Prunus avium "Sam" cv.	2% O_2 20% CO_2	LDPE film, 76.6 µm	Blue mold, altenaria rot, brown rot, rhizopus rot, gray mold, firmness	Ethanol and acetaldehyde 100 ppm	Control: 3–4 days MAP: 8 days at 25°C	Petracek et al. (2001)
Prunus avium "Burlat" cv.	3% O_2 12% CO_2	LDPE film, 50 µm	Acidity, color, firmness, anthocyanin, rotting		Control: 5 days MAP: 21 days	Remon et al. (1999)

Spotts et al. (2001) implemented an integrated approach for the control of postharvest brown rot of sweet cherry fruit (*Prumus avium* and *Lambert* and *Lapin* cultivars). System components included a preharvest application of propiconazole, a postharvest application of a wettable dispersible granular formulation of the yeast *Cryptococcus infirmo-miniatus* (CIM) *Pfaff* and *Fell*. The postharvest treatments applied to 'Lapins' and 'Lambert' sweet cherry fruit were: *M. fructicola* alone (control) and CIM combined with *M. fructicola*. These treatments were applied to the fruit from each of the two preharvest treatment regimes (no spray and propiconazole).

Fruits were immersed for 30s in treatment suspensions and stored in MAP at 2.8°C for 20 days or −0.5°C for 42 days. Preharvest propiconazole and postharvest CIM were equally effective for control of brown rot. And a synergism was observed. MAP significantly reduced brown rot compared to air-stored fruit. Fifty wet fruits were transferred into the PE (3.781) bags. Half of these bags were placed into larger folded but unsealed PE bags and stored in air at −0.5°C and 2.8°C, whereas the other half were stored in MAP. Prior to storage at −0.5°C and 2.8°C, the bags were evacuated, injected with a 50–50 gas mixture of CO_2 and N_2, and sealed. Preharvest propiconazole decreased brown rot in 'Lambert' and 'Lapins' in both storage atmospheres and temperatures except for 'Lapins' in MAP at −0.5°C. 'Lapins' fruit stored in MAP displayed substantially less decay than air-stored fruit at both storage temperatures in both years (Spotts et al., 1998). This treatment alone ensured control of brown rot but improved considerably when combined with a postharvest treatment of CIM. It was found that the combination of propiconazole–CIM displayed synergism demonstrated in effective decay control, thereby providing a means of excellent control with low fungicide residues and without the need for postharvest fungicide application. Although high CO_2 suppresses *M. fructicola*, its effect is rather fungistatic than fungicidal. This is because as soon as fruits are returned to air, decay resumes rapidly. Application of MAP prolonged the shelf life of cherries and reached 20 days at 2.8°C and 40 days at −0.5°C.

Tian et al. (2003) conducted experiments based on responses of physiology and quality of sweet cherry fruit *Prunus avium* Lapis cultivar to different atmospheres in storage. Fruits were stored in MAP (PE film 40 μm) and controlled atmospheres (CA of 5% O_2 plus 10% CO_2; or 70% O_2 plus 0% CO_2 at 1°C) to determine the effects of different O_2 and CO_2 concentrations on physiological properties, quality attributes, and storability during storage periods of 60 days. The ethanol contents of sweet cherry fruits enhanced rapidly from 7.1 nmol/g at harvest to 22.1, 17.0, and 11.8 nmol/g in MAP, control atmosphere with 5% O_2 + 10% CO_2 and control atmosphere with 70% O_2 + 0% CO_2, respectively, after 10 days. Sweet cherries preserved at 70% O_2 level always kept the lowest level of ethanol compared with the fruits with other treatments. However, CA with 5% O_2 + 10% CO_2 was much more effective in limiting ethylene production than other treatments. Decay incidence reached 16% in MAP-stored fruits, and only 1.6% decay rate was shown in the fruits under CA with 5% O_2 + 10% CO_2 after 60 days of storage. Firmness of sweet cherries stored in MAP slightly decreased with storage time. Fruits in 70% O_2 + 0% CO_2 exhibited a rapid decrease in firmness after 40 days and showed browning at this time. Vitamin C contents decreased rapidly with storage time. Sweet cherries stored under 5% O_2 + 10%

CO_2 had a relatively higher vitamin C content than that in other treatments. However, high O_2 atmosphere exhibited a beneficial effect in maintaining vitamin C content only up to 10 days, after which it rapidly dropped.

Alique et al. (2003) studied the benefits of different wrap-films for Navalinda cultivar sweet cherry, packed in small format (consumer packages) with MAP conditions. The ultimate purpose of this research was to propose guidelines for optimal preparation of Navalinda cultivar sweet cherries packed with macro- and microperforated films for distribution and marketing in the European markets. MAP effectively retarded deterioration of certain cherry quality parameters and decay caused by fungal growth. The use of modified atmospheres for cherry preservation required strict temperature control over all stages such as storage, distribution, and sale. Exposure of the product to high temperatures enhanced respiration, thereby resulting in ideal conditions for anaerobic conditions and fermentation induction thus releasing off-flavors. A storage and distribution process entailing 8 days at 4°C plus 4 days at 20°C was simulated by packing cherries in punnets and film-wrapping with one macroperforated film and two microperforated films of 0.30 (MAP.30) and 0.55 (MAP.55) mol cm/cm^2 atm day of CO_2 permeability at 0°C. Prolonged low-temperature storage caused increased respiratory intensity at 20°C and signs of senescence in the fruit. The sweet cherries packed in MAP.30 and MAP.55 microperforated films maintained similar acidity levels throughout the storage process, whereas the acidity of the cherries packed in macroperforated film differed significantly from that packed in microperforated films. The fact that acidity was maintained in the fruits packed in microperforated film indicated that the MAP slowed down the senescence process and the degradation of fruit quality. It was shown that MAP with microperforated wrapping of Navalinda cherries effectively retarded the loss of acidity and firmness and darkening of color and generally considerably limited the loss of quality and deterioration of the cherries, thereby resulting in shelf-life prolongation of sweet cherries up to 16 days.

Petracek et al. (2001) investigated the effect of O_2, the CO_2 partial pressures ($p_{O_2}^{pkg}$ and $p_{CO_2}^{pkg}$, respectively), and temperature on sweet cherry (*Prunus avium* cultivar Sam) respiration in MA (LDPE) packages. Steady-state $p_{O_2}^{pkg}$ and $p_{CO_2}^{pkg}$ were reached within 2 (25°C) to 10 (0°C) days. Respiration rates were calculated based on polymer film permeability, thickness, surface area, steady-state $p_{O_2}^{pkg}$ O_2 and $p_{CO_2}^{pkg}$, and packed fruit mass in an attempt to determine the effect of $p_{O_2}^{pkg}$, $p_{CO_2}^{pkg}$, and temperature on sweet cherry respiration, fermentative volatile formation, and shelf life. The potential of MA packaging as a practical technique for extending the shelf life of sweet cherries was confirmed by the obtained results.

The shelf life of sweet cherries is considerably shortened by loss of firmness, discoloration of the stem, desiccation, and decays due to blue mold (*Penicillium expansum*), gray mold (*Botrytis cinerea*), Rhizopus rot (*Rhizopus* spp.), brown rot (*M. fructicola*), Cladosporium rot (*Cladosporium herbarum*), and Alternaria rot (*Alternaria alternate*). Previous studies on MA packaging of sweet cherry revealed that while fruit stored in PE film retained firmness, off-odors closely linked with anaerobic respiration accumulated. Therefore, it was suggested that sweet cherry shelf life may benefit from MA packaging, provided the low $p_{O_2}^{pkg}$ and/or high $p_{CO_2}^{pkg}$ can effectively suppress mold growth, respiration, and other metabolic processes without stimulating anaerobic respiration. From a practical viewpoint, strict control of temperature was found to be the best tool for prolonging the shelf life of sweet cherries (Meheriuk et al., 1995).

The effects of the degree of ripeness and packaging atmosphere on the quality of the early season cherry (*Prunus avium*, cultivar Burlat) were investigated by Remon et al. (1999). Cherries were classified into two groups depending on their stage of ripening and color as a maturity index (red and purple). After rapid chilling, cherries were sealed in 50 μm LDPE bags at four different MAs and stored at 2°C. The analysis conducted revealed that the packed red cherries were regarded as commercial during 3 weeks of storage, whereas the purple ones lasted only 2 weeks.

Although all MAP conditions used displayed a beneficial effect on retention of firmness, the best results were obtained with red cherries. In red cherries, L* values decreased over the second week and then increased to values found at harvest (MAP 2 and MAP 3) or slightly lower (other samples). In purple cherries, all MAP conditions displayed clearly a decrease in L*. Under all MAP conditions, TA increased in red cherries during the first week and decreased afterward.

The commercial shelf life of Red Burlat cherries stored at 2°C in LDPE bags was prolonged substantially (threefold) compared with unpacked cherries. In the case of riper (purple-colored) cherries, their shelf life doubled when stored under MAP. The changes in oxygen and carbon dioxide composition are displayed in Figure 10.4.

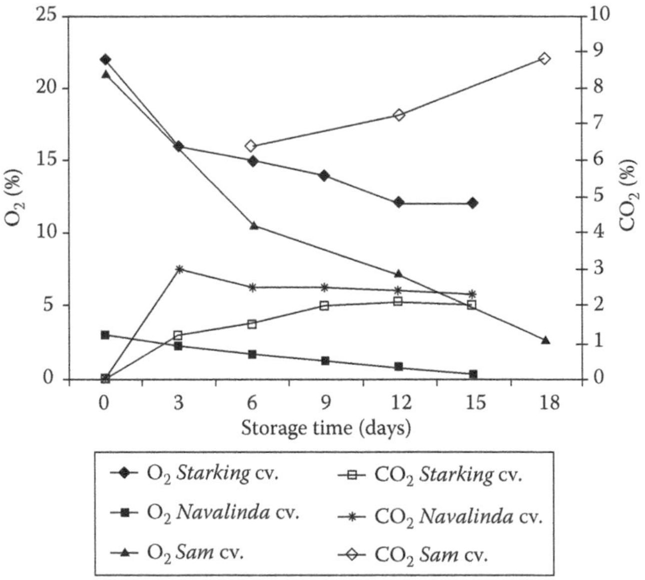

FIGURE 10.4 Oxygen percentage in MAP for cherry cultivars with storage time (days) at 5°C (From Alique, R. et al., *Eur. Food Res. Technol.*, 217, 416, 2003) and change in carbon dioxide composition in MAP for cherries stored at 5°C up to 18 days. (From Serrano, M. et al., *Innov. Food Sci. Emerg. Technol.*, 6, 115, 2004.)

10.1.3 Pineapple

The effect of packaging conditions on fresh-cut "Gold" pineapple shelf life was studied by Montero-Calderon et al. (2007), over 20 days of storage at 5°C. Fresh-cut fruit pieces were packed in PP trays and wrapped with 64 m PP film under active (high 40% or low oxygen, 11.4%) or passive MA (air or cut fruit coated with 1%, w/v alginate). Changes in headspace composition, TA, pH, SS content, juice leakage, color, texture, and microbial growth were evaluated over time. Although oxygen content decreased with storage time, it never dropped to levels below 2% inside packaging O_2, thereby avoiding the formation of off-odors. In the meantime, CO_2 concentration inside all packages displayed a steady rise over time up to 10.6%–11.7%. The high CO_2 levels recorded inside all PP containers after 8–11 days of storage at 5°C had no undesirable changes in quality attributes for over 15 days.

The fluctuation in ethylene content could be due to processing damages (initial rise), ethylene production of healthy tissue (considerable drop), and tissue deterioration after 12 days storage (increase in ethylene again).

Pineapple pieces stored under PP-HO displayed the least ethanol concentration (13 μL) up to the 20th storage day at 5°C. The average values for TA, soluble solids content (SS), and pH, which exhibited slight changes over storage, were 0.68 ± 0.02 g/100 g, 13.9% ± 0.2% SS, and 3.58 ± 0.04, respectively.

The liquid inside the package enhanced considerably over time for all packaging conditions. Fresh-cut pineapple stored in PP-ALG packaging conditions had less liquid leakage per fruit weight than those packed under PP-HO, PP-LO, and PP-AIR through time. Although no significant differences were recorded among packaging conditions for the microbial growth, considerable differences were reported throughout the storage time. Initial populations ranged from 3 to 4 log CFU/g for molds and yeasts at day 0 and reached 7–7.5 log CFU/g after 18 days of storage. A similar increase was reported for mesophilic and psychrophilic bacteria reaching values of 7–8.5 log CFU/g, respectively, after 18 days at 5°C. MAP allowed conservation of fresh-cut pineapples without undesirable changes in quality parameters during refrigerated storage. In fact, both fermentation and deterioration symptoms (ethanol concentration, off-odors, and off-flavors) were avoided during the first 2 weeks of storage, and the end of shelf life was apparent by mesophilic bacterial growth after 14 days storage.

The effects of pretreatment and MAP on the quality of fresh-cut pineapples stored at 4°C were evaluated for 7 days by Liu et al. (2006). The pretreatment was carried out by immersing the pineapple slices in a solution containing 0.25% ascorbic acid (AA) and 10% sucrose for 2 min. The applied MAP consisted of 4% oxygen, 10% carbon dioxide, and 86% nitrogen. Both pretreatment and MAP could decrease the respiration rate, ethylene production, textural and color deterioration, as well as the overall sensory degradation in fresh-cut pineapples. Over storage, the deterioration of color properties greatly affected the acceptance of fresh-cut fruits. Thus, both pretreatment and MAP could restrict the drop of luminosity in fresh-cut pineapples. Compared to the color properties, the textural properties were equally important for fresh-cut fruits. The cut surfaces allowed rapid growth of bacteria, thereby triggering deteriorations of texture, color, and even flavor properties of fresh-cut fruits. The total bacterial counts (TBC) increased rapidly during storage, and passive MAP-WP and passive MAP-P reached 10^7 and 10^6 CFU/g on day 7, respectively. However, the TBC in active MAP-WP and active MAP-P amounted to just 10^4 CFU/g. The obtained results indicated that MAP could suppress the growth of bacteria, but the pretreatment had a rather minor effect on the TBC. The sensory score for passive MAP-WP dropped dramatically over storage. Moreover, MAP-WP could sustain the overall sensory properties score of fresh-cut pineapples. The obtained results indicated that both pretreatment and MAP had a strong effect on the overall sensory properties (acceptable range up to 5 days) of fresh-cut pineapples.

The effect of storage temperature and modified O_2 and CO_2 concentrations on the atmosphere on the post-cutting life and quality of fresh-cut pineapple (*Ananas comosus*) was investigated by Marrero and Kader (2005). Changes in preservation conditions, pH, firmness, and quality parameters were evaluated versus time. Temperature was the main factor affecting post-cutting life, which ranged from 4 days at 10°C to over 14 days at 2.2°C and 0°C. The end of post-cutting life was evident with a sharp increase in CO_2 release followed by an enhancement in ethylene production. The main effect of decreased (8 kPa or lower) O_2 levels was better maintenance of the yellow color of the pulp pieces, as shown in higher final chroma values, while elevated (10 kPa) CO_2 levels resulted in a reduction in browning (higher L* values). MAP allowed preservation of pulp pieces for over 2 weeks at 5°C or lower without undesirable changes in quality parameters.

Enhanced (10%) CO_2 levels in conjunction with low (8% or lower) O_2 concentrations led to higher final values of luminosity (L*) and acidity of pulp wedges, whereas reduced O_2, with or without increased CO_2, improved the retention of color, as measured by the chroma values, compared to wedges kept under humidified air for 15 days at 5°C. The post-cutting life of pulp pieces varied from 4 days at 10°C to over 2 weeks at 0°C.

Quality changes (firmness, color, total soluble solids [TSS], TA, sensory quality, and microbial safety) in fresh-cut pineapples were evaluated by Chonhenchob et al. (2005). The effect of chemical treatments on reduction of browning, firmness loss, and decay of fresh-cut tropical fruits was investigated. The most effective agents for fresh-cut pineapples were 0.1 M AA 0.2 M and 0.2 M AA + 0.2 M calcium chloride (CALC), respectively. Fresh-cut tropical fruits were packed in various rigid containers (PET, OPS, and OPLA). Fresh-cut pineapple treated with AA gave the highest scores for color and overall acceptance, whereas CALC had no effect on the firmness of fresh-cut fruits. In-package gas compositions at equilibrium of pineapples in PET were 6% O_2 and 14% CO_2. An MA of 6% O_2 and 14% CO_2 in PET extended the shelf life of fresh-cut pineapples from 6 days to 13 days. The results revealed that fresh-cut pineapples packed under 4% O_2, 10% CO_2, and 86% N_2 had the longest shelf-life as compared with those in air and VP. L* values decreased slightly in fresh-cut fruits over storage due to the effectiveness of AA (0.2 M) in browning inhibition. The TSS of fresh-cut fruits augmented with storage, whereas the TA decreased. The results revealed that pineapples packaged in PET had a significantly lower TSS/TA than those packaged in OPS and OPLA. All treatments displayed a sharp drop in sensory quality over storage. The major indicators of quality loss were surface slime and visible fungi. Less than 10^2 CFU/g yeasts and mold formation were reported in market acceptable fresh-cut fruits. Passive MA with reduced O_2 and elevated CO_2 in rigid container packaging could be beneficial to maintaining the quality and prolonging shelf life of fresh-cut pineapples provided an appropriate package was selected (Table 10.3).

Postharvest physiology and quality maintenance of sliced pear were investigated by Rosen and Kader (1989). Partially ripe pears (cv. 'Bartlett') were dipped in various solutions (citric acid (CITA), AA, and CALC) and stored in air or in CA for 7 days at 2.5°C followed by 1 day at 20°C. 'Bartlett' pears were used for various tests after storage at −1°C for 2–12 weeks. Pears were sorted to eliminate damaged or

TABLE 10.3
Impact of MAP (Gas Composition, Packaging Material) on Shelf Life of Pineapple Cultivars

Fruit	MAP Composition	Packaging Material	Shelf-Life Parameters Studied	Usage of Other Parameters	Shelf Life	References
Ananas comosus Gold pineapple	HO: 38%–40% O_2 LO: 10%–12% O_2, 1% CO_2, AIR: 20.9% O_2	PP film, 64 µm	Gas composition, TA, pH, SS, color, texture, microbial growth, juice leakage	Alginate coating, AA, CITA	Control: 2 days MAP: 30 days	Montero-Calderon et al. (2007)
Ananas comosus Gold pineapple	4% O_2 10% CO_2 86% N_2	Polystyrene (PS) container	Respiration rate, ethylene production, textural and color deterioration, sensory quality	AA, sucrose, sodium hypochlorite (NaClO)	Control: 2 days MAP: 30 days	Liu et al. (2006)
Ananas comosus Gold pineapple	8% O_2 10% CO_2	Oriented PS cups	Conservation conditions, pH, firmness, quality evaluation	AA, sodium hypochlorite (NaClO)	Control: 2 days MAP: 30 days	Marrero and Kader (2005)
Ananas comosus Gold pineapple	6% O_2 14% CO_2	Oriented polylactide container (PLLA), Polyethylene terephthalate (PET), Oriented PS	Firmness, color, SS, TA, sensory quality, microbial safety	AA, CALC	Control: 2 days MAP: 30 days	Chonhenchob et al. (2005)

defective fruits. They were then partially ripened by exposure to 10 ppm ethylene at 20°C for 48 h and sorted by color. CA treatments for pears consisted of 0.5%, 1%, or 2% O_2 and 12% CO_2. The sliced pears respired at a higher rate than whole pears throughout the 8 day storage period. The whole fruit consistently produced more ethylene than the sliced fruit throughout the 8 day storage period. CA storage considerably reduced respiration and ethylene production rates of sliced fruits. The firmness of strawberry and pear slices was maintained by storage in air, 12% CO_2 and in a 0.5% O_2 atmosphere, respectively, or by dipping in 1% CALC. These treatments resulted in lighter colored pear slices than the air control treatment.

Geeson et al. (1991b) investigated the effect of MAP on unripe green Conference pears with initial firmness values of 46–55 N sealed in MA LDPE (30 μm) pillow packs. Changes in pack atmosphere composition, skin chlorophyll content, flesh firmness, and sensorial quality were monitored during 14–20 days simulated shelf life at 20°C. In MA packs, equilibrated atmospheres contained 5%–9% CO_2 and <5% O_2. The concentration of CO_2 in MA packs increased to between 5% and 9%, and that of O_2 decreased to 5% or even less during the first 3 days. In control packs, the chlorophyll content declined steeply, but in MA packs there was a negligible change over the course of the experiment. In MA packs, the pears softened much more slowly over the first 4 days, but thereafter the firmness readings declined at a faster rate to reach mean values of chlorophyll. MA packaging initiated only partial inhibition of flesh softening and, even after perforation, the pears failed to develop the same texture as that of the control fruit. In MA packs, ripening was very slow, and the pears remained quite firm with little flavor for at least 14 days. The failure of the MA-packed fruit to develop the aromatic and textural characteristics of normal ripe Conference pears in conjunction with high pack variability rather precludes the use of MA packaging for extending the market life of this particular variety.

The application of high O_2 atmospheres (HOA) of 70 kPa O_2, balance N_2, for active MAP of fresh-cut 'Flor de Invierno' pears was evaluated by Oms-Oliu et al. (2007) as an alternative to conventional low O_2 atmosphere (LOA), active MP (2.5 kPa O_2+7 kPa CO_2, balance N_2), and to traditional passive atmosphere (PA) packaging. Gas exchange, color, firmness, and microbiological stability were determined throughout 28 days storage at 4°C. Although HOA did not prevent the production of acetaldehyde and ethanol during storage of fresh-cut pears, their accumulation was promoted under anoxic conditions. After being peeled and their core removed, the remaining tissue of pears was cut into wedges. Pear wedges were dipped for 1 min in an aqueous solution of N-acetylcysteine at 0.75% (w/v) and glutathione at 0.75% (w/v). The amount of O_2 inside packages stored under HOA decreased over storage, while the accumulation of CO_2 under HOA was similar to that of packages stored under PA. For fresh-cut pears stored under HOA, the O_2 concentrations inside the packages remained higher than 50 kPa throughout storage. However, CO_2 increased to levels higher than 20 kPa by the end of storage. Ethylene concentrations in package headspaces were considerably affected by storage atmosphere. It was shown that the storage temperature greatly affected the ethylene concentrations in package headspaces and the changes in acetaldehyde concentrations inside packages. Acetaldehyde started to accumulate in packages stored under HOA or

LOA after 1 week of storage, and it continuously increased throughout time. The changes in color and firmness of fresh-cut 'Flor de Inviero' pears dipped into 0.75% of N-acetylcysteine and 0.75% of glutathione solution and stored under different MA conditions were investigated. Browning on pear wedges stored under HOA could be visually assessed after 2 weeks of storage. However, HOA best maintained the initial firmness of fresh-cut pears. The drop in firmness reported in fresh-cut pears stored under LOA or PA, especially after 10–14 days of storage, may be related to excessively low O_2 and high CO_2, and ethanol accumulation in packages. The growth rate for aerobic psychrophilic microorganisms was much higher in pears stored under PA than under both LOA and HOA. The most important spoilage microorganisms isolated from fresh-cut pears packaged under MA conditions were *Candida parapsilosis* and *Rhodotorula mucilaginosa*.

In an attempt to determine which was the most effective treatment against the enzymatic browning of pears (Williams, Conference, Passacrassana), two antioxidant treatments were tested (treatment 1: 2% AA + 1% CITA + 1% $CaCl_2$; treatment 2: 2% AA + 0.01% 4-hexylresorcinol [4-HR] + 1% $CaCl_2$) (Arias et al., 2006). An MAP was designed based on two previous tests: measurement of respiratory activity of the peeled and cut pear at three temperatures (4°C, 15°C, and 25°C); and evaluation of fruit tolerance at three different atmospheric compositions (21% O_2 + 10% CO_2, 2% O_2 + 0% CO_2, 2% O_2 + 10% CO_2) per storage duration (0, 3, 6, and 9 days at 4°C). Browning potential was evaluated over 9 days. *Passacrassana* pears consistently showed greater absorbance than the other two varieties. Initial polyphenol oxidase (PPO) activity and phenolic content were determined per variety. A greater degree of inhibition to browning was reached by the treatment based on AA, 4-HR, and $CaCl_2$. The samples packed in an MA with a composition of 10% O_2 + 10% CO_2 + 80% N_2 atmospheres with a high concentration of CO_2 (21% O_2 + 10% CO_2 and 2% O_2 + 10% CO_2) were found to be the most appropriate to preserve the product at low temperatures.

Guevara et al. (2000) tested the effect of packaging nopalitos (*Opubtia ficus-indica*, *Milpa Alta* cultivar of pears) at 5°C under MA. Changes in texture, weight, chlorophyll content, crude fiber content, color, and microbial growth were evaluated over storage time. The atmosphere modification that occurred was probably due to the high respiration rate of the cactus stems. The oxygen concentration in MAP decreased and reached about 8.6 kPa after 30 days in storage, while the CO_2 concentration increased to about 6.9 kPa in the same period. The in-package O_2 concentration underwent a decrease of 35% after 30 days, whereas the in-package CO_2 increased up to 100% after 30 days. The weight loss increased considerably with storage time extension. The top quality of packed produce in MA was recorded in cladodes packaged in MA, followed by those maintained at high RH, and was the least in cladodes that were kept without packaging. The overall quality of two cladodes preserved in MA was still high (with scores of over 6) after 30 days in storage. MAP considerably limited the brown discoloration of the cut surface and decreased the mucilage loss as well. Crude fiber content and texture of cladodes held in MAP decreased only slightly. There has been a close relation between the changes in texture and crude fiber contents, thereby suggesting that the retention in texture might be due to the positive effects of MA on preventing crude fiber losses.

It is possible that the decrease in crude fiber content is due to increase in the activity of enzymes such as celluloses, hemicellulases, and pectinases. The decrease in total chlorophyll after 30 days was about 250% higher for the cladodes in bulut compared to the ones in MA. Cladodes held in MAP displayed an enhanced level of anaerobic mesophiles, which reached the same level as in cladodes held at high RH after 30 days storage.

Although some retardation in ripening of Doyenne du Comice pears was reported when unripe fruits were held in MA packs at 20°C, softening proceeded at a more rapid rate than chlorophyll loss so that the appearance of the fruit failed to match its internal condition. Gas composition, color, firmness, aroma, and sensory quality were assessed versus time by Geeson et al. (1991a). Experiments were conducted with four Doyenne du Comice pears packed in a preformed foam tray of 30 μm LDPE, with atmosphere equilibrated to 6%–8% CO_2 and 2%–3% O_2. Multiperforated control packs were already slightly over ripe 4 days after packing. In packs with four m-p (microperforations), pears were sweet, juicy, and eating-ripe after 4 or 7 days but had become overripe with some internal browning after 10 days and with one m-p pears ripened more slowly and remained firm after 10 days. However, all pears in micro-perforated MA packs were as ripe as the controls after 5 days and all were similarly dry, over-ripe with internal browning and breakdown after 8 days. Sensory differences between MA-packed and control fruit corresponded to those found by instrumental methods. Thus, the MA pears were significantly firmer and, by day 12, less juicy and aromatic (one m-p) than the control fruit. In fact, the obtained results revealed that MA packaging retarded (the less the microperforation number, the greater the delay) the rates of degreening and softening of conditioned, part-ripe Comice similarly to unripe Conference pears.

However, provided the skin color had already reached an acceptable stage of "yellow color" prior to packaging, the slight retardation in the rate of softening in the one m-p MA packs seemed to offer a better opportunity for extending ambient shelf life by several days without adversely affecting their quality. Internal atmosphere, quality attributes, and sensory evaluation of MA packaged fresh-cut Conference pears were investigated by Soliva-Fortuny et al. (2005). Conference pear cubes processed at partially ripe maturity preserved an acceptable sensory quality during 3 weeks storage. On the contrary, CO_2 internal levels enhanced dramatically from the beginning of storage under all tested conditions, reaching values of 38.9–49.1 kPa at 21 day. These high concentrations of CO_2 suggest that anaerobic respiration could be triggered specially during the second and third weeks of storage. The product underwent a progressive but slight loss of flavor, color, and firmness. Proper processing and storage managed to ensure a shelf life of 3 weeks for fresh-cut Conference pear cubes obtained from adequately ripened fruits. The presence of O_2 in the storage atmosphere had a strong effect on the sensorial shelf life of the product. Storage under a low amount of oxygen was detrimental both to flavor perception and to the fruit tissue in conjunction with high CO_2 concentrations, releasing off-flavors and triggering the production of fermentative metabolites beyond 3 weeks. The changes in yeast and molds of two cultivars stored at 5°C versus storage time (28 days) and change in mesophilic aerobic microorganisms in pears stored at 5°C versus time (20 days) are displayed in Figure 10.5.

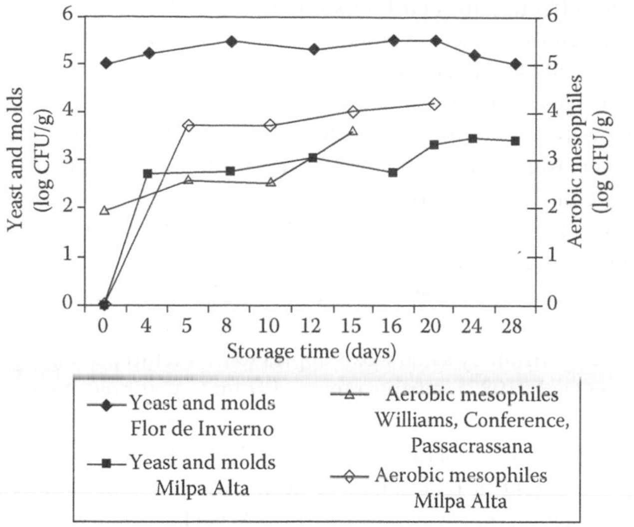

FIGURE 10.5 Changes in yeast and molds in pears for cultivars Flor de Invierno and Milpa Alta (From Guevara, J.C. et al., *Lebensm. Wiss. Technol.*, 34, 445, 2000) and changes in mesophilic aerobic microorganisms found at pear cultivars versus storage time. (From Soliva-Fortuny, R. et al., *Int. J. Food Sci. Technol.*, 42, 208, 2005.)

10.1.4 Apple

Apple Bravo de Esmolfe originating from the parish of Esmolfe is cultivated in specific regions of Portugal. This cultivar, although it requires very low temperatures, is extremely sensitive to frost. Rocha et al. (2002) assessed the effect of MA on the quality of apple (cv. *Bravo de Esmolfe*) during cold storage. Fruits were harvested and stored at 2°C and 85% RH in air and MA packages over 6.5 months and assessed at different periods of storage. Two different plastic films were used: perforated bags to allow normal atmosphere (NA), plastic 1, and plastic 2–PP 100 μm at 2°C (MA). Although the initial concentrations of gases were 13.9% and 2.3% for O_2 and CO_2, respectively, after 130 days O_2 concentration dropped to 10% and the CO_2 concentration increased to 5%. The weight loss of apples packed in MA amounted to less than 2% after 6.5 months. Water loss can greatly affect the deterioration by causing losses in appearance (wilting), texture (softening, flaccidity, loss of juiciness), and nutritional quality. Apples stored in MA packs exhibited better color than fruits stored in air. Apples stored under MA storage for 6.5 months at 2°C were firmer than fruits stored under air. The absence of any significant differences in the pH and SS content of apples stored in air and MA after 6.5 months at 2°C implied that the MA had no effect on these parameters.

Smith et al. (1998) investigated the effects of MA on physiological maturity of Discovery apples to retail packaging. The apples were packed in MA (30 μm ethylene vinyl acetate [EVA]) or perforated control packs at marketing conditions (10°C or 20°C). The effects of MA packaging on fruit ripening changes were greatly affected by harvest date and related to the respiration rate of fruit when packed.

MA packs were effective in retarding softening and skin color changes in immediately pre-climacteric and early climacteric fruit. The concentration of CO_2 rose rapidly, whereas that of O_2 decreased equally fast, over the first 24 h at both temperatures. These changes were more marked at 20°C than at 10°C. At 10°C, CO_2 and O_2 concentrations were similar for MA packed fruit from all harvest dates, but at 20°C, O_2 concentrations were lower in packs coming from the later harvests. The weight loss was consistently higher in control packs than in MA packs at both temperatures. At all harvests, the flesh of the MA-packed fruit was found to be firmer, tougher (woodier), and less mealy than that of the control fruit. After 14 days, the number of significant sensorial differences in textural attributes augmented at all harvests. Some tainting or off-flavor development was reported in MA-packed fruit.

The responses of Bramley's and Cox's apples to MA retail packaging were investigated by Geeson et al. (1987) in fruits removed from CA storage at monthly intervals throughout the marketing season, packed in MA or perforated control packs and held under simulated marketing conditions at 15°C. Moreover, differential effects of MA packaging on the extent of retardation of flesh softening and skin yellowing were found as the storage period was prolonged. MA packs made from LDPE (30 μm) were effective in delaying changes in firmness and skin color of Bramleys, previously stored for less than 22 weeks in 8% CO_2 + 13% O_2 or for less than 30 weeks in 6% CO_2 + 3% O_2. Similarly, during a 2-week marketing period at 15°C the concentrations of CO_2 in both LDPE and EVA (30 μm) packs reached the highest peak after 1–2 days and thereafter showed a gradual decline, while the concentrations of oxygen decreased rapidly for 2 days and then gradually increased. Weight losses both in Bramley's and Cox's were consistently approximately 0.1% per week in MA packs than in perforated control packs (0.4%–1.1% and 0.2%–0.4% per week) at all packing dates. As with Bramley's, the firmness of Cox's after 2 weeks in all types of packs declined progressively through the storage season. The skin color of fruit held for 2 or 4 weeks in MA packs was significantly greener than that in control packs per packing date. Although, the skin of Bramley's in MA packs remained dry, the fruit in control packs became very greasy, particularly after longer storage or holding periods. Some samples packed in both types of MA pack were significantly firmer, tougher, and more acid than the corresponding controls.

MAP was used for collecting data for apple (*Malus × domestica Borkh.*) slices at 0°C, 5°C, 10°C, and 15°C (Lakakul et al., 1999). Researchers evaluated the activation energy of oxygen, the respiration rate, and the permeability of oxygen uptake. They used 76.2 and 101.6 mm thick LDPE film and a polyolefin plastomer (POP) film made from a proprietary resin. The respiratory rate of apple slices was two to three times that reported for the whole fruit, due to wounding. Since diffusive resistance of whole fruit resulted in an O_2 partial pressure gradient of only 2–5 kPa, the interior O_2 level was high enough to very nearly saturate the respiratory response. The suggested model could be used for designing a package to minimize fermentation and predict the headspace atmosphere composition for sliced apples in packages consisting of any polymer film of known permeability, thickness, area, and fruit mass. Decay was observed on the apple slices in packages within 5 days at 15°C and 8 days at 10°C and browning was considered unacceptable within 24 h of packaging at all temperatures (Table 10.4).

TABLE 10.4
Effect of MAP (Gas Composition, Packaging Material) in Conjunction with Other Parameters on the Shelf Life of Apple Cultivars

Fruit	MAP Composition	Packaging Material	Shelf-Life Parameters Studied	Usage of Other Parameter	Shelf Life	References
Malus domestica "Bravo de Esmolfe" cv.	13.9% O_2 2.5% CO_2	Plastic 2-PP film, 100 μm	Gas composition, weight loss, color, pH, SS, firmness		Control: 2 months MAP: 6.5 months	Rocha et al. (2002)
Malus domestica "Discovery" cv.		EVA film, 30 μm	Gas composition color, firmness, sensory analysis, respiration rate, maturity measurement		Control: 6 days MAP: 14 days	Smith et al. (1998)
Cox's Orange Pippin apples	13% O_2 8% CO_2 or 6% O_2 3% CO_2	EVA film, 30 μm	Gas composition, weight loss, color, pH, sensory analysis, firmness	Benomyl, diphenylamine	Control: 6 days MAP: 2–4 weeks	Geeson et al. (1987)
Malus domestica "Borkh" cv.	20.7% O_2	LDPE film, 76.2 μm or POP film 101.6 μm	Activation energy of O_2, respiratory rate, permeability O_2 update	Ethanol	Control: 6 days MAP: 20 days	Lakakul et al. (1999)
Malus domestica "Borkh" cv.	2.5% O_2 7% CO_2		SS, color, pH, sensory analysis, firmness, acidity	AA, CALC, ethanol, acetaldehyde	Control: 6 days MAP: 21 days	Soliva-Fortuny et al. (2004)

The organoleptic shelf life of Golden Delicious fresh-cut apples, dipped in 10 g/L AA and 5 g/L CALC and packaged under different MA conditions, was assessed by Soliva-Fortuny et al. (2003) over refrigerated storage. Fruits were packed in plastic bags under 2.5 kPa O_2 and 7 kPa CO_2. Oxygen concentrations in the fresh-cut tissue were found to decrease exponentially. Simultaneously, carbon dioxide increased during 3 weeks of storage. Ethylene concentrations augmented dramatically, thus inducing wounding response leading to maximum concentrations between the first and the second weeks of storage. Acetaldehyde and ethanol were released during the storage period in low concentration. Although ethanol accumulation was insignificant over the first 3 days of storage, it enhanced thereafter. Application of a dipping treatment of 10 g/L AA and 5 g/L CALC was shown to be very effective in preserving the initial color of fresh-cut Golden Delicious apple cubes. The product was microbiologically stable during the first 3 weeks of storage. Apples packaged under O_2-restrictive conditions (0 kPa O_2 and plastic material) kept their initial firmness without significant changes during 3 weeks. The packaged apple texture, SS, and total acidity and the amounts of glucose and fructose did not change significantly or very slightly over storage.

10.1.5 Banana

An integrated strategy to control postharvest decay of Embul banana was put forward by Ranasinghe et al. (2004). Mature Embul (Musa, AAB) bananas were treated with emulsions of either cinnamon bark or leaf (*Cinnamomum zeylanicum*) or clove (*Syzygium aromaticum*) oils to control postharvest diseases, packed under MA using (0.075 mm, LDPE) bags, and stored in a cold room (14°C, 90% RH) or at ambient temperature (28°C). Treatment with emulsions of cinnamon oils (crown rot control) combined with MAP can be recommended as a safe, cost-effective method for prolonging the storage life of Embul bananas up to 21 days in a cold room and 14 days at 28°C ± 2°C without any effect on the organoleptic and physicochemical properties. Essential oil treatments hardly had any effect on the flesh firmness, TSS, pH, titratable acids, carbohydrates, and reducing sugar content of bananas after 14 days at 28°C or 21 days at 14°C ± 1°C. Apart from odor, flavor, taste, and overall acceptability, oil-treated banana exhibited considerable differences compared with the benomyl treatment when stored at 28°C.

Senescent spotting of banana peel was investigated by Choehom et al. (2003). Banana fruits (*Musa cavendishii* [*Musa acuminata*] AA Group cv. Sucrier) were placed in trays and held at 29°C–30°C. Covering the trays with "Sun wrap" polyvinyl chloride (PVC) film prevented the early senescent peel spotting, typical for this cultivar. CO_2 and ethylene concentrations within the packages rose, but inclusion of CO_2 scrubbers or ethylene absorbents had no effect on spotting. The positive effect of MAP on peel spotting was accompanied with reduced in vitro phenylalanine ammonia lyase (PAL) activity in the peel, in conjunction with an increase of in vitro PPO activity (Table 10.5).

The effect of MAP on chilling-induced peel browning in banana was examined by Bich et al. (2003). Sucrier bananas (*Musa acuminata*) stored at 10°C resulted in chilling injury (CI). Fruit was placed in packages with and without MA. Oxygen levels in the MA packages were approximately 12% while CO_2 concentrations amounted to 4%. The peel of control bananas displayed grayish brown discoloration after 6 days of storage. It is noteworthy that after 12 and 18 days of storage, scores for softness, sweetness, and flavor were higher in MA-packed fruit. Over storage, the total free phenolics content in the peel of control bananas was considerably lower than that of bananas with MA packaging. Pulp softness, sweetness, and flavor of MA-packed fruit were better than in control fruit, thereby confirming that MA reduced CI symptoms. PAL and PPO activities may be causally related to CI-induced peel browning.

TABLE 10.5
Effect of MAP (Gas Composition, Packaging Material) on the Shelf Life of Several Banana Cultivars

Fruit	MAP Composition	Packaging Material	Shelf-Life Parameters Studied	Usage of Other Parameters	Shelf Life	References
Musa acuminate Embul cv.	8% O_2 11% CO_2	LDPE film, 75 μm	Organoleptic properties, pathological properties, physicochemical parameters	Cinnamon leaf or bark, clove oils, fungicide	Control: 14 days MAP: 21 days	Ranasinghe et al. (2004)
Musa cavendishii Sucrier cv.	3.1% O_2 6.8% CO_2	PVC film, 12.5 μm	Gas composition, ethylene content, pH	Phenylalanine ammonium lyase, PPO	Control: 6 days MAP: 20 days	Choehom et al. (2003)
Musa cavendishii Sucrier cv.	12% O_2 4% CO_2	Nonperforated PE film, 11 μm	Gas composition, softness, color, pH, flavor, browning	Phenylalanine ammonium lyase, PPO	Control: 6 days MAP: 30 days	Bich et al. (2003)
Pachbale variety	3% O_2 5% CO_2 92% N_2	PE pouches, 100 gouge	Penetration, shear, force-relaxation, sensory analysis	Ethylene scrubber	Control: 8 days MAP: 21 days	Chauhan et al. (2006)

Textural properties of MAP banana (var. *Pachbale*), stored at 13°C ± 1°C, followed by Ethrel-induced ripening at 30°C ± 1°C were studied by Chauhan et al. (2006). MA included active as well as passive types involving flushing of PE pouches with specific gas mixture (3% O_2 + 5% CO_2 + 92% N_2) at partial vacuum (52.63 kPa). All the textural parameters investigated displayed a decline except for adhesiveness with the progress in ripening. In the case of partial VP samples, a threshold low-temperature duration of 30 days was found to be optimal to avoid abnormal ripening in terms of texture.

10.1.6 Litchi Cultivars

The effect of MAP and postharvest treatments on quality retention of litchi cv. *Mauritius* was studied by Sivakumar and Korsten (2005). Sulfur dioxide (SO_2) fumigation is a commonly adopted commercial treatment to retain the postharvest quality of litchi (*Litchi chinensis* Sonn.) fruit. The fruits of litchi cv. *Mauritius* were dipped for 2 min at 8°C in postharvest treatment solutions of ethylenediamine tetraacetic acid (EDTA), calcium disodium salt hydrate (EDTA) (0.1%), phosphoric acid (0.1%), or 4-HR (0.1%). Then, fruits were packed in three types of BOPP packaging: (1) BOPP-1, (2) BOPP-2, and (3) BOPP-3, heat-sealed to modify the atmosphere around the fruits and placed in commercial cardboard cartons. Fruits were first held at 2°C, 95% RH for 34 days to simulate refrigerated sea shipment conditions and then, at market conditions of 14°C for 2 days. The MA created (17.0% O_2 and 6.0% CO_2) inside the BOPP-3 and the high RH around the fruit minimized the transpiration rate, thereby preventing the weight loss and deterioration of fruit quality. Multivariate canonical variate analysis (CVA) of the results indicated that fruit packed in BOPP-3 retained color and excellent eating qualities during long-term storage. The BOPP-3 packaging can be recommended as a safe, cost-effective alternative for extending the storage life of litchi cv. *Mauritius* during sea shipment.

Postharvest browning of litchi fruit caused by water loss was investigated by Jiang and Fu (1998) in relation to anthocyanin content, PPO activity, pH value, and membrane permeability. Pericarp desiccation resulted in decreased anthocyanin concentration. Fruit stored at 90% RH displayed the lowest loss of total anthocyanins (TA) over the storage period, whereas TA content of fruit stored at 60% and 70% RHs declined considerably. PO activity with 4-methylcatechol (4-MC) as a substrate enhanced with reduced storage RH, but no activity toward TA was detected. Moreover, in the presence of 4-MC, the oxidation of anthocyanins by PPO was reported. The pH value was initially low and then increased with pericarp desiccation, being related to increased PPO activity. It was shown that oxidation of both phenolic and anthocyanins by PPO affected the response of litchi fruit to water loss in terms of browning and suggested that substrate–enzyme contact should be emphasized as this could promote enzymatic reaction leading to browning. Storage at 13°C under CA (3%–5% O_2 transpiration and 3%–5% CO transpiration) at 90% RH led to satisfactory good browning control and fruit quality maintenance.

Litchi (*Litchi chinensis* Sonn. cv. Heiye) fruits were stored in air, MAP, and CA at 3°C to determine the effects of different O transpiration and CO transpiration atmospheres on physiology, quality, and decay during the storage periods. Tian et al. (2004) reported that CA conditions were effective in reducing total phenol content, delaying anthocyanidin decomposition and preventing pericarp browning. PPO, peroxidase (POD), anthocyanin, and total phenols were involved in cellular browning. High O_2 treatment considerably limited ethanol production of litchi flesh in the early period of storage. The fruit stored in CA conditions for 42 days preserved good quality without any off-flavor. Color deterioration in litchi pericarp was due to browning reactions, and active PPO and POD were reported. It was found that CA conditions were more effective in delaying pericarp browning and reducing fruit decay than MAP.

The effects of MAP (BOPP-1 or BOPP-2) in conjunction with antimicrobial agents *Bacillus subtilis*, 10^7 CFU/mL, EDTA, calcium disodium salt hydrate (0.1%), or 4-HR (0.15%) on postharvest decay control and quality retention of litchi cv. *McLean's Red* were assessed by Sivakumar et al. (2007) as possible substitutes for commercial SO_2 fumigation. *B. subtilis* was shown to survive effectively in BOPP-1 (16% O_2, 6% CO_2, 90% RH), but its survival was adversely affected in BOPP-2 (5% O_2, 8% CO_2, 93% RH). *A. alternata* and *Cladosporium* spp. were the major decay-causing fungi in BOPP-1 treatments, and *Candida*, *Cryptococcus*, and *Zygosaccharomyces* were the predominant yeasts in BOPP-2 treatments. Combination treatments of EDTA, 4-HR, or *B. subtilis* in BOPP-1 inhibited PPO and significantly reduced pericarp browning and severity. Among the combination treatments, *B. subtilis* + BOPP-1 had the best potential to control decay, maintain the color, and the overall litchi fruit quality during a marketing chain of 20 days.

10.2 CONCLUSIONS

It should be noted that one can hardly claim that all strawberries being sold in the future should be packaged, not even the cultivars responding well to MAP, as it is not of use if the produce is sold within a day after harvest. However, packaging of strawberries might be a very promising alternative for producers who would like to export their produce (longer transportation time).

The cherry industry is interested in decay minimization by nonchemical means and preservation of fruit quality during long-term storage and distant transport. On many occasions, slower surface transportation is used rather than more expensive air transport. Implementation of the preharvest iprodione–postharvest yeast-MAP system would improve the decay control over long-term storage and transport of sweet cherry fruit.

It was found that Burlat cherries can be kept in an excellent commercial state for as long as 3 weeks by adopting these recommendations: (i) harvesting the cherries at the red color stage, (ii) quick chilling, (iii) packaging in air using LDPE, and (iv) keeping the temperature during the storage period at 2°C. These conditions allow for successful long-distance transportation of this highly perishable fruit.

MAP is a very promising technique particularly in certain areas of novel processed foods for the prolongation of shelf life and control of the growth of food pathogens at food refrigeration temperatures. However, it should not be considered a substitute for refrigeration. Although CO_2 has bacteriostatic properties, there is little evidence that it represents a significantly greater hazard than packaging in air. A ratio of spoilage to pathogenesis can be applied to determine the best MA conditions for the simultaneous control of spoilage and pathogenic microorganisms for a given product. Further research is required to widen our understanding of the effects of MAs on food pathogens and in particular clostridis. The development of spoilage relative to pathogenesis should be studied for more non-seafood products. A determination of the mechanism by which CO_2 inhibits microorganisms would allow its more effective use. Although researchers are beginning to determine the effects of carbon dioxide on certain pathogenic organisms, there is a potential area for research about how MA conditions can affect toxin production. An in-depth study understanding this area is bound to assist scientists to identify the ultimate requirements of MAs.

ABBREVIATIONS

4-HR	4-hydroxy resorcinal
4-MC	4-methyl catechol
a_w	water activity
b	blue to yellow
BOPP	bioriented polypropylene

CA	controlled atmosphere
CFU	colony form units
CI	chilling injury
CIM	*Cryptococcus infirmo-miniatus*
CK	check experiment
CS	cellophane sheets
CVA	canonical variate analysis
EA	ethylene absorber
EDTA	ethylenediaminetetracetic acid
EMA	equilibrium modified atmosphere
EVA	ethylene vinyl acetate
HOA	high-oxygen atmosphere
L	luminosity
LDPE	low-density polyethylene
MAP	modified atmosphere packaging
MAP-P	modified atmosphere packaging-pretreatment
MAP-WP	modified atmosphere packaging-without pretreatment
MDA	malondiadelhyde
NA	normal atmosphere
N-OPP	nonoriented polypropylene
OPS	oriented polystyrene
p_{O_2}	partial pressure
PAL	phenylalanine ammonia lyase
PB	polyethylene bags
PET	polyethylene terephthalate
POD	peroxidase
POP	polyolefin plastomer
PP	polypropylene
PP-ALG	polypropylene alginate
PP-HO	polypropylene high oxygen
PP-LO	polypropylene low oxygen
PPO	polyphenol oxidase
PS	polystyrene
PVC	polyvinylchloride
r_{O_2}	respiration rate
RH	relative humidity
RQ	respiration quote
SS	soluble solids
TA	titratable acidity
TSS	total soluble solids

REFERENCES

Alique, R., Martinez, M.A., and Alonso, J. (2003). Influence of the modified atmosphere packaging on shelf life and quality of Navalinda sweet cherry. *European Food Research and Technology*, 217: 416–420.

Arias, E., Gonzalez, J., Lopez-Buesa, P., and Oria, R. (2006). Optimization of processing of fresh-cut pear. *Journal of the Science of Food and Agriculture*, 88: 1755–1763.

Ben-Yehoshua, S., Fishman, S., Fang, D., and Rodov, V. (1989). New developments in modified atmosphere packaging and surface coatings for fruits, pp. 250–260. http://aciar.gov.au/files/node/2248/PR050%20part%2012.pdf (accessed on February 18, 2012).

Bich, T., Nguyen, T., Ketsa, S., and van Doorn, W.G. (2003). Effect of modified atmosphere packaging on chilling-induced peel browning in banana. *Postharvest Biology and Technology*, 31: 313–317.

Chauhan, O.P., Raju, P.S., Dasgupta, D.K., and Bawa, A.S. (2006). Instrumental textural changes in banana (var. *Pachbale*) during ripening under active and passive modified atmosphere. *International Journal of Food Properties*, 9: 237–253.

Choehom, R., Ketsa, S., and van Doorn, W.G. (2003). Senescent spotting of banana peel is inhibited by modified atmosphere packaging. *Postharvest Biology and Technology*, 31: 167–175.

Chonhenchob, V., Chantarasomboon, Y., and Singh, S.P. (2005). Quality changes of treated fresh-cut tropical fruits in rigid modified atmosphere packaging containers. *Packaging Technology Science*, 20: 27–37.

Geeson, J.D., Genge, P.M., Sharples, R.O., and Smith, S.M. (1991a). Limitations to modified atmosphere packaging for extending the shelf-life of partly ripened Doyenne du Comice pears. *International Journal of Food Science and Technology*, 26: 225–231.

Geeson, J.D., Genge, P.M., Smith, S.M., and Sherples, S.O. (1991b). The response of unripe Conference pears to modified atmosphere retail packaging. *International Journal of Food Science and Technology*, 26: 215–223.

Geeson, J.D., Smith, S.M., Everson, H.P., Genget, P.M., and Browne, K.M. (1987). Responses of CA-stored Bramley's seedling and Cox's orange pippin apples to modified atmosphere retail packaging. *International Journal of Food Science and Technology*, 22: 659–668.

Guevara, J.C., Yahia, E.M., and Brito de la Fuente, E. (2000). Modified atmosphere packaging of prickly pear cactus stems (*Opuntia* spp.). *Lebensmittel Wissenschaft und Technology*, 34: 445–451.

Hertog, M.L.A.T.M., Boerrigter, H.A.M., van den Boogaard, G.J.P.M., Tijskens, L.M.M., and van Schaik, A.C.R. (1998). Predicting keeping quality of strawberries (cv. 'Elsanta') packed under modified atmospheres: An integrated model approach. *Postharvest Biology and Technology*, 15: 1–12.

Jiang, Y.M. and Fu, J.R. (1998). Postharvest browning of litchi fruit by water loss and its prevention by controlled atmosphere storage at high relative humidity. *Lebensmittel Wissenschaft und Technology*, 32: 278–283.

Lakakul, R., Beaudry, R.M., and Hernandez, R.J. (1999). Modeling respiration of apple slices in modified atmosphere packages. *Journal of Food Science*, 64(1): 5–110.

Lambert, R.J., Skandamis, P.N., Coote, P.J., and Nychas, G.J. (2001). A study of the minimum inhibitory concentration and mode of action of oregano essential oil, thymol and carvacrol. *Journal of Applied Microbiology,* 91: 453–462.

Liu, C.L., Hsu, C.K., and Hsu, M.M. (2006). Improving the quality of fresh-cut pineapples with ascorbic acid/sucrose pretreatment and modified atmosphere packaging. *Packaging Technology and Science*, 20: 337–343.

Marrero, A. and Kader, A.A. (2005). Optimal temperature and modified atmosphere for keeping quality of fresh-cut pineapples. *Postharvest Biology and Technology*, 39: 163–168.

Meheriuk, M., Girard, B., Moyls, L., Beveridge, H.J.T., McKenzie, D.L., Harrison, J., Weintraub S., and Hocking, R. (1995). Modified atmosphere packaging of 'Lapins' sweet cherry. *Food Research International,* 28: 239–244.

Montero-Calderon, A., Rojas-Graó, M.A., and Martin-Belloso, O. (2007). Effect of packaging conditions on quality and shelf-life of fresh-cut pineapple (*Ananas comosus*). *Postharvest Biology and Technology*, 50: 182–189.

Nielsen, T. and Leufven, A. (2007). The effect of modified atmosphere packaging on the quality of Honeoye and Korona strawberries. *Food Chemistry*, 107: 1053–1063.

Oms-Oliu, G., Soliva-Fortuny, R., and Martın-Belloso, O. (2007). Physiological and microbiological changes in fresh-cut pears stored in high oxygen active packages compared with low oxygen active and passive modified atmosphere packaging. *Postharvest Biology and Technology*, 48: 295–301.

Petracek, P.D., Joles, D.W., Shirazi, A., and Cameron, A.C. (2001). Modified atmosphere packaging of sweet cherry (*Prunus avium* L., cv. 'Sams') fruit: Metabolic responses to oxygen, carbon dioxide, and temperature. *Postharvest Biology and Technology*, 24: 259–270.

Picon, A., Martinez-Jaivega, J.M., Cuquerella, J., Del Rio, M.A., and Navarro, P. (1992). Effects of precooling, packaging film, modified atmosphere and ethylene absorber on the quality of refrigerated Chandler and Douglas strawberries. *Food Chemistry*, 48: 189–193.

Ranasinghe, L., Jayawardena, B., and Abeywickrama, K. (2004). An integrated strategy to control post-harvest decay of Embul banana by combining essential oils with modified atmosphere packaging. *International Journal of Food Science and Technology*, 40: 97–103.

Remon, S., Ferrer, A., Marquina, P., Burgos, J., and Oria, R. (1999). Use of modified atmospheres to prolong the postharvest life of Burlat cherries at two different degrees of ripeness. *Journal of the Science of Food and Agriculture*, 80: 1545–1552.

Renault, P., Houal, L., Jacquemin, G., and Chambroy, Y. (1994). Gas exchange in modified atmosphere packaging. 2: Experimental results with strawberries. *International Journal of Food Science and Technology*, 29: 379–394.

Rocha, A.M.C.N., Barreiro, M.G., and Morais, A.M.M.B. (2002). Modified atmosphere package for apple *Bravo de Esmolfe*. *Food Control*, 15: 61–64.

Rosen, J.C. and Kader A.A., (1989). Postharvest physiology and quality maintenance of sliced pear and strawberry fruits. *Journal of Food Science*, 54(3): 656–659.

Serrano, M., Martınez-Romero, D., Castillo, S., Guillen, F., and Valero, D. (2004). The use of natural antifungal compounds improves the beneficial effect of MAP in sweet cherry storage. *Innovative Food Science and Emerging Technologies*, 6: 115–123.

Silva, F.M., Chau, K.V., Brecht, J.K., and Sargent, S.A. (1999). Modified atmosphere packaging for mixed loads of horticultural commodities exposed to two postharvest temperatures. *Postharvest Biology and Technology,* 17: 1–9.

Siro, I., Devlieghere, F., Jacxsens, L., Uyttendaele, M., and Debevere, J. (2003). The microbial safety of strawberry and raspberry fruits packaged in high-oxygen and equilibrium-modified atmospheres compared to air storage. *International Journal of Food Science and Technology*, 41: 93–103.

Sivakumar, D., Arrebola, E., and Korsten, L. (2007). Postharvest decay control and quality retention in litchi (cv. *McLean's Red*) by combined application of modified atmosphere packaging and antimicrobial agents. *Crop Protection*, 27: 1208–1214.

Sivakumar, D. and Korsten, L. (2005). Influence of modified atmosphere packaging and postharvest treatments on quality retention of litchi cv. *Mauritius*. *Postharvest Biology and Technology*, 41: 135–142.

Smith, S.M., Geeson, J.D., and Genge, P.M. (1998). The effect of harvest date on the responses of Discovery apples to modified atmosphere retail packaging. *International Journal of Food Science and Technology*, 23: 81–90.

Soliva-Fortuny, R., Alos-Saiz, N., Espachs-Barroso, A., and Martin-Belloso, O. (2004). Influence of maturity at processing on quality attributes of fresh-cut Conference pears. *Journal of Food Science,* 69(7): 290–294.

Soliva-Fortuny, R., Ricart-Coll, M., Elez-Martınez, P., and Martın-Belloso, O. (2005). Internal atmosphere, quality attributes and sensory evaluation of MAP packaged fresh-cut Conference pears. *International Journal of Food Science and Technology*, 42: 208–213.

Soliva-Fortuny, R.C., Ricart-Coll, M., and Martın-Belloso, O. (2003). Sensory quality and internal atmosphere of fresh-cut Golden Delicious apples. *International Journal of Food Science and Technology*, 40: 369–375.

Spotts, R.A., Cervantes, L.A. and Facteau, T.J., (2001). Integrated control of brown rot of sweet cherry fruit with a preharvest fungicide, a postharvest yeast, modified atmosphere packaging, and cold storage temperature. *Postharvest Biology and Technology*, 24: 251–257.

Spotts, R.A., Cervantes, L.A., Facteau, T.J., and Chand-Goyal, T. (1998). Control of brown rot and blue mold of sweet cherry with preharvest Iprodione, postharvest *Cryptococcus infirmo-miniatus*, and modified atmosphere packaging. *Plant Disease*, 82(10): 1158–1160.

Tano, K., Kouame, E.A., Nevry, R.K., and Oule, M.K. (2008). Strawberries (*Fagaria Xananassa Duch.*) stored under temperature fluctuation conditions. *European Journal of Scientific Research,* 21(2): 353–364.

Tian, S.P., Jiang, A.L., Xu, Y., and Wang, Y.S. (2003). Responses of physiology and quality of sweet cherry fruit to different atmospheres in storage. *Food Chemistry*, 87: 43–49.

Tian, S.P., Li, B.Q., and Xu, Y. (2004). Effects of O_2 and CO_2 concentrations on physiology and quality of litchi fruit in storage. *Food Chemistry*, 91: 659–663.

Van der Steen, C., Jacxsens, L., Devlieghere, F., and Debevere, J. (2001). Combining high oxygen atmospheres with low oxygen modified atmosphere packaging to improve the keeping quality of strawberries and raspberries. *Postharvest Biology and Technology*, 26: 49–58.

Zhang, M., Xiao, G., and Salokhe, V.M. (2005). Preservation of strawberries by modified atmosphere packages with other treatments. *Packaging and Technology Science*, 19: 183–191.

11 Bakery Products

Ioannis S. Arvanitoyannis and Konstantinos Bosinas

CONTENTS

11.1 INTRODUCTION

Baking has been many cultures' favorite technique for the production of snacks, desserts, and side dishes to meals for many years. Nowadays, baking is well known as the method for producing sweets and all sorts of wondrous appetizing pastries. In the old days, the first incident of baking occurred when humans took wild grass grains, soaked them in water, and mixed everything together, turning it into a kind of broth-like paste. The paste was then cooked by pouring it onto a flat, hot rock, resulting in a bread-like substance. Later, this paste was roasted on hot embers, which facilitated considerably the bread-making process. According to existing records, the Egyptians already had bread since 2500 BC and most probably learned the process from the Babylonians. The Greek Aristophanes, around 400 BC, recorded information that tortes with patterns and honey flans existed in Greek cuisine. Dispyrus, a Greek sweet prepared by the Greeks around that time, resembling a donut made from flour and honey, shaped like a ring and soaked in wine, was eaten when hot (Tanis).

The production of bread and other bakery products gradually evolved from a primitive, artisanal industry into a large-scale, modern manufacturing industry, generating billions of dollars in revenue and employing many thousands of personnel and resulting in huge profits of many thousands of Euros world wide (Smith et al., 2004).

Hasan (1997) claimed that global consumption of bakery products and cereals has augmented at an average rate of 25% per annum since 1970 and they are consumed more than any other food source in the daily diet. This sustained growth has been driven by consumer demand for convenient, premium baked goods that are fresh, nutritious, conveniently packaged, and shelf stable. Furthermore, this enhanced demand is being addressed by various new processing and packaging technologies, such as modified atmosphere packaging (MAP), which has considerably augmented the availability and prolonged the shelf life of a great variety of bakery products. An increase in in-store bakeries and a new interest in "organic," ethnic, and artisan-type bakery products have recently been reported (Smith et al., 2004).

The western European bread industry produces 25 million tonnes of bread on an annual basis, of which the industrial or plant sector's share is 8 million tonnes. Germany and the United Kingdom are the main operators with 60% of plant sector production—France, the Netherlands, and Spain produce another 20% (The federation of Bakers).

The most common raw materials used are flour, water, yeasts (basically *saccharomyces cerevisiae*), sugar and glucose, vegetable fats and oils, eggs, starch, milk and milk products, gluten, emulsifiers, improvers, conditioners, etc. (Kotsianis et al., 2002).

A wide variety of bakery products can be found on supermarket shelves, such as breads, cakes, bagels, biscuits, unsweetened rolls and buns, doughnuts, meat pies, dessert pies, pizza, crackers, cookies, and other products as given in Table 11.1.

11.2 CLASSIFICATION OF BAKERY PRODUCTS

Several methods can be applied toward bakery product classification. The latter can be based on product type, that is, bread or sweet goods; the method of leavening such as biological, chemical, or unleavened; or on the basis of their pH, moisture content, or water activity (a_w) (Table 11.2).

The three groups of bakery products classified on the basis of pH are the following:

1. High acid with $pH<4.6$
2. Low acid with $pH>4.6$ but <7
3. Neutral products with $pH>7$ (Smith et al., 2004)

According to another classification based on their a_w, the categories are as follows (Smith and Simpson, 1995):

1. Low-moisture bakery products ($a_w<0.6$)
2. Intermediate-moisture bakery products (a_w between 0.6 and 0.85)
3. High-moisture bakery products ($a_w>0.85$ and generally between 0.95 and 0.99)

Bakery products, similar to the rest of processed foods, can undergo physical, chemical, and microbiological spoilage. Moreover, classification of products on the basis of their pH and a_w can assist on recognizing the spoilage and safety potential of bakery products. Although chemical spoilage problems can considerably limit the shelf life of low- and intermediate-moisture bakery products, there is a major hazard of microbiological spoilage in intermediate- and high-moisture products. Several high-moisture, unfilled, and filled, bakery products have also been involved in outbreaks of foodborne illness and, therefore, pose safety concerns (Smith et al., 2004).

TABLE 11.1
Categories of Bakery Products Found on Supermarket Shelves

Categories of Bakery Products	Types within Each Category
Unsweetened goods	Bread: sliced, crusty, par-baked, ethnic
	Rolls: soft, crusty
	Crumpets
	English muffins
	Croissants
	Pizza base
	Raw pastry
Sweet goods	Large cakes: plain, fruited
	Pancakes
	Doughnuts
	Waffles
	Cookies
	Biscuits
	American muffins
	Buns
	Wafers
Filled goods	Tarts: fruit, jam
	Pies: meat, fruit
	Sausage rolls
	Pasties
	Cakes: cream, custard
	Pizza
	Quiche

Source: Adapted from Smith, J.P. et al., *Crit. Rev. Food Sci. Nutr.*, 44, 19, 2004.

11.3 SPOILAGE CONCERNS OF BAKERY PRODUCTS

Spoilage stands for any change in food condition that makes it less appetizing at consumption time. The spoilage problems of bakery products could be further classified or subdivided into the following:

1. Physical spoilage (moisture loss, staling)
2. Chemical spoilage (rancidity)
3. Microbiological spoilage (yeast, mold, bacterial growth) (Hasan, 1997)

The predominant spoilage problem is greatly affected by interrelated factors, such as storage temperature, relative humidity, preservatives, content pH, packaging

TABLE 11.2
Classification of Bakery Products Based on Moisture Content and Water Activity

Product Type	Water Activity (a_w)
Low moisture content	
Cookies	0.20–0.30
Crackers	0.20–0.30
Intermediate moisture content	
Cake donut	0.85–0.87
Chocolate-coated donuts	0.82–0.83
Danish pastries	0.82–0.83
Cream-filled snack cakes	0.78–0.81
Pound cake	0.84–0.86
Banana cake	0.84–0.86
Sofi cookies	0.50–0.78
Breadjam cake	0.85
High moisture content	
Bread	0.96–0.98
Egg bread	0.90
Pumpernickel bread	0.90
Pita bread	0.90
Yeast raised donuts	0.96–0.98
Fruit pies	0.95–0.98
Soy bean pie	0.93
Carrot cake	0.94–0.96
Custard cake	0.92–0.94
Cheese cake	0.91–0.95
Butter cake	0.90
Pizza crust	0.94–0.95
Butter cake	0.99

Source: Adapted from Hasan, S., Methods to extend the mold free shelf life of pizza crusts, MSc thesis, McGill University, Montreal, QC, Canada, 1997.

material, and gas atmosphere surrounding the product, and, most importantly, by the moisture content (m_c) and water activity (a_w) (Smith et al., 2004).

11.3.1 Physical Spoilage

The physical spoilage of bakery products is often closely linked with moisture loss or gain leading to texture deterioration or mold growth. The occurrence of such a problem can be effectively prevented with the use of moisture-impermeable wrapping materials, (LDPE, PP), which are high barrier to moisture (Smith and Simpson, 1996; Hasan, 1997). Unfortunately, application of such films may lead to conditions favorable to mold growth, especially in high-moisture bakery products (Smith et al., 2004).

The more serious physical spoilage problem in bakery products is *staling*. The latter has been defined as "almost any change, short of microbiological spoilage, which occurs in bread or other products, during the postbaking period, making it less acceptable to the consumer" (Zobel and Kulp, 1996). Staling is accompanied by many changes among which the most important ones are the following:

Loss of crispiness (water absorption or movement)
Increase in crumb firmness (moisture loss/or starch retrogradation)
Loss of crumb firmness (moisture migration from fillings to cakes)
Increases in crumbliness (loss of cohesion)
Loss or change of taste and aroma (Cauvain, 1998)

The staling mechanism has been the subject of many investigations. Bread staling falls into two categories: crust staling and crumb staling. Crust staling is due to moisture transfer from the crumb to the crust and is often less objectionable than crumb staling (Gray and Bemiller, 2003). Products of a higher moisture content, that is, bread and cakes, stale more rapidly than intermediate- or low-moisture products, such as cookies or crackers (Smith et al., 2004). On the other hand, crumb staling is more complicated and important, and less understood.

It is generally believed that the retrogradation/recrystallization of starch, and in particular, the short amylopectin side chains, plays a major role in bread firming, after the initial cooling process. The extent of crystallization over storage does not necessarily correlate well with crumb firming, particularly in amylase-supplemented breads. Gallagher et al. (2003) reported that changes in the firming rate of bread are mainly due to the formation of hydrogen bonding between gluten and starch granules. In one study, breads baked from lower (10.4%) protein flours staled at a more rapid rate than those from higher (13.1%) protein flours. It was concluded that the gluten was the main flour component responsible for the shelf life of bakery products. Moreover, apart from starch crystallization, water migration is considered to be a very important aspect of firming; changes in water distribution between the biopolymers in the crumb are often neglected in bread firming models. Despite a great amount of research conducted, the crumb firming and the induced antifirming properties of amylases have not been properly comprehended at a molecular level (Goesaert et al., 2009).

Staling is of great economic importance to the bakery industry. Various commercial methods are applied to delay the staling of bakery products, including reformulation with lipids and shortenings, surfactants, emulsifiers, gums, and mono- and diglycerides. It has been shown that maltodextrin displays antistaling action in chapattis and the combination of SSL + α amylase was shown to exhibit a synergistic effect (Shaikh et al., 2008). Zhou et al. (2007) indicated that trehalose could enhance the quality and retard the staling of bread effectively. In another study by Tian et al.

(2009) it was demonstrated that β-cyclodextrin (β-CD) has significant impact on the staling of crust and crumb.

While staling is usually delayed with the addition of chemical additives, the application of a CO_2-enriched atmosphere as a means of retarding staling is quite questionable and conflicting results have been reported in the literature. Rasmussen and Hansen (2001) found no significant effects (measurements with differential scanning calorimetry) of MAP during storage of bread for 7 days at 20°C compared with control bread. Similar reports were published from Doerry (1985), Black et al. (1993), and Brody (1989) (cited in Smith et al., 2004) regarding the ineffectiveness of MAP on staling for various bakery products. However, in contrast to the aforementioned studies, others have shown that MAP decreased the rate of firming (Knorr and Tomlins, 1985; Avital et al., 1990).

11.3.2 Chemical Spoilage

Bakery products, in particular the ones with a high fat content, are subject to chemical spoilage or rancidity. Rancidity is equivalent to lipid degradation, thereby leading to off-odors, which make products unappetizing and limit their shelf life (Smith et al., 2004). Crackers and cookies are conducive to rancidity, due to their high fat content and their prolonged shelf life (Mcfeaters and Cassone, 1999).

The occurring rancidity problems can be distinguished into two types: oxidative and hydrolytic. The liberation of odorous products results due to oxidative rancidity breakdown of unsaturated fatty acids, thus leading to the release of aldehydes, ketones, and shorter chain fatty acids (Hui et al., 2004). During hydrolytic rancidity, which is the result of the hydrolysis of triglycerides, free fatty acids, such as capric, lauric, and myristic acids, are produced, which have very strong off-flavors (Wekell and Barnett, 1999).

According to Nanditha and Prabhasankar (2009), "chemical spoilage can be often effectively limited with the addition of antioxidants. The latter include tocopherols and derived compounds in baked and fried products and vegetable oils such as β-carotene in butter, butterfat, coconut oil, and corn oil. BHA (Butylated Hydroxy Anisole); BHT (Butylated Hydroxy toluene), in bakery products and candies; Tertiary-butyl hydroquinone (TBHQ), in oils, fats, and meat products; Gallates in frying fats and oils; Rosmarinic acid, in the form of the herb rosemary and Italian seasoning mixtures in naturally or minimally processed foods." An alternative might be the replacement of O_2 with gas packaging in 100% N_2 to delay rancidity in low-moisture bakery products, in case microbiological hazards are not at stake (Smith et al., 2004).

11.3.3 Microbiological Spoilage

Microbiological spoilage is frequently the major hazard limiting the shelf life of high- and intermediate-moisture bakery products.

The major bacterial problem in bread is "rope." Ropy spoilage is due mainly to *Bacillus subtilis* and *Bacillus licheniformis*, the spores of which contaminate raw materials (flour, bread conditioners, yeast) and survive at baking temperatures. Ropy spoilage in bread is first detected by the release of an odor similar to that of pineapple. Later, the crumb becomes discolored, soft, and sticky to the touch, thereby making the bread both unpalatable and inedible. The deterioration of bread texture is due to formation of slime because of a combined action of the proteolytic and amylolytic enzymes produced by some *Bacillus* strains (Valerio et al., 2008). Application of chemical or natural preservatives (propionates, and acetic acid, respectively) can effectively overcome rope problems.

Unfilled pastries can follow a deterioration pattern similar to that of bread. However, when pastries are filled, they are subject to other types of microbial spoilage. Most fillings favor the growth of food spoilage microorganisms, in particular the ones including egg and dairy products (i.e., *Bacillus cereus* and *Staphylococcus aureus*). This latter pathogen has also been implicated in foodborne poisoning outbreaks from cream-filled bakery products. Other bakery potential hazardous ingredients include chocolate, desiccated coconut, and cocoa powder, which have been accused for outbreaks of Salmonella food poisoning from bakery products. At this instance, it is worth noting that bacterial food poisoning outbreaks in filled bakery products in both the United States and Canada are very rare (Smith and Simpson, 1995).

11.3.3.1 Yeast Spoilage

In general, there are two kinds of yeast spoilage of bakery products and ingredients: The visible growth of yeasts on the surface of products (white, cream, or pink patches).

The fermentative spoilage of a wide range of products and ingredients is manifested by means of the release of alcoholic, estery, or other odors and/or visible evidence of bubbles in jams and fondants or expansion of flexible packaging.

Visible yeast growth most often occurs in high a_w short shelf-life products, whereas fermentative spoilage is more frequently linked with low a_w long shelf-life products. Some yeasts are resistant to low a_w without causing any spoilage until they get adapted to the new conditions, or the conditions in the product are modified, thereby allowing their growth (Legan, 1993).

11.3.3.2 Mold Spoilage

Mold spoilage of bakery products is of serious economic concern. Losses due to mold spoilage vary between 1% and 5% of products depending on season, type of product, and method of processing. By far the most widespread, and important, with regard to biodeterioration, are the species of *Eurotium*, *Aspergillus*, and *Penicillium*. The most important molds isolated from bakery products are *Aspergillus* spp., xerophilic species of *Penicillium* and, mainly, species of *Eurotium* (Abellana et al., 2000).

Further to the economic losses related to bakery products, another serious concern is the possibility of mycotoxin production. *Eurotium* species are most often the first fungi to colonize improperly dried, stored commodities. In fact, when they do grow, the level of available water is enhanced, thereby allowing other species to thrive as well. *Eurotium* spp. do not produce any significant mycotoxins, but it is important to be aware of the conditions under which species of *Aspergillus* and *Penicillium* can grow and spoil bakery products, since several species can potentially produce mycotoxins. Toxigenic *Aspergillus flavus* has been isolated from 3 out of 15 home-stored bakery products, and toxigenic *Penicillium* spp. have been isolated from wheat flour and bread in the United States. Several species of *Penicillium* including *Penicillium chrysogenum* produce mycotoxins (Abellana et al., 2001).

Although freshly baked products are viable, vegetative molds- and mold spores-free products soon get contaminated as a result of postbaking cross-contamination by mold spores from the air, bakery surfaces and equipment, food handlers, and raw ingredients such as glazes, nuts, spices, and sugars (Hasan, 1997; Smith et al., 2004). Mold problems get more hazardous over the summer season, due to airborne contamination and warmer and more humid storage conditions (Smith, 1993). Several factors affect the types and numbers of mold spores occurring in bakery products, including product type and equilibrium, relative humidity, and season. The most common molds detected in bakery products are the following.

11.3.3.3 *Aspergillus* sp.

This mold produces greenish to black spore heads. The yellow pigment from the spore head spreads downward through the mycelium and can stain the product surface. At higher temperature, *Aspergillus* sp. accounted for 39% of molds that grow on bread (Hasan, 1997). Many *Aspergillus* species are xerophilic and able to grow on media containing high concentrations of salt or sugar (Marin et al., 2002). According to Suhr and Nielsen (2005) *A. flavus* displayed the highest capacity to grow at low residual O_2 level at increasing CO_2 on wheat bread as it was the only fungus growing practically in the absence of oxygen (0.03% O_2) at 75% CO_2. In the same study, it was reported that Miller and Golding (1949) also determined that *A. flavus* required less O_2 than other molds (*A. niger, P. expansum, P. notatum, P. roqueforti*) on malt agar.

11.3.3.4 *Penicillium* sp.

Penicillium is the most important mold limiting the shelf life of bakery products. Hasan (1997) reports that according to Legan (1993) 90%–100% of loaves each month were spoiled by *Penicillium* species. The predominance of *Penicillium* may be partially due to their ability to grow over a wide range of temperatures (with preference to cooler temperatures) and water contents and the production of spores ubiquitous in the atmosphere. However, at 22°C–24°C there is a 50% reduction in *Penicillium* contamination, and in warmer climates *Aspergillus* and *Eurotium* prevail (Magan et al., 2003). Suhr and Nielsen (2005) report that in their experiments *P. commune* exhibited the highest CO_2 tolerance with growth at 1% residual O_2 in 99% CO_2.

11.3.3.5 *Eurotium* sp.

Some of these molds are known to grow on the walls of bakeries, more often in areas where condensation is a problem. This mold can grow at lower a_w values than other molds. Therefore, *Eurotium* species, that is, *Eurotium glaucus* and *E. amstelodami*, are the major molds limiting the shelf life of low a_w bakery products (Hasan, 1997). The most common *Eurotium* species occurring in bakery products are very resistant to temperature. Vytrasova et al. (2002) reported that *E. amstelodami, E. rubrum,*

E. Herbariorum, and *E. chevalieri* were heated to 75°C–85°C for 60 min without reaching a thermal death point and that during production, the food is never exposed to elevated temperatures for such a long time. *Eurotium* species are potentially the most hazardous of all. They are often the first colonizers of improperly dried stored commodities, and their growth is followed with a_w favoring other species (like potentially toxigenic *Aspergilli* and *Penicillia*) to dominate. Although *Eurotium* species produce no important mycotoxins, they do release several secondary metabolites, thereby causing oxidative rancidity problems (Abellana et al., 1999a,b).

11.3.3.6 *Rhizopus* sp.

Rhizopus stolonifer is by far the most common species in foods occasionally referred to as "bread molds," and they produce watery soft rot of apples, pears, stone fruits, grapes, figs, and other fruits (Jay, 2000). Another mold, *R. nigricans*, is the common black bread mold. It has a very fluffy appearance of white cottony mycelium and black sporangia (Pateras, 2007).

11.3.3.7 *Monilia* sp.

Monilia sp., in particular, *Monilia sitophila* (*conidial stage of Neurospora intermedia*), induces bread spoilage (Hasan, 1997). This microorganism is a very difficult mold to eliminate in bakery products because its spores are capable of withstanding fairly high temperatures for a reasonable length of time. Its color is reddish, and it is found in bread stored at a high humidity or wrapped while still warm (Pateras, 2007).

11.3.3.8 *Mucor* sp.

This mold produces grey colonies (Pateras, 2007) and often infects bread products (Hasan, 1997; Pateras, 2007). According to a study by Odell (1983) cited in Legan (1993), mucorales were determined on 20% of loaves stored in sealed plastic bags for 5–6 days at 22°C.

11.3.4 Strategies to Control Microbiological Spoilage of Bakery Products

Microbiological spoilage is often the main factor limiting the shelf life of intermediate- and high-moisture bakery products. Therefore, the developments of methods to control microbiological spoilage, in particular mold growth, are of substantial economic importance to the bakery industry. Three basic strategies are available toward prolonging the microbiological shelf life of bakery products (Smith et al., 2004)

1. Prevention of postbaking contamination by packaging prior to or after baking preferably under aseptic conditions
2. Destruction of postbaking contaminants on the surface of products
3. Growth control of postbaking contaminants

11.3.4.1 Prevention of Postbaking Contamination

Mold spoilage of bakery products is due to postbaking contamination. Flour contains a substantial amount of mold, similar to other dry ingredients, and flour dust can spread widely throughout the bakery, especially in small bakeries where separation of different processes is much more difficult than in large plant bakeries. Therefore, airborne distribution of dust and mold spores is likely to give rise to contamination of bread (Legan, 1993). Extension of the mold-free shelf life of bakery products can be reached by packaging as soon as possible after cooling or slicing and wrapping under sterile conditions (Hasan, 1997).

Recently developed air filtration/circulation systems are able to remove all microorganisms from air by blowing sterile air into the bakery environment. While these systems may effectively prevent cross-contamination of processed foods, the major disadvantage of aseptic packaging conditions continues to be their too high cost (Smith et al., 2004).

11.3.4.2 Destruction of Postbaking Contaminants

Despite latest improvements in bakery design, cooling, and packaging systems, attempts to prevent postbaking cross-contamination of bakery products with mold spores did not enjoy a widespread success. Therefore, attention has focused on methods to destroy and control any postprocessing cross-contamination of products with mold spores. Several traditional and novel methods have been comparatively investigated with regard to destroying postbaking contaminants of high- and intermediate-moisture bakery products including ultraviolet light (UV), infrared radiation (IR), microwave (MW) heating, low irradiation dose, pulsed light technology, and ultra-high pressure (UHP) (Smith et al., 2004).

11.3.4.3 UV Light

UV radiation has a wavelength between 210 and 328 nm, with maximum antimicrobial activity between 240 and 280 nm. UV radiation has low energy and does not penetrate foods. Since it is absorbed by glass and plastics, it can be effectively used for surface disinfection (Evans, 1999). In bakeries, UV lamps find application on bread slicing machines to minimize the contamination from airborne molds and mold growth in the packages. UV irradiation does not generate heat capable of charring films or causing condensation problems (Begum et al., 2009). The poor penetrative capacities of UV light seriously restrict its food use to surface applications, where it may catalyze oxidative changes that result in rancidity and discoloration, among others (Jay, 2000). Other disadvantages of UV light that limit its extensive application as a method of increasing the mold-free shelf life of products refer to cost, mold contamination due to inadequate sealing of the package, and safety concerns (Smith et al., 2004).

11.3.4.4 IR Radiation

IR treatment can be applied to kill mold spores on bread by heating surfaces to the desired temperature of 75°C without adversely affecting the quality and appearance of the product or the integrity of the packaging material (Pateras, 2007). The time to reach this temperature greatly depends on the thickness of the packaging material, the nature of the product, and the distance between the IR projector and the product surface. The advantage of IR radiation over MW and UV is that it is required to heat only the exterior surfaces, thereby minimizing the occurrence of problems due to condensation or air expansion. One main disadvantage is its high cost for multisided products, which must either rotate between heaters or be treated in separate ovens (Pateras, 2007).

11.3.4.5 MW Heating

Most MW food research has been carried out at two specific frequencies, 915 and 2450 Mc/s (Jay et al., 2005). Experimental treatment of packaged bread with high-frequency MW energy for 45–60 s was shown to prolong the mold-free shelf life of bread (Hasan, 1997; Smith et al., 2004). MW heating of bread for 3–12 s resulted in 100% increase in the shelf life before the growth of both *A. niger* and *P. notatum* was evident in approximately 7 days (El-Khoury, 1999). MW baked cakes had some quality defects such as lack of color, high weight loss, very firm texture, and low volume (Sumnu et al., 2005). The two main reasons this technology has not found widespread acceptance are its high capital costs (equipment) and the wide temperature distribution range across a package (Coles et al., 2003).

11.3.4.6 Low-Dose Irradiation

Gamma irradiation is considered to be an alternative method for food preservation to prevent food spoilage and insect infestation and capable of reducing the microbial load. In one recent investigation, commercial Mexican bread made of wheat flour was irradiated at 1.0 kGy using a ^{60}Co γ-beam 651 PT irradiator facility. No changes were detected in moisture, protein, and ashes in γ-irradiated samples as compared with those of nonirradiated samples. The obtained results confirmed the effectiveness of γ-irradiation in reducing the microbial load in bread making wheat flour without undergoing any substantial change in the physicochemical and baking properties (Agundez et al., 2006).

11.3.4.7 Pulsed Light Technology

A new processing technology applies intense flashes of broad-spectrum white light to kill microorganisms on packaging and food surfaces and in air and water. The flashes have a duration of 100–300 millionths of a second and can effectively kill all microorganisms in less than a second. The process appears to be extremely effective for sterilization when the light can reach all surfaces (i.e., packaging materials) or the total volume of a product such as water. The process was initially marketed under the trade name of PureBright, by PurePulse Technologies in San Diego, California, in the 1990s (Ott, 1999).

Bread treated with pulsed light had a mold-free shelf life of more than 2 weeks at ambient temperature, while control loaves were visibly spoiled with mold growth after 1 week (Smith et al., 2004). However, there has been no report on the effect of pulsed light on the sensory properties of bread or on the packaging film. Nevertheless, pulsed light technology is promising as a novel method to prolong the mold-free shelf life of prepackaged bakery products.

11.3.4.8 High-Pressure Technology

Another novel and promising technology as a method of prolonging the shelf life and maintaining the quality of prepackaged food products is high-pressure processing (HPP), also known as UHP processing.

HPP in conjunction with refrigeration and ambient or moderate heating temperature makes possible the inactivation of pathogenic and spoilage microorganisms in foods with minimal changes in texture, color, and flavor compared with conventional technologies. Five decimal reductions in pathogens including *Salmonella typhimurium*, *Salmonella enteritidis*, *Listeria monocytogenes*, *S. aureus*, and *Vibrio parahaemolyticus* can be reached with HPP (Torres and Velazquez, 2005).

HPP disrupts the bacterial membranes, thus affecting transport phenomena involved in nutrient uptake and disposal of cell waste (Torres and Velazquez, 2005).

11.3.4.9 Controlling the Growth of Postbaking Contaminants

While the bakery industry has a widespread choice of traditional and novel methods to destroy postprocessing mold contaminants, they are not widely applied to prolong the mold-free shelf life of bakery products. The most practical, commonly applied, and cost-efficient approach adopted by the bakery industry to meet this target is to control the growth of any postbaking contamination in the packaged products. This can be done through reformulation to reduce product a_w or m_c or pH, freezing (Hasan, 1997), use of chemical preservatives (sorbates or propionates) either by incorporation or surface application MAP, using gas packaging or interactive packaging sachet technology (Smith et al., 2004).

11.3.4.10 Product Reformulation to Reduce a_w and pH

Reformulation involves a reduction in available water or pH. Reduction in product a_w can be achieved by dehydration, either through evaporation or freeze-drying or by using high osmotically active additives, for example, sugars and salts, incorporated directly into the food (Wagner and Moberg, 1989 cited in Hasan, 1997). The a_w reduction is of practical significance because the food is made nonperishable. The general strategy of lowering a_w below optimum is to enhance the length of the lag phase of growth and to minimize the growth rate and size of final population. This effect may result from adverse effects of lowering water on all metabolic activities because all chemical reactions of cells presuppose the presence of an aqueous environment (Jay, 2000). The RH of the storage environment is important both from the standpoint of a_w within foods and the microorganism growth at the surfaces. Foods likely to undergo surface spoilage from molds, yeasts, and certain bacteria should be stored under conditions of low RH (Jay, 2000) (Figure 11.1).

Reduction in pH can be achieved through the use of acidulants, (e.g., citric, lactic, acetic acids) or cultures of LAB (sour dough cultures) (Smith et al., 2004).

However, product reformulation to reduce pH and a_w may affect negatively the sensory properties, whereas, on the other hand, the use of preservatives is diminishing due to consumer concern (Sanguinetti et al., 2009).

11.3.4.11 Freezing

Freezing is maybe the oldest preservation method and has been effectively applied for long-term preservation of bakery products, particularly cream-filled products (Hasan, 1997).

Quick freezing processes have more advantages than slow freezing from the standpoint of overall product quality. With respect to crystal formation upon freezing,

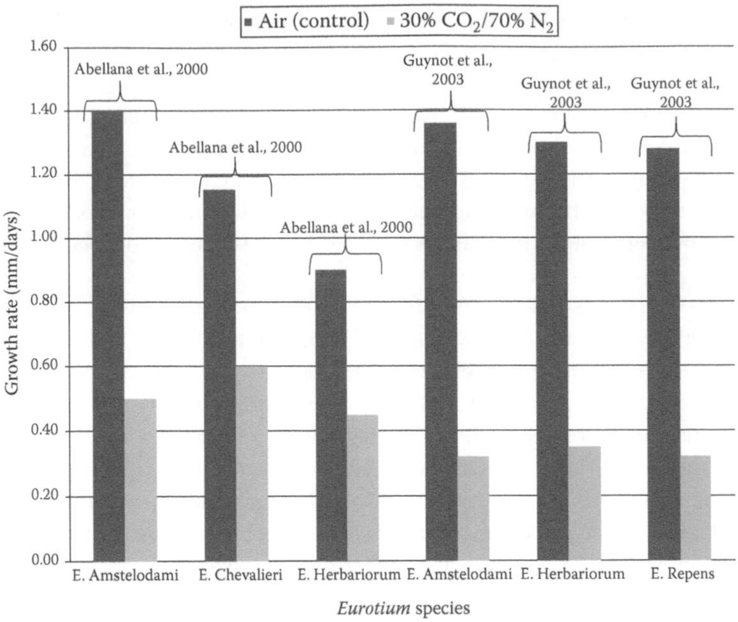

FIGURE 11.1 Growth rates of *Eurotium* species (20°C–25°C) under air (control) and MA (30% CO_2/70% N_2) atmosphere.

slow freezing favors the formation of large extracellular crystals, whereas rapid freezing induces the formation of much smaller intracellular ice crystals (Jay, 2000).

Cakes, cookies, short cake, and pancakes are commonly frozen and marketed in the frozen form. Bread has been held fresh for many months by storage at −22°C (Hasan, 1997). Bread quality can be improved by applying freezing temperatures to the breadmaking process. Freezing temperatures can be applied to bread dough interrupting the breadmaking prior to proofing. Moreover, freezing temperatures can also be used in partially baked bread (known as part-baked or prebaked bread). Freezing the prebaked bread is the easiest option to extend the shelf life of the bread, keeping its freshness (Barcenas et al., 2003).

11.3.5 Preservatives

11.3.5.1 Chemical Preservatives

The use of weak organic acids such as propionic, benzoic, and sorbic; investigation on the packaging material; or MAP in the last years have been the main approaches toward satisfying the market demands to prolong the shelf life of bakery products (Gutierrez et al., 2009).

11.3.5.2 Propionic Acid/Propionates

Although propionic acid and sodium, potassium, and calcium propionates are primarily used against molds, some yeasts and bacteria can also be inhibited. Propionic acid and propionates are used as antimicrobials in baked goods and cheeses. Propionates' direct addition to bread dough has no effect on the activity of baker's yeast (Davidson, 1999). These preservatives have two functions: (1) to prevent the bacterial spoilage of bread known as "rope" caused by certain *Bacillus* spp., notably *B. subtilis* and *B. licheniformis*, and (2) to delay the rate of mold development. Typical concentrations applied are 0.1% propionic acid or 0%–2% calcium propionate based on flour weight, much lower than the maximal permitted levels but adequate to prolong the mold-free shelf life of commercial British bread by approximately 1 day (Legan, 1993).

11.3.5.3 Benzoic Acid/Benzoates

The main application of benzoic acid and sodium benzoate is as antifungal agents. The inhibitory concentration of benzoic acid at pH less than 5.0 against most yeasts ranges from 20 to 700 µg/mL, whereas for molds it is 20–2000 µg/mL. However, there are several fungi including *Byssochlamys nivea, Pichia membranaefaciens, Talaromyces flavus, and Zygosaccharomyces bailii* resistant to benzoic acid. Although bacteria associated with food poisoning such as *B. cereus, L. monocytogenes, S. aureus,* and *V. parahaemolyticus* are inhibited by 1000–2000 µg/mL undissociated acid, the control of many spoilage bacteria demands much higher concentrations (Davidson, 1999).

11.3.5.4 Sorbic Acid/Sorbates

Sorbates are probably the most well characterized and among the most important of all food antimicrobials as to their spectrum of action and bacteria, yeasts, and molds inhibition at concentrations of 0.05%–0.3%. *Byssochlamys, Candida, Saccharomyces, Zygosaccharomyces, Aspergillus, Fusarium, Geotrichum,* and *Penicillium* are among the food-related yeasts and molds inhibited by sorbates. Sorbate can be applied to foods in various ways: direct addition, dipping, spraying, dusting, or incorporation into packaging. Baked goods, icing, fruit, and cream fillings can be protected from yeast and molds by means of 0.05%–0.10% potassium sorbate applied either as a spray after baking or by direct addition (Davidson, 1999). Several investigators concluded that PS is the most suitable preservative to be applied in bakery products. It has been demonstrated to be the most effective preservative to inhibit mycoflora, one of the most common sources of spoilage of bakery products (Guynot et al., 2004). Although sorbates are more effective at inhibiting mold growth than propionates, they also have greater adverse effects on yeast activity and dough rheology, thereby seriously decreasing the loaf volume and making dough sticky and difficult to process (Legan, 1993).

11.3.5.5 Natural Preservatives

The bakery industry has initially relied on chemical preservatives for the shelf-life extension of products. However, recently its attention has focused on natural preservatives. One such preservative is nisin, a bacteriocin produced by *Lactococcus lactis*. Nisin is an effective natural food preservative displaying a wide spectrum of activity against gram-positive bacteria, in particular LAB, and hence, it is employed to control spoilage by LAB in various food products (Cabo et al., 2001). Although nisin is most often used in dairy and meat products, it has also been investigated in high-moisture baked products, such as crumpets and pikelets, to effectively control the outgrowth of bacterial spores.

Two other commercially available biopreservatives are Alta-R 2341 and Perlac 1911, produced by Quest International in Montreal, Canada. Alta 2341 is a pediocin produced as a by-product of fermentation when *Pediococcus acidlactici* is grown on yeast extract, corn syrup, and vegetable protein. Perlac 1911 is produced by *L. lactis* and is isolated from the fermentation of glucose solids. They are multifunctional ingredients that enhance flavor, water retention, and shelf life of various foods (Smith et al., 2004). One natural preservative studied as a promising alternative to organic acids as mold inhibitors is Upgrade. The latter consists of 79% soy flour, 17.5% whey, and 3.5% calcium sulfate and was shown to be effective as a mold inhibitor in both white and whole wheat bread when used at levels of 1%–2%, (0.125%–0.25%) (Smith et al., 2004).

11.3.6 Early Detection of Food Spoilage

Early detection of food spoilage is crucial because of legislative and consumer pressure to minimize the use of preservatives, especially those based on organic acids, in bakery products of intermediate moisture (Keshri et al., 2002).

In the past, control of food safety has been carried out by testing of both raw materials and end products. The major problem with end (final) product testing is the great number of samples to be analyzed before one reaches a decision on the safety of a product batch, especially when the microorganisms are considered to be heterogeneously distributed within the batch. Over the last decade, the rapid advances in the development of electronic nose technology attracted great interest in applications for the fast detection of spoilage microorganisms (Needham et al., 2005).

Needham et al. (2005) reported that there is potential to distinguish between various types of bread spoilage by means of an electronic nose. Microbial spoilage due to bacteria, yeast, and fungi and enzymic spoilage initiated by lipoxygenase can be distinguished from one another and from unspoiled bread analogues, before any signs of spoilage are visible (Needham et al., 2005).

11.4 MODIFIED ATMOSPHERE PACKAGING

MAP has been defined as

> the enclosure of a food product in a high gas barrier film in which the gaseous environment has been changed or modified to slow respiration rate, reduce microbiological growth, and retard enzymatic spoilage with the intent of extending shelf life (Hasan, 1997)

MAP technology has been much more successful in European countries than in North America. The United Kingdom is the leader in MAP technology, accounting for approximately 50% of the European market, followed by France, which has approximately 25% of the MAP packaged food market (Smith et al., 2004).

11.4.1 Methods of Atmosphere Modification

11.4.1.1 Vacuum Packaging

The simplest form of MAP is vacuum packaging (VP). In VP, products are placed in a package made of films with low oxygen permeability, that is, high gas barrier, air is evacuated, and the package is sealed. Under appropriate packaging conditions, headspace O_2 is reduced to <1%. This very low level of oxygen helps extend the shelf life of products by inhibiting the two major spoilage agents that is oxidative reactions and aerobic microorganisms (Hasan, 1997). VP is not a suitable technology to prolong the mold-free shelf life of most soft bakery products, because of its crushability effect. It has been used, however, to prevent mold problems in flat breads (naan, pita) and pizza crusts and to avoid rancidity problems in shortbread, a high-fat product bearing a hard texture (Smith et al., 2004).

11.4.1.2 Gas Packaging

This technology is extensively used in both the United Kingdom and Europe, to prolong the shelf life of bakery products. Gas packaging involves packaging of products in high gas barrier films with the appropriate mixture of gases, followed by heat-sealing of the package. Gases in packaging mainly include N_2 and CO_2. In most cases, it will be the microbial growth that is inhibited (CO_2 is dissolved into the surface moisture of the product to form a weak acid, H_2CO_3 (carbonic acid), and the absence of oxygen prevents the growth of aerobic spoilage bacteria and molds), but in dried foods, the inception of rancidity and other chemical changes can be delayed. Most MAP foods are packaged in transparent film to allow the retail customer to view the food (Coles et al., 2003).

11.4.1.2.1 Antimicrobial Effect of Carbon Dioxide

The antimicrobial effect of CO_2 augments when applied under CO_2 pressure (Erkmen, 1997) because molds, yeasts, and highly aerobic spoilage bacteria are highly susceptible to CO_2. Facultative bacteria may or may not be inhibited by CO_2 while LAB and anaerobes are highly resistant. The bacterial inhibition by CO_2 comprises the extension of the lag phase and the generation time, thereby leading to a drop in growth rate and spoilage delay (Genigeorgis, 1985) (Figure 11.2).

11.4.1.2.2 Factors Influencing the Antimicrobial Effect of CO_2

11.4.1.2.2.1 Types of Organisms The number and type of microorganisms present in a food product affect the antimicrobial effect of CO_2. Microorganisms differ considerably in their sensitivity to CO_2. In general, the gram-negative bacteria are more sensitive to CO_2 inhibition than gram-positives, with *pseudomonads* being among the most sensitive and clostridia the most resistant (Jay et al., 2005). However, certain gram-positive bacteria (e.g., LAB and certain *Bacillus* species) are very resistant to CO_2 and can tolerate and even grow in atmospheres containing 75%–100% CO_2 (Smith et al., 2004). Anaerobic bacteria, such as *Clostridium botulinum*, are not inhibited by elevated levels of CO_2 in the package atmosphere. In fact, if pathogenic species of *Bacillus* and *Clostridium* are present, they could possibly increase

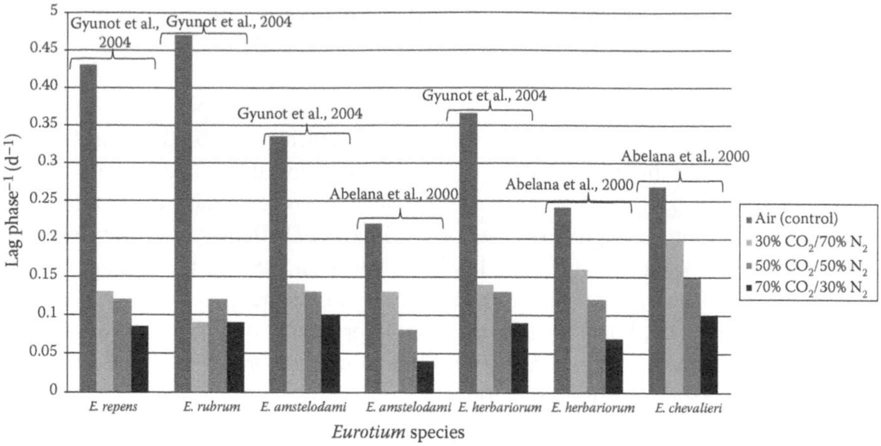

FIGURE 11.2 Inverse phases of *Eurotium* species (20°C–25°C) under air (control) and MAs (CO_2/N_2: 30/70, 50/50, and 70/30).

to hazardous levels in high-moisture, MAP bakery products, particularly at high temperature or abuse storage conditions (Smith et al., 2004).

Mold species also vary in their sensitivity to the inhibitory effects of CO_2. There is evidence that raising the level of CO_2 in the atmosphere above 70%–80% greatly decreases the growth rate of most molds, with some notable exceptions, such as *P. roqueforti* or *B. nivea* (Ottaviani and Ottaviani, 2003).

The age of the microbiological population also affects the inhibitory effect of CO_2. It has been shown that, as bacteria go from the lag phase to the log phase, the inhibitory effects of CO_2 are reduced. Therefore, the earlier the product is gas packaged, the more effective CO_2 will be (Jay, 2000).

11.4.1.2.2.2 Concentration of CO_2 Early experiments demonstrated that the success in controlling the growth of aerobic spoilage microorganisms in foods was not simply due to the elimination of O_2 but, rather, to a definite requirement for CO_2 in the gas atmosphere (Jay, 2000). Moreover, the growth of aerobic bacteria and molds can be inhibited at low concentrations of CO_2. Carbon dioxide at 20%–60% has bacterio- and fungi-static properties and will delay the growth of mold and aerobic bacteria by prolonging the lag phase and generation time of susceptible microorganisms (Coles et al., 2003; Daifas et al., 1999). However, the concentration required amounted to 50% CO_2 for complete inhibition of mold growth on bread (Smith et al., 2004). *A. flavus*, *A. ochraceus* group, *P. patulum*, species of *Eurotium*, and *Fusarium sporotrichioides* are among the molds that were inhibited only when the CO_2 levels exceeded 50% (Abellana et al., 2000).

From these studies, it becomes evident that the concentration of CO_2 in the gas mix is critical, if the anticipated prolongation in shelf life of a product is to be reached. For most products, a minimum of 20%–30% CO_2 (v/v) is required to inhibit most aerobic spoilage microorganisms, whereas for longer shelf-life prolongation, a concentration of 60% should be used (Smith et al., 2004). The inhibitory effect of

CO_2 was shown to increase linearly with increasing concentration. However, CO_2 at concentrations above 50%–60% had no further effect on microorganisms (Abellana et al., 2000).

Although there is little or no enhanced antimicrobial effect or prolongation on product shelf life at concentrations of CO_2 above 50%–60%, slightly greater concentrations (70%–80%) have been employed to make up for losses of CO_2 through packaging films or its eventual absorption by products (Jay, 2000).

11.4.1.2.2.3 Temperature The growth inhibitory effect of CO_2 is heavily related to temperature. The inhibitory effect of a certain partial pressure of carbon dioxide in the gas phase increases with decreasing temperature. According to Molin (2000) "this increased inhibitory effect at lower storage temperatures has been attributed to the greater dissolution of CO_2 in the aqueous phase of products and resulting changes in the intracellular pH and enzymatic activities of microorganisms. The solubility of carbon dioxide in water increases with decreasing temperature and the vegetative microorganisms, mainly consisting of water, and enclosed by the water phase, are exposed to a higher CO_2 concentration when the temperature is low." Smith et al. (1988) applied response surface methodology (RSM) and displayed that if the temperature in the retail environment of crumpets rises, the same product will have to be packed in higher concentration of CO_2 environment to reach similar shelf life.

Temperature control is very critical if gas packaging is employed with any meat- or cream-filled baked products. Several studies showed that *Salmonella* and *Staphylococcus* species can multiply under anaerobic conditions at 10°C–12.5°C, that is, conditions of moderate temperature abuse (Smith et al., 2004). Other studies have confirmed that the maximal growth of *A. flavus* occurred at 33°C–35°C and dropped as storage temperature decreased (Ellis et al., 1993).

11.4.1.3 Package Material Permeability

The success or failure of MAP for respiring and nonrespiring foods mainly depends on both the O_2 and CO_2 permeability of package materials to maintain the correct gas mixture in the package headspace. Moreover, membranes applied in gas packaging should have low water vapor transmission rate (WVTR) to prevent moisture loss or moisture gain, that is, to maintain the proper WV concentration because it takes part in the MAP reaction. Polymers commonly used for MAP of food include polyester, polyamide (PA), polypropylene (PP), polyvinylidene chloride (PVDC), ethylene vinyl alcohol (EVOH), ethylene vinyl acetate (EVA), and polyethylene (PE) (Table 11.3). Since all the desired characteristics of a package material, such as, structural strength, permeability, and heat sealability, are very difficult to be found in one polymer, individual polymers may be co-extruded or laminated to one another to produce films endowed with the required characteristics for MAP (Brody, 1999).

11.4.1.3.1 Membranes Commonly Used in Bakery Products

In case the objective for a gas-packaged bakery product, such as bread, is short shelf life (i.e., 2–3 days), LDPE/HDPE bags are appropriate. However, if a longer shelf life is anticipated, individual polymers are laminated to one another, since all expected characteristics of a packaging film for MAP applications, that is, strength,

TABLE 11.3
Evaluation of Suitability of Gas Packaging for Some Bakery Products

Product	a_w	pH	Moisture (%)	Swelling	CO_2 Level	Mold	Yeasts	Bacteria	Problem
Dough or batter									
Cake donut	0.82	6.60	17.85	—	—	—	—	+	T, C
Crumpet	0.97	6	47–52.41	—	+++	—	—	+++	B, S
Crusty roll	0.95	5.60	29.28	—	—	+ (14 days)	—	++	M
Yeast donut	0.91	6.40	27.98	+	++	—	+	+++	B, Y, S (14 days)
Waffle	0.94	7.20	65.79	—	+	—	—	++	Not much
Cake/pastry									
Chocolate danish	0.83	6.28	22.54	—	—	—	—	—	None
Carrot muffin	0.91	8.70	35.14	—	—	—	—	+	
Butter tart	0.78	5.70	19.95	—	—	—	—	+	
Cake (layer)									
Strawberry layer cake	0.90	6.66	37.24	+	+	—	++	++	Y, B, S (21 days)
Cherry cream cheese cake	0.94	4.51	49.90	++	++	—	++	—	Y, S (14 days)
Pies									
Blueberry pie	0.94	3.78	40.20	—	—	—	—	—	T, C
Apple turnover	0.94	4.60	35.12	+	++	—	+++	+++	Y, B, S (14 days)
Apple pie (baked)	0.95	4.21	47.82	—	—	—	+		B, Y, T, C
Apple pie (unbaked)	0.96	4.25	54.75	+++	+++	—	++	+++	Y, B, S (7 days)

Source: Adapted from Ooraikul, B., Technological consideration in modified atmosphere packaging (Chapter 3) and Modified atmosphere packaging of bakery products (Chapter 4), in *Modified Atmosphere Packaging of Foods.* Ooraikul, B. and Stiles M.E. (Eds.), Ellis Horwood, Chichester, U.K., 1991, pp. 26–48, and pp. 49–117.

T, Texture; C, Color; B, Bacteria; Y, Yeast; M, Mold; S, Swelling of package.

impermeability, and heat sealability, hardly occur in one polymer (Smith et al., 2004). Examples of laminated structures for gas packaging include PA/PE, nylon/PVDC/PE, or PA/EVOH/PE. These composite structures have all the anticipated characteristics of a packaging film for gas-packaging applications: strength, provided by the outermost layer of PA; gas and moisture vapor impermeability, provided by EVOH or PVDC; and heat sealability, provided by LDPE or EVA or an ionomer (Surlyn) (Smith et al., 2004).

The important attributes of laminated films for MAP foods are high bond strength, consistent and uniform thickness, consistent seal strength, and consistent barrier to O_2 and moisture vapor (Smith et al., 1983). The latter attribute is crucial, since the O_2 level in the package headspace may soon reach concentrations of 1% or even use high barrier films. Several studies have revealed that molds can tolerate and grow in quite low concentrations of headspace O_2, even in the presence of elevated levels of CO_2 (Smith et al., 1986, 1987, 2004; Ellis et al., 1993).

Packaging films commonly used with bakery products are barrier films such as PA/PE and PVDC-coated PP laminated or extrusion coated with PE or ionomer (Jay, 2000).

11.4.1.4 Active Packaging

Active packaging refers to the incorporation of certain additives into packaging film or within packaging containers with the aim of maintaining and extending product shelf life. Packaging may be termed *active* if it carries out some desired role in food preservation apart from providing an inert barrier to external conditions (Day, 1989). Active packaging includes additives or *freshness enhancers* capable of scavenging oxygen; adsorbing CO_2, moisture, ethylene, and flavor or odor taints; releasing ethanol, sorbates, antioxidants, and other preservatives; and maintaining temperature control (Coles et al., 2003).

Recent studies on sliced bread (Rodríguez et al., 2008) revealed a strong antimicrobial activity of cinnamon-based active paper packaging against *R. stolonifer* food spoilage. Other studies confirmed that active packaging is more effective than MAP from a microbiological point of view, for maintaining the original texture and sensory attributes of products (Sanguinetti et al., 2009).

Although many active packaging additives, such as oxygen scavengers, have been used in sachets placed into the package, the additives can also be included in the polymeric package structure. The packagers are thus profiting from easier handling and improving consumer safety by eliminating the potential of accidentally consuming a sachet (Markarian, 2006).

11.4.1.4.1 Oxygen Scavengers

The most extensively used active packaging technologies for foods are those developed as oxygen scavengers since they were the first to be commercialized in the late 1970s by Japan's Mitsubishi Gas Chemical Company (Ageless). O_2 scavengers are capable of eliminating the oxygen contained both in the packaging headspace and in the product permeating through the packaging material during storage. Oxygen scavengers are extensively employed toward retarding or preventing deterioration due to product component oxidation and growth of microorganisms (Charles et al., 2006).

The most well-known oxygen scavengers are small sachets containing various iron-based powders (iron or iron oxide powder that can be still oxidized) combined with a suitable catalyst. These chemical systems often react with water provided by the food to produce a very reactive hydrated metallic reducing agent that effectively scavenges oxygen within the food package and irreversibly turns it to a stable oxide. The iron powder is isolated from the food by storing it in a small, highly oxygen-permeable sachet labeled "*Do not eat*." The greatest advantage of such oxygen scavengers is that they are able to reduce oxygen levels to less than 0.01%, which is much lower than the typical 0.3%–3.0% residual oxygen levels obtained by application of MAP (Coles et al., 2003).

The use of oxygen scavengers has proved to be more advantageous than gas-packaging methods as more scavengers than normally required can be included, to allow for any ingress of oxygen through faulty seals. The color of indicator tablets is altered according to the amount of oxygen present. The major drawback of this system is its high cost and the potential consumer objection to the presence of a sachet inside the pack (Kotsianis et al., 2002). A consumer survey within the frame of a pilot study comprising personal interviews was carried out to determine the consumer attitudes toward oxygen absorbers used in food packages in three Helsinki area supermarkets. The greatest percentage overall of consumers (72%) were positively inclined to accept the use of oxygen absorbers, whereas 23% could not decide (Mikkola et al., 1997). Oxygen scavengers can be used alone or in combination with MAP. The common practice is removal of most of the atmospheric oxygen by MAP followed by insertion of a relatively small and low-cost scavenger to mop up the residual oxygen remaining within the food package (Coles et al., 2003).

The potential of oxygen absorbents to prolong the mold-free shelf life of bakery products has been investigated in several experiments. Experiments with crusty rolls showed that they remained mold free for more than 60 days, when Ageless absorber was placed in the package, compared with 16–18 days for those packaged without oxygen absorber (Smith et al., 1986) (Figures 11.3 and 11.4).

The shelf life of white bread packaged in PP film can be prolonged from 5 to 45 days at room temperature by incorporating into the package an O_2 absorbent sachet (Smith et al., 2004). Pizza crust, which molds in 2–3 days at 30°C, can also remain mold free for more than 14 days at this temperature when an appropriate O_2 absorbent is present (Smith et al., 1983) (Figure 11.5).

11.4.1.4.2 Antimicrobial Packaging

Antimicrobial packaging is a bold and extremely challenging technology that could effectively prolong shelf life and improve food safety in both synthetic polymers and edible films. The market volume for antimicrobial use in polyolefins is anticipated to rise from 3300 ton in 2006 to 5480 ton in 2012 (McMillin, 2009).

Research in the area of antimicrobial food packaging materials has considerably augmented over the last few years as an alternative method for controlling microbial contamination of foods either through the incorporation of antimicrobial substances or as coatings on the packaging materials. A great deal of research is oriented to the design of antimicrobial packaging containing natural antimicrobial agents for specific or broad microbial inhibition depending on the nature of the agents used or on

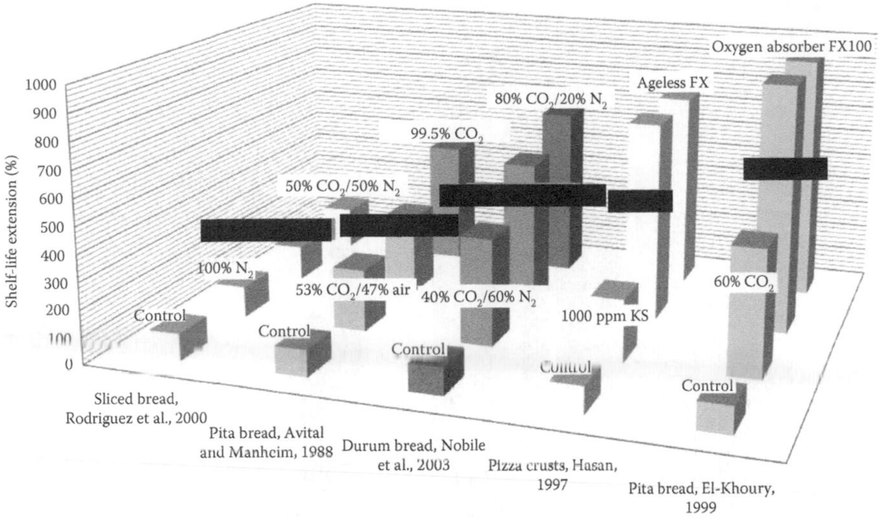

FIGURE 11.3 Percentage of shelf-life extension of bakery products under modified atmosphere and use of additives.

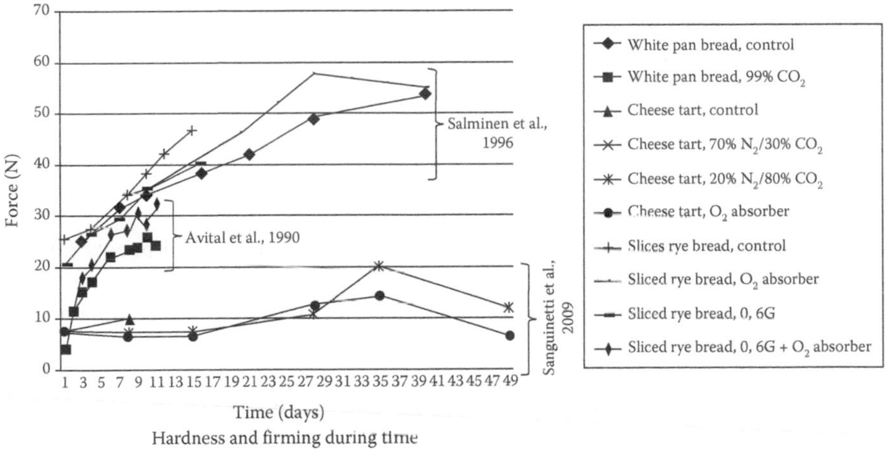

FIGURE 11.4 Effect of time storage on hardness and firmness (Force [N]) of white pan bread, cheese tart, and sliced rye bread under several MAs (CO_2/N_2: 99/1, 30/70, 80/20) in conjunction with O_2 absorber.

their concentration. Various types of antimicrobial delivery systems and packaging materials for food combinations have been developed to optimize the effectiveness of the system (Dainelli et al., 2008).

Antimicrobial packaging is a typical form of active packaging because it interacts with the product or the headspace between the package and the food system, to obtain the best possible result. Likewise, antimicrobial food packaging can potentially decrease, inhibit, or delay the growth of microorganisms present in the packed

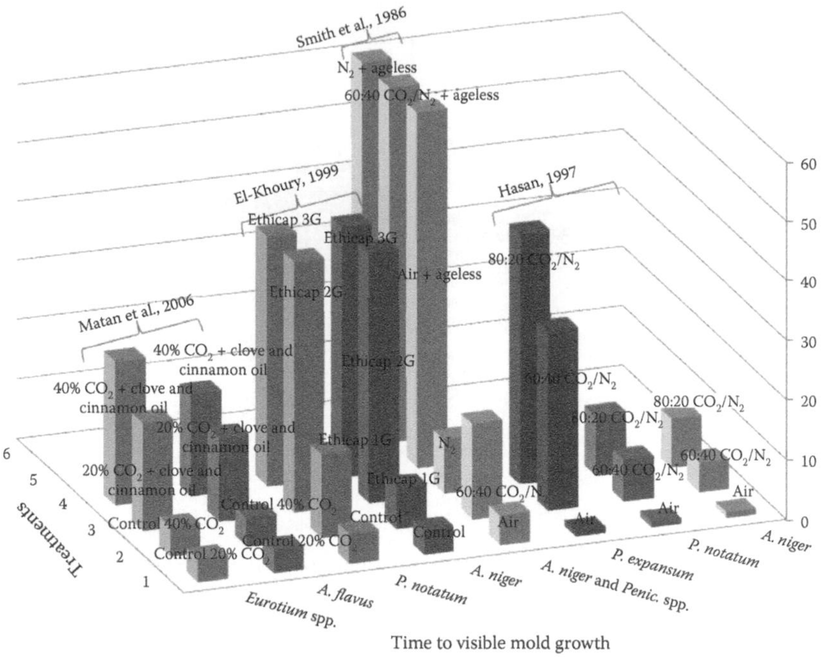

FIGURE 11.5 Mold growth of various bakery products packed under MAP in the presence or absence of active compounds versus time (days).

food or packaging material itself. Antimicrobial packaging may be in the following forms:

1. Inclusion of sachets or pads containing volatile antimicrobial agents into packages
2. Direct incorporation of volatile and nonvolatile antimicrobial compounds into polymers
3. Antimicrobials adsorption onto polymer surfaces
4. Immobilization of antimicrobials on inert substrate by means of covalent or ion bonds
5. Application of inherently antimicrobial polymers (Appendini and Hotchkiss, 2002)

Gutierrez et al. (2009) claimed that the use of a cinnamon-based active package proved to extend more than three times the product shelf life of a complex bakery product with least change in the packaging and no supplementary processing steps. Both quality and property of the product were not affected by the application of this concept. The shelf life increased from 3 to 10 days with maximum quality and safety. However, additional work has to be carried out for potential industrial scale-up. Rodriguez et al. (2008) assessed the antimicrobial activity of a new active paper package including cinnamon essential oil to solid wax paraffin as an active coating. The antifungal activity of the active paper was tested against *R. stolonifer*, and the obtained results demonstrated that 6% (w/w) of the essential oil in the active coating formulation completely inhibits the growth of *R. stolonifer*, while 4% still has strong antimicrobial activity in in vitro conditions. After 3 days of storage, approximately complete inhibition was reached with 6% cinnamon essential oil.

11.4.1.5 Use of Gas Packaging for Shelf-Life Extension of Bakery Products

The concept of altering the gas phase inside a package of bread to prolong the product shelf life is not novel. As early as 1933, it was shown that the storage of bread in atmospheres containing at least 17% carbon dioxide considerably retarded the appearance of mold. At a CO_2 concentration of 50%, the mold-free shelf life of bread doubled under storage conditions favoring mold development.

Research on the use of CO_2 for prolonging the shelf life of bakery products was undertaken by Seiler in the United Kingdom in the 1960s. In detailed studies with bread and cake stored at 21°C and 27°C and CO_2 concentrations of 0%–60%, it was shown that the mold-free shelf life increased with rising CO_2 concentrations, in conjunction with lower temperatures. Subsequent studies with mixtures of CO_2 and N_2 and with 100% CO_2 confirmed the demand for CO_2 in the package headspace, where simply displacing headspace O_2 with N_2 alone was inadequate to prevent mold growth (Robertson, 2005). Investigations on British bread suggested a shelf-life extension of 100%, 200%, and 300% when packed in atmospheres of 40%, 60%, or 90% carbon dioxide, respectively, by volume. The effect of CO_2 on the mold-free shelf life of products is RH dependent. In bakery products such as cakes, with RH values of 85% or below, it is possible to obtain 400% extension in their shelf lives in atmospheres of 75% or above of carbon dioxide. In bread-type products, of RH higher than 90%, when a high content of carbon dioxide is present, the obtained prolongation is about 250%. Utilization of a mixture of two gases, carbon dioxide and nitrogen, is suggested to avoid package collapse, which can occur since carbon dioxide is absorbed into the product if it is used alone. The most appropriate composition was found to be a gas with 60% carbon dioxide and 40% nitrogen (Pateras, 2007).

According to Smith et al. (1986) "packaging of bakery products such as crumpets in a combination of CO_2 and N_2 (60:40) has been found to be effective against microbial spoilage for several weeks under ambient temperature. They also determined the minimum oxygen level in the CO_2/N_2 (60:40) atmosphere required for visible mold growth. Those experiments revealed that all gas-packaged inoculated Potato dextrose Agar (PDA) plates with oxygen ranging from <0.05% to 10% displayed growth after 4–6 days, while those packaged in air (control) exhibited growth only after 1–1.5 days."

Investigation of gas-packaged crusty rolls revealed that although mold growth was obvious in all air-packaged crusty rolls after 5–6 days, growth was retarded for further 4–5 days and 11–12 days in rolls packaged in 100% N_2 and CO_2/N_2 (60:40), respectively (Smith et al., 1986). Smith et al. (1988) also looked into the effectiveness of CO_2:N_2 (60:40) gas mixture to prolong the shelf life of English style crumpets by applying an RSM.

Though the uses of MAs have advantageous effects on the shelf-life prolongation of bakery products, minor or major problems have been occasionally recorded. El-Khoury (1999) reported that in all MAP products packages of less than 60% CO_2 (balance N_2) the duration of sensory shelf life was approximately 7 days lesser than microbiological shelf life due to changes in odor, texture, and flavor. After day 28, sensory analysis displayed a certain staling of the bread accompanied with a slight acidic flavor and odor. Daifas et al. (2000) reported that alcoholic odors were frequently determined in sensory analysis of crumpets used in that study irrespective of the amount of ethanol (Ethicap) included in the package. In the same study it was stated that Seiler (1978) exhibited that cake and bread treated with 1% (w/w) of ethanol were sensorially acceptable but when treated with >2% (w/w), alcoholic flavors were too apparent and products were rejected on the basis of flavor and odor. Salminen et al. (1996) also reported that panelists detected the flavor of ethanol in very rare cases in experiments where rye bread was packed under MAs.

Most of the major problems occurring are attributed to the inability of the gaseous environment to maintain headspace oxygen below 0.4%, as described by Smith et al. (1986). It was shown that "anaerobic environment" containing enhanced levels of CO_2 and/or N_2 is not always possible to prevent the mold growth; in fact, complete mold growth prevention is feasible only if the headspace oxygen is reduced and maintained at levels less than 0.4%. Guynot et al. (2003a) reported that the inhibitory effect of CO_2 on microorganisms in a culture medium or in a food system depends on partial CO_2 pressure, O_2 content, headspace gas volume, temperature, acidity, and a_w. According to Suhr and Nielsen (2005) the MAP applied in their experiments hardly inhibited the growth of the spoilage yeast *Endomyces fibuliger* on wheat bread. On rye bread *E. fibuliger* displayed growth on day 2 in all 1% O_2 treatments, while *P. commune* could grow under 99% CO_2 environment in the presence of 1% residual oxygen. Moreover, *A. flavus* exhibited the highest capacity to grow under low O_2, as it could grow when no O_2 was added (0.03%) at 75% CO_2. The effect of MAP in conjunction with other treatments on shelf life of various bakery products is given in Table 11.4.

11.4.1.6 Advantages and Disadvantages of Gas Packaging of Bakery Products

The most important advantages associated with MAP of bakery products are as follows

1. Shelf-life prolongation
2. Retardation (no consensus by all authors)
3. Improved external appearance of product
4. Easier and less frequent transportation and distribution
5. Less additives (safer and more inexpensive product)

TABLE 11.4
Effect of MAP and Other Treatments on Shelf-Life Extension of Bakery Products

Product	Combination of Gases	Film Type	Parameter	Shelf-Life Extension	Other Treatment	Source
Wheat flour	Air		Growth *A. flavus*	2 days fungal growth delay determined by the «Gompertz» model after 3, 4, and 5 days	Phenyllactic acid (PLA) (7.5 mg/mL)	Lavermicocca et al. (2003)
Bread	>>		Growth *P. chrysogenum*	34.5 h after 3 days, 39.7 h after 4 days, and 48.3 h after 5 days determined by the «Gompertz» model		
>>	>>		Growth *P. polonicum*	155.2 h after 3 days, 190 h after 4 days, and 225 h after 5 days determined by the «Gompertz» model		
Sponge cake	>>	PA and PE co-extrusion mix film (FontPack, Commercial del Paper Font Spain), with a thickness of 90 µm Transmission rates: O_2, 19.913 cm^3/m^2/day/atm; CO_2, 164.903 cm^3/m^2/day/atm; and H_2O, 2.60 cm^3/m^2/day/atm	Colony diameter *E. amstelodami*	After 27 days, the colony diameter was 80% of the control	ATCO LH100 (100 mL a.c)	Guynot et al. (2003b)

(continued)

TABLE 11.4 (continued)
Effect of MAP and Other Treatments on Shelf-Life Extension of Bakery Products

Product	Combination of Gases	Film Type	Parameter	Shelf-Life Extension	Other Treatment	Source
Sponge cake	100% N_2	>>	Colony diameter *E. amstelodami*	After 27 days, the colony diameter was 37% of the control		
>>	30% CO_2, 70% N_2	>>	>>	>>		Guynot et al. (2003a)
>>	Air	>>	Colony diameter *E. amstelodami, E. rubrum, E. herbariorum, E. repens*	After 27 days, no fungal growth was detected.	ATCO LH210 (210 mL a.c)	
>>	30% CO_2, 70% N_2	>>	>>	>>	ATCO oxygen scavenger	
>>	100% N_2	>>	>>	>>	>>	
Spanish sponge cake	Air	OPP20/(PELD/EVOH/PELD)45/PELLD30 with a total thickness of 95 µm. Transmission rates: O_2: 2 cm^3/m^2/day/atm (at 50% RH and 23°C), H_2O: 1g/m^2/day/atm ((at 75% RH and 25°C)	Colony diameter *E. amstelodami, E. herbariorum, E. chevalieri*	No isolate growth was detected	a_w 0.75–0.80	Abellana et al. (2000)
>>	>>	>>	>>	Reduced growth rate with similar behavior of all isolates	a_w 0.85	
>>	>>	>>	>>	Rapid growth rate	a_w 0.90	

>>	>>	>>	>>	Lag phase between 2 and 4 days	a_w 0.90	Abellana et al. (2000)
>>	>>	>>	>>	Lag phase >30 days	a_w 0.80 and 0.75	
>>	100% CO_2	>>	Colony diameter *E. amstelodami, E. herbariorum, E. chevalieri*	No isolate growth was detected		
>>	>>	>>	>>	Lag phase was >30 days irrespective of the values of O_2		
>>	50% CO_2	>>	>>	No isolate growth was detected	a_w 0.85	
>>	>>	>>	>>	>>	>>	
High-moisture, high-pH bakery products	Air	>>	Growth of *B. Cereus*	No growth throughout the experiment	Potassium sorbate (KS) >1000 ppm, pH = 5	Koukoutsis et al. (2004)
>>	>>	>>	>>	>>	SHA 2000 ppm, pH = 9	
>>	>>		>>	1 day inhibition of growth	SHA (Sorbic hydroxamic acid) 2000 ppm, pH = 7	
>>	>>		Growth of *B. Subtilis and S. enteritidis*	No growth throughout the experiment	SHA 2000 ppm, pH = 5 and 9	
>>	>>		Growth of *S. enteritidis*	>>	SHA 2000 ppm, pH = 7	
>>	>>		Growth of *L. monocytogenes*	Inhibition of growth for over 28 days	KS 1500–2000 ppm, pH = 5	
>>	>>		Growth of *S. cerevisiae* and *P. notatum*	No growth throughout the experiment	SHA 1000 ppm, pH = 9	

(*continued*)

TABLE 11.4 (continued)
Effect of MAP and Other Treatments on Shelf-Life Extension of Bakery Products

Product	Combination of Gases	Film Type	Parameter	Shelf-Life Extension	Other Treatment	Source
>>	>>		Growth of *S. cerevisiae* and *P. notatum*	>>	SHA 1000 ppm, pH = 9	
>>	>>	Cryovac® bags	Growth *B. Cereus, B. subtilis, S. enteritidis*	1 day inhibition of growth	Water–ethanol emitters	
>>	>>	>>	Growth of *S. cerevisiae, P. notatum*	2 days inhibition of growth	>>	
>>	>>	Metalized bags	Growth of *B. Cereus, B. subtilis, S. enteritidis*	>>	>>	
>>	>>	>>	Growth of *S. cerevisiae, P. notatum*	4 days inhibition of growth	>>	
>>	>>	Cryovac bags	Growth of *B. Cereus, B. subtilis, S. enteritidis*	11 days inhibition of growth	Mastic-ethanol emitters	
>>	>>	>>	Growth of *S. cerevisiae, P. notatum*	13 days inhibition of growth	>>	
>>	>>	Metalized bags	Growth of *B. Cereus, B. subtilis, S. enteritidis*	>28 days inhibition of growth	>>	
	>>	>>	Growth of *S. cerevisiae, P. notatum*	>>	>>	
Sponge cake	100% N_2	PA–PE co-extrusion mix film (total thickness 90 μm), transmission rates: O_2: 19.913 cm^3/m^2/day/atm, CO_2: 164.903 cm^3/m^2/day/atm, H_2O: 2.60 cm^3/m^2/day/atm)	Growth of *E. repens* and *E. rubrum*	No growth	pH = 6, a_w = 0.80	Guynot et al. (2004)

>>	>>	>>	Growth of *E. amstelodami, E. herbariorum*	>>	pH=6, a_w=0.80, 0.05% PS (potassium sorbate)
>	>>	>>	Growth of *E. repens* and *E. rubrum*	>>	pH=6, a_w=0.85–0.90, 0.05% PS
>>	>>	>>	Growth of *E. amstelodami, E. herbariorum*	>>	pH=6, a_w=0.85, 0.2% PS
>>	50% CO_2, 50% N_2	>>	Growth of *E. amstelodami*	>>	pH=6, a_w=0.85–0.90, 0.1% PS
>>	70% CO_2, 30% N_2	>>	>>	>>	pH=6, a_w=0.85–0.90, 0.05% PS
>>	50% CO_2, 50% N_2	>>	Growth of *E. herbariorum*	>>	pH=6, a_w=0.85, 0.1% PS
>>	70% CO_2, 30% N_2	>>	>>	>>	pH=6, =a_w=0.90, 0.1% PS
>>	>>	PA–PE co-extrusion mix film (total thickness 90 μm), transmission rates: O_2: 19.913 cm^3/m^2/day/atm, CO_2: 164.903 cm^3/m^2/day/atm, H_2O: 2.60 cm^3/m^2/day/atm)	>>	>>	pH=6, a_w=0.85
>>	100% CO_2	>>	>>	>>	pH=6, a_w=0.90
>>	70% CO_2, 30% N_2	>>	Growth of *E. amstelodami*	>>	pH=7.5, a_w=0.90, PS=0.05%
>>	>>	>>	Growth of *E. herbariorum, E. repens,* and *E. rubrum*	>>	pH=7.5, a_w=0.90, PS=0.2%

(continued)

TABLE 11.4 (continued)
Effect of MAP and Other Treatments on Shelf-Life Extension of Bakery Products

Product	Combination of Gases	Film Type	Parameter	Shelf-Life Extension	Other Treatment	Source
>>	30% CO_2, 70% N_2	>>	Growth of *E. herbariorum*	>>	pH=7.5, a_w=0.85, PS=0.1%	
>>	<50% CO_2	>>	>>	>>	pH=7.5, a_w=0.85, PS=0.05%	
>>	100% N_2	>>	Growth of *E. amstelodami, E. repens*, and *E. rubrum*	>>	pH=7.5, a_w=0.80	
>>	>>	>>	Growth of *E. herbariorum*	>>	pH=7.5, a_w=0.80, PS=0.05%	
African steamed bread			Growth of microorganisms	1 week growth inhibition	Hurdle technology (salt, glycerol, fat, heat)	
Agar model system	80% CO_2, 20% N_2	High gas barrier Cryovac bags (O_2: 3–6 cc/m^2/day/atm, 4.4°C, 0% RH)	Growth *A. niger* PDA	52 days growth inhibition	1240 ppm KS, pH=5.5	Hasan (1997)
>>	>>	>>	>>	40 days growth inhibition	1240 ppm KS, pH=6	
>>	>>	>>	>>	27 days growth inhibition	1240 ppm KS, pH=6.5	
>>	>>	>>	>>	>52 days growth inhibition	500–1000 ppm KS, pH=5.5	
>>	40% CO_2, 60% N_2	High gas barrier Cryovac bags (O_2: 3–6 cc/m^2/day/atm, 4.4°C, 0% RH)	>>	18 days growth inhibition	1240 ppm KS, pH=5.5	

>>	60% CO_2, 40% N_2	>>	>>	31 days growth inhibition	>>
>>	80% CO_2, 20% N_2	>>	>>	52 days growth inhibition	>>
Pizza crust (without inoculation)		>>	>>	9 days growth inhibition	1000 ppm KS (direct incorporation into the dough), pH = 5.5
>>		>>	>>	15 days growth inhibition	2000 ppm KS (direct incorporation into the dough), pH = 5.5
>>		>>	>>	5 days growth inhibition	1000 ppm KS (direct incorporation into the dough), pH = 6
>>		>>	>>	13 days growth inhibition	2000 ppm KS (direct incorporation into the dough), pH = 6
		>>	>>	5 days growth inhibition	1000 ppm KS (direct incorporation into the dough), pH = 5.5
>>		>>	>>	8 days growth inhibition	2000 ppm KS (direct incorporation into the dough), pH = 5.5
>>		High gas barrier Cryovac bags (O_2: 3–6 cc/m^2/day/atm, 4.4°C, 0% RH)	>>	1 day growth inhibition	>>
>>		>>	>>	>>	>>
>>		>>	>>	6 days growth inhibition	1000 ppm KS (with surface spraying), pH = 5.5

(continued)

TABLE 11.4 (continued)
Effect of MAP and Other Treatments on Shelf-Life Extension of Bakery Products

Product	Combination of Gases	Film Type	Parameter	Shelf-Life Extension	Other Treatment	Source
>>		>>	>>	12 days growth inhibition	2000 ppm KS (with surface spraying) pH = 5.5	
>>		>>	>>	4 days growth inhibition	1000 ppm KS (surface spraying onto the crust), pH = 6	
>>		>>	>>	>>	2000 ppm KS (surface spraying onto the crust) pH = 6	
>>		>>	>>	3 days growth inhibition	1000 ppm KS (surface spraying onto the crust), pH = 5.5	
>>		>>	>>	6 days growth inhibition	2000 ppm KS (surface spraying onto the crust) pH = 5.5	
>>		High gas barrier Cryovac bags (O_2: 3–6 cc/m^2/day/atm, 4.4°C, 0% RH)	>>	1 day growth inhibition	1000 ppm KS (surface spraying onto the crust), pH = 6	
>>		>>	>>	>>	2000 ppm KS (surface spraying onto the crust) pH = 6	
>>		>>	>>	>>	1000 ppm KS (impregnation of packaging material), pH = 5.5	

>>	>>	>>	>>	2000 ppm KS (impregnation of packaging material) pH = 5.5
>>	>>	>>	>>	1000 ppm KS (impregnation of packaging material) pH = 6
>>	>>	>>	>>	2000 ppm KS (impregnation of packaging material) pH = 6
>>	>>	>>	>>	1000 ppm KS (impregnation of packaging material), pH = 5.5
>>	>>	>>	>>	2000 ppm KS (impregnation of packaging material) pH = 5.5
>>	>>	>>	>>	1000 ppm KS (impregnation of packaging material) pH = 6
>>	High gas barrier Cryovac bags (O_2: 3–6 cc/m^2/day/atm, 4.4°C, 0% RH)	>>	>>	2000 ppm KS (impregnation of packaging material) pH = 6

(*continued*)

TABLE 11.4 (continued)
Effect of MAP and Other Treatments on Shelf-Life Extension of Bakery Products

Product	Combination of Gases	Film Type	Parameter	Shelf-Life Extension	Other Treatment	Source
>>	60% CO_2, 40% N_2	>>	>>	21 days growth inhibition		
>>	80% CO_2, 20% N_2	>>	>>	25 days growth inhibition		
>>		>>	>>	No growth after 42 days	Ageless (FX-100)	
>>	60% CO_2, 40% N_2	>>	>>	11 days growth inhibition		
>>	80% CO_2, 20% N_2	>>	>>	15 days growth inhibition		
>>		>>	>>	No growth after 42 days	Ageless (FX-100)	
>>	60% CO_2, 40% N_2	>>	>>	>>	1000 ppm PS, pH=6	
>>	80% CO_2, 20% N_2	>>	>>	>>	>>	
>>	60% CO_2, 40% N_2	>>	>>	>>	>>	
>>	80% CO_2, 20% N_2	>>	>>	>>	>>	
>>	Air	>>	Mold growth	Mold growth after 4 days	pH=6	
>>	60% CO_2, 40% N_2	>>	>>	No growth after 42 days	>>	
>>		>>	>>	>>	pH=6, Ageless FX	

>>	Air	>>	>>	Mold growth after 7–9 days	pH=6, 1000 ppm KS	
>>	60% CO_2, 40% N_2	High gas barrier Cryovac bags (O_2: 3–6 cc/m^2/day/atm, 4.4°C, 0% RH)	>>	No growth after 42 days	pH=6, 1000 ppm KS	
>>		>>	>>	>>	pH=6, 1000 ppm KS, Ageless FX	
>>	Air	>>	>>	Mold growth after 14 days	pH=6, 2000 ppm KS	
>>	60% CO_2, 40% N_2	>>	>>	No growth after 42 days	>>	
>>		>>	>>	>>	pH=6, 2000 ppm KS, Ageless FX	
Crusty rolls	Air	PE-coated PA film. Average permeabilities per m^2/24 h at 25°C and 100% RH: Oxygen 40 cc, N_2 14 cc, CO_2 155 cc water vapor 11 g	>>	growth inhibition for 5–6 days		Smith et al. (1986)
>>	60% CO_2, 40% N_2	>>	>>	Growth inhibition for 11–13 days		
>>	100% N_2	>>	>>	Growth inhibition for 4–6 days		
>>	Air	>>	>>	Growth inhibition for >60 days	Ageless	
>>	60% CO_2, 40% N_2	>>	>>	>>	>>	
>>	100% N_2	>>	>>	>>	>>	

(continued)

TABLE 11.4 (continued)
Effect of MAP and Other Treatments on Shelf-Life Extension of Bakery Products

Product	Combination of Gases	Film Type	Parameter	Shelf-Life Extension	Other Treatment	Source
Brown rice muffin	Air	Co-extrusion film (oriented polypropylene [OPP] and cast polypropylene [CPP]), 45 µm thickness	Microbial spoilage and sensory evaluation	1 day extension		Keawmanee and Haruenkit (2007)
>>	>>	>>	>>	3 days extension	oxygen absorber (Ageless PR50)	
>>	60% CO_2, 40% N_2	>>	>>	>>		
>>	Air	>>	Firmness (staling)	6 days observation		
>>	With oxygen absorber	>>	>>	No significant difference was observed		
>>	60% CO_2, 40% N_2	>>	>>	>>		
>>	60% CO_2, 40% N_2 with oxygen absorber	>>	>>	>>		
Wheat bread	Air	Laminate with ethylene vinyl alcohol as barrier layer. 95 µm thickness. Permeability (at 75% relative air humidity and 25°C) of O_2 and CO_2 was	Bread firmness (staling)	15 days observation		Rasmussen and Hansen (2001)

		less than 2 and 2.3 mL/m^2/24 h, respectively. Water permeability <g/m^2/24 h. Moisture loss about 0.1 g/24 h per bread packing.			
>>	50% CO_2, 50% N_2	>>	>>	No significant difference was observed	
>>	100% CO_2	>>	>>	>>	
>>	Air	>>	Melting enthalpy of retrograded starch	8 days observation	
>>	100% CO_2	>>	>>	No significant difference was observed	
Whole wheat flour bread, white breads, and biscuits	Air	Flexible package	Compressibility tests	13–15 days for bread and 7–15 days for biscuits	Knorr and Tomlins (1985)
>>	100% N_2	>>	>>	No significant differences	
>>	100% CO_2	>>	Compressibility tests	Significantly lower regression coefficients ($P<0.01$)	
Cheese cake	Air	Multistrate (EVOH/OPET/PE) gas and water barrier film with a thickness of 54 μm	48 days storage. YPD (yeast peptone dextrose agar) growth, puncturing test	Spoilage after 7 days	Sanguinetti et al. (2009)

(*continued*)

TABLE 11.4 (continued)
Effect of MAP and Other Treatments on Shelf-Life Extension of Bakery Products

Product	Combination of Gases	Film Type	Parameter	Shelf-Life Extension	Other Treatment	Source
>>	30% CO_2, 70% N_2	>>		Spoilage after 14 days, significant hardness increase		
>>	80% CO_2, 20% N_2	>>		Spoilage after 34 days, significant hardness increase		
>>	O_2 absorber	>>		No spoilage, insignificant hardness increase		
Sliced bread	Cinnamon-based active paper	Double coated film with active paraffin formulation in between, containing the appropriate amount of cinnamon essential oil as an active agent	Inoculation on potato dextrose agar for 5 days	Complete inhibition after 3 days with 6% cinnamon essential oil		Rodríguez et al. (2008)
>>	>>	>>	>>	60%–80% inhibition after 3 days with 4% cinnamon essential oil		
>>	>>	>>	>>	No inhibition after 3 days with 2% cinnamon essential oil		

Bread	100% N_2	High barrier plastic bag (laminate of OPP20/EVOH/PELD45/PELLD30)	Fungi growth on Czapek yeast extract agar (CYA) (*A. flavus, P. commune, P. roqueforti*, and *E. fibuliger*)	*E. fibuliger*: 1.5 mm (14% of the control)	Oxygen absorber	Nielsen and Rios (2000)
>>	>>	>>	>>	*E. fibuliger*: 4 mm (36% of the control)	Without oxygen absorber	
>>	Air	>>	>>	*E. fibuliger*: 11 mm		
>>	100% N_2	>>	>>	Most common bread fungi: no growth	Oxygen absorber	
>>	50% N_2, 50% CO_2	>>	>>	>>	>>	
>>	100% CO_2	>>	>>	>>	>>	
>>	>>	>>	>>	Most common bread fungi: 18%–29% of the control	0.02%–0.03% residual O_2	
>>	50% N_2, 50% CO_2	>>	>>	>>	>>	
>>	100% CO_2	>>	>>	Most common bread fungi: 1%–6% of the control	>>	
>>	>>	>>	>>	Most common bread fungi: 4%–11% of the control	1.00% residual O_2	
Apple turnovers	Air	PE/PA film	Yeast growth	Evident growth after 1 day	a_w 0.95	Smith et al. (1987)
>>	>>	>>	>>	Evident growth after 5 days	a_w 0.90	

(continued)

TABLE 11.4 (continued)
Effect of MAP and Other Treatments on Shelf-Life Extension of Bakery Products

Product	Combination of Gases	Film Type	Parameter	Shelf-Life Extension	Other Treatment	Source
>>	>>	>>	>>	Evident growth after 14 days	a_w 0.85	
>>	Air + Ethicap (ethanol vapor)	>>	>>	Heavy growth after 3 days	a_w 0.95, Ethicap size 4	
>>	>>	>>	>>	>>	a_w 0.95, Ethicap size 6	
>>	>>	>>	>>	Medium growth after 7 days, heavy growth after 14 days	a_w 0.90, Ethicap size 2	
>>	>>	>>	>>	Slight growth	a_w 0.90, Ethicap size 4	
>>	>>	>>	>>	>>	a_w 0.85, Ethicap size 1 and 2	
>>	Air	>>	>>	Growth after first day of storage	a_w=0.93	
>>	60% CO_2, 40% N_2	>>	>>	>>	>>	
>>	Air + Ethicap (ethanol vapor)	>>	>>	No growth	>>	
>>	60% CO_2, 40% N_2 + Ethicap	>>	>>	>>	>>	

Crumpets	60% CO_2, 40% N_2	12/75 polyester/polyethylene top web and 100/100 nylon/polyethylene bottom web, with average permeabilities, per m^2/24 h, at 25°C and 100% RH, of: O_2, 40 mL; N_2, 14 mL; CO_2, 155 mL; water vapor, 11 g	Volume measurements, aerobic plate counts (APC)	(In all cases, package volume and headspace CO_2 decreased during the first week of storage.) Remained stable or gradually decreasing	20°C	Smith et al. (1983)
>>	>>	>>	>>	Remained stable or gradually decreasing	22°C	
>>	>>	>>	>>	The volume did not start to increase until after 25 days	24°C	
>>	>>	>>	>>	Volume increase after 12 days due to production of CO_2 and other metabolites by microorganisms. APC and metabolites were higher in the product stored at 30°C for 19 days than in those stored at 20°C for 1 month	30°C	
Yeast leavened crumpets	Air with ethanol vapor (2G Ethicap) generator		Neurotoxin production (*C. botulinum*), 30 days storage (inoculation)	No detection		Daifas et al. (2003)

(*continued*)

TABLE 11.4 (continued)
Effect of MAP and Other Treatments on Shelf-Life Extension of Bakery Products

Product	Combination of Gases	Film Type	Parameter	Shelf-Life Extension	Other Treatment	Source
Chemical leavened crumpets	>>		>>	10 days delay of detection		
Yeast leavened crumpets	100% CO_2 (25°C)		>>	Neurotoxin detection after 5 days		
Chemical leavened crumpets	>>		>>	>>		
Crumpets	Air	High gas barrier bags (ethanol transmission rate [ETR] 0.21 g/m^2/day, 25°C).	Neurotoxin production (*C. botulinum*) (inoculation)	Toxin detection after 5 days		
>>	Air with ethanol vapor generators (Ethicap 2, 4, or 6G)	>>	>>	Ethicap 2G delayed toxicity for 10 days while complete inhibition (>21 days) was observed in all crumpets packaged with 4 or 6G Ethicap		

>>	Air with cotton wool pads saturated with 2, 4, or 6 g of 95% food grade ethanol and stored at 25°C	>>	>>	Complete inhibition		
>>	60% CO_2, 40% N_2		>>	Toxin detection after 4 days		
>>	100% CO_2		>>	Toxin delay from 1.5 (low pH crumpets)—3 days (high pH crumpets)		
Whole grain rye bread	Air-packed control	HDPE containers (500 mL)	Microbial growth, 42 days	First microbial growth observed after 3 days	Air-packed control	Salminen et al. (1996)
>>	Ethanol emitter		>>	>>	0.6G	
>>	>>		>>	>>	1G	
>>	>>		>>	First microbial growth observed after 26–27 days	2G	
>>	>>		>>	>>	3G	
>>	Oxygen absorber		>>	No growth observed	Oxygen absorber	
>>	1% (v/v) ethanol-containing gases		>>	First microbial growth observed after 8 days		

However, the application of MAP entails several disadvantages among which the most important are the following:

1. Secondary fermentation problems followed by the potential growth of molds resistant to high CO_2 content and, eventually, growth of *C. botulinum* (inconclusive evidence) (Katsarou et al., 2002). Although it was initially believed that the presence of O_2 could inhibit the growth of *C. botulinum*, this assumption has been recently refuted by several studies that proved the inclusion of O_2 provides no benefit but, on the contrary, it can even enhance the toxin production (Smith et al., 2004).
2. Relatively high cost of packaging equipment and materials.
3. Nonuniform distribution of gases in the package.
4. Possible collapse of packaging in case the amount of CO_2 employed is high.

11.4.2 MAP Techniques and Machinery

The MAP machinery used for bakery products can be distinguished based on the applied technique in the following three categories: thermoforming systems, preformed container machines, and horizontal or vertical form-fill-seal machine systems (Kotsianis et al., 2002). Opting for a packaging machine depends on various factors, mainly determined by the nature of the product and the market requirements. For example, the chamber type of machine will result in minimum residual oxygen content in the pack whenever this is required by the process of evacuation prior to back flushing with gas, compared with the continuous flushing systems. However, this method is not applicable to handling fragile products (Hastings, 1998).

In the continuous flow wrap/gas flushing technique, the machine forms a tube of flexible material that encloses the product either by itself or on a carrier tray. The influx of the opted gas mixture is carried out in a continuous countercurrent flow into the package to force the air out, the ends of the package are heat sealed, and the packages are cut from each other. The greatest advantage of this technique is its high production rate, which can reach up to 120 packages/min (Jay, 2000). These systems are by far the most popular MAP machinery for bakery products in view of their flexibility and high speed (Ooraikul, 1991).

In both preformed tray and thermoform/fill/gas flush/seal techniques, a compensated vacuum method is applied to insert the gas mixture. In this method, the product is placed into a thermoformed tray and a vacuum applied to remove most of the air. The vacuum is stopped by the appropriate gas mixture and the package heat sealed with a top web of film. One considerable advantage of the in-line thermoform method of gas packaging is the high efficiency of oxygen removal to residual levels of <1%. The in-line thermoforming type of equipment may also be used for VP or vacuum skin packaging (Jay, 2000).

11.4.3 Conclusions

Bakery products were among the first to be used both in MAP and active packaging. The gas compositions most frequently employed are CO_2/N_2: 100/0, 80/20,

70/30, 60/40, and 0/100. The main objective of MAP and active packaging is the growth inhibition of yeasts and *A. niger, E. amstelodami, E. herbariorum, E. repens, E. rubrum, B. cereus, B. subtilis, P. notatum, S. enteriditis, S. cerevisiae,* and *E. chevalieri.* As regards the active packaging components, the most widely used ones in bread packaging are Ageless (Mitsubishi Ltd, Japan), oxygen scavengers, and ethanol emitters.

REFERENCES

Abellana, M., Magrí, X., Sanchis, V. and Ramos, A.J. 1999a. Water activity and temperature effects on growth of *Eurotium amstelodami, E. chevalieri and E. herbariorum* on a sponge cake analogue. *International Journal of Food Microbiology,* 52: 97–103.

Abellana, M., Benedí, J., Sanchis, V. and Ramos, A. J., 1999b. Water activity and temperature effects on germination and growth of *Eurotium amstelodami*, E. *chevalieri* and E. *herbariorum* isolates from bakery products. *Journal of Applied Microbiology,* 87: 371–380.

Abellana, M., Sanchis, V. and Ramos, A. J. 2001. Effect of water activity and temperature on growth of three *Penicillium* species and *Aspergillus flavus* on a sponge cake analogue. *International Journal of Food Microbiology*, 71: 151–157.

Abellana, M., Sanchis, V., Ramos, A. J. and Nielsen, P. V. 2000. Effect of modified atmosphere packaging and water activity on growth of *Eurotium amstelodami*, *E. chevalieri* and *E. herbariorum* on a sponge cake analogue. *Journal of Applied Microbiology*, 88(4): 606–616.

Agundez-Arvizu, Z., Fernandez-Ramırez, M. V., Arce-Corrales, M. E., Cruz-Zaragoza, E., Melendrez, R., Chernov, V. and Barboza-Flores, M. 2006. Gamma radiation effects on commercial Mexican bread making wheat flour. *Nuclear Instruments and Methods in Physics Research B*, 245: 455–458.

Appendini, P. and Hotchkiss, J. H. 2002. Review of antimicrobial packaging. *Innovative Food Science & Emerging Technologies*, 3: 113–126.

Avital, Y., Mannheim, C. H. and Miltz, J. 1990. Effect of carbon dioxide atmosphere on staling and water relations in bread. *Journal of Food Science*, 55: 413–418.

Barcenas, M. E., Haros, M., Benedito, C. and Rosell, C. M. 2003. Effect of freezing and frozen storage on the staling of bre-baked bread. *Food Research International*, 36: 863–869.

Begum, M., Hocking, A. D. and Miskelly, D. 2009. Inactivation of food spoilage fungi by ultra violet (UVC) irradiation. *International Journal of Food Microbiology*, 129: 74–77.

Black, R. G., Quail, K. J., Reyes, V., Kuzyk M. and Ruddick L. 1993. Shelf life extension of pita bread by modified atmosphere packaging. *Food Australia,* 45: 387–391.

Brody, A. L. 1989. *Controlled/Modified Atmosphere/Vacuum Packaging of Foods*. Trumbull, CT: Food and Nutrition Press.

Brody, A. 1999. Controlled/modified atmosphere/vacuum packaging of foods. In *Wiley Encyclopedia of Food Science and Technology*, F. J. Francis (Ed.), New York: John Wiley & Sons.

Cabo, M. L., Pastoriza, L., Bernandez, M. and Herrera, J. J. R. 2001. Effectiveness of CO_2 and Nisaplin on increasing shelf-life of fresh pizza. *Food Microbiology*, 18: 489–498.

Cauvain, S. P. 1998. Improving the control of staling in frozen bakery products. *Trends in Food Science & Technology*, 9: 56–61.

Charles, F., Sanchez, J. and Gontard, N. 2006. Absorption kinetics of oxygen and carbon dioxide scavengers as part of active modified atmosphere packaging. *Journal of Food Engineering*, 72: 1–7.

Coles, R., Mcdowell, D. and Kirwan, M. J. 2003. *Food Packaging Technology*. London, U.K.: Blackwell Publishing, p. 63.

Daifas, D. P., Smith, J. P., Blanchfield, B. and Austin, J. W. 1999. Effect of pH and CO_2 on growth and toxin production by *Clostridium botulinum* in English-style crumpets packaged under modified atmospheres. *Journal of Food Protection*, 62(10): 1157–1161.

Daifas, D. P., Smith, J. P., Blanchfield, B., Cadieux, B., Sanders, G. and Austin, J. W. 2003. Challenge studies with proteolytic *Clostridium botulinum* in yeast and chemically leavened crumpets packaged under modified atmospheres. *Journal of Food Safety*, 23(2): 107–125.

Daifas, D. P., Smith, J. P., Tarte, I., Blanchfield, B. and Austin, J. W. 2000. Effect of ethanol vapor on growth and toxin production by *Clostridium botulinum* in a high moisture bakery product. *Journal of Food Safety*, 20(2): 111–125.

Dainelli, D., Gontard, N., Spyropoulos, D., Zondervan van den Beuken, E. and Tobback, P. 2008. Active and intelligent food packaging: Legal aspects and safety concerns. *Trends in Food Science & Technology*, 19: 103–112.

Davidson, P. M. 1999. Antimicrobial compounds. In *Wiley Encyclopedia of Food Science and Technology*, F. J. Francis (Ed.), New York: John Wiley & Sons.

Day, B. P. F. 1989. Extension of shelf-life of chilled foods. *European Food and Drink Review*, 4: 47–56.

Doerry, American Institute of Baking. 1985. Packaging bakery foods in controlled atmospheres, *Technical Bulletin* 7(4). https://www.aibonline.org/perl/techbulletins.pl?term=doerry (accessed on February 18, 2012).

Ellis, W. O., Smith, J. P., Simpson, B. K., Khanizadeh, S. and Oldham, J. H. 1993. Control of growth and aflatoxin production by *Aspergillus flavus* under modified atmosphere packaging (MAP) conditions. *Food Microbiology*, 10: 9–16.

El-Khoury, A. A. 1999. Shelf life extension studies on pita bread. MSc thesis, McGill University, Montreal, Quebec, Canada.

Erkmen, O. 1997. Antimicrobial effect of pressurized carbon dioxide on *Staphylococcus aureus* in broth and milk. *Lebensmittel Wissenschaft und Technologie*, 30(8): 826–829.

Evans, D. A. 1999. Disinfectants. In *Wiley Encyclopedia of Food Science and Technology*, F. J. Francis (Ed.), New York: John Wiley & Sons.

Gallagher, E., Kunkel, A., Gormley, T. R. and Arendt, E. K. 2003. The effect of dairy and rice powder addition on loaf and crumb characteristics, and on shelf life (intermediate and long-term) of gluten-free breads stored in a modified atmosphere. *European Food Research Technology*, 218: 44–48.

Genigeorgis, C. A. 1985. Microbial and safety implications of the use of modified atmospheres to extend the storage life of fresh meat and fish. *International Journal of Food Microbiology*, 1(5): 237–251.

Goesaert, H., Slade, L., Levine, H. and Delcour, J. A. 2009. Amylases and bread firming—An integrated review. *Journal of Cereal Science*, 50(3): 345–352, doi:10.1016/j.jcs.2009.04.010

Gray, A. and Bemiller, J. N. 2003. Bread staling: Molecular basis and control. Institute of Food Technologists. *Comprehensive Reviews in Food Science and Food Safety*, 2: 1–21.

Gutierrez, L., Sanchez, C., Battle, R. and Nerin, C. 2009. New antimicrobial active package for bakery products. *Trends in Food Science & Technology*, 20: 92–99.

Guynot, M. E., Marın, S., Sanchis, V. and Ramos, A. J. 2003a. Modified atmosphere packaging for prevention of mold spoilage of bakery products with different pH and water activity levels. *Journal of Food Protection*, 66(10): 1864–1872.

Guynot, M. E., Marın, S., Sanchis, V. and Ramos, A. J. 2004. An attempt to minimize potassium sorbate concentration in sponge cakes by modified atmosphere packaging combination to prevent fungal spoilage. *Food Microbiology*, 21: 449–457.

Guynot, M. E., Sanchis, V., Ramos, A. J. and Marín, S. 2003b. Mold-free shelf-life extension of bakery products by active packaging. *Journal of Food Science*, 68(8): 2547–2552.

Hasan, S. 1997. Methods to extend the mold free shelf life of pizza crusts. MSc thesis, McGill University, Montreal, QC, Canada.

Hastings, M. J. 1998. MAP machinery. In *Principles and Applications of Modified Atmosphere Packaging of Foods*, B. A. Blakistone (Ed.), 2nd edn., London, U.K.: Blackie Academic and Professional, pp. 39–62.

Hui, Y. H., Lim, M. H., Nip, W. K., Smith, J. S. and Yu, P. H. F. 2004. Principles of food processing. In *Food Processing: Principles and Applications*, S. J. Smith and Y. H. Hui (Eds.), Malden, MA: Wiley-Blackwell.

Jay, J. M. 2000. *Modern Food Microbiology*, 6th edn., Gaithersburg, MD: Aspen.

Jay, J. M., Loesnner, M. J. and Golden, D. A. 2005. Food protection with modified atmospheres. In *Modern Food Microbiology*. J. M. Jay, M. J. Loessner, and D. A. Golden (Eds.), New York: Springer, pp. 351–370.

Katsarou, E., Kolintza, S. and Lougovois, V. 2002. Modified atmosphere packaging of fish—A review. *Technological Educational Institute of Athens*. (http://www.srcosmos.gr/srcosmos/showpub.aspx?aa=8299).

Koukoutsis, J., Smith, J. P., Phillips D. D., Yayalan, V., Cayouette, D., Ngadi, M. and El khoury, W. 2004. *In vitro* studies to control the growth of microorganisms of spoilage and safety concern in high-moisture, high-pH bakery products. *Journal of Food Safety*, 24(3): 211–230.

Keawmanee and Haruenkit. 2007. *Shelf Life of Brown Rice Muffin under Modified Atmosphere Packaging*. http:/iat.sut.ac.th/food/FIA2007/FIA2007/paper/P2-11-NC.pdf (accessed on October 09, 2011).

Keshri, G., Voysey, P. and Magan, N. 2002. Early detection of spoilage moulds in bread using volatile production patterns and quantitative enzyme assay. *Journal of Applied Microbiology*, 92: 165–172.

Knorr, D. and Tomlins, R. 1985. Effect of carbon dioxide modified atmosphere on the compressibility of stored baked goods. *Journal of Food Science*, 50: 1172–1179.

Kotsianis, I. S., Giannou, V. and Tzia, C. 2002. Production and packaging of bakery products using MAP technology. *Trends in Food Science & Technology*, 13: 319–324.

Lavermicocca, P., Valerio, F. and Visconti, A. 2003. Antifungal activity of phenyllactic acid against molds isolated from bakery products. *Applied and Environmental Microbiology*, 69: 634–640.

Legan, L. D. 1993. Mould spoilage of bread—The problem and some solutions. *International Biodeterioration and Biodegradation*, 32: 33–53.

Magan, N., Arroyo, M. and Aldred, D. 2003. Mould prevention in bread. In *Bread Making: Improving Quality*, S. P. Cauvain (Ed.), Boca Raton, FL: CRC.

Marın, S., Guynot, M. E., Neira, P., Bernadó, M., Sanchis, V. and Ramos, A. J. 2002. Risk assessment of the use of sub-optimal levels of weak-acid preservatives in the control of mould growth on bakery products. *International Journal of Food Microbiology*, 79: 203–211.

Markarian, J. 2006. Consumer demands push growth in additives for active packaging. *Plastics, Additives and Compounding*, 8(5): 30–33.

Matan, N., Rimkeeree, H., Mawson, A. J., Chompreeda, P., Haruthaithanasan, V. and Parker, M. 2006. Antimicrobial activity of cinnamon and clove oils under modified atmosphere conditions. *International Journal of Food Microbiology*, 107: 180–185.

Mcfeaters, R. R. and Cassone, D. R. 1999. Biscuit and cracker technology. In *Wiley Encyclopedia of Food Science and Technology*, F. J. Francis (Ed.), New York: John Wiley & Sons.

McMillin, K. W. 2009. Where is MAP going: A review and future potential of modified atmosphere packaging for meat. *Meat Science*, 80: 43–65.

Mikkola, V., Lahteenmaki, L., Hurme, E., Heinio, R. L., Kaariainen, T. J. and Ahvenainen, R. 1997. Consumer attitudes towards oxygen absorbers in food packages. *Technical Research Centre of Finland ESPOO*. http:/www.vtt.fi/inf/pdf/tiedotteet/1997/T1858.pdf (accessed on October 09, 2011).

Miller, D. D. and Golding, N. S. 1949. The gas requirements for molds. V. The minimum oxygen requirements for normal growth and for germination of six mold cultures. *Journal of Dairy Science,* 32: 191–210.

Molin, G. 2000. Modified atmospheres. In *The Microbiological Safety and Quality of Food*, B. M. Lund, T. C. Baird-Parker, and G. W. Gould (Eds.) Gaithersburg, MD: An Aspen Publication.

Nanditha, B. and Prabhasankar, P. 2009. Antioxidants in bakery products: A review. *Critical Reviews in Food Science and Nutrition*, 49(1): 1–27.

Needham, R., Williams, J., Beales, N., Voysey, P. and Magan, N. 2005. Early detection and differentiation of spoilage of bakery products. *Sensors and Actuators B: Chemical*, 106(1): 20–23.

Nielsen, P. V. and Rios, R. 2000. Inhibition of fungal growth on bread by volatile components from spices and herbs, and the possible application in active packaging, with special emphasis on mustard essential oil. *International Journal of Food Microbiology*, 60(2–3): 219–229.

Odell, D. E. 1983. Fungi isolated from incubated commercial wheat germ bread produced in the UK during 1982–83. *FMBRA Bulletin,* No. 6. December: 281–286.

Ooraikul, B. 1991. Technological consideration in modified atmosphere packaging (Chapter 3) and Modified atmosphere packaging of bakery products (Chapter 4). In: *Modified Atmosphere Packaging of Foods,* B. Ooraikul and M. E. Stiles (Eds.), Chichester, U.K.: Ellis Horwood, pp. 26–48, and 49–117.

Ott, T. M. 1999. Pulsed light processing. In *Wiley Encyclopedia of Food Science and Technology*. F. J. Francis (Ed.), New York: John Wiley & Sons.

Ottaviani, F. and Ottaviani, M. G. 2003. Spoilage molds in spoilage. In *Encyclopedia of Food Sciences and Nutrition*, B. Caballero, L. C. Trugo, and P. C. Finglas (Eds.), Oxford, U.K.: Academic Press, pp. 5522–5530.

Pateras, M. C. I. 2007. Bread spoilage and staling. In *Technology of Breadmaking*, S. P. Cauvain and L. S. Young (Eds.), New York: Springer, p. 282.

Rasmussen, P. H. and Hansen, A. 2001. Staling of wheat bread stored in modified atmosphere. *Lebensmittel Wissenschaft und Technologie*, 34: 487–491.

Robertson, G. L. 2005. *Food Packaging: Principles and Practice*, 2nd edn., (Food Science and Technology). Boca Raton, FL: CRC.

Rodríguez, M., Medina, L. M. and Jordano, R. 2000. Effect of modified atmosphere packaging on the shelf life of sliced wheat flour bread. *Nahrung/Food*, 44(4): 247–252.

Rodríguez, A., Nerín, C. and Batlle, R. 2008. New cinnamon-based active paper packaging against *Rhizopus stolonifer* food spoilage. *Journal of Agricultural and Food Chemistry*, 56(15): 6364–6369.

Salminen, A., Latva-Kala, K., Randell, K., Hurme, E., Linko, P. and Ahvenainen, R. 1996. The effect of ethanol and oxygen absorption on the shelf-life of packed sliced rye bread. *Packaging Technology and Science*, 9(1): 29–42.

Sanguinetti, A. M., Secchi, N., Del Caro, A., Stara, G., Roggio, T. and Piga, A. 2009. Effectiveness of active and modified atmosphere packaging on shelf life extension of a cheese tart. *International Journal of Food Science and Technology*, 44(6): 1192–1198.

Seiler, D. A. L. 1978. The microbial content of wheat and flour. *FMBRA Bulletin,* February, 26–36.

Shaikh, I. M., Ghodke, S. K. and Ananthanarayan, L. 2008. Inhibition of staling in chapati (Indian unleavened flat bread). *Journal of Food Processing and Preservation*, 32(3): 378–403.

Smith, J. P. 1993. Bakery products. In *Principles and Application of Modified Atmosphere Packaging of Food.* R. T. Parry (Ed.) Glasgow, U.K.: Blackie Academic and Professional, pp. 134–169.

Smith, J. P., Daifas, D. P., El-Khoury, W., Koukoutsis, J. and El-Khoury, A. 2004. Shelf life and safety concerns of bakery products—A review. *Critical Reviews in Food Science and Nutrition*, 44: 19–55.

Smith, J. P., Jackson, E. D. and Ooraikul, B. 1983. Storage study of a gas-packaged bakery product. *Journal of Food Science*, 48: 1370–1375.

Smith, J. P., Khanizadeh, S., van de Voortl, F. R., Hardin, R., Ooraikul, B. and Jackson, E. D. 1988. Use of response surface methodology in shelf life extension studies of a bakery product. *Food Microbiology*, 5: 163–176.

Smith, J. P., Ooraikul, B., Koersen, W. J., Jacksont, E. D. and Lawrence, R. A. 1986. Novel approach to oxygen control in modified atmosphere packaging of bakery products. *Food Microbiology*, 3: 315–320.

Smith, J. P., Ooraikul, B., Koersen, W. J., van de Voortl, F. R., Jackson, E. D. and Lawrence, R. A. 1987. Shelf life extension of a bakery product using ethanol vapor. *Food Microbiology*, 4: 329–337.

Smith, J. P. and Simpson, B. K. 1995. Modified atmosphere packaging of bakery and pasta products. In *Principles of Modified-Atmosphere and Sous Vide Product Packaging*, J. M. Farber and K. Dodds (Eds.), Lancaster, PA: Technomic Publications.

Smith, J. P. and Simpson, B. K. 1996. Modified atmosphere packaging. In *Baked Goods Freshness*, R. E. Hebeda and H. F. Zobel (Eds.) New York: Marcel Dekker, Inc., pp. 205–237.

Suhr, K. I. and Nielsen, P. V. 2005. Inhibition of fungal growth on wheat and rye bread by modified atmosphere packaging and active packaging using volatile mustard essential oil. *Journal of Food Science*, 70(1): M37–M44.

Sumnu, G., Sahin, S. and Sevimli, M. 2005. Microwave, infrared and infrared-microwave combination baking of cakes. *Journal of Food Engineering*, 71: 150–155.

Tanis, P. 2012. Lecture on processing fillings in bakery products. In *Biscuits & Snack Symposium,* Sweden, 2012. ZDS, Academy of Sweets, Solingen, Germany.

Tian, Y. Q., Li, Y., Jin, Z. Y., Xu, X. M., Wang, J. P., Jiao, A. Q., Yu, B. and Talba T. 2009. β-Cyclodextrin (β-CD): A new approach in bread staling. *Thermochimica Acta*, 489(1–2): 22–26.

Torres, J. A. and Velazquez, G. 2005. Commercial opportunities and research challenges in the high pressure processing of foods. *Journal of Food Engineering*, 67(1–2): 95–112.

Valerio, F., Bellis, P., Lonigro, S. L., Visconti, A. and Lavermicocca, P. 2008., Use of *Lactobacillus plantarum* fermentation products in bread-making to prevent *Bacillus subtilis* ropy spoilage. *International Journal of Food Microbiology*, 122(3): 328–332.

Vytrasova, J., Pribanova, P. and Marvanova, L. 2002. Occurrence of xerophilic fungi in bakery gingerbread production. *International Journal of Food Microbiology*, 72: 91–96.

Wekell, J. C. and Barnett, H. J. 1999. Seafood: Flavors and quality. In *Wiley Encyclopedia of Food Science and Technology*, F. J. Francis (Ed.) New York: John Wiley & Sons.

Zhou, J.-C., Peng, Y. A.-F. and Xu, N. 2007. Effect of trehalose on fresh bread and bread staling. *Cereal Foods World*, 52(6): 313–316.

Zobel, H. F. and Kulp, K. 1996. The staling mechanism. In *Baked Goods Freshness* (Food Science and Technology), R. S. Hebeda and H. F. Zobel (Eds.), Boca Raton, FL: CRC.

The Federation of Bakers. http://www.bakersfederation.org.uk/europe.aspx (accessed on September 01, 2011).

Smith, J.P., Daifas, D. P., El-Khoury, W., Koukoutsis, J. and El-Khoury, A. 2004. Shelf-life and safety concerns of bakery products—A review. *Critical Reviews in Food Science and Nutrition* 44: 19–55.

[illegible] *Journal of Food Science* [illegible]

[illegible]

[illegible]

Part V

Other Applications of MAP

12 Ready-to-Eat Foods

Ioannis S. Arvanitoyannis and Maria Andreou

CONTENTS

12.1 INTRODUCTION

At present, there is a great need to develop techniques to maintain the natural qualities of cooked ready-to-eat (RTE) foods without using chemical preservatives, for example, vacuum packaging or modified atmosphere packaging (MAP) (Day, 1998). Foods must be prepared at relatively high temperatures (around 100°C) and then packed in vacuum or modified atmosphere (MA). The process is finalized with a rapid cooling step followed by refrigerated storage. The final product should be reheated at approximately 70°C for 2 min before consumption (Sallares, 1995).

Consumer interest in convenience foods with preferably long shelf life has focused attention on MAP (Young et al., 1988). RTE food products have emerged as one of the most rapid-growing categories of the retail market all over the world (http://www.iip-in.com/foodservice/13_readytoeat.pdf).

Application of MAP to RTE meals in fact revealed substantial practical difficulties. For example, a tray may have several partitions to keep the ingredients separate in a complex dish, required both for an appealing, neat appearance and for preventing the product from the mixing of various incompatible food components. These partitions have also the disadvantage they can act as barriers to the flow of gas across and though the product and make application of MAP ineffective and impractical on packaging machines. Moreover, the vacuum cycle also lowers the production line speed significantly (Tucker, 2005).

12.1.1 Convenience Foods

Convenience foods or tertiary processed foods are foods designed to save consumers time in the kitchen and reduce costs due to spoilage. Minimum preparation is required for these foods (only heating). Such foods are packaged for an extended shelf life with negligible loss of flavor and nutrients with time. They were developed to preserve the oversupply of agricultural products available at the time of harvest in order to stabilize the food markets in developed countries (http://www.nationmaster.com/encyclopedia/Convenience-food).

Convenience foods can be mainly grouped into two categories:

- Shelf-stable convenience foods
- Frozen convenience foods or table convenience food. The latter are further classified as follows
- RTE and ready-to-serve (RTS) foods, for example, meats such as precooked sausages, ham, and chicken products
- Ready to cook (RTC) foods, for example, instant mixes such as cake mixes, ice cream mix, jelly mix, and pudding mix and pasta products such as noodles, macaroni, and vermicelli, etc. (http://www.iip-in.com/foodservice/13_readytoeat.pdf)

Shelf-stable convenience foods are commercially prepared foods targeting ease of consumption. Products indicated as convenience foods are often preprepared food stuffs sold as hot, RTE dishes; as room temperature, shelf-stable products; or as refrigerated or frozen products requiring minimal preparation, typically just heating, by the consumer (http://en.wikipedia.org/wiki/Convenience food).

12.1.2 RTE Foods

RTE foods play a very important role in the daily diet of the consumer (Shah and Nath, 2006). RTE meals can be generally defined as a complex assemblage of pre-cooked foodstuffs, packaged together and sold through the refrigerated retail chain in order to provide the consumer with a rapid meal solution (Tucker, 2005).

12.1.3 Cook-Chill Foods

Retail refrigerated RTE meals have gained popularity because of their convenience and freshness. The majority of large-scale food services also apply the cook-chill technology and often produce tonnes of products at a time (Rybka-Rodgers, 2000).

The food is stored at 0°C–3°C before being reheated to at least 70°C for 2 min prior to consumption. According to the Department of Health (1989) guidelines for cook-chill, the maximum recommended shelf life for such products is 5 days. At this instance, it is worth noting that these guidelines only refer to cook-chill products to have longer permitted shelf life if packaged under vacuum or MAP by food manufacturers for the retail market (Blakistone, 1999).

12.1.4 Sous-Vide (under Vacuum)

Sous-vide (SV) foods depend strongly on temperature control (refrigeration) and often do not dispose safety margins in case of a breakdown in the cold chain. Absence of visual indicators of temperature abuse intensifies the hazard. Potential safety risks remain a restraining factor for wider acceptance of the cook-chill technology. Furthermore, manufacturers are often confronted with a quandary between applying sufficient heat treatment and preserving the quality of heat-sensitive ingredients, such as eggs, cheese, and poultry. A potential solution to this problem could be the combination of subtle inhibitory factors (E:\Sous-Vide-Fresh Cooking.mht).

Opting for an approach depends on available scientific information, effect on the product's quality, availability of application techniques, costs, and legislative and marketing considerations. Since many of the described preservation methods are novel and only a few applied commercially in Australia, their legal status has not been established yet (Rybka-Rodgers, 2001).

Double heating, irradiation, addition of lactate, salt, and pH lowering are either affecting the product sensorial/structural/nutritional quality or perceived as "unnatural" by the consumers. MAP and hydrostatic pressure were shown to be ineffective against spore-formers. Relying on the application of natural antimicrobial substances is of interest only in specific foods. Although bacteriocins and protective cultures do not dispose the aforementioned disadvantages, the loss of the inhibitory effect and lack of practical applications make necessary the need for further research in this field (Rybka-Rodgers, 2001).

The applications of MAP to RTE products are given in Table 12.1.

TABLE 12.1
Application of Modified Atmosphere Packaging to Ready-to-Eat Products

RTE Commodity	MAP Atmosphere Composition	MAP Material	Shelf Life	Microbiological Quality	Quality	Reference
Precooked chicken	MAP_1: 76% CO_2: 13.3% N_2: 10.7% O_2 MAP_2: 80% CO_2: 20% N_2: 0% O_2	Nylon/ polyethylene barrier bags rated with a maximum OTR of 9 cc/m^2 in 24 h. Bags containing samples for air storage (Al were heat sealed without evacuation or gas flush. Bags containing samples for MA storage were evacuated to −950 mbar, backflushed with a commercial gas mix to +200 mbar, and heat sealed using a Multivac model gas packaging machine	18 days	*L. monocytogenes* 10^3–10^8 *P. fluorescens* $10^{4.8}$–$10^{8.1}$	Many of the proteolytic enzymes of the genus *Pseudomonas* are heat stable and can survive mild heat treatments. MAP generally had no effect on the interaction between the two bacteria	Marshall et al., 1992
Precooked chicken	MAP_1: 30% CO_2:70% N_2 MAP_2: 60% CO_2:40% N_2 MAP_3: 90% CO_2:10% N_2	Low-density polyethylene/PA/ low-density polyethylene (LDPE/ PA/LDPE) barrier pouches (1 fillet/pouch), 75 μm in thickness, having an oxygen permeability of 52.2 mL/m^2 day atm at 60% RH/251°C and water vapor permeability of 2.4 g/m^2 day at 100% RH/251°C	20 days	Total viable counts (TVC), 10^2–10^6 LAB, 10^4–10^8 *B. thermosphacta* 10^2, *Pseudomonads* 10^2, yeasts and molds, and *Enterobacteriaceae*	Precooked chicken meat was better preserved under M_2 and M_3 mixtures maintaining acceptable odor/taste attributes even on final day of storage	Patsias et al., 2006

(continued)

TABLE 12.1 (continued)
Application of Modified Atmosphere Packaging to Ready-to-Eat Products

RTE Commodity	MAP Atmosphere Composition	MAP Material	Shelf Life	Microbiological Quality	Quality	Reference
Cooked poultry	MAP: 40% CO_2:60% N_2	Internal film liner tray had a moisture vapor transmission rate (MVTR) of 13.85 gm/m^2 in 24 h at 37.8°C and 100% RH and an OTR of 5.5 cc/m^2 in 24 h at 22.8°C and 0% RH	20 days	*Lactococcus raffinolactis* *Carnobacterium divergens* *Carnobacterium piscicola* *Lactococcus garvieae* *Lactococcus lactis* *Enterococcus faecalis* *Carnobacterium, L. raffinolactis, L. lactis,* and *L. garvieae*	Only *C. piscicola* isolates had an inhibitory substance active against other LAB and against several *Listeria* spp. Species-specific PCR primers were used for the differentiation of *Carnobacterium, L. raffinolactis, L. lactis,* and *L. garvieae* strains associated with the MAP poultry products	Bakarat et al., 2000
Cooked minced pork meat	MAP_1: 5% O_2: 95% N_2 MAP_2: 3% O_2: 97% N_2 MAP_3: 1% O_2: 99% N_2 MAP_4: 100% N_2	Packed in sealed PE bags with a high O_2 transmission rate (OTR = 3800 cm^3/m^2/24 h/bar)	10 days	Thiobarbituric acid reactive substances (TBARS) $10^{0.5}$–10^2	The antioxidative effect of individual spices depends on the substrate or the nature of the product. Most investigations have used an oil-based medium or lard for determination of the antioxidative effect of different spices	Huisman et al., 1994
Minced turkey	MAP: 21% O_2: 79% N_2	Meat balls (a total of approximately 120), each	9 days	TBARS	The combination of a high phospholipid fraction and a	Bruun-Jensen et al., 1994

		weighing 34.0 + 0.5 g packed separately in polyethylene bags			low level of natural tocopherols in turkey meat with the high iron content in dark thigh muscles make the meat sensitive to the presence of O_2, especially after grinding and heating. On reheating such products, even after a few hours of storage, a characteristic off-flavor, known as warmed-over flavor (WOF) develops	
Smoked turkey	MAP_1: 30% CO_2: 70% N_2 MAP_2: 50% CO_2: 50% N_2	PET//LDPE/EVOH/LDPE barrier pouches (200 g/pouch) having an oxygen permeability of 2.32 mL/m^2	30 days	Aerobic plate count 10^2–10^9 LAB $10^{1.8}$–10^8 *Pseudomonas* spp. 10^1–$10^{2.5}$ *Enterobacteriaceae* 10^1–10^3	The lowest levels of all BAs determined were obtained for the samples stored in skin packages, possibly due to the most effective removal of air from the product environment	Ntzimani et al., 2008
Cooked frankfurter-type sausages and sliced cooked cured pork shoulder	MAP: 80% CO_2, 20% N_2	Cryovac-type bags with low oxygen permeability (35 cm^3/m^2/day at 22°C, 65% RH)	28 days	*Leuconostoc mesenteroides* and the bacteriocin from *L. mesenteroides* pathogenic nonpathogenic *Staphylococci enterococci*,	D-Lactate is a parameter of the bacterial contamination of vacuum or MAP meat and meat products	Metaxopoulos et al., 2002

(continued)

TABLE 12.1 (continued)
Application of Modified Atmosphere Packaging to Ready-to-Eat Products

RTE Commodity	MAP Atmosphere Composition	MAP Material	Shelf Life	Microbiological Quality	Quality	Reference
				B. thermosphacta $10^{2.5}$–10^{6} *Pseudomonads* $10^{2.5}$, *Enterobacteria* 10^{1} *Micrococci* 10^{3} TVS $10^{2.5}$–$10^{7.5}$, LAB $10^{2.5}$–10^{6}		
Sliced turkey breast fillets	MAP_1: 80% CO_2/20% N_2 MAP_2: 60% CO_2/20% O_2/20% N_2 MAP_3: 0.4% CO/80% CO_2/N_2 MAP_4: 1% CO/80% CO_2/N_2 MAP_5: 0.5% CO/24% O_2/50% CO_2/rest N_2; MAP_6: 100% N_2	Henco Vac 1900 packaging machine and a wv-impermeable film (type V 40–2, 40 mm thick). The OTR of the film was <35 cm^3/m^2/24 h/atm at 20°C and 65% RH	21 days	TVC 10^{4}–10^{8}, LAB $10^{2.8}$–$10^{8.8}$	Use of vacuum is better than MAP for sliced breast turkey fillets and pork–piroski sausages	Pexara et al., 2002
"Piroski" (pork sausage)	MAP_1: 80% CO_2/20% N_2	Henco Vac 1900 packaging machine and a wv-impermeable	21 days	TVC, LAB	The use of vacuum instead of MAs for sliced breast turkey	Pexara et al., 2002

	MAP_2: 60% CO_2/20% O_2/20% N_2 MAP_3: 0.4% CO/80% CO_2/N_2 MAP_4: 1% CO/80% CO_2/N_2 MAP_5: 0.5% CO/24% O_2/50% CO_2/rest N_2; MAP_6: 100% N_2	film (type V 40–2, 40 mm thick). The OTR of the film was <35 cm3/m2/24 h/atm at 20°C and 65% RH			fillets and pork–piroski sausages	
Turkish pastirma	MAP package containing 50% or 80% CO_2	Bags (80% PE and 20% PA). The plastic film in the bags had a WVP of 100 g/m²/24 h at 23°C and OP of 40–50 mL/m²/24 h at 23°C	120 days	TBARS 1.30–1.36	Pastirmas packaged under MAP yielded the lowest TBARS	Veli Gök et al., 2007
Pastirma	MAP: 50% N_2, 50% CO_2	Gas-impermeable bag OPAEVOH/PE (water vapor transmission rate 15 g/m²/24 h at 38°C, 90% RH, 1 atm; O_2 transmission rate 5 cm³/m²/24 h at 23°C, 50% RH, 1 atm; N^2 transmission rate 1 cm³/m²/24 h at 23°C, 50% RH, 1 atm; CO_2 TR 23 cm³/m²/24 h at 23°C, 50% RH, 1 atm)	150 days	LAB and *Micrococcus*/*Staphylococcus*	The inhibition of the tested MAs was higher when the gaseous mixture contained 50% CO_2, while higher concentrations of carbon dioxide had no retarding effect on microbial growth	Aksu and Kaya, 2005
Chilled, cooked, and peeled shrimps	MAP: 50% CO_2, 30% N_2, 20% O_2	Packaging film (NEN 40 HOB/LLPDE) with low gas permeability (0.45 ± 0.15 cm³/m²	4 months	*L. monocytogenes* $10^{1.5}$–10^8 *B. thermosphacta*	This RTE product is mildly preserved and frozen; retail distribution prior to chilled distribution in supermarkets	Mejlholm et al., 2005

(continued)

TABLE 12.1 (continued)
Application of Modified Atmosphere Packaging to Ready-to-Eat Products

RTE Commodity	MAP Atmosphere Composition	MAP Material	Shelf Life	Microbiological Quality	Quality	Reference
		for O_2 and 1.8 ± 0.6 cm^3/m^2 for CO_2)		10^3–10^9 LAB 10^1–10^9	could overcome potential shelf-life problems. Frozen storage has been reported to inhibit or inactivate both spoilage and pathogen microorganisms	
Precooked fish	MAP: 50% CO_2:50% N_2	Laid on polystyrene trays (6–8 pieces per tray), which were packed in multilayer co-extruded film bags. A mixture of gases consisting of 50% CO_2:50% N_2 was injected to half of each group (MAP-samples) at a product:gas ratio of 1:2 (w/v)	30 days	TVC $10^{0.5}$–10^8 *Streptococcus agalactiae* *L. plantarum*	Significant differences in odor and flavor scores between MAP samples with and without lauric acid after 3 weeks of storage, and the latter was rejected. Combination of lauric acid and MAP is appropriate for precooked fish products	Pastoriza et al., 2002
Cold-smoked salmon	Vacuum-packaged 30–40 cm^3/m^2/day/bar for O_2 (23°C, 75% RH), 90 cm^3/m^2/day/bar for CO_2 (23°C, 75% RH), 2.5 g	PA/PE 30 Am/70 with low TRs: 30–40 cm^3/m^2/day/bar for O_2 (23°C, 75% RH), 90 cm^3/m^2/day/bar for CO_2 (23°C, 75% RH), 2.5 g/m^2/day for H_2O (20°C, 85% RH)	55 days	Mesophilic *L. monocytogenes*	Physiochemical characteristics of cold-smoked salmon, the contents in salt and phenolic compounds affect the growth rates of *L. monocytogenes*	Cornu et al., 2006

	m^2/day for H_2O (23°C, 85% RH)					
Precooked red claw crayfish (Cherax quadricarinatus)	MAP: 80% CO_2/10% O_2/10% N_2	VP trays were placed inside vacuum bags (2.2 mil, 3–6 cm^3/m^2/24 h OTR at 23°C); packaged using a vacuum machine	21 days	2-Thiobarbituric acid (TBA)	Combination of MAP gases is effective in inhibiting microbial growth while minimizing oxidative and textural changes in precooked, shell-less red claw crayfish tails stored at refrigerator temperature	Gong and Xiong, 2007
RTC spotted wolffish (Anarhichas minor)	MAP: CO_2 60%:N_2 40%	HDPE semirigid trays	15 days	*Shewanella putrefaciens* psychrotrophic bacteria 10^3–10^8	MAP in combination with superchilling gave a shelf life with high to moderate sensory quality for 15 days. MA was effective in reducing the growth and activity of bacteria	Rosnes et al., 2006
Fish salad	MAP_1: 1% O_2: 7% CO_2: N_2 12% MAP_2: CO_2/N_2: 3%/7%	Gas-barrier film (O_2 permeability, 6.89 mL/m^2; CO_2, 5.42 mL/m^2; N_2, 2.48 mL/m^2 at 4°C; and VP, 7.86 mL/m^2 at 37.8°C ± 1°C, 90% ± 2 RH g/m^2 days atm)	14 days	Coliform bacteria Yeasts and molds *Staphylococcus aureus* Aerobic bacteria TVB-N (mg/100 g fish)	The initial quality of fish must be good for packaging to be effective MAP was effective in extending the shelf life of fish salads at 4°C	Suhendan et al., 2002
Hard-boiled eggs	10% H_2, 10% CO_2, and 80% N_2	Placed in 210 × 210 mm^2 Cryovac bags (O_2 transmission rate = 3–6 cm^3 m^2, 24 h, atmosphere at 4.4°C, 0% RH)	21 days	*C. botulinum* *L. monocytogenes* $10^{2.5}$–10^9	Monitoring and control of this important critical control point (CCP) at all stages of the processing and distribution chain is essential to prevent the	Claire et al., 2004

(continued)

TABLE 12.1 (continued)
Application of Modified Atmosphere Packaging to Ready-to-Eat Products

RTE Commodity	MAP Atmosphere Composition	MAP Material	Shelf Life	Microbiological Quality	Quality	Reference
					growth of proteolytic strains of *C. botulinum* in case they are present in the final produce	
Cottage cheese	MAP: 100% CO_2	Permeability of PS containers to O_2, as measured on the Mocon, was 1.7 mL/(day atm)/container The permeability of the containers to CO_2 was estimated to be 8.5 mL/(day atm)/container	24 days	*Pseudomonas* spp. and *L. monocytogenes Pseudomonads, Alcaligenes, Proteus, Aerobacter or Aeromonads, Listeria innocua, Clostridium sporogenes, and L. monocytogenes*	Sensory evaluation of cheese in high- barrier trays showed no difference in taste between cottage cheese with or without addition of CO_2 during the first 9 days of storage. On the 13th day, both cheeses were slightly sour and control cheese also had a putrid odor	Mannheim and Soffer, 1996
Greek cheese "Anthotyros"	MAP_1: 30% CO_2: 70% N_2 MAP_2: 70% CO_2: 30% N_2	LDPE/PA/LDPE barrier pouches of 75 mm thickness, with an oxygen permeability of 52.2 cm^3/m^2/day/atm at 75% RH, at 25°C, and a WVP of 2.4 g/m^2/day at 100% RH at 25°C	37 days	LAB 10^3–10^8 TVC *Enterobacteriaceae* 10^1–10^8	Whey cheeses in the Mediterranean basin are popular, and are consumed as table cheeses due to their nutritional value, low fat and salt contents, and good sensory characteristics	Papaioannou et al., 2007
Cheddar cheese	MAP: 73% CO_2: 27% N_2	A laminate film (control film), consisting of Bx Nylon/LLDPE/	30 days	*Penicillium thomii Aspergillus puniceus*	The 73% CO_2/27% N_2 atmosphere resulted in the	Oyugi and Buys, 2007

		LDPE/LLDPE It had an OTR < 20 mL/m^2/24 h/atm at 22°C and 75 RH		*Sclerophoma Penicillium expansum*	cheese with the best microbiological qualities	
Cheese	MAP$_1$: 20% CO_2/1% O_2 MAP$_2$: 20% CO_2 /5% O_2 MAP$_3$: 40% CO_2/1% O_2 MAP$_4$: 40% CO_2/5% O_2	In barrier plastic bags PP/EVA/PP (polypro-pylene/ethylene–vinyl alcohol/polypropylene)	30 days	*Mucor plumbeus, Fusarium oxysporum, Byssochlamys fulva, B. nivea, Penicillium commune, P. roqueforti, Aspergillus flavus,* and *Eurotium chevalieri*	Mycotoxin production greatly decreased but not totally inhibited in 20% or 40% CO_2 and 1% or 5% O_2 in comparison with production in air	Taniwaki et al., 2001
Cheese "Myzithra Kalathaki"	MAP$_1$: 20% CO_2/80% N_2, MAP$_2$: 40% CO_2/60% N_2 MAP$_3$: 60% CO_2/40% N_2	200 g of cheese was placed into plastic pouches composed of co-extruded LDPA/PA/LDPE 75 mm in thickness having an OTR of 52.2 mL/(m^2 day atm) and a WVTR of 2.4 g/(m^2 day), measured using the Oxtran 2/20 and Permatran 3/31 permeability testers (MOCON, MN, the United States)	35 days	LAB, *Enterobacteriaceae* mesophilic bacteria psychrotrophic bacteria	Lipolysis, proteolysis, and lipid oxidation were inhibited due to the presence of CO_2-containing atmospheres	Dermiki et al., 2008
Semihard cheeses	MAP$_1$: 100% N_2 MAP$_2$: 20% CO_2 80% N_2	A thermoformed transparent laminate of PA and PE (80 mm PA/120 mm PE, with an oxygen permeability of 13 cm^3/m^2/24 h at 23°C, 50% RH, and 1 atm) was used for the experiment	21 days	1-Pentanol, 1-octanol, 2-ethyl-1-butanol, 2-butanol, 2-heptanol, 2-pentanol, 2-nonanol, benzaldehyde,	Volatile compounds found in concentrations below the odor threshold values are not expected to play a major role in cheese flavor	Jurica et al., 2003

(continued)

TABLE 12.1 (continued)
Application of Modified Atmosphere Packaging to Ready-to-Eat Products

RTE Commodity	MAP Atmosphere Composition	MAP Material	Shelf Life	Microbiological Quality	Quality	Reference
				2-butanone, 2-nonanone, dimethyl trisulfide, and tetrahydrofuran		
Graviera hard cheese	MAP_1: 100% CO_2, MAP_2: 100% N_2, and MAP_3: 50% CO_2/50% N_2	In LDPE/PA/LDPE barrier pouches, 80 g per pouch, 75 mm in thickness having an OTR of 52.2 mL/m^2/day/atm and a WVTR of 1.29 g/m^2/day	63 days	TBARS	The best cheese protection was achieved in MAs containing either 100% N_2 or 50% CO_2/50% N_2, attaining a shelf life of at least 9 weeks when stored in the dark	Trobetas et al., 2008
Salad	MAP: CO_2 20%: O_2 80%	Rinsed in potable water with 100 mg/L active chlorine at 2°C then centrifuged and mixed in the above ratio. After weighing, wrapping in 290 × 155 mm PP film (CO and O_2 TRs of 142 cm^3/m^2/24 h/atm and 44 cm^3/m^2/24 h/atm, respectively) and stored at 4°C	10 days	*Pseudomonas* spp. 10^5–10^8 LAB 10^3–10^8	Browning of the cut lettuce pieces was minimal. Cutting damage and off-odors were the most noticeable defects	Garcia-Gimeno and Zurera-Cosano, 1997

Flour bread	MAP_1: 100% CO_2 MAP_2: 50% CO_2 50% N_2 MAP_3: 20% CO_2/80% N_2	Cryovacuum BB4L bag (601 m thick with permeability at O_2, CO_2, and steam of 35, 150 cm^3/24 h N m^2 for bar and 20 g/24 h for m^2, respectively)	37 days	*Penicillium Aspergillus*	Continual bacterial growth, producing a small amount of CO_2, replaces the amount of CO_2 lost through the packaging film and ensures the continued reduction of O_2 and the restoration of CO_2 levels, thus prolonging mold-free shelf life	Rodrıguez et al., 2002
Pita bread	MAP of 99% CO_2 or 73% CO_2 with 27% N_2	A high-barrier Plastic laminate (Hy-bar 502/1 made by BXL England), was used The barrier layer was EVOH (Ethylene-Vinyl-Alcohol), and its permeability was found to be O_2, 0.45 mL/m^2/24 h, CO_2, 2.3 mL/m^2/24 h, H_2O, 3.5 g/m^2/24 h. The total thickness of the laminate was 100 pm	14 days	*Aspergillus, Rhizopus,* and *Penicillium*	Shelf-life extension of MAP pita at room temperature to 14 days by means of CO_2 and N_2/CO_2 combinations MAP. The shelf-life extension was reached by staling of pita in the presence of CO_2 Two weeks old MAP pita could be refreshened by heating to yield an acceptable product, very similar to a frozen-defrosted pita	Avital and Mannheim, 1988

12.2 PARAMETERS RELATED TO COOKED MEATS PRIOR TO STORAGE UNDER MAP

12.2.1 Meat

MAP is one of several processes applied by many manufacturers to maintain the quality and extend the shelf life of cooked meats. The resulting gaseous mixture within the package is different from that of the atmosphere. The gases most often employed are mixtures of O_2, CO_2, and N_2 and are selected according to product type or bacterial risk. Further to the gaseous atmosphere composition, time, and temperature are other important factors that affect bacterial growth (Williamson et al., 2002).

A disadvantage of MAP cooked meats is keeping the product stationary within the package. The meat product often touches the package top and leaves isolated meat juice. This juice warms more rapidly when separated from the cold meat mass, thereby providing a hospitable environment for microbial growth. Therefore, foods are often double packaged, (first in vacuum pack [VP], which is then placed in a sealed tray), thereby resulting in excess packaging costs for consumers (http://www.fass.org/fass01/pdfs/Rourke.pdf).

12.2.1.1 Cooked Meat

Adequate cooking of meat is necessary to inactivate microbial pathogens. This is particularly important for ground meat products and some specific meats where pathogens can be present internally. Although consumers are advised on appropriate temperature to be reached, it is likely that they assess cooking status by the color of the meat or juice. Many factors can prolong the pink "uncooked" color in meat, including high pH, MAP, rapid thawing, low fat content, nitrite, and irradiation. Such factors may result in overcooking and loss of food quality and consumer rejection. On the other hand, some factors cause "premature browning" of meat, where the interior of the product looks cooked but a microbiologically safe temperature has not been reached, are food safety issues. Pale and soft exudative meats can prematurely brown, as can meats packaged under oxygenated conditions, frozen in bulk, or thawed over long periods, or those that have had salts or lean finely textured beef added. Summarizing, the color of cooked meat is not necessarily a good indicator of sufficient cooking, and the use of a proper thermometer for foods is recommended (King and Whyte 2006).

12.2.1.2 Growth of Aeromonas hydrophila in MAP Cooked Meat Products

Aeromonas hydrophila was shown to multiply very rapidly at refrigerated temperatures. At high water activity ($a_w \geq 0.992$), a generation time of 13.6h was reported at 4°C. The developed models clearly showed, however, that proliferation of *A. hydrophila* could be prevented by the use of CO_2 in the packaging atmosphere in conjunction with a lower water activity ($a_w < 0.985$). The a_w of cured cooked meat products varies in the range of 0.980–0.960. Gas-packed cured cooked meat products will not sustain the growth of *A. hydrophila* when kept at refrigerated temperatures (5°C–7°C) (Devlieghere et al., 2000a,b).

12.2.1.3 *Listeria monocytogenes* and *Pseudomonas fluorescens* on Precooked Chicken

In the presence of *Pseudomonas fluorescens*, the growth of *Listeria monocytogenes* was stimulated at 3°C in air relative to growth alone. A rather negligible effect was reported on the growth of *P. fluorescens* in the presence of *L. monocytogenes* at lower temperatures. However, the growth of the spoilage bacterium was inhibited in mixed culture. Thus, both bacteria were capable of growing in the presence of each other. Slight differences were noted between growth of the bacteria either alone or in mixed culture at 11°C. MAP generally had no effect on the interaction between the two bacteria (Marshall et al., 1992).

The growth of *L. monocytogenes* against time (days) for chicken-based products is shown in Figure 12.1.

12.2.1.4 Chilled Precooked Chicken

Although the combination of precooking treatment and subsequent aerobic storage of chicken results in a good-quality poultry product with acceptable shelf life (14–15 days), as judged both by sensory and microbiological analyses, the use of MAP combined with precooking treatment may further increase the shelf life of the precooked chicken product by at least 6 days (Patsias et al., 2006).

12.2.1.5 Chicken Nuggets

Chicken nuggets were prepared from spent hen meat and roasted pea flour at 5% and 10% levels in an attempt to reduce the cost. Chicken nuggets showed higher emulsion stability with a progressive improvement in the cooking yield up to 10% levels of pea flour inclusion. Shear force value also displayed a significant ($p<0.05$) decrease at 10% pea flour level. Incorporation of 10% pea flour resulted in lower moisture content than the rest of formulations and a decrease in both protein content. RTE

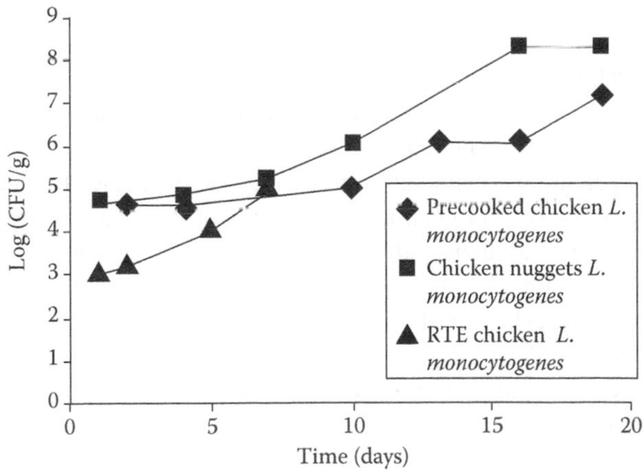

FIGURE 12.1 Growth of *L. monocytogenes* chicken products (precooked, nuggets, and RTE) versus storage time. (From Marshall, D.L. et al., *Food Microbiol.*, 9, 303, 1992; From Bakarat, R.K. and Harris, J. *Appl. Environ. Microbiol.*, 65, 42, 1999; From Elliot, S.C. et al., *J. Food Sci.*, 63, 1080, 1998.)

chicken nuggets prepared by replacing spent hen minced meat with 10% pea flour packed in two-ply laminated aluminum foil (0.124 mm) under nitrogen gas could be stored for 30 days under refrigeration condition (4°C ± 1°C) without any loss of sensory attributes and increase in microbial load (Singh et al., 2008).

12.2.1.6 Cooked Cured Meat (Lactic Acid Bacteria)

Lactic acid bacteria (LAB) are the major group associated with the spoilage of refrigerated vacuum or MA-packaged cooked cured meat products. Storage under MA (CO_2 plus N_2) can prolong considerably the shelf life of the cooked cured meat products compared with vacuum packaging, because the growth rate of the LAB is strongly limited under CO_2 atmosphere (Metaxopoulos et al., 2002).

12.2.1.7 Piroski

The hygienic quality of the tested industrially manufactured smoked turkey fillets and "piroski" pork sausage was shown to be satisfactory. Only the high levels of LAB were of great concern for the product shelf life. To increase the shelf life of vacuum- or MA-packed cooked meats, the first measure was to reduce postheating contamination, mainly with LAB. The quality of the hygiene in the slicing and packaging room is very important since the higher the contamination at this stage is, the shorter the shelf life will be, irrespective of the applied storage conditions (Pexara et al., 2002).

12.2.1.8 Pastirma

MAP is a better method than aerobic/vacuum to retard the growth of food pathogens. Thus, the growth of *Enterobacteriaceae* is restricted in MA (CO_2+N_2). The CO_2 in MAP had an antibacterial effect on Gram-negative bacteria, the *Enterobacteriaceae* counts in the high CO_2 atmospheres were heavily decreased compared with VPs and MAP, and the effect of temperature was substantial in vacuum and 50% CO_2 + 50% N_2 (Aksu et al., 2005).

12.2.1.9 Turkish Pastirma

MAP was found to be the most effective packaging system at preserving pastirma quality, especially cured meat color, over the course of 120 days of storage. The rest of the applied packaging treatments did not manage to delay the deterioration of microbiological, sensory, and chemical properties of pastirmas with storage time. Very high correlation coefficients (all >0.90) existed between sensory properties and TBARS, hexanal content, L*, a*, and b*, proving that sensory panel could determine the quality changes over storage time as precisely as instrumentally determined parameters could. MAP must become the preferred choice of packaging for pastirma producers in case they wish to extend their share in the pastirma market (Gok et al., 2008).

12.2.1.10 Cooked Beef

Ferioli et al. (2007) evaluated the effect of an oxygen-enriched MAP on the development of cholesterol and susceptibility to lipid oxidation in raw and cooked minced beef. MAP led to a gradual rise in cholesterol oxidation in raw beef over a 15 days refrigerated storage. After 8 and 15 days at 4°C, cholesterol oxidation products (COPs) were detected at levels two (rise of 196%) and five times (rise of 483%) higher with respect to muscle stored for 1 day after packaging. Pan frying hardly displayed any effect on cholesterol oxidation in raw fresh beef. However, higher amounts of COPs were found in cooked burgers with respect to the corresponding raw minced muscle. Cooking resulted in consistent COP increase (rise of 71%) in muscle after 15 days storage. Therefore, the connection between the progress of lipid peroxidation and the formation of cholesterol oxides was clearly shown.

12.2.1.11 Beef Kebab

Kebab is a convenient RTE beef product, commercially prepared. Hence, the shelf-life extension and hygienic quality are very important for economic concern of traders and the safety of consumers involved. Irradiation could be effectively used for the shelf-life extension and improving the safety of such produces. It was shown that a dose of 5 and 2.5 kGy extended the shelf life of beef kebab more than 4 and 2 weeks, respectively, compared with the unirradiated samples stored at refrigeration temperature (Hussain et al., 2006).

12.3 DELICATESSEN PRODUCTS

For safe preservation without risk of spoilage and color changes and for innovative and appetizing presentation of deli meats, MAP offers safety guarantees. Nitrogen-only atmospheres consisting exclusively of N_2 (absolutely excluding oxygen from the air) or with slight concentrations of carbon dioxide can effectively change the aspect and preservation of all different types of deli meats (http://www.vakuumverpacken.de/GB /fachberichte/vacmap-informat.pdf).

12.3.1 Meats

Retail slicing of cooked, cured, vacuum-packaged bologna, ham, and pastrami increased the bacterial numbers in the meats without augmenting the public health risk of food poisoning at 48°C storage temperature. Products had a refrigerated shelf life of 21 days at 4°C, but shelf life could not be defined with regard to pH, fermentable carbohydrate remaining, or bacteria numbers present after 21 days of storage. LAB developed more rapidly at 8°C than at 4°C and were abundant at meat surface–package interfaces, thereby making the development of bacterial guidelines for these products impossible. Vacuum-packed products should be considered, from a legislative perspective, as a specific category of semidry fermented sausages. In fact, it is not practical to state the maximum for bacterial numbers since manufacturing of these products is based on LAB food fermentation under specific refrigeration conditions and vacuum (low O_2) (Halley et al., 1996).

12.3.2 Sausage "salchichon"

It has been concluded that manufacturing of dry fermented sausages with high monounsaturated fatty acid (MUFA) or polyunsaturated fatty acid (PUFA) content,

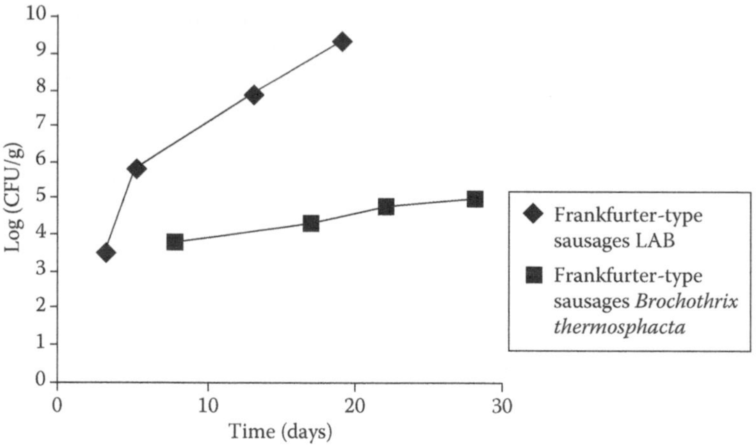

FIGURE 12.2 Growth of LAB and *B. thermosphacta* on frankfurter-type sausages. (From Rubio, B. et al., *Meat Sci.*, 76, 128, 2007; From Gok, V. et al., *Meat Sci.*, 80, 335, 2008; From Aksu, M.I. et al., *J. Muscle Food.*, 16(3), 206, 2005.)

using as raw material fat and meat from pigs fed on diets enriched with either sunflower oil or soya oil, is feasible. Rubio et al. (2002) reported that the modification of fatty acid percentages and the packaging method did not involve substantial changes over the microbiological, physical, and texture parameters. Since storage time was found to be the major factor of modification of sausage characteristics, a storage time longer than 150 days should not be applied. Focusing on sensory characteristics, the different sliced salchichons packed under vacuum and with 20%/80% CO_2/N_2 MA, after long storage, revealed a loss of spicy odor and taste, an increase in acid taste, and only a minor increase in hardness.

The growth of LAB and *Brochothrix thermosphacta* versus time for sausages is shown in Figure 12.2.

12.4 CHEESES

Many cheeses during aging and preservation initiate a physiological production of carbon dioxide and a moderate consumption of oxygen, exactly the same as in breathing. Therefore, an adequate atmosphere (10%–30% CO_2 with respect to N_2) does not affect dairy products nor does it present a risk to the consumer since such gas is a normal product of ageing. Packaging under MA is effective toward protecting numerous dairy products from molding and other disagreeable odor and taste modifications (http://www.vakuumverpacken.de/GB/fachberichte /vacmap-informat.pdf).

12.4.1 Dairy Products

Considerable changes in distribution patterns and increasing demand for increased food quality have resulted in improved shelf life of nonsterile dairy products.

Refrigerated shelf-life prolongation requires reductions in the growth rate of spoilage microorganisms and ensuing product deterioration. Reduction in initial bacterial loads, improved pasteurization regimes, and reduction of postprocessing contamination have been employed with reasonable success. The employment of antimicrobial additives has been discouraged primarily due to labeling requirements and reported toxicity risks. Carbon dioxide is a naturally occurring milk component playing an inhibitory role toward selected dairy spoilage microorganisms. Addition of CO_2 by means of MAP or direct injection as a cost-effective shelf-life extension strategy is commercially employed worldwide for several dairy products and is being considered for others as well. New CO_2 technologies are currently being developed aiming at improvements in the shelf life, quality, and yield of a diversity of dairy products, including raw and pasteurized milk, cheeses, cottage cheese, yogurt, and fermented dairy beverages (Hotchkiss et al., 2006).

12.4.2 Cottage Cheese

Mannheim and Soffer (1996) found that the addition of pure CO_2 to commercial-sized headspaces of cottage cheese packages could effectively serve as an additional hurdle against spoilage bacteria, yeasts, and molds. Addition of CO_2 can only be promising in conjunction with good hygiene practices and low-temperature storage. CO_2 coming from the headspace of packages dissolves in the cottage cheese, thereby contributing to an extension of the lag phase of coliforms, yeasts, molds, and Gram-negative spoilage bacteria. Flushing the headspace with pure CO_2 hardly affected the taste and texture of the product. The presence of carbon dioxide did not inhibit LAB and thus does not prevent an increase in cheese acidity and a lowering of its pH. In order to maintain the desired level of CO_2 and cause its dissolution into the product, a high-barrier package must be used.

12.4.3 Parmigiano Reggiano Cheese

By the end of storage time, the MA conditions did not induce any considerable changes in the flavor intensity, but over the first and second months an increase in sour and sharp tastes was observed, followed by a steep reduction in both attributes by the end of storage period. The early increase in proteolysis index reported in the MA-packed cheeses could be related to the contemporary increase in the "sharpness" and "intensity." The VP Parmigiano Reggiano (PR) did not show any modification in sensory terms, even after 90 days of storage when the highest values of maturation coefficient had been reached. A possible explanation to this phenomenon could be the partial removal of the aromatic components determined by the occurring oil stripping phenomena. A distinct sharp taste was perceived in MA-packed PR cheese after the first and second months of storage. This fact could be linked to the high percentage of N_2 in the gas atmosphere. Other studies, conducted on MA-packed portioned PR cheese (Romani et al., 1999) revealed that CO_2/N_2 MA, particularly rich in N_2, showed an increase in sharp taste perception over storage related to microbial growth and the loss of CO_2 antimicrobial effect (Romani et al., 2002).

12.4.4 Greek Cheese (Anthotyros)

Based primarily on sensory evaluation, the use of MAP extended the shelf life of fresh Anthotyros stored at (4°C) by ca. 10 days (M_1=30%/70% CO_2/N_2 gas mixture) and 20 days (M_2=70%/30% CO_2/N_2 gas mixture) compared with VP. The shelf-life extension of cheese at 12°C was almost 2 days (M_1) and 4 days (M_2), with cheese preserving satisfactory sensory characteristics. It must be made clear that these recommendations on MAP packaging conditions for Anthotyros corresponded to samples from a single dairy plant, and thus, their general application has to be tested further. In view of the limited information available on the use of MAP, more studies are required with regard to preservation of whey cheeses. Establishment of optimal CO_2 concentrations toward prolonging the shelf life and preservation of good sensorial characteristics is one of the problems to be solved (Papaioannou et al., 2007).

12.4.5 Cheese (Myzithra Kalathaki)

Based on sensory analysis, it can be concluded that the MA containing 40% CO_2/60% N_2 (M_1) resulted in a shelf-life extension of the whey cheese "Myzithra Kalathaki" by 14–16 days, while that containing 60% CO_2/40% N_2 (M_2) extended cheese shelf life by 18–20 days. Moreover, lipolysis, proteolysis, and lipid oxidation were inhibited due to the presence of CO_2-containing atmospheres. It should also be pointed out that there are concerns regarding the safety of foods packaged under MA since the absence of oxygen may lead to the growth of anaerobic pathogens (Farber, 1991). Therefore, further work must be carried out to determine the effect of various MAs on the growth of particular pathogens (Dermiki et al., 2008).

12.4.6 Cheddar Cheese

The film with oxygen scavengers was more effective than the control film against mold growth, whereas the 73% CO_2/27% N_2 atmosphere resulted in cheese production with the best microbiological qualities. The three MAs and the packaging film strongly affected the mycoflora of shredded cheddar cheese, as the mold species isolated initially differed from those isolated at 16 weeks in the six treatments. It was shown that O_2 scavengers were very effective in controlling the growth of molds on shredded Cheddar cheese (Oyugi and Buys, 2007).

The growth of *Enterobacteriaceae* versus time for cheese products is displayed in Figure 12.3.

12.4.7 Semihard Cheeses

Jurica et al. (2003) reported that light exposure resulted in significantly decreased yellowness and increased redness of sliced Samso cheese packaged in 0%, 20%, or 100% CO_2 during storage. Cheese stored in 100% CO_2 and exposed to light had significantly lower L* values (lightness) than cheeses stored under the other storage

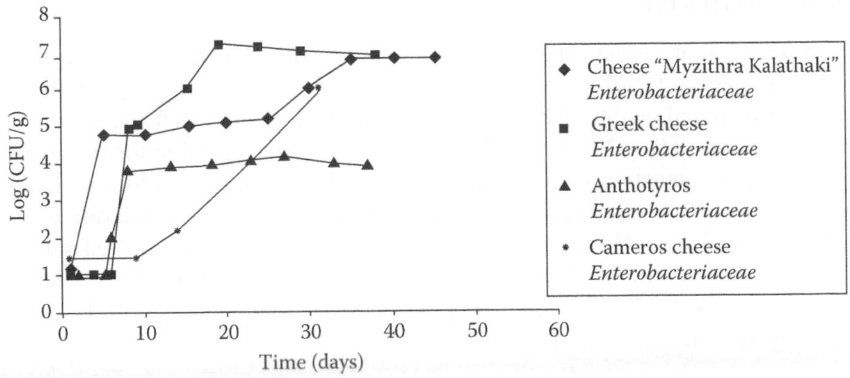

FIGURE 12.3 Growth of *Enterobacteriaceae* on four soft cheeses (three Greek and one Spanish) versus storage time. (From Papaioannou, G. et al., *Int. Dairy J.*, 17, 358, 2007; From Oyugi, E.A. and Buys, E.M., *Int. J. Dairy Tech.*, 60(2), 85, 2007; From Dermiki M. et al., *Lebensmittel Wissenschaft und Technologie,* 41, 284, 2008.)

and packaging combinations. Moreover, storage under 100% CO_2 had a drying-out effect on the cheese. According to Jurica et al. (2003), "the sensory panel characterized cheeses stored in the dark in 0% and 20% CO_2 as most buttermilk-like (odor and taste) and buttery (odor). The diacetyl concentration of these samples was above the odor threshold value and hexanal at a concentration close to the threshold. Cheese stored under 100% CO_2 and exposed to light were tasted by the sensory panel and characterized as rancid (taste and odor) and of a dry/crumbly texture. These cheeses contained 2-butanol and dimethyl trisulfide at concentrations above the odor threshold values and benzaldehyde at a concentration close to the threshold. Furthermore, the formation of primary and secondary alcohols was reported in cheeses packaged in 100% CO_2 and light exposed with storage time."

12.4.8 Mold Cheese from Goat Milk

The material for experiments comprised four batches of the Rokpol-type mold cheese manufactured at 25 day intervals. The raw material for the cheese manufacturing was goat milk of extra class total bacterial count (TBC)—less than $100 \times 10^3/cm^3$ and number of somatic cells—$400 \times 10^3/cm^3$. The experimental Rokpol cheese was packed in aluminum foil as well as in MA and under vacuum. The packaging material used to pack cheeses in MA was a barrier foil (oriented polyethylene polyamide [OPE-PA]). The microbiological investigations conducted revealed that, in the case of packaging of the experimental cheese in aluminum foil, proliferation of the *Escherichia coli* type of bacteria occurred. In the remaining cheeses packed in MA, the number of bacteria from the *E. coli* group remained stable from the moment of their production up to the end of their shelf life, irrespective of the type of the applied gas mixture. The performed organoleptic evaluation displayed that cheeses packed in aluminum foil were characterized by the highest consumer overall acceptability (Pikul Danków et al., 2006).

12.4.9 Dried Milk

Twenty samples of regular and instant nonfat dry milk (NFDM) stored up to 29 years in residential locations at ambient conditions in cans were analyzed for sensory and nutritional quality. Although the overall acceptability hedonic scores varied considerably (2.9–6.2), all samples were considered acceptable for use in case of an emergency situation by the majority of panelists. A 23 year old sample with low oxygen was not significantly different from fresh samples in overall acceptability hedonic scores. No difference in overall acceptability was reported between regular and instant NFDM. Amounts of available lysine (1.4–2.6 mg/100 g), thiamin (2.7–4.0 μg/g), and riboflavin (12.7–20.2 μg/g) were not significantly correlated with sample age (Lloyd et al., 2004).

12.4.10 Dried Milk Products

NFDM and powdered whey beverages are available at the retail level, packaged in cans under reduced oxygen atmosphere to prolong shelf life. In the 10 brands tested, wide variations were reported in headspace oxygen, can seam quality, sensory quality, and vitamin A (with 6 of 10 brands entirely lacking the vitamin). Manufacturers of dried milk products packaged in cans for long-term storage need to pay more attention to can seam quality, product labeling, and vitamin fortification. Consumers must be equally aware to evaluate several brands of dried milk products prior to purchasing the most convenient one (Lloyd et al., 2004).

12.4.11 Malai Peda

Sharma et al. (2003) studied "malai peda, a traditional sweet prepared from milk condensed to 25% of its volume and subsequently sweetened for its shelf life at room and refrigerated temperatures under controlled conditions after packaging in flexible packaging material (Poster paper/A1 foil/LDPE). The samples were tested for several parameters such as moisture content, peroxide value, titratable acidity, and overall acceptability after the specified time period. The overall acceptability of peda decreased considerably at 6 days of storage at room temperature. All the packaged samples were found to be of satisfactory quality even after 31 days of storage at 11°C ± 1°C. Malai peda samples packed under vacuum-nitrogen and stored under refrigeration were found to be of superior quality compared with the conventional ones."

12.5 SEAFOOD

RTE fish has a relatively short shelf life (12 days) under refrigerated conditions, not presenting hygienic quality hazards when properly packaged (Oetterer, 1999). There has been a recent interest in prolonging RTE fish shelf life due to the increase in demand for fresh products, which has led to a greater variety of products being packaged under MA. Such increase in RTE fish shelf life brings about great industrial advantages, because of less losses both in distribution and in display of the product at the retail stores, thereby stabilizing the supply at reasonable prices (Oetterer et al., 2003).

12.5.1 Cold-Smoked Salmon

Physicochemical characteristics of cold-smoked salmon, especially the contents in salt and phenolic compounds, strongly affect the growth rates of *L. monocytogenes*. Secondary models can be effectively used to model these effects and, among the four tested models, the secondary model proposed by Devlieghere et al. (2001) and modified by Giménez and Dalgaard (2004) appeared the most appropriate one. However, it was obvious that the studied factors, including the phenolic content, were not sufficient to describe the whole variability of the behavior of *L. monocytogenes* in cold-smoked salmon. Further sources of uncertainty and variability affecting the growth rate such as the between-strain variability and the between-product variability, which is not explained by the measured physicochemical factors, must be considered (Cornu et al., 2006).

12.5.2 RTE Shrimps (*Pandalus borealis*)

Cooked deep-water shrimp, *Pandalus borealis*, is a popular food product in Europe and is sold RTE as whole, peeled in brine, or brined packaged in MA. Due to the biological and physiological characteristics, shrimps are highly vulnerable to spoilage reactions (Yamagata and Low, 1995; Mendes et al., 2002), and a shelf life between 4 and 7 days when stored in ice has been reported for raw (Goncalves et al., 2003) and cooked shrimp (Sivertsvik et al., 1997). The main parameters to determine shrimp quality are sensory odor and appearance (Goncalves et al., 2003), further to microbiological characteristics. For thawed, cooked, and peeled MAP shrimp (*P. borealis*), the microorganisms *B. thermosphacta* and *Carnobacterium maltaromaticum* were identified to be responsible for the sensory spoilage (Mejlholm et al., 2005).

12.5.3 Shrimps (*Pandalus borealis*)

Sivertsvik and Birkeland (2006) investigated the effects of storage time, MA (30% or 60% CO_2), soluble gas stabilization (SGS), and gas to product volume (*g*/*p*) ratio on the microbiological and sensory characteristics of cooked, peeled, and brined RTE deep-water shrimps (*P. borealis*). SGS treatment prior to packaging (2 h) was shown to increase the CO_2 content in the packaged shrimp and counteracted package collapse, even at low *g*/*p* ratios (0.66). SGS treatment reduced the aerobic plate count and psychotropic count. The recorded increase in CO_2 levels during MA packaging and the application of SGS enhanced ($P<0.01$) the sensory quality of the shrimps. The exudates in the packages (%) dropped significantly ($P<0.01$) when applying SGS treatment. Therefore, SGS treatment in conjunction with MA packaging can be applied successfully on RTE shrimps to reduce the package volume and to improve the microbiological and sensory characteristics.

12.5.4 Chilled, Cooked, and Peeled Shrimps (*Pandalus borealis*)

To prevent *L. monocytogenes* from becoming a safety problem, cooked and peeled MAP shrimps should be distributed at 2°C and with a maximum shelf

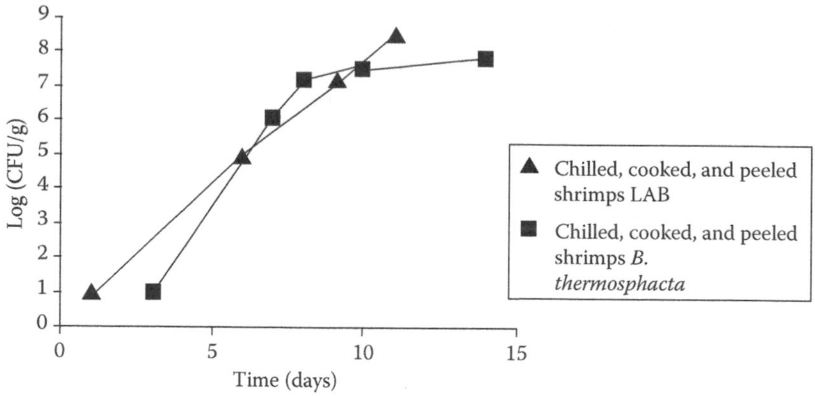

FIGURE 12.4 Growth of *B. thermosphacta* and LAB on chilled, cooked, and peeled shrimps. (From Patsias, I. et al., *Eur. Food. Res. Technol.*, 223, 683, 2006; From Mejlholm, O. et al., *J. Appl. Microbiol.*, 99, 66, 2005.)

life of 20–21 days. At higher temperatures, the shelf life is significantly reduced (Mejlholm et al., 2005).

The growth of *B. thermosphacta* and LAB versus time for shrimps is shown in Figure 12.4.

12.5.5 Desalted Cod Fillets

To meet up-to-date consumer demands, it is often important to offer ready-to-use (RTU) desalted cod in consumer packs with adequate keeping quality at chill temperatures. The aim of this work was to investigate. The effect of potassium sorbate (PS), citric acid, and MAP in varying combinations on the keeping quality of desalted cod fillets in consumer packs was investigated by Magnússon et al. (2006). After desalting, cod fillets were packed in trays and stored at 1.0°C ± 0.2°C for up to 33 days. The fillets were packed in open bags, in MA ($CO_2/N_2/O_2$:75/20/5) only, or in MA following a PS and a citric acid treatment. The rate of microbial growth was by far fastest in fillets in open bags. MAP alone dropped the growth rate substantially, and even further decrease was obtained in MAP fillets treated with citric acid and/or sorbate solutions. The synergistic effect of these treatments was quite distinct. Quantitative descriptive analysis (QDA) was used to assess cooked samples. During the first day of storage, the samples were described by sweet and butter odor, salt taste, clammy and rubber-like texture, which became less evident with increasing storage time. Both taste and texture differed by groups and the least in samples treated with sorbate and citric acid/sorbate solutions. Sensory spoilage attributes and total volatile bases (TVB-N)/trimethylamine (TMA) measurements could be related well with microbial counts. Employment of MAP prolonged the shelf life from 6–10 to 18–24 days, MAP and citric acid to 24–28 days, while the addition of sorbate to MAP fillets extended the shelf life to at least 33 days.

12.6 SALAD

Deli salads may contain *L. monocytogenes* due to contaminated ingredients, post-process contamination, or both. A preservation approach combining temperature and acidulants can be applied to control the survival and growth of *L. monocytogenes* in RTE foods. *L. monocytogenes* has been isolated and identified from many RTE foods, including luncheon meats and coleslaw (Schlech et al., 1983; CDC, 1999). Control of *L. monocytogenes* in foods is difficult because the microorganism is (i) halotolerant, (ii) ubiquitous in raw ingredients and the processing environment, (iii) capable of survival or growth over a broad pH range (pH 4.4–9.6) and a broad temperature range (<1°C–45°C), (iv) and capable of survival or growth in atmospheres of varied oxygen content (Farber and Peterkin, 2000).

The purpose of this investigation was to develop an upstream processing method to facilitate the detection of *L. monocytogenes* in RTE salads by polymerase chain reaction (PCR). Potato salad, a model RTE commodity, was seeded with *L. monocytogenes* and treated with two alternative upstream sample processing methods (designated one-step and two-step centrifugation), followed by DNA extraction amplification, and hybridization. The two-step method resulted in 1000-fold improvements in the PCR detection limit, from 10^6 CFU/g (no sample processing) to 10^3 CFU/g (http://cat.inist.fr/?aModele=afficheN&cpsidt=17722892).

12.6.1 RTE Vegetable Salad

The shelf life of RTE vegetable salads established by manufacturers is usually 7–14 days depending on the type of vegetable and is determined by loss in organoleptic qualities. The evolution of spoilage organisms in a mixed salad of red cabbage, lettuce, and carrot stored at 4°C, 10°C, and 15°C was monitored. Changes in carbon dioxide and oxygen concentrations and pH were also recorded. Predictive modeling was used to establish a theoretical shelf-life time as a function of temperature. LAB at levels of 10^6 CFU/g appeared to be related to both spoilage and theoretically predicted shelf-life values (Garcia-Gimeno and Zurera-Cosano, 1997).

12.6.2 Pesto

Fabiano et al. (2000) optimized the performance of MAP on the shelf life of the Italian vegetable sauce used for spaghetti and known as "pesto." A mathematical model was developed to predict gas composition inside the polymeric package (CO_2 concentration). Predictions of equilibration time and equilibrium gas composition exhibited moderately good agreement with experimental data. A useful extension of pesto shelf life, up to 120 days, was reached by adopting a target atmosphere containing, at steady state, 10% CO_2 and 90% N_2 at 5°C. Microbial growth was inhibited by the CO_2 levels but was not accompanied by significant changes in odor and color.

12.6.3 Kumagai

Guavas cv. Kumagai fruits were packed in plastic bags and stored at 10°C and 85%–90% RH during 7, 14, 21, and 28 days, followed by open exposure for 3 days at

25°C and 70%–80% RH to simulate commercial handling conditions. The packaging materials used were multilayer coextruded polyolefin (PO) film with selective permeability (polysaccharide peptide, PSP), LDPE, LDPE film with mineral incorporation (LDPEm), and heat-shrinkable polyolefin film (SHR). Guavas without plastic packages were used as control samples. The concentrations of O_2 and CO_2 in the package headspace and the sensorial characteristics of the guavas (skin color, pulp texture, off-flavor, and overall quality) were evaluated. The LDPE film had the higher gas barrier and promoted the lowest oxygen level (0.1%) and the highest CO_2 concentration (19%) inside the packages. Such passive MA induced both off-flavor and abnormal ripeness. The PO heat-shrinkable film provided poor modification in the package atmosphere, so the fruits displayed fast senescence (decay). The films including minerals resulted in an atmosphere of 3% O_2 and 4.5% CO_2 inside the packages, which maintained the fruit with good sensorial characteristics for 14 days. The coextruded PO films provided an atmosphere of 0.5% O_2 and 4.5% CO_2 sufficient to maintain the fruit with good sensorial characteristics for 28 days (Jacomino et al., 2001).

12.6.4 Fish Salad

MAP was shown to be effective in prolonging the shelf life of fish salads at 4°C. Gas packaging managed to delay both microbiological activity and sensory changes in the fish salads. Determination of shelf life and quality of the MAP products that both the treatments used for preparing the MAP groups were successful in extending the shelf life of fish salads (Suhendan et al., 2002).

12.7 BREAD AND BAKERY PRODUCTS

Ready leavened products, snack-foods, cookies, and many other bakery products are subject to the phenomenon of molding (with respect to their humidity level), in fatty ingredients (becoming rancid), and changes in their consistency (becoming stale). Modification from 100% nitrogen to 100% carbon dioxide according to the specific was found to be effective toward doubling and tripling the shelf life at ambient temperature of these packaged products (HIntlîan et al., 1986).

The shelf life of many perishable foods such as baked products is limited in the presence of atmospheric oxygen by three important factors: the chemical effect of atmospheric oxygen, the growth of aerobic spoilage microorganisms, and attack by insects and pests. The effect of all these factors was changes in color, flavor, odor, and overall deterioration in food quality. However, the main factor initiating deterioration in dry baked products is oxidation (Berenzon and Saguy, 1998).

12.7.1 Bakery Products

The total elimination of air represents a serious hurdle in MAP of bakery products, because of high spin-rates of the packaging lines and, particularly, because of the texture of bakery products, which retain large quantities of air inside their porous structure. Simulating the gas-flushing MAP with laboratory equipment and measuring the oxygen concentration directly inside bread rolls, the rate of atmosphere substitution was determined by evaluating the effects of different baking treatments (7, 12, and 23 min at 230°C) and the role of different gases (nitrogen, argon, helium, nitrous oxide, and carbon dioxide). Plotting the oxygen content inside the products versus time led to logistic "dose–response" curves for predictive purposes of residual oxygen concentration. The gas properties were found to strongly affect the rate of oxygen substitution. The less water-soluble the gas was, the faster the oxygen reduction was. On the other hand, the larger the gas molecule, the slower the process. Furthermore, baking time had some measurable effects on the rate of oxygen substitution, thereby providing necessary information for optimization of choice of proper gas mixture MAP (Piergiovanni and Fava, 1997).

12.7.2 Mold Spoilage of Bakery Products

A sponge cake analog was used to investigate the influence of pH, water activity (a_w), and carbon dioxide (CO_2) levels on the growth of seven fungal species commonly causing bakery product spoilage (*Eurotium amstelodami, Eurotium herbariorum, Eurotium repens, Eurotium rubrum, Aspergillus niger, Aspergillus flavus,* and *Penicillium corylophilum*). Water activity, CO_2, and their interactions were the main factors considerably affecting the fungal growth. Water activity at levels of 0.80–0.90 exhibited a significant effect on fungal growth, and determination of CO_2 was required to prevent cake spoilage. At an a_w level of 0.85, the lag phases increased up to 200% when the level of CO_2 in the headspace increased from 0% to 70%. Generally, no fungal growth was observed for up to 28 days of incubation at 25°C when samples were packaged with 100% CO_2, regardless of the a_w level. Partial least squares (PLS) projection to latent structure regression was applied to build a polynomial model to predict the sponge cake shelf life on the basis of the lag phases of all seven species tested. The model developed described quite well (R^2 = 79%) the growth of almost all species, which responded similarly to changes in tested factors. The importance of combining several hurdles, such as MAP, a_w, and pH, that have synergistic or additive effects on the inhibition of mold growth was greatly emphasized (Guynot et al., 2003).

12.7.3 Sliced Wheat Flour Bread

An increase of 130% was reported in the shelf life of common bread packed in CO_2, and 160% when packaging was modified with the introduction of a tray on which the product was placed, allowing for greater contact with the atmosphere in which the product was contained. The increase in shelf life of products packed in an atmosphere with a CO_2 concentration of approximately 50% reached 200%–250%. At CO_2 concentrations higher than 75%, a considerable higher increase in shelf life was recorded. In products with a_w < 0.90, the increase in shelf life obtained by packing under carbon dioxide varied from one test to another. In crumpets, fruitcakes, and bread products packed under high CO_2 concentrations, the increase amounted to 250%, whereas in other experiments with fermented products, the increase was only 150%. Such differences in shelf life were attributed more to the type of microbial population present (especially mold) than to the efficiency of CO_2, depending on the a_w of the packed product (Rodriguez et al., 2000).

12.7.4 Pita Bread

The shelf life of pita bread was only a few hours, mainly because of its large surface to volume ratio. Hardening, initiated by staling and drying, was shown to be the main factor influencing the shelf life. Extension of shelf life by packaging pita in a high-barrier laminate under MA of 99% CO_2 or 73% CO_2 with 27% N_2 was studied. A shelf life of 14 days, as determined by microbial spoilage, was obtained for MAP pita bread. Staling was delayed in MAP pita. Organoleptic hedonic comparisons of MAP pita versus frozen defrosted pita from the same batch revealed no differences. The appearance of microorganisms on bread indicated the termination of shelf life (Avital and Mannheim, 1988).

Shelf life of MAP pita at room temperature was prolonged by 14 days by means of CO_2 and N_2/CO_2 combinations MAP. The shelf-life extension was achieved by delaying microbial growth. CO_2 was also found to reduce the dehydration and staling of pita. Two-week-old MAP pita could be refreshened by heating to yield an acceptable product, very similar to a frozen-defrosted pita (Avital and Mannheim, 1988).

12.8 PIZZA

12.8.1 Prebaked Pizza

The tentative effect of various MAPs on the shelf life of prebaked pizza dough, with and without added calcium propionate, was investigated. Three packaging atmospheres were used: 20% CO_2:80% N_2, 50% CO_2:50% N_2, 100% CO_2, and air (control). Samples were examined daily for visible mold growth and analyzed after 2, 8, 17, and 31 days throughout storage (15°C–20°C and 54%–65% relative humidity, RH) for changes in gaseous composition, pH, and microbial populations (mesophilic aerobic and anaerobic bacteria, LAB, and yeasts and molds). Microbiological analysis revealed that molds displayed a greater sensitivity to CO_2 than bacteria and yeasts. Products containing calcium propionate displayed no mold growth throughout storage (31 days) when packaged in air or in CO_2-enriched atmospheres (20%, 50%, and 100%). However, in pizza dough without preservative (calcium propionate) the mold growth was evident after 7 days, except for 100% CO_2 atmosphere (13 days) regardless of the packaging atmosphere. It was thus concluded that the addition of calcium propionate had a strong effect on the shelf-life extension of prebaked pizza dough (Rodríguez et al., 2003).

12.8.2 Pizza

Acidification and CO_2 release resulting from fermentative metabolism of LAB and yeasts were the main factors for rejection of ham pizza stored under 20% CO_2 (commercially applied conditions). Ham and tomato paste were identified as major sources of fermentation. A rotatable factorial design with the aim at studying the effects of Nisaplin and MAP revealed that CO_2 reduced gas release and had a negligible effect on acidification. On the contrary, Nisaplin did not have a positive effect on gas release, despite the lowest acidification at high CO_2 concentrations. The combined use of Nisaplin and MAP lead to significant increases in shelf life with regard to commercially stored samples and were ascribed to complementary effects of nisin and CO_2 against LAB and yeasts, respectively. Subsequent studies aiming at optimizing the addition of Nisaplin displayed that optimum effectiveness in spoilage prevention could be reached only under highly CO_2-enriched atmospheres in combination with 100 mg/kg of Nisaplin added to the top of the pizza and mixed in the tomato paste (Cabo et al., 2001).

12.8.3 Pizza with a Color Indicator

Ahvenainen et al. (1997) studied the effects of leaking on the quality of MAP chilled minced meat pizza. Capillary-like leaks of various sizes were made experimentally in the sealing area by means of tungsten threads of diameters 50 and 100 μm. Test packages were stored for 5 weeks at 5°C (in darkness or under illumination) or at 10°C in darkness, and the microbial and sensory deterioration of pizzas packed at leaking packages was compared to that of intact packages. The main parameter affecting the deterioration of pizzas was the diameter of the capillary leakage. To be more specific, the higher storage temperature accelerated the deterioration of the quality of pizzas in all packages. On the other hand, illumination enhanced the rate of deterioration especially in leaking packages. The color indicator tested reacted to different leak sizes as anticipated and was found to be reliable in terms of leak detection.

12.9 CAKES

The combined effect of MAP and PS on the preservation of a typical Spanish cake (sponge cake) was analyzed. A full factorial experiment design including a_w (0.80, 0.85, and 0.90) and pH (6 and 7.5) was applied. *E. amstelodami, E. herbariorum, E. repens,* and *E. rubrum* colony radii were measured daily for 28 days. All combinations of PS concentration and MAP were more effective at pH 6 than at 7.5. As a_w increased, the amount of PS or CO_2 necessary to prevent or delay fungal growth increased. The antifungal activity of PS could be strongly improved by packaging in a suitable MA. At pH 6 inclusions of 0.1% PS in the presence of 50%–70% CO_2 effectively prevented mold spoilage up to 28 days over the entire a_w range. Another promising combination was the usage of 0.2% of PS in 30% N_2:70% CO_2 MA packages that effectively prevented spoilage of neutral cake analogues at all a_w levels. Control air-packaged bags with 0.2% of PS showed visible growth after 6 days, at pH.6 and 0.90 a_w (Guynot et al., 2003).

The effect of three different MAs (50% CO_2:50% N_2, 100% CO_2, and standard air as control) on the shelf life of sponge cake was investigated. The samples were examined daily for visible mold growth and analyzed periodically (after 2, 6, 13, 22, and 27 days) throughout storage (15°C–20°C and 51%–63% RH) for gaseous composition and microbial populations: of mesophilic aerobic and anaerobic bacteria, LAB, and yeasts and molds. It was found that the atmosphere CO_2:N_2 (50:50) extended the shelf life of the sponge cakes by 2–3 days with respect to packaging in standard air (Rodriguez et al., 2002).

Ramos Abellana et al. (2000) investigated the combined effects of water activity (a_w), oxygen (O_2), and carbon dioxide (CO_2) levels on growth variables of three species of *Eurotium* on a sponge cake analogue. The use of PLS for the analysis of data exhibited that the fungi behaved similarly in terms of their growth responses, to changes in the three tested factors. CO_2, a_w, and the interaction between CO_2 and O_2 were found to be the most significant factors affecting growth variables. Moreover, the model adopted was shown to have a good predictive power. No isolates were able to grow when CO_2 concentrations were higher than 60% under anaerobic conditions. At lower values of CO_2, a_w, and O_2 affected the growth variables considerably. Low levels of O_2 (0.02%–0.5%) did not affect the growth variables studied when levels of CO_2 in the bags were high. However, when the CO_2 concentration dropped, the different O_2 levels strongly affected the growth variables.

12.10 BREAKFAST

Breakfast has been described as the most important meal of the day. Over the years, there has been considerable evidence that individuals who have breakfast, including RTE cereals, display better overall nutrition profiles, show improvements in cognitive functioning, and might be less likely to be overweight (Rampersaud et al., 2005).

12.10.1 Quality of Breakfast Cereals Marketed to Children

Although specific claims were generally justified by the nutritional content of the product, cereals with such claims do not have better overall nutrition profiles. These findings are particularly important in the light of recent evidence that nutrition- and health-related claims can lead consumers to perceive foods as more effective health wise than they are and to ignore other nutrition information (e.g., the fact that low fat does not mean low energy). Only in a few cases, cereals with low-sugar claims provided a disclaimer indicating that the cereal is not a reduced-energy product. However, this disclaimer was hardly readable because of its much smaller font than the low-sugar claim (Schwartz, 2008).

12.11 RTE EGGS

12.11.1 Hard-Boiled Eggs

Benoıt et al. (2004) determined the sensory shelf life and microbiological safety of hard-boiled eggs packaged under various gas atmospheres, challenged with *L. monocytogenes* and proteolytic strains of *Clostridium botulinum.* It was found that the

> growth of *L. monocytogenes* occurred in all inoculated egg samples stored at 4°C, 8°C, and 12°C with counts increasing from ~10^2 to >10^6 CFU/g after 3–20 days, depending on the packaging atmosphere and storage temperature. The obtained results revealed

that packaging eggs, even under elevated CO_2 concentrations (80%), had a rather limited inhibitory effect on the growth of this psychrotrophic pathogen. In challenge studies with proteolytic strains of *C. botulinum*, botulinum neurotoxin was not detected in any sample after 21 days at 12°C. However, neurotoxin was detected in all inoculated eggs that had been initially stored at 12°C, then transferred to 25°C for further 7 days, regardless of the packaging atmosphere. It was shown that MAP cannot be regarded as an adequate barrier to control the growth of *L. monocytogenes* in hard-boiled eggs, even at refrigerated storage temperatures. However, refrigerated temperatures were essential to extend the sensory shelf life and prevent the growth of proteolytic strains of *C. botulinum* in these packaged products

12.11.2 Chicken Liver Egg Pâté

Soffer et al. (1994) investigated the shelf life of pâté (salad in Israel), made by blending fried chicken liver and onions with hard-boiled eggs, from 6 days in air to 14 days by applying MAP, using high CO_2/low O_2 mixtures. Aerobic microbial growth, mainly *Bacillus subtilis*, and lipid oxidation (as TBA values) were largely inhibited, during 14 days, and odor up to 8 days (not tested beyond). It is noteworthy that although the microbial growth inhibition was due to high CO_2 levels, oxidation was prevented because of microbial inhibition and not because of low O_2 content.

The growth of *L. monocytogenes* against time (days) for egg products is shown in Figure 12.5.

12.12 FRESH PASTA

Whatever the production method used for fresh pasta (handmade or industrial, pasteurized or sterilized, packaged or loose) an MA with a medium low content of CO_2

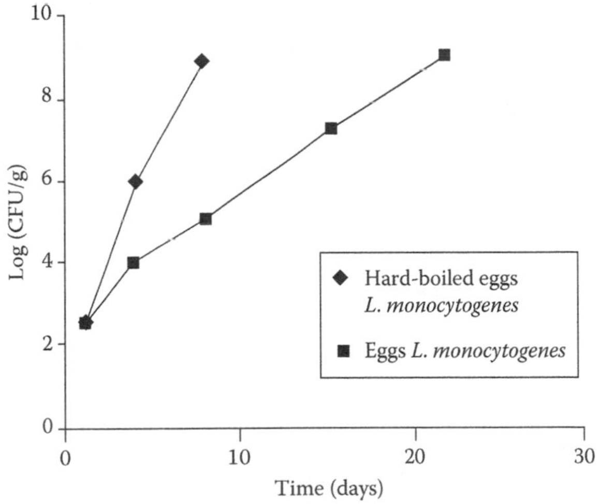

FIGURE 12.5 Growth of *L. monocytogenes* on eggs versus storage time. (From Claire, B. et al., *Food Microbiol.*, 21, 141, 2004; From Soffer, T. et al., *Int. J. Food Sci. Technol.*, 29, 161, 1994.)

(10%–30%) with respect to N_2 allows inhibiting the growth of possible contaminating microorganisms, thereby avoiding the undesirable appearance of mold and increasing the life of the product at refrigerated temperatures. This allows for greater flexibility in both production and distribution of fresh pasta (http://www.vakuumverpacken.de/GB/fachberichte/vacmap-informat.pdf).

12.13 FRUITS AND VEGETABLES

Consumer's demand for minimally processed fruits and vegetables free of additives and with high overall quality and safety has shown a marked rise over the recent years. Fresh cut is a branch of the food industry with the highest increasing pace and potential for growth due to its convenience, healthiness, and attractive appearance and flavor. Consequently, the variety and quantity of fruits and vegetables prepared and conditioned for immediate consumption (RTE) have greatly increased (Artés et al., 2008).

12.14 MICROORGANISMS IN RTE FOODS UNDER MAP

The extended shelf life of MAP, RTE, and refrigerated foods has increased concern about the growth of facultative anaerobic psychrotrophic pathogens (Barakat, 1998).

12.14.1 *Clostridium botulinum* in Convenience Foods

Detection of *C. botulinum* spores in 1 of 400 samples examined demonstrated a low incidence rate. However, the isolation of type B spores from vacuum-packed frankfurters emphasized the need to refrigerate and to prevent abuse of VP meats to eliminate a possible health hazard to the consumer. Reports on the risk of toxin development in VP foods vary. According to Insalata et al. (1969), the risk of toxin formation is mainly dependent on the product itself and not the packing. In a study of smoked ciscoes inoculated with type E spores and held at 30°C, it was found that packages open to the air became toxic as rapidly as those in plastic, airtight packages. It was also reported that anaerobic packaging of foods may increase the food poisoning hazard by allowing the production of botulinum toxin at refrigerator temperatures (Insalata et al., 1969).

12.14.2 *Listeria monocytogenes*

L. monocytogenes in RTE food products presents serious food safety concerns to food processors, regulatory agencies, and consumers (http://scholar.lib.vt.edu/theses/available/)

L. monocytogenes is a Gram-positive bacterium, in the division Firmicutes, named after Joseph Lister. Motile via flagella at 30°C and below but usually not at 37°C, *L. monocytogenes* can instead move within eukaryotic cells by explosive polymerization of actin filaments (known as *comet tails* or *actin rockets*).

Pseudomonas is capable of hydrolyzing proteins in foods resulting in the release of stimulatory factors for growth of *L. monocytogenes*. If MAP inhibits the growth of aerobic *pseudomonads*, *L. monocytogenes* would not grow more rapidly under MAP compared with air conditions in the presence of *Pseudomonas* sp. (Marshall et al., 1991). *L. monocytogenes* grows in RTE meat products under refrigeration without exhibiting any apparent signs of spoilage (Barakat, 1998).

12.14.3 *Pseudomonas fluorescens*

P. fluorescens has many flagella and is endowed with an extremely versatile metabolism. It occurs both in soil and water. Although it is an aerobe, some of its strains are capable of using nitrate instead of oxygen as a final electron acceptor during cellular respiration. Optimal temperatures for the growth of *P. fluorescens* vary from 25°C up to 30°C heat-stable lipases and proteases. The latter initiate milk spoilage, by causing bitterness, casein breakdown, and ropiness because of slime production and proteins coagulation.

12.14.4 Lactic Acid Bacteria

LAB are Gram-positive, acid tolerant, generally nonsporulating, nonrespiring rod or cocci that are associated by their common metabolic and physiological characteristics. These bacteria often occur either in decomposing plants or lactic products and produce lactic acid as the end product of carbohydrate fermentation. The latter linked LAB with food fermentations because acidification inhibits the growth of spoilage agents. Proteinaceous bacteriocins are also produced by several LAB strains and stand for an additional hurdle for spoilage and pathogenic microorganisms. Furthermore, LAB and other metabolic products play a major role in the sensorial and textural profile of food items. The industrial importance of LAB is further strengthened by their generally regarded as safe (GRAS) status, in view of their ubiquitous appearance in food and their contribution to the healthy microflora of human mucosal surfaces. The LAB comprise the genera *Lactobacillus*, *Leuconostoc*, *Pediococcus*, *Lactococcus*, and *Streptococcus*. LAB also include the more peripheral *Aerococcus*, *Carnobacterium*, *Enterococcus*, *Oenococcus*, *Sporolactobacillus*, *Teragenococcus*, *Vagococcus*, *and Weisella*; which belong to the order *Lactobacillales*. (www.encyclopedia.com/doc/1G1-15139266.html - 44k -).

12.14.5 *Brochothrix thermosphacta*

B. thermosphacta is the predominant spoilage organism in chilled meats and processed meat products stored aerobically or under MA. Spoilage is greatest in depleted aerobic conditions, often aided by increased CO_2 levels. Such conditions are common in VP products.

As a facultative anaerobe, *B. thermosphacta* is well suited to growing under MA environments. The successful spoilage of chilled products is due mainly to psychrotrophic nature. Its growth range is from 0°C up to 30°C, with an optimum at 20°C–25°C. Refrigeration temperatures selectively favor its growth. Its pH growth range (pH 5–9) falls well within most meat products. These factors, along with their ability to tolerate low a_w values and remain relatively resistant to curing agents, increase its ability to outgrow many of the other food spoilage microflora (http://www.arrowscientific.com.au/Brochothrix_thermosphacta.htm).

12.14.6 *Shigella*

Shigella, the causative agent of shigellosis or "bacillary dysentery," was first discovered over 100 years ago by Kiyoshi Shiga, a Japanese scientist. *Shigellae*, members of the family *Enterobacteriaceae*, are nearly identical genetically to *E. coli* and are closely related to *Salmonella* and *Citrobacter* spp. (Lampel and Shapira, 2001). *Shigellae* are Gram-negative, facultatively anaerobic, nonspore forming, nonmotile rods that typically do not ferment lactose. Moreover, they are lysine-decarboxylase, acetate, and mucate negative and do not produce gas from glucose (Warren et al., 2006).

12.14.7 *Clostridium botulinum*

C. botulinum is a Gram-positive, rod-shaped bacterium that produces the neurotoxin botulin, which causes the flaccid muscular paralysis reported in botulism. It is also the main paralytic agent in botox. It is an anaerobic spore-former, which produces oval, subterminal endospores and is commonly found in soil. *C. botulinum* is a rod-shaped microorganism. It is an obligate anaerobe, meaning that oxygen is poisonous to the cells. However, it tolerates very small traces of oxygen due to an enzyme called superoxide dismutase (SOD), which is an important antioxidant defense in nearly all cells exposed to oxygen. Under unfavorable circumstances, it is able to form endospores that allow it to survive in a dormant state until exposed to conditions that can support its growth. In laboratory, the microorganism is usually isolated in Tryptose Sulfite Cycloserine (TSC) growth media, always in an anaerobic environment with less than 2% of oxygen. This is feasible by using commercial kits that employ a chemical reaction to replace O_2 with CO_2 (E.J. GasPak System). *C. botulinum* is a lipase-negative microorganism, growing between pH values of 4.8 and 7, and cannot use lactose as a primary carbon source (Doyle, 2007).

12.14.8 *Enterobacteriaceae*

Enterobacteriaceae is a family of Gram-negative bacilli containing more than 100 species of bacteria that normally inhabit the intestines of humans and animals. *Enterobacteriaceae*, commonly part of the normal intestinal tract flora, are referred to as coliforms. Members of the *Enterobacteriaceae* are relatively small, nonspore forming bacilli and frequently resistant to common antibiotics. They ferment a variety of different carbohydrates. *Enterobacteriaceae* are differentiated and classified based on the patterns of their carbohydrate fermentation. Apart from human and animal intestines, *Enterobacteriaceae* are found in soil, water, and decaying matter. Some pathogenic strains also produce exotoxins, while others produce exotoxins known as "enterotoxins" because they specifically affect the intestinal tract, causing diarrhea and body fluid loss (http://www.innvista.com/health/microbes/bacteria/entero.htm).

12.14.9 *LACTOBACILLUS PLANTARUM*

Lactobacillus plantarum is a Gram-positive bacteria producing lactic acid and lives in various environments, including some foods and the human gastrointestinal tract. The possibility of using the microbe to deliver vaccines and other therapeutic compounds to patients is currently under study.

LAB are extensively used in the dairy industry for the manufacturing of cheese and yoghurt. Certain LAB, often members of the *Lactobacillus* genus, are believed to have a positive effect on consumers' health. Those probiotic strains are still viable during passage through the gastrointestinal tract. In the intestine the presence or activity of the strain can provide possible benefits (www.bodyecology.com/07/08/16/lactobacillus_plantarum_benefits.php - 19k -).

12.15 CONCLUSIONS

The shelf life of all RTE products (irrespective of the nature of food; plant or animal origin) and convenience foods was shown to be able to be extended by means of MAP. The most predominant gas compositions were 100% N_2, 100% CO_2, and intermediate compositions in the presence of low oxygen content. It is interesting that, even in the case of cooked fish and seafood, which are generally regarded as more susceptible to spoilage, their shelf life was prolonged up to 21 days, thereby providing a longer time for food transportation to more remote destinations.

ABBREVIATIONS

Abbreviation	Full Names
A. hydrophila	*Aeromonas hydrophila*
a_w	water activity
C. botulinum	*Clostridium botulinum*
CO_2	carbon dioxide
COPs	cholesterol oxidation products
E. herbariorum	*Eurotium herbariorum*
E. repens	*Elytrigia repens*
E. rubrum	*Etheostoma rubrum*
E. coli	*Escherichia coli*
GRAS	generally regarded as safe
L. monocytogenes	*Listeria monocytogenes*
LAB	lactic acid bacteria
LDPE	low-density polyethylene
LDPEm	LDPE film with mineral incorporation
MA	modified atmosphere
MAP	modified atmosphere packaging
MUFA	monounsaturated fatty acid

NFDM	nonfat dry milk
P. fluorescens	*Pseudomonas fluorescens*
Pandalus borealis	deep-water shrimps
PCR	polymerase chain reaction
PLS	partial least squares
PO	polyolefin
PS	potassium sorbate
PSP	polysaccharide peptide
PUFA	poly-unsaturated fatty acid
QDA	quantitative descriptive analysis
RH	relative humidity
PR	parmigiano Reggiano
RTC	ready to cook
RTE	ready to eat
RTS	ready to serve
RTU	ready to use
SGS	soluble gas stabilization
SHR	heat-shrinkable
SOD	superoxide dismutase
TBC	total bacterial count
TMA	trimethylamine
TSC	tryptose sulfite cycloserine
TVB	total volatile bases
VP	vacuum packed

REFERENCES

Aksu, M.I., Kaya, M. and Ockerman, H.W. (2005). Effect of modified atmosphere packaging and temperature on the shelf life of sliced pastirma produced from frozen/thawed meat. *Journal of Muscle Foods*, 16(3): 192–206.

Aksu, M.I. and Kaya, M. (2005). Effect of storage temperatures and time on shelf-life of sliced and modified atmosphere packaged Pastirma, a dried meat product, produced from beef. *Journal of the Science and Food Agriculture,* 85: 1305–1312.

Ahvenainen, R., Eilamo, M. and Hurme, E. (1997). Detection of improper sealing and quality deterioration of modified-atmosphere-packed pizza by a colour indicator. *Food Control*, 8(4): 177–184.

Artés, F., Gómez, P.A. and Artés-Hernández, F. (2008). Physical, physiological and microbial deterioration of minimally fresh processed fruits and vegetables. *Food Science and Technology International*, 13(3): 177–188.

Avital, Y. and Mannheim, C.H. (1988). Modified atmosphere packaging of pita (pocket) bread. *Packaging Technology and Science*, 1: 7–23.

Barakat, R.K. (1998). Shelf life, microbial ecology, and safety of a cooked modified atmosphere packaged refrigerated poultry product. PhD thesis, University of Guelph, Guelph, Ontario, Canada.

Barakat, R.K. and Harris, J. (1999). Growth of *Listeria monocytogenes* and *Yersinia enterocolitica* on cooked modified atmosphere-packaged poultry in the absence of a naturally occurring microbiota. *Applied and Environment Microbiology*, 65: 342–345.

Barakat, R.K., Griffiths, M.W. and Harris, L.J. (2000). Isolation and characterization of *Carnobacterium, Lactococcus* and *Enterococcus spp*. from cooked, modified atmosphere packaged, refrigerated, poultry meat. *International Journal of Food Microbiology*, 62: 83–94.

Blakistone, B.A. (1999). Meats and Poultry. In: *Principles and Applications of Modified Packaging of Foods*, 2nd edn., Gaithersburg, MD: Aspen Publication, pp. 240–290.

Benoıt, C., Smith, J.P., El- Khoury, W., Cayouettea, B., Ngadi, M., Blanchfield, B. and Austin, J.W. (2004). Challenge studies with *Listeria monocytogenes* and proteolytic *Clostridium botulinum* in hard-boiled eggs packaged under modified atmospheres. *Food Microbiology*, 21(2): 131–141.

Berenzon, S. and Saguy, I.S. (1998). Oxygen absorbers for extension of crackers shelf-life. *Lebensmittel-Wissenschaft und-Technology*, 31: 1–5.

Bruun-Jensen, L., Skovgaard, M., Skibsted, L.H. and Bertelsen, G. (1994). Antioxidant synergism between tocopherols and ascorbyl palmitate in cooked, minced turkey. *Zournal Lebensmittel Untersuchung and Forschung*, 199: 210–213.

Cabo, M.L., Pastoriza, L., Bernárdez, M. and Herrera, J.J.R. (2001). Effectiveness of CO_2 and Nisaplin on increasing shelf-life of fresh pizza. *Food Microbiology*, 18(5): 489–498.

Centers for Disease Control and Prevention (CDC). (1999). Outbreak of *Salmonella serotype* Muenchen infections associated with unpasteurized orange juice-United States and Canada, June 1999. *Morbility and Mortality Weekly Report*, 48: 582–585.

Claire, B., Smith, J.P., El-Khoury, W., Cayoutte, B., Ngadi, M., Blanchfield, B. and Austin, J.W. (2004). Challenge studies with *Listeria monocytogenes* and proteolytic *Clostridium botulinum* in hardboiled eggs packaged under modified atmospheres. *Food Microbiology*, 21: 131–141.

Cornu, M., Beaufort, A., Rudelle, S., Laloux, L., Bergis, H., Miconnet, N., Serot, T. and Delignette-Muller, M.L. (2006). Effect of temperature, water-phase salt and phenolic contents on *Listeria monocytogenes* growth rates on cold-smoked salmon and evaluation of secondary models. *International Journal of Food Microbiology*, 106: 159–168.

Day, B.P.F. (1998). *Campden and Chorleywood Food Research Association Chilled Food Packaging*. In ftp://166.111.30.161/incoming/new_book/Food%20Science/Chilled%20 Foods%20(2nd%20Edition)/34990wp_06.pdf

Dermiki, M., Ntzimani, A., Badeka, A., Savvaidis, I.N. and Kontominas, M.G. (2008). Shelf-life extension and quality attributes of the whey cheese "Myzithra Kalathaki" using modified atmosphere packaging. *Lebensmittel Wissenschaft und Technologie*, 41: 284–294.

Devlieghere, F., Geeraerd, A.H., Versyck, K.J., Bernaert, H., Van Impe, J.F. and Debevere, J. (2000a). Shelf life of modified atmosphere packed cooked meat products: Addition of Na-lactate as a fourth shelf life determinative factor in a model and product validation. *International Journal of Food Microbiology*, 58: 93–106.

Devlieghere, F., Gil, M.I. and Debevere, J. (2001). *Modified Atmosphere Packaging* (MAP). In ftp://166.111.30.161/incoming/new_book/Food%20Science/Nutrition%20 Handbook%20for%20Food%20Processors/36659_16.pdf

Devlieghere, F., Lefevere, I., Magninand, A. and Debevere, J. (2000b). Growth of *Aeromonas hydrophila* in modified-atmosphere-packed cooked meat products. *Food Microbiology*, 17: 185–196.

Doyle, M.P. (2007). *Food Microbiology: Fundamentals and Frontiers*. ASM Press. In http:// en.wikipedia.org/wiki/Clostridium_botulinum

Eliot, S.C., Vuillemard, J.C. and Emond, J.P. (1998). Stability of shredded mozzarella cheese under modified atmospheres. *Journal of Food Science*, 63: 1075–1080.

Fabiano, B., Perego, P., Pastorino, R. and Del Borghi, M. (2000). The extension of the shelf-life of "pesto" sauce by a combination of modified atmosphere packaging and refrigeration. *International Journal of Food Science and Technology*, 35(3): 293–303.

Farber, J.M. (1991). Microbiological aspects of modified-atmosphere packaging technology—A review. *Journal of Food Protection*, 54: 58–70.

Farber, J.M. and Peterkin, P.I. (2000). *Listeria monocytogenes*. In: *The Microbial Safety of Food*, vol. 2. Lund B.M., Baird-Parker T.C. and Gould G.W. (eds.), Gaithersburg, Maryland: Aspen, pp. 1178–1232.

Ferioli, F., Caboni, F.M. and Dutta, P.C. (2007). Evaluation of cholesterol and lipid oxidation in raw and cooked minced beef stored under oxygen-enriched atmosphere. *Meat Science*, 80(3): 681–685.

Garcia-Gimeno, R.M. and Zurera-Cosano, G. (1997). Determination of ready-to-eat vegetable salad shelf-life. *International Journal of Food Microbiology*, 36: 31–38.

Gok, V., Obuz, E. and Akkaya, L. (2008). Effects of packaging method and storage time on the chemical, microbiological, and sensory properties of Turkish pastirma –A dry cured beef product. *Meat Science*, 80: 335–344.

Gonçalves, A., Lopez-Caballero, M. and Nunes, M.L. (2003). Quality changes of deepwater rose shrimp (*Parapenaeus longirostris*) packed in modified atmosphere. *Journal of Food Science*, 68: 2586–2590.

Gong, C. and Xiong, Y.L. (2007). Shelf-stability enhancement of precooked red claw crayfish (*Cherax quadricarinatus*) tails by modified $CO_2/O_2/N_2$ gas packaging. *LWT—Food Science and Technology*, 41(8): 1431–1436.

Guynot, M.E., Marín, S., Sanchis, V. and Ramos, A.J. (2003). Modified atmosphere packaging for prevention of mold spoilage of bakery products with different pH and water activity levels. *Journal of Food Protection*, 66(10): 1864–1872.

Halley, R.A., Doyon, G., Fortin, J., Rodrigueb, N. and Carbonnea, M. (1996). Post-process, packaging-induced fermentation of delicatessen meats. *Food Research International*, 29(1): 35–48.

Hotchkiss, J.H., Werner, B.G. and Lee, E.Y.C. (2006). Addition of carbon dioxide to dairy products to improve quality: A comprehensive review. *Comprehensive Reviews in Food Science and Food Safety*, 5(4): 158–168.

Huisman, M., Madsen, H.L., Skibsted, L.H. and Bertelsen, G. (1994). The combined effect of rosemary (*Rosmarinus officinalis L.*) and modified atmosphere packaging as protection against warmed over flavour in cooked minced pork meat. *Zournal Lebensmittel Untersuchung and Forschung*, 198: 57–59.

Insalata, N.F., Witzeman, S.J., Fredericks, G.J. and Sunga, F.C.A. (1969). Incidence study of spores of *Clostridium botulinum* in convenience foods. *Applied Microbiology*, 62(6): 479–490.

Jacomino, A.P., Grígoli De Luca Sarantóoulos, C.I., Monteiro Sigrist, J.M., Do Alfredo Kluge, R. and Minami, K. (2001). Sensorial characteristics of "Kumagai" guavas submitted to passive modified atmosphere in plastic packages. *Journal of Plastic Film and Sheeting*, 17(1): 6–21.

Jurica, M., Bertelsena, G., Mortensenb, G. and Agerlin Petersen, M. (2003). Light-induced colour and aroma changes in sliced, modified atmosphere packaged semi-hard cheeses. *International Dairy Journal*, 13: 239–249.

King, N.J. and Whyte, R. (2006). Does it look cooked? A review of factors that influence cooked meat color. *Journal of Food Science*, 71(4): 31–40.

Lampel, J. and Shapira, Z. (2001). Judgmental errors, interactive norms, and the difficulty of detecting strategic surprises. *Organization Science*, 12(5): 599–611.

Lloyd, M.A., Zou, J., Farnsworth, H., Ogden, L.V. and Pike, O.A. (2004a). Quality at time of purchase of dried milk products commercially packaged in reduced oxygen atmosphere. *Journal of Dairy Science*, 87(8): 2337–2343.

Lloyd, M.A., Zou, J., Ogden, L.V. and Pike, O.A. (2004b). Sensory and nutritional quality of nonfat dry milk in long-term residential storage. *Journal of Food Science*, 69(8): 326–331.

Magnússon, H., Sveinsdóttir, K., Lauzon, H.L., Thorkelsdóttir, Á., Martinsdóttir, E. and Keeping (2006). Quality of desalted cod fillets in consumer packs. *Journal of Food Science*, 71(2): 69–76.

Mannheim, C.H. and Soffer, T. (1996). Shelf-life extension of cottage cheese by modified atmosphere packaging. *Lebensmittel-Wissenschaft und-Technology*, 29: 767–771.

Marshall, D.L., Wiese-Lehigh, P.L., Wells, J.H. and Farr, A.J. (1991). Comparative growth of *Listeria monocytoyenes* and *Pseudomonas fluorescens* on precooked chicken nuggets stored under modified atmospheres. *Journal of Food Protection*, 54: 841–843.

Marshall, D.L., Andrews, L.S., Wells, J.H. and Farr, A.J. (1992). Influence of modified atmosphere packaging on the competitive growth of *Listeria monocytogenes* and *Pseudomonas fluorescens* on precooked chicken. *Food Microbiology*, 9: 303–309.

Mejlholm, O., Boknæs, N. and Dalgaard, P. (2005). Shelf life and safety aspects of chilled cooked and peeled shrimps (*Pandalus borealis*) in modified atmosphere packaging. *Journal of Applied Microbiology*, 99: 66–76.

Mendes, R., Huidobro, A. and Caballero, E.L. (2002). Indole levels in deepwater pink shrimp (*Parapenaeus longirostris*) from the Portuguese coast. Effects of temperature abuse. *European Food Research and Technology*, 214: 125–130.

Metaxopoulos, J., Mataragas, M. and Drosinos, E.H. (2002). Microbial interaction in cooked cured meat products under vacuum or modified atmosphere at 4°C. *Journal of Applied Microbiology*, 93: 363–373.

Ntzimani, A.G., Paleologos, E.K., Savvaidis, I.N. and Kontominas, M.G. (2008). Formation of biogenic amines and relation to microbial flora and sensory changes in smoked turkey breast fillets stored under various packaging conditions at 4°C. *Food Microbiology*, 25: 509–517.

Oetterer, M. (1999). Agroindústrias beneficiadoras de pescado cultivado: unidades modulares e polivalentes para implantação, com enfoque nos pontos críticos e higiênicos e nutricionais. Tese (livre-docência)—Escola Superior de Agricultura Luiz de Queiroz. Universidade de São Paulo, Piracicaba, Brazil, p.198

Oetterer, M., Perujo, S.D., Gallo, C.R., Arruda, L.F., Borghesi R. and Cruz, A.M.P. (2003). Monitoring the Sardine (Sardinella brasiliensis) fermentation process to obtain Anchovies, *Brasil. Scientia Agricola*, 60: 511–517.

Oyugi, E.A. and Buys, E.M. (2007). Microbiological quality of shredded Cheddar cheese packaged in modified atmospheres. *International Journal of Dairy Technology*, 60(2): 89–85

Papaioannou, G., Chouliara, I., Karatapanis, A.E., Kontominas, M.G. and Savvaidis, I.N. (2007). Shelf-life of a Greek whey cheese under modified atmosphere packaging. *International Dairy Journal*, 17: 358–364.

Pastoriza, L.M.L., Cabo Bernárdez Sampedro, M. and Juan, J.R. (2002). Herrera combined effects of modified atmosphere packaging and lauric acid in the stability of pre-cooked fish products during refrigerated storage. *European Food Research Technology*, 215: 189–193.

Patsias, A., Chouliara, I., Paleologos, E.K., Savvaidis, I. and Kontominas, M.G. (2006). Relation of biogenic amines to microbial and sensory changes of precooked chicken meat stored aerobically and under modified atmosphere packaging at 4°C. *European Food Research and Technology*, 223: 683–689.

Pexara, S., Metaxopoulos, J. and Drosinos, E.H. (2002). Evaluation of shelf life of cured, cooked, sliced turkey fillets and cooked pork sausages "piroski"-stored under vacuum and modified atmospheres at +4 and +10°C. *Meat Science*, 62: 33–43.

Piergiovanni, L. and Fava, P. (1997). Minimizing the residual oxygen in modified atmosphere packaging of bakery products. *Food Additives and Contaminants*, 14(6–7): 765–773.

Pikul Danków, R.J., Wójtowski, J. and Cais-Sokolińska, D. (2006). Effect of packaging systems on the quality and shelf-life of the Rokpol type mould cheese from goat milk. *Archives of Animal Breeding*, 49(SPEC): 214–218.

Ramos Abellana, M., Sanchis, A.J. and Nielsen, P.V. (2000). Effect of modified atmosphere packaging and water activity on growth of *Eurotium amstelodami*, *E. chevalieri* and *E. herbariorum* on a sponge cake analogue. *Journal of Applied Microbiology*, 88(4): 606–616.

Rampersaud, G., Pereira, M., Girard, B., Adams, J. and Metzl, J. (2005). Breakfast habits, nutritional status, body weight, and academic performance children and adolescents. *Journal of American Dietecians Association*, 105: 743–760.

Rodrıguez, M., Medina, L.M. and Jordano, R. (2000). Effect of modified atmosphere packaging on the shelf life of sliced wheat flour bread. *Nahrung*, 44(4): 247–252.

Rodriguez, M.V., Medina, L.M. and Jordano, R. (2002). Prolongation of shelf life of sponge cakes using modified atmosphere packaging. *Acta Alimentaria*, 31(2): 191–196.

Rodríguez, V., Medina, L.M. and Jordano, R. (2003). Influence of modified atmosphere packaging on the shelf life of prebaked pizza dough with and without preservative added. *Nahrung*, 47(2): 122–125.

Romani, S., Sacchetti, G., Pittia, P., Pinnavaia1, G.G. and Dalla Rosa, M. (2002). Physical, chemical, textural and sensorial changes of portioned *Parmigiano reggiano* cheese packed under different conditions. *Food Science and Technology International*, 8: 203–212.

Romani S., Sacchetti G., Vannini L., Pinnavaia G. G., Dalla Rosa M. and Corradini C. (1999). Stabilita`in conservazione del Parmigiano Reggiano confezionato in porzioni. *Scienza e Tecnica Lattiero-Casearia,* 50(4): 273–290.

Rosnes, J.T., Kleiberg, G.H., Sivertsvik, M., Lunestad, B.T. and Lorentzen G. (2006). Effect of modified atmosphere packaging and superchilled storage on the shelf-life of farmed ready-to-cook spotted wolf-fish (*Anarhichas minor*). *Packaging Technology and Science,* 19: 325–333.

Rubio, B., Martínez, B., Sánchez, M.J., Dolores, M., García-Cachán, J.R. and Isabel, J. (2007). Study of the shelf life of a dry fermented sausage "salchichon" made from raw material enriched in monounsaturated and polyunsaturated fatty acids and stored under modiWed atmospheres. *Meat Science*, 76: 128–137.

Rybka-Rodgers, S. (2001). Improvement of food safety design of cook-chill foods. *Food Research International*, 34: 449–455.

Rybka-Rogers, S. (2000). Improvement of food safety, designing of cook–chill foods., Centre for Advanced Food Research, University of Western Sydney, Penrith, New South Wales, Australia, pp. 6–9.

Sallares, E. (1995). Principios y aplicaciones de la tecnica del vacıo en hostelerıa. Ed. Enrique Sallares, Espana.

Schlech, W.F., Lavigne, P.M., Bortolussi, R.A., Allen, A.C., Haldane, E.V., Wort, A.J., Hightower, A.W. et al. (1983). Epidemic listeriosis evidence for transmission by food. *The New England Journal of Medicine*, 308: 203–206.

Schwartz, B. (2008). Examining the nutritional quality of breakfast cereals marketed to children. *Journal of American Dieticians Association*, 108: 702–705.

Shah, N.S. and Nath, N. (2006). Effect of calcium lactate, 4-hexyl resorcinol and vacuum packing on physico-chemical, sensory and microbiological qualities of minimally processed litchi (*Litchi chinensis Sonn.*) *International Journal of Food Science and Technology*, 41: 1073–1081.

Sharma, H.K., Singhal, R.S. and Kulkarni, P.R. (2003). Effect of modified atmosphere packaging on the keeping quality of Malai peda. *Journal of Food Science and Technology*, 40(5): 543–545.

Singh, O.P., Singh, J.N., Bharti, M.K. and Soni, K. (2008). Refrigerated storage stability of chicken nuggets containing pea flour. *Journal of Food Science and Technology*, 45(5): 460–462.

Sivertsvik, M. and Birkeland, S. (2006). Effects of soluble gas stabilisation, modified atmosphere, gas to product volume ratio and storage on the microbiological and sensory characteristics of ready-to-eat shrimp (*Pandalus borealis*). *International Food Science and Technology*, 12(5): 445–454.

Sivertsvik, M., Rosnes, J.T. and Bergslien, H. (1997). Shelf-life of whole cooked shrimp (*Pandalus borealis*) in carbon dioxide-enriched atmosphere. In: *Seafood from Producer to Consumer, Integrated Approach to Quality*. Luten J.B., Børresen T. and Oehlenschläger J. (eds), Amsterdam, the Netherlands: Elsevier Science, B.V., pp. 221–230.

Soffer, T., Pinhas, M. and Mannheim Chaim, H. (1994). Shelf-life of chicken liver/egg pate in modified atmosphere packages. *International Journal of Food Science and Technology*, 29: 161–166.

Suhendan, M., Erkan, N., Baygar, T. and Ozden, O. (2002). Modified atmosphere packaging of fish salad. *Fisheries Science*, 68: 204–209.

Taniwaki, M.H., Hocking, A.D., Pitt, J.I. and Fleet, G.H. (2001). Growth of fungi and mycotoxin production on cheese under modified atmospheres. *International Journal of Food Microbiology*, 68: 125–133.

Trobetas, A., Badeka, A. and Kontominas, M.G. (2008). Light-induced changes in grated Graviera hard cheese packaged under modified atmospheres. *International Dairy Journal*, 18: 1133–1139.

Tucker, G (2005). Thermal processing of ready meals. In: *Thermal Food Processing: New Technologies and Quality Issues*, Da-Wen Sun (ed.), Boca Raton, FL: CRC Press, pp. 363–385.

Warren, B.R., Parish, M.E. and Schneider, K.R. (2006). *Shigella* as a foodborne pathogen and current methods for detection in food. *Critical Reviews in Food Science and Nutrition*, 46(7): 551–567.

Williamson, K., Allen, G. and Bolton, F.J., PHLS North West FESL—Preston PHL. (2002). Report of the Greater Manchester/Lancashire/Preston PHL Liaison Group Survey on the microbiological quality of modified atmosphere packaged ham.

Yamagata, M. and Low, L.K. (1995). Banana shrimp, *Penaeus merguiensis*, quality changes during iced and frozen storage. *Journal of Food Science*, 60: 721–726.

Young, L.L., Reverie, R.D., and Cole, A.B., (1988). Fresh red meats: A place to apply modified atmospheres. *Food Technology* 42: 64–66, 68–69.

ELECTRONIC REFERENCES

http://www.iip-in.com/foodservice/13_readytoeat.pdf
http://www.nationmaster.com/encyclopedia/Convenience-food
http://en.wikipedia.org/wiki/Convenience food
E:\Sous-Vide-Fresh Cooking.mht
http://www.fass.org/fass01/pdfs/Rourke.pdf
http://www.vakuumverpacken.de/GB /fachberichte/vacmap-informat.pdf
http://www.vakuumverpacken.de/GB/fachberichte /vacmap-informat.pdf
http://cat.inist.fr/?aModele=afficheN&cpsidt=17722892
http://www.vakuumverpacken.de/GB/fachberichte/vacmap-informat.pdf
http://scholar.lib.vt.edu/theses/available/
www.encyclopedia.com/doc/1G1-15139266.html - 44k -
http://www.arrowscientific.com.au/Brochothrix_thermosphacta.htm
http://www.innvista.com/HEALTH/MICROBES/bacteria/entero.htm
www.bodyecology.com/07/08/16/lactobacillus_plantarum_benefits.php - 19k -

13 Miscellaneous Foods (Coffee, Tea, Beer, Snacks)

Ioannis S. Arvanitoyannis

CONTENTS

13.1 INTRODUCTION

There is a continuous search for improved methods of transporting food products from producers to consumers. It is well known that the preservative effect of chilling can be considerably increased in combination with modified atmosphere packaging (MAP) over storage. Such methods have been used commercially for more than 100 years for the bulk storage and transport of fresh meat and fruits and are referred to as controlled atmosphere storage (CAS). In fact, the latter, in view of the widespread availability of polymeric packages, was extensively applied to consumer packs and given the name modified atmosphere packaging (MAP) because the atmosphere surrounding the food is modified but continuously not controlled (Robertson, 2005).

MAP is a form of packaging where the removal of air from the pack can occur either with vacuum application (VA) or with gas flushing. The gas mixture used is dependent on the type of product. It should be borne in mind that the gaseous atmosphere is continuously altered over the storage period due to factors such as respiration of the packed product, biochemical changes, and slow permeation of gases through the container (Blakistone, 1999).

The aim of this chapter is to review the effects of MAP and vacuum packaging (VP) on *coffee, tea, corn curls, almonds, peanuts, beer, and crisps* shelf life, physicochemical and organoleptic properties, and the survival and growth of spoilage and pathogenic microorganisms as synoptically given in Table 13.1.

MAP goes back to 1821 when Jacques Etienne Bernard, a professor at the School of Pharmacy at Montpellier in France, published a report on the application of MAP in fruit ripening. It was reported that ripening of fruits was delayed when placed in an atmosphere with less O_2. However, no commercial usage of this technique was reported for almost 100 years (Dilley, 1978).

One of the most important applications of CAS took place in 1865 in Cleveland (Ohio) when Benjamin Nyce constructed a reasonably airtight store that used ice for cooling and a special paste for filtering the atmosphere to remove CO_2. This store was operated for several years but did not allow others to use his patented procedures (Dilley, 1978).

In 1903, Thatcher and Booth of Washington State University were the first American scientists to study the controlled atmosphere (CA) storage for 2 years and show that it was a promising technique. Later on, between 1907 and 1915, research personnel at the U.S. Department of Agriculture and Cornell University investigated the behavior of several fruits to both lower O_2 and higher CO_2 levels in storage atmospheres (Kader et al., 1989).

TABLE 13.1
Historical Milestones in MAP Development

Year	Technique of MAP of Fruits	Researcher	References
1821	MAP on fruits	J.E. Berard	Dilley (1978)
1865	CAS	B. Nyce	Dilley (1978)
1903	MAP on fruits	Thatcher, Booth	Kader et al. (1989)
1907 and 1915	Behavior of several fruits to both lower O_2 and higher CO_2	Thatcher, Booth	Kader et al. (1989)
1930	High levels of CO_2 in holding rooms	—	Church and Parsons (1995)
1970	MAP followed an upward trend	—	Church and Parsons (1995)
In the late 1970s	The inhibitory effects of CO_2 on microbial growth	—	Sivertsvik et al. (2002), Ben-Yehoshua et al. (1985)

CAS was already used in the 1930s when ships transporting fruits were found to have high levels of CO_2 in their holding rooms, thereby increasing the shelf life of the product. In the 1970s, modified atmosphere (MA) packages reached the stores when bacon and fish were sold in retail packs in the United Kingdom. Since then, the development of MAP followed an upward trend due to consumer demand. This has led to advances, for example, in the design and manufacturing of polymeric films. New techniques were engineered, like the employment of an antifogging layer to improve product visibility. Equilibrium modified atmosphere packaging (EMAP) is a recently developed technique based on MAP. The growing popularity of EMAP of vegetables and fruits could be attributed to the modern consumers' demands for fresh vegetables and fruits of a long shelf life without the use of preservatives (Church and Parsons, 1995).

The effective and widespread commercialization of MAP started in the late 1970s following the elapse of more than 150 years of scientific research on the inhibitory effects of CO_2 on microbial growth, as well as the effect of gaseous atmospheres on respiring produce. It required the convergence of scientific knowledge, polymeric films, gas flushing/VP equipment, and cold distribution chains to achieve the commercial success it currently enjoys. It is worth noting that MAP has already been well ahead of the more widely publicized canning, freezing, and aseptic packaging in terms of the volume of preserved food. Although shelf-life prolongation is the most apparent advantage of MAP, there are also several other advantages (as well as disadvantages) as given in Table 13.2 (Sivertsvik et al., 2002; Ben-Yehoshua et al., 1985).

13.2 MAP ON MISCELLANEOUS FOODS

A miscellaneous food is a generic term referring to all foods that do not fall in one of the classical categories (i.e., meat, fish, dairy, fruits, vegetables, etc.). A common characteristic of these products is that they are processed foods and of high added value. The miscellaneous foods in this chapter could not be distinguished in foods of animal and plant origin since all of them are of plant origin and processed as well.

13.2.1 Coffee

Coffee is the second most valuable commodity in the world market after oil. The United States consumes 20% of the entire world's coffee, making it by far the largest consumer in the world. There are many varieties of coffee beans, grown in different parts of the world. Coffee beans are shipped from the countries of origin in green state, the processing of the beans being carried out by companies that convert them into final products such as roasted whole beans, roasted ground beans, instant coffee, and coffee extracts (Robertson, 1997).

The various types of raw beans are cleaned, blended, and roasted by heating to obtain the expected flavor and aroma characteristic to the product. Since bean cultivars may vary in terms of their chemical composition, appropriate roasting conditions will have to be selected. Over roasting, there is a substantial loss of moisture from the beans, and chemical reactions are initiated. These reactions release aroma and flavor volatiles from nonvolatile components (fatty acids or lipids) of the beans,

TABLE 13.2
Advantages and Disadvantages of MAP

Advantages	Disadvantages
Shelf life increases by possibly 50%–400%	High costs for gases, packaging materials, and machinery
Reduced economic losses due to longer shelf life	Temperature control necessary
Decreased distribution costs, longer distribution distances, and fewer deliveries required	Different gas formulations per product type
Provides a high-quality product	Special equipment and training required
Easier separation of sliced products	Potential growth of food-borne pathogens due to temperature abuse by retailers and consumers
Centralized packaging and portion control	Increased pack volume adversely affects transport costs and retail display space
Improved presentation, clear view of product, and all-around visibility	Loss of benefits once the pack is opened or leaks
Little or no need for chemical preservatives	CO_2 dissolving into the food could lead to package collapse and increased drip
Sealed packages are barriers against product recontamination and drip from package	Additional cost for gas composition determination
Odorless and convenient packages	Requirement of additional investment in machinery and labor in the packaging line
Reduction of weight loss, desiccation, and shriveling	Risks of spoiled produce due to improper packaging or temperature abuse
Delay of ripening	Possible occurrence of new risks of microbiological safety due to possible development of anaerobic pathogenic flan
Alleviation of chilling injury	Plastic films may be environmentally undesirable unless effective recycling is in place
Semicentralized manufacturing options	
Reduction of labor and waste at the retail level	
Quality advantages such as color, moisture, flavor, and maturity retention	
Excellent branding options	
Reduction of handling and distribution of unwanted or low-grade produce	
Expanded radius of distribution systems	
Quality advantages transferred to the consumer	

Sources: Sivertsvik, M. et al., Modified atmosphere packaging, in: *Minimal Processing Technologies in the Food Industry*, T. Ohlsson and N. Bengtsson (Eds.), Woodhead Publishing Ltd, Cambridge, England, U.K., 2002; Ben-Yehoshua, S. et al., *Plant Physiol.*, 79, 1048, 1985.

which break down and react together in very complex reactions finally resulting in CO_2 release (Brado, 2004). The temperature at which roasted coffee beans are stored and its fluctuation have a direct effect on coffee staling rate. The presence of moisture heavily affects and promotes nonenzymatic browning and lipid oxidation (http://www.theqarrcoffee.com/packaginghtml).

13.2.1.1 Roast and Ground Coffee

Substantial quantities of CO_2 are released on grinding especially with the finer grinds. For example, estimates have been given that 45% is released within 5 min of grinding for a fine ground coffee, while others have determined that 30% is lost within 5 min of grinding to an average particle size of 1000 μm and 70% at 500 μm.

Oxygen is a prime determinant of the shelf life of coffee. There are currently three main ways of lowering its concentration inside a package. The first method is to apply a high vacuum immediately after filling into the package and then sealing it. The second is to flush the roasted and ground coffee and package with an inert gas immediately prior to sealing. The third is to place into the sachet compounds that will act as O_2 scavengers inside the package.

When used by consumers, the package of roast and ground coffee will be opened and closed frequently. In such situations, the shelf life of the product is practically reduced in view of the O_2 intrusion. A maximum figure for satisfactory shelf life in a commercial pack might be 20 μm of O_2 per gram of coffee (including that which has been absorbed), which corresponds to 1% oxygen content in a high vacuum pack immediately after desorption. In a recent development, to avoid the softening of the package, a CO_2 sachet containing a sorbent was applied.

13.2.1.2 Whole Beans

During roasting, CO_2 is released and becomes trapped within the beans. However, the amount of CO_2 produced depends heavily on several factors including the bean variety and roasting conditions. If the roasted beans remain intact, the CO_2 is slowly released from the bean (Barbera, 1967). If beans, however, have been sealed in packages directly after roasting, the CO_2 released from the beans builds up within the pack, eventually triggering its explosion. However, the roasted beans cannot be left unpackaged because beans may absorb moisture and oxygen thereby leading to flavor deterioration (Clarke, 1987). Investigation revealed that roasted whole beans exposed to air managed to maintain the quality equivalent to that of the freshly roasted product for only up to 10–12 days. After 40 days, flavor alteration became apparent and after 70 days, the beans were sensorially unacceptable (Wasserman et al., 2002).

Whole beans must be thoroughly degassed to remove CO_2 prior to packaging. The degassing process frequently comprises the application of vacuum. Packaging is an important consideration as the barrier properties of the materials considerably affect the product shelf life. Whole beans are by far more stable to oxidation than their ground form; as a result regular bags may be sufficient provided the product is to be sold in the short term. In view of the fact that dark roasted coffees can release oil, the package of this product needs to be grease resistant (Clarke, 1987).

VP must be effectively applied to minimize the presence of O_2 within the packs, because the latter induces staling. VP can maintain the beans in an acceptable condition for a maximum period of 18 months. An alternative to this method is inert gas (N_2, Ar, Xe) flushing, which favors a low residual level of O_2 inside the packaging.

Coffee freshness is heavily dependent on the time elapsed since roasting, in conjunction with the harvesting time of beans. Coffee beans can be kept for years, provided the conditions are appropriate, until they are roasted. It is noteworthy however, that, after roasting, the flavorsome coffee oils of the bean begin to deteriorate and will eventually turn rancid. Moreover, after roasting, coffee beans release gases (mostly carbon dioxide) for around a week, with the majority of the gases being generated within the first 2–3 days (http://www.coffee-makers-cafe.com/coffee-bean-storage-keep-fresh.html).

13.2.1.3 Instant Coffee

Instantized coffee production comprises the extraction of water-soluble components; the latter are filtered and dried, commonly with spray drying or freeze drying. The oil coming from the pressed ground coffee is added back and mixed again with the dried solids to improve aroma (Clarke, 1987). Although the CO_2 release is hardly a problem to consider in instant product packaging, moisture absorption, oxidation, and staling are important problems to solve, particularly due to the extended surface area and porous structure of the particles. Another crucial issue for instant coffee packs, similarly to ground coffee, is that they both have a low residual O_2 level. Therefore, gas flushing with inert gas has been put forward as a promising solution for ensuring shelf-life prolongation of instant products.

Instant products are most often packaged in tins and glass jars, these packs being gas flushed before the lid is attached. The diaphragm is made of waxed paper or aluminum foil to improve the moisture and gas barrier properties. A current competitor to the glass jar is the plastic jar, considered to be advantageous because of being lighter in weight, not fragile, and of lower cost.

Flushing with inert gases such as argon and xenon singly or in mixtures of more than one gas proved to be highly effective in reducing the oxidation of coffee for more than a year. The performance of these gases is regarded to be better than that of N_2 and CO_2 (Spencer and Humphreys, 2002).

13.2.2 Snacks

Nowadays, potato chip packaging systems are designed to keep the oxygen and water partial pressures in the package headspace as low as possible during storage. This is because the quality of this product depends on the extent of lipid oxidation and the amount of sorbed water (the latter is inversely proportional related to the crispness of the product) (Quast et al., 1972). To meet the aforementioned requirements, potato chips are currently packed under nitrogen as inert gas and polymeric films of low permeability to oxygen and water vapor. It is worth reporting that an increase in the initial value of the oxygen partial pressure in the package headspace induces an increase in the lipid oxidation rate, thereby resulting in a decrease in the shelf life of the product. On the other hand, an increase in the initial value of the water vapor partial pressure in the package headspace can cause either an increase in the product's shelf life or a decrease, depending on the factors responsible for its unacceptability. In fact, the shelf life is prolonged when the unacceptability of the product is due to rancidity.

Impact assessment of storage temperature and headspace package atmosphere, on induction of potential lipid oxidation in packed potato crisps, is of great importance for the corresponding industry (Tabee, 2008). The main reason for using MA inside the package container is to produce a gas composition that will most effectively avoid or, in the worst case, delay lipid oxidation over storage (Berry, 2003).

Snack foods are considered as "junk foods" or "empty calories." Their convenience to consume resides in the fact that they require neither preparation time nor special effort by the consumer. Snacks are consumed not only to satisfy hunger or supply the nutrients to our body but also for social reasons as well. Snack items and processed foods usually have a high sodium content, which has a disastrous effect on human health. Teenage boys are most likely to have a low intake level of calcium, iron, magnesium, and vitamin B6, and teenage girls consume these nutrients at even lower levels. The most frequently consumed foods by both sexes between meals are cakes, cookies, pies, candy, and desserts (Bawa and Sidhu, 2003).

The impact of MAP (gas composition and packaging material) on the shelf life of crisps is summarized in Table 13.3.

13.2.2.1 Corn Curls

Many extruded and puffed snack foods are packaged in similar material to that reported earlier for fried snack foods. However, since the loss of crispness is the major deterioration mode, a package providing a good barrier to water vapor is the primary requirement. Some extruded and puffed snacks are comparatively less sensitive to O_2 than fried snack foods, and the oxygen barrier requirements for these packages can consequently be less stringent (Bawa and Sidhu, 2003; FAO, 1992).

13.2.2.2 Almonds

Flexible plastic pouches of low permeability to the main MAP gases CO_2, O_2, N_2, and H_2O vapor were used to pack the kernels. The pouches were fitted with a septum in order to check the atmosphere. A partial vacuum was carried out to remove the air before packing with nitrogen. The gaseous content was determined just after packing the almonds. The initial O_2 and CO_2 content in the pouches with a plain air atmosphere was 20.6% and 0.04%, respectively. In the pouches with a nitrogen atmosphere, the O_2 content was always lower than 0.25%. A periodic measurement of the atmosphere inside the packages was carried out, and any change in the atmosphere composition due to a lack of air tightness led to the rejection of samples (García-Pascual et al., 2003).

No significant differences were recorded for the samples stored at 8°C for 9 months and the others stored at 36°C for 4 months. The inert atmosphere had no effect on the changes in the α-tocopherol content. Aflatoxins were not detected in concentrations above 0.5 mg/kg in any sample. However, the unshelled almonds stored at ambient

TABLE 13.3
Effect of MAP Composition/Packaging on Shelf Life of Crisps

Foods	Gas Composition	Packaging	Shelf Life	Other Parameters Studied	Other Processes Involved	Results	References
Potato crisps	N_2 with a mixture of N_2 and water vapor RH: 0.1% → 32%	Polymeric films of low permeability to oxygen and water vapor	Prolongation		Different types of packaging, how they react to the shelf life	Mathematical modeling for optimal headspace gas composition	
Chips	Flushed with nitrogen until the O_2 concentration <1%	Levered film metalized polyester	5 months	a and b carotene vitamin A	PE pouches <1% O_2 (a) 0°C–1°C → 94%–98% RH (b) 22°C–23°C → 31%–45% RH	The optimal long-term storage conditions for maximizing provitamin A activity and carotenoids for carrot chips were 22°C–23°C and 31%–45% RH	Sulaeman et al. (2003)
Potato chips	Nitrogen as inert gas	Polymeric films (low permeability to O_2 and H_2O)	More than 3 months	Oxygen and water removal	40°C in dark	—	Silva et al. (2004)
Crisps	N_2	Clear OPP (oriented polypropylene) film (without N_2 in clear plastic bags)	120 days (without N_2 55°C–65°C)	Reduce the level of O_2	Flushed and then stored at 23°C and 40°C at 90% RH in the dark	Crisps packaged in metalized OPP retained their freshness for a longer period in the presence of N_2. The use of metalized films is expected to grow with the use of gas flushing	Robertson (2005)

temperature for 9 months did not undergo any change in their initial fat content, the peroxide value (PV), or the α-tocopherol content. The explanation for this (atmosphere inside the shell, effective barrier against light or gaseous exchanges) remains, at present, unclear (Kazantzis et al., 2003).

Two flexible films, comprising a polyamide (PA) or an ethylene-vinyl alcohol (EVOH) copolymer layer as a barrier to low molecular weight compounds, were investigated. Products were packed alternatively under nitrogen or in the presence of oxygen scavengers. Samples were stored at 37°C and analyzed at regular internals. The obtained results proved that the loss of moisture, and the consequent hardening of samples, affected greatly both quality and shelf life of almond paste (Baiano and Del Nobile, 2005).

After, storage of shelled almonds for 6 months at either 5°C or 20°C, the results were lower moisture content, higher oil content, similar oil quality and composition, similar sugar content, and small changes in sugar composition compared with freshly harvested almonds. Almonds stored at 5°C retained higher kernel weight and moisture content than almonds stored at 20°C.

The effect of MAP composition in conjunction with other techniques on the shelf life of almonds is shown in Table 13.4.

In developed countries, moisture and oxygen scavengers were among the first types of active and intelligent packaging to be developed and successfully applied for improving food quality and prolonging shelf life (i.e., for delicatessen, cooked meats etc.). Apart from these, numerous other concepts such as ethanol emitters (e.g., for bakery products), ethylene absorbers (e.g., for climacteric fruits), carbon dioxide emitters/absorbers, and time/temperature and oxygen indicators were developed as well. Packaging interacts with foods for altering their texture and organoleptic properties or retarding microbial spoilage (Dainelli et al., 2008).

Almonds (*Prunus amygdalus Batsch*) belonging to the *Rosaceae* family are one of the most popular tree nuts on a worldwide basis and rank number one in tree nut production. They are widely used both as snack foods and as ingredients in a wide range of processed foods, especially in bakery and confectionery products (Sang et al., 2002a). Almonds, when incorporated in the diet, have been reported to reduce colon cancer risk in rats (Davis and Iwahashi, 2001) and increase high-density lipoprotein (HDL) cholesterol reducing at the same time low-density lipoprotein (LDL) cholesterol levels in humans (Hyson et al., 2002).

Another outlet for almonds and peanuts has been their application in confectionery products, such as chocolates in the United States, whereas hazelnuts have been more extensively used in Europe (Nattress et al., 2004). The formation of primary oxidation products was recorded by determining the PV. Changes in PV, hexanal content, fatty acid composition, volatile compounds, color parameter L*, odor and taste of dark chocolate with hazelnuts as a function of active or MA, packaging material oxygen permeability, and time of storage in the dark were reported at 20°C. Most of the determined compounds had the flavor profile of cocoa beans, formed either over fermentation or cocoa roasting (Hoskin and Dimick, 1984; Jinap et al., 1998). After 12 months of storage, an increase in concentration of aldehydes, ketones, alcohols, and alkanes with a parallel decrease in pyrazines was reported particularly in the least protected products after 6 and 12 months of storage.

TABLE 13.4
Effect of MAP Composition/Packaging on Shelf Life of Almonds

Foods	Gas Composition	Packaging	Shelf Life	Other Parameters Studied	Other Processes Involved	Results	References
Almonds	Air and N_2/O_2 20%–6% CO_2 0%–0.4%	Flexible plastic pouches	4–9 months	Aflatoxins	Raw and roasted 8°C → 9 months 36°C → 4 months	The unshelled almonds stored at ambient temperature for 9 months did not undergo any change in their initial fat content, the peroxide value, or the α-tocopherol content	García-Pascual et al. (2003)
Almonds	N_2	Metal cans	6 months 1 year at 20°C	Oil and sugar composition	In shell 5°C and 80% RH versus 20°C and 60% RH	Almonds stored at 5°C retained higher kernel weight and moisture content than almonds stored at 20°C	Kazantzis et al. (2003)
Almond (paste)	Nitrogen or in the presence of oxygen scavengers	Polymeric films, PA or ethylene-vinyl aluminum vessels	(a) 4.5 months under air (b) 5 months in the presence of oxygen scavengers (c) 4 months under nitrogen (d) 3.5 months with nylon multilayer	*Aspergillus* spp. *Bacillus cereus* *Escherichia coli*	Stored at 37°C	The loss of moisture, and the consequent hardening of samples, affected both quality and shelf life of almond paste	Baiano and Del Nobile (2005)

According to Mexis and Kontominas (2010),

> the effect of active packaging, nitrogen flushing, container oxygen barrier and storage conditions on quality retention of raw whole unpeeled almonds. Almond kernels were packaged in: (a) polyethylene terephthalate/low-density polyethylene (PET/LDPE), and (b) low-density polyethylene/ethylene vinyl alcohol/low-density polyethylene (LDPE/EVOH/LDPE) pouches under N_2, with or without an oxygen scavenger, heat-sealed and stored for a period of 12 months. PV varied between 0.17 for fresh almonds and 9.22 meq O_2/kg oil for almonds packaged in PET/LDPE pouches under N_2 exposed to light at 20°C after 12 months of storage. The obtained values for hexanal were lower than 28.5 mg/kg and 4.88 mg/kg, respectively. Polyunsaturated fatty acids (PUFA) and saturated fatty acids (SFA) displayed an increasing trend with a parallel decrease of monounsaturated fatty acids (MFA) after 12 months of storage in all treatments. Volatile compounds (aldehydes, ketones, alcohols, alkanes and aromatic hydrocarbons) augmented thereby indicating enhanced lipid oxidation. Employment of the oxygen absorber provided a shelf life of at least 12 months for all samples irrespectively of container and holding conditions

The package of walnut kernels consisted of (Mexis et al., 2009a) (a) LDPE, 55 lm in thickness in air, (b) PET/polyethylene (PET/PE), 70 lm in thickness under N_2, and (c) PETSiOx/PE pouches, 62 lm in thickness under N_2. Samples were stored either under fluorescent light or in the dark at 4°C or 20°C for a period of 12 months. Hexanal values for walnut kernels under pouches were lower than 28.5 mg/kg and 36.0 mg/kg and for tetrabutylammonium (TBA) ca. 0.2 and 11 mg malondialdehyde (MDA)/kg. The obtained values for odor ranged between 0.2 for fresh walnut kernels and 5.7 for walnut kernels packaged in PE exposed to light after 12 months of storage at 20°C. Respective values for taste were 0.7 and 6.8. Based on shelf life (taste) values and PV data, it was suggested that "the upper limit value for PV is close to 10.0 meq O_2/kg walnut oil. Respective limit values for hexanal are 1–2 mg hexanal/kg walnut and for TBA they are 1–2 mg MDA/kg walnut. Walnuts maintained acceptable quality for ca. 2 months in PE-air, 4–5 months in PET/PE-N_2 and at least 12 months in PET-SiOx/PE-N_2 pouches at 20°C, with samples stored in the dark retaining slightly higher quality than those exposed to light." Figures 13.1 through 13.3 show the effect of storage time on hexanal content, peroxide values (POV), and chromatometric parameters of MAP-packaged almonds and walnuts.

Antioxidant effects of electron beam–irradiated almond skin powder (ASP) in raw minced chicken breasts (MCB) over refrigerated and frozen storage were studied by Teets and Were (2008). MCB samples were treated with butylated hydroxytoluene (BHT), nonirradiated ASP (0 kGy), and irradiated ASP (10, 20, and 30 kGy) and compared to MCB without antioxidants. Color was determined on initial and final day of analysis, while conjugated dienes (CD), POV, thiobarbituric acid reactive substances (TBARS), and hexanal content were recorded periodically over a period of 12 days of refrigerated storage and 7 months of frozen storage, respectively. ASP addition decreased considerably the L* values compared with MCB without ASP or BHT. During refrigerated storage, MCB containing ASP decreased the formation of lipid oxidation products ranging from 0% to 66%, 7% to 24%, 0% to 37%, and 4% to 71% reduction in POV, CD, TBARS, and hexanal content, respectively, as compared with

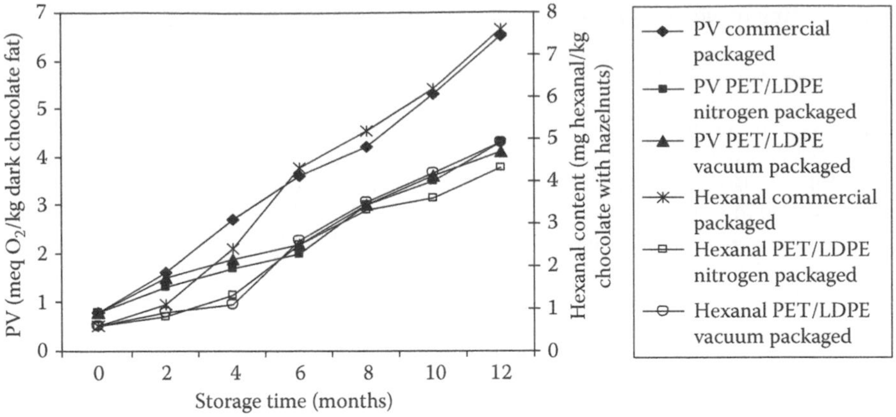

FIGURE 13.1 PV and hexanal content of dark chocolate with hazelnuts as a function of active and MAP, packaging material oxygen permeability, and storage time in the dark at 20°C. (From Mexis, S.F. et al., *Innov. Food Sci. Emerg. Technol.*, 10, 580, 2009a; From Mexis, S.F. et al., *Food Control*, 20, 743, 2009b.)

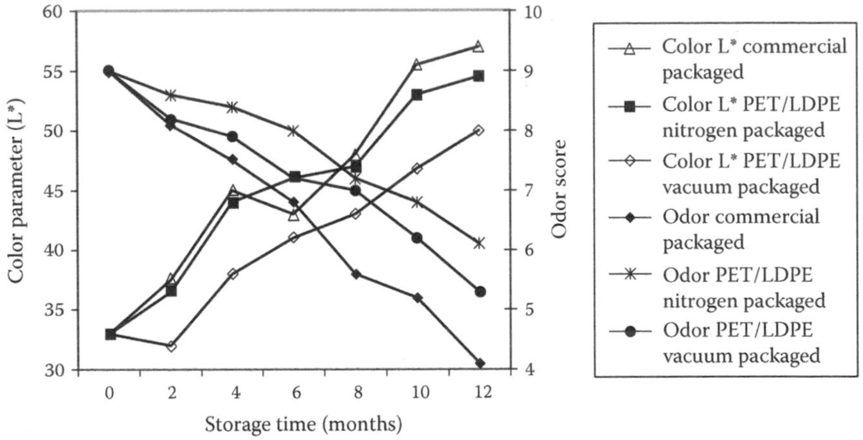

FIGURE 13.2 Changes in L* color parameter and odor of dark chocolate with hazelnuts as a function of active and MAP, packaging material oxygen permeability, and storage time in the dark at 20°C. (From Mexis, S.F. et al., *Innov. Food Sci. Emerg. Technol.*, 10, 580, 2009a; From Mexis, S.F. et al., *Food Control*, 20, 743, 2009b.)

MCB without antioxidants over the duration of study. The effect of storage on hexanal content in walnuts and color parameter (L*) is displayed in Figure 13.4.

Sánchez-Bel et al. (2008) investigated the effects of irradiation with accelerated electrons (0, 3, 7, and 10 kGy) on the chemical composition (water content, proteins, neutral detergent fiber, sugars, lipid content, organic acids, and color) and sensorial properties (rancidity, sweetness, off-flavors and odors, texture, and whiteness) of the shelled almond cultivar Guara, packaged under air atmosphere and stored for

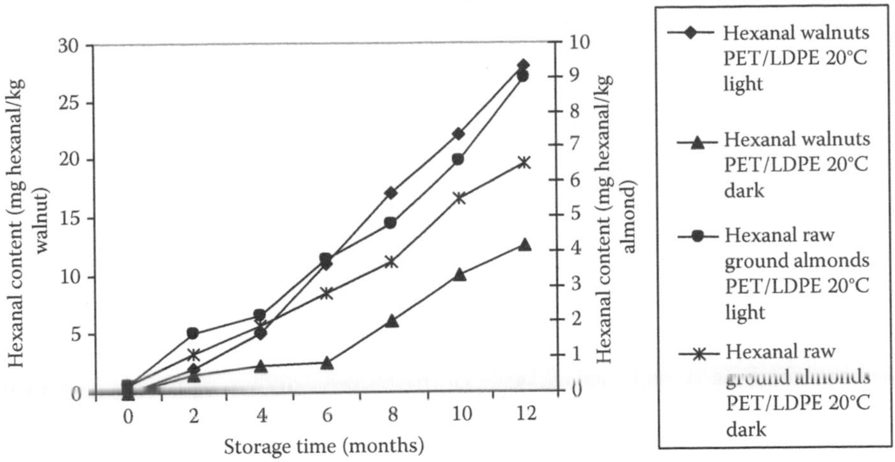

FIGURE 13.3 Hexanal content in walnuts (From Mexis, S.F. et al., *Innov. Food Sci. Emerg. Technol.*, 10, 580, 2009a.) and almonds packaged with PET/LPDE against storage time. (From Mexis, S.F. et al., *Food Control*, 20, 743, 2009b.)

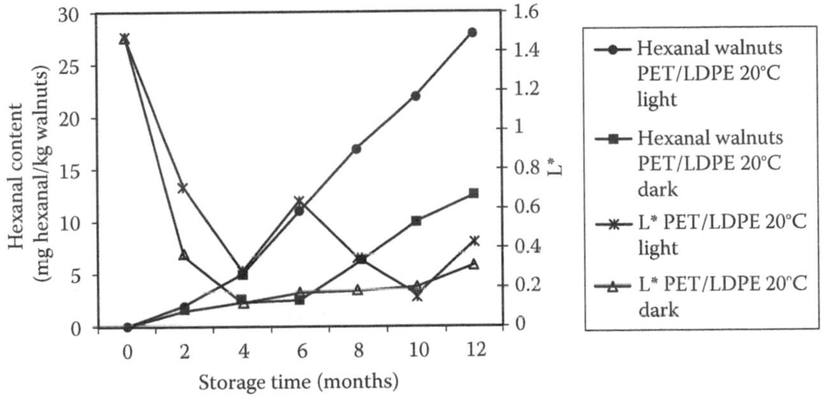

FIGURE 13.4 Hexanal content in walnuts (From Mexis, S.F. et al., *Innov. Food Sci. Emerg. Technol.*, 10, 580, 2009a.) and color parameters (L*) of raw unpeeled almonds as a function of irradiation dose and storage time in the light and dark at 20°C. (From Mexis, S.F. and Kontominas, M.G., *Food Sci. Technol.*, 43, 1, 2010.)

5 months. Changes observed were a decrease of glucose in samples treated at all irradiation doses. A substantial increase in citric acid, at doses above 3 kGy, and then a decrease in values similar to those of the control were recorded. With respect to the sensorial analysis, there was no treatment effect on the sweetness, texture, or color; but there was a marked rancidity in the samples treated with 10 kGy that decreased the overall acceptance of the samples.

Figure 13.5 exhibits clearly the strong impact of storage time on protein content of irradiated almonds and peroxide value in shelled and roasted marcona almonds.

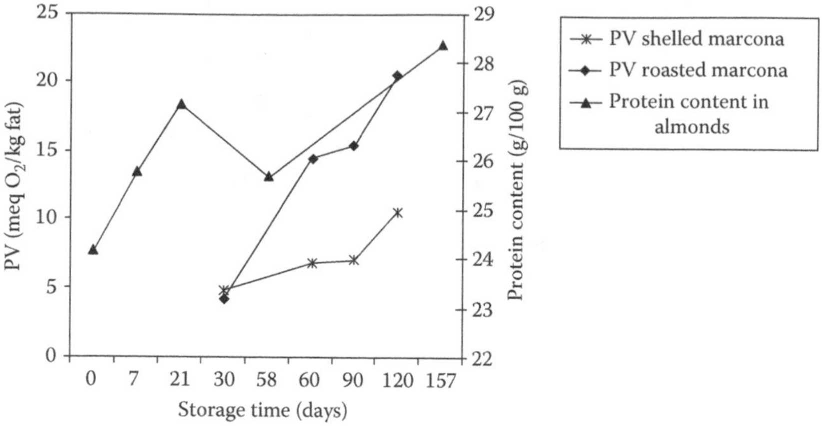

FIGURE 13.5. PV content in shelled and roasted marcona against storage time (From García-Pascual, P. et al., *Biosyst. Eng.*, 84(2), 201, 2003.) and protein content of irradiated almonds over storage time. (From Sánchez-Bel, P. et al., *Sci. Direct*, 41, 442, 2008.)

13.2.2.3 Peanuts

Peanuts were packaged in two atmospheres: (i) under air and (ii) under CO_2:N_2 (65%:35% v/v). Packaging under air was carried out by placing plates in 210 × 210mm bags in each of three packaging films and sealing with a pulse heat sealer. Packaging and sealing of packages in a CO_2 enriched atmosphere was conducted using a Multipack vacuum/gas packaging unit. Each package was filled with a 65:35 ratio of CO_2:N_2 gas mixture. A proportional mixer was employed to provide the appropriate ratio of CO_2 and N_2 in each package. Packaged plates were incubated at 25°C and 30°C in a precision gravity convection incubator and at 20°C in an incubator for 15 days.

MAP involving a CO_2:N_2 (65:35) gas mixture was effective in controlling the production of aflatoxin B by *Aspergillus flavus* in peanuts to levels below the regulatory limit of 20ng/g. However, the antimycotic effect of CO_2 enriched atmospheres is dependent on the barrier characteristics of the packaging films, especially at higher storage temperatures. Should peanuts be gas packaged in a high barrier film to both O_2 and CO_2, their safety is ensured. Further removal of oxygen (residual) could take place by incorporating oxygen absorbent to scavenge the gas-packaged product.

Estimation of the shelf life of a food is directly related to its packaging (e.g., peanuts in MAP); both product conditions and the package as well as gas mixture composition should be considered and any other problems/failures this particular product has been related to, as shown in Table 13.5.

13.2.3 Tea

MAP was shown to be very useful in maintaining the fresh quality of tea. Tea accounts for nearly half of all drinks consumed in the United Kingdom and is the most popular beverage in the country. Research carried out on black tea revealed

TABLE 13.5
Effect of MAP Composition/Packaging on the Shelf Life of Peanuts

Foods	Gas Composition	Packaging	Shelf Life	Other Parameters Studied	Other Processes Involved	Results	References
Peanuts	CO_2/N_2 65%/35%	High medium barrier films (ASI and ASII)	15 days	Aflatoxin production (*A. flavus*)	Storage at 20°C, 25°C, 30°C	Peanuts should be gas packaged in a high barrier film to both O_2 and CO_2, e.g., film ASI. Inclusion of an oxygen absorbent to scavenge any residual O_2 in the gas packaged product was beneficial to the product	Ellis et al. (1994a,b)
Peanuts	(a) 0%–1% O_2 100%–99% N_2 (b) 1%–3% O_2 99%–97% N_2 (c) 3%–6% O_2 97%–94% N_2	Multilayer laminated materials, cans	Longest at 0%–1% O_2 100%–99% N_2	Oxygen content	Roasting and salting Storage for 13 months 18°C and RH 75%	It was clearly shown that food shelf life is greatly affected by package, gas mixture, and RH	Ucherek (2004)
Peanuts	10% O_2 60/40 CO_2/N_2 25°C	High gas barrier bags	9 months	a_w, storage temperature, headspace O_2, CO_2 on the growth of aflatoxin production by *A. flavus*		Packaging/storage conditions (a_w, headspace O_2, temperature) affect both mold growth and the level of aflatoxin detected in gas-packaged peanuts	Ellis et al. (1993)

(continued)

TABLE 13.5 (continued)
Effect of MAP Composition/Packaging on the Shelf Life of Peanuts

Foods	Gas Composition	Packaging	Shelf Life	Other Parameters Studied	Other Processes Involved	Results	References
Roasted peanuts	Flushing with nitrogen	High-density barrier bags made of 0.75 PA and 2.25 PE	Cracker-coated peanuts (CCP): 78, 56, and 32 days Roasted peanuts (RP): 116, 105, and 94 days	Different temperatures in roasted and cracker coated against development of oxidized flavors	23°C, 30°C, 40°C, roasted peanut Freezer −19°C or oven 23°C	Sensory profiles predicted shelf-life RP and CCP better than hexanal content. Quantification of sensory attribute profile and hexanal content of RPs and CCPs under various storage conditions of time and temperature also identified critical measurements over storage to determine shelf life	Grosso et al. (2007)

that postharvest processing produces most of the aroma volatiles in the product. However, changes in the aroma quality of the tea were found to occur during storage due to oxidative deterioration, thereby reducing the quality of the product (Springett et al., 1994). These changes often occurred during the first 10 weeks of storage of the picked and fired tea.

Following a great amount of research, a U.K. company coined a technique to capture the true freshly picked flavor and aroma of the tea by protecting the tea against oxidation from the plantation all the way to the consumer. According to Darrington (1991) the first stage of the operation involved VP the fresh tea in large blocks in aluminum foil/polyester/polyethylene laminate at the tea estate. Upon its arrival at the U.K. factory, the tea was blended and held in stainless steel to the bins, continuously flushed with N_2. The tea was then filled into tea bags and the bags were sealed inside foil pouches after N_2 flushing.

The gas-flushed tea had both milder taste and stronger aroma than the conventional product and targeted tea drinkers who would be willing to pay the difference in price for the better quality.

13.2.3.1 Black and Green Tea

It has been reported that at 32% relative humidity (RH) and 20°C under conditions excluding light, black tea could be stored for a period of 300 days without loss of tea character. Exclusion of light is clearly important because photooxidation of lipids and nonenzymic browning reactions, both of which contribute to quality loss in black tea, are accelerated by light (Robertson, 2005).

Potential deterioration in the quality of green tea over storage is recognized by determining the following parameters: (1) reduction in ascorbic acid content; (2) change in color from bright green to olive green and then brownish green; (3) change in leafy and refreshing odor to dull and heavy odor; (4) changes from a well-balanced, complex taste to a flat taste lacking in characteristic briskness. All the aforementioned reactions are accelerated by moisture, oxygen, elevated temperatures, and exposure to light similarly to black tea.

The effect of MAP composition in conjunction with other conditions/parameters on the shelf life of tea, beer, and coffee is given in Table 13.6.

13.2.3.2 Tea Packaging

Loose tea is packaged in a multitude of different shapes, sizes, and types of materials, the most common being a paperboard carton with either an aluminum foil liner or an overwrap of PP. Metal containers with snap-on lids are also employed for specific premium products. Nowadays, tea bags have become the most popular form of retail packaging, and considerable development occurred toward improving the tissue paper used for this type of package, with porous wet-strength paper being required. Once filled, the tea bags must be placed inside a package with an adequate barrier to moisture vapor such as paperboard cartons lined with PP or RCF. Since the initial moisture content of tea amounts to 3%–4%, there is a strong need for a good moisture vapor barrier to prevent a relatively rapid rise in moisture content.

TABLE 13.6
Effect of MAP Composition/Packaging on Shelf Life of Tea, Beer, and Coffee

Foods	Gas Composition	Packaging	Shelf Life	Other Parameters Studied	Other Processes Involved	Results	References
Green tea	Nitrogen gas flushing	Paperboard carton: a. aluminum foil liver; b. overwrap of PP RCF Metal containers with snap-on lids	300 days	Ascorbic acid	Heat (6–8 h), saturated humidity (3 h), heating 82°C–93°C, exposure to light	During storage, 1. Reduction in ascorbic acid 2. Change in color from bright green to olive green and brownish green 3. Change in leafy and refreshing odor to dull and heavy odor 4. Changes from a well-balanced complex taste to flat	Yamanishi (1986)
Beer	CO_2, removal of O_2	Glass, metal, plastics (PVdC) copolymer coated	80–120 days	Microbiological stability killing beer spoilage organisms	Held at 70°C → 20 s before cooling and packaging		Robertson (2005)
Coffee (roasted whole beans)	Flushing with inert gas either CO_2 or more typically N_2	Plastic films, one-way valves	12 months	The evolution of CO_2	Dark roasted beans	MAP is commercially used to replace VP of whole beans. Packaging under N_2 in packs, removes the need for the long storage after roasting to degas the beans with one-way valve. Easy opening and closing of the packs with peel seal	Anonymous (1994)
Coffee (instant)	Oxygen content is lowered to less than 4%	In flexible laminates of PET–Al foil–LDPE or metalized PET–LDPE. Metal containers and glass jars	18 months	Evolution of CO_2, moisture absorption, and oxidation		i. Extraction of water-soluble components ii. Filtering and drying, with spray drying or freeze drying	Clarke (1985)

However, even when tea has gained a large amount of moisture, its market value is still similar to tea protected from moisture, leading to the conclusion that moisture content alone does not result in deterioration of tea. A reduction in the levels of various chemical constituents is more likely to contribute to the end of shelf life of tea than an increase in moisture content. The volatile fraction of black tea clearly displays an overall decline during storage accelerated by moisture uptake and, to some extent, by storage at elevated temperatures (in http://books.google.gr/books?id=cFzIphx7CUQC&pg=PA598&lpg=PA598&dq=RCF+in+Tea+Packaging&source=bl&ots=sc3xBMvC5k&sig=F8PAzFNBSfta0BcOoqgkUVXsfiw&hl=el&ei=ebjTSu6pN8TJ_gaO4tzZAg&sa=X&oi=book_result&ct=result&resnum=1&ved=0CAsQ6AEwAA#v=onepage&q=RCF%20in%20Tea%20Packaging&f=false).

The storage stability of green tea is the lowest among various teas including black tea, oolong tea, and pouching tea. The most effective protection of the quality of green tea over storage required the usage of nitrogen gas flushing or VP in order to avoid any changes in the moisture content and ascorbic acid content of green tea over storage (Yamanishi, 1986)

13.2.4 Beer

Beer is an alcoholic beverage prepared by brewing and fermentation from cereals (usually malted barley) and flavored with hops to impart a bitter flavor. Two major technological steps are required to complete the transformation of the raw material into the finished product: (i) controlled germination (i.e., malting), which allows the ultimate production of a fermentable extract through the activities of enzymes formed during seedling growth, and (ii) fermentation.

Beer is drunk all over the world. In some countries, such as parts of Germany, it stands for the drink of choice for accompanying food. Beer is probably the most important drink of relaxation and moderation. The alcohol range in "Normal" beers varies from 2.5% up to 13%. Although the non- and low-alcohol beers (NAB/LABs) can be classified in many ways, the most applied approach is to classify beers containing less than 0.05% or less than 2% alcohol (by volume), respectively. The sweetness of beers is attributed to residual sugars that have not been fermented into alcohol. The Brewery often adds sugar ("primings") to the beer prior to packaging (Bamforth, 2003).

Microbial degradation is not considered a problem with beer because of its low pH (around 4.0). The two main parameters minimizing the probability of wild yeast growths are pasteurization and aseptic cold filtration. Unfortunately, even over storage, beer can still undergo severe changes leading to the appearance of haze, development of off-flavors, and darker color mainly due to oxidation. One of the major oxidative reactions is the oxidation of linoleic acid (introduced into the wart from the malt) the product of which is 2-trans-nonenal, which imparts to beer a cardboard-like flavor (flavor threshold: 0.1 ppd). It is noteworthy that flavor loss occurs more rapidly in the presence of light in conjunction with certain metal ions.

Depletion of flavor compounds from beer is possible by means of sorption by the packaging material. Some of the nonpolar flavor compounds in beer are readily soluble in polymers such as the polyvinyl chloride (PVC) liners in crown corks than

TABLE 13.7
Milestones in Beer Canning

Year	Individual/Company
1933	Test market (New York, New Jersey)
1935	2000 cans by Kruegers Finest Beer and Kruegers Cream Ale
1935	36 U.S. breweries canned beer
1960	Schlitz used Aluminum
1965	Detectable pull-tab Ale
1975	Patent for a can end with an inseparable tear strip to avoid litter

in water. The degree of partitioning into such materials is closely linked with the polarity of the molecule in question, the volume and chemical composition of the polymer or other lining material, the volume and alcohol content of the beer, and the temperature. For several bottled beers, loss of hop flavor to packaging material does occur and could potentially be a major source of hop losses in flavor loss (Macleod and Macleod, 1968).

A pronounced rise in oxygen scavengers is clear because the growth rate is estimated at more than 50% annually for beer crowns alone; bottles for other beverages, fruit juices, sport drinks, and case-ready meat also head the list. Other markets include trays and lidding stock for home replacement meals and composite cans. Three billion packages employing oxygen absorbers were used in 2004 by North America, and more than six billion worldwide. Considerable activity on the enhanced employment of oxygen scavengers was reported with PET bottles for beer and other beverages. One of the most popular products included is "Amsorb DFC" based on binding protein (BP) chemical (http://pffc-online.com/mag/paper_packaging_really_works/).

The milestones in beer packaging are summarized in Table 13.7.

13.2.4.1 Packaging of Beer

Beer can be packaged in glass, metal, or plastic recipients. The research and development of new food packaging is a quite challenging endeavor. Generally, one has to distinguish between (a) passive packaging, (b) active and intelligent packaging, and (c) new techniques for the design and the presentation of a package. In the first case of new materials, in most cases composite materials are manufactured. The advantage of composite materials is that they display properties that cannot be reached with single materials such as brown PET bottles in which beer was sold for some years. Moreover, intelligent packaging becomes increasingly important. Another issue is the proper labeling that informs the user about the freshness of the product he is about to consume (Stubenrauch, 2005).

In combination with multilayer injection technology, the process made possible the production of a PET bottle with effective O_2 and CO_2 barriers to protect the beer and keep it fresh. Another marketing bonus was that beer in a plastic bottle always "wears" its brand name while beer in a plastic cup usually does not. In fact, the consumer receives solely the branded image along with the product (http://www.packagingdigest.com/article/CA6607575.html).

Innovative packaging has started to play a central role in a company's effort to differentiate itself in a competitive marketplace. In the European beer market, Heineken is the company that has adopted creative packaging in an attempt to stand out in the crowd. In 1998, Heineken began offering several of its most important European beer brands, such as "33" Export in France, Dreher in Italy, and Cruzcampo in Spain, in 0.5L, 660mL, and 1L plastic bottles (http://www.packagingdigest.com/article/CA6580713.html).

PET beer bottles have been a major driver for oxygen scavengers, particularly in Japan and central and eastern Europe. The growing predominance of major supermarket chains in developed countries and the development of a more advanced retail infrastructure in developing markets are anticipated to be key drivers for MAP and VP. The trend for consumers to demand more convenient packaged foods is anticipated to result in even greater sales of MAP and active packaging, including the use of moisture scavengers, self-venting films, and microwave (MW) apparatus (http://www.pirainternational.com/businessintelligence/GrowthOpportunitiesinActiveandModifiedAtmospherePackaging.aspx).

Nishida et al. (2005) developed several physical techniques to fractionate a malt kernel into several components. The most recently developed malt fractionation technique involved abrading and polishing the malt kernel from the outside using a stone grinder and then separating the fractions by means of a sieve shaker. By applying this technique, a malt kernel is separated into the inner fraction, the outer layer fraction, and the husk fraction.

The final processing stage, prior to storage of the beer, is to fill it into the intended package. In the United States in 2000 the balance of packaging was 9% on draft, 51% into cans, 38% into nonreturnable glass bottles, and the remainder in glass bottles that are returned to the Brewer for washing and refilling. In France and Italy, though, much of the beer is retailed in nonreturnable glass, while Sweden is the country that sells the highest proportion of beer in cans. As regards the United States and Canada, there is a strong preponderance of nonreturnable glass in the United States, whereas in Canada, the bulk of bottled beer is in returnable glass (Bamforth, 2003).

13.2.4.2 Packaging in Glass and Metal

The traditional packaging media for beer is the glass bottle sealed with a crown corn. Pasteurization of the beer in the bottle after sealing is the most common means of ensuring microbiological stability. The aim is to heat the beer to a high enough temperature and hold it there long enough to kill any beer spoilage organisms. The brewing industry has developed its own standard measure of the effectiveness of the pasteurization process and uses the term pasteurization units (PUs) where one PU is equivalent to holding beer at 60°C for 1 min. Application of 10 PUs is regarded as a suitable heat treatment for most bottled beers produced under good manufacturing practices.

In an attempt to prolong the shelf life of bottled beer, oxygen-absorbing materials have been incorporated into the lining material of crown corks to remove any residual oxygen present in the headspace at the time of filing, as well as absorbing oxygen permeating into the bottle through the crown closure. The linings containing the O_2 absorber can be used in metal crowns, as well as roll-on metal and screw-on plastic caps. They have also found extensive applications in the fruit juice and carbonated beverage industries. The most important milestones in beer canning starting almost 80 years ago are summarized in Table 13.7.

One of the great beer genres (viz. the English cask ale) appeared in the frame of the "natural" clarification process rooted in a protein preparation called isinglass. Isinglass is a very pure form of collagen extracted from the dried swim bladders of certain warm-water fish, amongst them the catfish, jewfish, threadfish, and croaker (Bamforth, 2006).

Glass bottles used for holding beer come in diverse shapes and sizes and colors. The glass may be brown or black, green, or clear. Marketing people would want beer packaged only in brown because of the protection it gives from light (Bamforth, 2003).

13.2.4.3 Packaging in Plastics

Although the use of poly-vinylidene dichloride (PVdC) copolymer coated PET bottles for the packaging of beer commenced in the early 1980s in the United Kingdom, it was not widely adopted by brewers in other countries. The PVdC copolymer coating provided an acceptable barrier to oxidation and prevented flavor degradation in the beer. It also lowered the CO_2 permeability of the bottle. Taking into account the surface area to volume relationship and the O_2 permeability of the bottle material, most brewers tended to use larger sized bottles (typically 2 L) to obtain a satisfactory shelf life. A clean filling technique was employed to handle the sterile filtered beer. However, this bottle never enjoyed widespread commercialization and alternative barrier coatings were developed.

Nowadays, a small but increasing quantity of beer is packaged in PET bottles with an O_2 barrier to provide acceptable shelf life. Various barriers are commercially available for multilayer laminated polymers. Shelf lives of up to 9 months are now being claimed for some of these bottles.

13.3 CONCLUSIONS

The application of MAP to miscellaneous products is of great importance because, under this term, there are products of high added value. In fact, products like coffee, tea, snacks (crisps, corn curls), nuts, and beer had attracted research interest both in terms of quality and safety. MAP proved to be an excellent approach in this field especially in the case of snacks, coffee, and almonds, and only secondarily for tea and beer. Optimization of the internal atmosphere in conjunction with the packaging material showed that the shelf life could be effectively prolonged by three to four times. Furthermore, the quality of the aforementioned foods was substantially better because the oxidation problems were eliminated and both the appearance and the taste of the product improved considerably.

ABBREVIATIONS

Abbreviation	Full Names
ASP	Almond Skin Powder
BHT	Butylated Hydroxytoluene
BP	Binding Protein
CA	Controlled Atmosphere
CAS	Controlled Atmosphere Storage
CD	Conjugated Dienes
DFC	Data Flow Control
EMAP	Equilibrium Modified Atmosphere Packaging
EVOH	Ethylene Vinyl Alcohol
HDL	High-Density Lipoprotein
HDPE	High-Density Polyethylene
LABs	Low-Alcohol Beers
LDL	Low-Density Lipoprotein
LDPE	Low-Density Polyethylene
MA	Modified Atmosphere
MAP	Modified Atmosphere Packaging
MCB	Minced Chicken Breasts
MDA	Malondialdehyde
MFA	Monounsaturated Fatty Acids
MW	Microwave
NAB	Nonalcohol Beer
PE	Polyethylene
PET	Polyethylene Terephthalate
PUFA	Polyunsaturated Fatty Acids
Pus	Pasteurization Units
PV	Peroxide Value
PVC	Polyvinyl Chloride
PVdC	Poly-Vinylidene Dichloride
RH	Relative Humidity
SFA	Saturated Fatty Acids
TBA	Tetrabutylammonium
TBARS	Thiobarbituric Acid Reactive Substance
VP	Vacuum Packaging

REFERENCES

Anonymous. (1994). Permeable plastics film for respiring food produce. *Food Cosmetics Drug Pack*, 17: 7.

Baiano, A. and Del Nobile, M.A. (2005). Shelf life extension of almond paste pastries. *Journal of Food Engineering*, 66: 487–495.

Bamforth, C.W. (2003). *Tap into the Art and Science of Brewing*. Oxford University Press Inc., Madison Avenue, New York.

Bamforth, C.W. (2006). *Brewing New Technologies*. 1st edn, Woodhead Publishing Limited, Cambridge, U.K.

Barbera, C.E. (1967). Gas-volumetric method for determination of internal non-odourous atmosphere of coffee beans, *Proceedings of Third International Colloquium on ASIC*, Paris, France, pp. 436–442.

Bawa, A.S. and Sidhu, J.S. (2003). Snack foods, *Encyclopedia of Food Science and Technology*, 2nd edn., Vol. 8, Eds. B. Caballero, I.C. Trugo and P.M. Finglas, Academic Press, New York, pp. 5322–5332.

Ben-Yehoshua, S., Burg, S.P. and Young, R. (1985). Resistance of citrus fruit to mass transport of water vapor and other gases. *Plant Physiology*, 79: 1048–1053.

Berry, D. (2003). Fats Change in Food Product Design, in http://www.foodproductdesign.com/articles/2003/06/fats146....change.aspx

Blakistone, B.A. (Ed). (1999). *Principles and Applications of Modified Packaging of Foods*. 2nd edn, An Aspen publication, Gaithersburg, MD, pp. 172–185.

Brado, C.H.J. (2004). Harvesting and green coffee processing, in *Coffee: Growing, Processing, Sustainable Production*. Ed. J.N. Witgens. Wiley-VCH, Weinheim, Germany, pp. 604–715.

Church, I.J. and Parsons, A.L. (1995). Modified atmosphere packaging technology: A review. *Journal of the Science of Food and Agriculture*, 67(2): 143–152.

Clarke, R.J. (1985). Water and mineral contents, *Coffee: Chemistry*, Vol. 1., Eds. R.J. Clarke and R. Macrae. Elsevier Applied Science, London, U.K., pp. 42–82.

Clarke, R.J. (1987). Packing of roast and instant coffee, in *Coffee: Technology*. Eds. R.J. Clarke. and R. Macrae. Elsevier Applied Science, Amsterdam, the Netherlands.

Dainelli, D., Gontard, N., Spyropoulos, D., Zondervan-van den Beuken, E. and Tobback, P. (2008). Active and intelligent food packaging: Legal aspects and safety concerns. *Trends in Food Science and Technology*, 19: 103–112.

Darrington, H. (1991). The Darwin project. *Food Manufacturing*, 2: 52–54.

Davis, P.A. and Iwahashi, C.K. (2001). Whole almonds and almond fractions reduce aberrant crypt foci in a rat model of colon carcinogenesis. *Cancer Letters*, 165: 27–33.

Dilley, D. (1978). Approaches to maintenance of postharvest integrity. *Journal of Food Biochemistry*, 2: 235–242.

Ellis, W.O., Smith, J.P., Simpson, B.K., Khanizadeh, S. and Oldham, J.H. (1993). Control of growth and aflatoxin production of *Aspergillus flavus* under modified atmosphere packaging (MAP) conditions. *Food Microbiology*, 10: 9–21.

Ellis, W.O., Smith, J.P., Simpson, B.K. and Ramaswamy, H. (1994a). Effect of gas barrier characteristics of films on aflatoxin production by *Aspergillus flavus* in peanuts packaged under modified atmosphere packaging (MAP) conditions. *Food Research International*, 27: 505–512.

Ellis, W.O., Smith, J.P., Simpson, B.K., Ramaswamy, H. and Doyon, G. (1994b). Growth of and aflatoxin production by *Aspergillus flavus* in peanuts stored under modified atmosphere packaging (MAP) conditions. *International Journal of Food Microbiology*, 22: 173–187.

FAO. (1992). Snack foods, in *Small-Scale Food Processing—A Guide For Appropriate Equipment*. Eds. P. Fellows. and A. Hampton. Intermediate Technology Publications in Association with CTA, London, U.K.

García-Pascual, P., Mateos, M., Carbonell, V. and Salazar, D.M. (2003). Influence of storage conditions on the quality of shelled and roasted almonds. *Biosystems Engineering*, 84(2): 201–209.

Grosso, N.R., Resurreccion, A.V.A., Walker, G.M. and Chinnan, M.S. (2007). Sensory profiles and hexanal content of cracker-coated and roasted peanuts stored under different temperatures. *Journal of Food Processing and Preservation*, 32(1): 1–23.

Hoskin, J. and Dimick, P. (1984). Role of sulfur compounds in the development of chocolate flavors—A review. *Process Biochemistry*, 19: 150–156.

Hyson, D., Schneeman, B.O. and Davis, P.A. (2002). Almonds and almond oil have similar effects on plasma lipids and LDL oxidation in healthy men and women. *Journal of Nutrition*, 132: 703–707.

Jinap, S., Wan, R., Russly, A.R. and Norsin, L.M. (1998). Effect of roasting time and temperature on volatile component profiles during roasting of cocoa beans (*Theobroma cacao*). *Journal of the Science of Food and Agriculture*, 77(4): 441–448.

Kader, A.A., Zagory, D. and Kerbel, E.L. (1989). Modified atmosphere packaging of fruits and vegetables. *Critical Reviews in Food Science and Nutrition,* 28: 1–30.

Kazantzis, I., Nanos, G. and Stavroulakis, G. (2003). Effect of harvest time and storage conditions on almond kernel oil and sugar composition. *Journal of the Science of Food and Agriculture*, 83(4): 354–359.

Macleod, A.J. and Macleod, G. (1968). Volatiles of cooked cabbage. *Journal of the Science of Food and Agriculture,* 19: 273–277.

Mexis, S.F., Badeka, A.V. and Kontominas, M.G. (2009a). Quality evaluation of raw ground almond kernels (*Prunus dulcis*): Effect of active and modified atmosphere packaging, container oxygen barrier and storage conditions. *Innovative Food Science and Emerging Technologies*, 10: 580–589.

Mexis, S.F., Badeka, A.V., Riganakos, K.A., Karakostas, K.X. and Kontominas, M.G. (2009b). Effect of packaging and storage conditions on quality of shelled walnuts. *Food Control*, 20: 743–751.

Mexis, S.F. and Kontominas, M.G. (2010). Effect of oxygen absorber, nitrogen flushing, and packaging material oxygen transmission rate and storage conditions on quality retention of raw whole unpeeled almond kernels (*Prunus dulcis*). *Food Science and Technology*, 43: 1–11.

Nattress, L.A., Ziegler, G.R., Hollender, R. and Peterson, D.G. (2004). Influence of hazelnut paste on the sensory properties and shelf-life of dark chocolate. *Journal of Sensory Studies*, 19(2): 133–148.

Nishida, Y., Tada, N., Inui, T., Kageyama, N., Furukubo, S., Takaoka, S. and Kawasaki, Y. (2005). Innovative control technology of malt components by use of a malt fractionation technique, *Proceedings of the 30th European Brewery Convention Congress*, Prague, Czech Republic.

Quast, D.G., Karel, M. and Rand, W.M. (1972). Development of a mathematical model for oxidation of potato chips as a function of oxygen pressure, extent of oxidation, and equilibrium relative humidity. *Journal of Food Science,* 37: 673–678.

Robertson, G.L. (2005). *Food Packaging Principles and Practice*. Taylor & Francis, Boca Raton, FL.

Sánchez-Bel, P., Egea, I., Romojaro, F. and Martínez-Madrid, M.C. (2008). Sensorial and chemical quality of electron beam irradiated almonds (*Prunus amygdalus*). *Science Direct*, 41: 442–449.

Sang, S., Lapsley, K., Jeong, W.S., Lachence, P.A., Ho, C.T. and Rosen, R.T. (2002a). Antioxidative phenolic compounds isolated from almond skins (*Prunus amygdalus Batsch*). *Journal of Agriculture and Food Chemistry*, 50: 2459–2463.

Sivertsvik, M., Rosnes, J.T. and Bergslien, H. (2002). Modified atmosphere packaging, in *Minimal Processing Technologies in the Food Industry*. Eds. T. Ohlsson and N. Bengtsson. Woodhead Publishing Ltd, Cambridge, England, U.K.

Spencer, K.C. and Humphreys, D.J. (2002). Argon packaging and processing preserves and enhances flavor, freshness, and shelf life of foods. *Freshness and Shelf Life of Foods,* Vol. 836, Eds. K. R. Cadwallader and H. Weenen, ACS Symposium Series, Washington, DC, pp. 270–291.

Springett, M.B., Williams, B.M. and Barnes, R.J. (1994). The effect of packaging conditions and storage time on the volatile composition of Assam black leaf tea. *Food Chemistry*, 49: 393–398.

Stubenrauch, C. (2005). Neue verpackungen für lebensmittel. *Chemie Unserer Zeit*, 39: 310–316.

Sulaeman, A., Keeler, L., Giraud, D.W., Taylor, S.L. and Driskell, J.A. (2003). Changes in carotenoid, physicochemical and sensory values of deep-fried carrot chips during storage. *International Journal of Food Science and Technology*, 38: 603–613.

Tabee, E. (2008). Lipid and phytosterol oxidation. In *Vegetable Oils and Fried Potato Products*. Doctoral Thesis, Swedish University of Agricultural Sciences. Uppsala, Sweden.

Teets, A.S. and Were, L.M. (2008). Inhibition of lipid oxidation in refrigerated and frozen salted raw minced chicken breasts with electron beam irradiated almond skin powder. *Meat Science*, 80: 1326–1332.

Ucherek, M. (2004). An integrated approach to factors affecting the shelf life of products in modified atmosphere packaging (MAP). *Food Reviews International*, 20(3): 297–307.

Wasserman, G.S., Bradbury, A., Cruz, A. and Penson, S. (2002). Coffee, in *Kirk-Othmer Encyclopedia of Chemical Technology*, Volume 7. Wiley & Sons, Inc., New York, pp. 250–272.

http://aem.asm.org/cgi/content/abstract/65/1/342

http://pffc-online.com/mag/paper_packaging_really_works/

http://scholar.lib.vt.edu/theses/available/etd-08252004-160303/unrestricted/Arrittdissertation04.pdf

http://www.coffee-makers-cafe.com/coffee-bean-storage-keep-fresh.html

http://www.foodproductdesign.com/articles/2003/06/fats146….cnange.aspx

http://www.google.com/books?hl=el&lr=&id=rI7b34Pwjv8C&oi=fnd&pg=PA208&dq=Devlieghere,+F.,+Gil,+M.I.+and+Debevere,+J.+Ghent+University+Modified+atmosphere+packaging+(MAP)+in+www&ots=0v0vGSr5ic&sig=zayN4y00CF8GyO-YXqbtHuRFzLg#v=onepage&q=Devlieghere%2C%20F.%2C%20Gil%2C%20M.I.%20and%20Debevere%2C%20J.%20Ghent%20University%20Modified%20atmosphere%20packaging%20(MAP)%20in%20www&f=false

http://www.packagingdigest.com/article/CA6580713.html

http://www.packagingdigest.com/article/CA6607575.html

http://www.pirainternational.com/businessintelligence/GrowthOpportunitiesinActiveandModifiedAtmospherePackaging.aspx

Part VI

Active Packaging and Its New Trends

14 Active and Intelligent Packaging

Ioannis S. Arvanitoyannis and Georgios Oikonomou

CONTENTS

14.1 INTRODUCTION

14.1.1 Role of Food Packaging

Packaging is of essential importance for achieving the primary objective of the food supply chain, feeding large numbers of consumers, each with individual consumption and shopping habits. One of the most important functions of packaging in food products is the function of preservation. Various methodologies of incorporating food preservation technologies in packaging technology have been developed over the years. This is of essential importance. Today's modern packaging technology has reduced losses of valuable food products during distribution to levels less than 1%. However, in developing countries due to poor packaging and storage, food losses can still reach levels of up to 50% losses that the community itself cannot afford, given the scarcity of food (Sonneveld, 2000).

Packaging has a significant role in the food supply chain, and it is an integral part both of the food processes and the whole food supply chain. Food packaging has to perform several tasks as well as fulfilling many demands and requirements. Traditionally, a food package makes distribution easier. It has protected food from environmental conditions, such as light, oxygen, moisture, microbes, mechanical stresses, and dust. Other basic tasks have been to ensure adequate labeling for providing information for example, to the customer, and a proper convenience to the consumer, for example, easy opening, reclosable lids, and a suitable dosing mechanism (Ahvenainen, 2003). The interactions between package, food, and environment are shown in Figures 14.1 through 14.3.

14.1.2 History of Packaging

Very early in time, food was consumed where it was found. Families and villages were self-sufficient, making and catching what they used. When containers were needed, nature provided gourds, shells, and leaves to use. Later, containers were fashioned from natural materials, such as hollowed logs, woven grasses, and animal organs.

Fabrics descended from furs were used as primitive clothing. Fibers were matted into felts by plaiting or weaving. These fabrics were made into garments, used to wrap products, or formed into bags. With the weaving process, grasses, and later reeds, were made into baskets to store food surpluses. Some foods could then be saved for future meals and less time was needed for seeking and gathering food.

As ores and compounds were discovered, metals and pottery were developed, leading to other packaging forms (Hook and Heimlich in http://ohioline.osu.edu/cd-fact/0133.html.).

Packaging is used for several purposes:

- To contain products, defining the amount the consumer will purchase
- To protect products from contamination, from environmental damage, and from theft
- To facilitate transportation and storing of products
- To carry information and colorful designs that make attractive displays.

(Berger in http://edis.ifas.ufl.edu/pdffiles/AE/AE20600.pdf)

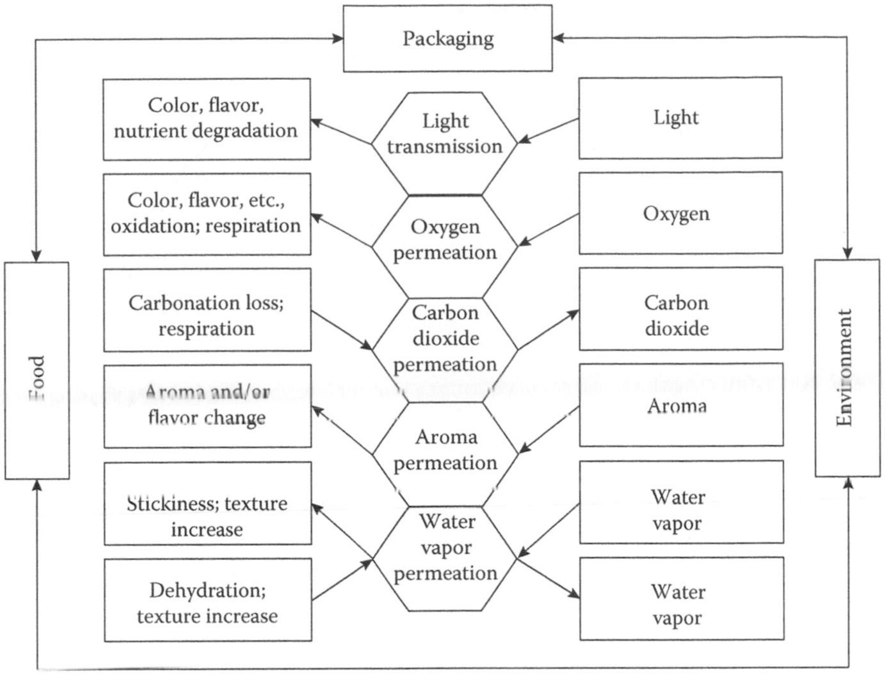

FIGURE 14.1 Interactions between package, food, and environment.

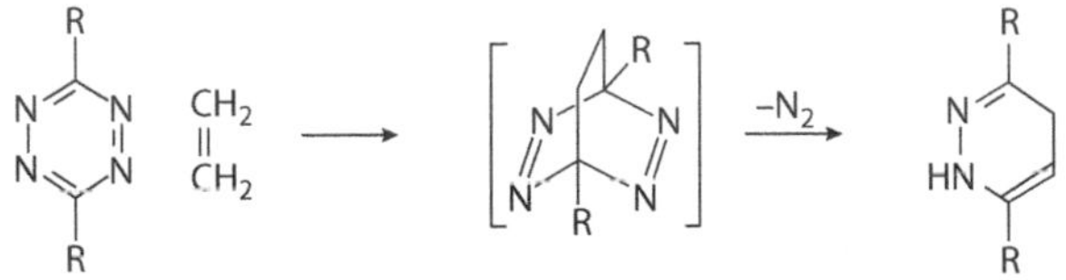

FIGURE 14.2 Reaction scheme for ethylene-scavenging (Food Science Australia).

An approximate chronology of food packaging is as follows:

- 20,000 years ago—modified natural materials—grass, reeds, skins
- 8,000 years ago—ceramics, amphorae—developed in the Middle East
- 5,000 years ago—wood, barrels, boxes, crates—wooden boxes found in Egyptian tombs
- 3,500 years ago—mass-produced ceramics, pottery—invention of the pottery wheel
- 2,500 years ago—glass containers—glass blowing developed by the Phoenicians and Syrians
- 2,000 years ago—paper and cellulose fibers—not true paper

The last 1000 years have seen many changes and advances in packaging as a result of huge social change.

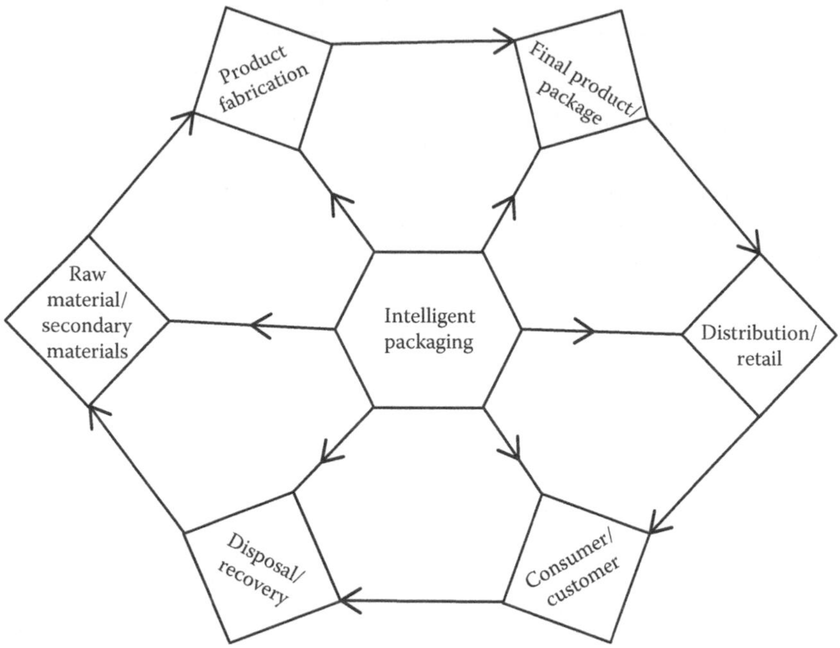

FIGURE 14.3 Material and information flow in the food supply chain cycle.

Notable advances in packaging in the twentieth century include

- Aluminum foil containers—1950s
- Aluminum cans—1959
- Cellulose packaging—1950s
- Heat shrinkable plastic films—1958
- Styrene foam—1930 to 1950s
- PETE (polyethylene terephthalate) containers—1977

(http://www.reducepackaging.com/history.html)

The history of food packaging can be used to study the development of civilization. At any time, food packaging reflects evolution in science, technology, art, psychology, sociology, politics, and law. The earliest packaging consisted of leaves, shells, gourds, animal skins and bladders, and even human skulls to contain and transport foods short distances from harvest. Woven baskets, leather bags, clay pots, glass containers, wood barrels, cloth sacks, and paper wraps reflect later development of human skills, tools, and discoveries about materials. These later packages not only still functioned mainly to contain and transport food but also provided increased protection (Krochta, 2006a,b).

14.1.3 Packaging Today

Nowadays, there is strong increase in new technologies for food packaging, which offer many advantages to the products, like safety and quality. The most important types of package are

- Active packaging
- Intelligent or smart packaging
- MAP

The emphasis of this chapter is on active and intelligent packaging.

Active and intelligent packaging systems can provide several benefits to the quality and safety of food. The active systems aim at shelf-life extension of the food products by keeping their quality for longer (e.g., oxygen absorbers in nuts to prevent rancidity, caused by fat oxidation). The intelligent systems are aiming to monitor the quality of the food product or its surrounding environment to predict or measure the safe shelf life better than a best-before date (De Jong et al., 2005).

Recent technological developments have allowed the food industry to create active packaging to prolong food quality and shelf life. Active packaging interacts with food to reduce oxygen levels or add flavorings or preservatives. Intelligent packaging can monitor the food and transmit information on its quality (http://74.125.39.104/search?q=cache:D-l-tvbE3uwJ:europa.eu/rapid/pressReleasesAction.do%3Freference%3DIP/03/1554%26format%3DPDF%26aged%3D1%26language%3DEN%26guiLanguage%3Den+IP/03/1554&hl=el&ct=clnk&cd=2&gl=gr).

According to the definitions of the Actipak project, active and intelligent packagings have the following characteristics (Ahvenainen, 2003):

- Active packaging changes the condition of the packed food to extend the shelf life or to improve the safety or sensory properties, while maintaining the quality of the packaged food.
- Intelligent packaging systems monitor the condition of packaged foods to provide information about the quality of the packaged food during transport and storage.

14.2 ACTIVE PACKAGING

The recognition of active packaging as a generic approach is a relatively new occurrence, as evidenced by the earliest reviews bringing together the concepts, even if using different descriptions—such as "smart" (Sacharow, 1988; Labuza and Breene, 1989). The field has been developing largely as a series of niche markets owing to the current approach of the package converting and resin industries of viewing it in terms of a series of market opportunities. The user industries, typified by the food industry, have presented these opportunities in isolation, and this may continue for some years. The approach of considering a range of packaging options (both passive and active) as a whole is not yet common practice.

Many of the developments have been logical consequences of earlier commercial products or of noncommercial research publications. These are, however, some concepts that appear to have established new lines of investigation or commercial development. Any choice of this type is necessarily subjective (Rooney, 2005).

Active packaging is defined as an intelligent or smart system that involves interactions between package or package components and food or internal gas atmosphere and complies with consumer demands for high quality, fresh-like, and safe products.

Active packaging extends the shelf life of foods, while maintaining their nutritional quality, inhibiting the growth of pathogenic and spoilage microorganisms, preventing and/or indicating the migration of contaminants, and displaying any package leaks present, thus ensuring food safety (Ozdemir and Floros, 2004).

Active packaging performs some desired role other than providing an inert (passive) barrier to external conditions. Active packaging concepts thus enhance the performance of the package, by changing the condition of the packaged food to improve quality and shelf life. With consumer interest in ever higher quality and safety in foods, active packaging is a field of high interest and development.

14.2.1 Protective Active Packaging

Most active packaging concepts enhance the protective function of food packaging. Protective active packaging approaches include oxygen-scavenger sachets, labels, closure liners, and films that complement the oxygen-barrier property of the package. Such concepts could increase the protection of oxygen-sensitive frozen foods.

14.2.2 Convenience Active Packaging

A number of active packaging concepts enhance the convenience of packaged frozen foods. Packaging that is stable to the conditions of a microwave or conventional oven (dual-ovenable) can serve as a convenient container for preparation, service, and consumption of frozen foods. Incorporation of susceptors in microwavable packaging allows crisping and browning of the food (Sun, 2006).

Active packaging includes components of packaging systems that are capable of acting as (Ozdemir and Floros, 2004) the following:

- Oxygen scavengers
- Carbon dioxide scavengers/emitters
- Moisture absorbers
- Ethylene and flavor/odor taints
- Releasing carbon dioxide
- Ethanol emitters
- Antioxidants and other preservatives
- Maintaining temperature control compensating for temperature changes
- Microwave

Some examples of active packaging systems are given in Table 14.1.

14.2.3 Antimicrobial Packaging

Antimicrobial films can be classified into two types:

1. Those that contain an antimicrobial agent that migrates to the surface of the food
2. Those that are effective against the surface growth of microorganisms without migration

TABLE 14.1
Examples of Active Packaging Systems

Active Packaging System	Mechanisms	Purpose	Food Applications
Oxygen scavengers	Iron based Metal/acid Nylon MXD6 Metal (e.g., platinum) catalyst Ascorbate/metallic salts Enzyme based	Reduction/preventing of mold, yeast, and aerobic bacteria growth Prevention of oxidation of fats, oils, vitamins, colors Prevention of damage by worms, insects, and insect eggs	Bread, cakes, cooked rice, biscuits, pizza, pasta, cheese, cured meats and fish, coffee, snack foods, dried foods, and beverages
Carbon dioxide	Calcium hydroxide, sodium hydroxide, or potassium hydroxide Calcium oxide/activated charcoal Ferrous carbonate/metal halide	Removing of carbon dioxide formed during storage in order to prevent bursting of a package	Coffee, fresh meats and fish, nuts, and other snack food products and sponge cakes
Ethylene scavengers	Potassium permanganate Activated carbon Activated clays/zeolites	Prevention of too fast ripening and ripening and softening	Fruit, vegetables, and other horticultural products
Preservative releasers	Organic acids Silver zeolite Spice and herb extracts BHA/BHT antioxidants Vitamin E antioxidant Chlorine dioxide/sulfur dioxide	Oxidation	Cereals, meats, fish, bread, cheese, snack foods, fruit, and vegetables
Ethanol emitters	Encapsulated ethanol	Sterilization	Pizza crusts, cakes, bread, biscuits, fish, and bakery products
Humidity absorbers (drip-absorbent sheets, films, sachets)	Polyacrylates (sheets) Propylene glycol (film) Silica gel (sachet) Clays (sachet)	Control of excess moisture in packed food Reduction of water activity on the surface of food in order to prevent the growth of molds, yeast, and spoilage bacteria	Meat, fish, poultry, bakery products, and cuts of fruits and vegetables

(continued)

TABLE 14.1 (continued)
Examples of Active Packaging Systems

Active Packaging System	Mechanisms	Purpose	Food Applications
Moisture absorbers	PVA blanket Activated clays and minerals Silica gel	Prevention of microbial growth Removal of dripping water Prevention of fogging	Fish, meats, poultry, snack foods, cereals, dried foods, sandwiches, fruit, and vegetables
Flavor/odor absorbers	Cellulose acetate film containing naringinase enzyme Ferrous salt and citric or ascorbic acid (sachet) Specially treated polymers	Reduction of bitterness in grapefruit juice Improving the flavor of fish and oil-containing food	Fruit juices, fish, oil-containing foods such as potato chips, biscuits, and cereal products, Beer
Temperature control packaging	Nonwoven plastics Double-walled containers Hydrofluorocarbon gas Quicklime/water Ammonium nitrate/water Calcium chloride/water Super corroding alloys/salt water Potassium permanganate/glycerin	Temperature stability	Ready meals, meats, fish, poultry, and beverages
UV-light absorbers	Polyolefins like polyethylene and polypropylene doped the material with a UV-absorbent agent	Restricting light-induced oxidation	Light-sensitive foods such as ham, drinks
Lactose remover	Crystallinity modification of nylon 6 UV stabilizer in polyester bottles Immobilized lactase in the packaging material	Serving milk products to the people suffering lactose intolerance	Milk and other dairy products
Cholesterol remover	Immobilized cholesterol reductase in the packaging material	Improving the healthiness of milk products	Milk and other dairy products

Source: Ahvenainen, R., *Novel Food Packaging Techniques*, Woodhead Publishing Limited, Cambridge, U.K., 2003; Day, B.P.F., Active packaging of food, in *Smart Packaging Technologies for Fast Moving Consumer Goods*, J. Kerry and P. Butler, eds., John Wiley & Sons, Ltd, New York, 2008, pp. 1–18.

TABLE 14.2
Patents for Oxygen-Scavenging Systems

Patent Number	Title	Issue Date	Reference
4536409	Oxygen scavenger	Aug 20, 1985	Farrell and Tsai (1985)
4702966	Oxygen scavenger	Oct 27, 1987	Farrell and Tsai (1987)
5284871	Oxygen removal	Feb 8, 1994	Graf (1994)
5364555	Polymer compositions containing salicylic acid chelates as oxygen	Nov 15, 1994	Zenner et al. (1994)
5648020	Oxygen-scavenging composition for low-temperature use	Jul 15, 1997	Speer and Roberts (1997)
5667863	Oxygen-absorbing label	Sep 16, 1997	Cullen et al. (1997)
5744246	Oxygen-scavenging ribbons and articles employing the same	Apr 28, 1998	Ching (1998)
5766473	Enzyme-loaded hydrophilic porous structure for protecting oxygen	Jun 16, 1998	Strobel and Gagnon (1998)
5798055	Oxygen-scavenging metal-loaded ion-exchange compositions	Aug 25, 1998	Blinka et al. (1998)
5806681	Article for scavenging oxygen from a container	Sep 15, 1998	Frisk (1998)
WO/1999/048963	Oxygen scavengers with reduced oxidation products for use in plastic films and beverage and food containers	Sep 9, 1999	Ching et al. (1999)
5952066	Transparent package with aliphatic polyketone oxygen scavenger	Sep 14, 1999	Schmidt et al. (1999)
6063503	Oxygen-absorbing multilayer film and method for preparing the same	May 16, 2000	Hatakeyama and Takahashi (2000)
6083585	Oxygen-scavenging condensation copolymers for bottles and packaging	Jul 4, 2000	Cahill and Chen (2000)
6086786	Oxygen-scavenging metal-loaded ion-exchange composition	Jul 11, 2000	Blinka et al. (2000)
6254804	Oxygen scavengers with reduced oxidation products for use in plastic	Jul 3, 2001	Matthews and Depree (2001)
6391403	Zeolite in packaging film	May 21, 2002	Blinka et al. (2002)
6423776	Oxygen-scavenging high-barrier polyamide compositions for packaging	Jul 23, 2002	Akkapeddi et al. (2002)
6544611	Oxygen-scavenging PET-based polymer	Apr 8, 2003	Schiraldi et al. (2003)
7078100	Oxygen scavenger compositions derived from isophthalic acid	Aug 28, 2003	Ebner et al. (2003)
6689437	Oxygen-absorbing material	Feb 10, 2004	Ubara et al. (2004)

(continued)

TABLE 14.2 (continued)
Patents for Oxygen-Scavenging Systems

Patent Number	Title	Issue Date	Reference
6709724	Metal catalyzed ascorbate compounds as oxygen scavengers	Mar 23, 2004	Teumac et al. (2004)
6793994	Oxygen-scavenging polymer compositions containing ethylene vinyl alcohol copolymers	Sep 21, 2004	Tsai and Akkapeddi (2004)
7288586	Polyester-based cobalt concentrates for oxygen-scavenging compositions	Oct 30, 2007	Stewart et al. (2007)
7452601	Oxygen scavenger compositions derived from isophthalic acid or terephthalic acid monomer or derivatives thereof	Nov 18, 2008	Ebner et al. (2008)

For both concepts, an intense contact between the food surface and the packaging film is required, and therefore, potential food applications are skin- and vacuum-packaged food products. An exception is made for volatile compounds that need no contact with the food. The major potential food applications for antimicrobial films include meat, fish, poultry, bread, cheese, fruits, and vegetables. Food packages can be made antimicrobial by incorporation and immobilization of antimicrobial agents or by surface coating. Some polymers are inherently antimicrobial, such as chitosan film. Another possible application of antimicrobial packaging is sachets that are enclosed or attached to the interior of a package, such as ethanol vapor generators (Heirlings, 2003).

Antimicrobial packaging is one of many applications of active packaging (Floros et al., 1997). Active packaging is the packaging system that possesses attributes beyond basic barrier properties, which are achieved by adding active ingredients in the packaging system and using actively functional polymers (Han and Rooney, 2002). Antimicrobial packaging is a packaging system that is able to kill or inhibit spoilage and pathogenic microorganisms that contaminate foods. The new antimicrobial function can be achieved by adding antimicrobial agents in the packaging system and using antimicrobial polymers that satisfy conventional packaging requirements. When the packaging system acquires antimicrobial activity, the packaging system (or material) limits or prevents microbial growth by extending the lag period and reducing the growth rate or decreases live counts of microorganisms (Han, 2000).

The primary goals of an antimicrobial packaging system are (Han, 2003) as follows:

- Safety assurance
- Quality maintenance
- Shelf-life extension

14.2.4 Oxygen Scavengers

Oxygen scavenging is today the most important type of active packaging from a commercial point of view. If one looks at the feasible commercial solutions for oxygen scavenging, the following approaches have been used until now:

- The presence of additional sachets containing oxygen-scavenging material, similar, for example, to the familiar sachets of desiccant in packaging containing electronic goods
- The presence of internal labels, or the packaging materials themselves, which contain the O_2 scavengers in very layers (Goldhan, 2003)

The main cause of most food spoilage is oxygen, since its presence allows a myriad of aerobic food-spoiling microorganisms to grow and thrive. Oxygen also spoils many foods through enzyme-catalyzed reactions, as in the browning of fruit and vegetables, destruction of ascorbic acid, and the oxidation of a wide range of flavors.

Thus, keeping food chilled in an ambient atmosphere that is low in oxygen features strongly in most current, popular methods of food packaging and storage (Mills, 2005).

A patent assigned to Grace and Company (1969), an early participant in oxygen-scavenging package material structures, described the incorporation of lipid antioxidant *n*-propyl gallate into both flexible and thermoformable plastic package materials. The development was for polyolefins such as polyethylene and polypropylene and especially for polyvinylidene chloride (PVdC) (saran) film. The antioxidant could be incorporated into two-ply or three-ply structures. An example would be two layers of PVdC sandwiching the antioxidant. Furthermore, cited as antioxidants were butylated hydroxyl anisole (BHA), BHT, and dihydroguaretic acid, among others, with the preference being for *n*-propyl gallate (Brody et al., 2001).

The presence of air or oxygen increases the corrosiveness of a system, for example, drilling fluids or air drilling systems used in drilling oil and gas wells, brines employed in the secondary recovery of petroleum by water flooding and in the disposal of waste water and brines from oil and gas wells, steam generating systems, water circulating systems, automobile radiators, diesel locomotive engines, boiler water, sea water ship ballast, etc. Infact, in any system, the presence of oxygen or air causes or increases corrosion. Therefore, it is highly desirable to remove oxygen from such systems. The practice of removing dissolved oxygen from such systems is so well known that the agent employed to achieve this result is known as an oxygen scavenger (Redmore and Mo, 1973).

An oxygen scavenger comprises an oxygen-absorbent composition and an oxygen-permeable film covering the oxygen-absorbent composition and including an asymmetric porous membrane, whose outer surface portion in the thickness direction of the asymmetric porous membrane is formed as a dense skin layer (Yamada et al., 1992).

The oxygen-scavenging reaction may be summarized as follows (Redmore and Mo, 1973):

$$2H_2S + O_2 \xrightarrow[Na_3VO_4]{AQ} 2S + 2H_2O \qquad (14.1)$$

$$H_2NNH_2 + O_2 \xrightarrow[Na_3VO_4]{AQ} N_2 + 2H_2O \qquad (14.2)$$

By the use of an O_2 scavenger, which absorbs the residual O_2 after packaging, quality changes of O_2-sensitive foods can often be minimized. In general, existing O_2-scavenging technologies utilize one or more of the following concepts (Vermeiren et al., 1999):

- Iron powder oxidation
- Ascorbic acid oxidation
- Photosensitive dye oxidation
- Enzymatic oxidation (e.g., glucose oxidase and alcohol oxidase)
- Unsaturated fatty acids (e.g., oleic acid or linolenic acid)
- Immobilized yeast on a solid material

The majority of currently commercially available O_2 scavengers are based on the principle of iron oxidation:

$$Fe \rightarrow Fe^{2+} + 2e^- \qquad (14.3)$$

$$\frac{1}{2}O_2 + H_2O + 2e^- \rightarrow 2OH^- \qquad (14.4)$$

$$Fe^{2+} + 2OH^- \rightarrow Fe(OH)_2 \qquad (14.5)$$

$$Fe(OH)_2 + \frac{1}{4}O_2 + \frac{1}{2}H_2O \rightarrow Fe(OH)_3 \qquad (14.6)$$

In enzymatic oxygen-scavenging systems, an enzyme reacts with a substrate to scavenge oxygen. These systems are more expensive than iron-based systems, due to the cost of enzymes used for the oxygen-scavenging purpose. Enzymatic oxidation systems are also usually very sensitive to temperature, pH, water activity, and solvent/substrate present in the sachet, thus limiting the widespread use of these enzyme-based systems.

Oxygen-scavenging sachets are not appropriate for liquid foods, because the direct contact of the liquid with the sachet usually causes the spillage of sachet contents. In addition, sachets may cause accidental consumption with the food or may be ingested by children (Ozdemir and Floros, 2004).

A list of representative patents for oxygen-scavenging systems is given in Table 14.3.

TABLE 14.3
Commercial Carbon Dioxide Generators

Patent Number	Title	Issue Date	Reference
3348922	Carbon dioxide generator	Oct 1967	Bose et al. (1967)
D220454	Generator for compressing carbon dioxide	Apr 1971	Klopp (1971)
4506473	Carbon dioxide generator insect attractant	Mar 26, 1985	Waters (1985)
D293249	Carbon dioxide compressing generator for dry ice manufacture	Dec 15, 1987	Milden (1987)
4786519	Delayed reaction carbon dioxide generator package	Nov 22, 1988	Gupta (1988)
D298754	Carbon dioxide compressing generator for dry ice manufacture	Nov 29, 1988	Broeils (1988)
5267455	Liquid/supercritical carbon dioxide dry cleaning system	Dec 7, 1993	Dewees et al. (1993)
5412958	Liquid/supercritical carbon dioxide/dry cleaning system	May 9, 1995	Iliff et al. (1995)
5669251	Liquid carbon dioxide dry cleaning system having a hydraulically	Sep 23, 1997	Townsend and Purer (1997)
5783082	Cleaning process using carbon dioxide as a solvent and employing	Jul 21, 1998	DeSimone et al. (1998)
5904737	Carbon dioxide dry cleaning system	May 18, 1999	Preston and Turner (1999)
6004400	Carbon dioxide cleaning process	Dec 21, 1999	Bishop and Harrover (1999)
6117190	Removing soil from fabric using an ionized flow of pressurized gas	Sep 12, 2000	Chao et al. (2000)
6129451	Liquid carbon dioxide cleaning system and method	Oct 10, 2000	Rosio and Fulton (2000)
6148644	Dry cleaning system using densified carbon dioxide and a surfactant adjunct	Nov 21, 2000	Jureller et al. (2000)
6182318	Liquified gas dry-cleaning system with pressure vessel temperature compensating compressor	Feb 6, 2001	Roberts and Keqler (2001)
6200393	Carbon dioxide cleaning and separation systems	Mar 13, 2001	Romack et al. (2001)
6216302	Carbon dioxide dry cleaning system	Apr 17, 2001	Preston and Turner (2001)
6233772	Carbon dioxide cleaning apparatus with rotating basket and external drive	May 22, 2001	McClain and Schrebe (2001)
6264753	Liquid carbon dioxide cleaning using agitation enhancements at low temperature	Jul 24, 2001	Chao et al. (2001)
6299652	Method of dry cleaning using densified carbon dioxide and a surfactant	Oct 9, 2001	Jureller and Kerschner (2001)
6442980	Carbon dioxide dry cleaning system	Sep 3, 2002	Preston et al. (2002)

(continued)

TABLE 14.3 (continued)
Commercial Carbon Dioxide Generators

Patent Number	Title	Issue Date	Reference
6711773	Detergent injection methods for carbon dioxide cleaning apparatus	Mar 30, 2004	DeYoung et al. (2004)
7445602	Carbon dioxide sensor and airway adapter incorporated in the same	Nov 4, 2008	Yamamori et al. (2008)

Selecting the right type of oxygen scavenger: Oxygen scavengers must satisfy several requirements. They must

1. Be harmless to the human body. Though the oxygen scavengers themselves are neither food nor food additives, they are placed together with food in a package, and there is therefore the possibility of accidental intake by consumers.
2. Absorb oxygen at an appropriate rate. If the reaction is too fast, there will be a loss of oxygen absorption capacity during introduction into the package. If it is too slow, the food will not be adequately protected from oxygen damage.
3. Not produce toxic substances or unfavorable gas or odor.
4. Be compact in size and are expected to show a constant quality and performance.
5. Absorb a large amount of oxygen.
6. Be economically priced (Nakamura and Hoshino, 1983; Abe et al., 1994; Rooney, 1995).

14.2.5 Carbon Dioxide Scavengers

CO_2 scavengers, another active packaging technology, used not as much as O_2 scavengers, are commercialized to avoid gas pressure buildup inside rigid packaging or volume expansion in flexible packaging by absorbing CO_2 produced by fermented or roasted foods (Lee et al., 2001). CO_2 scavengers can be composed either of a physical absorbent (zeolite or an active carbon powder) or of a chemical absorbent (calcium hydroxide, Na_2CO_3, $Mg(OH)_2$, etc.) (Charles et al., 2006).

Carbon dioxide is known to suppress microbial activity. Relatively high CO_2 levels (60%–80%) inhibit microbial growth on surfaces and, in turn, prolong shelf life. Therefore, a complementary approach to O_2 scavenging is the impregnation of a packaging structure with a CO_2-generating system or the addition of the latter in the form of a sachet. Table 14.3 lists the main commercial CO_2 generators (Suppakul et al., 2003).

The reactant commonly used to scavenge CO_2 is calcium hydroxide, which, at a high enough water activity, reacts with CO_2 to form calcium carbonate (Vermeiren et al., 2003):

$$Ca(OH)_2 + CO_2 \rightarrow CaCO_3 + H_2O$$

A disadvantage of this CO_2-scavenging substance is that it scavenges carbon dioxide from the package headspace irreversibly and results in depletion of CO_2, which is not always desired. In the case of packaged kimchi, depletion of CO_2 in the kimchi juices causes loss of the product's characteristic fresh carbonic taste. Therefore, reversible absorption or adsorption by physical sorbents such as zeolites and active carbon may be an alternative (Lee et al., 2001).

Since the permeability of CO_2 is three to five times higher than that of O_2 in most plastic films, it must be continuously produced to maintain the desired concentration within the package. High CO_2 levels may, however, cause changes in the taste of products and the development of undesirable anaerobic glycolysis in fruits. Consequently, a CO_2 generator is only useful in certain applications such as fresh meat, poultry, fish, and cheese packaging (Floros et al., 1997).

A list of patents for carbon dioxide scavengers is given in Table 14.4.

14.2.6 Ethylene Scavengers

Ethylene is a growth hormone that functions in the sprouting of plant seedlings, the growth of plants, and the growth of fruit. It helps to accelerate ripening in fruit, followed by aging and ultimately death. Ethylene production is a biochemical process, independent of respiration that occurs in each living cell for the purpose of producing energy (Brody et al., 2001).

Packaging technologies designed to scavenge or absorb ethylene from the surrounding environment of packaged produce have also been developed. The most widely used ethylene-scavenging packaging technology today is in the form of a sachet containing potassium permanganate immobilized on an inert porous support, such as alumina and silica, at a level of about 5% w/w. The ethylene is scavenged through an oxidation reaction with the potassium permanganate to form carbon dioxide and water. Although these permanganate-based ethylene-scavenging sachets are effective at removing ethylene, their use is sometimes accompanied by undesirable effects. These include possible migration of the potassium permanganate from the sachet onto the produce, lack of specificity to ethylene resulting in desirable aromas being scalped, and a general lack of user enthusiasm for the use of sachets. Other types of sachet-based ethylene-scavenging technologies utilize activated carbon with a metal catalyst (e.g., palladium), such as SendoMate from Mitsubishi Chemicals and Neupalon from Sekisui Jushi Ltd (Scully and Horsham, 2007).

Many C_2H_4-adsorbing substances have been described in the patent literature, but those that have been commercialized are based on potassium permanganate ($KMnO_4$), which oxidizes C_2H_4 in a series of reactions to acetaldehyde and then acetic acid, which, in turn, can be further oxidized to CO_2 and H_2O:

$$3C_2H_4 + 2KMnO_4 + H_2O \rightarrow 2MnO_2 + 3CH_3CHO + 2KOH \quad (14.7)$$

$$3CH_3CHO + 2KMnO_4 + H_2O \rightarrow 3CH_3COOH + 2MnO_2 + 2KOH \quad (14.8)$$

TABLE 14.4
Patents of UV Absorbers

Patent Number	Title	Issue Date	Reference
3328491	UV- light-absorbing copolymers of acryl-oxymethyl benzoates and dihydroxy-benzophenone derivatives	May 1, 1964	Fertig et al. (1964)
4301267	UV light stable copolymer compositions comprising monomers that are.alpha., 62-unsaturated dicarboxylic acid half-esters of 2-hydroxy, alkoxy, methylolbenzophenones, and styrene-butadiene comonomers	Nov 17, 1981	Barabas et al. (1981)
4528311	UV-absorbing polymers comprising 2-hydroxy-5-acrylyloxyphenyl-2H-benzotriazoles	Jul 9, 1985	Beard et al. (1985)
4716234	UV-absorbing polymers comprising 2-(2′-hydroxy-5′-acryloyloxyalkoxyphenyl)-2H-benzotriazole	Dec 29, 1987	Dunks et al. (1987)
4845180	UV- light-absorbing compounds, compositions and methods for making same	Jul 4, 1989	Henry and Reich (1989)
5133745	UV-absorbing hydrogels	Jul 28, 1992	Falcetta et al. (1992)
5459222	UV-absorbing polyurethanes and polyesters	Oct 17, 1995	Rodgers et al. (1995)
5806834	UV-absorbing polymer film	Sep 15, 1998	Yoshida (1998)
6244707	UV-blocking lenses and material containing benzotriazoles and benzophenones	Jun 12, 2001	Faubl (2001)
6242597	Trisaryl-1,3,5-triazine UV light absorbers	Jun 5, 2001	Gupta et al. (2001)
6252032	UV-absorbing polymer	Jun 26, 2001	Van Antwerp and Yao (2001)
6468609	UV-absorbing film and its use as protective sheet	Oct 22, 2002	Mariën and Moeyersons (2002)
6773104	UV filter coating	Aug 10, 2004	Cornelius and Torii (2004)
6872766	UV light filter element	Mar 29, 2005	Schunk et al. (2005)
7014797	Low-color UV absorbers for high UV wavelength protection	Mar 21, 2006	Danielson et al. (2006)
7381762	UV-absorbing compounds and compositions containing UV-absorbing compounds	Jun 3, 2008	Xia and Moore (2008)

$$3CH_3COOH + 8KMnO_4 \rightarrow 6CO_2 + 8MnO_2 + 2H_2O \tag{14.9}$$

$$\text{Overall}: 3C_2H_4 + 12KMnO_4 \rightarrow 12MnO_2 + 12KOH + 6CO_2 \tag{14.10}$$

As ethylene in the atmosphere surrounding the produce increases, the plant's respiration rate increases. If the ethylene level of the surrounding environment is maintained low, respiration slows. The effect on the respiration rate depends primarily on the relative concentration compared to the ethylene emission of the plant, rather than on the absolute ethylene concentration.

When ethylene is removed from the fresh fruit or vegetable environment, the ripening and deterioration processes of plant products are slowed, and so the storage life in extended (Brody et al., 2001).

Ethylene accelerates the respiration rate, resulting in shorter shelf life; therefore, storing fruits and vegetables in refrigeration temperatures is recommended to lower the respiration rate. Ethylene production is also enhanced with increasing amounts of oxygen and it is important to reduce oxygen, but not to levels that inhibit fruit and vegetable respiration or produce anaerobic conditions. Carbon dioxide may inhibit the production of ethylene because it binds ethylene-binding sites of specific enzymes, thus preventing ethylene-induced conformational changes of the tissue. Consequently, reduced oxygen and ethylene concentrations, combined with increased CO levels, may avoid the accumulation of ethylene and in turn prevent accelerated deterioration of the plant tissues (Yuan, 2003).

14.2.7 Flavor Odor/Absorbers

As far as food aromas are concerned, plastics are usually considered to have a negative impact on food quality. Flavor scalping, that is, sorption of food flavors by polymeric packaging materials, may result in loss of flavor and taste intensities and changes in the organoleptic profile of foods. However, flavor sorption could be used in a positive way to selectively absorb unwanted odors or flavors. Odor removers have the potential to scavenge the malodorous constituents of both oxidative and nonoxidative biochemical deterioration (Vermeiren et al., 2003). Many foods such as fresh poultry and cereal products develop very slight but nevertheless detectable deterioration odors during product distribution, e.g., such as sulfurous compounds and amines from protein/amino acid breakdown or aldehydes and ketones from lipid oxidation or anaerobic glycolysis. These odors are trapped within gas-barrier packaging so that, when the package is opened, they are released and detected by consumers. Another reason for incorporating odor removers into packages is to obviate the effect of odors developed in the package materials themselves (Vermeiren et al., 1999; Brody et al., 2001).

Flavor enhancers are among the most controversial areas of active packaging due to concerns regarding their ability to mask product taint. In some applications, off-odors are the only by-products of food product spoilage, and their removal does impair the consumers' ability to recognize the point at which a food has become a safety hazard. These devices have been used to mask the impact on food flavor of plastic and other packaging materials by the release of neutralizing substances from the active element. Such devices have been applied to the positive effect of reinforcing and improving the flavors of foods, for example, to the packaging of vitamins with activated carbon used to absorb naturally occurring off-odors. Although much interest has been shown in this technology, relatively little R&D has been undertaken and this situation is likely to remain until the principle of flavor/odor absorption becomes more widely accepted (Sun, 2006).

14.2.8 Microwave

Microwave susceptors have seen significant commercial opportunities for many years. The active nature of these packages is that they become functional during the heating of the food and package in the microwave oven and can create greater uniformity of microwave heating and the possibility of browning and crisping the surface of microwave-heated foods. Due to the nature of microwave heating, these active packaging devises are particularly well suited to frozen foods.

14.2.9 Ethanol Emitters

Another approach to lengthen shelf life has been to spray ethanol or another alcohol on the surface of a food. The earliest work published was a Japanese patent (72JP-073439) issued to Showa Tansen Co. in which both ethanol and propylene glycol were sprayed onto the surface of bakery goods. Nippon Polycello Co. of Japan built machinery to do this using at least 2% ethanol in the gas space (78JP-016118). Another Japanese scientist in fact patented a process for sterilization of foods at reduced temperature by using ethanol, hydrogen peroxide, or propylene glycol in the sterilizing steam. Two U.S. companies researched this extensively in the 1970s (Keebler and Creative Crust) for application to shelf stable intermediate moisture pizza crusts with a_ws of 0.85–0.90. Since this constituted a new food additive application, they had to petition FDA (Labuza and Breene, 1989).

Ethanol has been used as a preservative for centuries. At high concentrations, ethanol denatures the proteins of molds and yeasts and, although the action is not as severe, it exhibits antimicrobial effects even at low levels. Spraying the foodstuff with ethanol prior to packaging can be used to obtain the desired effect, but in some cases a more practical option can be to use sachets generating ethanol vapor. A product called Ethicap consists of an ethanol/water mixture adsorbed onto a silicon dioxide powder, contained in a sachet of a laminate of paper and ethyl vinyl acetate copolymer. The odor of alcohol can be masked by adding traces of flavors, such as vanilla, to the sachet. Ethicap acts by absorbing moisture from the food and releasing ethanol vapor into the packaging headspace. The size of the sachet used depends on the water activity of the food and the desired shelf life on the product (Ohlsson and Bengtsson, 2002).

The size and capacity of the ethanol emitting sachet used depends on the weight of food the a_w of the food and the desired shelf life required. When food is packed with an ethanol-emitting sachet, moisture is absorbed by the food and ethanol vapor is released and diffuses into the package headspace. Ethanol emitters are used extensively in Japan to extend the mold-free shelf life of high-ratio cakes and other high-moisture bakery products by up to 2000% (Rooney, 1995; Day, 2003). Research has also shown that such bakery products packed with ethanol emitting sachets did not get as hard as the controls, and results were better than those using an oxygen scavenger alone to inhibit mold growth. Hence, ethanol vapor also appears to exert an antistaling effect in addition to its antimold properties. Ethanol emitting sachets are also widely used in Japan for extending the shelf life of semimoist and dry fish products (Rooney, 1995; Day, 2003).

14.2.10 UV Absorbers

Ultraviolet (UV) light blockers are often used in clear or tinted packaging to protect personal care, food, and beverage products from UV light degradation coming from sunlight or in-store fluorescent lighting. "UV light can degrade not only the color but also the flavor and nutritional value of beverages packaged in clear PET bottles. Fruit juices, tea, and sports drinks are particularly susceptible," explains Mark Jordan, vice president of marketing at Techmer PM. "As clear and transparent packaging materials gain more popularity and use, protecting products from UV damage becomes of critical importance," adds Ciba Specialty Chemicals. Ciba® Shelfplus® UV 1100 blocks 90% of UV light up to 390 nanometers (nm). Techmer PM recently introduced Techspere™ PTM 12125P UV light blocker for PET, which is specially formulated to minimize haze when processed at less than 288°C, says the company. ColorMatrix is commercializing new, unique, UV-blocking chemistries for PET bottles and films. The new chemistry is much stronger, blocking UV light over 390 nm, compared to the typical benzotriazole chemistry that blocks effectively up to 370–380 nm (http://www.sciencedirect.com/science?_ob=ArticleURL&_udi=B6VPY-4BYYGS0-X&_user=83475&_rdoc=1&_fmt=&_orig=search&_sort=d&view=c&_acct=C000059672&_version=1&_urlVersion=0&_userid=83475&md5=8526474d4c5fe666589a7b49b7fba0ad).

Some patents for UV absorbers are summarized in Table 14.5.

14.2.11 Temperature Control Packaging

The sensorial appreciation of a food product is highly dependent on its serving temperature. If the product is to be consumed directly from the packaging, it is desirable to use packages that help the product obtain its optimum temperature.

14.2.11.1 Self-Heating

Beverage packages with self-heating functions have been described by Katsura (1989). There is also a growing demand in the outdoor market for food packaging that cooks or prepares the food via built-in healing mechanisms. The principle of heating is based on the theory that heat is generated when certain chemicals arc mixed. Ready meals can be heated, without the aid of a conventional or a microwave oven, by mixing iron, magnesium, and salt water. The metals are supplied in a PET bag placed on a heat-resistant tray. Salt water, which is supplied in a separate bag, is added to the metal-containing bag and the actual food package is then placed on the tray. Within 15 min the food has reached a temperature of 60°C and is ready. Other

TABLE 14.5
Temperature Control Patents

Patent Number	Title	Issue Date	Reference
3273634	Self-sustaining temperature control package	Sep 1966	Snelling (1966)
3971876	Temperature control apparatus	Jul 27, 1976	Witkin and Bowles (1976)
4066868	Temperature control method and apparatus	Jan 3, 1978	Witkin and Bowles (1978)
4744408	Temperature control method and apparatus	May 17, 1988	Pearson and Cordino (1988)
5181214	Temperature stable solid-state laser package	Jan 19, 1993	Berger et al. (1993)
6009712	Temperature controller of optical module package	Jan 4, 2000	Ito and Funakawa (2000)
6038865	Temperature-controlled appliance	Mar 21, 2000	Watanabe et al. (2000)
6489793	Temperature control of electronic devices using power following feedback	Dec 3, 2002	Jones et al. (2002)
6976364	Temperature regulation of a sprayed fluid material	Dec 20, 2005	Bengtsson (2005)

containers with healing functions are based on the reaction between lime and water (Hurme et al., 2002).

A list of temperature control patents is given in Tables 14.6 and 14.7.

14.2.12 Cholesterol Removal

Several methods to reduce cholesterol include extraction with solvents (Larsen and Froning, 1981), supercritical fluid extraction (Arul et al., 1988), and adsorption with saponin to form complexes (Micich, 1990). However, a major disadvantage of solvent use is that other lipid-soluble components may be extracted along with cholesterol, and proteins may be denatured. Additionally, most solvents are relatively nonselective, remove flavor and nutritional components, and are costly (Ahn and Kwak, 1999).

For cholesterol removal, several methods are used (Roczniak et al., 1991):

- A first method for reducing the amount of cholesterol in foods is using a procedure known as steam stripping.
- Another method for removing cholesterol from edible fats is to extract the cholesterol containing fat with supercritical carbon dioxide.
- A further method for removing cholesterol comprises using a charcoal adsorption process wherein the fat is passed through a bed of charcoal that functions as an adsorbent material.
- Still another method for removing cholesterol is to add enzymes or microorganisms to the fat to convert the cholesterol to another compound.

TABLE 14.6
Cholesterol Removal Patents

Patent Number	Title	Issue Date	Reference
5045242	Removal of cholesterol from edible fats	Sep 3, 1991	Roczniak et al. (1991)
5061505	Process for the removal of cholesterol and/or cholesterol esters from foodstuffs	Oct 29, 1991	Cully et al. (1991)
5063077	Process for the removal of cholesterol and cholesterol esters from egg yolk	Nov 5, 1991	Vollbrecht et al. (1991)
WO/1991/016824	Cholesterol removal	Nov 14, 1991	Oakenfull et al. (1991)
5091203	Method for removing cholesterol from eggs	Feb 25, 1992	Conte et al. (1992)
WO/1992/005710	Enhanced cholesterol extraction from egg yolk	Apr 4, 1992	Lombardo (1992)
5128162	Method for removing cholesterol from edible oils	Jul 7, 1992	Wrezel et al. (1992)
5292546	Process for the removal of cholesterol from egg yolk	Mar 8, 1994	Cully and Vollbrecht (1994)
5302405	Method for removing cholesterol and fat from egg yolk by chelation and reduced-cholesterol egg product	Apr 12, 1994	Hsieh et al. (1994)
5316780	Method for extracting cholesterol from egg yolk	May 31, 1994	Stouffer et al. (1994)
5326579	Process to remove cholesterol from dairy products	Jul 5, 1994	Richardson and Jimenez-Flores (1994)
WO/1995/004473	Removal of cholesterol from edibles	Feb 16, 1995	Garti (1995)
5468511	Method for removal of cholesterol and fat from liquid egg yolk with recovery of free cholesterol as a by-product	Nov 21, 1995	Zeidler (1995)
5496637	High efficiency removal of low-density lipoprotein–cholesterol from whole blood	Mar 5, 1996	Parham et al. (1996)
5498437	Process for the removal of cholesterol derivatives from egg yolk	Mar 12, 1996	Kohlrausch et al. (1996)
6093434	Enhanced cholesterol extraction from egg yolk	Jul 25, 2000	Kijowski and Lombardo (2000)
6110517	Method for removing cholesterol from milk and cream	Aug 29, 2000	Kwak et al. (2000)
6933291	Cholesterol lowering supplement	Aug 23, 2005	Qi et al. (2005)

TABLE 14.7
Time–Temperature Indicators Patents

Patent Number	Title	Issue Date	Reference
3942467	Time–temperature indicator	Mar 9, 1976	Witonsky (1976)
3965741	Time–temperature indicator device and method	Jun 29, 1976	Wachtell and Jones (1976)
3999946	Time–temperature history indicators	Dec 28, 1976	Patel et al. (1976)
4042336	Time–temperature integrating indicator	Aug 16, 1977	Larsson (1977)
4195056	Vapor permeation time–temperature indicator	Mar 25, 1980	Patel (1980)
4339207	Temperature-indicating compositions of matter	Jul 13, 1982	Hof and Ulin (1982)
4488822	Time–temperature indicator	Dec 18, 1984	Brennan (1984)
4509449	Temperature–time limit indicator	Apr 9, 1985	Chalmers (1985)
4812053	Activatable time–temperature indicator	Mar 14, 1989	Bhattacharjee (1989)
5085801	Temperature indicators based on polydiacetylene compounds	Feb 4, 1992	Thierry and Moigne (1992)
5120137	Time–temperature-indicating device	Jun 9, 1992	Ou-Yang (1992)
5159564	Thermal memory cell and thermal system evaluation	Oct 27, 1992	Swartzel et al. (1992)
5182212	Time–temperature indicator with distinct end point	Jan 26, 1993	Jalinski (1993)
5529931	Time–temperature indicator for establishing lethality of high	Jun 25, 1996	Narayan (1996)
5637475	Time–temperature method for establishing lethality of high temperature	Jun 10, 1997	Narayan (1997)
5662419	Time–temperature monitor and recording device and method for using the same	Sep 2, 1997	Lamagna (1997)
5709472	Time–temperature indicator device and method of manufacture	Jan 20, 1998	Prusik et al. (1998)
5779364	Temperature-sensitive device for medicine containers	Jul 14, 1998	Cannelongo and Cugini (1998)
6103351	Time–temperature integrating indicator device	Aug 15, 2000	Ram et al. (2000)
6158381	Time–temperature indicator	Dec 12, 2000	Bray (2000)
6176197	Temperature indicator employing color change	Jan 23, 2001	Thompson (2001)
6214623	Time–temperature indicator devices	Apr 10, 2001	Simons and Weldy (2001)
6244208	Time–temperature integrating indicator device with barrier material	Jun 12, 2001	Qiu et al. (2001)
6435128	Time–temperature integrating indicator device with barrier material	Aug 20, 2002	Qiu et al. (2002)
6514462	Time–temperature indicator devices	Feb 4, 2003	Simons (2003)
6524000	Time–temperature indicators activated with direct thermal printing and methods for their production	Feb 25, 2003	Roth (2003)

TABLE 14.7 (continued)
Time–Temperature Indicators Patents

Patent Number	Title	Issue Date	Reference
6544925	Activatable time–temperature indicator system	Apr 8, 2003	Prusik et al. (2003)
6614728	Time–temperature integrating indicator	Sep 2, 2003	Spevacek (2003)
6737274	Comparator for time–temperature indicator	May 18, 2004	Wright (2004)
6916116	Time- or time–temperature-indicating articles	Jul 12, 2005	Diekmann and Bommarito (2005)
7290925	Full history time–temperature indicator system	Nov 6, 2007	Skjervold et al. (2007)
7360946	Time–temperature indicators linked to sensory detection	Apr 22, 2008	Tester and Al-Ghazzewi (2008)

14.3 INTELLIGENT PACKAGING

14.3.1 Definition of Intelligent Packaging

According to the American Heritage Dictionary, the word "intelligent" is defined as "showing sound judgment and rationality" and as "having certain data storage and processing capabilities." A prerequisite of making sound decisions is effective communication—the ability to acquire, store, process, and share information—and this is where IP can make a significant contribution.

To date, four types of applications of intelligent packaging systems are used to

1. Improve product quality and product value (quality indicators, temperature and time–temperature indicators (TTIs), and gas concentration indicators)
2. Provide more convenience (quality, distribution, and preparation methods)
3. Change gas permeability properties
4. Provide protection against theft, counterfeiting, and tampering (Rodrigues and Han, 2003)

Examples of intelligent packaging include (Brody, 2007)

- Time–temperature and other indicators that can imply/signal the user about the quality of the packaged product (reported as far back as the 1960s)
- A biosensor, in theory, which can inform the user of the growth of microorganisms or even a specific microorganism in the package
- A bar code to help communicate information for more precise reheating or cooking of the contained food in an appliance
- An ethylene sensor, probably for the ripeness of fresh fruit
- Nutritional attributes of the contained food
- Gas concentrations on modified atmosphere packages (coincided with the advent of oxygen scavengers during the late 1980s)

Intelligent or smart packaging systems monitor the condition of the packaged foods to give information about the quality of the packaged food during transport and storage. Intelligent packaging systems give information on product quality directly (freshness indicators), the package and its headspace gases (leak indicators), and the storage conditions of the package (TTIs). In a broader sense, an intelligent packaging can also provide information on the product and its origin as such and reveal, for example, rough physical handling of the package and protect the product from tampering and pilferage (Smolander, 2003).

Intelligent packaging is defined as a packaging system that is capable of carrying out intelligent functions (such as detecting, sensing, recording, tracing, communicating, and applying scientific logic) to facilitate decision making to extend shelf life, enhance safety, improve quality, provide information, and warn about possible problems. It is believed that the uniqueness of IP lies in its ability to communicate: since the package and the food move constantly together throughout the supply chain cycle, the package is the food's best companion and is in the best position to communicate the conditions of the food (Yam et al., 2005).

Intelligent (or smart) packaging can be divided into two types. "Simple" intelligent packaging contains components that sense the environment and communicate information important to proper handling of the food product. "Interactive" or "responsive" intelligent packaging has additional capability allowing response to environmental change and, thus, prevention of food deterioration.

There are several reasons for the bright future of intelligent packaging (Ahvenainen, 2003); the significance of freshness and safety will increase, demands of consumers will increase, globalization and expansion of the marketing area make logistic chains longer thereby placing more demands on traceability, and the facilitation of in-house control for industry and retailing in the complete food supply chain. Intelligent packaging can also monitor product quality and trace the critical points in the food supply chain.

Several intelligent packaging concepts involve sensors that provide information related to food quality. One category includes temperature sensors that indicate whether frozen food package has been exposed to temperatures above a critical limit. TTIs that provide time-integrated information about the temperature history of the product are also available. TTIs are often self-adhesive color-changing labels that respond gradually and irreversibly to the cumulative exposure of the product to temperature. TTIs can be matched to the specific shelf-life characteristics of each product. Such indicators allow more accurate assessment of the remaining product shelf life. However, the Arrhenius-type temperature behavior of TTIs does not take into account the concentration effect, ice-crystal growth, and glass transition phenomena of frozen foods.

Another category of intelligent packaging includes components that range from bar codes to radio frequency transmitters that allow accurate tracking of product for improved supply chain management and rapid traceability.

Intelligent packaging has been proposed for a future smart kitchen. The cooking appliance system would read a bar code that includes information on optimum cooking conditions and appropriately adjust the oven. The system could also read a TTI to alert the consumer to spoiled food (Sun, 2006).

The techniques of intelligent packaging with the greatest commercial value are (a) TTIs and (b) O_2 indicators.

14.3.2 Time–Temperature Indicators

This appliance should indicate any temperature abuse and, ideally, estimate the remaining shelf life of the foodstuff. There are two types of temperature indicators: those providing the entire temperature history to indicate the cumulative time–temperature exposure above a critical temperature (TTI) and those providing information as to whether the indicator has been exposed above or below a critical temperature (temperature indicators [TI]). Temperature indicators labeled on a package can inform of the heat load to which the package surface has been subjected in the distribution chain, usually expressed as a visible response in the form of mechanical deformation, color change, or color movement. Hundreds of patents for TIs and for TTIs have been accepted, but they have found little use in the marketplace. However, once the concerns have been addressed there is an enormous potential for the use of TTIs for foods. Their use would advantageously lead to increased quality, safety, and integrity of the packaged product (Hurme et al., 2002).

A TTI can be defined as a simple, inexpensive device that can show an easily measurable, time–temperature-dependent change that reflects the full or partial temperature history of a food product to which it is attached (Taoukis and Labuza, 1989). The principle of TTI operation is a mechanical, chemical, electrochemical, enzymatic, or microbiological irreversible change usually expressed as a visible response, in the form of a mechanical deformation, color development, or color movement. The rate of change is temperature dependent, increasing at higher temperatures. The visible response thus gives a cumulative indication of the storage conditions that the TTI has been exposed to. The extent to which this response corresponds to a real time–temperature history depends on the type of the indicator and the physicochemical principles of its operation. Indicators can thus be classified according to their functionality and the information they convey (Taoukis and Labuza, 2003).

TTI was brought forward 40 years ago, which was applied to biological bacterin, but it was not until the last century that TTI was applied to the food industry. TTI was mainly applied to reflect the time–temperature history of the chilled, frozen food that was sensitive to temperature, such as fresh milk, frozen fish, meat, and seafood. Moreover, TTI was also applied to assess the sterilization process, such as thermally processed milk, and to estimate the remaining shelf life of foodstuff and might be used to control the distribution chain of horticultural products (Yan ct al., 2008).

TTIs allow such control down to product unit level. TTI can show an easily measurable, time- and temperature-dependent change that cumulatively reflects the time–temperature history of the food product. A TTI-based management system aiming to improve both quality and safety in the food chill chain can be developed by applying the state of the art in TTI technology in conjunction with predictive models for microbial growth and risk evaluation. In order to use a TTI-based system accurately, mathematical models are needed that describe the effect of temperature on the behavior of fish spoilage bacteria assessing effects of dynamic storage conditions. In addition, a full kinetic study of the TTI response is needed. Based on reliable models of the shelf life and the kinetics both of the product and the TTI response, the integral effect of temperature can be monitored, and quantitatively translated to food quality, from production to the point of consumption (Tsironi et al., 2008).

The TTIs can be used as an alternative to conventional temperature probes. TTIs are particles that contain thermally labile species that undergo irreversible changes during passage through a thermal process. Compared to thermocouples, TTIs are small, can be made neutrally buoyant, and can be manufactured from materials with the same thermal conductivity as food particles. The time–temperature history of the product is not needed to determine the impact of thermal treatments when detailed kinetic information is available (Mehauden et al., 2008).

Different substances can be encapsulated inside TTIs:

- Microbiological TTIs are based on the quantification of the destruction of a target microorganism.
- Chemical and physical TTIs are based on the detection of a change in a physical or chemical property.
- Enzymatic TTIs are based on the quantification of enzyme activity after thermal treatment.

Among all TTI types, the microbial TTI edges out the others in that its TTI response is directly related to microbial food spoilage, as it reflects the bacterial growth and metabolism that occurs in the TTI system itself. To date, three types of commercial microbial TTIs, Traceo, Traceo restauration, and eO, which are based on the growth and metabolic activities of patent microbial strains, have been developed by Cryolog (Cryolog S.A., Quimper, France). However, details of the kinetic properties (e.g., the E_ and endpoint options) of these TTIs are not available (Vaikousi et al., 2008).

Most TTI systems can be designed to have a useful response time matching or correlating to the shelf life at a target constant temperature. On the other hand, the temperature dependence of the response expressed in kinetic terms as activation energy) can only be set at certain limited values. A difference in sensitivity between TTI response and food spoilage can result in an accumulating error in the translation of the response to actual quality loss of the food under the variable temperature conditions of the chill chain (Taoukis and Labuza, 1992).

A list of patents for TTIs is given in Table 14.8.

14.3.3 Oxygen Indicators

Oxygen indicators relate to packages of food, pharmaceuticals, chemical materials, clinicals, and similar materials to detect and demonstrate whether there has been any tampering. Various methods for ascertaining the integrity of a package have been previously described using certain chemical and physical tests to indicate a change in the environment within the package. Thus, a change in moisture content, exposure to light, loss of packaging gas, and color changes resulting from the presence of oxygen have all been suggested (Perlman and Linschitz, 1985).

Some packaging indicator patents are displayed in Table 14.9.

TABLE 14.8
Oxygen Indicator Patents

Patent Number	Title	Issue Date	Reference
4169811	Oxygen indicator	Oct 2, 1979	Yoshikawa et al. (1979)
4349509	Oxygen indicator adapted for printing or coating and oxygen-indicating device	Sep 14, 1982	Yoshikawa et al. (1982)
4526752	Oxygen indicator for packaging	Jul 2, 1985	Perlman and Linschitz (1985)
5293866	Oxygen flow meter indicator	Mar 15, 1994	Padula (1994)
5358876	Oxygen indicator	Oct 25, 1994	Inoue et al. (1994)
6703245	Oxygen-detecting composition	Mar 9, 2004	Sumitani et al. (2004)
WO/2007/018301	Ink composition and oxygen indicator	Feb 15, 2007	Hurme et al. (2007)
WO/2007/059901	Oxygen scavenger/indicator	May 31, 2007	Langowski and Wanner (2007)

TABLE 14.9
Problems and Solutions Encountered with Introducing New Products Using Active and/or Intelligent Packaging Techniques

Problem/Fear	Suggested Solution
Consumer attitude	Consumer research: education and information
Doubts about performance	Storage tests before launching. Consumer education and information
Increased packaging costs	Use in selected, high-quality products. Marketing tool for increased quality and quality assurance
False sense of security, ignorance of date markings	Consumer education and information
Mishandling and abuse	Active compound incorporated into label or packaging film. Consumer education and information
False complaints and return of packs with color indicators	Color automatically readable at the point of purchase
Difficulty of checking every color indicator at the point of purchase	Barcode labels: intended for quality assurance for retailers only

Source: Adapted from Hurme, E. and Ahvenainen, R., Active and smart packaging of ready-made foods, in *Minimal Processing and Ready Made Foods,* T. Ohlsson, R. Ahvenainen, and T. Mattila-Sandholm (eds.), SIK, Goteborg, Sweden, pp. 169–182.

An ideal oxygen indicator should have the following properties (Mills, 2005):

- Being inexpensive and not add significantly to the overall cost of the package
- Comprising nontoxic, non-water-soluble components that have food contact approval, since the indicator will be placed inside the food package

- Having a very long shelf-life under ambient conditions and only be activated as an oxygen indicator when the package has been sealed and is largely or wholly oxygen-free
- Being tunable with respect to oxygen sensitivity
- Exhibiting an irreversible response toward oxygen
- Being easily incorporated into the food package

14.4 CONCLUSIONS

Packaging has always played a very important role in food preservation by prolonging food shelf life. Over the last 40 years, the innovations introduced in food packaging led to the coining of another, nowadays quite popular, term "active packaging," whereas the classical packaging was characterized as passive packaging. Active packaging, also known as intelligent and smart packaging, aims at (1) enhancing the shelf life of packaged foods by removing the residual oxygen or water, or by restricting microorganism or mold growth; (2) improving food properties, for example, by cholesterol and lactose removal; and (3) monitoring the temperature/time abuse by applying time temperature indicators/integrators. Although active packaging appeared to be a very promising technique in terms of shelf life prolongation, food properties improvement, and consumer protection, its applicability is unfortunately quite limited compared to the tremendous amount of research and patents published. Globalization is anticipated to play a crucial role in further applying this technology because there will be a strong need for transporting goods to even longer distances and requirement for longer shelf life. The latter, in conjunction with the continuously stricter legislation requirements for food safety, is anticipated to be the major factor that will push forward the active, smart, and intelligent packaging applications.

REFERENCES

Abe, Y., A., Ahvenainen, R., Nattila-Sandholm, T., and Ohlsson, T. (1994). Active packaging with oxygen absorbers. *Minimal Processing of Foods, VTT Symposium 142*, Espoo, Finland, pp. 209–233.

Ahn, J. and Kwak, H.S. (1999). Optimizing cholesterol removal in cream using β-cyclodextrin and response surface methodology. *Journal of Food Science*, 64(4): 629–632.

Ahvenainen, R. (2003). *Novel Food Packaging Techniques*. Woodhead Publishing Limited, Cambridge, U.K.

Akkapeddi, M.K., Kraft, T.J., and Socci, E.P. (2002). Oxygen scavenging high barrier polyamide compositions for packaging. U.S. Patent No. 6423776.

Arul, J.A., Boudreau, A., Marhlouf, J., Tardif, R., and Grenier, B. (1988). Distribution of cholesterol in milk fraction. *Journal of Dairy Research*, 55: 361–371.

Barabas, E.S., Mallya, P., and Gromelski, S.J. (1981). Ultraviolet light stable copolymer compositions comprising monomers which are alpha, 62-unsaturated dicarboxylic acid half-esters of 2-hydroxy, alkoxy, methylolbenzophenones and styrene-butadiene comonomers. U.S. Patent No. 4301267.

Beard, C.D., Yamada, A., and Doddi, N. (1985). Ultraviolet absorbing polymers comprising 2-hydroxy-5-acrylyloxyphenyl-2H-benzotriazoles. U.S. Patent No. 4528311.

Bengtsson, B.G. (2005). Temperature regulation of a sprayed fluid material. U.S. Patent No. 6976364.

Berger, J., Mick, D., and Kleefeld, J. (1993). Temperature stable solid-state laser package. U.S. Patent No. 5181214.

Bhattacharjee, H.R. (1989). Activatable time-temperature indicator. U.S. Patent No. 4812053.

Bishop, P.W. and Harrover, A.J. (1999). Carbon dioxide cleaning process. U.S. Patent No. 6004400.

Blinka, T.A., Edwards, F.B., Miranda, N.R., Speer, D.V., and Thomas, J.A. (2002). Zeolite in packaging film. U.S. Patent No. 6391403.

Blinka, T.A., Speer, D.V., and Feehley, A. (1998). Oxygen scavenging metal-loaded ion-exchange compositions. U.S. Patent No. 5798055.

Blinka, T.A., Speer, D.V., and Feehley, A. (2000). Oxygen scavenging metal-loaded ion-exchange composition. U.S. Patent No. 6086786.

Bose, R.N., Tibbitts, W.I., and Ranum, R.I. (1967). Carbon dioxide generator. U.S. Patent No. 3318022.

Bray, A.V. (2000). Time temperature indicator. U.S. Patent No. 6158381.

Brennan, T.A. (1984). Time/temperature indicator. U.S. Patent No. 4488822.

Brody, A.L., Strupinsky, E.R., and Kline, L.R. (2001). *Active Packaging for Food Applications*. CRC Press, New York, pp. 31–64.

Broeils, J. (1988). Carbon dioxide compressing generator for dry ice manufacture. U.S. Patent No. D298754.

Cahill, P.J. and Chen, S.Y. (2000). Oxygen scavenging condensation copolymers for bottles and packaging. U.S. Patent No. 6083585.

Cannelongo, J.F. and Cugini, C.D. (1998). Temperature sensitive device for medicine containers. U.S. Patent No. 5779364.

Chalmers, S.P. (1985). Temperature/time limit indicator. U.S. Patent No. 4509449.

Chao, S.C., Purer, E.M., and Sorbo, N.W. (2000). Liquid carbon dioxide cleaning using agitation enhancements at low temperature. U.S. Patent No. 6264753.

Chao, S.C., Sorbo, N.W., and Purer, E.M. (2001). Removing soil from fabric using an ionized flow of pressurized gas. U.S. Patent No. 6117190.

Charles, F., Sanchez, J., and Gontard, N. (2006). Absorption kinetics of oxygen and carbon dioxide scavengers as part of active modified atmosphere packaging. *Journal of Food Engineering*, 72: 1–7.

Ching, T.Y. (1998). Oxygen scavenging ribbons and articles employing the same. U.S. Patent No. 5744246.

Ching, T.Y., Cai, G., Depree, C., Galland, M.S., Goodrich, J.L., Leonard, J.P., Matthews, A., Russell, K.W., and Yang, H. (1999). Oxygen scavengers with reduced oxidation products for use in plastic films and beverage and food containers. WO/1999/048963.

Conte, J.A, Johnson, B.R., Hsieh, R.J., and Ko, S.S. (1992). Method for removing cholesterol from eggs. U.S. Patent No. 5091203.

Cornelius, L.E. and Torii, T. (2004). Ultraviolet filter coating. U.S. Patent No. 6773104.

Cullen, J.S., Idol, R.C., and Powers, T.H. (1997). Oxygen-absorbing label. U.S. Patent No. 5667863.

Cully, J. and Vollbrecht, H.-R. (1994). Process for the removal of cholesterol from egg yolk. U.S. Patent No. 5292546.

Cully, J., Vollbrecht, H.R., and Schütz, E. (1991). Process for the removal of cholesterol and/or cholesterol esters from foodstuffs. U.S. Patent No. 5061505.

Danielson, T.D., Zhao, X.E., Mason, M.E., Connor, D.M., Stephens, E.B., Sprinkle, J.D., and Xia, J. (2006). Low-color ultraviolet absorbers for high UV wavelength protection applications. U.S. Patent No. 7014797 B2.

Day, B.P.F. (2003). Active packaging. In: *Food Packaging Technologies* (R. Coles, D. McDowell, and M. Kirwan, eds.). CRC Press, Boca Raton, FL, pp. 282–302.

Day, B.P.F. (2008). Active packaging of food. In: *Smart Packaging Technologies for Fast Moving Consumer Goods* (J. Kerry and P. Butler, eds.). John Wiley & Sons, Ltd, New York, pp. 1–18.

De Jong, A.R., Boumans, H., Slaghek, T., Van Veen, J., Rijk, R., and Van Zandvoort, M. (2005). *Active and Intelligent Packaging for Food: Is It the Future?* Taylor & Francis, London, U.K.

DeSimone, J.M., Romack, T., Betts, D.E., and McClain, J.B. (1998). Cleaning process using carbon dioxide as a solvent and employing. U.S. Patent No. 5783082.

Dewees, T.G., Knafelc, F.M., Mitchell, J.D., Taylor, R.G., Iliff, R.J., Carty, D.T., Latham, J.R., and Lipton, T.M. (1993). Liquid/supercritical carbon dioxide dry cleaning system. U.S. Patent No. 5267455.

DeYoung, J.P., Romack, T.J., and McClain, J.B. (2004). Detergent injection methods for carbon dioxide cleaning apparatus. U.S. Patent No. 6711773.

Diekmann, T.J. and Bommarito, G.M. (2005). Time or time-temperature indicating articles. U.S. Patent No. 6916116.

Dunks, G.B., Yamada, A., Beard, C.D., and Doddi, N. (1987). Ultraviolet absorbing polymers comprising 2-(2′-Hydroxy-5′-acryloyloxyalkoxyphenyl)-2H-benzotriazole. U.S. Patent No. 4716234.

Ebner, C.L., Matthews, A.E., and Millwood, T.O. (2003). Oxygen scavenger compositions derived from isophthalic acid. U.S. Patent No. 7078100.

Ebner, C.L., Matthews, A.E., and Millwood, T.O. (2008). Oxygen scavenger compositions derived from isophthalic acid or terephthalic acid monomer or derivatives thereof. U.S. Patent No. 7452601.

Falcetta, J.J., Park, J., and Smith, C.G. (1992). Ultraviolet absorbing hydrogels. U.S. Patent No. 5133745.

Farrell, C.J. and Tsai, B.C. (1985). Oxygen scavenger. U.S. Patent No. 4536409.

Farrell, C.J. and Tsai, B.C. (1987). Oxygen scavenger. U.S. Patent No. 4702966.

Faubl, H. (2001). UV blocking lenses and material containing benzotriazoles and benzophenones. U.S. Patent No. 6244707.

Fertig, J., Goldberg, A.I., and Skoultchi, M. (1964). UV light absorbing copolymers of acryloxymethyl benzoates and dihydroxy-benzophenone derivatives. U.S. Patent No. 3328491.

Floros, J.D., Dock, L.L., and Han, J.H. (1997). Active packaging technologies and applications. *Food Cosmetics and Drug Packaging,* 20(1): 10–17.

Frisk, P. (1998). Article for scavenging oxygen from a container. U.S. Patent No. 5806681.

Garti, N. (1995). Removal of cholesterol from edibles. WO/1995/004473.

Goldhan, G. (2003). Oxygen scavenging packaging concepts. The future of food safety research in the European Union: Fosare seminar series 3. Food safety in relation to novel packaging technologies.

Graf, E. (1994). Oxygen removal. U.S. Patent No. 5284871.

Gupta, A.S. (1988). Delayed reaction carbon dioxide generator package. U.S. Patent No. 4786519.

Gupta, R.B., Jakiela, D.J., and Haacke, G. (2001). Trisaryl-1,3,5-triazine ultraviolet light absorbers. U.S. Patent No. 6242597 B1.

Han, J. (2000). Antimicrobial food packaging. *Food Technology*, 54(3): 56–65.

Han, J.H. (2003). Time–temperature indicators (TTIs). In: *Novel Food Packaging Techniques* (R. Ahvenainen, ed.). CRC Press, Boca Raton, FL.

Han, J.H. and Rooney, M.L. (2002). Personal communications. In: *Active Food Packaging Workshop, Annual Conference of the Canadian Institute of Food Science and Technology (CIFST),* May 26, 2002.

Hatakeyama, H. and Takahashi, H. (2000). Oxygen-absorbing multi-layer film and method for preparing same. U.S. Patent No. 6063503.

Henry, J.C. and Reich, C.J. (1989). Ultraviolet light absorbing compounds, compositions and methods for making same. U.S. Patent No. 4845180.

Hof, C.R. and Ulin, R.A. (1982). Temperature indicating compositions of matter. U.S. Patent No. 4339207.

Hsieh, R.J., Snyder, D.P., and Ford, E.W. (1994). Method for removing cholesterol and fat from egg yolk by chelation and reduced-cholesterol egg product. U.S. Patent No. 5302405.

Hurme, E. and Ahvenainen, R. (1996). Active and smart packaging of ready-made foods. In: *Minimal Processing and Ready Made Foods* (T. Ohlsson, R. Ahvenainen, and T. Mattila-Sandholm, eds.). SIK, Goteborg, Sweden, pp. 169–182.

Hurme, E., Sipilainen-Malm, T., Ahvenainen, R., and Nielsen, T. (2002). Active and intelligent packaging. In: *Minimal Processing Technologies in the Food Industry* (T. Ohlsson and N. Bengtsson, eds.). Woodhead Publishing Limited, Cambridge, U.K., pp. 87–123.

Hurme, E., Sipila, I-M.T., Ruskeepaa, A-L., Kawashima, M., and Kowsaka, K. (2007). Ink composition and oxygen indicator. WO/2007/018301.

Iliff, R.J., Mitchell, J.D., Carty, D.T., Latham, J.R., and Kong, S.B. (1995). Liquid/supercritical carbon dioxide/dry cleaning system. U.S. Patent No. 5412958.

Inoue, Y., Hatakeyama, H., and Yoshino, I. (1994). Oxygen indicator. U.S. Patent No. 5358876.

Ito, A. and Funakawa, S. (2000). Temperature controller of optical module package. U.S. Patent No. 6009712.

Jalinski, T.J. (1993). Time temperature indicator with distinct end point. U.S. Patent No. 5182212.

Jones, T.P., Turner, J.E., and Malinoski, M.F. (2002). Temperature control of electronic devices using power following feedback. U.S. Patent No. 6489793.

Jureller, S.H. and Kerschner, J.L. (2001). Method of dry cleaning using densified carbon dioxide and a surfactant. U.S. Patent No. 6299652.

Jureller, S.H., Kerschner, J.L., and Murphy, D.S. (2000). Dry cleaning system using densified carbon dioxide and a surfactant adjunct. U.S. Patent No. 6148644.

Katsura, T. (1989). Present state and future trend of functional packaging materials attracting considerable attention. *Packaging Japan,* September, pp. 21–26.

Kijowski, M. and Lombardo, S.P. (2000). Enhanced cholesterol extraction from egg yolk. U.S. Patent No. 6093434.

Klopp, E.M. (1971). Generator for compressing carbon dioxide. U.S. Patent No. D220454.

Kohlrausch, U., Cully, J., and Schmid, B. (1996). Process for the removal of cholesterol derivatives from egg yolk. U.S. Patent No. 5498437.

Krochta, J.M. (2006a). Food packaging. In: *Handbook of Food Engineering* (D.R. Heldman and D.B. Lund, eds.). CRC Press, New York, pp. 847–928.

Krochta, J.M. (2006b). Introduction to frozen food packaging. In: *Handbook of Frozen Food Processing and Packaging* (D.-W. Sun, ed.). CRC Press, New York, pp. 615–640.

Kwak, H.S., Ahn, J.J., and Lee, D.K. (2000). Method for removing cholesterol from milk and cream. U.S. Patent No. 6110517.

Labuza, T. P. and Breene, W. M. (1989). Applications of "active packaging" for quality of fresh and extended improvement of shelf-life and nutritional shelf-life foods. *Journal of Food Processing and Preservation*, 13: 1–69.

Lamagna, D.J. (1997). Time-temperature monitor and recording device and method for using the same. U.S. Patent No. 5662419.

Langowski, H-C. and Wanner, T. (2007). Organic oxygen scavenger/indicator. WO/2007/059901.

Larsen, J.E. and Froning, G.W. (1981). Extraction and processing of various components from egg yolk. *Poultry Science*, 60: 160–167.

Larsson, R.P. (1977). Time temperature integrating indicator. U.S. Patent No. 4042336.

Lee, D.S., Shin, D.H., Lee, D.U., Kim, J.C., and Cheigh, H.S. (2001). The use of physical carbon dioxide absorbents to control pressure buildup and volume expansion of kimchi packages. *Journal of Food Engineering*, 48: 183–188.

Lombardo, S.P. and Kijowski, M. (1992). Enhanced cholesterol extraction from egg yolk. WO/1992/005710.

Mariën, A. and Moeyersons, B. (2002). UV-absorbing film and its use as protective sheet. U.S. Patent No. 6468609.

Matthews, A.E. and Depree, C. (2001). Oxygen scavengers with reduced oxidation products for use in plastic. U.S. Patent No. 6254804.

McClain, J.B. and Schrebe, G. (2001). Carbon dioxide cleaning apparatus with rotating basket and external drive. U.S. Patent No. 6233772.

Mehauden, K., Bakalis, S., Cox, P.W., Fryer, P.J., and Simmons, M.J.H. (2008). Use of time temperature integrators for determining process uniformity in agitated vessels. *Innovative Food Science and Emerging Technologies*, 9: 385–395.

Micich, T.J. (1990). Behaviors of polymer supported digitonin with cholesterol in the absence and presence of butter oil. *Journal of Agricultural and Food Chemistry*, 38: 1839–1843.

Milden, M. (1987). Carbon dioxide compressing generator for dry ice manufacture. U.S. Patent No. D293249.

Mills, A. (2005). Oxygen indicators and intelligent inks for packaging food. *Chemical Society Reviews*, 34: 1003–1011

Nakamura, H. and Hoshino, J. (1983). *Techniques for the Preservation of Food by Employment of an Oxygen Absorber*. Mitsubishi Gas Chemical Co., Tokyo, Japan, Ageless Division, pp. 1–45.

Narayan, K.A. (1996). Time-temperature indicator for establishing lethality of high. U.S. Patent No. 5529931.

Narayan, K.A. (1997). Time-temperature method for establishing lethality of high temperature. U.S. Patent No. 5637475.

Oakenfull, D.G., Sidhu, G.S., and Rooney, M.L. (1991). Cholesterol removal. WO/1991/016824.

Ohlsson, T. and Bengtsson, N. (2002). *Minimal Processing Technologies in the Food Industries*. CRC Press, Boca Raton, FL.

Ou-Yang, D.T. (1992). Time and temperature indicating device. U.S. Patent No. 5120137.

Ozdemir, M. and Floros, J. D. (2004). Active food packaging technologies. *Critical Reviews in Food Science and Nutrition*, 44: 185–193.

Padula, J. (1994). Oxygen flow meter indicator. U.S. Patent No. 5293866.

Parham, M.E., Duffy, R.L., and Nicholson, D.T. (1996). High efficiency removal of low density lipoprotein-cholesterol from whole blood. U.S. Patent No. 5496637.

Patel, G.N. (1980). Vapor permeation time-temperature indicator. U.S. Patent No. 4195056.

Patel, G.N., Preziosi, A.F., and Baughman, R.H. (1976). Time-temperature history indicators. U.S. Patent No. 3999946.

Pearson, W.K. and Cordino, C.E. (1988). Temperature control method and apparatus. U.S. Patent No. 4744408.

Perlman, D. and Linschitz, H. (1985). Oxygen indicator for packaging. U.S. Patent No. 4526752.

Preston, A.D. and Turner, J.R. (1999). Carbon dioxide dry cleaning system. U.S. Patent No. 5904737.

Preston, A.D. and Turner, J.R. (2001). Carbon dioxide dry cleaning system. U.S. Patent No. 6216302.

Preston, A.D., Turner, J.R., and Svoboda, C. (2002). Carbon dioxide dry cleaning system. U.S. Patent No. 6442980.

Prusik, T, Arnold, R.M., and Fields, S.C. (1998). Time-temperature indicator device and method of manufacture. U.S. Patent No. 5709472.

Prusik, T., Arnold, R.M., and Piechowski, A.P. (2003). Activatable time-temperature indicator system. U.S. Patent No. 6544925.

Qi, C., De Bont, H.B.A., Van Der Zee, L., Lansink, M., and Van Norren, K. (2005). Cholesterol lowering supplement. U.S. Patent No. 6933291.

Qiu, J., Noyola, J.M., and Yarusso, D.J. (2001). Time-temperature integrating indicator device with barrier material. U.S. Patent No. 6244208.

Qiu, J., Noyola, J.M., Yarusso, D.J., and Green, K.R. (2002). Time-temperature integrating indicator device with barrier material. U.S. Patent No. 6435128.

Ram, A.T., Manico, J.A., Gisser, K.R., Cowdery-Corvan, P.J., and Weaver, T.D. (2000). Time temperature indicator device. U.S. Patent No. 6103351.

Redmore, D. and Mo, M. (1973). Oxygen scavenger and use thereof. U.S. Patent No. 3764548.

Richardson, T. and Jimenez-Flores, R. (1994). Process to remove cholesterol from dairy products. U.S. Patent No. 5326579.

Roberts, J.L. and Kegler, A. (2001). Liquified gas dry cleaning system with pressure vessel temperature compensating compressor. U.S. Patent No. 6182318.

Roczniak, S., Hill, J.B., and Erickson, R.A. (1991). Removal of cholesterol from edible fats. U.S. Patent No. 5045242.

Rodgers, J., Borsody, I., Karydas, A., Falk, R.A., Mueller, K.F., and Kovaleski, M. (1995). UV-absorbing polyurethanes and polyesters. U.S. Patent No. 5459222.

Rodrigues, E.T. and Han, J. H. (2003). Intelligent packaging. In: *Encyclopaedia of Agricultural, Food and Biological Engineering* (D.R. Heldman, ed.). Marcel Dekker, New York, pp. 528–535.

Romack, T.J., McClain, J.B., Stewart, G.M., and Givens, R.D. (2001). Carbon dioxide cleaning and separation systems. U.S. Patent No. 6200393.

Rooney, M.L. (1995), Active packaging in polymer films. In: *Active Food Packaging* (M.L. Rooney, ed.). Blackie Academic and Professional, London, U.K., pp. 74–110.

Rooney, M.L. (2005). Introduction to active food packaging technologies. In: *Innovations in Food Packaging* (J.H. Han, ed.). Academic Press, Oxford, U.K., pp. 63–79.

Rosio, L.R. and Fulton, Ed.D. (2000). Liquid carbon dioxide cleaning system and method. U.S. Patent No. 6129451.

Roth, J.D. (2003). Time-temperature indicators activated with direct thermal printing and methods for their production. U.S. Patent No. 6524000.

Sacharow, S. (1988). Freshness enhancers: The control in controlled atmosphere packaging. *Prepared Foods,* 157: 121–122.

Schiraldi, D.A., Sekelik, D.J., and Smith, B.L. (2003). Oxygen scavenging PET based polymer. U.S. Patent No. 6544611.

Schmidt S.L., Collette W.N., Coleman E.A., and Krishnakumar S.M. (1999). Transparent package with aliphatic polyketone oxygen scavenger. U.S. Patent No. 5952066.

Schunk, T.C., Schroeder, K.M., Appell, C.H., and Linehan, D.T. (2005). Ultraviolet light filter element. U.S. Patent No. 6872766.

Scully, A.D. and Horsham, M.A. (2007). Active packaging for fruits and vegetables. In: *Intelligent and Active Packaging for Fruits and Vegetables* (C.L. Wilson, ed.). CRC Press, Boca Raton, FL, pp 57–72.

Simons, M.J. and Weldy, J. A. (2001). Time-temperature indicator devices. U.S. Patent No. 6214623.

Simons, M.J. (2003). Time-temperature indicator devices. U.S. Patent No. 6514462.

Skjervold, P.O., Salbu, B., Heyerdahl, P.H., and Lien, H. (2007). Full history time-temperature indicator system. U.S. Patent No. 7290925.

Smolander, M. (2003). Intelligent packaging systems as quality and safety indicating tools. The future of food safety research in the European Union: Fosare seminar series 3. Food safety in relation to novel packaging technologies.

Snelling, C.D. (1966). Self-sustaining temperature control package. U.S. Patent No. 3273634.

Sonneveld, K. (2000). *What Drives (Food) Packaging Innovation? Centre for Packaging, Transportation and Storage*. Victoria University of Technology, Melbourne City, Victoria, Australia.

Speer, D.V. and Roberts, W.P. (1997). Oxygen scavenging composition for low temperature use. U.S. Patent No. 5648020.

Spevacek, J.A. (2003). Time-temperature integrating indicator. U.S. Patent No. 6614728.

Stewart, M.E., Sharpe, E.E., Gamble, B.B., Stafford, S.L., Estep, R.N., Williams, J.C., and Clark, T.R. (2007). Polyester based cobalt concentrates for oxygen scavenging compositions. U.S. Patent No. 7288586.

Stouffer, S.C., Majeres, L.J., and Charintranond, W. (1994). Method for extracting cholesterol from egg yolk. U.S. Patent No. 5316780.

Strobel, J.M. and Gagnon, D.R. (1998). Enzyme loaded hydrophilic porous structure for protecting oxygen. U.S. Patent No. 5766473.

Sumitani, M., Inoue, H., and Sugito, K. (2004). Oxygen-detecting composition. U.S. Patent No. 6703245.

Sun, D.W. (2006). *Handbook of Frozen Food Processing and Packaging*. CRC Press, Boca Raton, FL.

Suppakul, P., Miltz, J., Sonneveld, K., and Bigger, S.W. (2003). Active packaging technologies with an emphasis on antimicrobial packaging and its applications. *Journal of Food Science*, 68(2): 408–420.

Swartzel, K.R., Ganesan, S.G., Kuehn, R.T., Hamaker, R.W., and Sadeghi, F. (1992). Thermal memory cell and thermal system evaluation. U.S. Patent No. 5159564.

Taoukis, P.S. and Labuza, T.P. (1989). Applicability of time temperature indicators as shelf-life monitors of food products. *Journal of Food Science*, 54: 783–788.

Taoukis, P.S. and Labuza, T.P. (1992). Assessing the food quality monitoring ability of a time-temperature indicator. In: *Annual IFT Meeting*. IFT, New Orleans, LA, p. 210.

Taoukis, P.S. and Labuza, T.P. (2003). Time-temperature indicators (TTIs). In: *Novel food packaging Techniques* (R. Ahvenainen, ed.). CRC Press, Boca Raton, FL.

Tester, R. and Al-Ghazzewi, F. (2008). Time temperature indicators linked to sensory detection. U.S. Patent No. 7360946.

Teumac, F.N., Zenner, B.D., Ross, B.A., Deardurff, L.A., and Rassouli, M.R. (2004). Metal catalyzed ascorbate compounds as oxygen scavengers. U.S. Patent No. 6709724 B1.

Thierry, A. and Moigne, J.L. (1992). Temperature indicators based on polydiacetylene compounds. U.S. Patent No. 5085801.

Thompson, G.M. (2001). Temperature indicator employing color change. U.S. Patent No. 6176197.

Townsend, C.W. and Purer, E.M. (1997). Liquid carbon dioxide dry cleaning system having a hydraulically. U.S. Patent No. 5669251.

Tsai, M.L. and Akkapeddi, M.K. (2004). Oxygen scavenging polymer compositions containing ethylene vinyl alcohol copolymers. U.S. Patent No. 6793994.

Tsironi, T., Gogou, E., Velliou, E., and Taoukis, P.S. (2008). Application and validation of the TTI based chill chain management system SMAS (Safety Monitoring and Assurance System) on shelf life optimization of vacuum packed chilled tuna. *International Journal of Food Microbiology*, 128: 108–115.

Ubara, H., Kodama, T., Mizuno, N., Sakaguchi, Y., Takasugi, K., Kurahara, S., Matsuda, K., Gyobu, S., Shimizu, Y., and Yoshida, H. (2004). Oxygen-absorbing material. U.S. Patent No. 6689437.

Vaikousi, H., Biliaderis, C.G., and Koutsoumanis, K.P. (2008). Development of a microbial time/temperature indicator prototype for monitoring the microbiological quality of chilled foods. *Applied and Environmental Microbiology*, 74: 3242–3250.

Van Antwerp, W.P. and Yao, L. (2001). UV absorbing polymer. U.S. Patent No. 6252032.

Vermeiren, L., Devlieghere, F., Van Beest, M., De Kruijf, N., and Debevere, J. (1999). Developments in the active packaging of foods. *Trends in Food Science and Technology*, 10: 77–86.

Vermeiren, L., Heirlings, L, Devlieghere, F., and Debevere, J. (2003). Oxygen, ethylene and other scavengers. In: *Novel Food packaging Techniques* (R. Ahvenainen, ed.). CRC Press, Boca Raton, FL.

Vollbrecht, H.R., Cully, J., and Wiesmüller, J. (1991). Process for the removal of cholesterol and cholesterol esters from egg yolk. U.S. Patent No. 5063077.

Wachtell, G.P. and Jones, E.W. (1976). Time temperature indicator device and method. U.S. Patent No. 3965741.

Watanabe, H., Sakai, M., and Tezuka, H. (2000). Temperature-controlled appliance. U.S. Patent No. 6038865.

Waters, J. (1985). Carbon dioxide generator insect attractant. U.S. Patent No. 4506473.

Witkin, D.E. and Bowles, A.G. (1976). Temperature control apparatus. U.S. Patent No. 3971876.

Witkin, D.E. and Bowles, A.G. (1978). Temperature control method and apparatus. U.S. Patent No. 4066868.

Witonsky, R.J. (1976). Time temperature indicator. U.S. Patent No. 3942467.

Wrezel, P.W., Krishnamurthy, R.G., and Hasenhuettl, G.L. (1992). Method for removing cholesterol from edible oils. U.S. Patent No. 5128162.

Wright, B.B. (2004). Comparator for time-temperature indicator. U.S. Patent No. 6737274.

Xia, J. and Moore, S.C. (2008). UV-absorbing compounds and compositions containing UV-absorbing compounds. U.S. Patent No. 7381762 B2.

Yam, K.L., Takhistov, P.T., and Miltz, J. (2005). Intelligent packaging: Concepts and applications. *Journal of Food Science*, 70(1): R1–R10.

Yamada, S., Sakuma, I., Himeshima, Y., Aoki, T., Uemura, T., and Shirakura, A. (1992). Oxygen scavenger. U.S. Patent No. 5,143,763.

Yamamori, S., Ono, Y., Takatori, F., Dainobu, H., Inoue, M., and Todokoro, N. (2008). Carbon dioxide sensor and airway adapter incorporated in the same. U.S. Patent No. 7445602.

Yan, S., Huawei, C., Limin, Z., Fazheng, R., Luda, Z., and Hengtao, Z. (2008). Development and characterization of a new amylase type time–temperature indicator. *Food Control*, 19: 315–319.

Yoshida, T. (1998). Ultraviolet-absorbing polymer film. U.S. Patent No. 5806834.

Yoshikawa, Y., Nawata, T., Goto, M., and Kondo, Y. (1979). Oxygen indicator. U.S. Patent No. 4169811.

Yoshikawa, Y., Nawata, T., Goto, M., and Kondo, Y. (1982). Oxygen indicator adapted for printing or coating and oxygen-indicating device. U.S. Patent No. 4349509.

Yuan, J.T.C. (2003). Modified atmosphere packaging for shelf-life extension. In: *Microbial Safety of Minimally Processed Foods.* (I.S. Novak, G.M. Sapesrs, and U.K. Uneja, eds.). CRC Press, Boca Raton, FL, pp. 205–219.

Zeidler, G. (1995). Method for removal of cholesterol and fat from liquid egg yolk with recovery of free cholesterol as a by-product. U.S. Patent No. 5468511.

Zenner, B.D., Teumac, F.N., Deardurff, L.A., and Ross, B.A. (1994). Polymer compositions containing salicylic acid chelates as oxygen. U.S. Patent No. 5364555.

Electronic References

http://74.125.39.104/search?q=cache:D-l-tvbE3uwJ:europa.eu/rapid/pressReleasesAction.do%3Freference%3DIP/03/1554%26format%3DPDF%26aged%3D1%26language%3DEN%26guiLanguage%3Den+IP/03/1554&hl=el&ct=clnk&cd=2&gl=gr

http://edis.ifas.ufl.edu/pdffiles/AE/AE20600.pdf

http://ohioline.osu.edu/cd-fact/0133.html.)
http://www.ipexl.com/patents/USPTO_7078100.html
http://www.ipexl.com/patents/USPTO_7381762.html
http://www.patentgenius.com/patent/5293866.html
http://www.patentgenius.com/patent/7445602.html
http://www.patentstorm.us/patents/5292546.html
http://www.patentstorm.us/patents/5316780/description.html
http://www.patentstorm.us/patents/6004400.html
http://www.patentstorm.us/patents/6117190.html
http://www.patentstorm.us/patents/6129451.html
http://www.patentstorm.us/patents/6148644.html
http://www.patentstorm.us/patents/6182318.html
http://www.patentstorm.us/patents/6216302.html
http://www.patentstorm.us/patents/6233772.html
http://www.patentstorm.us/patents/6264753.html
http://www.patentstorm.us/patents/6299652.html
http://www.patentstorm.us/patents/6442980.html
http://www.patentstorm.us/patents/6703245.html
http://www.patentstorm.us/patents/6711773/fulltext.html
http://www.pharmcast.com/Patents100/Yr2005/Aug2005/082305/6933291_Cholesterol082305.htm
http://www.reducepackaging.com/history.html
http://www.sciencedirect.com/science?_ob=ArticleURL&_udi=B6VPY-4BYYGS0-X&_user=83475&_rdoc=1&_fmt=&_orig=search&_sort=d&view=c&_acct=C000059672&_version=1&_urlVersion=0&_userid=83475&md5=8526474d4c5fe666589a7b49b7fba0ad
http://www.wipo.int/pctdb/en/wo.jsp?IA=AU1990000490&DISPLAY=STATUS
http://www.wipo.int/pctdb/en/wo.jsp?IA=JP2006315982&DISPLAY=STATUS
http://www.wipo.int/pctdb/en/wo.jsp?IA=US1991006942&DISPLAY=STATUS
http://www.wipo.int/pctdb/en/wo.jsp?wo=1995004473
http://www.wipo.int/pctdb/en/wo.jsp?wo=1999048963&IA=US1999006379&DISPLAY=STATUS
http://www.wipo.int/pctdb/en/wo.jsp?wo=2007059900

15 Adaptations of Food Packaging Trends via Nanotechnology

Muhammad Imran, Anne-Marie Revol-Junelles, and Stéphane Desobry

CONTENTS

15.1 NANOTECHNOLOGY AS AN INSPIRATION FROM NATURE

Above and beyond the copyright issues of ideas, a large portion of nanoscience innovations is an attempt to imitate what has evolved in Nature. Living organisms are not just a compilation of nanoscale objects: Atoms and molecules are organized in hierarchical structures and dynamic systems that are the outcomes of millions of years of Mother Nature's experiments (Weiss et al., 2006). Of course, all physiological processes on some level occur at the *submicron*, *nanometer*, and even smaller *picometer* scale. The essential biomolecules such as sugars, amino acids, hormones, and DNA are in the nanometer range. Tenth-nanometer diameter ions such as potassium and sodium generate nerve impulses. Most protein and polysaccharide molecules have nanoscale dimensions. Every living organism on earth exists owing to the presence, absence, concentration, location, and interaction of these nanostructures (Weiss et al., 2006; Ravichandran, 2010).

Nanoscience is currently enabling evolutionary changes in several technology areas, but new paradigms will eventually have a much wider and revolutionary impact (Imran et al., 2010). Nanoscience is "the study of phenomena and manipulation of materials at atomic, molecular, and macromolecular scales (0.2–100 nm), where properties differ significantly from those at a larger scale," whereas nanotechnology is "the design, characterisation, production, and application of structures, devices, and systems by controlling the shape and size at the nanometer scale" (RSRAE, 2004). The U.S. definition is that "nanotechnology is the understanding and control of matter at dimensions of roughly 1–100 nm, where unique phenomena enable novel applications." Encompassing nanoscale science, engineering, and technology, nanotechnology involves imaging, measuring, modeling, and manipulating matter at this length scale (NNI, 2001).

With the increased funding opportunities and interest in this field, the term *nano* is more frequently and often liberally used (Nel et al., 2006). The Royal Commission on Environmental Pollution has published a report on "Novel materials in the environment: The case of nanotechnology," which examined issues related to innovation in the materials sector and made important recommendations on how to deal with ignorance and uncertainty in this area. The small size of nanomaterials gives them specific or enhanced physicochemical properties, compared with the same materials at the macroscale, which generate great interest in their potential for development for different uses and products. Nevertheless, a good working definition of a nanomaterial is one that is between 1 and 100 nm in at least one dimension and which exhibits novel properties (Figure 15.1) (RCEP, 2008).

15.2 FOOD NANOTECHNOLOGY APPLICATIONS

Even though successful applications of nanotechnology to foods are still limited, several fundamental concepts based on nanoscale have been well established. Nanotechnology is important because it is cheap, relatively safe, clean, and the financial rewards are very high. Nanotechnology touches or will touch every aspect of our lives (El Naschie, 2006). The vital nanotechnological benefits in the food regime have been presented in Figure 15.2. In the food industry, new functional materials, micro- and nanoscale processing, product development, and the design of methods and instrumentation for food safety and biosecurity are the significant advancements by nanotechnology (Moraru et al., 2003). The immense opportunities of nanoscience are possible in the areas of (i) food safety and biosecurity, (ii) food processing, (iii) food packaging, and (iv) ingredient technologies (Bugusu and Lubran, 2007). One can speculate that understanding the unique possessions of edible stuff of nanometer size will result in innovative, safer, healthier, and tastier foods. Nanotechnology, a new frontier of this century, has relatively recent applications as compared with its use in drug delivery and pharmaceuticals. Smart delivery of nutrients, bioseparation of proteins, rapid sampling of biological and chemical contaminants, and nanoencapsulation of nutraceuticals are some of the emerging topics of nanotechnology for foods (Garcia et al., 2010).

Food nanotechnology has the potential to alter nutrient intake by broadening the number of enriched and fortified food products (Nickols-Richardson, 2007). Nanotechnology provides a good opportunity to improve the solubility of active (carotenoids, phytosterols, essential fatty acids, natural antioxidants) ingredients and to increase their bioavailability (Moraru et al., 2003). Omega-3 present in marine

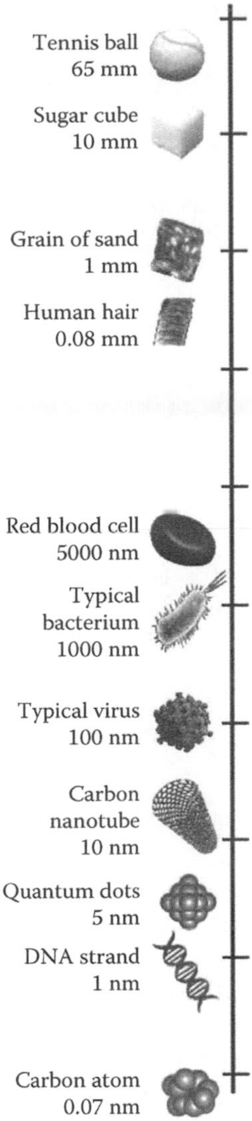

FIGURE 15.1 Length scale showing the nanometer in context. One nanometer (nm) is equal to one-billionth of a meter, 10^{-9} m. Most structures of nanomaterials that are of interest are below 100 nm. (From RCEP, Available at [www.rcep.org.uk/reports/27], 2008.)

oils is responsible for their numerous beneficial effects on the retina, the cardiovascular system, the nervous system, etc. The problem with their direct incorporation into food products is their ease of oxidation and, therefore, the corresponding losses. For this reason, they must be micro/nanoencapsulated in most food applications (Kolanowski et al., 2006, 2007; Luff, 2007; Shahidi, 2000; Yep et al., 2002). In developing nations, malnutrition contributes to more than half of the deaths of children under 5 years. Several inexpensive agricultural and food applications of

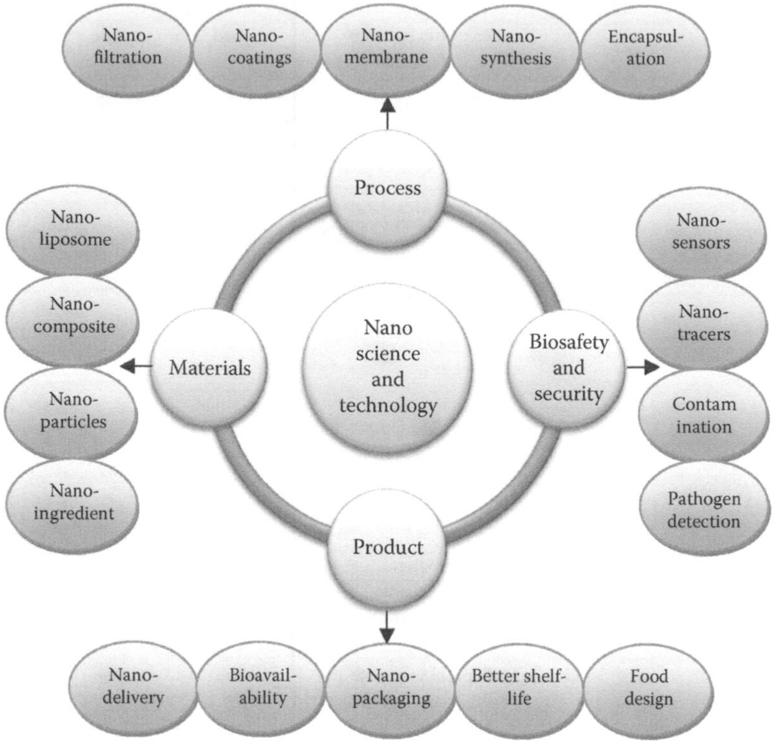

FIGURE 15.2 The impacts and needs of nanotechnology applications in foods.

nanotechnology have the potential to decrease malnutrition, and thus infant mortality (Court et al., 2005). Flavored waters and milk fortified with vitamins, minerals, and other functional ingredients via nanoemulsion technology have gained a lot of importance (Tan and Nakajima, 2005). Particle size is a determinant of iron (major malnutrition element) assimilation from feebly soluble Fe compounds. Decreasing the particle size of metallic Fe and ferric pyrophosphate added to foods increases Fe absorption (Rohner et al., 2007). Thus, nanoingredient food technology has the potential to get rid of major malnutrition dilemmas.

More than one-third of the population of rural areas in Africa, Asia, and Latin America has no clean water. More than two million children die each year from water-related diseases. Inexpensive, easily transportable, and easily cleanable systems like nanomembranes (filtration technologies) and nanoclays purify, detoxify, and desalinate water more efficiently than conventional systems (Court et al., 2005).

Food nanobioprocessing will achieve the goal of bioprocessing (utilize natural biological processes to generate a required compound from a specific waste/feedstock) with greater efficiency. The developments of devices that allow rapid identification of microbes present in feedstock are examples of research at the nanoscale that will increase the efficiency of bioprocessing (Scott and Chen, 2002). Nanosensors for the detection of pathogens and contaminants will possibly make manufacturing, processing, and consignment of food products more secure. Particular nanodevices may perhaps enable precise tracking and recording of the environmental conditions and shipment history of a specific product (Fonseca et al., 2007). In contrast, the use of nanotechnology for intelligent inks in modified atmosphere packaging (MAP) is gaining popularity.

Packaging that incorporates nanomaterials can be "smart/intelligent," which means that it can respond to environmental conditions or repair itself or alert a consumer to contamination and the presence of pathogens. Self-healing packaging materials use nano/microencapsulated repairing agents. Small amounts of an encapsulated "healing agent" will be released by crack propagation or other triggering mechanisms, which have been incorporated into polymeric coatings (Andersson et al., 2009). The security of the food supply may possibly be enhanced by making "pathogen-repulsive" surfaces or packaging materials that change color in the incidence of injurious microorganisms or toxins (Moraru et al., 2003). Nanotechnology for food packaging had grown from a $66 million business in 2003 to a $360 million business in 2008, with an average annual increase of 40% (Brody, 2006). There is an increase in the number of nanotechnology developments in either type of packaging, active and intelligent (Nachay, 2007).

15.3 POTENTIAL ADVANTAGES OF NANOTECHNOLOGY: WHY SIZE MATTERS

Nanotechnology applications are expected to bring a range of paybacks to active agents' uses in food sector, including new tastes, textures and sensations, less use of fat, enhanced absorption of nutrients, improved packaging, traceability, and security of food products (Chaudhry et al., 2008). Applications of novel micro- and nanotechnologies to food structuring are likely to bring large benefits to the food/health industry (Table 15.1). In due course, micro- and nanoencapsulation has brought a revolution because of the fact that by utilizing these techniques, the control against food-borne pathogens is not only more effective but also long-lasting, thus enabling the researchers and industrialists to provide a healthier, safer, and an enhanced shelf-life food to the consumer. It is true that micro- or nanoentrapment in liposomes enhances nisin stability, availability, and distribution, which may improve the control of undesirable bacteria in foods stored for long periods. It can be proposed that allergenicity and product labeling concerns are expected to be minimal as liposomes used in micro- and nanotechnology are generally formed from lipids that are naturally occurring in various food staples. Therefore, it is very crucial to induce further work in the nanoactive packaging, where natural antimicrobials would be used with, or instead of, conventional preservatives utilizing micro- or more preferably nanotechnology to improve the shelf life and safety of perishable foods.

15.3.1 Nanoactive Packaging

Nanotechnology approaches are being broadened in food science, especially in packaging with high performances and low concentrations and prices, so this category

TABLE 15.1
Prospective Advantages of Micro- or Nanotechnology for Novel Trends

Prospective Advantages	References
Protection	
To reduce or prohibit bacteriocins affinity to food components	Laridi et al. (2003)
Safeguard antimicrobial peptide from inhibitors or unfavorable conditions in food matrix	Laridi et al. (2003)
Stability	
Act as long-term preservative in foods for long periods of time	Laridi et al. (2003)
Avoid interference with lactic starter growth during fermentation	Laridi et al. (2003)
Relatively stable to pasteurization protocols	Laridi et al. (2003)
Improved viability of probiotics for future use	Taylor et al. (2007)
Cost-effective	
Economical as relatively lower dose of antimicrobial is required as compared with free form	Picot and Lacroix (2004)
Functionality	
Comparatively enhanced activity and bioavailability	Roberts and Zottola (1993)
Isolation/immobilization	
No or lesser harmful effects on organoleptically important bacteria, e.g., LAB for flavor	Roberts and Zottola (1993)
Use of active ingredients in low- or high-pH foods, e.g., nisin, a pH-sensitive compound	Buyong et al. (1998)
Controlled release	
Release of active agents can be triggered by physical or chemical stress	Taylor et al. (2007)
Effective and easier lactose peroxidase system provision	Lee and Rosenberg (2000)
Controlled release of active ingredients	Jacquot et al. (2000)
Nutrition	
Biopolymers for nanoencapsulation are usually edible, hence nutritive	Sorrentino et al. (2007)
Structurization	
High mechanical and barrier properties with nanocomposite	Thompson and Singh (2006)
Nanoparticles disperse and act as reservoirs of active ingredients	Rhim et al. (2006)
Sensotextural	
Certain masking effect without destroying texture is possible	Rhim and Ng (2007)

(continued)

TABLE 15.1 (continued)
Prospective Advantages of Micro- or Nanotechnology for Novel Trends

Prospective Advantages	References
Optimization	Nachay (2007)
Integration of nanosensors for detection of pathogen	Nachay (2007)
Application of nanotechnology promise to expand the use of edible and biodegradable packaging	Chen et al. (2006), Rhim and Ng (2007)

of nanoresearch is estimated to be revolutionary in food packaging. The linkage of a 100% bioriginated material and nanomaterials opens new windows for answering environmental and health concerns (Jamshidian et al., 2010). It is believed that the first area that nanotechnology will impact upon in the food industry is food packaging. Nanotechnology is likely to be engaged for effective and efficient amendments of food products by bioactive and smart nanopackaging technology.

Nanocomposites packaging to create molecular barrier: In food packaging, a key issue is the development of high barrier properties against the migration of oxygen, carbon dioxide, flavor compounds, and water vapor. The nanoscale plate morphology of clays and other fillers helped the development of gas barrier properties. Thus, the development of nanocomposites may offer a new approach to improve mechanical strength, thermal stability, and gas barrier properties (Arora and Padua, 2010). Numerous biopolymers have been used to develop materials for eco-friendly food packaging. However, the use of biopolymers has been restricted due to their usually poor mechanical and barrier properties, which may be improved by adding reinforcing compounds (fillers), forming composites. Nanoparticles have proportionally larger surface area than their microscale counterparts, which favors the filler–matrix interactions and the performance of the resulting material (Azeredo, 2009). Incorporation of chitosan nanoparticles in the hydroxypropyl methylcellulose (HPMC) films improved their mechanical properties significantly, while also improving film barrier properties significantly. The chitosan poly(methacrylic acid) (CS-PMAA) nanoparticles tend to occupy the empty spaces in the pores of the HPMC matrix, inducing the collapse of the pores and thereby improving tensile and barrier properties of film (De Moura et al., 2008).

Nanoactive packaging for controlled release of additives: Hydrophilic or hydrophobic active agents can be encapsulated into liposomes and then dispersed in the temperature-reversible chitosan-glycerophosphate hydrogel. The liposomes provided excellent sustained drug release from chitosan matrix when compared with that of free drugs, especially for hydrophilic drugs. The negative charge of the liposomes was complexed with the positive charge of the chitosan's protonized amine group, which results in controlling drug release (Chiang et al., 2009).

Nanotechnology to improve micronutrition: The ferrous glycinate liposomes will possibly be a category of capable iron fortifier. Ferrous glycinate liposomes might be obtained with a high encapsulation efficiency (EE) of 84.80%. The stability of micro/nanoencapsulated ferrous glycinate in a strong acid environment was greatly improved by protecting it from the extracapsular environment by a lipid bilayer. The bioavailability of ferrous glycinate, as the iron source for biological activity including hemoglobin formation, may be increased (Ding et al., 2009).

To enhance the curcumin absorption by oral administration, liposome-encapsulated curcumin (LEC) was prepared from commercially available lecithins. High bioavailability of curcumin was evident in the case of oral LEC; a faster rate and better absorption of curcumin were observed as compared to the other forms. Thus, liposome encapsulation of ingredients such as curcumin may be used as a novel nutrient delivery system (Takahashi et al., 2009).

Nanoactive packaging for improved shelf life: The nanopackaging might provide an attractive alternative to improve preservation quality of the strawberry fruits during extended storage. A novel nanopacking material was synthesized by blending polyethylene with nanopowder (nano-Ag, kaolin, TiO). This nanopackaging was able to maintain the sensory, physicochemical, and physiological quality of strawberry fruits at a higher level compared with the normal packing (polyethylene bags) at 4°C. Additionally, nanopacking has the advantages of simple processing and feasibility to be industrialized in contrast with other storages (Yang et al., 2010).

Nanosensors and bioswitches: In effect, the necessity to generate fast, reliable, and precise information on the quality and security of foodstuffs and food industry has resulted in an intensive search for more selective and sensitive analytical methods. Chemical sensor and biosensor technology for use in this area has taken advantage of the unique merits of nanomaterials rather early. Various nanobased sensing approaches for exogenous compounds (e.g., pesticides, toxic anions, ripening gases, or vitamin supplements) and endogenous compounds (from microorganisms to vitamins) in food are under development (Valdes et al., 2009). Researchers are working on nanoparticle films and other packaging with embedded sensors that will detect food pathogens. With the electronic tongue technology, the sensors can detect substances in parts per trillion (PPT) and would trigger a color change in the packaging to alert the consumer about food contamination or spoilage.

Nanotechnology to develop antimicrobial packaging: The silver-based nanoclay showed strong antimicrobial activity against Gram-negative *Salmonella* spp. The active agent dispersed well throughout the PLA matrix to a nanoscale, yielding nanobiocomposites. The films were transparent with improved water barrier and strong antimicrobial properties. The migration levels of silver were within the specific migration levels referenced by the European Food Safety Agency (EFSA), and antimicrobial activity displayed supported the potential application of this biocidal additive in nanoactive food-packaging applications (Busolo et al., 2010). Nanocomposite low-density polyethylene films containing Ag and ZnO nanoparticles were prepared to preserve orange juice. The microbial growth rate significantly reduced as a result of using this nanocomposite packaging material. Packaging made from nanocomposite film containing nanosilver showed more pronounced antimicrobial effects, as compared with nano-ZnO during 112 days storage of inoculated orange juice (Emamifar et al., 2010).

15.3.2 Active Agent Nanoencapsulation: Basis of Nanoactive Packaging

A key approach for superior delivery of active ingredients is "micro- and nanoencapsulation." In the following section, we have attempted to discuss some micro- and nanotechnology trends to fabricate the nanoparticles to be incorporated eventually in active food packaging.

15.3.2.1 Example of Antimicrobial Nanoencapsulation

Nisin: Excessive nisin amounts are required for guaranteeing effective pathogen growth inhibition because nisin is structurally unstable in food due to its deprivation by interaction with food and cell matrices and the development of tolerant and resistant *Listeria* strains (Chi-Zhang et al., 2004). On the other hand, proteolytic enzymes in the food systems, especially in fresh meat products, are responsible for the inactivation of bacteriocin and thus decreasing antimicrobial efficacy (Degnan et al., 1993). In order to improve bioavailability, the practical way reported is encapsulation of bacteriocins, which limits the degree of its degradation in a food model system. The higher stability of encapsulated nisin may be attributed to its maintenance at a high concentration and purity inside nanovesicles or immobilized on vesicle membranes (Laridi et al., 2003).

The ability of liposomes to withstand exposure to the environmental and chemical stresses, typically encountered in foods and food processing operations, was analyzed by the EE, ζ-potential, and particle size distribution. Liposomes consisting of distearoylphosphatidylcholine and distearoylphosphatidylglycerol, with nisin entrapped, retained 70%–90% EE despite exposure to elevated temperatures (25°C–75°C) and a range of pH (5.5–11.0). Results suggest that liposomes may be an appropriate candidate for nisin entrapment in low- or high-pH foods with light heat treatment tolerance (Taylor et al., 2007).

Benech et al. (2002) investigated whether the encapsulation of nisin Z in liposomes can provide a powerful tool to improve nisin stability and inhibitory action. The inhibition of *Listeria innocua* in cheddar cheese was evaluated during 6 months and compared by in situ production of nisin Z by *Lactococcus lactis* subsp. *lactis biovar diacetylactis* UL719. Immediately after cheese production, 3- and 1.5-log-unit reductions in viable counts of *L. innocua* were obtained in cheeses with encapsulated nisin and the nisinogenic starter, respectively. After 6 months, cheeses made with encapsulated nisin contained less than 101 CFU of *L. innocua* per gram and 90% of the initial nisin activity, compared with 10^4 CFU/g and only 12% of initial activity in cheeses made with the nisinogenic starter (Benech et al., 2002).

Pediocin: In a meat model system, the entrapment of pediocin AcH in liposomes (18% entrapment efficiency) made from phosphatidylcholine improved the antilisterial activity of pediocin compared with free pediocin. In case of direct incorporation, a decrease in pediocin activity (12%–54% recovery of original activity) occurred. Higher pediocin activity (29%–62% increase; average overall concentrations) was recovered from the model food system containing encapsulated bacteriocin as compared with free pediocin AcH (Degnan and Luchansky, 1992). The additional recovery of pediocin activity provided by liposomes decreased the potential for the direct application of biopreservatives.

Bacteriocin-loaded micro-/nanoparticles seem to be promising formulations to achieve long-lasting antimicrobial activity. These polymeric micro-/nanocolloids are physically stable and can be easily formulated with a variety of materials obtaining the controlled release rate of the active agent (Salmaso et al., 2004).

Lactoferrin: Lactoferrin has been used as a potential antimicrobial in coatings, but the cations including Na^+, Ca^2, and Mg^2 interfere with its activity (Al-Nabulsi and Holley, 2005; Franklin et al., 2004). To avoid this microencapsulation and controlled release, technology has found broad applications. Paste-like microcapsules were incorporated in edible whey protein isolates (WPI) packaging film to test the antimicrobial activity of LF against a meat spoilage organism *Carnobacterium viridans*. The film was applied to the surface of bologna after its inoculation with the organism and stored under vacuum at 4°C or 10°C for 28 days. The growth of *C. viridans* was delayed at both temperatures, and microencapsulated LF had greater antimicrobial activity than when unencapsulated (Al-Nabulsi et al., 2006).

Lysozyme: Lysozyme is of interest for use in food systems since it is a naturally occurring enzyme with antimicrobial activity. Entrapment efficiency, the mean average size, and the stability of the commercially available form that contains 2.5% nisin and lysozyme are influenced by lipid composition. Encapsulation of commercial nisin extract and lysozyme in PC-, PG-, and cholesterol-containing liposomes was achieved. The highest concentration of antimicrobials was encapsulated in 100% PC liposomes which also resulted in higher leakage. The antimicrobial loading was decreased by the addition of cholesterol and PG, but cholesterol decreased the leakage of PC liposomes (Gregoriadis and Davis, 1979; Hsieh et al., 2002). Application of nisin and lysozyme affected liposome stability; nevertheless, the intact encapsulated liposomes were physically stable for 2 weeks (Were et al., 2003). We can assume that for microbiological stabilization of food products, stable nanoparticulates of polypeptide antimicrobials can be achieved by selecting suitable lipid–antimicrobial combinations.

Silver: The use of nanoparticle metallic silver particles as an antimicrobial agent in polyurethane coatings has been achieved, which are applied on particular parts of food packaging machines and also on food handling robots to reduce the risk of bacterial contamination (Wagener, 2006). The effectiveness of Ag^+-based antimicrobial film in inhibiting the growth of an *A. acidoterrestris* strain in acidified malt extract broth and apple juice was investigated. The results indicate a 2-log comparative reduction in viable microbial count of thermal resistance microorganisms in acidic beverages (Vermeiren et al., 2002). Concerns relating to the amount of silver used in edible films should not be neglected during the production of such active films.

Silver nanoparticle thin layers were deposited onto medical and food-grade silicone rubber, stainless steel, and paper surfaces. The antimicrobial properties of the silver-coated surfaces were demonstrated by exposing them to *Listeria monocytogenes*. No viable bacteria were detected after 12–18 h on silver-coated silicone rubber surfaces; thus, 4–5 log reduction was achieved (Jiang et al., 2004).

These results depict that silver is one of the strongest bactericides, and it becomes more effective as nanoparticles.

Chitosan: Recently, a study has been carried out to develop biodegradable antimicrobial bionanocomposite films with acceptable properties for applications in food packaging using biopolymers such as chitosan, as well as nanosilver, silver zeolite, and nanoscale-layered silicates. Tensile strength increased by 7%–16%, whereas water vapor permeability decreased by 25%–30% depending on the nanoparticle material tested. The silver-containing chitosan-based nanocomposite films depicted a potential of antimicrobial activity (Al-Nabulsi and Holley, 2006). The chitosan films containing silver nanocomposites depicted an excellent potential of antimicrobial activity.

15.4 BIOACTIVE DELIVERY MECHANISMS AT MICRO OR NANOSCALE

To date, the literature reveals that relatively low attention has been given to the release mechanism of diverse active agents from liposome at micro- or nanoscale. The research studies related to the application of encapsulation concept in the food are mainly focused on methodology, liposomal composition, EE, and prolonged stability. However, little emphasis has been placed on the release phenomenon occurring at molecular or membrane level.

The question arises, what type of mechanisms are involved in the binding of active agents to the liposomal membrane. Is this binding really irreversible and just caused by their lipo- or hydrophilicity? Do other factors participate in leakage kinetics from the liposomal bilayer membrane? The appreciation of the observed reversible character of the active agent binding is of paramount importance, since the term "encapsulated" is often overinterpreted as being "irreversibly encapsulated" (Fahr et al., 2005).

In the case of the active agents capable of forming pores, like antimicrobial bacteriocins, different postulates of the mechanism of factors of release include the following: liposomes are "kinetically stable" for a defined period of time; "fusion" between liposome and pathogen outer membrane; core (antimicrobial) material induced leakage; liposome membrane permeability, which, for instance, had been reduced by cholesterol; diffusion due to low molecular weight of the antimicrobials; certain antimicrobials like nisin induce leakage by changing membrane structure; interaction between liposome and fat globule membrane may result in destabilization of liposomal membrane and subsequent release of active compounds and bivalent ions Ca^{++}/Mg^{++} induce de-stability of liposomes, especially in the food products containing higher bivalent ions (Al-Nabulsi and Holley, 2006; Anal and Singh, 2007; Fahr et al., 2005; Jiang et al., 2004; Laridi et al., 2003; Taylor et al., 2005; Vermeiren et al., 2002; Wagener, 2006; Were et al., 2003, 2004).

Table 15.2 summarizes the critical features of food active agents (antioxidant, antimicrobials, enzymes, amino acids, vitamins, enzymes, fortifiers, essential fatty

TABLE 15.2
Micro- or Nanoencapsulation of Various Bioactive Agents in Food Systems, Their Key Features, and Mechanisms of Release (In Vitro/In Vivo)

Bioactive Agent	Particle/Vesicle Composition	Encapsulation Objective	Encapsulation Method	Size (nm)	Entrapment Efficiency (%)	Mechanisms of Release H: Hypothetical; P: Proved	References
Antioxidant							
Curcumin	Commercial soya lecithin liposome	To improve gastrointestinal absorption	Microfluidization	220	68	H: The mechanisms in uptake include steps of the diffusion of particles through mucus and accessibility to an enterocyte surface, epithelial interaction, cellular trafficking, exocytosis, and systemic dissemination	Takahashi et al. (2009)
Quercetin	(SLP-PC70/white) Eudragit® E nanoparticles	Solubility	Precipitation method	82–473	95–99.9	H: The dissolution increased due to particle size reduction and high-energy amorphous state formation	Wu et al. (2008)
Bacteriocin							
Bacteriocin-like substance (BLS)	Phosphatidyl-choline (PC) liposomes	Slow release, lesser amount	Reverse phase evaporation	570	26	H: Weak hydrophobic interaction between BLS and PC results in slow release	Teixeira et al. (2008)
Nisin	Phospholipon mixture liposomes	Long-term preservation, nisin protection	Mozafari method	190–284	12–54	H/P: Fusion of liposome with bacterial membrane proved with electron microscope and lipid mixing assay, but intracellular nisin delivery is not yet proved	Colas et al. (2007)

Nisin	Proliposome	Longer stability	Heating method	725–800	9.5–47	H: Extraliposomal environmental conditions, bivalent ions like Ca^{2+} and Mg^{2+}, fat concentration in medium	Laridi et al. (2003)
Nisin	PC: Phosphatidyl-glycerol (PG) liposome	Stability against pH	Reverse phase evaporation	100–240	72–89 (Calcein dye)	H: Nisin rapidly creates pores in membranes that contain high levels of anionic lipids H: Brownian motion inducing rearrangement of liposome and thus reducing its stability	Taylor et al. (2007)
Nisin and lysozyme (LZ)	PC:PG:Cholesterol liposome	Microbiological safety of food products	Reverse phase evaporation	85–145 161–174	54–71 (Nisin) 60–61 (LZ) (Calcein dye)	H: Unstable pore formation due to binding of nisin to negatively charged head groups of phospholipids H: Insertion of antimicrobial into liposome membrane due to its low molecular weight	Were et al. (2003)
Enzyme							
Flavourzyme (FZ)	Chitosan: Alginate capsules	Controlled proteolysis of cheese	Extrusion	Microscale	7–84	H: Enzyme release by the pressure applied during pressing the cheese curd	Anjani et al. (2007)
Chymotrypsin	Proliposome	Cheese accelerated ripening	Agitation/ hydration	—	63	H: The lipid exchange between milk fat globule membranes and liposomes might explain the early release of enzymes	Laloy et al. (1998)

(continued)

TABLE 15.2 (continued)
Micro- or Nanoencapsulation of Various Bioactive Agents in Food Systems, Their Key Features, and Mechanisms of Release (In Vitro/In Vivo)

Bioactive Agent	Particle/Vesicle Composition	Encapsulation Objective	Encapsulation Method	Size (nm)	Entrapment Efficiency (%)	Mechanisms of Release H: Hypothetical; P: Proved	References
Glucose oxidase	Egg PC: Cholesterol liposome	Sustained release of hydrogen peroxide	Dehydration rehydration (DRV) method	—	24	H: The liposomal membrane was considered as semipermeable, thus enzyme release by permeation	Rodriguez-Nogales (2004), Rodriguez-Nogales et al. (2004)
Lipase/protease/FZ	Proliposome VPF	Cheese texture and flavor improvement	Agitation/hydration	Microscale	20–36	H: Rupturing of liposome during cheese making H: Release of encapsulated agents due to liposome membrane collapse by lipase action	Kheadr et al. (2003)
Fortifier							
Iron (Ferrous glycinate)	Egg lecithin liposome	To improve bio-availability	Reverse phase evaporation	207–559	66–84.8	P: The leakage is attributed to low pH resulting in instability of liposome related to permeation of protons P: Bile salts are surfactants that seem to break down the bilayer membranes	Ding et al. (2009)

Nutrient							
PUFA + vitamin E + flavonoid	Marine lecithin emulsion	PUFA stability against oxidation	High-pressure homogenization	160–207	Emulsion based	H: Release due to thermodynamic instability of emulsions affected by temperature, fat, and lecithin ration	Belhaj et al. (2010)
Miscellaneous							
Free amino acids (FAA)	Soy lecithin liposome	To enhance nutritional value	Reverse phase evaporation and rehydration of freeze-dried	2000–8000	42.6	H: Release tested under in vitro simulating condition but no mechanism of release investigated	Barr and Helland (2007)
Curcuminoids	Polybutylcyano-acrylate	To improve bioavailability and stability against photo degradation	Solvent evaporation method	173–281	66–82	H/P: Release due to environment changes in the cell, like, acidic value; however, what phenomenon occurs on membrane level due to low pH is nct proved	Mulik et al. (2009)

PUFA, Polyunsaturated fatty acid.

acids) by encapsulation in the micro- or nanoparticles (liposome), with release mechanisms/factors under in vitro or in vivo conditions. A few research studies had proved the actual phenomenon underlying the release from liposome. In the case of antimicrobial release for minimizing food-borne illness, transmission electron microscope reveals the fusion of liposome with the bacterial cell envelope to discharge the encapsulated nisin in liposome (Colas et al., 2007). Other interesting facts had revealed the effects of pH and bile salts for increasing the permeation of iron (ferrous glycinate) through liposomal membrane due to bilayer instability and surfactant action, respectively (Ding et al., 2009). Various logical hypotheses had been formulated for different active agent release from liposomal encapsulation, but they are yet to be proved (Table 15.2).

15.4.1 Release Efficiency of Bioactives at Micro/Nanoscale

Encapsulation and targeting the bioactive agents—including nutrients, drugs, vaccines, and cosmetics—and their protection from degradation and inactivation have been investigated extensively using microencapsulation systems (Mozafari et al., 2006, 2008). The complex term *release* includes two major phenomena: diffusion of active agents from liposome core through uni/multilamellar bilayers and desorption from liposome into the medium where its bioavailability is required (Champagne and Fustier, 2007). A recent study has investigated the first phase of the release process of lipophilic agents in aqueous liposomal dispersions. The sequential steps included (1) active agent dissolved in the lipid domain of the membrane, (2) departure of the drug from the membrane into the aqueous phase, (3) association of the drug component in the aqueous phase with the acceptor membrane, followed by (4) dissolving of the drug in the acceptor membrane. These steps may differ at high phospholipid concentrations, where it is believed that collision between the lipid vesicles is the main transferring mechanism (Fahr et al., 2005). Different mechanisms of active agent/drug delivery from nanoparticles to the target cell are illustrated in Figure 15.3.

Owing to either the lack of advanced tools to study the nanostructured materials or the novelty of this field, a study of the literature shows no significant work done to elaborate the release process of food active agents from liposomes. An exciting work related to antimicrobial release from liposomes had analyzed this phenomenon by using calcein (fluorescent dye) (Were et al., 2003). However, the major complication to imitate the nisin release in this manner is its capability to form pores in membrane models (liposome), and thus, results of its release will be quite different from fluorescent dye, as a high concentration of this antimicrobial bioactive would disrupt the membrane instead of slow inside-out migration.

Table 15.3 summarizes the different approaches utilized till date for the quantification of the release efficiency of encapsulated active agents being employed in the food sector. More than half of the research data comprise an indirect evaluation of bioactive agents through residual enzymatic or antimicrobial activity measurement. It can be foreseen that with improvements in manufacturing technologies, new strategies for stabilization of fragile nutraceuticals, and development of novel approaches

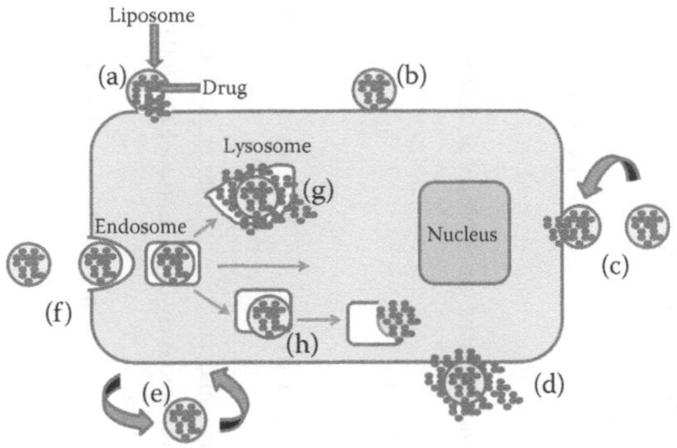

FIGURE 15.3 Drug-loaded liposomes can specifically (a) or nonspecifically (b) adsorb onto the cell surface. Liposomes can also fuse with the cell membrane (c), and release their contents into the cell cytoplasm, or can be destabilized by certain cell membrane components when adsorbed on the surface (d) so that the released drug can enter cell via micropinocytosis. Liposome can undergo the direct or transfer-protein-mediated exchange of lipid components with the cell membrane (e) or be subjected to a specific or nonspecific endocytosis (f). In the case of endocytosis, a liposome can be delivered by the endosome into the lysosome (g) or, en route to the lysosome, the liposome can provoke endosome destabilization (h), which results in drug liberation into the cell cytoplasm. (From Torchilin, V.P., *Nat. Rev. Drug Discov.*, 4, 145, 2005.)

to site-specific carrier targeting, encapsulation carriers will play an important role in increasing the efficacy of functional foods (Chen et al., 2006).

At the present time, greater fundamental understanding of polymer–polymer and polymer–active agent interactions at the molecular level is still required to ensure the design of ideal nutraceutical carriers for use in the food industry. The labeling of the active agents with colored dyes or radioactive probes may result as an exciting approach toward the direct quantification of bioactive bioavailability and its mechanism of release.

15.5 NANOACTIVE PACKAGING: CONCLUDING NOTES

Thus far at its infancy stage, nanoactive packaging is an innovative cutting edge approach to enhance food safety and security. The nanoactive packaging strengthens the preexisting principal objectives of packaging, which includes physical protection (mechanical parameters), molecular barrier (oxygen, water vapor, dust, or chemicals), sensory attributes (transparency, homogenous topographical attributes), high quality food (biosensors, smart nanosystems), and extended shelf life or food safety (nanoactive particles, controlled release of antimicrobials, and target delivery). The need-based evolution toward nanoactive packaging for superior packaged food quality is illustrated in Figure 15.4. Furthermore, the nanoactive concept complements the eco-friendly biodegradable polymers for environmental conservation. The adaptations toward the bioactive, biodegradable,

TABLE 15.3
Different Approaches of Quantification of the Release Efficiency of Encapsulated Active/Fortification Agents

Bioactive	Release Efficiency	Quantification D: Direct; ID: Indirect	References
Antioxidant			
Quercetin	Higher solubility of nanosystem, 74-fold higher release	D: Partition coefficient analysis based on UV-spectrometer ID: Antioxidant activities assays	Wu et al. (2008)
Bacteriocin			
BLS	The encapsulated BLS remained with approximately 90% its initial activity for 30 days, as compared with 14 days for free BLS	ID: Residual antimicrobial activity against *L. monocytogenes*	Teixeira et al. (2008)
Nisin	Indirectly correlated to stability of liposome	ID: Size stability and image analysis	Colas et al. (2007)
Nisin	After 18 days storage 39.2%–78.2% (of initially encapsulated) nisin was released in different food medium	D: Competitive enzyme immunoassay (c-EIA)-based quantification	Laridi et al. (2003)
Nisin and lysozyme	During 35 min, 5%–20% release was observed for a mixture of nisin and lysozyme for different PC:PG:Cholesterol liposome	ID: Fluorescent assay based on calcein dye release	Were et al. (2003)
Enzyme			
Flavourzyme	During 4–16 h under simulated cheese block pressing, only 30% of enzyme was released from microcapsules and thus beneficial for initial stages of cheese maturation	ID: Enzyme quantification analyzed indirectly by its activity against substrate L-leucine-p-nitroanilide expressed as leucine amino peptidase units (LAPU) using spectrophotometer	Anjani et al. (2007)

Chymotrypsin	During the first 3 weeks, the liposome enzyme treated cheese had little decreased in residual activity as compared with >20% decrease for free enzyme treated cheese model	ID: Residual enzymatic activity in cheese was characterized against fluorescein isothyocyanate–casein (FITC)	Laloy et al. (1998)
Fortifier			
Iron	Bile salt micelles and vesicles can solubilize phospholipid bilayer membranes. A small amount of ferrous glycinate was released from liposomes in the first 4 h in the medium of pH 1.3. However, after 20 h, more than 50% ferrous glycinate was released	D: The in vitro (simulated gastrointestinal conditions) release of ferrous glycinate from liposomes was measured by bathophenanthrolin Colorimetry and dialysis method	Ding et al. (2009)
Miscellaneous			
Curcuminoids	Initial burst release obtained for first 4–5 h attributed to the drug adsorbed on the surface of the nanoparticles. After 24 h, 77% and 85% cumulative release was observed for neutral and acidic medium, respectively	D: HPLC method using UV–vis detector and compared with standard curcuminoid solution	Mulik et al. (2009)

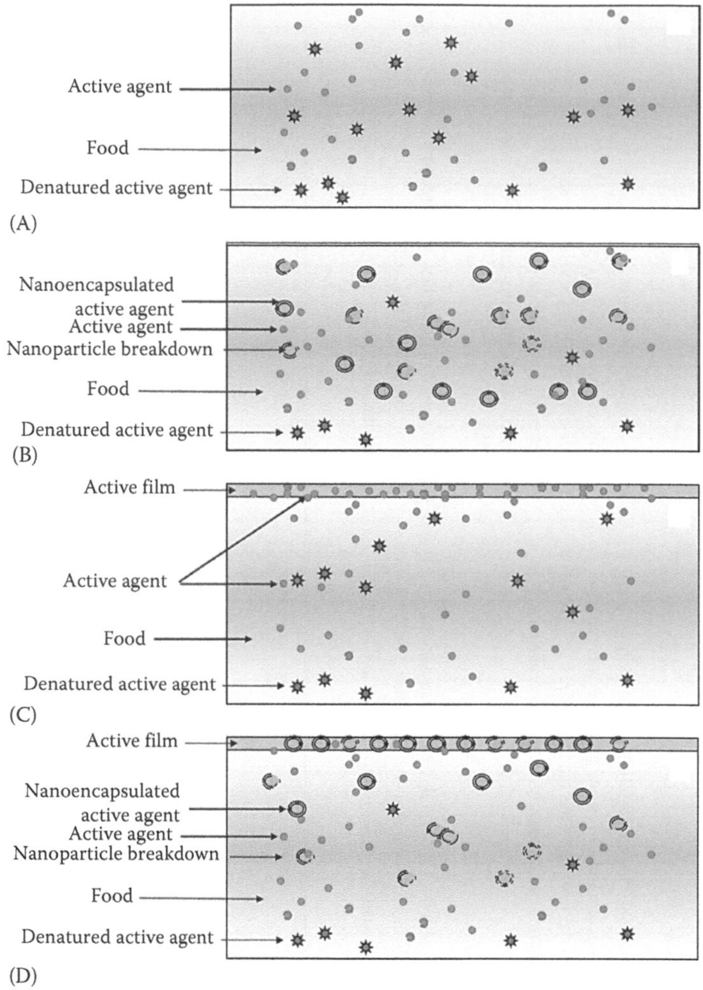

FIGURE 15.4 Illustration of different approaches opted for food security. (A) Direct incorporation of active agent, i.e., free form into food; major disadvantages—degradation/denaturation of active agent by enzymes/ions/food matrices interaction/fat, no protection at food surface where microbial/contamination load is maximum, costly as excessive quantity of active agent required, and low shelf-life of food. (B) Nanoencapsulation of active and subsequent incorporation in food; major disadvantage—inadequate quality due to contamination and microbial spoilage at food surface. (C) Active packaging, incorporation of active agent in food coatings for controlled release; major disadvantage—inactivation of additive by complex food system; however, sustained desorption/diffusion of active agent from package toward food. (D) Nanoactive packaging, fusion of all the aforementioned approaches to eliminate the respective disadvantages for longer food shelf life.

and bionanocomposite concepts in nanoactive packaging are most likely to be the smartest development yet to be seen in modern packaging innovations to provide healthier food.

REFERENCES

Al-Nabulsi, A.A., Han, J.H., Liu, Z., Rodrigues-Vieira, E.T. and Holley, R.A. (2006). Temperature-sensitive microcapsules containing lactoferrin and their action against *Carnobacterium viridans* on Bologna, *Journal of Food Science*, 71: M208–M214.

Al-Nabulsi, A.A. and Holley, R.A. (2005). Effect of bovine lactoferrin against *Carnobacterium viridans, Food Microbiology*, 22: 179–187.

Al-Nabulsi, A.A. and Holley, R.A. (2006). Enhancing the antimicrobial effects of bovine lactoferrin against *Escherichia coli* O157:H7 by cation chelation, NaCl and temperature, *Journal of Applied Microbiology*, 100: 244–255.

Anal, A.K. and H. Singh, H. (2007). Recent advances in microencapsulation of probiotics for industrial applications and targeted delivery, *Trends in Food Science and Technology*, 18: 240–251.

Andersson, C., Järnstöm, L., Fogden, A., Mira, I., Voit, W., Zywicki, S. and Bartkowiak, A. (2009). Preparation and incorporation of microcapsules in functional coatings for self-healing of packaging board, *Packaging Technology and Science*, 22: 275–291.

Anjani, K., Kailasapathy, K. and Phillips, M. (2007). Microencapsulation of enzymes for potential application in acceleration of cheese ripening, *International Dairy Journal*, 17: 79–86.

Arora, A. and Padua, G.W. (2010). Review: Nanocomposites in food packaging, *Journal of Food Science*, 75: R43–R49.

Azeredo, H.M.C.D. (2009). Nanocomposites for food packaging applications, *Food Research International*, 42: 1240–1253.

Barr, Y. and Helland, S. (2007). A simple method for mass-production of liposomes, in particular large liposomes, suitable for delivery of free amino acids to filter feeding zooplankton, *Journal of Liposome Research*, 17: 79–88.

Belhaj, N., Arab-Tehrany, E. and Linder, M. 2010. Oxidative kinetics of salmon oil in bulk and in nanoemulsion stabilized by marine lecithin. *Process Biochemistry* 45 (2) 187–195.

Benech, R.O., Kheadr, E.E., Lacroix, C. and Fliss, I. (2002). Antibacterial activities of nisin Z encapsulated in liposomes or produced in situ by mixed culture during Cheddar cheese ripening, *Applied and Environmental Microbiology*, 68: 5607–5619.

Brody, A.L. (2006). Nano and food packaging technologies converge, *Food Technology*, 60: 92–94.

Bugusu, B. and Lubran, M. (2007). International food nanoscience conference, *Food Technology*, 61: 121–124.

Busolo, M.A., Fernandez, P., Ocio, M.J. and Lagaron, J.M. (2010). Novel silver-based nanoclay as an antimicrobial in polylactic acid food packaging coatings, *Food Additives and Contaminants—Part A Chemistry, Analysis, Control, Exposure and Risk Assessment*, 27: 1617–1626.

Buyong, N., Kok, J. and Luchansky, J.B. (1998). Use of a genetically enhanced, pediocin-producing starter culture, Lactococcus lactis subsp. lactis MM217, to control Listeria monocytogenes in cheddar cheese, *Applied and Environmental Microbiology*, 64: 4842–4845.

Champagne, C.P. and Fustier, P. (2007). Microencapsulation for the improved delivery of bioactive compounds into foods, *Current Opinion in Biotechnology*, 18: 184–190.

Chaudhry, Q., Scotter, M., Blackburn, J., Ross, B., Boxall, A., Castle, L., Aitken, R. and Watkins, R. (2008). Applications and implications of nanotechnologies for the food sector, *Food Additives and Contaminants—Part A Chemistry, Analysis, Control, Exposure and Risk Assessment*, 25: 241–258.

Chen, L., Remondetto, G.E. and Subirade, M. (2006). Food protein-based materials as nutraceutical delivery systems, *Trends in Food Science and Technology*, 17: 272–283.

Chiang, H., Huang, Y.C., Yeh, H.Y., Yeh, S.Y. and Huang, Y.Y. (2009). The drug release from liposomal carrier within the chitosan matrix, *Biomedical Engineering—Applications, Basis and Communications*, 21: 107–114.

Chi-Zhang, Y., Yam, K.L. and Chikindas, M.L. (2004). Effective control of Listeria monocytogenes by combination of nisin formulated and slowly released into a broth system, *International Journal of Food Microbiology*, 90: 15–22.

Colas, J.C., Shi, W., Rao, V.S.N.M., Omri, A., Mozafari, M.R. and Singh, H. (2007). Microscopical investigations of nisin-loaded nanoliposomes prepared by Mozafari method and their bacterial targeting, *Micron*, 38: 841–847.

Court, E.B., Daar, A.S., Persad, D.L., Salamanca-Buentello, F. and Singer, P.A. (2005). Tiny technologies for the global good, *Materials Today*, 8: 14–15.

Degnan, A.J., Buyong, N. and Luchansky, J.B. (1993). Antilisterial activity of pediocin AcH in model food systems in the presence of an emulsifier or encapsulated within liposomes, *International Journal of Food Microbiology* 18: 127–138.

Degnan, A.J. and Luchansky, J.B. (1992). Influence of beef tallow and muscle on the antilisterial activity of pediocin AcH and liposome-encapsulated pediocin AcH, *Journal of Food Protection* 55: 552–554.

De Moura, M.R., Avena-Bustillos, R.J., McHugh, T.H., Krochta, J.M. and Mattoso, L.H.C. (2008). Mattoso, Properties of novel hydroxypropyl methylcellulose films containing chitosan nanoparticles, *Journal of Food Science*, 73: N31–N37.

Ding, B., Xia, S., Hayat, K. and Zhang, X. (2009). Preparation and pH stability of ferrous glycinate liposomes, *Journal of Agricultural and Food Chemistry*, 57: 2938–2944.

El Naschie, M.S. (2006). Nanotechnology for the developing world, *Chaos, Solitons and Fractals*, 30: 769–773.

Emamifar, A., Kadivar, M., Shahedi, M. and Soleimanian-Zad, S. (2010). Effect of nanocomposite packaging containing Ag and ZnO on inactivation of Lactobacillus plantarum in orange juice, *Food Control*, 22: 408–413.

Fahr, A., Hoogevest, P.V., May, S., Bergstrand, N. and Leigh, M.L.S. (2005). Transfer of lipophilic drugs between liposomal membranes and biological interfaces: Consequences for drug delivery, *European Journal of Pharmaceutical Sciences*, 26: 251–265.

Fonseca, L., Cane, C. and Mazzolai, B. (2007). Application of micro and nanotechnologies to food safety and quality monitoring, *Measurement and Control*, 40: 116–119.

Franklin, N.B., Cooksey, K.D. and Getty, K.J.K. (2004). Inhibition of Listeria monocytogenes on the surface of individually packaged hot dogs with a packaging film coating containing nisin, *Journal of Food Protection*, 67: 480–485.

Garcia, M., Forbe, T. and Gonzalez, E. (2010). Potential applications of nanotechnology in the agro-food sector, *Ciencia e Tecnologia de Alimentos*, 30: 573–581.

Gregoriadis, G. and Davis, C. (1979). Stability of liposomes in vivo and in vitro is promoted by their cholesterol content and the presence of blood cells, *Biochemical and Biophysical Research Communications*, 89: 1287–1293.

Hsieh, Y.F., Chen, T.L., Wang, Y.T., Chang, J.H. and Chang, H.M. (2002). Properties of liposomes prepared with various lipids, *Journal of Food Science*, 67: 2808–2813.

Imran, M., Revol-Junelles, A.M., Martyn, A., Tehrany, E.A., Jacquot, M., Linder, M. and Desobry, S. (2010). Active food packaging evolution: Transformation from micro- to nanotechnology, *Critical Reviews in Food Science and Nutrition*, 50: 799–821.

Jacquot, M., Revol-Junelles, A.M., Milliere, J.B., Miclo, A. and Poncelet, D. (2000). Anti Listeria monocytogenes activity of the lactoperoxidase system (LPS) using encapsulated substrates, *Minerva Biotecnologica*, 12: 301–304.

Jamshidian, M., Tehrany, E.A., Imran, M., Jacquot, M. and Desobry, S. (2010). Poly-Lactic Acid: Production, applications, nanocomposites, and release studies, *Comprehensive Reviews in Food Science and Food Safety*, 9: 552–571.

Jiang, H., Manolache, S., Wong, A.C.L. and Denes, F.S. (2004). Plasma-enhanced deposition of silver nanoparticles onto polymer and metal surfaces for the generation of antimicrobial characteristics, *Journal of Applied Polymer Science*, 93: 1411–1422.

Kheadr, E.E., Vuillemard, J.C. and El-Deeb, S.A. (2003). Impact of liposome-encapsulated enzyme cocktails on cheddar cheese ripening, *Food Research International*, 36: 241–252.

Kolanowski, W., Jaworska, D., Laufenberg, G. and Weißbrodt, J. (2007). Evaluation of sensory quality of instant foods fortified with omega-3 PUFA by addition of fish oil powder, *European Food Research and Technology*, 225: 715–721.

Kolanowski, W., Ziolkowski, M., Weißbrodt, J., Kunz, B. and Laufenberg, G. (2006). Microencapsulation of fish oil by spray drying—Impact on oxidative stability. Part 1, *European Food Research and Technology*, 222: 336–342.

Laloy, E., Vuillemard, J.C., Dufour, P. and Simard, R. (1998). Release of enzymes from liposomes during cheese ripening, *Journal of Controlled Release*, 54: 213–222.

Laridi, R., Kheadr, E.E., Benech, R.O., Vuillemard, J.C., Lacroix, C. and Fliss, I. (2003). Liposome encapsulated nisin Z: Optimization, stability and release during milk fermentation, *International Dairy Journal*, 13: 325–336.

Lee, S.J. and Rosenberg, M. (2000). Whey protein-based microcapsules prepared by double emulsification and heat gelation, *LWT—Food Science and Technology*, 33: 80–88.

Luff, J. (2007). Omega-3 and micro-encapsulation technology—Making functional foods taste better for longer, *Food Science and Technology*, 21: 30–31.

Morais, H.A., Da Silva Barbosa, C.M., Delvivo, F.M., Mansur, H.S., De Oliveira, M.C. and Silvestre, M.P.C. (2004). Comparative study of microencapsulation of casein hydrolysates in lipospheres and liposomes, *Journal of Food Biochemistry*, 28: 21–41.

Moraru, C.I., Panchapakesan, C.P., Huang, Q., Takhistov, P., Liu, S. and Kokini, J.L. (2003). Nanotechnology: A new frontier in food science, *Food Technology*, 57: 24–29.

Mozafari, M.R., Flanagan, J., Matia-Merino, L., Awati, A., Omri, A., Suntres, Z.E. and Singh, H. (2006). Recent trends in the lipid-based nanoencapsulation of antioxidants and their role in foods, *Journal of the Science of Food and Agriculture*, 86: 2038–2045.

Mozafari, M.R., Johnson, C., Hatziantoniou, S. and Demetzos, C. (2008). Nanoliposomes and their applications in food nanotechnology, *Journal of Liposome Research*, 18: 309–327.

Mulik, R., Mahadik, K. and Paradkar, A. (2009). Development of curcuminoids loaded poly(butyl) cyanoacrylate nanoparticles: Physicochemical characterization and stability study, *European Journal of Pharmaceutical Sciences*, 37: 395–404.

Nachay, K. (2007). Analyzing nanotechnology, *Food Technology*, 61: 34–36.

Nel, A., Xia, T., Madler, L. and Li, N. (2006). Toxic potential of materials at the nanolevel, *Science*, 311: 622–627.

Nickols-Richardson, S.M. (2007). Nanotechnology: Implications for food and nutrition professionals, *Journal of the American Dietetic Association*, 107: 1494–1497.

NNI, National Nanotechnology Initiative. (2001). Available at http://www.nano.gov/ (accessed on December 14, 2010).

Picot, A. and Lacroix, C. (2004). Encapsulation of bifidobacteria in whey protein-based microcapsules and survival in simulated gastrointestinal conditions and in yoghurt, *International Dairy Journal*, 14: 505–515.

Ravichandran, R. (2010). Nanotechnology applications in food and food processing: Innovative green approaches, opportunities and uncertainties for global market, *International Journal of Green Nanotechnology: Physics and Chemistry*, 1: 72–96.
RCEP, Royal Commission on Environmental Pollution. (2008). Novel materials in the environment: The case of nanotechnology. Available at http://www.rcep.org.uk/reports/27 (accessed on December 10, 2010).
Rhim, J.W., Hong, S.I., Park, H.M. and Ng, P.K.W. (2006). Preparation and characterization of chitosan-based nanocomposite films with antimicrobial activity, *Journal of Agricultural and Food Chemistry*, 54: 5814–5822.
Rhim, J.W. and Ng, P.K.W. (2007). Natural biopolymer-based nanocomposite films for packaging applications, *Critical Reviews in Food Science and Nutrition*, 47: 411–433.
Roberts, R.F. and Zottola, E.A. (1993). Shelf-life of pasteurized process cheese spreads made from cheddar cheese manufactured with a nisin-producing starter culture, *Journal of Dairy Science*, 76: 1829–1836.
Rodriguez-Nogales, J.M. (2004). Kinetic behaviour and stability of glucose oxidase entrapped in liposomes, *Journal of Chemical Technology and Biotechnology*, 79: 72–78.
Rodriguez-Nogales, J.M., PÃ©rez-Mateos, M. and Busto, M.D. (2004). Application of experimental design to the formulation of glucose oxidase encapsulation by liposomes, *Journal of Chemical Technology and Biotechnology*, 79: 700–705.
Rohner, F., Ernst, F.O., Arnold, M., Hilbe, M., Biebinger, R., Ehrensperger, F., Pratsinis, S.E., Langhans, W., Hurrell, R.F. and Zimmermann, M.B. (2007). Synthesis, characterization, and bioavailability in rats of ferric phosphate nanoparticles, *Journal of Nutrition*, 137: 614–619.
RSRAE, Royal Society and Royal Academy of Engineering. (2004). Nanoscienceand nanotechnologies: Opportunities and uncertainties. Available at http://www.nanotec.org.uk/finalreport. pdf (accessed on December 14, 2010).
Salmaso, S., Elvassore, N., Bertucco, A., Lante, A. and Caliceti, P. (2004). Nisin-loaded poly-L-lactide nano-particles produced by CO_2 anti-solvent precipitation for sustained antimicrobial activity, *International Journal of Pharmaceutics*, 287: 163–173.
Scott, N.R. and Chen, H. (2002). Nanoscale science and engineering for agriculture and food systems, in *National Planning Workshop*. Washington DC, November 18–19, 2002.
Shahidi, F. (2000). Antioxidants in food and food antioxidants, *Nahrung—Food*, 44: 158–163.
Sorrentino, A., Gorrasi, G. and Vittoria, V. (2007). Potential perspectives of bio-nanocomposites for food packaging applications, *Trends in Food Science and Technology*, 18: 84–95.
Takahashi, M., Uechi, S., Takara, K., Asikin, Y. and Wada, K. (2009). Evaluation of an oral carrier system in rats: Bioavailability and antioxidant properties of liposome-encapsulated curcumin, *Journal of Agricultural and Food Chemistry*, 57: 9141–9146.
Tan, C.P. and Nakajima, M. (2005). [beta]-Carotene nanodispersions: preparation, characterization and stability evaluation, *Food Chemistry*, 92: 661–671.
Taylor, T.M., Davidson, P.M., Bruce, B.D. and Weiss, J. (2005). Liposomal nanocapsules in food science and agriculture, *Critical Reviews in Food Science and Nutrition*, 45: 587–605.
Taylor, T.M., Gaysinsky, S., Davidson, P.M., Bruce, B.D. and Weiss, J. (2007). Characterization of antimicrobial-bearing liposomes by zeta potential, vesicle size, and encapsulation efficiency, *Food Biophysics*, 2: 1–9.
Teixeira, M.L., dos Santos, J., Silveira, N.P. and Brandelli, A. (2008). Phospholipid nanovesicles containing a bacteriocin-like substance for control of Listeria monocytogenes, *Innovative Food Science and Emerging Technologies*, 9: 49–53.
Thompson, A.K. and Singh, H. (2006). Preparation of liposomes from milk fat globule membrane phospholipids using a microfluidizer, *Journal of Dairy Science*, 89: 410–419.
Torchilin, V.P. (2005). Recent advances with liposomes as pharmaceutical carriers, *Nature Reviews Drug Discovery*, 4: 145–160.

Valdes, M.G., Gonzalez, A.C.V., Calzon, J.A.G. and Diaz-Garcia, M.E. (2009). Analytical nanotechnology for food analysis, *Microchimica Acta*, 166: 1–19.

Vermeiren, L., Devlieghere, F. and Debevere, J. (2002). Effectiveness of some recent antimicrobial packaging concepts, *Food Additives and Contaminants*, 19: 163–171.

Wagener, M. (2006). Antimicrobial coatings, *Polymers Paint Colour Journal*, 196: 34–37.

Weiss, J., Takhistov, P. and McClements, D.J. (2006). Functional materials in food nanotechnology. *Journal of Food Science*, 71: R107–R116.

Were, L.M., Bruce, B.D., Davidson, P.M. and Weiss, J. (2003). Size, stability, and entrapment efficiency of phospholipid nanocapsules containing polypeptide antimicrobials, *Journal of Agricultural and Food Chemistry*, 51: 8073–8079.

Were, L.M., Bruce, B., Davidson, P.M. and Weiss, J. (2004). Encapsulation of nisin and lysozyme in liposomes enhances efficacy against *Listeria monocytogenes*, *Journal of Food Protection* 67: 922–927.

Wu, T.H., Yen, F.L., Lin, L.T., Tsai, T.R., Lin, C.C. and Cham, T.M. (2008). Preparation, physicochemical characterization, and antioxidant effects of quercetin nanoparticles, *International Journal of Pharmaceutics*, 346: 160–168.

Yang, F.M., Li, H.M., Li, F., Xin, Z.H., Zhao, L.Y., Zheng, Y.H. and Hu, Q.H. (2010). Effect of nano-packing on preservation quality of fresh strawberry (fragaria ananassa duch. cv fengxiang) during storage at 4°C, *Journal of Food Science*, 75: C236–C240.

Yep, Y.L., Li, D., Mann, N.J., Bode, O. and Sinclair, A.J. (2002). Bread enriched with microencapsulated tuna oil increases plasma docosahexaenoic acid and total omega-3 fatty acids in humans, *Asia Pacific Journal of Clinical Nutrition*, 11: 285–291.

16 Bioactive Packaging Technologies with Chitosan as a Natural Preservative Agent for Extended Shelf-Life Food Products

Vasiliki I. Giatrakou and Ioannis N. Savvaidis

CONTENTS

16.1 INTRODUCTION

Consumers, food companies, and governments all want to raise the quality level of food products to a higher extent. The new generation of food products needs to be more appealing, tastier, made with fresher ingredients, but most importantly healthy and safe. Therefore, the food industry aims to develop products that have a high nutritional density, are based on high-quality and fresh ingredients, contain natural antioxidants like spice and herb extracts, and contain less salt and synthetic preservatives. To meet the growing consumer demand for safer and better quality food products with a fresh and natural (green) image, new and novel packaging technologies or materials have been and continue to be developed (Cutter, 2006). Bioactive packaging technologies for extended shelf-life raw or precooked products have become one of the major areas of research in food packaging. Of these active packaging systems, the antimicrobial version is of great importance (Coma, 2008).

Antimicrobial active bio-based packaging materials could be developed and used because they can inhibit or kill the microorganisms and thus extend the shelf life of perishable products and enhance the safety of packaged products (Coma, 2008). For example, chitosan (CH)-based antimicrobial films have shown potential to be used as packaging material for the preservation of quality of a variety of foods, as CH is a natural biopolymer that can be formed into fibers, gels, sponges, beads, or even nanoparticles (Dutta et al., 2009). Moreover, it has the ability to dissolve and create film-forming solutions to provide edible protective coating to the food products (dipping, spraying, or blending).

CH is an edible and biodegradable polymer derived from chitin, the most abundant natural polymer available, next to cellulose (Shahidi et al., 1999). Chitin is the major structural component of the exoskeleton of crustaceans found in marine environment like lobsters, crabs, shrimp, prawn, and krill, as a composite with proteins, lipids, and calcium carbonate. It is also found in abundance in the complex carbohydrate cell wall of fungi and yeasts (Mathur and Narang, 1990). The name "chitin" is derived from the Greek word "chiton," meaning "coat of mail" (Shahidi et al., 1999). Full (or partial) deacetylation of chitin using hot alkali (concentrated NaOH) or enzymatic hydrolysis produces CH. However, commercial CH still contains N-acetyl groups. From a chemical point of view, CH is a linear polysaccharide composed mainly of β-1 4-2-deoxy-2-amino-D-glucopyranose and of β-1,4–2-deoxy-2-acetamido-D-glucopyranose (chitin) residues to a lesser extent (Fernandes et al., 2008).

CH is a natural antimicrobial that offers real potential for applications in the food industry, fulfilling the growing consumer demand for foods without chemical preservatives, due to its particular physicochemical properties, short-time biodegradability, biocompatibility with human tissues, and antibacterial activity toward spoilage and pathogenic foodborne microorganisms (Gram-positive and Gram-negative bacteria) (Aider, 2010). It is also effective against several fungi and yeasts and has revealed significant antioxidant properties on muscle foods (Jo et al., 2001; Lee et al., 2003; Georgantelis et al., 2007a,b; Soultos et al., 2008). The main applications of chitosan (CH) on food preservation, based on definitions for bioactive packaging proposed by Coma (2008), could be summarized as follows:

1. Application as a bioactive natural additive, either directly applied onto the food in the absence or presence of other antimicrobial and/or antioxidant agents (spice extracts, tocopherol, sulfites, nitrites). CH or its modified form (oligomer, glutamate salts, etc.) is used in its initial powder form (insoluble) and incorporated into the core of the foodstuff and, thus, is mainly applied on minced meat mixtures used for the preparation of meatballs, patties, sausages, etc. (Darmadji and Izumimoto, 1994; Roller et al., 2002; Juneja et al., 2006; Georgantelis et al., 2007a,b; Soultos et al., 2008).
2. Application as a bioactive edible coating or soluble additive (CH solutions) directly applied onto the food by dipping, spraying, or blending. This application is based on the fact that CH has the ability to be dissolved in aqueous acidic solutions (acetic, lactic, propionic, and citric acids are the most frequently used), creating film-forming solutions that are used as edible coatings or liquid additives for foodstuffs (Roller and Covill, 2000; Lin and Chao, 2001; Sagoo et al., 2002; Beverlya et al., 2008; Chhabra et al., 2006; Tsai et al., 2006, Kok and Park, 2007; Waimaleongora-Ek et al., 2008). Nowadays, new promising technologies integrating edible CH-based solutions or bioactive coatings in combination with natural antimicrobial substances (essential oils [EOs], spices, lysozyme, nisin, etc.) have been successfully applied for preservation of fish, meat, poultry, or vegetable-based products, as well as on inactivation of foodborne pathogenic bacteria (Inatsu et al., 2005; Kanatt et al., 2008; Kim et al., 2008; Ponce et al., 2008; Giatrakou, 2010; Giatrakou et al., 2010b; Ojagh et al., 2010). Moreover, CH solutions in combination with modified atmosphere/vacuum packaging (with or without EOs) have been successfully used for prolonging the shelf life of meat, poultry, dairy products, and fresh homemade pasta (Yingyuad et al., 2006; Del Nobile, 2009a,b; Duan et al., 2010; Giatrakou et al., 2010a,b)
3. Application as a polymer matrix not directly applied onto the product surface (as described earlier) but used as carrier for controlled release of other antimicrobials or flavoring substances (EOs or other antimicrobials). In that case, edible (or not in some cases) dry CH-based films are easily prepared by evaporating from dilute acid conditions using a casting technique (Ouattara et al., 2000; Coma et al., 2002; Pranoto et al., 2005; Zivanovic et al., 2005; Hosseini et al., 2008; Fernandez-Saiz et al., 2010, Sanchez-Gonzalez et al., 2010).

This review surveys the utilization of CH as a direct additive (in powder form), an edible coating, or antimicrobial biopolymer for active packaging of perishable food products. The potential of these technologies is evaluated in terms of their capacity to extend the shelf life and also to ensure the safety and preservation of food products.

16.2 ANTIMICROBIAL ACTIVITY OF CHITOSAN

CH has shown strong antimicrobial activity both in vivo and in vitro studies, including foodborne, spoilage, and pathogenic microorganisms: *Aeromonas hydrophila, Bacillus cereus, Brochothrix thermosphacta, Clostridium perfringens, Enterobacter aeromonas, Escherichia coli, Lactobacillus sakei, Listeria monocytogenes, Photobacterium phosphoreum, Pseudomonas fluorescens, Pseudomonas aeruginosa, Salmonella* spp., *Shigella dysenteriae, Staphylococcus aureus, Yersinia enterocolitica, Vibrio parahaemolyticus, Vibrio cholerae*, as well as on fungi and yeasts like *Saccharomyces cerevisiae, Candida lambica, Candida albicans, Candida parapsilosis, Rhodotorula glutensi, Botrytis cinerea,* and *Rhizopus stolonifer* (Shahidi et al., 1999; Helander et al., 2001; Devlieghere et al., 2004; Wang et al., 2004; Inatsu et al., 2005; Chung and Chen, 2008; Fernandes et al., 2008; Marques et al., 2008).

Minimum inhibitory concentrations (MICs, defined as the lowest concentration of an antimicrobial that will inhibit the visible growth of a microorganism after overnight incubation) recorded for CH of various molecular weight (MW)/deacetylation degree (D.D.) as a natural antimicrobial substance against several bacterial species are presented in Table 16.1.

16.2.1 Mechanism of Antimicrobial Action of Chitosan toward Gram-Negative and Gram-Positive Bacteria

The exact mechanism of the antimicrobial action of CH is not yet fully understood. However, several theories have been proposed and it is very likely that the antimicrobial mechanism of CH is attributed to a combination of several different actions. It is generally assumed that protonation of amine groups ($-NH_2$ groups) of CH chain in acidic conditions makes CH positively charged ($-NH_3^+$ groups); the cationic nature of CH is a fundamental factor for its antibacterial activity since interaction between positively charged CH molecules and negatively charged microbial cell membranes may cause membrane disruption and leakage of intracellular constituents leading to impairment of vital activities and eventually to lysis and death (Raafat et al., 2008).

The cell wall of Gram-negative bacteria is composed of three different sections: an outer membrane, 1–2 thin layers of peptidoglycan, and an inner membrane consisting of lipids, phospholipids, and structural proteins. The surface of the outer membrane consists of lipopolysaccharides (LPS) and proteins. LPS have three different regions: O-specific chain, core, and Lipid-A. O-specific chain comprises glycose, galactose including uronic acid, and some sugars, which are negatively charged. These negatively charged groups get involved in ionic-type of binding with $-NH_3^+$ groups of CH

TABLE 16.1
Minimum Inhibitory Concentrations of CH of Various Molecular Weight and Deacetylation Degree as a Natural Antimicrobial Substance, as Reported in Several Studies In Vitro

Bacterium	Type of CH (MW/D.D.)	MIC	Reference
S. aureus	MW = 35 kDa D.D. = not reported	0.005% w/v	Chang et al. (1989)
	MW = not reported D.D. = 69%	0.01% w/v	Chen et al. (1998)
	MW = 107 kDa D.D. = 80%–85%	0.20% w/v	Fernandes et al. (2008)
	MW = 591 kDa D.D. = 80%–85%	0.20% w/v	Fernandes et al. (2008)
E. coli	MW = not reported D.D. = 92%	0.0075% w/v	Simpson et al. (1997)
	MW = not reported D.D. = 69%	0.01% w/v	Chen et al. (1998)
	MW = 107 kDa D.D. = 80%–85%	0.20% w/v	Fernandes et al. (2008)
	MW = 591 kDa D.D. = 80%–85%	0.25% w/v	Fernandes et al. (2008)
L. monocytogenes	MW = not reported D.D. = 69%	0.01% w/v	Chen et al. (1998)
	MW = 43 kDa D.D. = 94%	0.006%–0.01% w/v	Devlieghere et al. (2004)
V. parahaemolyticus	MW = not reported D.D. = 69%	0.01% w/v	Chen et al. (1998)
S. typhimurium	MW = not reported D.D. = 69%	>0.2% w/v	Chen et al. (1998)
S. enterica	MW = Medium D.D. = 75%–85%	0.03% v/v	Marques et al. (2008)
P. fluorescens	MW = 43 KDa D.D. = 94%	0.008% w/v	Devlieghere et al. (2004)
E. aeromonas	MW = 43 KDa D.D. = 94%	0.006% w/v	Devlieghere et al. (2004)
B. thermosphacta	MW = 43 KDa D.D. = 94%	0.008% w/v	Devlieghere et al. (2004)
B. cereus	MW = 43 KDa D.D. = 94%	0.006% w/v	Devlieghere et al. (2004)
	MW = 470 kDa D.D. = not reported	0.03% w/v	No et al. (2002)
	MW = 1671 kDa D.D. = not reported	0.05% w/v	No et al. (2002)

(*continued*)

TABLE 16.1 (continued)
Minimum Inhibitory Concentrations of CH of Various Molecular Weight and Deacetylation Degree as a Natural Antimicrobial Substance, as Reported in Several Studies In Vitro

Bacterium	Type of CH (MW/D.D.)	MIC	Reference
	MW=224kDa D.D.=not reported	0.05% w/v	No et al. (2002)
	MB=not reported D.D.=69%	0.1% w/v	Chen et al. (1998)
	MW=59kDa D.D.=not reported	>0.1% w/v	No et al. (2002)
Bacillus spp.	MW=1671KDa D.D.=not reported	0.005% w/v	No et al. (2002)
	MW=470kDa D.D.=not reported	0.005% w/v	No et al. (2002)
	MW=224kDa D.D.=not reported	0.005% w/v	No et al. (2002)
	MW=59kDa D.D.=not reported	0.005% w/v	No et al. (2002)

with concomitant release of LPS from the outer membrane, Ca^{+2} and Mg^{+2} drawn from the cell wall, exposure of cell-wall and cell membrane to osmotic shock, and thus spillage of the cytoplasm (Vishu Kumar et al., 2007; Chung and Chen, 2008). In addition, membrane-bound energy generation pathways, due to impairment of the proper functional organization of the electron transport chain, interfere with proper oxygen reduction and thereby force cells to shift to anaerobic energy production (Raafat et al., 2008).

In Gram-positive bacteria, the cell wall is not surrounded by an outer membrane as in the case of Gram-negative ones, but it is made up of 30–40 layers of peptidoglycans, which contain repeating units of N-acetylglycosamine and N-acetylmuramic acid, amino acids, and teichoic acids (polymers of ribitol and sorbitol). Positively charged $-NH_3^+$ groups could bind on the negatively charged surface components of the cell wall, resulting in hydrolysis of the peptidoglycan layer and exudation of metallic ions and cytoplasmic contents of low MW such as proteins, nucleic acid, glycose, and lactate dehydrogenase (Vishu Kumar et al., 2007; Goy et al., 2009).

CH could also act as a chelating agent that selectively binds trace metals, thereby inhibiting production of toxins and microbial growth (Shahidi et al., 1999). According to Wang et al. (2005), this specific antibacterial mechanism occurs mainly at higher pH values than the one of electrostatic interaction. In particular, at pH=7–9 amine groups of CH are not protonated ($-NH_2$ groups) and thus the electron pair of amine nitrogen is available for donation to metal ions. In other words, metallic ions act as an electron receptor that binds to CH chains via $-NH_2$ (electron donors) and by forming bridges to hydroxyl groups (Wang et al., 2005;

Goy et al., 2009). Moreover, CH acts as a water-binding agent and inhibits various enzymes (Shahidi et al., 1999) or surrounds bacterial cells by forming a polymer layer that eventually leads to the cell death by preventing the uptake of nutrients. However, this mechanism seems to affect only growth of vegetative bacterial cells, as bacterial spores can survive for extended periods without nutrients (Fernandes et al., 2009).

Finally, CH may form complexes with extracellular DNA of bacterial cells, which bind to receptors of cell membrane and consequently passively transported to the inner cell regions. Penetration of CH–DNA complexes to the nuclei of the microorganisms may result in inhibition of mRNA and protein synthesis, via interference of the former complex with enzymes regulating these specific functions (Shahidi et al., 1999; Goy et al., 2009).

To conclude, it is very possible that the antibacterial action of CH is a combination of all aforementioned different mechanisms, as also proposed by Chung and Chen (2008).

16.2.2 Factors Affecting Antimicrobial Activity of Chitosan

According to data presented in Table 16.1, the antimicrobial activity of CH (referred to as MIC) is not always the same, but it varies according to its MW and D.D. However, there are also other factors influencing the antimicrobial action of CH, such as origin (marine animals, fungi, algae), type of organic solvent used for the preparation of CH solutions, the species of the microbial contaminant, and bacterial cell age and growth phase (lag, exponential, or stationary phase). If the antibacterial activity of CH is evaluated in real food systems, pH and composition of food matrix, as well as the length and temperature of storage influence its effectiveness (Chhabra et al., 2006). Some of the most important factors affecting the antimicrobial activity of CH are further analyzed as follows:

16.2.2.1 Molecular Weight of Chitosan

CH is commercially available with >85% deacetylated units and MW ranging between 100 and 1000 kDa (Goy et al., 2009). Recent studies have revealed the antimicrobial potential of CH to be dependent on its MW. However, the results reported to date sometimes lead to contradictory results. High MW of native CH causes high viscosity and low solubility in acid-free aqueous media, resulting in a limited application on food preservation (Fernandes et al., 2009). However, recent studies have focused attention toward conversion to chitooligosaccharides (COS), which are water-soluble due to their shorter chain length (MW of COS $\leq$ 10 kDa). Some studies reported that CH is more effective in inhibiting growth of bacteria than COS, whereas others claimed that CH of high MW may sometimes lead to a decrease in its activity (Eaton et al., 2008). Most important findings of recent studies focusing on antimicrobial characteristics of high, medium, and low MW CH as well as of CH COS are highlighted as follows:

Fernandes et al. (2009) investigated how CH with different MW acts differently upon *B. cereus* vegetative cells and spores. CH with an MW of 628 kDa (high MW) reduced the number of *B. cereus* vegetative cells by almost 3 log CFU/g, whereas CH of 100 kDa (low MW) and COS (≤3 kDa) led to lower reduction levels of the initial cell population (1–2.5 log CFU/g). Authors suggested that the longer polymer chains of 628 kDa CH gave stronger antibacterial action. However, analyzing the results for *B. cereus* spores, neither of the CHs used showed any effect upon them, while for COS the reduction achieved was 1-log cycle (ca. 90%). In general, molecules of high and low MW CH surrounded cells of *B. cereus* by forming a polymer layer functioning as a barrier for the uptake of nutrients, while COS provoked more visible damages in the vegetative cells; thus, their small cell size enabled penetration into the bacterial cell.

The relationship between MW of CH and their antimicrobial activity upon Gram-positive and negative bacteria was also studied by Fernandes et al. (2008). The authors stated that CH with MW of 107 (low MW), 591 (medium MW), and 628 (high MW) presented almost the same antibacterial action against growth of *E. coli*. However, COS (MW < 5 kDa) were more effective to reduce *E. coli* initial population in vitro. On the contrary, COS did not reveal any antimicrobial action toward *S. aureus*, unlike low, medium, and high MW CHs. Fernandes et al. (2008) concluded that CH of low MW had a more pronounced antibacterial action toward Gram-negative bacteria, and the reverse in the case of Gram-positive ones. This exact conclusion was also reported by Eaton et al. (2008), who studied antibacterial action of low and high MW CH against the same pathogens. In addition, Vishu Kumar et al. (2007) reported that very low MW CH (8.5–9.5 kDa) showed better growth inhibitory effects toward both Gram-negative and Gram-positive bacteria (*B. cereus, E. coli, Y. enterocolitica*, and *Bacillus licheniformis*) than native CH of high MW. Moreover, Devlieghere et al. (2004) reported that CH of low MW (43 kDa) was very effective against Gram-negative bacteria (MIC ≤ 0.006% w/v), whereas the sensitivity of Gram-positive bacteria was highly variable (MIC > 0.05% w/v).

In general, the antimicrobial activity of CH is dependent on the type of target microorganism and the MW of CH, being stronger for lower MW in the case of Gram-negative bacteria. This is probably attributed to the fact that CH of low MW and especially COS may easily penetrate the outer cell wall of Gram-negative bacteria, due to their small molecular chain length, as compared to CH with medium or high MW (Fernandes et al., 2008). On the other hand, CH with higher MW may prove to be more effective toward Gram-positive bacteria, as they posses the ability to form a thick polymer layer surrounding cells, thus preventing nutrient absorption (No et al., 2002; Eaton et al., 2008; Fernandes et al., 2009).

Besides the type of target microorganism, pH of the medium appears to influence the differential action of low and high MW CH. Tsai et al. (2006) investigated the effect of high or low MW CH on *B. cereus*, inoculated into nutrient broth and cooked rice. Results of the study showed that low MW CH had a greater inhibitory effect on *B. cereus* as compared to high MW CH, at pH = 6.0. However, when pH of the medium increased up to 7.0, the antibacterial action of high MW CH was almost negligible, whereas the antibacterial activity of low MW CH was not affected by that change in pH. This suggests that the pKa and solubility of low MW CH is higher than the one of native (high MW) CH, contributing to a better antibacterial activity at neutral pH (Tsai et al., 2006).

As a conclusion, prior to application of low (including COS), medium, or high MW CH on real food systems as potential antimicrobial agents, composition/pH of the matrix, storage temperature, and packaging conditions, which all affect the type of bacterial species dominating the microflora (and thus determine the type of spoilage), must be taken into account. For example, COS lost their antibacterial activity against Gram-negative and Gram-positive bacteria when added in milk, after 4–8 h, in contrast to medium and high MW CH. However, the addition of medium and high MW CH to apple juice led to the formation of several off-odors and off-flavors, whereas COS maintained desirable sensory characteristics of the product (Fernandes et al., 2008). Moreover, CH oligomer did not reduce microbial populations of pork sausages during refrigerated storage (Jo et al., 2001), whereas Lin and Chao (2001) reported that total viable counts (TVC) and lactic acid bacteria (LAB) populations were lowest in reduced-fat Chinese-style sausages containing low MW CH (150 kDa) in comparison to high (1250 kDa) and medium MW (600 kDa) CHs. Moreover, sausages with low MW and medium MW CH were higher in sensory acceptability (Lin and Chao, 2001).

16.2.2.2 Effect of pH

In general, the antibacterial activity of CH seems to increase as the pH of the matrix decreases. This is attributed to the fact that at low acidity, more amino groups of CH are protonated ($-NH_3^+$ groups), which leads to more interaction with negatively charged surfaces of bacteria and thus better disruption of cell wall/membrane structure (Devlieghere et al., 2004). CH can be considered a strong base as it possesses primary amine groups with a pKa value of 6.3. The presence of the amine groups indicates that pH substantially alters the charged state and properties of CH. At acidic environments, these amines become protonated and the electron pair of nitrogen amine groups become more easily adapted by other molecules. However, at pH above 6, amine groups of CH become deprotonated (more NH_2 groups) and the polymer loses its charge (Pillai et al., 2009). As a result, interaction with negatively charged molecules of cell wall/membranes of bacteria is weakened and the polymer becomes insoluble. Helander et al. (2001) stated that the antibacterial action of CH toward Gram-negative bacteria is more intense at pH < pKa < 6.3 owing to a large number of CH's amine groups being positively charged. However, Tsai et al. (2006) showed that low MW CH has the potential to maintain its antibacterial action at neutral pH = 7, in contrast to native (high MW) CH, which showed a complete loss of its antibacterial ability against *B. cereus*. Chung et al. (2004) reported that the pKa of CH varies between 6.3 and 7.0. These results verify that the positive charge on the amine groups is not the sole factor resulting in antimicrobial activity. As a conclusion, CH and COS may be effective food preservatives in low pH food products like apple juice, but when applied to food products with higher pH, low MW CH may be more efficient to inhibit the growth of spoilage and pathogenic microorganisms. However, it is worth mentioning that acidic environments do not always enhance the efficiency of CH, as Roller and Covill (2000) reported that in mayonnaise containing CH combined with lemon juice, although the pH of the medium was relatively low (4.5), its antibacterial activity toward three inoculated microorganisms (*Salmonella*, *Lactobacillus fructivorans*, *Zygosaccharomyces bailii*) was relatively weak. On the contrary, the combination of CH with acetic acid instead of lemon juice enhanced the antibacterial activity toward the three aforementioned microorganisms, indicating that not only the pH but also the type of acidic environment also influences the antimicrobial action of CH.

In another study, it was observed that the antibacterial activity of the N-alkylated CH derivatives against *E. coli* increased as the pH increased from 5.0 and reached a maximum level around the pH of 7.0–7.5 (Yang et al., 2005). This may be attributed to the fact that electrostatic interaction is not the only mechanism controlling antibacterial action of CH. In fact, CH has excellent metal-binding capacities and not only the amine groups but also hydroxyls in the CH molecules are responsible for the uptake of metal cations by chelation (Goy et al., 2009). According to the aforementioned mechanism, CH deprives nutrients of microorganisms that are essential for their growth. This type of antibacterial mechanism is more efficient at high pH where positive ions are bounded to CH, since the amine groups are deprotonated and the single electron pair on the amine nitrogen is highly available for donation to metal ions (Goy et al., 2009). At $pH<6$, the complexation involves only one $-NH_2$ group and three hydroxyls (–OH) or H_2O molecules, while at $pH>6.7$, the complexation is likely to involve two $-NH_2$ groups involved in the complex formation, thus showing better antimicrobial activity (Goy et al., 2009).

16.2.2.3 Effect of Deacetylation Degree

It is generally accepted that highly deacetylated CH has shown more antimicrobial activity than CH with higher proportion of acetylated amino groups (Goy et al., 2009; Aider, 2010). This is due to the fact that as D.D. increases, more number of reactive sites ($-NH_3^+$ groups) become available for electrostatic interaction with negatively charged molecules on the surface of bacterial cell wall/membranes. Furthermore, D.D. is a determinant factor in the solubility of CH (Goy et al., 2009; Pillai et al., 2009).

16.2.2.4 Composition of the Food Matrix

Antimicrobial activity of CH is influenced to a great extent when being examined in real food products instead of models in vitro. This is due to the fact that reactive sites of CH may eventually interact with nutrient components of foods, thus minimizing its antimicrobial effect toward several microorganisms. For example, Ausar et al. (2002) reported that the antibacterial action of CH against lactic acid fermenting bacteria was limited when applied to milk instead of in vitro study. Devlieghere et al. (2004) thoroughly studied the effect of different food components (starch, whey proteins, NaCl, and oil) on the antimicrobial properties of CH. According to their findings, the antimicrobial activity of CH was negatively affected by increasing starch and whey protein content, but it did not seem to be affected by the oil content of foods. As regards the effect of NaCl, results seem to be controversial. Devlieghere et al. (2004) stated that adding NaCl to the medium had a negative effect on antimicrobial activity, whereas Chung et al. (2003) stated the opposite, as they observed that increasing ionic strength by adding salt contributes to a higher solubility and antimicrobial activity of CH.

16.2.2.5 Cell Age and Concentration of Target Microorganisms

The antimicrobial action of CH seems to be affected by the growth phase of target microorganisms. Tsai et al. (2006) reported that the susceptibility of four strains of *B. cereus* cells varied in relation to their growth phase. Specifically, cells of *B. cereus* in the middle and late exponential were the most sensitive to low MW CH and cells in the stationary phase were the least sensitive. In addition, Tsai and Su (1999) found that cells of *E. coli* in the late exponential phase were the most sensitive to the action of CH. The differences in susceptibility of two aforementioned pathogens toward the action of CH may be attributed to the fact that the electronegativity of the cell surface of these pathogens appeared to be the highest when the cells were in the late exponential phase and lowest on stationary-phase cells, with the latter being most resistant to electrostatic interactions with positively charged active sites of CH molecules (Tsai et al., 2006).

The initial cell concentration of bacteria has proved to influence the antibacterial action of CH, as Fernandes et al. (2008) reported that the antibacterial action of CH and COS toward *E. coli* and *S. aureus* cells was closely related to the inoculum's level. Although CH revealed a significant bactericidal action toward *E. coli* (initial inoculum of 10^3 cells/mL) when applied at several concentrations (0.1%–0.5% w/v), CH's antibacterial activity was negatively affected at a higher inoculum (10^7 cells/mL) while COS maintained their activity. Interestingly, when the same procedure was repeated using a high inoculum of *S. aureus*, high MW CH managed to maintain its antibacterial activity to a better degree than low MW CH.

16.3 APPLICATIONS OF CHITOSAN ON FOOD PRESERVATION

16.3.1 Utilization of Chitosan as a Food Additive Directly Incorporated into Meat Formulations

Direct incorporation of CH as a bactericidal agent for protection of meat-based products from microbial deterioration may result in partial inactivation of the active substances by the food constituents and is therefore expected to have only a limited effect on the surface flora (Coma et al., 2002). However, the use of CH in its initial powder form, without being dissolved in any organic solvent, may offer a good potential for the preservation of minced meat products (e.g., patties, meatballs, etc.) or meat mixtures used for formulation of emulsion-type or traditional (British or Greek style) sausages, as described in Section 16.3.2 (Table 16.2).

Soultos et al. (2008) investigated the effect of CH powder (MW: 4.9×10^5) added individually or in combination with nitrites (150 ppm) on microbiological, physicochemical, and sensory properties of fresh pork Greek-style sausages stored at 4°C for 28 days. Sausages of this traditional type were made using pork meat and fat, chopped and thoroughly mixed with salt and seasonings. The mixture was treated with two levels of CH (Dalian Xindie Chitin Co., Dalian, China) in powder form (0.5% and 1% w/w) alone or in combination with nitrites (150 ppm) and finally stuffed into natural casings from the clean intestine of pigs using a filling machine. CH resulted in significant inhibition on microbial growth, while nitrites did not seem

TABLE 16.2
Overview of Studies Testing the Antibacterial Activity of CH (in Powder Form) Directly Incorporated into Meat Formulations

Food Product	Application of CH (± Other Agents)	Concentration Applied	Storage Temperature/Period/Other Notes on Experimental Setup	Bacterial Species	Results: Reduction of Final Population (log CFU/g)	Results: Shelf-Life Extension (Based on Criteria Applied)	Other Observations/Comments	Reference
Greek-style pork sausages	CH of high MW ± nitrites	1% w/w ± 150 ppm	4°C/28 days	Natural microflora	1.0–1.7	+14 days (threshold limit of TVC = 7.0 log CFU/g)	(a) CH–nitrite combination improved sensory characteristics (b) Lipid oxidation lower in CH–nitrite combination (c) Nitrites did not protect from microbial spoilage	Soultos et al. (2008)
Greek-style pork sausages	CH of high MW + rosemary extract	1% w/w + 260 mg/kg	4°C/20 days	Natural microflora	1.5–2.4	+ more than 10 days (threshold limit of *Pseudomonas*, yeasts–molds count = 7.0 log, rancidity level)	CH–rosemary combination showed the best antimicrobial and antioxidant effect (possible synergistic action)	Georgantelis et al. (2007a)

British-style pork sausages	CH glutamate + sulfites	0.6% w/w + 170 ppm	4°C, 24 days	Natural microflora	3.0–4.0	+ more than 12 days (threshold limit of TVC = 7.0 log CFU/g, yeasts–molds count = 6.0 log CFU/g)	CH–sulfites combination improved sensory characteristics (odor, appearance)	Roller et al. (2002)
Emulsion-type pork sausages	CH oligomer	0.2% w/w	4°C/21 days/ aerobic or vacuum packaging	Natural microflora	No effect	—	CH reduced lipid oxidation	Jo et al. (2001)
Fresh minced beef	CH (MW not reported)	1% w/w	4°C/10 days	Natural microflora	1.0–2.7	—	CH improved sensory attributes, maintained redness, and reduced lipid oxidation and putrefaction	Darmadji and Izumimoto (1994)
Cooked ground beef and turkey	CH glutamate	3% w/w	Abusive cooling from 54.4°C to 7.2°C after (a) 12, 15, 18 h (b) 21 h	*C. perfringens* spore cocktail	(a) 4.0–5.0 (b) 2.0	Not determined	—	Juneja et al. (2006)

to protect sausages from microbial spoilage. Specifically, CH at a concentration of 1% w/w in the meat mixture reduced TVC, *Pseudomonas*, *Enterobacteriaceae*, *B. thermosphacta*, and LAB by 1.0–1.7 log CFU/g, after 28 days of storage at 4°C. The authors concluded that Gram-positive bacteria were more sensitive than Gram-negative bacteria to the antimicrobial action of CH. As regards sensory evaluation, samples containing CH–nitrites combination at any level were judged as more acceptable than other treatments.

The effect of high MW CH in powder form (1% w/w) on microbial parameters of Greek-style pork sausages was also studied by Georgantelis et al. (2007a), with the exception that the authors combined CH (Dalian Xindie Chitin Co., Dalian, China) with rosemary extract and α-tocopherol. CH addition in the sausage formulation resulted in significant inhibition of microbial growth during storage at 4°C. In fact, CH 1% w/w reduced TVC, *Enterobacteriaceae*, and yeasts–molds by ca. 1.5–1.9 log CFU/g after 20 days of storage at 4°C, as compared with the control. Interestingly, CH had a more pronounced effect on Gram-positive LAB, as their counts were reduced by ca. 2.4 log CFU/g, during the same period, as compared with the control. Counts of *Pseudomonas* spp. reached level of 7 log CFU/g on day 9 for control, whereas samples containing CH 1% w/w slightly exceeded 6 log CFU/g during the entire storage period. Finally, samples containing the combination of CH and rosemary extract (260 mg/kg) exhibited the lowest microbial counts, indicating a possible synergistic effect.

Roller et al. (2002) developed a novel preservation system for the chill preservation of British-style pork sausages based on combinations of CH glutamate (Pronova, Drammen, Norway), carnocin, and low concentrations of sulfite. Sausages were prepared using minced pork and fat. Minced meat was mixed with a mixture of dry ingredients, containing rusk, seasoning, 2% skim milk powder, as well as CH glutamate (0.6%) and sodium metabisulfite as additives. CH and low sulfite (170 ppm) combination proved to be the most effective to reduce TVC, LAB, and yeast–molds by 3.0–4.0 log CFU/g, especially during the first 7–10 days of storage at 4°C. In conclusion, the authors stated that CH and low sulfite combination extended the shelf life of sausages nearly three times, based on maximum acceptable limits for total and yeast counts established for meat (IFST, 1999). Finally, sensory evaluation showed that CH–sulfite treatments deteriorated less rapidly and were acceptable for a longer period than other batches by the panelists, based on appearance and odor attributes.

Jo et al. (2001) prepared emulsion-type sausages with the addition of a CH oligomer (MW: 5000, Shinyoung CH Co, Seoul, Korea). Sausages were prepared using a mixture of ground pork meat and fat, salt, ice water, sugar, nitrites, phosphates, and a spice mix. CH oligomer (in powder form) was incorporated into the meat mixture at a final concentration of 0.2% w/w. Sausages were stuffed with the former mixture into a casing, dried, and smoked. The combination of CH oligomer with aerobic or vacuum packaging did not affect microbial counts during storage of sausages at 4°C for 21 days. Moreover, the addition of CH oligomer in the emulsion-type sausages did not influence flavor, texture, and overall acceptance, according to sensory analysis.

Finally, Darmadji and Izumimoto (1994) studied the effect of CH (in powder form) on quality attributes of fresh minced beef. CH (Katokichi Co., Japan) was incorporated into the formulation of minced beef in its initial powder form, at final concentrations of 0%, 0.2%, 0.5%, and 1.0% w/w. According to their findings, CH at a final concentration of 1.0% w/w reduced TVC, *Staphylococci,* and coliforms by ca. 1–1.4 log CFU/g, as compared to the control, during storage of the product for 10 days at 4°C. Moreover, CH addition (1.0% w/w) resulted in a significant reduction of Gram-negative bacteria, *Micrococci*, and *Pseudomonas* counts by ca. 2.0–2.7 log CFU/g at the end of the storage period. Finally, CH-treated samples resulted in better sensory attributes and maintained the redness of the surface of the product in favorable levels.

Juneja et al. (2006) were the first to study the effect of CH glutamate on controlling spore germination and outgrowth of *C. perfringens* during storage of cooked ground beef and turkey, during abusive cooling from 54.4°C to 7.2°C in predetermined time intervals (hours). CH glutamate (containing 44% glutamic acid, Pronova Biopolymer, Drammen, Norway) was incorporated into ground beef or turkey meat formulations to final concentrations of 0.5%–3.0% w/w. CH-treated and control samples were inoculated with *C. perfringens* spore cocktail of 3.0 log CFU/g, cooked, and chilled. Results of the study suggest that 3% w/w CH reduced *C. perfringens* spore germination and outgrowth by 4–5 log CFU/g during exponential cooling of beef and turkey in 12, 15, or 18 h time intervals, but higher chilling time (21 h) resulted in a lower reduction by ca. 2.0 log CFU/g.

All the aforementioned studies suggest that CH or CH glutamate, both in powder form, can be utilized at concentrations as low as 1.0% w/w, direct additives incorporated into the formulation of food products made of raw ground meat (sausages, patties, meatballs, cocktail salami, etc.), in order to control microbial spoilage and to extend their shelf life during refrigerated storage. Furthermore, the fact that CH/low sulfite/spice extract treatments revealed a possible synergistic effect on delaying microbial spoilage or sensory deterioration of this kind of meat products suggests that combined application of CH with the aforementioned antimicrobial agents allows the use of the latter in lower concentrations than the ones required to achieve the same target if individual additives were used alone. However, CH oligomers added in sausages did not prove to be effective in inhibiting microbial deterioration and improving sensory characteristics and, thus, may not be a promising technology to maintain the microbial quality of such meat products. Interestingly, CH glutamate at concentrations of 3.0% w/w may reduce the potential risk of spore germination and outgrowth of pathogenic bacteria that remain a major cause of foodborne illness (like *C. perfringens*) during improper storage and inadequate cooling practices of cooked ground meat products, in retail food operations.

16.3.2 Application of Chitosan Solutions as Bioactive Edible Coating or Liquid (Acid-Soluble) Additive Applied onto the Food (Dipping, Spraying, or Blending)

CH solutions have been investigated as possible means of extending the shelf life and enhancing the safety of perishable food products. Till the year 2000, little work had been reported on antagonistic properties of CH against pathogenic and spoilage microorganisms important in foods; however, over the last 10 years CH has attracted the attention of many researchers in order to be exploited as a natural shelf-life extender of meat- or fish-based products. Some of the work focused on its possible role in food preservation is presented in the following section (Table 16.3).

Sagoo et al. (2002) developed a novel preservation system for chilled raw skinless sausages and an unseasoned minced pork mixture, using CH glutamate solutions. Antimicrobial solutions were prepared by dissolving CH glutamate (Pronova Co., Drammen, Norway) in deionized water. Dipping of skinless sausages in CH solution (1.0% w/v) or blending CH solution (1.5%, 3.0% w/v) with the formulation of unseasoned minced pork mixture at a final concentration of 0.3% and 0.6% w/w reduced TVC, yeast–molds, and LAB populations by up to 3 log CFU/g, after storage for 18 days at 7°C and 4°C, respectively, as compared with control samples. Furthermore, based on the upper acceptability limit of 7.0 log CFU/g established for TVC (IFST, 1999), CH dipping resulted in a microbiological shelf-life extension of pork sausages by 8 days, as compared with control.

In another study, Kok and Park (2007) evaluated the potential of two different CH edible coatings to extend the shelf life of a surimi-based product, called "set fish balls," during refrigerated storage (4°C). Acid-soluble CH of low MW (Vanson Halosource, Inc., Redmond, WA) was dissolved in 1% acetic acid solution, to a final concentration of 1% w/v. Water-soluble CH lactate of low MW (Kyowa Technos Co. Ltd., Chiba, Japan) was prepared in water to a final concentration of 0.5% w/v. Although fish balls dipped in water-soluble CH solution had (by ca. 5.0 log CFU/g) lower aerobic plate counts as compared to the uncoated samples on the third storage day, prolonged storage resulted in a significant loss of antibacterial activity. On the other hand, dipping fish balls in 1% acid-soluble CH led to an important antibacterial effect, as TVC remained below 1 log CFU/g during the entire storage period of 21 days, whereas control uncoated samples exceeded 7.0 log CFU/g between 15 and 18 days of storage.

The potential of CH solutions to prolong shelf life and reduce spoilage microorganisms of meat products was also studied by Lee et al. (2003). The authors dissolved CH of low MW (30 and 120 kDa) as well as a CH oligomer (5 kDa) in lactic acid solution 0.5% (w/v) and water, respectively, to achieve a final concentration of 0.1%–1.0% w/v. Pork samples were dipped in these solutions for 1 min and then stored at 10°C for 8 days. The microbiological shelf life of control samples, based on TVC limit of 7.0 log CFU/g, was ca. 6 days, whereas treatments coated with CH solution of low MW (1.0% w/v) had TVC counts lower than 6.0 log CFU/g at the end of the storage period. Interestingly, water-soluble CH oligomer (5 kDa) was not effective in reducing TVC and thus prolonging the shelf life of pork samples, whereas CH of MW 30 and 120 kDa (1.0% w/v) reduced final TVC counts by up to 1.5 and 2.0 log CFU/g, respectively, as compared with control samples.

The use of a natural antimicrobial coating such as CH combined with vacuum or modified atmosphere packaging (MAP) may result in foods with better organoleptic and microbiological quality than those of conventional packaged raw or precooked

TABLE 16.3
Overview of Studies Testing the Antibacterial Activity of CH in Solution Form, Applied onto Food Products via Dipping or Spraying onto Product Surface or via Blending with Product Formulation

Food Product	Storage Temperature/Period/Other Notes on Experimental Setup	Type of CH Solution	Method of Application onto the Food Product	Bacterial Species	Results: Reduction of Final Population (log CFU/g)	Results: Shelf-Life Extension (Based on Criteria Applied)	Other Observations/Comments	References
(a) Raw skinless sausages	(a) 7°C/18 days	(a) CH glutamate in water (1.0% w/v)	(a) Dipping	(a) Natural microflora	(a) 1.0–3.0	(a) +8 days (threshold limit of TVC = 7.0 log CFU/g)	—	Sagoo et al. (2002)
(b) Minced pork mixture	(b) 4°C/18 days	(b) CH glutamate in water (1.5%, 3.0% w/v)	(b) Blending with meat mixture (i) 0.3% (ii) 0.6% (w/w)	(b) Natural microflora	(b) 1.0–3.0	(b) (i) +9 (ii) +15 days (threshold limit of TVC = 7.0 log CFU/g)	—	
Surimi-based set fish balls	4°C/21 days	CH of low MW in acetic acid (1.0% w/v)	Dipping	Natural microflora	5.0–7.4	TVC remained too low to determine	CH treatment retained TVC below detection limit (<1.0 log CFU/g) throughout 21 days of storage	Kok and Park (2007)

(continued)

TABLE 16.3 (continued)
Overview of Studies Testing the Antibacterial Activity of CH in Solution Form, Applied onto Food Products via Dipping or Spraying onto Product Surface or via Blending with Product Formulation

Food Product	Storage Temperature/Period/Other Notes on Experimental Setup	Type of CH Solution	Method of Application onto the Food Product	Bacterial Species	Results: Reduction of Final Population (log CFU/g)	Results: Shelf-Life Extension (Based on Criteria Applied)	Other Observations/Comments	References
Grilled pork	2°C/28 days/vacuum packaging	Commercial grade CH in acetic acid (2.0%, 2.5% w/v)	Dipping	Natural microflora	2.5–5.0	+ more than 14 days (threshold limit of TVC = 6.0 log CFU/g)	CH–VP combination minimized lipid oxidation and improved sensory characteristics (odor, color, overall acceptability)	Yingyuad et al. (2006)
Pork	10°C/8 days	(a) CH of low MW in lactic acid (1.0% w/v	Dipping	(a) Natural microflora	(a) 1.5–2.0	(a) + more than 2 days (TVC remained below threshold limit of 7.0 log CFU/g during the storage period)	CH reduced lipid oxidation and maintained redness	Lee et al. (2003)
		(b) CH oligomer in water (1.0% w/v)		(b) Natural microflora	(b) No effect	(b) No effect		

Amaranth-based fresh homemade pasta	4°C/54 days/ (a) Aerobic packaging (b) MAP (30 N_2/70% CO_2)	CH in lactic acid	Blending with pasta dough (2000, 4000 mg/kg)	(a) Natural microflora (b) Natural microflora		(a) +27–30 days (threshold limit of mesophilic counts = 6.0 log CFU/g) (b) Mesophilic counts never exceeded threshold limit	Additive antimicrobial effect between MAP and CH observed	Del Nobile et al. (2009a)
RTE roast beef cubes	4°C/28 days	(a) CH of low MW in acetic acid (0.5%, 1.0% w/v) (b) CH of low MW in lactic acid (0.5%, 1.0% w/v)	Dipping	*L. monocytogenes*	(a) 2.3–3.3 (b) 1.0–1.2	Not determined	Acetic acid–CH coating more effective than lactic acid–CH coating	Beverlya et al. (2008)
Bovine meat pâté	4°C/6 days	CH of high MW in acetic acid (0.5% w/v)	Blending with meat formulation (5 mg/g)	*L. monocytogenes*	4.0	Not determined	Some negative influence of CH on flavor and taste was found	Albuquerque Bento et al. (2011)
Raw oysters	4°C/12 days	Commercial food grade CH in HCl (0.5%–2.0% w/v)	Dipping	(a) *S. typhimurium* (b) *S. aureus*	(a) No influence (effect not clear) (b) 2.3–4.1	Not determined	—	Chhabra et al. (2006)

food products. From this point of view, Yingyuad et al. (2006) investigated the combined effect of CH coating solutions and vacuum packaging on maintaining the quality of refrigerated grilled pork. CH dipping solutions were prepared by dissolving commercial grade CH (Fisher Scientific, the United States) in aqueous acetic acid solution of 1% w/v, to a final CH solution concentration of 2.0% and 2.5% w/v. Grilled pork was immersed into the solutions for 1 min and then vacuum packaged (VP). CH coating (2.0% or 2.5% w/v) in combination with VP reduced TVC by ca. 5.0 and 3.0 log CFU/g as compared with air and vacuum uncoated samples, respectively, after 14 days of storage at 2°C. Interestingly, TVC of CH-coated samples remained below 4 log CFU/g till the end of the storage period (28 days). TVC of air and vacuum uncoated samples reached 6.85 and 6.3 log CFU/g on the 14th and 28th day of storage, respectively, while the quality of the product was unacceptable when TVC exceeded 6.0 log CFU/g.

More recently, a study on the combined effects of CH and MAP to improve the microbiological quality of a freshly prepared homemade product (amaranth-based fresh pasta) stored under refrigeration was conducted by Del Nobile et al. (2009a). Working active solutions were prepared by dissolving CH (Danisco, Braband, Denmark) in lactic acid solution (1.38% v/v). These solutions were mixed with the dough of pasta separately, to obtain final concentrations of 2000 and 4000 mg/kg of pasta. The samples were packaged under aerobic or MAP (80:20, 0:100, and 30:70 N_2:CO_2 combinations) conditions and stored at 4°C. Control treatment (no CH added, aerobic packaging) exceeded the legislation threshold limit of mesophilic bacteria for fresh pasta (6 log CFU/g) after ca. 20 days of storage, whereas treatments with CH (2000 and 4000 mg/kg) exceeded the aforementioned limit after 47–50 days (4°C). In the three MAP treatments, mesophilic counts never exceeded the proposed limit of 6 log CFU/g. Interestingly, the combined application of MAP (80:20, 0:100 N_2:CO_2) and CH resulted in lower mesophilic counts than MAP alone, by ca. 1.12–1.55, at the end of the storage period (54th day). Moreover, after the first 30 days of storage, all CH-air samples reduced *Staphylococcus* spp. counts by at least 3.0 log CFU/g, as compared with untreated control samples. The study concluded that MAP and CH, irrespective of the concentration tested, can act in a synergic mode in controlling the microbial quality loss of fresh pasta during refrigerated storage and thus be utilized as a valid alternative to more expensive food thermal treatments commonly used to prolong the shelf life of these kinds of product.

To conclude, all previously described research studies suggest that CH–organic acid solutions (mainly lactic or acetic acid) at concentrations of 1.0%–2.5% w/v could be potentially used as effective means of extending the shelf life of various perishable food products, including surimi-based seafood products, fresh homemade pasta, raw or grilled meat products, etc., during refrigerated storage, without any negative impact on sensorial acceptability. Acid-soluble CH applied on food products by dipping or blending proved to significantly reduce TVC by 2–6 log CFU/g, during refrigeration storage (2°C–10°C), and its effectiveness was concentration dependant. Physical characteristics of CH (e.g., MW) seem to be an important factor in controlling its antimicrobial properties on real food products. For example, although water-soluble CH glutamate solutions (1.0%–3.0% w/v) managed to reduce

TVC, LAB, and yeast–mold populations of raw skinless sausages and minced pork mixture stored under refrigeration by up to 3.0 log CFU/g, other water-soluble CH oligomers (0.5%–1.0% w/v) significantly lost their antibacterial properties against spoilage bacteria of seafood or raw pork after 2–4 days of refrigerated storage (4°C). Furthermore, acid-soluble (acetic/lactic acid) CH was successfully used in combination with VP or MAP to extend the shelf life of raw and precooked foods and resulted in products with better sensorial and microbiological quality than aerobic packaging, suggesting that CH–MAP or VP can act in a synergic mode in controlling microbial and sensorial quality loss.

Apart from these studies investigating the ability of CH solutions to prolong the shelf life of refrigerated food products, many authors have tried successfully to evaluate its potential to enhance food safety during chill storage. The great concern of health-conscious society for better quality and improved safety of food products was the main motive for the following studies to examine the effect of CH solutions on several foodborne pathogenic bacteria inoculated into fresh or precooked food products, being stored under refrigeration. *L. monocytogenes*, *S. aureus*, and *Salmonella enterica* are three of a wide range of pathogenic bacteria, implicated in serious foodborne outbreaks caused by the consumption of raw or precooked products (when contaminated with any of these bacteria); and thus, research has focused on the potential of CH as a natural preservative to inhibit their growth.

Refrigerated ready-to-eat (RTE) food products have a great risk of being contaminated and support the growth of *L. monocytogenes*, a psychrotrophic and ubiquitous pathogen, presenting a high mortality rate among some high-risk groups, including the newborn, the aged, and people with compromised immune systems. Motivated by the great concern of consumers about the potential harmful effects of synthetic food preservatives traditionally used for preventing pathogen growth, Beverlya et al. (2008) and Albuquerque Bento et al. (2011) evaluated the antimicrobial activity of CH solutions against the growth of this specific pathogen, after being inoculated into RTE popular meat products (roast beef, bovine pate) stored at 4°C.

Beverlya et al. (2008) evaluated the antimicrobial activity of high and low MW CH against *L. monocytogenes* inoculated in RTE roast beef cubes. The low MW CH (Keumco Chemical, Seoul, Korea) and high MW CH (Premix Ingredients, Avaldnes, Norway) were dissolved in acetic or lactic acid solution (1% w/v) to a final concentration of 0.5% or 1% w/v. Roast beef cubes, previously inoculated with *L. monocytogenes* at an initial concentration of 6.65 log CFU/g, were dipped once in the solutions for 30 s. The most effective CH coating was low MW CH in acetic acid (0.5% or 1% w/v), reducing pathogen counts by ca. 2.3–3.3 log CFU/g, as compared with control uncoated samples, after 28 days of storage at 4°C.

Albuquerque Bento et al. (2011) aimed at evaluating the efficacy of CH in inhibiting *L. monocytogenes* in bovine meat pâté stored at 4°C. CH originated from *Mucor rouxii* (fungi isolated from mangrove Sediment), and its production was performed by the authors themselves. The obtained CH presented MW of 2.60×10^4 g/mol and D.D. of 85%. The CH solution was prepared by dissolving 50 mg CH in 100 mL acetic acid 1% (v/v) and was added to bovine pâté samples to achieve a final concentration of 5 mg CH/g pâté. The addition of CH solution to bovine meat pâté decreased the counts of *L. monocytogenes* from approximately 7 to 3 log CFU/g after 6 days of storage at 4°C, whereas in pâté without CH added (control) pathogen counts were over 7 log CFU/g already after 2 days of storage. Sensory valuation suggested that addition of CH in pâté would be acceptable to consumers, although some negative influence on flavor and taste was found.

S. enterica, especially serotype enteritidis, is a well-known pathogenic bacterium responsible for the most frequently reported zoonotic disease in many countries (World Health Organization, 2002). Although relatively much in vitro work has been done on antimicrobial properties of CH against *S. enterica*, very little work (to our knowledge) has focused on its ability to inhibit growth of this pathogen on real food products, stored under low refrigeration temperatures. Results of in vitro study of Marques et al. (2008) performed at 10°C indicated that *S. enterica* required concentrations of 0.03%–0.05% for an inhibitory or bactericidal effect, respectively, whereas at 20°C MICs were greater than 0.10%.

The antimicrobial potential of CH solutions sprayed or coated onto fish products (oysters, shrimp-based salad) inoculated with *S. enterica* was investigated by Chhabra et al. (2006) and Roller and Covill (2000).

Chhabra et al. (2006) investigated the fate of *S. enterica* ser. typhimurium in raw oysters treated with CH. Commercial food-grade CH (T.C. Union Co, Thailand) was dissolved in 0.5% (v/v) HCl (0.6 M) to achieve final concentrations of 0.5%–2.0% w/v. Fresh shucked oysters were dipped in CH solution for 1 h at 4°C. CH-treated and control uncoated samples had no significant differences among them as regards *S. typhimurium* counts, over the 12 day storage period (4°C); however, pathogen counts were reduced from their initial high populations (7.2 log CFU/g) irrespective of treatment. In another study, CH glutamate solution (30 g/L, Pronova, Norway) prepared in water was tested against the growth of *S. enterica* ser. enteritidis, inoculated onto shrimps (9 mg of CH per g of shrimp) stored at 5°C and 25°C for up to 8 days (Roller and Covill, 2000). CH solution was sprayed over the surface of shrimps (9 mg of CH per g of shrimp). At both temperatures, CH did not reveal any antimicrobial effect toward pathogen counts, inoculated in high initial numbers (5–7 log CFU/g). Both aforementioned studies indicate that CH coating may be ineffective to control *Salmonella* growth in seafood, especially if contamination with this pathogen is at high levels. However, further studies focusing on antimicrobial properties of CH in foods contaminated with low levels (1–4 log CFU/g) of this specific pathogen are needed, to ensure its potential to control *Salmonella* growth.

S. aureus is a spherical Gram-positive bacterium (coccus), some strains of which are capable of producing a highly heat-stable protein toxin that causes illness in humans. Foods that require considerable handling during preparation and that are kept at slightly elevated temperatures before consumption are frequently involved in staphylococcal food poisoning. Although food handlers are usually the main source of food contamination in staphylococcal food poisoning outbreaks, equipment and environmental surfaces can also be sources of contamination with *S. aureus* (FDA, 2009). Various researchers have demonstrated the antimicrobial activity of CH against *S. aureus* in suspension studies, and it was found that CH was inhibitory at minimum concentrations of 0.005%–0.20%

w/v (Chang et al., 1989; Darmadji and Izumimoto, 1994; Fernandes et al., 2008). However, its antimicrobial potential toward this specific pathogen has only been sparingly investigated in real food systems, with few exceptions. For example, Chhabra et al. (2006) inoculated fresh oysters with *S. aureus* and observed that dipping in CH solution of 0.5%, 1.0%, and 2.0% w/v (see previous paragraph for preparation of CH solution) reduced pathogen counts by 3.5, 2.3, and 4.0 log CFU/g, respectively, as compared with untreated samples, after 12 days of storage at 4°C. The antimicrobial effect of CH solutions against *S. aureus* inoculated into semiskimmed milk was studied by Fernandes et al. (2008). In this study, CH of low, medium, and high MW was dispersed in 1.0% (v/v) acetic acid solution at a final concentration of 0.5% w/v., while COS (Nicechem, Shangai, China) in deionized water was used at a final concentration of 2.5% w/v. Results showed that after an initial positive inhibitory effect of COS, *S. aureus* viable counts increased afterward, due to possible trapping of their molecules in the milk protein network. However, the highest MW CH prevented loss of antibacterial effect against the same pathogen.

The results of the aforementioned studies focusing on antibacterial activity of CH solutions against Gram-positive pathogenic bacteria, inoculated on real food systems, suggest that CH (especially of low MW) could be considered as a possible alternative agent compound for controlling the growth of *L. monocytogenes* and *S. aureus* in RTE meat products, fresh seafood, or milk during storage at refrigeration temperatures. Although the antibacterial activity of CH against *Salmonella* spp., proved to be significant in vitro, its effectiveness on real food products seems to be questionable, as acid or water-soluble CH had no influence on the reduction of *S. typhimurium* or *enteritidis* during refrigerated storage of seafood (raw oysters and shrimp salad). It is worth mentioning that the majority of studies on the inhibitory effect of CH against food-related microorganisms has been carried out in laboratory media, so more research work using real food systems needs to be carried out in the future, to assess the efficacy of CH in inhibiting the survival of pathogens such as *L. monocytogenes*, *S. aureus*, or *Salmonella* spp.

Nowadays, new promising technologies integrating edible CH-based solutions or bioactive coatings enriched with natural antimicrobial substances (EOs, lysozyme, nisin, etc.) have been successfully applied for the preservation of fish, meat, poultry, or vegetable-based products, as well as in inactivation of foodborne pathogenic bacteria.

16.3.2.1 Application of Edible Chitosan Solutions or Bioactive Chitosan Coatings in Combination with Essential Oils on Preserving Food Quality and Safety

Recently, several researchers have revealed the beneficial effects of natural antimicrobials like CH and EOs individually and in combination when applied on food systems. EOs possess antibacterial, antiviral, antifungal, and antioxidant properties (Burt, 2004; Holley and Patel, 2005). Generally, among EOs, thyme and oregano oil have increasingly attracted the interest of researchers as potential "natural" antimicrobials to be used in the food industry, because they have revealed strong antibacterial activity against both foodborne pathogens and spoilage organisms due to a high percentage of phenolic compounds such as thymol, p-cymene, carvacrol, and γ-terpinene (Burt, 2004; Holley and Patel, 2005). In recent years, highly promising in vitro and in vivo studies on combined application of active CH-based solutions/bioactive coatings with thyme oil (Giatrakou, 2010; Giatrakou et al., 2010a,b), rosemary, oreganum, capsicum, and garlic oleoresin (Ponce et al., 2008), allyl isothyocyanate-hop extract (Wasaouro EXT®, Inatsu et al., 2005), mint (Kanatt et al., 2008), and cinnamon oil (Duan et al., 2010; Ojagh et al., 2010) have been reported.

Giatrakou et al. (2010a) studied the combined application of CH solution, thyme oil, and MAP on preservation of a fresh ready-to-cook (RTC) poultry product (chicken-pepper kebab). Stock CH solution 2% (w/v) was prepared by dissolving low MW CH (from crab shells) in powder form (Aldrich Co, Athens, Greece) in glacial acetic acid aqueous solution 1% (w/v). The antimicrobials were added to the RTC poultry product, either singly, or sequentially; thyme oil was applied using a micropipette (0.2% v/w), whereas CH solution was sprayed directly onto the product (1.5% v/w). All samples were stored at 4°C, under MAP (30% CO_2/70% N_2). MAP-CH (M-CH) and MAP-thyme (M-T) treatments significantly affected TVC, LAB, *Pseudomonas* spp., *B. thermosphacta*, *Enterobacteriaceae*, and yeast–molds, whereas combined application of MAP with CH and thyme oil (M-CH-T) had a more pronounced effect, indicating possible additive or synergistic antimicrobial effect between natural agents. For example, M-CH-T treatment reduced the aforementioned spoilage microorganisms of the RTC product by 3.0–4.5 log CFU/g, after storage for 14 days at 4°C, as compared with control samples. Moreover, M-CH-T treatment extended the shelf life of the product by 8–9 days, based on microbiological threshold limit for TVC = 7.0 log CFU/g and sensory analysis (taste attribute).

The same authors investigated the combined potential of MAP and CH thyme to inhibit the growth of three inoculated pathogens (*L. monocytogenes*, *S. enterica* ser. Montevideo, *B. cereus*) onto the aforementioned RTC product during refrigerated storage at 4°C (Giatrakou, 2010, unpublished results). According to their findings, thyme oil (M-T treatment) reduced *L. monocytogenes* counts by 1.0 log CFU/g, as compared with the control (M), while the anti-*Listerial* activity of CH when applied alone (M-CH) decreased progressively over time (Figure 16.1). However, combined CH–thyme treatment (M-CH-T) had a bacteriostatic effect on *L. monocytogenes*, during storage at 4°C for 8 days. In conclusion, CH and thyme oil revealed a possible synergistic effect on inhibiting growth of this specific pathogen.

During storage at 4°C and for a period of 8 days, *Salmonella* and *B. cereus* counts were reduced by 0.6–1.0 and 0.5 log CFU/g respectively, in the presence of thyme oil or CH, when applied individually onto the product. Combined application of the antimicrobials onto the RTC product (M-CH-T) decreased final population of *Salmonella* and *B. cereus* by 1.2 and 0.7 log CFU/g, indicating negligible synergistic action between CH and thyme oil toward these species at this specific temperature (Giatrakou, 2010, unpublished results).

The combined effect of CH solution and thyme oil on inoculated *L. monocytogenes* was also investigated during storage of the RTC poultry product at 8°C and

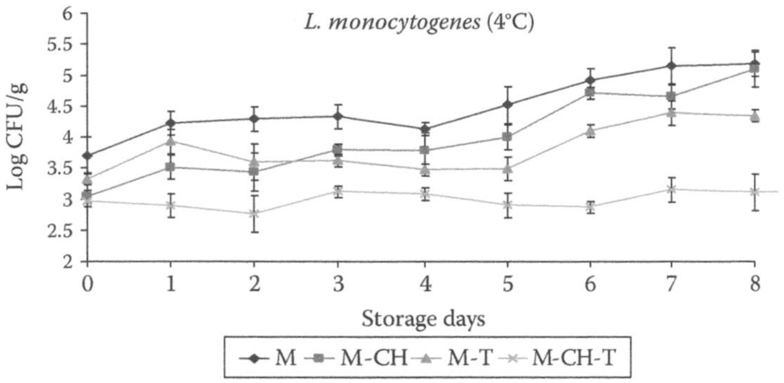

FIGURE 16.1 Fate of *L. monocytogenes* inoculated on an RTC poultry product (chicken-pepper kebab) during storage at 4°C under MAP (treatment M), with thyme oil (M-T), with CH (M-CH), and with CH plus thyme oil (M-CH-T). (From Giatrakou, V., Unpublished results, Ph.D. Thesis., Study of the Quality and Hygiene of a Raw Poultry Product (Chicken-pepper Kebab-"Souvlaki") Stored Under Modified Atmosphere Packaging-Combined Effect of Treatment with Thyme Oil and CH. PhD Thesis., Department of Chemistry, University of Ioannina, Ioannina, Greece, 2010.)

for a period of 8 days (Giatrakou, 2010, unpublished results). Control treatment (M) was stored under MAP with a gas atmosphere of 30% CO_2/70% N_2. In control samples (M) during the first 6 days of storage at abuse temperature (8°C), *L. monocytogene* growth was slow, showing an increase of ca. 1.30 log CFU/g from their initial values (Figure 16.2).

However, after the sixth day of storage at 8°C, the pathogen grew faster, reaching levels of ca. 6.2 log CFU/g (increase of ca. 3.0 log CFU/g from the initial value).

M-CH treatment decreased pathogen counts by 0.6–0.8 log CFU/g, during the first 4 days of storage. However, from the fourth day and till the end of the storage period (8th) addition (spraying) of CH alone did not decrease *L. monocytogenes* significantly, as compared to the control. This indicates that antibacterial effect of CH against the pathogen significantly weakened over time, probably attributed to the fact that the amine groups of CH bind to components of the bacterial cell surface and, therefore, are no longer available to attach to new *Listeria* cells (Coma et al., 2002). On the contrary, thyme oil (M-T treatment) maintained its antibacterial action against the pathogen to a better degree than CH, resulting in a final reduction of ca. 1.0 log CFU/g, as compared to control (M). Combined application of CH and thyme oil (M-CH-T) reduced pathogen counts by 1.2 log CFU/g, right from the initial day of storage. During the first 4 days of storage, treatments M-CH, M-T, and M-CH-T revealed no significant changes among them, the latter, however, was the only one to retain *Listeria* counts below 3.0 log CFU/g until the fifth day of storage (8°C) (bacteriostatic effect). Moreover, treatment M-CH-T produced lower counts of the pathogen in the poultry product resulting in ca. 1.5–1.6 and 0.7 log cycles as compared to treatments M, M-CH, and M-T, respectively. However, at the end of the storage period, the addition of thyme oil

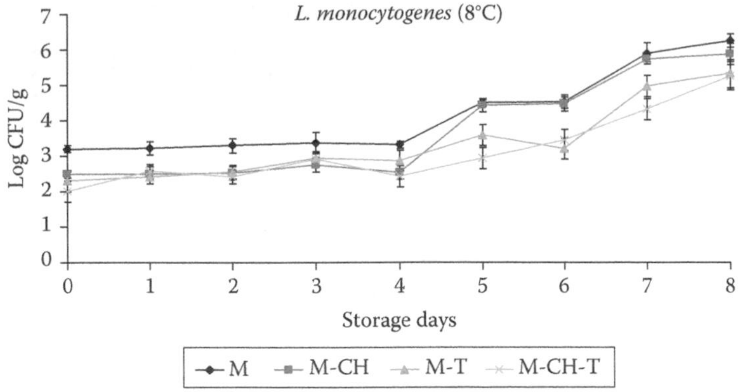

FIGURE 16.2 Fate of *L. monocytogenes* inoculated on an RTC poultry product (chicken-pepper kebab) during storage at 8°C under MAP (treatment M), with thyme oil (M-T), with CH (M-CH), and with CH plus thyme oil (M-CH-T). (From Giatrakou, V., Unpublished results, Ph.D. Thesis., Study of the Quality and Hygiene of a Raw Poultry Product (Chicken-pepper Kebab-"Souvlaki") Stored Under Modified Atmosphere Packaging-Combined Effect of Treatment with Thyme Oil and CH. PhD Thesis., Department of Chemistry, University of Ioannina, Ioannina, Greece, 2010.)

alone or with CH (M-T and M-CH-T) reduced viable counts of *Listeria* by the same magnitude (1.0 log CFU/g). Overall, although the combined application of CH and thyme oil (under MAP conditions) reduced *Listeria* counts by ca. 2.0 log CFU/g during storage at 4°C, indicating a possible synergistic effect between two antimicrobials, at 8°C, a different behavior was noted with a reduction in final counts of the pathogen by 1.0 log CFU/g, indicating that the synergistic effect of CH and thyme oil alone was less pronounced.

Additionally, the combined effect of CH and thyme oil on *S. enterica* ser. Montevideo inoculated onto the RTC poultry product stored at 8°C, for a period of 8 days, was investigated (Figure 16.3).

According to Figure 16.3 in control (M) samples *Salmonella* grew fast from an initial level of 3.3 log CFU/g to high final counts of 5.8 log CFU/g, after 8 days of storage at 8°C. CH (M-CH) did not inhibit the growth of the pathogen, as *Salmonella* counts did not differ significantly from the control treatment, during the entire storage period, except from day 4 when a slight difference was observed. Interestingly, thyme oil (M-T) had a more pronounced effect on *Salmonella* during the first 6 days of storage, reducing pathogen counts by ca. 1.0 log CFU/g, as compared to the control. However, on final day 8 of storage treatments M-T and M (control) had almost equal counts, indicating that the antimicrobial effect of thyme oil was progressively weakened. The combination of CH and thyme oil (M-CH-T) was the most effective treatment in inhibiting *Salmonella* growth, as compared to all other treatments during the first 6 days of storage at 8°C. Population counts of the pathogen in M-CH-T were by 2.0 log CFU/g lower as compared with M on the sixth day of storage, whereas extended storage resulted in a reduction of a lower magnitude (only by ca. 0.7 log CFU/g lower than M

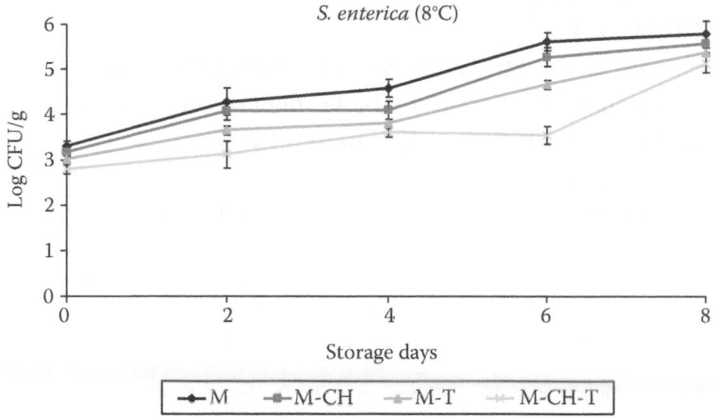

FIGURE 16.3 Fate of *S. enterica* inoculated on an RTC poultry product (chicken-pepper kebab) during storage at 8°C under MAP (treatment M) with thyme oil (M-T), with CH (M-CH), and with CH plus thyme oil (M-CH-T). (From Giatrakou, V., Unpublished results, Ph.D. Thesis., Study of the Quality and Hygiene of a Raw Poultry Product (Chicken-pepper Kebab-"Souvlaki") Stored Under Modified Atmosphere Packaging-Combined Effect of Treatment with Thyme Oil and CH. PhD Thesis., Department of Chemistry, University of Ioannina, Ioannina, Greece, 2010.)

on day 8). Interestingly, during the first 6 days of storage at 8°C, M-CH-T had significantly lower *Salmonella* counts than M-CH and M-T, indicating that CH and thyme oil may potentially exhibit a synergistic or (additive) effect against the pathogen (Giatrakou, 2010, unpublished results). In the same series of experiments, the combined effect of CH and thyme oil on a *B. cereus* cocktail mixture of psychrotrophic and mesophilic strains (Table 16.4), inoculated onto the poultry RTC product was examined during storage under MAP at 8°C, for a period of 8 days (Figure 16.4).

According to Figure 16.4 during storage at abuse temperature (8°C), under M treatment (control) *B. cereus* counts reached ca. 4.5 log CFU/g. Treatments with CH or thyme oil (M-CH and M-T) did not seem to affect growth of the pathogen significantly, during the first 5 days of storage (Giatrakou, 2010, unpublished results). However, toward the end of the storage period (days 6–8) these treatments resulted in lower counts of the pathogen (ca. 1.0 log CFU/g) as compared with the control (M). Interestingly, the combined application of CH and thyme oil (M-CH-T) revealed a more pronounced effect on these species, reducing its viability by almost 2.0 log CFU/g, indicating a possible additive antimicrobial effect between these two antimicrobial agents. In addition, *B. cereus* counts in treatment M-CH-T remained below 6.0 log CFU/g during the entire storage period, whereas M (control), M-CH, and M-T treatments exceeded this level right from the fifth day. This observation could lead to the assumption that a combined antimicrobial treatment involving MAP, CH, and an EO (e.g., thyme) could be used as a means of an active packaging/antimicrobial technology, for preservation of fresh poultry products, also guaranteeing safety from a likelihood of a *Bacillus* spp. toxin formation, as it is known that

TABLE 16.4
***Bacillus cereus* Strains Used in the Cocktail Mixture, Inoculated onto a Poultry RTC Product during Storage under MAP at 8°C, for a Period of 8 Days**

Type Strain and Reference of Source	Description and Origin
[a]ATCC 10987	Mesophilic, isolated from a spoiled cheddar cheese
ATCC 14579	Mesophilic, isolated from the air in a cow shed
PAL 22 (Z4222, [b]INRA)	Psychrotrophic, isolated from cooked chilled food
PAL 25 ([c]NCTC 11143)	Mesophilic, emetic outbreak

Source: Giatrakou, V., Unpublished results, Ph.D. Thesis., Study of the Quality and Hygiene of a Raw Poultry Product (Chicken-pepper Kebab-"Souvlaki") Stored Under Modified Atmosphere Packaging-Combined Effect of Treatment with Thyme Oil and CH. PhD Thesis., Department of Chemistry, University of Ioannina, Greece, 2010.

[a] ATCC, American Type Culture Collection, Manassas, VA.

[b] INRA, Institut National de la Recherche Agronomique, Avignon France.

[c] NCTC, National Collection of Type Cultures, Central Public Health Laboratory, London, the United Kingdom.

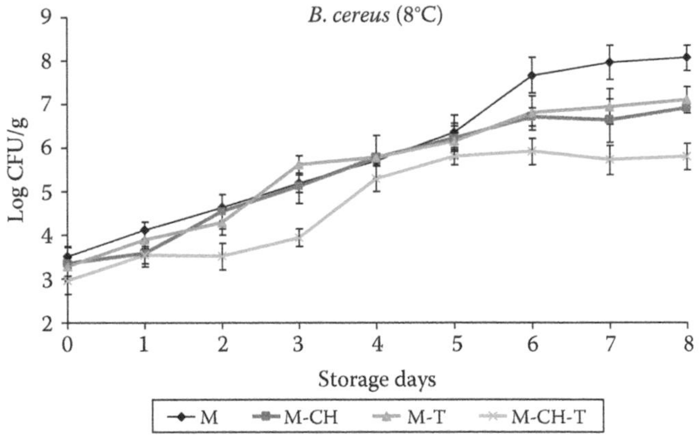

FIGURE 16.4 Fate of *Bacillus* spp. inoculated on an RTC poultry product (chicken-pepper kebab) during storage at 8°C under MAP (treatment M) with thyme oil (M-T), with CH (M-CH), and with CH plus thyme oil (M-CH-T). (From Giatrakou, V., Unpublished results, Ph.D. Thesis., Study of the Quality and Hygiene of a Raw Poultry Product (Chicken-pepper Kebab-"Souvlaki") Stored Under Modified Atmosphere Packaging-Combined Effect of Treatment with Thyme Oil and CH. PhD Thesis., Department of Chemistry, University of Ioannina, Greece, 2010.)

concentrations of 6.0 log CFU/g are required for enterotoxin formation (Grant et al., 1993; Beuchat et al., 1997; FSAI, 2007; FDA, 2009). Such likelihood was not investigated in the aforementioned series of experiments (Giatrakou, 2010, unpublished results).

It is worth mentioning that during storage of the poultry product at 8°C, the possible synergistic or additive effect between CH and thyme oil was more pronounced in the case of *Salmonella* and *B. cereus* than for *Listeria*, whereas at 4°C the opposite effect was observed. Results demonstrate that CH loses its antimicrobial action toward *Salmonella* and *Listeria* over time; however, combination with thyme oil may be a potential solution for this particular problem, as the combined application of these antimicrobials resulted in a bacteriostatic effect against these bacteria for 8 and 5–6 days during storage at 4°C and 8°C, respectively (Giatrakou, 2010, unpublished results).

With regard to other studies, Inatsu et al. (2005) evaluated the antibacterial effect of CH and allyl-isothiocyanate (AIT) against growth of *E. coli*, *Salmonella enteritidis*, *S. aureus*, and *L. monocytogenes*, inoculated onto Chinese cabbage. These authors prepared a CH (CH 10, Wako Pure Chemical, Osaka, Japan) solution in distilled water (5% w/v) and added a proper amount of the former to cups containing Chinese cabbage in order to obtain a final CH concentration of 0.1%. Moreover, an emulsion liquid containing AIT and hop extract (AIT-hop, Wasaouro EXT) at a final concentration of 0.2% was added singly or in combination with the aforementioned CH solution. According to their findings, the combination of CH and AIT-hop exhibited a slightly greater bactericidal effect against mesophilic and coliform bacteria compared with the two natural antimicrobials applied individually (ca. 2.0 log CFU/g decrease as compared with the control) during storage of the lightly fermented cabbage (10°C, 4 days). Moreover, CH alone or with AIT-hop reduced viable counts of *E. coli* and *S. enteritidis* by 1.2 and 0.7 log CFU/g, respectively. The combination of CH/AIT-hop slightly reduced counts of *E. coli* when compared with CH only, but this was not observed in the case of and *S. enteritidis*. Finally, this combination enhanced bactericidal activity toward inoculated *L. monocytogenes*, with a reduction of ca. 1–2 log CFU/g in counts, lower than the singly applied antimicrobials by the end of the storage period.

Ponce et al. (2008) evaluated the potential antibacterial and antioxidant benefits of film-forming solutions of CH enriched with oleoresins, in vivo and in vitro. Experiments in vitro showed that a pure edible coating solution of CH, prepared by dissolving 20 g of CH in 1% acetic acid and 1% glycerol solution, did not show any significant antimicrobial properties on butternut squash native microflora and *L. monocytogenes*. The authors attributed the limited anti-listerial activity of CH film-forming solution to the high number of the initial inoculum (ca. 10^6 and 10^7 CFU/petri dish), thereby exceeding CH inhibition activity. However, enrichment of the CH film-forming solution with 1% rosemary oleoresin was very effective against *L. monocytogenes* and the squash native microflora (in vitro). However, the use of CH coatings enriched with rosemary and olive oleoresins did not produce a significant antimicrobial effect when applied to butternut squash during in vivo study.

In another study, Kanatt et al. (2008) developed a novel natural preservative system, consisting of CH and mint mixture with antimicrobial and antioxidant properties. CH (Mahatani CH Pvt. Ltd., India) solution was made in 1% acetic acid and mixed with mint extract solution. CH–mint mixture solution (CM) was incorporated in the meat formulation for the preparation of pork salami, as well as in minced lamb meat, in order to achieve a final concentration of 0.1%. Mint alone did not reveal any antimicrobial action, whereas the antimicrobial action of CM was similar to that of CH. CM (0.1%) reduced counts of Gram-negative (*E. coli*, *Pseudomonas*, *S. typhimurium*) and Gram-positive (*B. cereus, S. aureus*) bacteria, inoculated into the minced lamb meat by 1.0 and 2–3 log CFU/g, respectively, after 28 days of storage at 0°C–3°C. Moreover, the same study evaluated the effect of CM on the shelf life of pork cocktail salami. Control salamis spoiled in less than 2 weeks, whereas salamis formulated with CM displayed a shelf life of 3 weeks (0°C–3°C) without any positive or negative effect on sensory characteristics (color, flavor, taste, and texture).

Ojagh et al. (2010) studied the effect of CH coating enriched with cinnamon oil on the quality of refrigerated rainbow trout, stored under refrigeration (4°C) for a period of 16 days. CH solution was prepared with 2% w/v CH of medium MW (Aldrich Chemical Co., Milwaukee, WI), and glycerol (0.75 mL/g) was added as a plasticizer. Cinnamon oil (1.5%) was dissolved in Tween 80 and then mixed with the aforementioned CH solution, to obtain a final film-forming solution. Fillet samples of rainbow trout were immersed twice in each of the aforementioned CH coating solutions and were left to drain at 10°C for 5 h. Final counts of Gram-negative psychrotrophic bacteria were reduced by 1.63 and 1.75 log CFU/g, as compared with the control, in the presence of CH and CH–cinnamon oil solutions, respectively, after 16 days of refrigerated storage. In addition, the results of the sensory evaluation showed that adding cinnamon oil into the CH coating significantly enhanced the beneficial effects on color and overall acceptability of raw fish fillets during the final days of storage, whereas CH coating with or without the EO added did not produce undesirable sensory properties on cooked fish fillet.

Duan et al. (2010) investigated the combined effect of CH–cinnamon–krill oil coating and MAP on the storability of cold-stored lingcod (*Ophiodon elongates*) fillets (Table 16.5). CH of low MW (Primex ehf) was dissolved in acetic acid solution 1% with the addition of glycerol 25% (w/w of CH) and finally added at a final concentration of 3.0% w/v. Krill oil (20% w/w) was mixed into the CH solution with the addition of Tween 80, and cinnamon oil was then mixed into the krill oil-incorporated CH combined solution at a concentration of 0–0.1 μL/mL. The solution mixtures were homogenized at 3000 r.p.m. for 1 min and used for coating fish fillets right after sample preparation. The lingcod fillets were cut into small pieces, coated using vacuum impregnation procedure to achieve uniform coatings, vacuum or modified atmosphere (MA) (60% CO_2 + 40% N_2) packaged, and then stored at 2°C for up to 21 days. The combined CH coating and vacuum or MA packaging reduced lipid oxidation, chemical spoilage as reflected in TVBN, and reduced TVC by 2.2–4.2 log reductions during storage. The addition of cinnamon oil in coating did not provide additional reduction in TVC of CH-coated samples, and no difference was observed between samples with vacuum and MA packaging. However, TVC of all CH-coated samples remained at very low levels (less than 5.0 log CFU/g) during the entire

TABLE 16.5
Overview of Studies Testing the Combined Effect of CH Solutions/Coatings and Essential Oils/Plant Extracts or Lysozyme on Preserving Food Quality and Safety

Food Product	Storage Temperature/ Period	Type of Active Packaging Technology Applied	Bacterial Species	Results: Reduction of Final Population (log CFU/g)	Results: Shelf-Life Extension (Based on Criteria Applied)	Other Observations/ Comments	Reference
Lightly fermented Chinese cabbage	10°C/4 days	CH solution (in water) + allyl-isothiocyanate-hop extract (AIT-hop, Wasaouro Ext)	(a) Natural flora	(a) 2.0	Not determined	(a) CH and AIT-hop combination exhibited slightly greater bactericidal effect against mesophilic bacteria and coliforms, but not against LAB	Inatsu et al. (2005)
			(b) Inoculated pathogens (i) *E. coli* (ii) *S. enteritidis* (iii) *S. aureus* (iv) *L. monocytogenes*	(b) (i) 1.0 (ii) 0.7 (iii) 1.0–1.3 (iv) 2.5–3.0		(b) Synergistic effect between CH and AIT-Hop only against *L. monocytogenes*	
Minced lamb meat	0°C–3°C/28 days	CH–mint mixture solution	(i) *P. fluorescens, E. coli, S. typhimurium* (ii) *B. cereus, S. aureus*	(i) 1.0 (ii) 2.0–3.0	Not determined	Mint extract alone had poor antimicrobial activity, no synergistic effect between CH and mint extract observed	Kanatt et al. (2008)

(*continued*)

TABLE 16.5 (continued)
Overview of Studies Testing the Combined Effect of CH Solutions/Coatings and Essential Oils/Plant Extracts or Lysozyme on Preserving Food Quality and Safety

Food Product	Storage Temperature/Period	Type of Active Packaging Technology Applied	Bacterial Species	Results: Reduction of Final Population (log CFU/g)	Results: Shelf-Life Extension (Based on Criteria Applied)	Other Observations/Comments	Reference
Pork cocktail salami	0°C–3°C/3 weeks	CH–mint mixture solution	Natural microflora	>1.0	+ 1 week (based on TVC, oxidative rancidity)	CH–mint mixture solution before cooking process minimized oxidative rancidity	Kanatt et al. (2008)
Rainbow trout	4°C/16 days	(a) CH–cinnamon oil solution	Natural microflora	(a) 1.75	+ more than 6 days (TVC remained below threshold limit of 7.0 log CFU/g during the storage period)	CH–cinnamon oil combination improved sensory characteristics of raw and cooked fish and reduced chemical spoilage	Ojagh et al. (2010)
Lingcod fish fillets	2°C/21 days	CH–cinnamon–krill oil solution in combination with MAP or VP	Natural microflora	2.2–4.2	+ more than 16 days (TVC remained below threshold limit of 7.0 log CFU/g during the storage period)	(a) Cinnamon did not provide additional TVC reduction than CH only (b) CH–cinnamon combination improved sensory characteristics (less fishy aroma)	Duan et al. (2010)

						(c) Combination retarded chemical spoilage and lipid oxidation	
RTC chicken-pepper kebab	4°C/14 days	CH solution + thyme oil in combination with MAP	Natural microflora	3.0–4.5	+ 8–9 days (threshold limit TVC = 7.0 log CFU/g, taste attribute)	(a) Synergistic or additive antimicrobial effect between CH and thyme oil was observed (b) CH–thyme oil combination improved sensory characteristics (c) CH–thyme oil combination reduced lipid oxidation and maintained redness	Giatrakou et al. (2010a)
RTC chicken-pepper kebab	4°C/12 days	CH solution + thyme oil in combination with aerobic packaging	Natural microflora	1.0–3.0	+ 3 days (threshold limit TVC = 7.0 log CFU/g)	(a) CH–thyme oil combination enhanced natural freshness of the product (b) CH–thyme oil combination retained acceptable taste–odor attributes over the storage period	Giatrakou et al. (2010b)

(continued)

TABLE 16.5 (continued)
Overview of Studies Testing the Combined Effect of CH Solutions/Coatings and Essential Oils/Plant Extracts or Lysozyme on Preserving Food Quality and Safety

Food Product	Storage Temperature/ Period	Type of Active Packaging Technology Applied	Bacterial Species	Results: Reduction of Final Population (log CFU/g)	Results: Shelf-Life Extension (Based on Criteria Applied)	Other Observations/ Comments	Reference
RTC chicken-pepper kebab	(a) 4°C/8 days	CH solution + thyme oil in combination with MAP	Inoculated pathogens (a) 4°C: (i) *L. monocytogenes* (ii) *Salmonella* spp. (iii) *B. cereus*	(a) 4°C: (i) 2.0 (ii) 1.2 (iii) 0.7	Not determined	(a) CH–thyme oil combination exhibited bacteriostatic effect on pathogen growth	Giatrakou (2010)
	(b) 8°C/8 days		(b) 8°C: (i) *L. monocytogenes* (ii) *Salmonella* spp. (iii) *B. cereus*	(b) 8°C: (i) 1.0 (ii) 0.7 (iii) 2.0		(b) Antimicrobial action of CH toward *Salmonella* and *Listeria* reduced over time at 8°C	
Hard boiled shell-on eggs	(a) 10°C/ 4 weeks	CL edible coating	(a) Inoculated pathogens: (i) *L. monocytogenes* (ii) *S. enteritidis*	(a) (i) 0.8 (ii) 4.0	(a) Not determined		Kim et al. (2008)
	(b) 10°C/10 weeks		(b) Natural microflora	(b) 1.5–2.6	(b) + 6 weeks (threshold limit TVC = 5 log CFU/g)	CL retarded moisture loss and color changes	

Hard boiled peeled eggs	10°C/6 weeks	CL edible coating in combination with VP	Natural microflora	2.0–2.3	+ more than 2 weeks (TVC remained below threshold limit of 5 log CFU/g)	CL controlled moisture loss and retarded color changes of egg yolk	Kim et al. (2008)
Fior di latte cheese	4°C/8 days	CH- alginate-lysozyme-EDTA and MAP	Natural microflora	2.0	+ 4 days (threshold limit of 6 and 5 log CFU/g for *Pseudomonas* and coliforms, sensorial parameters)	Synergic effect between active compounds and MAP was observed	Del Nobile et al. (2009b)

storage period (21 days), while respective counts of control uncoated samples (aerobic packaging) reached ca. 8.5 log CFU/g, on 14th storage day. Finally, according to sensorial analysis, consumers preferred the overall quality of CH-coated, cooked lingcod samples over the control, based on firm texture and aroma (less fishy).

From previously mentioned numerous research studies depicted in this section of the review, there is no doubt that CH can be effectively used either as a food preservative or as an edible coating material, in order to preserve quality and extend the shelf life of various food products. This potential of CH may be enhanced when combined with EOs (e.g., thyme, rosemary, mint, etc.) or allyl-isothiocyanate-hop extract. Based on the results reported in the studies of Duan et al. (2010) and Ojagh et al. (2010), cinnamon seemed to be the less effective among EOs to enhance the antibacterial effectiveness of CH coatings against spoilage bacteria but revealed desirable sensorial characteristics when applied on seafood. Combinations of CH solutions and EOs have the potential to give food products with enhanced microbiological quality, extended refrigerated storage, durability, and with improved sensory characteristics (poultry products, fish fillets, vegetable-based products).

Very often, either a possible or a synergistic/additive antimicrobial effect, compatible with the hurdle concept (which incorporates several antimicrobial measures to gain a synergistic antimicrobial net effect) between CH and active compounds of EOs is likely as demonstrated in the aforementioned studies (Inatsu et al., 2005; Giatrakou et al., 2010; Giatrakou, 2010). More importantly, CH–EO combinations may prove effective in controlling the growth of certain pathogenic bacteria (*L. monocytogenes*, *Salmonella* spp., *E. coli*, or *B. cereus*) likely to contaminate or be present in real food products that are stored under aerobic or VP/MAP conditions at chill or abuse temperatures (Inatsu et al., 2005; Kanatt et al., 2008; Ponce et al., 2008; Giatrakou, 2010).

In some cases, the antibacterial effect of CH against pathogenic bacteria may weaken over time, probably due to the fact that the amine groups of CH bind to cell debris or surface components of the bacteria and are no longer available to attach to other bacterial surfaces of new cells (Coma et al., 2002). However, combined application of CH–EOs may be an effective means of resolving this particular problem, maintaining its antibacterial action against pathogens.

In the majority of the aforementioned studies, Gram-negative pathogenic (e.g., *E. coli*, *Salmonella* spp.) or spoilage bacteria (e.g., *Pseudomonas*) revealed a greater resistance toward antibacterial action of CH than Gram-positive ones, probably due to the outer membrane surrounding their bacterial cells. Thus, EOs could enhance the antibacterial activity of CH, against Gram-negative bacteria, rendering the outer membrane of these species susceptible to the action of CH (Helander et al., 2001).

16.3.2.2 Application of Edible Chitosan Solutions or Bioactive Chitosan Coatings in Combination with Lysozyme in Preserving Food Quality and Safety

During the last decade there has been a great interest within the food industry in using enzymes naturally occurring in foods, such as lysozyme. Lysozyme is one of the most frequently used antimicrobial enzymes and shows antibacterial activity mainly on Gram-positive bacteria. Because of the protective outer membrane surrounding the peptidoglycan layer of Gram-negative bacteria, lysozyme does not show antibacterial activity against these species. However, when lysozyme is combined with other antimicrobial agents such as CH, EDTA, or EOs, the antibacterial spectrum may be widened and be effective against Gram-negative bacteria (Del Nobile et al., 2009b; Ntzimani et al., 2010).

Kim et al. (2008) developed a novel preservation system, consisting of CH and lysozyme for active edible coating of hard-boiled shell-on eggs, as well as on hard-boiled and peeled vacuum-packaged hard-boiled eggs. For preparing CH–lysozyme (CL) coating, shrimp-derived CH (Primex efh, Siglufjordur, Iceland) was dissolved in 1% acetic acid solution to obtain a final 3% CH solution. Glycerol 25% (w/w CH) was also added. Then, the CH solution was mixed with an appropriate amount of a previously prepared (stock) lysozyme solution 10% to achieve a final concentration of 60% (dry weight lysozyme/dry weight CH). Finally, CL edible coating was degassed using a vacuum pump. Dried hard-boiled eggs were twice coated by immersions in CL solution for 1 min and 10 s, respectively. Inoculated *L. monocytogenes* was able to grow in hard-boiled shell-on eggs, irrespective of being coated or not. However, CL coating reduced *L. monocytogenes* counts by 0.8 log CFU/g, as compared with the control, by the end of the 4-week storage period (10°C). On the other hand, CL coating revealed a stronger antimicrobial action against inoculated *Salmonella enteritidis*, reducing viable numbers by 4.0 log CFU/g compared with the control, during the same storage period. CL coating was also evaluated for controlling the multiplication of spoilage microorganisms of shell-on or peeled (vacuum-packaged) hard-boiled eggs. Total plate counts (TPC) in control hard-boiled eggs exceeded maximum level of TPC (5 log CFU/g for all egg products) after the 4-week storage, whereas shell-on or peeled (vacuum-packaged) eggs, coated with CL, did not reach this value even after 10 or 6 weeks of storage, respectively (reduction by ca. 2.3–2.6 log CFU/g).

Del Nobile et al. (2009b) combined CH with either alginate coating or active coating (lysozyme, ethylenediamine tetraacetic acid [EDTA], disodium salt) and MAP to prolong the shelf life of *Fior di latte* cheese. High MW CH (Sigma-Aldrich, Milan, Italy) was dissolved in lactic acid to achieve a final concentration of 1%. Then, a proper amount of the former solution was added into the working milk for cheese making, at a final CH concentration of 0.012%. Fior di latte cheese was dipped into sodium alginate solution (designed as active coating). CH–active coating MAP increased the shelf life in comparison with the traditional packaging from 1 to 5 days, due to the synergic effect between the active compounds and the atmospheric conditions in the package headspace (4°C). In fact, CH-active coating-MAP treatment maintained counts of *Pseudomonas* spp. below the threshold limit of 6 log CFU/g over the entire storage period of 8 days and reduced the growth of coliforms by ca. 2.0 log CFU/g, as compared with the control.

It needs to be stressed that to date, studies focusing on potential applications of CH with lysozyme on real foods are limited, in view of their applicability as a natural antimicrobial treatment combination, for either prolonging the shelf life or maintaining the safety of various foods, for example, fresh meat, poultry, or fish/seafood, including their products.

16.3.3 Application of Chitosan as an Active Packaging Material for Controlled Release of Active Compounds

Direct surface application of CH (via spraying, dipping) or incorporation into food formulations, as described in previous sections, may sometimes challenge some limitations because the active antimicrobial substance could be neutralized, evaporated, or diffused inadequately into the bulk of the food (Pranoto et al., 2005). The use of packaging films based on antimicrobial polymers could prove to be more efficient, by maintaining high concentrations on food surfaces, where the contamination is prevalent, with a low migration of active substances (Coma et al., 2002; Sanchez-Gonzalez et al., 2010). Dry edible or plastic CH-based films are easily prepared by evaporating from dilute acid conditions using a casting technique and possess the ability to retard moisture, oxygen, aromas, and solute transports (Ouattara et al., 2000; Coma et al., 2002; Pranoto et al., 2005; Zivanovic et al., 2005; Hosseini et al., 2008; Fernandez-Saiz et al., 2010; Sanchez-Gonzalez et al., 2010). Moreover, the development of new biodegradable or edible packaging material films has recently been undertaken for environmental aspects (Coma et al., 2002). Although CH has intrinsic antimicrobial activity, which is effectively expressed in aqueous system, as previously described, antimicrobial properties may become negligible when CH is in the form of insoluble films (Ouattara et al., 2000). Incorporation of EOs in CH dry films may not only enhance the film's antimicrobial properties but also reduce water vapor permeability and slow lipid oxidation of the product on which the film is applied (Zivanovic et al., 2005). It is worth mentioning that numerous studies have demonstrated that EOs are more effective in reducing microbial growth, when incorporated into a film or gel applied directly to the product, because of the active substances evaporating or diffusing into the medium (Sanchez-Gonzalez et al., 2010). In addition, the direct application of EOs to foods may pose an adverse effect on the sensory characteristics of the food; therefore, incorporation of EOs into edible films may have beneficial applications in food packaging (Hosseini et al., 2008).

In recent years, highly promising (a) in vitro and (b) in vivo studies on combined application of CH-based dry films and oregano oil (Zivanovic et al., 2005), garlic oil (GO) (Pranoto et al., 2005), thyme oil (Hosseini et al., 2008), clove oil (Hosseini et al., 2008), tea-tree oil (Sanchez-Gonzalez et al., 2010), or cinnamaldehyde (Ouattara et al., 2000) have been reported.

16.3.3.1 Studies on Food Models (In Vitro)

Pranoto et al. (2005) studied the antimicrobial effect of CH edible films incorporating GO against several foodborne pathogenic bacteria (*E. coli*, *S. aureus*, *S. typhimurium*, *L. monocytogenes*, and *B. cereus*). Edible films were prepared by dissolving shrimp CH of high MW in 1% acetic acid solution to achieve a final CH concentration of 1% w/v. GO was incorporated into the aforementioned CH solution, and final dry CH–GO films were prepared using a casting technique (drying at 40°C). Control CH films did not reveal any antimicrobial effect against all tested bacteria, probably due to the immobilization of CH molecules within the film. Interestingly, enrichment with GO at concentrations of at least 100 µL/g CH led to significantly high inhibitory zones for Gram-positive bacteria tested (*S. aureus*, *L. monocytogenes*, and *B. cereus*). *L. monocytogenes* was the most sensitive species against GO incorporated in the film. Interestingly, CH–GO films, in spite of reducing the growth of Gram-negative bacteria (*E. coli*, *S. typhimurium*) underneath film disks in direct contact with target microorganisms in agar, did not result in any clearing zone surrounding the same bacteria. The results of the study suggested that GO incorporated into CH film increased its antimicrobial efficacy without any effect on the mechanical and physical properties of the films.

A recent study conducted by Sanchez-Gonzalez et al. (2010) analyzed the antimicrobial, mechanical, optical, and barrier properties of CH-based film after enrichment with tea tree oil (TTO). Films were prepared by dissolving high MW CH (Sigma–Aldrich Quvmica, Madrid, Spain) in acetic acid solution (0.5% w/w) to a final concentration of 1% w/w. TTO was added to the CH (CH) solution at concentrations of 0%–2% w/w. Emulsions of CH–TTO were degasified with a vacuum pump, and final composite films were obtained by a casting procedure (drying at atmospheric conditions). *L. monocytogenes* inoculated in TSA plates showed a significant growth from 2 to 8 log CFU/cm^2 after incubation at 10°C, for 7 days. However, pure CH films (no TTO added) used as coating for TSA plates, retained *L. monocytogenes* counts at initial levels (2.0 log CFU/g) till the fifth day of storage at 10°C (bacteriostatic effect). Moreover, the same film had by ca. 2.0 log CFU/g lower counts of the pathogen at the end of the storage period (12 days). The incorporation of 2% w/w TTO in the CH film led to a further reduction of pathogen counts by 1.0–1.5 log CFU/g, and no significant effect was observed for lower TTO concentrations (0.5%, 1.0%). A partial loss of antimicrobial activity of CH or CH–TTO films was observed over time (after the seventh storage day), and the authors attributed this fact to a reduction of volatile compounds concentration (which also contribute to the total antimicrobial activity of EO) during the film drying process and the time required for the microbial experiments. This study concluded that CH is a good polymer matrix for entrapping TTO oil, which can be used in different applications.

Hosseini et al. (2008) evaluated the possible synergistic antibacterial effect of thyme and clove oil incorporated into CH-based edible films. CH-based edible film was prepared by dissolving practical grade CH from crab shells (Sigma Chemical Co., St. Louis, MO, the United States) in acetic acid solution 1% v/v to a final solution concentration of 2% w/v. Glycerol was mixed into the CH solution to a level of 0.50 mL/g, as a plasticizer. Thyme and clove EOs were added to CH solution to achieve a final concentration of 0.5%–1.5% per film. Final CH-based films were prepared by using a casting technique (drying at 25°C). Determination of the antimicrobial effect of CH-based edible films was done using the agar diffusion method. CH-based control films (no EOs added) did not show any inhibitory zone against several Gram-positive and Gram-negative pathogenic bacteria tested (*L. monocytogenes*, *S. aureus*, *S. enteritidis*, and *P. aeruginosa*), probably due to the immobilization of CH molecules within the film, resulting in a limited diffusion ability through the adjacent agar media. However, incorporation of thyme oil at a concentration of 1.0%–1.5% per CH film exhibited a clear inhibitory zone against all bacteria tested. CH films with 0.5% clove EO were not inhibitory toward the Gram-negative bacteria tested, but increasing concentration at 1.5% enhanced antimicrobial properties. The study indicates that the addition of clove and thyme EO has the potential of application in antimicrobial food packaging, both against Gram-negative and Gram-positive bacteria contaminating foods.

16.3.3.2 Studies on Real Food Systems (In Vivo)

A study evaluating the feasibility of using antimicrobial films, designed to slowly release bacterial inhibitors, to improve the preservation of vacuum-packaged processed meats during refrigerated storage was undertaken by Ouattara et al. (2000).

Simple CH-based antimicrobial films were prepared by dissolving technical grade CH from crab shells (Sigma, St. Louis, MO, the United States) in acetic or propionic acid solutions (1% v/v) at a final concentration of 2% w/v. In order to obtain translucent CH films, the casting technique was conducted (drying at 80°C). Cinnamaldehyde (active component of several EOs) was incorporated into the CH solution to reach a final concentration of 1.0% w/w, prior to casting. Antimicrobial CH-based dried films were placed on the upper surface of processed meat (bologna, ham, and pastrami). Application of CH-based film without cinnamaldehyde added, on vacuum-packaged bologna and pastrami, resulted in a reduction of *Enterobacteriaceae* counts by ca. 3.0 log CFU/g, as compared with control, after storage at 10°C for a period of 11 or 21 days, respectively. Interestingly, incorporation of cinnamaldehyde further decreased *Enterobacteriaceae* counts by 0.26–0.8 log CFU/g. On the contrary, CH-based films did not reveal any significant antimicrobial effect against LAB initially present on bologna and pastrami. Results of the study suggested that although CH-based films incorporating cinnamaldehyde could substantially delay growth of *Enterobacteriaceae*, successful application on vacuum-packaged processed meat is limited by the fact that the concept was not effective against LAB being responsible for spoilage of these refrigerated products. Toward this specific target, later studies focused on using better antimicrobial agents than cinnamaldehyde, which were active against a broader range of bacteria, and thus, improving antimicrobial properties of CH dry films and their results are discussed in the following paragraphs.

Zivanovic et al. (2005) studied the antimicrobial activity of CH-based dry films enriched with EOs in vitro and on processed meat. CH films were prepared as follows: Medium MW CH (Aldrich Chemical Co, Milwaukee, WI) was dissolved in 1.5% v/v acetic acid at a final concentration of 1.5% w/w. EOs were mixed with Tween 20 and then added to the aforementioned solution at various concentrations. Final film-forming CH–EO solution was poured into Petri dishes (10 mg CH/cm^2) and dried under vacuum (30°C, casting technique). CH films prepared with or without oregano EO were placed between two slices of bologna, previously inoculated with *L. monocytogenes* and *E. coli*. Pure CH films (no EOs added) reduced *L. monocytogenes* by 1–3 logs, whereas the films with 1% or 2% oregano EO decreased pathogen counts by ca. 3.6–4.0 logs, after 5 days of storage at 10°C, indicating a possible synergistic or additive effect of the two antimicrobial agents. The synergistic effect was more pronounced in the case of *E. coli*, inoculated on the bologna slices: pure CH films did not inhibit *E. coli* growth, while enrichment with 1% or 2% oregano EO increased the antimicrobial activity, resulting in a significant reduction of *E. coli* by ca. 3 logs. These authors concluded that CH–oregano oil films have the potential to be used for active packaging of processed meat, enhancing its safety.

Results of the aforementioned studies indicate that CH-based composite films have a great potential to improve their antimicrobial properties by incorporating antimicrobial agents like EOs. CH-based composite films with no EOs added have shown good, limited, or no antimicrobial activity depending on the procedure of dry film preparation, type of CH used, as well as target microorganism used. Three different types of CH have been used for dry antimicrobial film preparation in these studies: CH of technical (practical) grade, high MW, and of medium MW; low MW CH has not yet been used.

In vitro studies (Pranoto et al., 2005; Hosseini et al., 2008) reported that dry CH-based composite films did not exert any antimicrobial effect against several pathogenic Gram-positive and Gram-negative bacteria (*E. coli*, *S. aureus*, *S. typhimurium*, *S. enteritidis*, *L. monocytogenes*, *B. cereus*) due to immobilization of the CH molecules within the films, resulting in limited diffusion ability of the active substance through the adjacent medium (Pranoto et al., 2005; Hosseini et al., 2008). However, this was not always the case as recent in vivo (processed meat) and in vitro studies (agar diffusion method) showed good antimicrobial potential of CH-based composite films against *L. monocytogenes*, (reduction by 1–3 log CFU/g) (Zivanovic et al., 2005; Sanchez-Gonzalez et al., 2010) or *Enterobacteriaceae* (reduction by 3.0 log CFU/g, Ouattara et al., 2000). Incorporation of EOs (TTO, oregano, thyme, clove at concentrations of 1%–2.5% w/w or GO 100 μL/g) into CH films enhanced the inhibitory action against several pathogenic bacteria, as compared with control pure films, but the antimicrobial effect was more pronounced in the case of Gram-positive bacteria *L. monocytogenes* and *S. aureus* (Pranoto et al., 2005; Hosseini et al., 2008; Sanchez-Gonzalez et al., 2010).

Studies in vivo have used dry CH films enriched with EOs for active packaging of processed meat like bologna or pastrami, and the film was placed either on the upper surface (Ouattara et al., 2000) or between slices of the meat product (Zivanovic et al., 2005).

It is worth noting that additional research with the prospect of practical application on real foods is required with the view to explore the potential of films (singly or combined with natural antimicrobials) to be utilized as an active packaging solution for preservation of perishable food products, stored under refrigeration. So far, the main issues that need further research are the following: (a) study of the effect of EOs on mechanical and physical properties of the CH films, after incorporation, (b) study of the effect of CH–EO packaging films on organoleptic properties (odor, taste, color) of food products, and (c) study of the potential loss of antimicrobial effectiveness of the film due to instable diffusion of the active compounds from the film into the product.

ACKNOWLEDGMENTS

We thank the European Union for financial support of the project "DOUBLE FRESH" (Proposal. /Contract no.: PL 023182).

REFERENCES

Aider, M. (2010). Chitosan application for active-bio based films production and potential in the food industry: Review. *LWT-Food Science and Technology*, 43: 837–842.

Albuquerque Bento, R., Montenegro Stamford, T.L., Montenegro Stamford, T.C., Cardoso de Andrade, S.A. and Leite de Souza, E. (2011). Sensory evaluation and inhibition of *Listeria monocytogenes* in bovine pâté added of chitosan from *Mucor rouxii*. *LWT-Food Science and Technology*, 44: 588–591.

Ausar, S.F., Passalacqua, N., Castagna, L.F., Bianco, I.D. and Beitramo, D.M. (2002). Growth of milk fermentative bacteria in the presence of chitosan for potential use in cheese making. *International Dairy Journal*, 12: 899–906.

Beuchat, L.R., Rocelle, M., Clavero, S. and Jacquette, C.B. (1997). Effects of nisin and temperature on survival, growth, and enterotoxin production characteristics of psychrotrophic *Bacillus cereus* in beef gravy. *Applied and Environmental Microbiology*, 63: 1953–1958.

Beverlya, R., Janes, M.E., Prinyawiwatkula, W. and No, H.K. (2008). Edible chitosan films on ready-to-eat roast beef for the control of *Listeria monocytogenes*. *Food Microbiology*, 25: 534–537.

Burt, S. (2004). Essential oils: Their antibacterial properties and potential applications in foods—A review. *International Journal of Food Microbiology*, 94: 223–253.

Chang, D.S., Cho, H.R., Goo, H.Y. and Choe, W.K. (1989). A development of food preservation with the waste of crab processing. *Bulletin of Korean Fish Society*, 22: 70–78.

Chen, C., Liau, W. and Tsai, G. (1998). Antibacterial effects of N-sylfonated and N-sulfobenzoyl chitosan and application to oyster preservation. *Journal of Food Protection*, 61: 1124–1128.

Chhabra, P., Huang, Y.W., Frank, J.F., Chmielewski, R. and Gates, K. (2006). Fate of *Staphylococcus aureus*, *Salmonella enterica Serovar typhimurium*, and *Vibrio vulnificus* in raw oysters treated with chitosan. *Journal of Food Protection*, 69: 1600–1604.

Chung, Y.C. and Chen, C.Y. (2008). Antibacterial characteristics and activity of acid-soluble chitosan. *Bioresource Technology*, 99: 2806–2814.

Chung, Y.C., Su, Y.P., Chen, C.C., Jia, G., Wang, H.I., Wu, G.J.C. and Lin, J.G. (2004). Relationship between antibacterial activity of chitosan and surface characteristics of cell wall. *Acta Pharmacologica Sinica*, 25: 932–936.

Chung, Y.C., Wang, H.J., Chen, Y.M. and Li, S.L. (2003). Effect of abiotic factors on the antibacterial activity of chitosan against water borne pathogens. *Bioresource Technology*, 88: 179–184.

Coma, V. (2008). Bioactive technologies for extended shelf life of meat-based products. *Meat Science*, 78: 90–103.

Coma, V., Martial-Gros, A., Garreau, S., Copinet, A., Salin, F. and Deschamps, A. (2002). Edible antimicrobial films based on chitosan matrix. *Journal of Food Science*, 67: 1162–1169.

Cutter, C.N. (2006). Opportunities for bio-based packaging technologies to improve the quality and safety of fresh and further processed muscle foods. *Meat Science*, 74: 131–142.

Darmadji, P. and Izumimoto, M. (1994). Effect of chitosan in meat preservation. *Meat Science*, 38: 243–254.

Del Nobile, M.A., Cammariello, D., Conte, A. and Attanasio, M. (2009b). A combination of chitosan, coating and modified atmosphere for prolonging Flior di Latte cheese shelf life. *Carbohydrate Polymers*, 78: 151–156.

Del Nobile, M.A., Di Benedetto, N., Suriano, N., Conte, A., Corbo, M.R. and Sinigaglia, M. (2009a). Combined effects of chitosan and MAP to improve the microbial quality of amaranth homemade fresh pasta. *Food Microbiology*, 26: 587–591.

Devlieghere, F., Vermeulen, A. and Debevere, J. (2004). Chitosan: Antimicrobial activity, interactions with food components and applicability as a coating on fruit and vegetables. *Food Microbiology*, 21: 703–714.

Duan, J., Jiang, Y., Cherian, G. and Zhao, Y. (2010). Effect of chitosan-Krill oil coating and modified atmosphere packaging on the storability of cold stored lingcod (*Ophiodon elongatus*) fillets. *Food Chemistry*, 122: 1035–1042.

Dutta, P.K., Tripathi, S., Mehrotra, G.K. and Dutta, J. (2009). Perspectives for chitosan based antimicrobial films in food applications. *Food Chemistry*, 114: 1173–1182.

Eaton, P., Fernandes, J.C., Pereira, E., Pintado, M.E. and Malcata, F.X. (2008). Atomic force microscopy study of the antibacterial effects of chitosans on *Escherichia coli* and *Staphylococuus aureus*. *Ultramicroscopy*, 108: 1128–1134.

FDA (Food and Drug Administration). (2009). *Foodborne Pathogenic Microorganisms and Natural Toxins Handbook*, Office of the Commissioner, Center for Food Safety and Applied Nutrition, Department of Health and Human Services, College Park, MD.

Fernandes, J.C., Eaton, P., Gomes, A.M., Pintado, M.E. and Malcata, F.X. (2009). Study of the antibacterial effects of chitosans on *Bacillus cereus* (and its spores) by atomic force microscopy imaging and nanoidentation. *Ultramicroscopy*, 109: 854–860.

Fernandes, J.C., Tavaria, F.K., Soares, J.C., Ramos, O.S., Monteiro, M.J., Pintado, M.E. and Malcata, F.X. (2008). Antimicrobial effects of chitosan and chitooligosaccharide, upon *Staphylococcus aureus* and *Escherichia coli*, in food model systems. *Food Microbiology*, 25: 922–928.

Fernandez-Saiz, P., Soler, C., Lagaron, J.M. and Ocio, M.J. (2010). Effects of chitosan films on the growth of *Listeria monocytogenes*, *Staphylococcus aureus* and *Salmonella* spp. in laboratory media and in fish soup. *International Journal of Food Microbiology*, 137: 287–294.

FSAI (Food Safety Authority of Ireland), (2007). Resources and publications/factsheets. (http://www.fsai.ie/resources_and_publications /factsheets.html).

Georgantelis, D., Ambrosiadis, I., Katikou, P., Blekas, G. and Georgakis, S.A. (2007a). Effect of rosemary extract, chitosan and alpha-tocopherol on microbiological parameters and lipid oxidation of fresh pork sausages stored at 4°C. *Meat Science*, 76: 172–181.

Georgantelis, D., Blekas, G., Katikou, P., Ambrosiadis, I. and Fletouris, D. (2007b). Effect of rosemary extract, chitosan and alpha-tocopherol on lipid oxidation and colour stability during frozen storage of beef burgers. *Meat Science*, 75: 256–264.

Giatrakou, V. (2010). Unpublished results, Ph.D. Thesis. Study of the Quality and Hygiene of a Raw Poultry Product (Chicken-pepper Kebab-"Souvlaki") Stored Under Modified Atmosphere Packaging-Combined Effect of Treatment with Thyme Oil and Chitosan. Ph.D. Thesis. Department of Chemistry, University of Ioannina, Ioannina, Greece.

Giatrakou, V., Ntzimani, A. and Savvaidis, I.N. (2010a). Combined chitosan–thyme treatments with modified atmosphere packaging on a ready-to-cook poultry product. *Journal of Food Protection*, 73: 663–669.

Giatrakou, V., Ntzimani, A. and Savvaidis, I.N. (2010b). Effect of chitosan and thyme oil on a ready to cook chicken product. *Food Microbiology*, 27: 132–136.

Goy, R.C., de Britto, D. and Assis, O.B.G. (2009). A review of the antimicrobial activity of chitosan. *Polimeros: Ciencia e Tecnología*, 19: 241–247.

Grant, I.R., Nixon, C.R. and Patterson, M.F. (1993). Effect of low-dose irradiation on growth of and toxin production by *Staphylococcus aureus* and *Bacillus cereus* in roast beef and gravy. *International Journal of Food Microbiology*, 18: 25–36.

Helander, I.M., Nurmiaho-Lassila, E.L., Ahvenainen, R., Rhoades, J. and Roller, S. (2001). Chitosan disrupts the barrier properties of the outer membrane of Gram-negative bacteria. *International Journal of Food Microbiology*, 71: 235–244.

Holley, R.A. and Patel, D. (2005). Improvement in shelf-life and safety of perishable foods by plant essential oils and smoke antimicrobials. *Food Microbiology*, 22: 273–292.

Hosseini, M.H., Razavi, S.H., Mousavi, S.M.A., Shahidi, S.A. and Hasansaraei, A.G. (2008). Improving antibacterial activity of edible films based on chitosan by incorporating thyme and clove essential oils and EDTA. *Journal of Applied Sciences*, 8: 2895–2900.

IFST (Institute of Food Science and Technology). (1999). *Development and Use of Microbiological Criteria for Foods*. London, U.K.: Institute of Food Science and Technology, ISBN 0 905367 16 2.

Inatsu, Y., Bari, M.L., Kawasaki, S. and Kawamoto, S. (2005). Effectiveness of some natural antimicrobial compounds in controlling pathogen or spoilage bacteria in lightly fermented Chinese cabbage. *Journal of Food Science*, 70: 393–397.

Jo, C., Lee, J.W., Lee, K.H. and Byun, M.W. (2001). Quality properties of pork sausages prepared with water-soluble chitosan oligomer. *Meat Science*, 59: 369–375.

Juneja, V.K., Thippareddi, H., Bari, L., Inatsu, Y., Kawamoto, S. and Friedman, M. (2006). Chitosan protects cooked ground beef and turkey against *Clostridium perfrigens* spores during chilling. *Journal of Food Science*, 71: 236–240.

Kanatt, S.R., Chander, R. and Sharma, A. (2008). Chitosan and mint mixture: A new preservative for meat and meat products. *Food Chemistry*, 107: 845–852.

Kim, K.W., Daeschel, M. and Zhao, Y. (2008). Edible coatings for enhancing microbial safety and extending shelf life of hard-boiled eggs. *Journal of Food Science*, 73: 227–235.

Kok, T.N. and Park, J.W. (2007). Extending the shelf life of set fish ball. *Journal of Food Quality*, 30: 1–27.

Lee, H.Y., Park, S.M. and Ahn, D.H. (2003). Effect of storage properties of pork dipped in chitosan solution. *Journal of the Korean Society of Food and Nutrition,* 32: 519–525.

Lin, K.W. and Chao, J.Y. (2001). Quality characteristics of reduced-fat Chinese-style sausage as related to chitosan's molecular weight. *Meat Science*, 59: 343–351.

Marques, A., Encarnacao, S., Pedro, S. and Nunes, M.L. (2008). In vitro antimicrobial activity of garlic, oregano and chitosan against *Salmonella enterica. World Journal of Microbiology and Biotechnology*, 24: 2357–2360.

Mathur, N.K. and Narang, C.K. (1990). Chitin and chitosan, versatile polyssacharides from marine animals. *Journal of Chemical Education*, 67: 938–942.

No, H.K., Park, N.Y., Lee, S.H., Hwang, H.J. and Meyers, S.P. (2002). Antibacterial activities of chitosans and chitosan oligomers with different molecular weights on spoilage bacteria isolated from Tofu. *Journal of Food Science*, 67: 1511–1514.

Ntzimani, A.G., Giatrakou, V.I. and Savvaidis, I.N. (2010). Combined natural antimicrobial treatments (EDTA, lysozyme, rosemary and oregano oil) on semi cooked coated chicken meat stored in vacuum packages at 4°C: Microbiological and sensory evaluation. *Innovative Food Science and Emerging Technologies*, 11: 187–196.

Ojagh, S.M., Rezaei, M., Razavi, S.H. and Hosseini, S.M.H. (2010). Effect of chitosan coatings enriched with cinnamon oil on the quality of refrigerated rainbow trout. *Food Chemistry*, 120: 193–198.

Ouattara, B., Simard, R.E., Piette, G., Begin, A. and Holley, R.A. (2000). Inhibition of surface spoilage bacteria in processed meats by application of antimicrobial films prepared with chitosan. *International Journal of Food Microbiology*, 62: 139–148.

Pillai, C.K.S., Willi, P. and Sharma, C.P. (2009). Chitin and chitosan polymers: Chemistry, solubility and fiber formation. *Progress in Polymer Science*, 34: 641–678.

Ponce, A.G., Roura, S.I., del Valle, C.E. and Moreira, M.R. (2008). Antimicrobial and antioxidant activities of edible coatings enriched with natural plant extracts: *In vitro* and *in vivo* studies. *Postharvest Biology and Technology*, 49: 294–300.

Pranoto, Y., Rakshit, S.K. and Salokhe, V.M. (2005). Enhancing antimicrobial activity of chitosan films by incorporating garlic oil, potassium sorbate and nisin. *Lebensmittel-Wissenschaft und-Technologie*, 38: 859–865.

Raafat, D., Bargen, K., Haas, A. and Sahl, H.G. (2008). Insights into the mode of action of chitosan as an antibacterial compound. *Applied and Environmental Microbiology*, 74: 3764–3773.

Roller, S. and Covill, N. (2000). The antimicrobial properties of chitosan in mayonnaise and mayonnaise-based shrimp salads. *Journal of Food Protection*, 63: 212–209.

Roller, S., Sagoo, S., Board, R., O'Mahony, T., Caplice, E., Fitzgerald, G., Fogden, M., Owen, M. and Fletcher, H. (2002). Novel combinations of chitosan, carnocin and sulphite for the preservation of chilled pork sausages. *Meat Science*, 62: 165–177.

Sagoo, S., Board, R. and Roller, S. (2002). Chitosan inhibits growth of spoilage microorganisms in chilled pork products. *Food Microbiology*, 19: 175–182.

Sanchez-Gonzalez, L., Gonzalez-Martinez, C., Chiralt, A. and Chafer, M. (2010). Physical and antimicrobial properties of chitosan–tea tree essential oil composite films. *Journal of Food Engineering*, 98: 443–452.

Shahidi, F., Vidana Arachchi, J.K. and Jeon, Y.J. (1999). Food applications of chitin and chitosan. *Trends in Food Science and Technology*, 10: 17–51.

Simpson, B.K., Gagne, N., Ashie, I.N.A. and Noroozi, E. (1997). Utilization of chitosan for preservation of raw shrimp (*Pandalus borealis*). *Food Biotechnology*, 11: 25–44.

Soultos, N., Tzikas, Z., Abrahim, A., Georgantelis, D. and Amvrosiadis, I. (2008). Chitosan effects on quality properties of Greek-style fresh pork sausages. *Meat Science*, 80: 1150–1156.

Tsai, G.J. and Su, W.H. (1999). Antibacterial activity of shrimp chitosan against *Escherichia coli*. *Journal of Food Protection*, 62: 239–243.

Tsai, G.J., Tsai, M.T., Lee, J.M. and Zhong, M.Z. (2006). Effects of chitosan and a low-molecular-weight chitosan on *Bacillus cereus* and application in the preservation of cooked rice. *Journal of Food Protection*, 69: 2168–2175.

Vishu Kumar, A.B., Varadaraj, M.C., Gowda, L.G. and Tharanathan, R.N. (2007). Low molecular weight chitosan-preparation with the aid of pronase, characterization and their bactericidal activity towards *Bacillus cereus* and *Escherichia coli*. *Biochimica et Biophysica Acta*, 1770: 495–505.

Waimaleongora-Ek, P., Corredor, A.J.H., No, H.K., Prinyawiwatkul, W., King, J.M., Janes, M.E. and Sathivel, S. (2008). Selected quality characteristics of fresh-cut sweet potatoes coated with chitosan during 17-day refrigerated storage. *Journal of Food Science*, 73: 418–423.

Wang, X., Du, Y., Fan, L., Liu, H. and Hu, Y. (2005). Chitosan–metal complexes as antimicrobial agents: Synthesis, characterization, and structure-activity studies. *Polymer Bulletin*, 55: 105–113.

Wang, X.H., Du, Y.M. and Liu, H. (2004). Preparation, characterization and antimicrobialactivity of chitosan–Zn complex. *Carbohydrate Polymers*, 56: 21–26.

WHO (World Health Organization). (2002). Future trends in veterinary public health. Technical Report. WHO study group on future trends in veterinary public health, Geneva, Switzerland, vol. 907.

Yang, T.C., Chou, C.C. and Li, C.F. (2005). Antibacterial activity of N-alkylated disaccharide chitosan derivatives. *International Journal of Food Microbiology*, 97: 237–245.

Yingyuad, B.S., Ruamsin, S., Reekprkhon, D., Douglas, S., Pongamphai, S. and Siripatrawan, U. (2006). Effect of chitosan coating and vacuum packaging on the quality of refrigerated grilled pork. *Packaging Technology and Science*, 19: 149–157.

Zivanovic, S., Chi, S. and Draughon, A.A.E. (2005). Antimicrobial activity of chitosan films enriched with essential oils. *Journal of Food Science*, 70: 45–51.

Part VII

Consumer Behavior/Sensory Analysis and Legislation

17 Sensory Analysis and Consumer Search of MAP Acceptability

Ioannis S. Arvanitoyannis, Nikoletta Manti, and Nikoletta K. Dionisopoulou

CONTENTS

17.1 INTRODUCTION

Human beings consume food and drink both for acquiring energy and essential nutrients and for pleasure purposes (Chen, 2009). Food acceptability is considerably influenced by many factors, which may be related to the individual, the food, or the environment in which the food is consumed (Murray and Baxter, 2003; Guerra et al., 2008). Consumer opinion is determined in large part by the sensory perception of the food product (Wilkinson et al., 2000). Food scientists apply three distinct methods for tracing the causes and nature of various sensory quality problems in foods: (1) chemical procedures, (2) microbiological analyses, and (3) sensory evaluation (SE) techniques (Bodyfelt et al., 2008).

Several requirements regarding food quality and safety and healthier food have been put forward since the 1980s–1990. The modern consumer is accustomed to buying low-cost food products but does not like the idea of being at risk. In the 1990s, food safety became a key factor in food manufacturing as explicitly shown by the creation of the European Food Safety Authority (Cayot, 2007).

Enhanced competition and new opportunities stimulated by progressively vanishing trade barriers and expanding world markets have substantially speeded up the food industry's worldwide requirement for novel products, quality improvements, prolonged shelf life, greater productivity, and lesser production and distribution costs (Sidel and Stone, 1993).

If companies are to succeed in the market place, they need at least to be able to monitor the sensory properties of their products during production processes and product development. The companies of a competitive advantage will be the ones able to manipulate actively and control sensory properties. This can only take place if the relationship between food structure and sensory properties is well understood (Wilkinson et al., 2000).

The taste of foods is a very complicated issue in view of the many attributes involved therein. Awareness of the interactions of taste attributes leads to understanding of the development of new food products. After evaluating the taste interactions within a food, optimization of characteristics can occur, thus providing the most acceptable product to the consumer. Therefore, it would be beneficial to product developers to comprehend the combined perceptions of sensory attributes within a food (Duizer et al., 1997).

Sensory changes in food products are the result of intentional or unintentional interactions with packaging materials and the failure of materials in protecting product integrity or quality. On the other hand, sensory issues related to plastic food packaging involves knowledge provided by sensory scientists, materials scientists, packaging manufacturers, food processors, and consumers (Duncan and Webster, 2009). The migration of such additives or contaminants from polymeric food packaging materials into food may be separated into three different but closely interrelated stages. Once diffusion has occurred within the polymer, the next stages are dissolution at the polymer–food interface and dispersion into the bulk food (Kontominas et al., 2006).

Firmness studies have been conducted on many different foods using various kinds of objective tests. The most common is most likely the deformation test in which the food is squeezed comparatively gently so as not to cause permanent crushing or breaking. The distance to which the product is compressed under a standard force (or the force required to compress it to a standard distance) is recorded as an index of firmness (Szczesniak and Bourne, 1969). The qualitative aspects of a product include aroma, appearance, flavor, texture, aftertaste, and sound properties of a product, which distinguish it from others. Sensory panelists then quantify the product aspects to make easier the description of perceived product attributes. Several surveys have shown that the use and application of descriptive sensory testing has increased considerably (Murray et al., 2001).

Most foods and beverages stimulate a variety of sensory systems thus produce the complex variations in taste, smell, temperature, texture, and irritation. The combination of these attributes, especially taste, smell, and irritation, results in what people commonly refer to as a food or beverage's flavor. Two fundamental problems can be posed regarding the evaluation of flavor, one related to the integration of flavor components and the other to the analysis of a particular sensory dimension (Frank, 2002).

Perception of aroma, taste, and texture in foods is a dynamic, not a static, phenomenon (Cliff and Heymann, 1993). Among sensory features, aroma has attracted more attention in early times, thereby mirroring the importance of spices for foods and other aromatics widely added in perfumes and oils (York and Vaisey-Genser,

2003). Aroma suggests the flavor to be expected when consuming foods and thereby affecting the overall acceptance. Volatile compounds are generally responsible for aroma and vary from one food to another. Moreover, aroma is a promising indicator of product freshness (Murray and Baxter, 2003; Guerra et al., 2008).

In foods and beverages, odor is a considerable factor affecting both acceptance and preference. Unfortunately, the hedonic (degree of liking) and the qualitative (descriptive) aspects of food odors and flavors are not clearly understood (Stone et al., 1965). Less frequent modifiers of flavor were the hotness of spices such as ginger and the coolness of menthol. Then, there are the metallic, alkaline, and meaty tastes. Was one to accept the currently predominant view that there are only four true tastes: sweet, bitter, sour, and salt; then tastes like metallic and alkaline must presumably be regarded as modalities of the typical chemical senses (Bodyfelt et al., 2008).

The presence of food constituents such as texturing agents and flavorings has a strong effect on the dynamics of food breakdown and potential flavor release. Both from a fundamental food perception and, applied—product development, dynamic sensory methods are worthwhile studying and employing (Dijksterhuis and Piggott, 2001). In the case of fluid foods, the parameters affecting the nonoral evaluation of their textural properties will be the impressions triggered on tilting and shaking the container the food is held in and the amount of force required by the sample to get stirred with a spoon, fork, etc. Therefore, viscosity is bound to be among the most crucial textural properties of these foods (Shama et al., 1973). Rheological tests on foods are carried out for the quantitative evaluation of textural properties during, or after, processing, getting insight into the effect of micro- and macromolecular structures on textural properties, and possible correlation with oral and nonoral SEs by consumer or more specialized panels. The rheological properties of most foods display a clear dependence on the mechanical treatment they were subjected to during the test (Shama and Sherman, 1973). The SE includes several steps both outside and inside the mouth, starting with the first bite through mastication, swallowing, and finishing with residual feel in the mouth and throat (Szczesniak, 2002). Food oral processing is an indispensable procedure not only for the food consumption and digestion but also for the assessment and pleasure of food texture and flavor as well. The consumption of a food inside the mouth involves various oral operations, among which biting, chewing and mastication, transportation, bolus formation, and swallowing are the most important ones (Chen, 2009).

The methodology of SE is based on standard scales for the mechanical parameters that are also employed for selecting and training the panel members (Szczesniak, 2002). As early as 1930, rating scales were developed to assess the intensity of food qualities such as the meat tenderness. At the same time, other tests like paired comparison and rank order methods were introduced (York and Vaisey-Genser, 2003). The scoring methods suggested were either numerical intensity scales (frequently 0–7, with 0 denoting absence and 7 a very high intensity of a specific characteristic) or hedonic scales (ranging from "dislike extremely" to "like extremely"). The latter should not be used when the objective is to describe texture in terms of the characteristics present or to quantify their intensity (Bourne and Szczesniak, 2003).

With regard to foods and sensory stimuli, the SE field has not been able to move beyond its traditional link with product development and quality assurance.

Although some progress has been made in SE with the advent of e-nose, the information about how people evaluate foods, based on sensory properties, remains scarce. Researchers in the sensory field encountered substantial difficulties to model even simple combinations of tastes and odors (Meiselman, 1994).

The application of SE is a requirement for successful product development and quality control, and the food industry has embraced it by enhancing its demand for qualified sensory personnel (Risvik et al., 1989). Panelists need to participate in training for quality control sensory judges. The latter does not need to be long or involved. Companies with restricted resources are prone to use a limited number of participants in the SEs. As a result, it is of great importance to carry out a properly organized training program so that the participants get used to the commonly applied techniques how to evaluate foods (Munoz, 2002).

Sensory panels:

- Expert panel: Experts specialized in a specific technology evaluate typical products and define criteria of evaluation (Zeng et al., 2008).
- Trained panel: It comprises selected and trained assessors who have undergone sensory training and have experience in the methods to be applied and the ability to perceive the differences and describe sample products' sensory properties (Sinesio, 2005).
- Consumer panel: Consumer research is one of the key activities of consumer goods companies. Through this type of testing, companies determine consumer acceptance or liking, preference, and opinions on the products tested (Krishnamurthy et al., 2007). Selected nontrained consumers are invited to conduct evaluation in a laboratory under controlled conditions (Zeng et al., 2008). Their answers are subjective, and food acceptance is affected by numerous factors and parameters such as personal experience, sex, age, regular use of the product, eating habits, and social and nutritional aspects (Sinesio, 2005).

Once the method is defined and prior to starting the product assessment in a systematic way, several steps have to be conducted: assessor selection, basic and specific training, assessor qualification, and method validation. The applied method can be accredited according to ISO 17025 (ISO, 2005a) provided it gives reliable results (repeatability, reproducibility, and discrimination ability) (Etaio et al., 2010a).

Accreditation of analytical procedures based on ISO 17025 (ISO, 2005) has a history of approximately 20 years, although the accreditation of specific sensory methods is quite recent and the reported cases are limited compared to other kinds of analyses. Accreditation can be considered as an additional step with regard to official quality control, since it assures the technical competence by an external institution. It also enhances the guarantee and the quality image of the product. Accreditation is gradually gaining ground as an important tool in the food market, particularly in products bearing formal quality distinctive labels (Etaio et al., 2010b).

One element of the current drive toward harmonization is a general agreement that laboratories should be approved to ISO 17025—a laboratory standard equivalent to the more generic ISO 9000. However, ISO 17025 was a considerable advance from its predecessors (Maynard et al., 2003). Although the number of accredited laboratories with the new standard is growing, it is still small compared to the total number of functioning laboratories.

ISO/IEC 17025:1999 was the international standard initially introduced in 1999 for accreditation of both testing and calibration laboratories. ISO/IEC 17025 requirements were documented in two main sections: (i) the administrative requirements including purchasing, document control, corrective action, internal audits, and management review (Section 4) and (ii) the technical requirements including training, measurement uncertainly, proficiency testing, traceability, and reporting requirements (Section 5). The main difference between ISO/IEC 17025 and ISO 9000 focused not only on the technical requirements in Section 5 but also on most individual clauses (Bucher, 2004).

According to ISO 17025:2005, the laboratory should apply suitable methods and procedures for all tests and calibrations. This includes sampling, handling, transportation storage, and appropriate preparation of the items tested. The methods should include measurements uncertainly as well as statistical techniques for analysis of test and/or calibration data. Instructions and guidance related to handling and operating relevant equipment and all instructions should be available and familiar to the laboratory personnel (Sioblom and Keck, 2009).

The laboratory should properly arrange to ensure risk minimization of cross-contamination, in case these are critical to the type of test being performed. Within the frame of good laboratory practice (GLP), it is essential to separate locations, or clear designated areas, for (1) sample receipt and storage area, (2) sample preparation, (3) examination of samples, including incubation, (4) maintenance of reference organisms, (5) media equipment preparation, including sterilization, (6) sterility assessment, and (7) decontamination (Singer et al., 2005).

The competence of staff is ensured by training, evaluation of the acquired training followed by continual assessment of staff to perform their assigned duties. Internal training for a certain period of time is required depending on the experience of the person and the complexity of the task they will carry out. Frequently, the most efficient way of training new staff is to let the responsible technician train the new staff in the specific task. It is preferable to evaluate the effectiveness of training objectively, prior to approval of the person for a specific task (Thrane, 2008).

17.2 ISO FOR SENSORY ANALYSIS

The International Organization for Standardization (ISO) is a worldwide federation of national standards bodies (ISO member bodies). The work of preparing International Standards is normally carried out through ISO technical committees. International organizations, governmental and nongovernmental, in liaison with ISO, also participate in the work. ISO collaborates closely with the International Electrotechnical Commission (IEC) on all matters of electrotechnical standardization. International Standards are drafted in accordance with the rules clearly explained in the ISO/IEC Directives (http://www.mbu.cn/jpkc/UpFile/file/spapjyjs3/BS%20ISO%20412-2003.pdf).

When trying to describe the taste, smell, or feel of a product, a universe of terms and sensations will come to mind. This can be problematic when comparing products through sensory analysis. A new ISO standard (Table 17.1) will minimize this complexity by making sure that sensory tests and results are communicated in a consistent manner. Sensory analysis applies sight, smell, taste, touch, and hearing to test and compare consumer products. Food, cosmetic, fragrance, textile, and other

TABLE 17.1
Sensory Analysis–Related Standards Issued by TC 34/SC 12 Secretariat-ISO for Sensory Analysis

ISO 3591:1977	Sensory analysis—Apparatus—Wine-tasting class
ISO 3972:1991	Sensory analysis—Methodology—Method of investigating sensitivity of taste
ISO 4121:2003	Sensory analysis—Guidelines for the use of quantitative response scales
ISO 5492:2008	Sensory analysis—Vocabulary
ISO 5495:2005	Sensory analysis—Methodology—Paired comparison test
ISO 5496:2006	Sensory analysis—Methodology—Initiation and training of assessors in the detection and recognition of odors
ISO 6564:1985	Sensory analysis—Methodology—Flavor profile methods
ISO 6658:2005	Sensory analysis—Methodology—General guidance
ISO 8586-1:1993	Sensory analysis—General guidance for the selection, training and monitoring of assessors—Part 1: Selected assessors
ISO 8586-2:2008	Sensory analysis—General guidance for the selection, training and monitoring of assessors—Part 2: Expert sensory assessors
ISO 8587:2006	Sensory analysis—Methodology—Ranking
ISO 8588:1987	Sensory analysis—Methodology—"A"–"not A" test
ISO 8589:2007	Sensory analysis- General guidance for the design of test rooms
ISO 11035:1994	Sensory analysis—Identification and selection of descriptors for establishing a sensory profile by a multidimensional approach
ISO 11036:1994	Sensory analysis—Methodology—Texture profile
ISO 11037:1999	Sensory analysis—General guidance and test method for assessment of color of foods
ISO 11056:1999	Sensory analysis—Methodology—Magnitude estimation method
ISO 13299:2003	Sensory analysis—Methodology—General guidance for establishing a sensory profile
ISO 13300-1:2006	Sensory analysis—General guidance for the staff of a SE laboratory—Part 1: Staff responsibilities
ISO 13300-2:2006	Sensory analysis—General guidance for the staff of a SE laboratory—Part 2: Recruitment and training of panel leaders
ISO 13301:2002	Sensory analysis—Methodology—General guidance for measuring odor, flavor and taste detection thresholds by a three-alternative forced-choice (3-AFC)
ISO 13302:2003	Sensory analysis—Methods for assessing modifications to the flavor of foodstuffs due to packaging
ISO 16657:2006	Sensory analysis—Apparatus—Olive oil–tasting glass
ISO 16820:2004	Sensory analysis—Methodology—Sequential analysis
ISO 4120:2004	Sensory analysis—Methodology Triangle test

industries use it to evaluate both novel and established products, ensure that product quality is kept stable, assess shelf life, and investigate new products or revise and improve existing ones (http://www.iso.org/iso/pressrelease.htm?refid=Ref1176).

The International Standard ISO 3972:1991 is used to investigate taste sensitivity. It describes a set of tests to familiarize subjects with sensory analysis. It defines principles and different types of thresholds, for identification of tastes as well as familiarization with the different types of thresholds. Moreover, it advises reagents (water, stock solutions, and dilutions) and equipment (apparatus) and provides detailed principles for test conditions (Hoehl et al., 2010).

ISO 4121:2003 provides guidelines describing quantitative response scales (where the response obtained indicates the intensity of perception) and their use when assessing samples. It is applicable to all quantitative assessment, whether global or specific and whether objective or hedonic. It is intentionally limited to the most commonly used measurement scales for sensory assessment (http://www.iso.org/iso/catalogue_detail.htm?csnumber=33817).

ISO 5492:2008 defines the terms related to sensory analysis and is applied to all industries concerned with the evaluation of products by the sense organs. The terms are given under the following headings: (i) general terminology, (ii) terminology relating to the senses, (iii) terminology relating to organoleptic attributes, and (iv) terminology related to methods (http://www.iso.org/iso/catalogue_detail?csnumber=38051).

ISO 5495:2005 describes a procedure for determining whether there is a perceptible sensory difference or a similarity between samples of two products concerning the intensity of a sensory attribute. This test is sometimes also referred to as a directional difference test or a 2-AFC test (alternative forced choice). The paired comparison test is a forced choice test between two alternatives. The method is applicable irrespective of whether a difference exists in a single sensory attribute or in several. However, it does not provide any indication of the extent of that difference. The absence of difference for the attribute under investigation does not necessarily imply that there does not exist any difference between the two products. This method is only applicable if the products are relatively homogeneous. The method is effective for (i) determining whether a perceptible or nonperceptible difference exists (paired difference test), or paired similarity test when, for example, modifications are made in raw/packaging materials and processing operations, and (ii) selecting, training, and monitoring assessors. It is necessary to know, prior to carrying out the test, whether the test is a one-sided test (the test supervisor knows *a priori* the direction of the difference and the alternative hypothesis corresponding to the existence of a difference in the expected direction) or a two-sided test (http://www.iso.org/iso/catalogue_detail.htm?csnumber=31621, http://www.iso.org/iso/iso_catalogue/catalogue_tc/catalogue_detail.htm?csnumber=31621, http://www.iso.org/iso/iso_catalogue/catalogue_ics/catalogue_detail_ics.htm?csnumber=31621&ICS1=67&ICS2=240, http://www.complianceonline.com/ecommerce/control/product?product_id=808772&channel=rss_organic).

ISO 5496:2006 describes several methods for determining the aptitude of assessors and for training assessors to identify and describe odorous products. The methods described in ISO 5496:2006 are suitable for use by the agri-foodstuff industries employing olfactory analysis (e.g., perfumery, cosmetics, and aromatics) (http://www.iso.org/iso/catalogue_detail.htm?csnumber=44247).

ISO 6564:1985 specifies a family of methods for describing and assessing the flavor of food products by qualified and trained assessors. The method consists of procedures for describing and assessing in a reproducible way. The separate attributes contributing to the formation of the overall impression given by the product are identified and their intensity assessed in order to make up a holistic impression of the parameters affecting the flavor of the product (http://www.iso.org/iso/iso_catalogue/catalogue_tc/catalogue_detail.htm?csnumber=12966).

ISO 6658:2005 provides general guidance on the use of sensory analysis. Tests for the examination of foods by objective sensory analysis are given and information on the techniques to be used if statistical analysis of the results is required. In case a test is used for determining food preference, this is clearly indicated (http://www.iso.org/iso/iso_catalogue/catalogue_tc/catalogue_detail.htm?csnumber=36226).

ISO 8586-1:1993 supplements the information given in ISO 6658. The entire process consists of recruitment, preliminary screening, and initiation; training in general principles and methods; selection for particular purposes; monitoring performance; and possible training as expert assessors (http://www.iso.org/iso/catalogue_detail.htm?csnumber=15875, https://answers.complianceonline.com/ecommerce/control/product/~product_id=80902).

ISO 8586-2:2008 sets out requirements for expert sensory assessors to establish sensory profiles of products and materials by means of descriptors. However, specific knowledge of products or materials by expert sensory assessors is not required to fulfill these requirements. This standard supplements the information provided in ISO 6658 (http://www.iso.org/iso/iso_catalogue/catalogue_ics/catalogue_detail_ics.htm?csnumber=37389).

ISO 8587:2006 describes a method for SE to place a series of test samples in rank order. This method allows for assessing differences among several samples based on the intensity of a single attribute, or several attributes or an overall impression. It is used to identify whether differences are present but cannot determine the extent of difference between samples. ISO 8587:2006 is appropriate for the following cases: (a) evaluation of assessors' performance (training assessors and determining perception thresholds of individuals or groups) and (b) product assessment (pre-sorting of samples, determination of the effect on intensity levels of one or more parameters, and determination of the order of preference in a global hedonic test) (http://www.iso.org/iso/catalogue_detail.htm?csnumber=36172, http://www.iso.org/iso/iso_catalogue/catalogue_tc/catalogue_detail.htm?csnumber=36172).

The principle of ISO 8588:1987 consists in presentation to an assessor of a series of samples, some of which are composed of sample "A" while others are different from sample "A." The assessor has to determine whether each sample is or not identical to "A." A presupposition for this test is the assessor to have evaluated a known sample "A" before his exposure to test samples (http://www.iso.org/iso/catalogue_detail.htm?csnumber=15878).

General guidance for the design of test rooms intended for the sensory analysis of products is given in ISO 8589:2007. The requirements for setting up a test room comprising a testing area, a preparation area, and an office, specifying clearly the essential ones as opposed to those merely desirable. ISO 8589:2007 is not specific for any product or test type and does not address test facilities for the specialized examination of products in inspection or in-plant quality-control applications (http://www.iso.org/iso/catalogue_detail.htm?csnumber=36385).

A method for identifying and selecting descriptors that can be employed for drawing up the sensory profile of a product is described in ISO 11035:1994. The different stages of the process for setting up a test through which a complete description of the sensory attributes of a product (qualitatively and quantitatively) are thoroughly described. The qualitative approach resides in descriptors' perceptions whether they are able to distinguish one product from others of the same type. The quantitative analysis focuses an accessing the intensity of each descriptor. The intensity of each descriptor. This method can be used to define a production standard, to improve or develop products, to study the effect of the aging of products in conjunction with the conditions of storage and preservation, and to compare a product with others of the same type already available on the market (http://www.iso.org/iso/catalogue_detail.htm?csnumber=19015, http://www.techstreet.com/standards/ISO/11035_1994?product_id=714633).

ISO 11036:1994 provides a method for developing a textural profile of food products or nonfood products. It must be noted however that this specific method is just one approach to sensory textural profile analysis, whereas several other methods are available. A thorough description of the steps involved in the process of establishing a complete description of the textural attributes of a product is provided herein (http://www.iso.org/iso/catalogue_detail.htm?csnumber=19016, http://www.techstreet.com/standards/ISO/11036_1994?product_id=714634).

The overall process for sensory profile development is described in ISO 13299:2003. Sensory profiles can be established for products such as foods and beverages and can also be useful in studies of human cognition and behavior. Some representative applications of sensory profiling are as follows: (i) to develop or change a product; (ii) to define a product, production standard, or trading standard in terms of its sensory attributes; (iii) to study and improve shelf life; (iv) to define a reference fresh product for shelf-life testing; (v) to compare a product with a standard or with other similar products on the market or under development; and (vi) to map a product's perceived attributes for the purpose of relating them to factors such as instrumental, chemical, or physical properties, and/or to consumer acceptability, to characterize by type and intensity the off-odors or off-tastes in a sample of air or water (e.g., in pollution studies) (http://www.iso.org/iso/catalogue_detail.htm?csnumber=37227, http://www.nal.din.de/projekte/DIN+EN+ISO+13299/en/119351190.html).

This part of ISO 13300-1:2006 provides guidance on staff functions with regard to improving the organization of a SE laboratory, to optimizing the employment of personnel, and to improving the efficiency of sensory tests. This standard is applicable to any organization planning to establish a formal structure for SE. The main aspects to be considered are education, background, and professional competence of staff members and the responsibilities of staff members at three different functional levels: sensory manager, sensory analyst or panel leader, and panel technician (http://www2.nen.nl/nen/servlet/dispatcher.Dispatcher?id=BIBLIOGRAFISCHEGEGEVENS&contentID=225292). These guidelines are valid for all different types of SE laboratories, in particular those in industry, in R&D and service organizations, and in the field of official authorities dealing with product control. The SE laboratory is assumed to be able to conduct all types of sensory tests such as analytical tests related to discrimination tests and descriptive analysis (sensory profile), as well as consumer tests (hedonic tests). The individual profile of sensory activities of an organization determines the boundaries and conditions to be taken into account for designing and implementing the SE laboratory and its staff. The application of this guidance by the organization is quite flexible and depends greatly on the needs and possibilities within an organization. If, for example, personnel is not available for three levels of staff function the duties can be divided among staff accordingly. Moreover, in the case of a staff of two persons the technical/scientific functions can be shared between a person handling the administrative/management functions and the individual responsible for the operational functions (http://www.iso.org/iso/catalogue_detail.htm?csnumber=36387, http://www.iso.org/iso/iso_catalogue/catalogue_tc/catalogue_detail.htm?csnumber=36387, http://www.iso.org/iso/iso_catalogue/catalogue_ics/catalogue_detail_ics.htm?ics1=67&ics2=240&ics3=&csnumber=36387).

The guidelines for the recruitment and training of panel leaders are given in ISO 13300-2:2006. Moreover, this standard focuses on the main activities and responsibilities of a panel leader for sensory analysis (http://www.iso.org/iso/catalogue_detail.htm?csnumber=36388).

ISO 13301:2002 provides guidance on obtaining data on the detection of chemical stimuli that evoke responses to odor, flavor, and taste by a 3-AFC (three alternative forced choice) procedure; the processing of the data to estimate the value of a threshold and its error bounds; and other statistics related to the detection of the stimulus. The procedures are used in one of the following two modes: investigation of the sensitivity of assessors to specific stimuli and investigation of the ability of a chemical substance to stimulate the chemoreceptive senses. Examples of the first mode comprise studies of the differences among individuals or specified populations of individuals in sensitivities and of the effects of age, gender, physiological condition, disease, administration of drugs, and the effect of ambient conditions on sensitivity. Examples of the latter mode consist of (i) studies in flavor chemistry and the impact of specified chemicals on the flavor of foods, (ii) classification of chemicals (present in the environment) based on their impact on humans, (iii) studies on the relationship of molecular structure to capability of a chemical to act as a stimulant, (iv) quality assurance of gaseous effluents and of water, foods, and beverages, and (v) studies in the mechanism of olfaction (http://www.iso.org/iso/catalogue_detail.htm?csnumber=36791, http://www.complianceonline.com/ecommerce/control/product/~product_id=808568,).

ISO 13302:2003 focuses on methods for assessing the changes caused by packaging (migration) to the sensory attributes of foodstuffs or their simulants. This methodology can be used as initial selection to assess whether a packaging material is suitable or as subsequent acceptability screening of individual batches/production run. The standard is applicable to all materials usable for packaging foodstuffs (e.g., paper, cardboard, plastic, foils, and wood). Furthermore, the scope can be expanded to any objects to come into contact with foodstuffs (e.g., kitchen utensils, coatings, leaflets, or parts of equipment such as seals or piping) with the aim to controlling food compatibility from a sensory point of view in conformance with the legislation in force (http://www.iso.org/iso/iso_catalogue/catalogue_tc/catalogue_detail.htm?csnumber=33818&commid=47858, http://www.iso.org/iso/iso_catalogue/catalogue_ics/catalogue_detail_ics.htm?ics1=67&ics2=240&ics3=&csnumber=33818).

ISO 16657:2006 specifies the characteristics of a glass intended for use in the sensory analysis of the organoleptic attributes of odor, taste, and flavor of virgin olive oils, for the classification of such oils. The glass is not intended for the analysis of the color or texture of olive oils. Moreover, it describes a heating accessory used to reach and maintain the right temperature for this analysis (http://www.iso.org/iso/iso_catalogue/catalogue_ics/catalogue_detail_ics.htm?ics1=67&ics2=260&ics3=&csnumber=38031).

ISO 16820:2004 describes the procedure for statistically analyzing data from forced-choice sensory discrimination tests, such as the Triangle, Duo-Trio, 3-AFC, and 2-AFC, where after every trial of the discrimination test, the decision can be made to stop testing and either declare a difference or no difference, to keep on testing. The sequential method often facilitates a decision to be made after fewer trials of the discrimination test than would be required by conventional approaches that use predetermined numbers of assessments. The method is effective for (a) determining whether a perceptible difference results or not, or (b) selecting, training, and monitoring assessors (http://www.iso.org/iso/catalogue_detail.htm?csnumber=31085, http://www.iso.org/iso/iso_catalogue/catalogue_ics/catalogue_detail_ics.htm?ics1=67&ics2=240&ics3=&csnumber=31085).

A procedure for determining whether a perceptible sensory difference or similarity exists between samples of two products is very comprehensively described in ISO 4120:2004 and is a forced-choice procedure. The method applies whether a difference can exist in a single sensory attribute or in several attributes. Although the method is statistically more efficient than the duo-trio test, it still has restricted use with products that display strong carryover and lingering flavors. The method can be effectively applied even when the nature of the difference is unknown. The method is applicable only if the products are fairly homogeneous. The method is effective for (a) determining whether a perceptible difference results (triangle testing for difference) or not (triangle testing for similarity) when, for example, a change is made in ingredients, processing, packaging, handling, or storage; or (b) for selecting, training, and monitoring assessors (http://www.iso.org/iso/catalogue_detail?csnumber=33495, http://www.iso.org/iso/iso_catalogue/catalogue_ics/catalogue_detail_ics.htm?csnumber=33495, http://www.iso.org/iso/iso_catalogue/catalogue_tc/catalogue_detail.htm?csnumber=33495).

ISO 3591:1977 provides a "common standard" for wine tasting so that everybody uses the same type of glass to taste the wine. Should a series of different glasses be used, the judgment would be very different. The ISO wine tasting glass is strongly recommended by hundreds of professional wine tasters throughout the world (http://www.iso.org/iso/iso_cafe_sensory_analysis.htm). The tasting glass consists of a cup (an "elongated egg") supported by a stem resting on a base. The opening of the cup is narrower than the convex part so as to concentrate the bouquet. Several physical, dimensional, and special characteristics of glass are provided for clarification purposes. An annex comprises recommendations for use (http://www.iso.org/iso/

iso_catalogue/catalogue_ics/catalogue_detail_ics.htm?ics1=67&ics2=260&ics3=&csnumber=9002).

17.3 APPLICATIONS OF SENSORY EVALUATION

17.3.1 Meat

Balamatsia et al. (2007) used sensory analysis to assess the odor and taste of microwave cooked chicken. They found that both parameters decreased with time of refrigerated storage. All vacuum- and MA-packaged chicken samples received higher sensory scores than air-packaged samples, as judged by the aforementioned attributes. This trend became more apparent after day 3 and continued throughout the entire period of refrigerated storage. The limit of odor acceptability was reached approximately on day 7 for the air and VP samples and on day 9 for the MAP_1 (30%/65%/5% $CO_2/N_2/O_2$), whereas the limit of acceptability for odor was not reached for chicken samples packaged under MAP_2 (65%/30%/5% $CO_2/N_2/O_2$), even after 15 days of storage (Figure 17.1). Sensory scores for taste were only recorded for chicken samples that had not exceeded the microbiological limit value of 7 log CFU/g since these samples were unacceptable to taste. The limit of acceptability for taste was reached approximately on days 6–7 and 10 for chicken samples packaged in air and under vacuum, respectively. Similar to odor, the limit of acceptability for taste was not reached for samples packaged under MAP_2 even after 15 days of storage. The appearance scores for all chicken samples decreased at a slower rate than odor and taste scores and never reached the lower acceptability limit of 6 (Table 17.2).

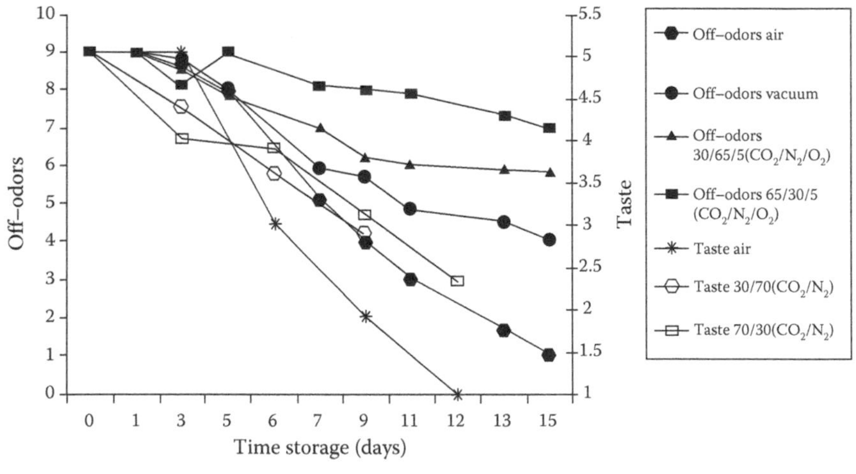

FIGURE 17.1 Changes in off-odor scores of chilled fresh chicken meat during storage in air, under VP, and under M_1 and M_2 gas mixtures (From Balamatsia, C.C. et al., *Food Chem.*, 104(4), 1622, 2007.) and effect of oregano essential oil and MAP on taste of chopped chicken meat stored at 4°C. (From Chouliara, E. et al., *Food Microbiol.*, 24(6), 607, 2007.)

TABLE 17.2
Effect of MAP on Foods of Animal Origin-Panel Type/Nr and Scale

Food	MAP Condition	Packaging Material	Panel: Type/Nr	Sensory Analysis: Scale	References
Chicken fillets	M_1: 30%/65%/5% ($CO_2/N_2/O_2$), M_2: 65%/30%/5% ($CO_2/N_2/O_2$)	1. Insulated polystyrene (PS) boxes, LDPE/PA/LDPE barrier pouches, 75 lm in thickness, oxygen permeability (OP) of 52.2 cm^3/m^2/day/atm at 75% RH, 23°C, carbon dioxide permeability of 191 cm^3/m^2/day/atm at 0% RH, 23°C, water vapor permeability (WVP) of 2.4/g^2/day at 100% RH, 23°C 2. PBI-Dansensor model mix 9000 gas mixer 3. BOSS model N48 vacuum sealer	Trained/7	0–9 9: excellent, 8: very good, 7: good, 6: acceptable, and 5: poor (lower limit of acceptability)	Balamatsia et al. (2007)
Fresh chicken breast meat	M_1: 30%/70% (CO_2/N_2) M_2: 70%/30% (CO_2/N_2)	1. LDPE/PA/LDPE barrier pouches, 75 mm in thickness having an OP of 52.2 cm^3/m^2/day/atm at 75% RH, 25°C and WVP of 2.4/g^2/day at 100% RH, 25°C 2. PBI-Dansensor model mix 9000 gas mixer 3. BOSS model N48 vacuum sealer	Trained/7	0–5 5: = most liked sample, 0: = least liked sample, and 3: = lower limit of acceptability	Chouliara et al. (2007)

(continued)

TABLE 17.2 (continued)
Effect of MAP on Foods of Animal Origin-Panel Type/Nr and Scale

Food	MAP Condition	Packaging Material	Panel: Type/Nr	Sensory Analysis: Scale	References
Beef	M_1: 50%/50% (O_2/CO_2) M_2: 80%/20% (O_2/CO_2) M_3: 80%/20% (CO_2/N_2)	M_1, M_2: Tray (13 × 18 × 4 cm) covered with transparent film, O_2 transmission rate: 10 $cm^3/m^2/24$ h/atm, vacuum: 5–10 mbar, 750 mbar gas filling M_3: Tray (13 × 18 × 4 cm) covered with transparent film, OTR: 2 $cm^3/m^2/24$ h/atm, vacuum: 5–10 mbar, 750 mbar gas filling, Convenience Food Systems (CFC) traysealer (CP)	Trained/8	0–15 15: very high intensity and 0: slight intensity	Clausen et al. (2009)
Galician chorizo Sausage	M_1: 100% CO_2, M_2: 100% N_2, M_3: 50/50 (CO_2/N_2)	M_1: Aluminum/PET/PE packs M_2: Aluminum/PET/PE packs M_3: Aluminum/PET/PE packs	Trained/11	1–7 7: intensity.	Fernandez et al. (2002)
Beef	M_1: 80%/20% (O_2/CO_2)	1. Packaged in oxygen-impermeable chub packaging using a Cartridge Pack 44 chub machine 2. Ross INPACK S45MA tray packaging machine	Untrained (Panel members were recruited by posting signs on the campus)/81, 72, and 76 judges for day 1, 6, and 10 panels, respectively	1–9 9: like extremely, 7: like moderately, 6: like slightly, 5: neither like nor dislike, 4: dislike slightly, and 1: dislike extremely	Jayasingh et al. (2002)

Smoked turkey breast fillets	M_1: 30%/70% (N_2/CO_2) M_2: 50%/50% (CO_2/N_2)	1. Packed in PET/LDPE/EVOH/LDPE barrier pouches (200 g/pouch) having an OP of 2.32 mL/m^2/day/atm at 23 1°C 2. Model mix 9000 gas mixer 3. BOSS model N48 vacuum sealer	Trained /5	1–5 5: extremely good, 4: good, 3: fair, 2: poor, 1: unacceptable quality, and 3.5: lower limit of acceptability	Ntzimani et al. (2008)
Precooked chicken	M_1: 30%/70% (CO_2/N_2) M_2: 60%/40 (CO_2/N_2) M_3: 90%/10% (CO_2/N_2)	1. LDPE/PA/LDPE barrier pouches (1 fillet/pouch), 75 mm in thickness having an OP of 52.2 mL/m^2 day atm at 60% RH/25°C and WVP of 2.4 g/m^2 day at 100% RH/25°C 2. PBI-Dansensor model 9000 gas mixer 3. BOSS model N48 vacuum sealer	Trained/7	1–9 9 ¼: no foreign flavor or like extremely, 1 ¼: extreme foreign flavor or dislike intensely, and 6: lower limit of acceptability	Patsias et al. (2006)
Beef	M_1: 20%/80% (CO_2/N_2) M_2: 80/20% (CO_2/N_2)	1. Plastic bags (PA/PE) with an OTR of 30–40 cm^3/m^2/24 h/bar at 23°C and 50% RH and a WVTR of 2.5 g/m^2/24 h at 23°C and 50% RH 2. Closed by heat sealing with a packer in a high barrier film with an OTR of 1.8 cm^3/m^2/24 h/bar at 20°C and 65% RH	Trained/8	1–5 5: excellent, 4: good, 3: acceptable, 2: fair, and 1: unacceptable	Rubio et al. (2007)
Sausage "Morcilla de Burgos"	M_1: 30%/70% (CO_2/N_2) M_2: 50%/50% (CO_2/N_2) M_3: 80%/20% (CO_2/N_2)		Trained/5	1–5 5: excellent, 4: good, 3: acceptable, 2: fair, and 1: unacceptable	Santos et al. (2005)

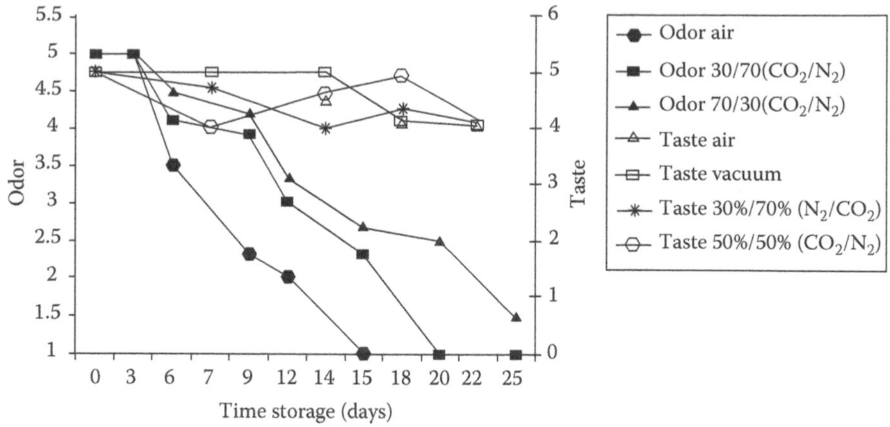

FIGURE 17.2 Changes in taste of smoked turkey breast fillets stored in air, under vacuum, skin, M_1 (30% CO_2/70% N_2), and M_2 (70% CO_2/50% N_2) at 4°C (From Ntzimani, A.G. et al., *Food Microbiol.*, 25(3), 509, 2008.) and effect of oregano essential oil and MAP on odor of chopped chicken meat stored at 4°C. (From Chouliara, E. et al., *Food Microbiol.*, 24(6), 607, 2007.)

Chouliara et al. (2007) reported on the effect of the addition of oregano oil on chopped chicken meat (Figure 17.2). However, when the concentration of oregano oil exceeded 1%, the samples were not sensorially evaluated because of the oregano oil strong flavor (1%). The lower acceptability score of 3.5 was reached for odor after 6 days for the air-packaged samples, 9–10 days for the air-packaged samples containing oregano oil 0.1%, 10–11 days for the samples under MAP_1, 13–14 days for the samples under MAP_1 containing oregano oil 0.1%, 11–12 days for the samples under MAP_2, and 13–14 days for the samples under MAP_2 containing oregano oil 0.1%. The lower acceptability limit for taste was reached after 6 days for the air-packaged samples, 8–9 days for the samples containing oregano oil 0.1%, 5–6 days for the samples under MAP_1, 10–11 days for samples under MAP_1 containing oregano oil 0.1%, 7–8 days for the samples under MAP_2, and 12 days for the samples under MAP_2 containing oregano oil 0.1%. Samples containing 0.1% oregano oil gave a characteristic desirable odor and taste to chicken meat, very compatible to cooked chicken flavor. Both odor and taste proved to be equally sensitive to sensory properties for chicken meat. Concentration of 0.1% of oregano oil had more or less the same effect as that of MAP on shelf-life prolongation of fresh chicken meat. Both MAP_1 (30%/70%, CO_2/N_2) and MAP_2 (70%/30%, CO_2/N_2) had the same effect on the organoleptic properties of chicken meat.

The highest cooking loss was found in steaks stored in high O_2 and CO_2 MA (12.0%–13.8%) and lowest in VSP-packed steaks (9.4%) and in steaks stored in N_2+air (2 days) (9.7%) (Clausen et al., 2009). The main reason for the increased cooking loss is the presence of CO_2 and not O_2. The results obtained revealed that a packaging system with high O_2 has a marked negative effect on several eating quality parameters. Both tenderness and juiciness in the cooked meat were considerably reduced and the flavor was more distinctly intense with a warmed-over/oxidized taint, in conjunction with the occurrence of a premature browning. Thus, the high-oxygen MAP systems cannot be recommended either to the meat industry or for retail meat distribution.

The most notable change reported was in external color, which became brownish red, particularly in samples stored under CO_2 (i.e., under 100% CO_2 or 50:50 CO_2/N_2). Some descriptors did not change significantly throughout storage or as a result of the packaging method considered. The descriptors used were degree of separation of skin, external smoky odor, ease of separation of skin, internal smoky odor, chewability, hot flavor, aftertaste intensity, and aftertaste persistence, displaying average values 2.17, 2.57, 4.36, 1.68, 4.82, 32.14, 3.61, and 3.50, respectively. The acid value was significantly affected by both storage time and packing method and by the interaction between the two (Fernandez-Fernandez et al., 2002).

Jayasingh et al. (2002) reported no significant difference in sensory scores for flavor between MAP and control samples on day 1. The flavor score of control samples was 6–7 over 10 day storage, where 6 = slightly and 7 = moderately. By day 6, the flavor score of MAP samples dropped to 4.8 (4 = dislike slightly, and 5 = neither like nor dislike). The 10 day old MAP samples were regarded as the least desirable of all the samples tested with a mean flavor score of 4.5. It is worth noting that the average sensory score for the stored MAP samples is on the "dislike" side of the scale. Further to lower flavor scores, the 6 and 10 day old MAP samples were the recipient of significantly lower scores than control samples for texture, juiciness, and overall acceptability.

According to Ntzimani et al. (2008) individual scores for odor and taste (Figure 17.3) revealed a similar decreasing trend for samples stored under all packaging treatments. Until day 7, all samples received a score of 5 and until day 18, all samples received a score higher than 4. On day 22 of storage, the first off-odors were

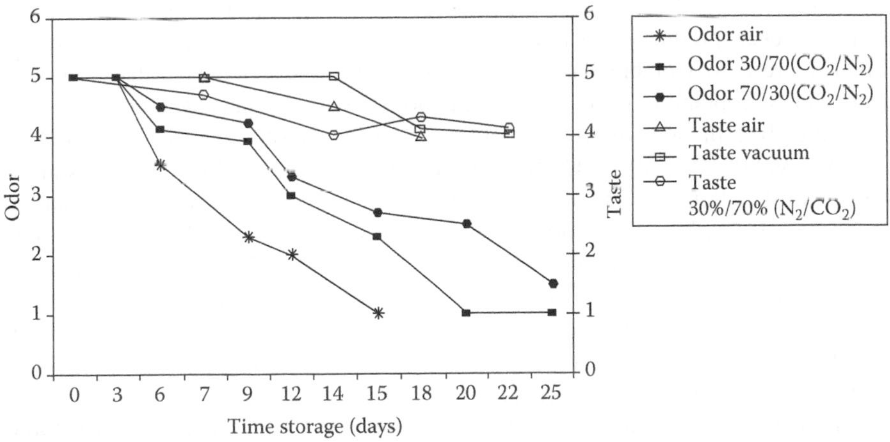

FIGURE 17.3 Effect of packaging type and storage time on taste morcillas (From Santos, E.M. et al., *Meat Sci.*, 71(2), 249, 2005.) and changes in odor of smoked turkey breast fillets stored in air, under vacuum, skin, M_1 (30% CO_2/70% N_2) and M_2 (70% CO_2/50% N_2) at 4°C. (From Ntzimani, A.G. et al., *Food Microbiol.*, 25(3), 509, 2008.)

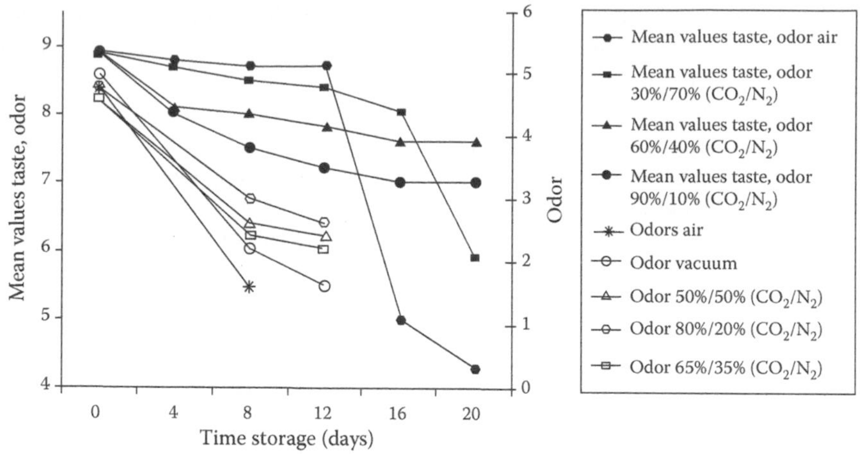

FIGURE 17.4 Changes in overall acceptability scores (mean values of taste and odor) of chilled precooked chicken meat packaged in air and under MAP (From Patsias, A. et al., *Food Microbiol.*, 23(5), 423, 2006.) and effects of MAP and storage time on odor of raw and cooked mussels. (From Caglak, E. et al., *Eur. Food Res. Technol.*, 226(6), 1293, 2008.)

developed in the air-packaged samples, while the rest of the samples maintained a large part of their original quality. By day 26 of storage, air-packaged samples were spoiled (score <3.5), whereas the rest of the samples were spoiled on day 30 of storage (score <3.5). Similar results were recorded for taste attributes. Based on sensory analysis (odor), the shelf life of all turkey samples, irrespective of MAP conditions, reached approximately 27–30 days with slightly better sensory characteristics obtained for skin-packaged samples. The shelf life of samples packaged in air amounted to 22–23 days.

Patsias et al. (2006): The results of the SE (odor and taste attributes) of the MW heated chicken product are shown as overall acceptability scores versus storage time (Figure 17.4). Combined scores for odor and taste revealed a similar pattern of decreasing acceptability (individual results not shown). Air-packaged chicken samples were given higher overall acceptability scores ($p<0.05$) than MAP_1 (30%/70% CO_2/N_2), MAP_2 (60%/40% CO_2/N_2), and MAP_3 (90%/10% CO_2/N_2) samples up to day 12, while, after this time period, this trend was reversed holding throughout the entire period of refrigerated storage. All chicken samples received high scores during the first 12 days, while after this period significant differences ($p<0.05$) were observed in sensory scores between air- and MA-packaged samples. The limit of overall acceptability (score 6) was reached just before day 16 (air samples) and day 20 (MAP_1 samples), while MAP_2 and MAP_3 samples never reached this limit within the time span of the experiment (Figure 17.4).

Rubio et al. (2007) carried out sensory analysis on Cecina de Leon (CL) stored under vacuum and MAP and reported that most sensory attributes evaluated did not vary significantly ($p>0.05$) throughout storage under vacuum and under 80%/20% CO_2/N_2, with the exception of the odor (Figure 17.5), which decreased significantly ($p<0.05$) at 150–210 days. When the CL portions were packed in 20%/80% CO_2/N_2,

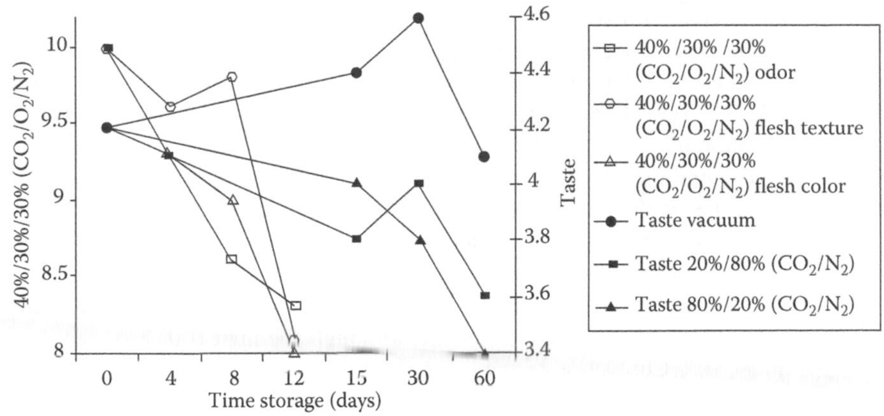

FIGURE 17.5 Effect of packaging systems and storage time on taste measured on "Cecina de Leon" portions stored at 6°C (From Rubio, B. et al., *Meat Sci.*, 75(3), 515, 2007.) and SE of raw sea bream stored at 4°C ± 0.5°C. (From Goulas, A.E. and Kontominas, M.G., *Food Chem.*, 100(1), 287, 2007.)

a reduction in odor, taste, hardness, and acceptability was recorded. The acceptability of the samples stored in 20%/80% CO_2/N_2 was lower than that in vacuum and under 80%/20% CO_2/N_2 packed samples at 210 days.

The effect of the method of packaging and storage time on sensory properties of morcillas was investigated by Santos et al. (2005). All sensory parameters dramatically dropped during storage, except for off-odors, which increased, especially in VP "morcilla" (Figure 17.6). Generally speaking, morcillas stored in 50% and 80% CO_2 presented better scores than morcillas VP and packaged under 30% CO_2, and these

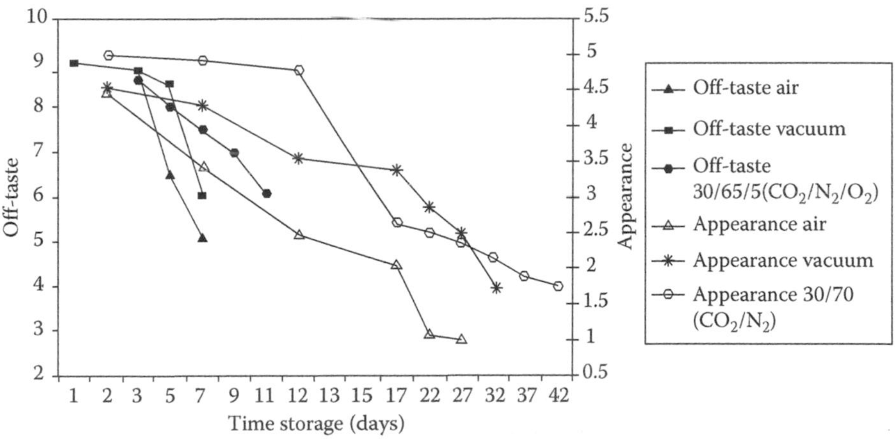

FIGURE 17.6 Changes in off-taste scores of chilled fresh chicken meat during storage in air, under VP, and under M_1 and M_2 gas mixtures (From Balamatsia, C.C. et al., *Food Chem.*, 104(4), 1622, 2007.) and effect of packaging type and storage time on appearance of morcillas. (From Santos, E.M. et al., *Meat Sci.*, 71(2), 249, 2005.)

presented better ($p<0.05$) scores than those stored in air. In the case of morcillas stored in air, the appearance was not acceptable after 12 days mainly due to the drying surface, and after 17 days the rest of the sensory attributes reached values below the acceptable level (less than 3 on the hedonic scale). These effects were by far more intense in VP morcillas than in MAP samples ($p<0.05$). In fact, the appearance of the VP samples and those with 30% CO_2 was below the acceptability limit after 22 and 17 days, respectively, because of vacuum loss, blowing of certain packs, and the presence of milky exudates and slime formation in the case of vacuum samples, and microbial colony spots on the surface of the gas-packaged products as well as the presence of slime. Samples packed in 50% and 80% CO_2 presented sporadically visible microbial colonies from the 27th day, although the appearance of sausages packaged under 80% CO_2 was acceptable until the 32nd day. Strong sour odors and tastes, as well as distinctive off-odors, when the packages were opened, were apparent from the 22nd day in the vacuum packages and those packaged with 30% CO_2. These effects were visible from the 32nd day in morcillas packaged under 50% and 80% CO_2. In general, no significant differences were found for the aforementioned parameters between sausages stored in 50% and 80% CO_2.

17.3.2 Fish

The effect of MAP and storage time on appearance and odor of raw and cooked mussel was studied by Caglak et al. (2008) and is shown in Figure 17.7. On day 8, the control sample (mussels in the air) was significantly less appealing and more odorous than the rest. On day 12, no significant differences in appearance and odor were recorded among all MAP samples except for VP ($p<0.05$) (Table 17.3).

Although the differences between Air and MAP_2 were found to be statistically significant ($p<0.05$) for odor, no significant difference was found in flavor between

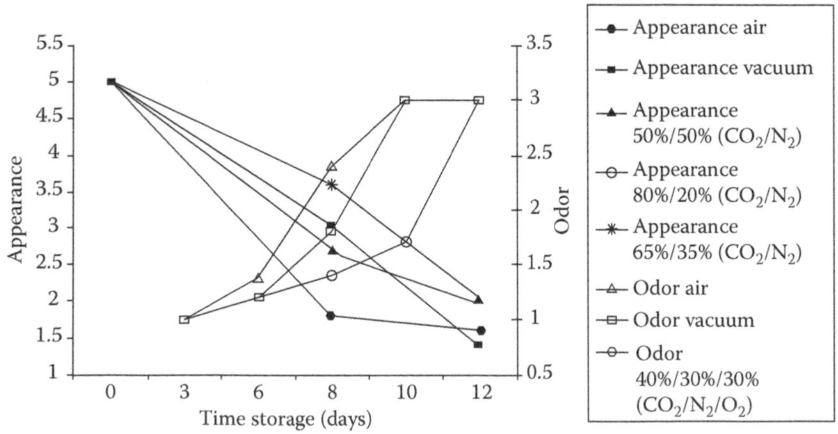

FIGURE 17.7 Effects of MAP and storage time on appearance of raw and cooked mussels (From Caglak, E. et al., *Eur. Food Res. Technol.*, 226(6), 1293, 2008.) and changes in odor of chilled fresh swordfish stored in air, vacuum, and under MAP. (From Pantazi, D. et al., *Food Microbiol.*, 25(1), 136, 2008.)

TABLE 17.3
Effect of MAP on Fish-Based Foods-Panel Type/Nr and Scale

Food	MAP Condition	Packaging Material	Panel: Type/Nr	Sensory Analysis: Scale	References
Mussels (*Mytilus galloprovincialis*)	M_1: 50%/50% (CO_2/N_2) M_2: 80%/20% (CO_2/N_2) M_3: 65%/35% (CO_2/N_2)	Packed on the PS tray, which was then placed inside the case (100 m PA/550 m PVC gas barrier) with thickness of 650 m	Trained/7	0–5 5: highly acceptable and 1: least acceptable	Caglak et al. (2008)
Fresh blue fish burger	M_1: 30%/40%/30% ($O_2/CO_2/N_2$) M_2: 50%/50% (O_2/CO_2) M_3: 5%/95% (O_2/CO_2)	Packed in PA/PE bags (95 μm) by means of S100-Tecnovac equipment. The bags were 170 × 250 mm long, with OP of 50.65 cm^3/m^2 day atm and WVTR of 1.64 g/m^2 day	Untrained (students and researchers)/10	0–5 5: excellent and 0: very poor	Nobile et al. (2009)
Cod (*Gadus morhua*)	M_1: 60%/30%10% ($CO_2/N_2/O_2$)	PE/PA/PE 20/70 pouches, 90 μm in thickness, using a Multivac model D8941 machine (Bury, Lancashire, the United Kingdom). At 85% RH and 23°C the film has an OP of 50 cm^3/m^2/24 h/1 bar, a carbon dioxide permeability of 150 cm^3/m^2/24 h/1 bar, a N_2P of 10 cm^3/m^2/24 h/1 bar, and a WVP of 2.6 g/m^2/24 h/1 bar	Trained/20	0–8 8: very pleasant, 4: neutral, and 0: very unpleasant	Fernández-Segovia et al. (2007)
Sea bream (*Sparus aurata*)	M_1: 40%/30%/30% ($CO_2/O_2/N_2$)	1. Packaged in LDPE/PA/LDPE barrier pouches 25 cm. 35 cm (2 fillets per pouch, weighing ca. 180 g) 75 lm in thickness having an OP of 52.2 cm^3/m^2/day/atm at 75% RH, 23°C, a carbon dioxide permeability of 191 cm^3/m^2/day/atm at 0% RH, 23°C and a WVP of 2.4 g/m^2/day at 100% RH, 23°C 2. PBI-Dansensor model 9000 gas mixer 3. BOSS model NE 48 vacuum sealer	Trained/7	0–10 10: most liked, 0: least liked, and <6: unacceptable	Goulas and Kontominas (2007)

(continued)

TABLE 17.3 (continued)
Effect of MAP on Fish-Based Foods-Panel Type/Nr and Scale

Food	MAP Condition	Packaging Material	Panel: Type/Nr	Sensory Analysis: Scale	References
Mussels (*Mytilus galloprovincialis*)	M_1: 50%/50% (CO_2/N_2) M_2: 80%/20% (CO_2/N_2) M_3: 40%/30%/30% ($CO_2/N_2/O_2$)	Packaged in LDPE barrier pouches (200 g/pouch) 75 µm in thickness having an OP of 52.2 cm^3/m^2/day/atm at 75% RH, 25°C and a WVP of 2.4 g/m^2/day at 100% RH, 25°C 2. HELMUT BOSS model N48 vacuum sealer.	Trained/7	0–10 10–9: excellent, 8–7: very good, and 6: lower limit of acceptability	Goulas et al. (2005)
Atlantic herring (*Clupea harengus*)	M_1: 60%/40% (CO_2/N_2)	Nylon-PE pouches (30×35 cm), the third lot VP and the fourth lot gas packed in a Multivac model A 300 VP machine. The OTR of pouches was 47 cc/m^2 24 h	Trained/6	0–3	Ozogul et al. (2000)
Sardines (*Sardina pilchardus*)	M_1: 60%/40% (CO_2/N_2)	Multivac model A 300 VP Machine, the OTR of pouches was 47 cm^3/m^2 24 h	Trained/6	Panelists were asked to state whether the fish were acceptable	Ozogul et al. (2004)
Swordfish (*Xiphias gladius*)	M_1: 40%/30%/30% ($CO_2/N_2/O_2$)	1. Packaged in LDPE/PA/LDPE barrier pouches (1 steak weighing approximately 150,710 g/pouch) 75 mm in thickness having an OP of 52.2 cm^3/m^2/day/atm at 75% RH, 25°C and a WVP of 2.4 g/m^2/day at 100% RH, 25°C 2. PBIDansensor model mix 9000 gas mixer 3. BOSS model N48 vacuum sealer	Trained/7	1–3 1: no off-flavors, equal to the reference sample; class 2: Slight off-flavors but not spoiled (initial decomposition change), and class 3: clearly recognizable off-flavors (unacceptable quality)	Pantazi et al. (2008)

Bass (*Dicentrarchus labrax*)	MAP_1: 70%/30% (CO_2/N_2) MAP_2: 20%/70%/10% ($O_2/CO_2/N_2$) MAP_3: 30%/60%/10% ($O_2/CO_2/N_2$) MAP_4: 40%/60% (O_2/CO_2) MAP_5: 30%/50%/20% ($O_2/CO_2/N_2$) MAP_6: 21%/79% (O_2/N_2)	Packed with PS tray laminated with a multilayer barrier film (V = 1000 cc) and sealed with a film of PA/EVOH/PE (PO_2 ¼ 1:5 cc/m^2 24 h atm; PH_2O ¼ 5:5 g/m^2 24 h atm; thickness = 55 lm)	Trained/7	0,5,10	Torrieri et al. (2006)

MAP_1 (50%/50% CO_2/N_2), MAP_3 (65%/35% CO_2/N_2), VP, Air, and MAP_2 (80%/20% CO_2/N_2). The mussels (A) were inappropriate for human consumption at day 8 from the microbiological and sensory aspect. By the end of the storage period (12 days), the differences between VP and MAP were found to be significant for appearance and odor (Figure 17.7).

The statistical analysis of the organoleptic assessment of the samples displayed no significant differences for any of the investigated attributes, among the panelists or during the time of storage, which implies that all samples evaluated (additives and blanched and packaged in vacuum) displayed high sensory stability throughout the storage time, with a high degree of agreement among the assessors. Application of one-way ANOVA revealed no significant differences between samples with and without additives for any of the attributes studied at any time or between stored and recently treated samples. All samples evaluated scored positively, since the scores of attributes for all samples were always higher than 4 on a scale from 0 to 8 (Fernández-Segovia et al., 2007).

All raw sea bream fillet samples investigated received acceptable sensory scores (significantly higher or similar to control samples than the lower acceptability limit of 6) during the first 15–16 days of storage. The salted samples remained acceptable (above the acceptability limit of 6) up to ca. 20–21 days while the MAP salted samples up to ca. 27–28 days of storage (Goulas and Kontominas, 2007).

Goulas et al. (2005) carried out SE of mussels (cooked and raw) and found that appearance scores for both cooked and raw samples dropped at a slower rate than odor and taste scores. In general, the appearance of the raw mussel samples received a lower score than appearance of cooked mussel samples. All mussel samples were graded "excellent" (9–10) or "very good" (7–8) during the first 5 days with regard to odor and taste and during the first 8 days with regard to appearance. After this period, significant differences ($p<0.05$) were recorded in sensory scores of VP, MAP, and control samples. The limit of acceptability (score 6) for odor and taste went over on day 11 (VP, two control samples) and on day 15 (MAP_1—50%/50% CO_2/N_2, MAP_2—80%/20% CO_2/N_2, and MP_3—40%/30%/30% $CO_2/N_2/O_2$) samples, respectively.

Although chemical and microbiological analyses indicated that CO_2 and VP extended the shelf life of herring, compared with storage in ice, sensory assessment revealed that the extension of shelf life was only with MAP (10 days) and VP (8 days) (Ozogul et al., 2000). It was found that 60% CO_2 treatments led to lower K values compared to those reported for aerobically held fish. The results disclosed considerable differences between ice and MAP storage conditions.

VP and MAP did significantly prolong the sensory shelf life of sardines as compared to storage in air. However, off-odors and drip losses in MAP and VP lowered the sensory quality of the sardines. The shelf life of sardine in MAP was longer than that in VP (Ozogul et al., 2004).

The individual scores for odor (Figure 17.8) and taste showed a similar pattern of decreasing acceptability for samples stored in air and under VP and MAP. Until day 6 of storage, all swordfish samples were given mean odor scores in the range 1.2–1.5. On day 8 of storage, first off-odors were developed in A (Air) samples; VP and MAP samples had acceptable quality. By day 10 of storage, Air and VP samples

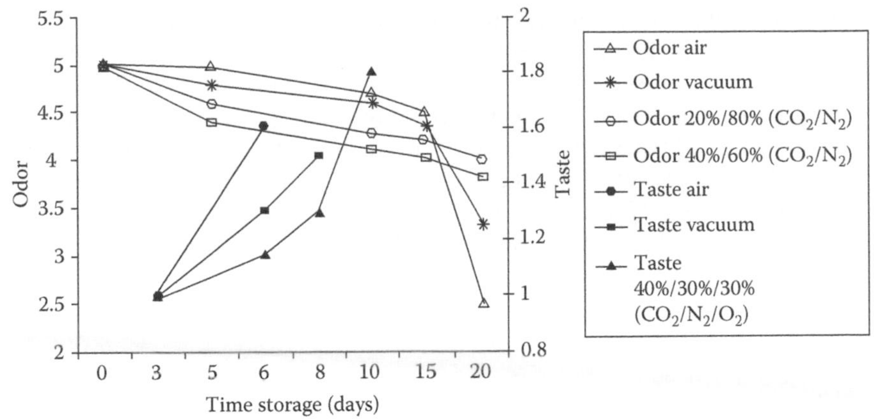

FIGURE 17.8 Changes in taste of chilled fresh swordfish stored in air, vacuum, and under MAP (From Pantazi, D. et al., *Food Microbiol.*, 25(1), 136, 2008.) and odor evaluation of Myzithra whey cheese packaged under various atmospheres during storage at 4°C. (From Dermiki, M. et al., *LWT*, 41(2), 284, 2008.)

were spoiled (score 3) while all MAP samples received a score of 1.7. Between days 10 and 16 of storage all samples were spoiled. The obtained data for taste generally were in satisfactory agreement with that for odor. Therefore, the MAP samples were the only samples given a score of less than 2 on day 10 of storage while Air and VP samples were spoiled (corresponding TVCs were higher than 7.0 log CFU/g) (Pantazi et al., 2008).

Torrieri et al. (2006): Time and atmosphere composition have a significant effect on all the attributes ($p<0.0001$) of chilled fresh swordfish. During storage, the odor (Figure 17.9) (evaluated as odor perceived at the opening of the package) changes

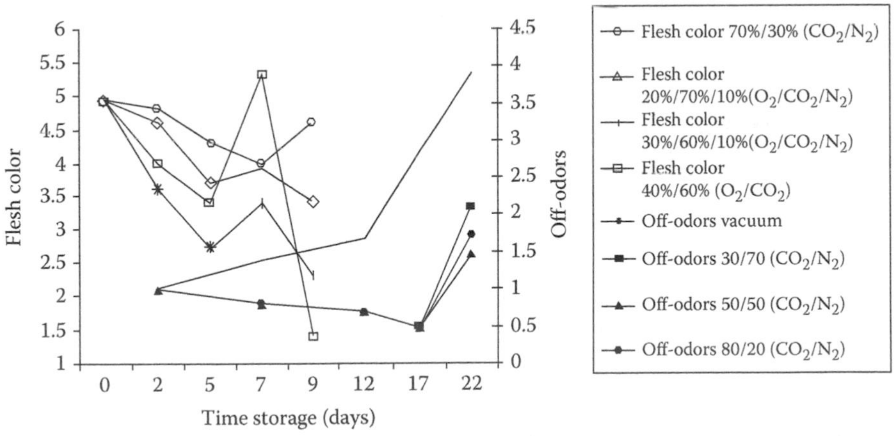

FIGURE 17.9 Effect of packaging type and storage time on off-odors of morcillas (From Santos, E.M. et al., *Meat Sci.*, 71(2), 249, 2005.) and flesh color of gutted bass in different MAP during storage at 3°C. (From Torrieri, E. et al., *J. Food Eng.*, 77(4), 1078, 2006.)

with increasing the storage time passing from "fresh fish odor" (day 0–2) for all the samples to "spoiled odor" as a result of the extensive spoilage occurring. Only the odor of the sample packed by using air as initial gas composition (F) is perceived as off-odor after 7 days of storage and after 9 days was given a score as low as 0.91. Among the other samples, those made by using 20% O_2–70% CO_2 (B) and 30% O_2–50% CO_2 (E) received after 9 days of storage a score higher than 5. The bass packed under 30% O_2–50% CO_2 (E) is the one awarded the best scores at each storage time. The color of fresh bass was ascribed as pink and corresponded to a score equal to 5.

17.3.3 Dairy

Dermiki et al. (2008): No significant differences ($p>0.05$) were recorded in odor (Figure 17.10) scores of Myzithra whey cheese for all packaging treatments for a given sampling day until day 15 of storage. By day 20 of storage, statistically significant ($p<0.05$) differences were reported among air-, vacuum-, and MA-packaged samples. On this day, cheeses packaged in air or UV were considered as unacceptable, while all MAP_1, MAP_2, and MAP_3 samples received a relatively high score. On day 26 of storage, all MAP samples received an odor score higher than 3.5, which was the lower acceptability limit. On day 30 of storage the only two samples receiving an acceptable score were those packaged under atmospheres M_2 and M_3. Odor characteristics, including "sour," "malty," and "facal," were recorded for cheese samples with scores lower than 3.5. As regards the taste scores, it was observed that up to day 5 of storage, there were no statistically significant differences ($p>0.05$) recorded among samples of different packaging treatments for a given sampling day. Statistically significant differences ($p<0.05$) were reported in the beginning with day 10 of storage and by day 15, samples packaged in air were regarded as unacceptable. All other samples maintained good sensory characteristics. Samples packaged under gas mixture MAP_1 kept their good sensory properties until day 20 of storage,

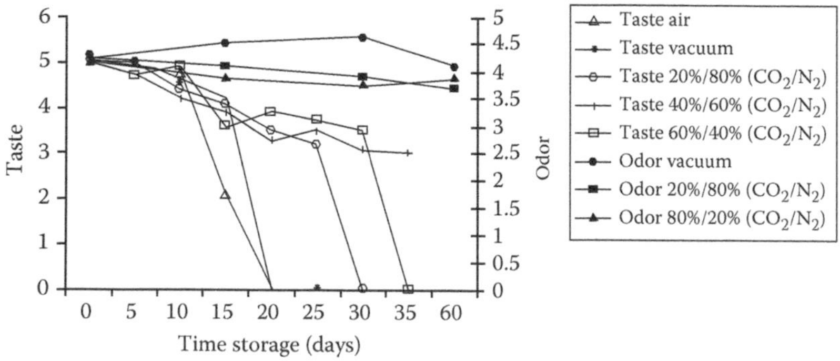

FIGURE 17.10 Effect of packaging systems and storage time on odor measured on "Cecina de Leon" portions stored at 6°C (From Rubio, B. et al., *Meat Sci.*, 75(3), 515, 2007.) and taste evaluation of Myzithra whey cheese packaged under various atmospheres during storage at 4°C. (From Dermiki, M. et al., *LWT*, 41(2), 284, 2008.)

TABLE 17.4
Effect of MAP on Dairy-Based Foods-Panel Type/Nr and Scale

Food	MAP Condition	Packaging Material	Panel: Type/ Nr	Sensory Analysis: Scale	References
Cheese "Myzithra Kalathaki"	M_1: 20%/80% (CO_2/N_2) M_2: 40%/60% (CO_2/N_2) M_3: 60%/40% (CO_2/N_2)	1. Plastic pouches composed of co-extruded LDPA/PA/LDPE 75 mm in thickness having an OTR of 52.2 mL/(m^2 day atm) and a WVTRof 2.4 g/(m^2 day), measured using the Oxtran 2/20 and Permatran 3/31 permeability testers 2. Dansensor, model Mix 9000, gas mixer and pouches were sealed using a Boss N 48 vacuum sealer 3. Dansensor model checkmate 9900 headspace analyzer 4. Rubber septum (Systech Instr. Ltd., the United Kingdom) was glued onto the surface of each pouch and pierced with a 23 gauge needle connected to the headspace analyzer	Trained/5	0–5 5: very good, 4: good, 3: fair, 2: poor, 1: very poor, and 0: unfit for human consumption	Dermiki et al. (2008)
Cheese (*Cameros cheese*)	M_1: 20%/80% (CO_2/N_2) M_2: 40%/60% (CO_2/N_2) M_3: 50%/50% (CO_2/N_2) M_4: 100% CO_2	1. Plastic films used were provided by Dixie with a CO_2 permeability of less than 13 $cm^3/m^2/24\,h$ at 1 atm and CO_2 permeability 5 $cm^3/m^2/24\,h$ at 1 atm	Trained/35	1–7 (from less to more)	Olarte et al. (2001)

MAP_2 samples until day 26, and MAP_3 samples until day 30 of storage. Between odor and taste, the latter proved to be a more sensitive sensory attribute for the determination of Myzithra Kalathaki whey cheese shelf life. In general, odor and taste scores were higher for samples packaged under a CO_2 atmosphere. This is due to the presence of CO_2, which inhibited a large portion of aerobic and psychrotrophic bacteria, usually responsible for the production of many metabolites leading to the development of off-flavors in foods (Table 17.4).

17.4 CONCLUSIONS

It was found by applying both microbiological analysis and sensory assessment that employment of MAP was effective at prolonging the shelf life of various high value-added products such as dairy (Myzithra whey cheese, Cameros cheese) and meat, and fish products as well. In fact, the usage of SE proved to be a powerful tool because it could effectively determine spoilage of food before microbiological analysis. Therefore, it is of great importance to undertake proper precautions with regard to both the infrastructure (premises) and the training panel to ensure that this "powerful tool" is practically unbiased and the obtained results reliable.

REFERENCES

Balamatsia, C.C., Patsias, A., Kontominas, M.G. and Savvaidis, I.N. (2007). Possible role of volatile amines as quality-indicating metabolites in modified atmosphere-packaged chicken fillets: Correlation with microbiological and sensory attributes. *Food Chemistry*, 104(4): 1622–1628.

Bodyfelt, F.W., Drake, M.A. and Rankin, S.A. (2008). Developments in dairy foods sensory science and education: From student contests to impact on product quality. *International Dairy Journal*, 18(7): 729–734.

Bourne, M.C. and Szczesniak, A.S. (2003). Texture, in *Encyclopedia of Food Science and Nutrition*. Eds. B. Caballero, P. Finglas, and L. Trugo, Academic Press, Amsterdam, the Netherlands, pp. 5167–5174.

Bucher, J.L. (2004). *The Metrology Handbook*. ASQ Quality Press, Milwaukee, WI, p. 109.

Caglak, E., Cakli, S. and Kilinc, B. (2008). Microbiological, chemical and sensory assessment of mussels (*Mytilus galloprovincialis*) stored under modified atmosphere packaging. *European Food Research and Technology*, 226(6): 1293–1299.

Cayot, N. (2007). Sensory quality of traditional foods. *Food Chemistry*, 10(1): 154–162.

Chen, J. (2009). Food oral processing—A review. *Food Hydrocolloids*, 23(1): 1–25.

Chouliara, E., Karatapanis, A., Savvaidis, I.N. and Kontominas, M.G. (2007). Combined effect of oregano essential oil and modified atmosphere packaging on shelf-life extension of fresh chicken breast meat, stored at 4°C. *Food Microbiology*, 24(6): 607–617.

Clausen, I., Jakobsen, M., Ertbjerg, P. and Madsen, N.T. (2009). Modified atmosphere packaging affects lipid oxidation, myofibrillar fragmentation index and eating quality of beef. *Packaging Technology Science*, 22: 85–96.

Cliff, M. and Heymann, H. (1993). Development and use of time-intensity methodology for sensory evaluation: A review. *Food Research International*, 26(5):375–385.

Del Nobile, M.A., Corbo, M.R., Speranza, B., Sinigaglia, M., Conte, A. and Caroprese, M. (2009). Combined effect of MAP and active compounds on fresh blue fish burger. *International Journal of Food Microbiology*, 135(3): 281–287.

Dermiki, M., Ntzimani, A., Badeka, A., Savvaidis, I.N. and Kontominas, M.G. (2008). Shelf-life extension and quality attributes of the whey cheese "*Myzithra Kalathaki*" using modified atmosphere packaging. *LWT*, 41(2): 284–294.

Dijksterhuis, G.B. and Piggott, J.R. (2001). Dynamic methods of sensory analysis. *Trends in Food Science & Technology*, 11(8): 284–290.

Duizer, L.M., Bloom, K. and Findlay, C.J. (1997). Dual-attribute time-intensity sensory evaluation: A new method for temporal measurement of sensory perceptions. *Food Quality and Preference*, 8(4): 261–269.

Duncan, S.E. and Webster, J.B. (2009). Sensory impacts of food–packaging interactions. *Advances in Food and Nutrition Research*, 56(8): 17–64.

Etaio, I., Albisu, M., Ojeda, M., Gil, P.F., Salmeron, J. and Perez Elortondo, F.J. (2010a). Sensory quality control for food certification: A case study on wine. Panel training and qualification, method validation and monitoring. *Food Control*, 21(4): 542–548.

Etaio, I., Albisu, M., Ojeda, M., Gil, P.F., Salmeron, J. and Perez Elortondo, F.J. (2010b). Sensory quality control for food certification: A case study on wine. Method development. *Food Control*, 21(4): 533–541.

Fernandez-Fernandez, E., Vazquez-Oderiz, M.L. and Romero-Rodriguez, M.A. (2002). Sensory characteristics of Galician chorizo sausage packed under vacuum and under modified atmospheres. *Meat Science*, 62(1): 67–71.

Fernández-Segovia, I., Escriche, I., Fuentes, A. and Serra, J.A. (2007). Microbial and sensory changes during refrigerated storage of desalted cod (*Gadus morhua*) preserved by combined methods. *International Journal of Food Microbiology*, 116(1): 64–72.

Frank, R.A. (2002). Response context affects judgments of flavor components in foods and beverages. *Food Quality and Preference*, 14(2): 139–145.

Goulas, A.E., Chouliara, I., Nessi, E., Kontominas, M.G. and Savvaidis, I.N. (2005). Microbiological, biochemical and sensory assessment of mussels (*Mytilus galloprovincialis*) stored under modified atmosphere packaging. *Journal of Applied Microbiology*, 98(3): 752–760.

Goulas, A.E. and Kontominas, M.G. (2007). Combined effect of light salting, modified atmosphere packaging and oregano essential oil on the shelf-life of sea bream (*Sparus aurata*): Biochemical and sensory attributes. *Food Chemistry*, 100(1): 287–296.

Guerra, S., Lagazio, C., Manzocco, L., Barnaba, M., and Cappuccio, R. (2008). Risks and pitfalls of sensory data analysis for shelf life prediction: Data simulation applied to the case of coffee. *LWT - Food Science and Technology*, 41: 2070–2078.

Hoehl. K., Schoenberger, G.U. and Busch-Stockfisch, M. (2010). Water quality and taste sensitivity for basic tastes and metallic sensation. *Food Quality and Preference*, 21(2): 243–249.

Jayasingh, P., Cornforth, D.P., Brennand, C.P., Carpenter, C.E. and Whittier, D.R. (2002). Sensory evaluation of ground beef stored in high-oxygen modified atmosphere packaging. *Sensory and Nutritive Qualities of Food*, 67(9): 3493–3496.

Kontominas, M.G., Goulas, A.E., Badeka, A.V. and Nerantzaki, A. (2006). Migration and sensory properties of plastics-based nets used as food-contacting materials under ambient and high temperature heating conditions. *Food Additives and Contaminants*, 23(6): 634–641.

Krishnamurthy, R., Srivastava, A.K., Paton, J.E., Bell, G.A. and Levy, D.C. (2007). Prediction of consumer liking from trained sensory panel information: Evaluation of neural networks. *Food Quality and Preference*, 18(3): 275–285.

Maynard, S., Foster, S. and Hall, D.J. (2003). ISO 17025 application within racing chemistry: A case study. *Technovation*, 23(9): 773–780.

Meiselman, H. (1994). Bridging the gap between sensory evaluation and market research. *Trends in Food Science & Technology*, 5(12): 396–398.

Munoz, A.M. (2002). Sensory evaluation in quality control: An overview, new developments and future opportunities. *Food Quality and Preference*, 13(6): 329–339.

Murray, J.M. and Baxter, I.A. (2003). Food acceptability and sensory evaluation, in *Encyclopedia of Food Sciences and Nutrition,* 2nd edn., Eds. B. Caballero, L. Trugo, and P. Finglas, Elsevier Science Ltd, Amsterdam, The Netherlands. pp. 5130–5136.

Murray, J.M., Delahunty, C.M. and Baxter I.A. (2001). Descriptive sensory analysis: Past, present and future. *Food Research International*, 34(6): 461–471.

Ntzimani, A.G., Paleologos, E.K., Savvaidis, I.N. and Kontominas, M.G. (2008). Formation of biogenic amines and relation to microbial flora and sensory changes in smoked turkey breast fillets stored under various packaging conditions at 4°C. *Food Microbiology*, 25(3): 509–517.

Olarte, C., Gonzalez-Fandos, E. and Sanz, S. (2001). A proposed methodology to determine the sensory quality of a fresh goat's cheese (Cameros cheese): Application to cheeses packaged under modified atmospheres. *Food Quality and Preference*, 12(3): 163–170.

Ozogul, F., Polat, A. and Ozogul, Y. (2004). The effects of modified atmosphere packaging and vacuum packaging on chemical, sensory and microbiological changes of sardines (*Sardina pilchardus*). *Food Chemistry*, 85(1): 49–57.

Ozogul, F., Taylor, K.D.A., Quantick, P. and Ozogul, Y. (2000). Chemical, microbiological and sensory evaluation of Atlantic herring (*Clupea harengus*) stored in ice, modified atmosphere and vacuum pack. *Food Chemistry*, 71: 267–273.

Pantazi, D., Papavergou, A., Pournis, N., Kontominasa, M.G. and Savvaidis, I.N. (2008). Shelf-life of chilled fresh Mediterranean swordfish (*Xiphias gladius*) stored under various packaging conditions: Microbiological, biochemical and sensory attributes. *Food Microbiology*, 25(1): 136–143.

Patsias, A., Chouliara, I., Badeka, A., Savvaidis, I.N. and Kontominas, M.G. (2006). Shelf-life of a chilled precooked chicken product stored in air and under modified atmospheres: Microbiological, chemical, sensory attributes. *Food Microbiology*, 23(5): 423–429.

Risvik, E., Poppert, R. and Rodgers, R. (1989). Expert systems and their application in sensory evaluation. *Food Quality and Preference*, 1(4–5): 183–184.

Rubio, B., Martinez, B., Gonzalez-Fernandez, C., Garcia-Cachan, M.D., Rovira, J. and Jaime, I. (2007). Effect of modified atmosphere packaging on the microbiological and sensory quality on a dry cured beef product: "Cecina de leon". *Meat Science*, 75(3): 515–522.

Santos, E.M., Diez, A.M., Gonzalez-Fernandez, C., Jaime, I. and Rovira, J. (2005). Microbiological and sensory changes in "Morcilla de Burgos" preserved in air, vacuum and modified atmosphere packaging. *Meat Science*, 71(2): 249–255.

Shama, F., Parkinson, C. and Sherman, P. (1973). Identification of stimuli controlling the sensory evaluation of viscosity. *Journal of Texture Studies*, 4(1): 102–110.

Shama, F. and Sherman, P. (1973). Identification of stimuli controlling the sensory evaluation of viscosity. *Journal of Texture Studies*, 4(1): 111–118.

Sidel, J.L. and Stone, H. (1993). The role of sensory evaluation in the food industry. *Food quality and Preference*, 4(1–2): 65–73.

Sinesio, F. (2005). Sensory evaluation. *Encyclopedia of Analytical Science*, Eds. P.J. Worsfold, A. Townshend, C.F. Poole, Academic Press, Cambridge, U.K. pp. 283–290.

Singer, D.C., Stefan, R.I. and Standen, J.F. (2005). *Laboratory Auditing for Quality and Regulatory Compliance*. Informa Healthcare, New York, p. 231.

Sioblom, C. and Keck, C. (2009). The ART laboratory in the era of ISO 1000 and GLP, in *Textbook of Assisted Reproductive Technologies*. Eds. D.K. Gardner, A. Weissman, C.M. Howles, and Z. Shoham. Informa Health Care, New York, p. 34.

Stone, H., Pangborn, R.M. and Ough, C.S. (1965). Techniques for sensory evaluation of food odors. *Advances in Food Research*, 14: 1–32.

Szczesniak, A.S. (2002). Texture is a sensory property. *Food Quality and Preference*, 13(4): 215–225.

Szczesniak, A.S. and Bourne, M.C. (1969). Sensory evaluation of food firmness. *Journal of Texture Studies*, 1(2): 52–64.

Thrane, C. (2008). Quality assurance in plant health diagnostics—The experience of the Danish Plant Directore, in *Sustainable Disease Management in a European Context.* Eds. D.B. Collinge, L. Munk, and B.M. Cooke. Copenhagen, Denmark, p. 341.

Torrieri, E., Cavella, S., Villani, F. and Masi, P. (2006). Influence of modified atmosphere packaging on the chilled shelf life of gutted farmed bass (*Dicentrarchus labrax*). *Journal of Food Engineering*, 77(4): 1078–1086.

Wilkinson, C., Dijksterhuis, G.B. and Minekus, M. (2000). From food structure to texture. *Trends in Food Science & Technology*, 11(12): 442–450.

York, R. and Vaisey-Genser, M. (2003). Sensory characteristics of human foods, in *Encyclopedia of Food Sciences and Nutrition,* 2nd edn., Eds. B. Caballero, L.Trugo, and P. Finglas, Elsevier Science Ltd, Amsterdam, The Netherlands, pp. 5125–5130.

Zeng, X., Ruan, D. and Koehl, L. (2008). Intelligent sensory evaluation: Concepts, implementations, and applications. *Mathematics and Computers in Simulation*, 77(5–6): 443–452.

https://answers.complianceonline.com/ecommerce/control/product/~product_id=809022

http://www.complianceonline.com/ecommerce/control/product/~product_id=808568

http://www.complianceonline.com/ecommerce/control/product?product_id=808772&channel=rss_organic

http://www.evs.ee/product/tabid/59/p-173919-iso-54962006.aspx

http://www.evs.ee/product/tabid/59/p-180617-iso-8586-11993.aspx

http://www.iso.org/iso/iso_cafe_sensory_analysis.htm

http://www.iso.org/iso/catalogu

http://www.iso.org/iso/catalogue_detail.htm?csnumber=15875

http://www.iso.org/iso/catalogue_detail.htm?csnumber=15878

http://www.iso.org/iso/catalogue_detail.htm?csnumber=19016

http://www.iso.org/iso/catalogue_detail.htm?csnumber=31085

http://www.iso.org/iso/catalogue_detail.htm?csnumber=31621

http://www.iso.org/iso/catalogue_detail.htm?csnumber=33817

http://www.iso.org/iso/catalogue_detail.htm?csnumber=36172

http://www.iso.org/iso/catalogue_detail.htm?csnumber=36385

http://www.iso.org/iso/catalogue_detail.htm?csnumber=36387

http://www.iso.org/iso/catalogue_detail.htm?csnumber=36791

http://www.iso.org/iso/catalogue_detail.htm?csnumber=37227

http://www.iso.org/iso/catalogue_detail?csnumber=38051

http://www.iso.org/iso/catalogue_detail.htm?csnumber=44247

http://www.iso.org/iso/iso_catalogue/catalogue_tc/catalogue_detail.htm?csnumber=12966

http://www.iso.org/iso/iso_catalogue/catalogue_ics/catalogue_detail_ics.htm?ics1=67&ics2=240&ics3=&csnumber=31085

http://www.iso.org/iso/iso_catalogue/catalogue_tc/catalogue_detail.htm?csnumber=31621

http://www.iso.org/iso/iso_catalogue/catalogue_ics/catalogue_detail_ics.htm?csnumber=31621&ICS1=67&ICS2=240

http://www.iso.org/iso/iso_catalogue/catalogue_ics/catalogue_detail_ics.htm?ics1=67&ics2=240&ics3=&csnumber=33818

http://www.iso.org/iso/iso_catalogue/catalogue_tc/catalogue_detail.htm?csnumber=36172

http://www.iso.org/iso/iso_catalogue/catalogue_tc/catalogue_detail.htm?csnumber=36226

http://www.iso.org/iso/iso_catalogue/catalogue_ics/catalogue_detail_ics.htm?ics1=67&ics2=240&ics3=&csnumber=36387

http://www.iso.org/iso/iso_catalogue/catalogue_ics/catalogue_detail_ics.htm?csnumber=37389

http://www.iso.org/iso/iso_catalogue/catalogue_ics/catalogue_detail_ics.htm?ics1=67&ics2=260&ics3=&csnumber=38031

http://www.iso.org/iso/iso_catalogue/catalogue_tc/catalogue_detail.htm?csnumber=33818&commid=47858

http://www.iso.org/iso/iso_catalogue/catalogue_ics/catalogue_detail_ics.htm?ics1=67&ics2=260&ics3=&csnumber=9002

http://www.iso.org/iso/pressrelease.htm?refid=Ref1176

http://www.nmkl.org/Publikasjoner/addendumprosedyre6.pdf

http://www.techstreet.com/standards/ISO/11035_1994?product_id=714633

http://www.techstreet.com/standards/ISO/11036_1994?product_id=714634

http://www2.nen.nl/nen/servlet/dispatcher.Dispatcher?id=BIBLIOGRAFISCHEGEGEVENS&contentID=225292

18 EU, U.S., and Canadian Legislation Related to Packaging Coming in Contact with Foods

Ioannis S. Arvanitoyannis and Persephoni Tserkezou

CONTENTS

18.1 INTRODUCTION

Food contact materials are all materials and articles intended to come into contact with foodstuffs. Many types of materials, such as polymers, regenerated cellulose, paper and board, glass and ceramics, elastomers (natural and synthetic rubbers), metals, wood, textile, waxes, etc., can be used for food packaging. (Simoneau, 2008).

The role of food packaging is to contain, maintain, protect, and provide convenience in the use of foodstuff (Robertson, 2006). A satisfactory packaging is designed to protect food from contamination by pathogens, foreign materials, and chemical substances (Kilcast, 1996; Robertson, 2006).

Food odor, flavor, texture, and characteristics are flexible. Packaging materials and foodstuffs may be transformed chemically and their characteristics could be altered. Reactions such as hydrolysis and oxidation can cause unpleasant odors and flavors and transform the texture and appearance of foodstuffs (Ayhan et al., 2001). These alterations may be created because of the packaging materials themselves, an interaction between packaging material and foodstuff, and poor packaging selection (Huber et al., 2002). Effects of materials on sensory characteristics and quality of foodstuffs can take place due to direct contact, as with a primary package intended for containment, or due to indirect means resulting from the environmental conditions (temperature, humidity, and presence of oxygen) as well as the characteristics of secondary packaging materials. Any material in contact with a foodstuff may have an effect (Duncan and Webster, 2009).

In the food industry, there is a possibility of interaction between the foodstuff and packaging material, which can cause alterations in the sensory characteristics of foodstuff. These interactions are related to a given chemical compound that adds a specific odor, flavor, and appearance. Thresholds for a given chemical compound differ according to the medium, food or drink, in which it is present, the temperature at which the product is maintained, the methodology used for the determination of the threshold, and individual sensitivity to the cause (Land, 1989; Kilcast, 1996; Meilgaard et al., 2007).

Novel packaging materials or modifications that improve product integrity, quality, and shelf life are important if product improvement is obvious by the consumer or end-user (Duncan and Webster, 2009). Over the last decades, one of the novel improvements in the field of food packaging is the "active and intelligent" packaging related to deliberate interactions with the food or its environment. Active packaging aims to enlarge the food shelf life, maintain its characteristics and properties, and develop its quality, while intelligent packaging aims to control the food freshness (Dainelli et al., 2008).

Active packages can be designed such that they contain compounds, like antioxidants, which can migrate into the food and improve the sensory quality by decreasing oxidation and increasing the shelf life and nutrition characteristics. Additives are usually used in polymer processing to reduce oxidative reactions (Robertson, 2006). Examples of active packaging systems include oxygen and ethylene scavengers, carbon dioxide scavengers and emitters, preservative releasers, ethanol emitters, moisture absorbers, and flavor/odor adsorbers. Novel technologies for these systems may aim to affect sensory characteristics of foodstuffs through microbial growth control, oxidation reactions, and moisture migration (Duncan and Webster, 2009). Intelligent packaging systems attached as labels, incorporated into, or printed onto a food packaging material offer improved possibilities to control product quality, mark the critical points, and give more detailed information all over the chain supply (Han et al., 2005).

As an example, Saint-Eve et al. (2008) indicated that the packaging material affected the sensory characteristics, specifically odor, of flavored stirred yogurts with 0% or 4% fat content. A trained sensory panel evaluated yogurt packaged in glass, polystyrene, or polypropylene over a 28 day refrigerated storage period. Evaluations, using an unstructured line scale with "weak" and "very intense" as the anchors, reflected 10 odor, 15 odor, 2 taste, and 3 texture-in-mouth characteristics. While odor intensity and profile changed for yogurts packaged in all materials over time, glass provided the best protection for aroma and flavor intensity as it had the best barrier properties.

Over the last years, several food safety crises occurred within the framework of the European Union (EU). Some of these crises resulted in great losses of human lives and money, whereas others simply made EU citizens more alert in terms of their everyday nutrition. As a result, the EU Commission tried to enhance the food safety level by either introducing new stricter directives/regulations or modifying the already existing ones (Arvanitoyannis et al., 2005). Based on deliberate interactions with the food or its environment, the main hazard is the migration of substances from the packaging to food. The EU laid down a number of directives and regulations with regard to materials and articles intended to come into contact with foodstuffs. The European Commission has laid down a new Regulation (EC) No 1935/2004, and this change will be soon achieved with a specific regulation entirely devoted to active and intelligent food contact materials.

The U.S. government was the first to introduce the meaning of food safety, which had a tremendous impact on everybody's life starting from the food and packaging companies up to consumers themselves. The rest of the nations simply followed the U.S. approach with a considerable delay, both in terms of legislation and implementation (Arvanitoyannis et al., 2006). In contrast to the United States, the saturation of active and intelligent packaging in the European market is limited. This interval compared to the United States and Canada was mostly attributed to an insufficient and inadequate flexible European legislation that could not keep up with technological improvements in the food packaging issues (Restuccia et al., 2010).

This chapter aims at providing an update of the current EU Regulations and Directives on issues related to food packaging materials. Initially, a brief presentation of EU, U.S., and Canadian legislation in terms of structure (horizontal, vertical) was attempted.

18.2 EU LEGISLATION RELATED TO FOOD PACKAGING

Directive 78/142/EEC concerns the presence of vinyl chloride monomer in, and possible migration from, materials and articles prepared with vinyl chloride polymers or copolymers, hereinafter called "materials and articles," which in their finished state are intended to come into contact with foodstuffs and are intended for that purpose. The level of vinyl chloride in materials and articles and the vinyl chloride released by materials and articles to foodstuffs are determined by means of gas-phase chromatography using the headspace method.

According to Directive 82/711/EEC, "plastics" shall mean the organic macromolecular compounds obtained by polymerization, polycondensation, polyaddition, or any other similar process from molecules with a lower molecular weight or by chemical alteration of natural macromolecules. Silicones and other similar macromolecular compounds shall also be regarded as plastics. Other substances or matter may be added to such macromolecular compounds. However, the following shall not be regarded as "plastics": (i) varnished or unvarnished regenerated cellulose film, (ii) elastomers and natural and synthetic rubber, (iii) paper and paperboard, whether modified or not by the addition of plastics, and (iv) surface coatings obtained from paraffin waxes, including synthetic paraffin waxes, and microcrystalline waxes and mixtures of the waxes listed in the first indent with each other and with plastics. This directive shall not apply to materials and articles composed of two or more layers, one or more of which do not consist exclusively of plastics, even if the one intended to come into direct contact with foodstuffs does consist exclusively of plastics. A decision on the application of this directive to the materials and articles referred to in the first subparagraph and on any adaptations to the directive that may become necessary shall be taken at a later date.

Directive 84/500/EEC concerns the possible migration of lead and cadmium from ceramic articles which, in their finished state, are intended to come into contact with foodstuffs, or which are in contact with foodstuffs, and are intended for that purpose. Where a ceramic article does not exceed the quantities given in Table 18.1 by more than 50%, that article shall nevertheless be recognized as satisfying the requirements of this directive if at least three other articles with the same shape, dimensions, decoration, and glaze are subjected to a test carried out under the conditions laid down in Annexes I and II, and the average quantities of lead and/or cadmium extracted from those articles do not exceed the limits set, with none of those articles exceeding those limits by more than 50%.

Directive 93/11/EEC concerns the release of N-nitrosamines and of substances capable of being converted into N-nitrosamines, hereinafter called "N-nitrosatable substances," from teats and soothers, made of elastomer or rubber. In Annex I of Directive 93/11/EEC, entitled basic rules for determining the release of N-nitrosamines and N-nitrosatable substances, release-test liquid (saliva test solution) is to obtain the release-test liquid, dissolve 4.2 g of sodium bicarbonate ($NaHCO_3$), 0.5 g of sodium chloride (NaCl), 0.2 g of potassium carbonate (K_2CO_3), and 30 mg of sodium nitrite ($NaNO_2$) in 1 L of distilled water or water of equivalent quality. The solution must have a pH value of 9. Test conditions where samples of material obtained from an appropriate number of teats or soothers are immersed in the release-test liquid for 24 h at a temperature of 40°C ± 2°C.

Directive 2002/72/EC shall apply to plastic materials and articles and parts thereof: (a) consisting exclusively of plastics or (b) composed of two or more layers of materials, each consisting exclusively of plastics, which are bound together by means of adhesives or by any other means, which, in the finished product state, are intended to come into contact or are brought into contact with foodstuffs intended for that purpose. For the purposes of this directive, "plastics" shall mean the organic macromolecular compounds obtained by polymerization, polycondensation, polyaddition, or any other similar process from molecules with a lower molecular weight or by chemical alteration of natural macromolecules. Other substances or matter may be added to such macromolecular compounds. However, the following shall not be regarded as "plastics": (a) varnished or unvarnished regenerated cellulose film, (b) elastomers and natural and synthetic rubber, (c) paper and paperboard, whether modified or not by the addition of plastics, (d) surface coatings obtained from (i) paraffin waxes, including synthetic paraffin waxes, and microcrystalline waxes, and (ii) mixtures of the waxes listed in the first indent with each other and with plastics, (e) ion-exchange resins, and (f) silicones. This directive shall not apply, until further action by the Commission, to materials and articles composed of two or more layers, one or more of which does not consist exclusively of plastics, even if the one intended to come into direct contact with foodstuffs does consist exclusively of plastics.

Annex I of Directive 2007/42/EC defined that regenerated cellulose film is a thin sheet material obtained from a refined cellulose derived from unrecycled wood or cotton. To meet technical requirements, suitable substances may be added either in the mass or on the surface. Regenerated cellulose film may be coated on one or both

TABLE 18.1
European Legislation Related to Food Packaging

Directive/ Regulation	Title	Main Points
Directive 78/142/ EEC (entry into force 01/02/1978)	On the approximation of the laws of the Member States relating to materials and articles that contain vinyl chloride monomer and are intended to come into contact with foodstuffs	Maximum vinyl chloride monomer level in materials and articles: 1 mg/kg in the final product
Directive 82/711/ EEC (entry into force 04/11/1982)	Laying down the basic rules necessary for testing migration of the constituents of plastic materials and articles intended to come into contact with foodstuffs	This directive shall apply to plastic materials and articles, that is to say to materials and articles and parts thereof: (a) consisting exclusively of plastics or (b) composed of two or more layers of materials, each consisting exclusively of plastics, which are bound together by means of adhesives or by any other means, which, in the finished product state, are intended to come into contact or are brought into contact with foodstuffs and are intended for that purpose
Directive 84/500/ EEC (entry into force 17/10/1984)	On the approximation of the laws of the Member States relating to ceramic articles intended to come into contact with foodstuffs	*Category 1*: Articles that cannot be filled and articles that can be filled, the internal depth of which, measured from the lowest point to the horizontal plane passing through the upper rim, does not exceed 25 mm (Pb: 0.8 mg/dm^2, Cd: 0.07 mg/dm^2) *Category 2*: All other articles that can be filled (Pb: 4.0 mg/L, Cd: 0.3 mg/L) *Category 3*: Cooking ware; packaging and storage vessels having a capacity of more than 3 L (Pb: 1.5 mg/L, Cd: 0.1 mg/L)
Directive 93/11/ EEC (entry into force 01/04/1994)	Concerning the release of the N-nitrosamines and N-nitrosatable substances from elastomer or rubber teats and soothers	0.01 mg in total of N-nitrosamines released/kg (of the parts of teat or soother made of elastomer or rubber) 0.1 mg in total of N-nitrosatable substances/kg (of the parts of teat or soother made of elastomer or rubber)
Directive 2002/72/ EC (entry into force 04/09/2002)	Relating to plastic materials and articles intended to come into contact with foodstuffs	Further provisions applicable when checking compliance with the migration limits—list of monomers and other starting substances that may be used in the manufacture of plastic materials and articles

(*continued*)

TABLE 18.1 (continued)
European Legislation Related to Food Packaging

Directive/ Regulation	Title	Main Points
Directive 2007/42/ EC (entry into force 20/07/2007)	Relating to materials and articles made of regenerated cellulose film intended to come into contact with foodstuffs	This directive shall apply to regenerated cellulose film within the meaning of the description given in Annex I that is intended to come into contact with foodstuffs or which, by virtue of its purpose, does come into such contact and which either (a) constitutes a finished product in itself or (b) forms part of a finished product containing other materials. This directive shall not apply to synthetic casings of regenerated cellulose
Regulation (EC) No 1935/2004 (entry into force 03/12/2004)	On materials and articles intended to come into contact with food and repealing directives 80/590/ EEC and 89/109/EEC	The purpose of this regulation is to ensure the effective functioning of the internal market in relation to the placing on the market in the community of materials and articles intended to come into contact directly or indirectly with food, whilst providing the basis for securing a high level of protection of human health and the interests of consumers
Regulation (EC) No 1895/2005 (entry into force 01/01/2006)	On the restriction of use of certain epoxy derivatives in materials and articles intended to come into contact with food	This regulation shall apply to materials and articles, including active and intelligent food contact materials and articles that are manufactured with or contain one or more of the following substances: (a) 2,2-bis (4-hydroxyphenyl) propane bis (2,3-epoxypropyl) ether, hereinafter referred to as "BADGE" (CAS No 001675-54-3), and some of its derivatives; (b) bis(hydroxyphenyl)methane bis(2,3-epoxypropyl)ethers, hereinafter referred to as "BFDGE" (CAS No 039817-09-9); (c) other novolac glycidyl ethers, hereinafter referred to as "NOGE"
Regulation (EC) No 2023/2006 (entry into force 18/01/2007)	On good manufacturing practice for materials and articles intended to come into contact with food	Detailed rules on good manufacturing practice—processes involving the application of printing inks to the nonfood contact side of a material or article
Regulation (EC) No 282/2008 (entry into force 17/04/2008)	On recycled plastic materials and articles intended to come into contact with foods and amending Regulation (EC) No 2023/2006	This regulation shall apply to the plastic materials and articles and parts thereof intended to come into contact with foodstuffs that contain recycled plastic (hereafter "recycled plastic materials and articles")

TABLE 18.1 (continued)
European Legislation Related to Food Packaging

Directive/ Regulation	Title	Main Points
Regulation (EC) No 450/2009 (entry into force 19/06/2009)	On active and intelligent materials and articles intended to come into contact with food	This regulation establishes specific requirements for the marketing of active and intelligent materials and articles intended to come into contact with food. These specific requirements are without prejudice to community or national provisions applicable to the materials and articles to which active or intelligent components are added or into which they are incorporated

sides. According to Article 2, the regenerated cellulose films shall belong to one of the following types: (a) uncoated regenerated cellulose film, (b) coated regenerated cellulose film with coating derived from cellulose, or (c) coated regenerated cellulose film with coating consisting of plastics. Regenerated cellulose films referred to in Article 2 shall be manufactured using only substances or groups of substances listed in Annex II of this directive subject to the restrictions set out therein.

Regulation (EC) No 1935/2004 shall apply to materials and articles, including active and intelligent food contact materials and articles (hereinafter referred to as materials and articles), which in their finished state (a) are intended to be brought into contact with food, or (b) are already in contact with food and were intended for that purpose, or (c) can reasonably be expected to be brought into contact with food or to transfer their constituents to food under normal or foreseeable conditions of use. This regulation shall not apply to (a) materials and articles that are supplied as antiques; (b) covering or coating materials, such as the materials covering cheese rinds, prepared meat products, or fruits, which form part of the food and may be consumed together with this food; or (c) fixed public or private water supply equipment. Materials and articles, including active and intelligent materials and articles, shall be manufactured in compliance with good manufacturing practice so that, under normal or foreseeable conditions of use, they do not transfer their constituents to food in quantities that could (a) endanger human health; (b) bring about an unacceptable change in the composition of the food; or (c) bring about a deterioration in the organoleptic characteristics thereof. The labeling, advertising, and presentation of a material or article shall not mislead the consumers.

Regulation (EC) No 1895/2005 shall not apply to containers or storage tanks with a capacity greater than 10,000 L or to pipelines belonging to or connected with them, covered by special coatings called "heavy-duty coatings." Materials and articles shall not release the substances listed in Annex I in a quantity exceeding the limits laid down in that Annex. The use and presence of BFDGE in the manufacture of materials and articles are prohibited. The use and/or presence of NOGE in the manufacture of materials and articles are prohibited.

Regulation (EC) No 2023/2006 shall apply to all sectors and to all stages of manufacture, processing, and distribution of materials and articles, up to but excluding the production of starting substances. The detailed rules set out in the Annex shall apply to the relevant individually mentioned processes, as appropriate. Printing inks applied to the nonfood-contact side of materials and articles shall be formulated and/or applied in such a manner that substances from the printed surface are not transferred to the food-contact side (a) through the substrate or (b) by set-off in the stack or the reel, in concentrations that lead to levels of the substance in the food that are not in line with the requirements of Article 3 of Regulation (EC) No 1935/2004. Printed materials and articles shall be handled and stored in their finished and semifinished states in such a manner that substances from the printed surface are not transferred to the food-contact side (a) through the substrate or (b) by set-off in the stack or reel, in concentrations that lead to levels of the substance in the food, which are not in line with the requirements of Article 3 of Regulation (EC) No 1935/2004. The printed surfaces shall not come into direct contact with food.

According to Regulation (EC) No 282/2008, the technical dossier shall contain the information specified in the guidelines for the safety assessment of a recycling process to be published by the authority at the latest 6 months after the date of publication of this regulation. The authority shall give an opinion within 6 months of receipt of a valid application as to whether or not the recycling process complies with the conditions laid down in Article 4. In the event of an opinion in favor of authorization of the evaluated recycling process, the opinion of the authority shall include the following: (a) a short description of the recycling process; (b) where appropriate, any recommendations on conditions or restrictions concerning the plastic input; (c) where appropriate, any recommendations on conditions or restrictions concerning the recycling process; (d) where appropriate, any criteria to characterize the recycled plastic; (e) where appropriate, any recommendations concerning conditions in the field of application of the recycled plastic; and (f) where appropriate, any recommendations concerning monitoring compliance of the recycling process with the conditions of the authorization.

Regulation (EC) No 450/2009 shall apply to active and intelligent materials and articles that are placed on the market within the community. For the purpose of this regulation, the following definitions shall apply: "active materials and articles" means materials and articles that are intended to extend the shelf life or to maintain or improve the condition of packaged food; they are designed to deliberately incorporate components that would release or absorb substances into or from the packaged food or the environment surrounding the food. "Intelligent materials and articles" means materials and articles that monitor the condition of packaged food or the environment surrounding the food. The community list shall specify (a) the identity of the substance(s); (b) the function of the substance(s); (c) the reference number; (d) if necessary, the conditions of use of the substance(s) or component; (e) if necessary, restrictions and specifications of use of the substance(s); and (f) if necessary conditions of use of the material or article to which the substance or component is added or into which it is incorporated. A summary of EU legislation for food packaging is given in Table 18.1.

18.3 U.S. LEGISLATION IN REGARD TO FOOD PACKAGING

The U.S. Food and Drug Administration produced the Federal Food, Drug and Cosmetic Act, which is based on the publication compilation of selected acts within the jurisdiction of the Committee on Energy and Commerce; Food, Drug, and Related Law, as amended through December 31, 2004.

According to Tolerances for Poisonous Ingredients in Food (Sec. 406 [21 USC §346]), any poisonous or deleterious substance added to any food, except where such substance is required in the production thereof or cannot be avoided by good manufacturing practice, shall be deemed to be unsafe for purposes of the application of clause (2)(A) of section 402(a); but when such substance is so required or cannot be so avoided, the secretary shall promulgate regulations limiting the quantity therein or thereon to such extent as he finds necessary for the protection of public health, and any quantity exceeding the limits so fixed shall also be deemed to be unsafe for purposes of the application of clause (2)(A) of section 402(a). While such a regulation is in effect limiting the quantity of any such substance in the case of any food, such food shall not, by reason of bearing or containing any added amount of such substance, be considered to be adulterated within the meaning of clause (1) of section 402(a). In determining the quantity of such added substance to be tolerated in or on different articles of food, the secretary shall take into account the extent to which the use of such substance is required or cannot be avoided in the production of each such article, and the other ways in which the consumer may be affected by the same or other poisonous or deleterious substances.

18.4 CANADIAN LEGISLATION FOR FOOD PACKAGING MATERIALS

According to Food and Drug Regulation from the Canadian Department of Justice, Division 23 is associated with food packaging materials. This division defines that no person shall sell any food in a package that may yield to its contents any substance that may be injurious to the health of a consumer of the food. No person shall sell any food in a package that has been manufactured from a polyvinyl chloride formulation containing an octyltin chemical. A person may sell food, other than milk, skim milk, partly skimmed milk, sterilized milk, malt beverages, and carbonated nonalcoholic beverage products, in a package that has been manufactured from a polyvinyl chloride formulation containing any or all of the octyltin chemicals, namely, di(n-octyl) tin S,S′-bis(isooctylmercaptoacetate), di(n-octyl)tin maleate polymer, and (n-octyl) tin S,S′,S″-tris(isooctylmercaptoacetate) if the proportion of such chemicals, either singly or in combination, does not exceed a total of 3% of the resin, and the food in contact with the package contains not more than one part per million total octyltin. Di(n-octyl)tin S,S′-bis(isooctylmercaptoacetate) shall be the octyltin chemical made from di(n-octyl)tin dichloride and shall contain 15.1%–16.4% of tin and 8.1%–8.9% of mercapto sulfur. For the purposes of this Division, di(n-octyl)tin dichloride shall be the chemical having an organotin composition of not less than 95% di(n-octyl) tin dichloride and shall contain no more than (a) 5% total of n-octyltin trichloride or tri(n-octyl)tin chloride or both, (b) 0.2% total of other eight (8) carbon isomeric alkyltin derivatives, and (c) 0.1% total of the higher and lower homologous alkyltin derivatives. Di(n-octyl)tin maleate polymer shall be the octyltin chemical made from di(n-octyl)tin dichloride and shall have the formula $((C_8H_{17})_2\ SnC_4H_2O_4)_n$ (where n is between 2 and 4 inclusive), and a saponification number of 225–255, and shall contain 25.2%–26.6% of tin.

REFERENCES

Arvanitoyannis, I.S., Choreftaki, S. and Tserkezou, P. (2005). An update of EU legislation (Directives and Regulations) on food-related issues (safety, hygiene, packaging, technology, GMOs, additives, radiation, labelling): Presentation and comments. *International Journal of Food Science and Technology*, **40**(10), 1021–1112.

Arvanitoyannis, I.S., Tserkezou, P. and Varzakas, T. (2006). An update of US food safety, food technology, GM food and water protection and management legislation. *International Journal of Food Science and Technology*, **41**(1), 130–159.

Ayhan, Z., Yeom, H.W., Zhang, Q.H. and Min, D.B. (2001). Flavor, color, and vitamin C retention of pulsed electric field processed orange juice in different packaging materials. *Journal of Agricultural and Food Chemistry*, **49**, 669–674.

Dainelli, D., Gontard, N., Spyropoulosm, D., Zondervan-van den Beuken, E. and Tobback, P. (2008). Active and intelligent food packaging: Legal aspects and safety concerns. *Trends in Food Science and Technology*, **19**, S103–S112.

Department of Justice Canada. Food and Drug Regulations, Division 23: Food Packaging materials. http://laws-lois.justice.gc.ca/eng/regulations/C.R.C.,_c._870/index.html (last access at 09/07/2011).

Directive 78/142/EEC on the approximation of the laws of the member states relating to materials and articles which contain vinyl chloride monomer and are intended to come into contact with foodstuffs. http://eur-lex.europa.eu/LexUriServ/LexUriServ.do?uri=OJ:L:1978:044:0015:0017:EN:PDF (last access at 22/06/2011).

Directive 82/711/EEC laying down the basic rules necessary for testing migration of the constituents of plastic materials and articles intended to come into contact with foodstuffs. http://eur-lex.europa.eu/LexUriServ/LexUriServ.do?uri=CELEX:31982L0711:EN:HTML (last access at 22/06/2011).

Directive 84/500/EEC on the approximation of the laws of the member states relating to ceramic articles intended to come into contact with foodstuffs. http://eur-lex.europa.eu/LexUriServ/LexUriServ.do?uri=CONSLEG:1984L0500:20050520:EN:PDF (last access at 22/06/2011).

Directive 2002/72/EC relating to plastic materials and articles intended to come into contact with foodstuffs. http://eur-lex.europa.eu/LexUriServ/LexUriServ.do?uri=OJ:L:2002:220:0018:0018:EN:PDF (last access at 22/07/2011).

Directive 2007/42/EC relating to materials and articles made of regenerated cellulose film intended to come into contact with foodstuffs. http://eur-lex.europa.eu/LexUriServ/LexUriServ.do?uri=OJ:L:2007:172:0071:0082:EN:PDF (last access at 22/06/2011).

Duncan, S.E. and Webster, J.B. (2009). Sensory impacts of food-packaging interactions (Chapter 2). In: *Advances in Food and Nutrition Research* (S.L. Taylor, ed.), Elsevier, Virginia Polytechnic Institute and State University (Virginia Tech), Blacksburg, VA, pp. 17–64.

Han, J.H., Ho, C.H.L. and Rodrigues, E.T. (2005). Intelligent packaging. In: *Innovations in Food Packaging* (J.H. Han, ed.), Elsevier Academic Press, Oxford, U.K.

Huber, M., Ruiz, J. and Chastellain, F. (2002). Off-flavour release from packaging materials and its prevention: A foods company's approach. *Food Additives & Contaminants*, **19**, 221–228.

Kilcast, D. (1996). Organoleptic assessment (Chapter 4). In: *Migration from Food Contact Materials* (L.L. Katan, ed.), Blackie Academic and Professional, New York, pp. 51–76.

Land, D.G. (1989). Taints—Causes and prevention (Chapter 2). In: *Distilled Beverage Flavour—Recent Developments* (J.R. Piggott and A. Paterson, eds.), Ellis Horwood, Chichester, West Sussex, U.K.

Meilgaard, M.C., Civille, G.V. and Carr, B.T. (2007). *Sensory Evaluation Techniques*, CRC Press, Boca Raton, FL, p. 448.

Regulation (EC) No 1935/2004 on materials and articles intended to come into contact with food and repealing Directives 80/590/EEC and 89/109/EEC. http://eur-lex.europa.eu/LexUriServ/LexUriServ.do?uri=OJ:L:2004:338:0004:0017:en:PDF (last access at 22/06/2011).

Regulation (EC) No 1895/2005 on the restriction of use of certain epoxy derivatives in materials and articles intended to come into contact with food. http://eur-lex.europa.eu/LexUriServ/LexUriServ.do?uri=OJ:L:2005:302:0028:0028:EN:PDF (last access at 22/06/2011).

Regulation (EC) No 2023/2006 on good manufacturing practice for materials and articles intended to come into contact with food. http://eur-lex.europa.eu/LexUriServ/LexUriServ.do?uri=OJ:L:2006:384:0075:0075:EN:PDF (last access at 22/06/2011).

Regulation (EC) No 282/2008 on recycled plastic materials and articles intended to come into contact with foods and amending Regulation (EC) No 2023/2006. http://eur-lex.europa.eu/LexUriServ/LexUriServ.do?uri=OJ:L:2008:086:0009:0018:EN:PDF (last access at 22/06/2011).

Regulation (EC) No 450/2009 on active and intelligent materials and articles intended to come into contact with food. http://eur-lex.europa.eu/LexUriServ/LexUriServ.do?uri=OJ:L:2009:135:0003:0011:EN:PDF (last access at 22/06/2011).

Restuccia, D., Gianfranco Spizzirri, U., Parisi, O.I., Cirillo, G., Curcio, M., Iemma, F., Puoci, F., Vinci, G. and Picci, N. (2010). New EU regulation aspects and global market of active and intelligent packaging for food industry applications. *Food Control*, **21**, 1425–1435.

Robertson, G.L. (2006). *Food Packaging Principles and Practice*, 2nd edn., CRC Press, Boca Raton, FL, p. 550.

Saint-Eve, A., Levy, C., le Moigne, M., Ducruet, V. and Souchon, I. (2008). Quality changes in yogurt during storage in different packaging materials. *Food Chemistry*, **110**, 285–293.

Simoneau, C. (2008). Food contact materials (Chapter 21). In: *Comprehensive Analytical Chemistry* (D. Barcelo, ed.), Elsevier, Amsterdam, the Netherlands, pp. 733–734.

US Food and Drug Administration. (2004). Federal Food, Drug, and Cosmetic Act, Chapter IV: Food, Sec. 406. [21 USC §346] Tolerances for Poisonous Ingredients in Food. http://www.fda.gov/regulatoryinformation/legislation/federalfooddrugandcosmeticactfdcact/default.htm (last access at 09/07/2011).

19 Conclusions and New Trends

Ioannis S. Arvanitoyannis

CONTENTS

19.1 CURRENT STATE OF THE ART AND STATISTICS OF APPLICATIONS BOTH IN TERMS OF PRODUCTION AND RESEARCH ARTICLES

Packaging helps to protect products against deteriorative effects, contain the product, communicate to the consumer, and provide consumers with ease of use and convenience (Yam et al., 2005). Shelf life is an important characteristic of all food products and related to its packaging; both product conditions and the package as well as gas mixture composition should be considered (Singh et al., 2011). The key to effective packaging is the selection of proper materials and designs that best respond to the needs of product's characteristics, marketing considerations, environmental and waste management issues, and cost (Marsh and Bugusu, 2007).

19.1.1 Consumers and Packaging

There is no doubt that consumers demand foods with greater convenience. There are numerous examples in the food industry that have addressed the consumer needs for convenience (Eilert, 2005). One typical example is the soup that does not require a can opener nor does it require a spoon, which in the past was a necessary requirement. Another example is the recent introduction into the market of a can within a can. Between layers of this package is crushed limestone. When the consumer wishes to heat the beverage, a button is pushed which releases water into the limestone and causes a thermal reaction to take place, resulting in the can heating. Nowadays, new food packaging concepts have been developed to meet the current consumption trends toward mildly preserved and fresh convenient food products. Packaging fresh-cut vegetables under an equilibrium modified atmosphere (MA) is one of the new applied food packaging technologies offering a prolonged shelf life to respiring products by suppression of their respiration rate (Jacxsens et al., 2002). By matching film permeability for O_2 and CO_2 to the respiration rate of the packaged fresh-cut produce, an equilibrium MA can be made to occur inside the package consisting of a decreased O_2 concentration and increased CO_2 concentration (Day, 1996).

Possible future developments in packaging may include self-venting microwave packs in which a vent opens at a predetermined temperature and closes on cooling, respiring trays that respond to changing atmospheric conditions to optimize respiration. Another possibility are materials that produce electrical pulses during movement, with which their electrical conductivity may change when exposed to light or their opacity may change or potentially can obtain softer or harder texture when subjected to small electrical charges (Fellows, 2000).

19.1.2 Environmental Impact and Food Packaging

Another important aspect of packaging is its environmental impact. The waste has initiated social, political, and research interest in several options, including source reduction (thinner gauges of materials, alternate materials, or reusable containers), recycling, composting, combustion or incineration, and land filling (Marsh and Bugusu, 2007). Nowadays, consumers demand environmentally friendly materials. This fact in combination with the increased price of petroleum-based packaging makes the use of biodegradable materials a promising alternative. However, this transition from petroleum-based materials to biomaterials is hindered, in many cases, by their limited functionality (Arvanitoyannis, 1999).

19.1.3 Food Packaging Production Trends

The global market for food and beverages of active and intelligent coupled with controlled/modified atmosphere packaging (CAP/MAP) increased from $15.5 billion in 2005 to $16.9 billion by the end of 2008 and is anticipated to reach $23.6 billion by 2013 with a compound annual growth rate of 6.9%. The global market is divided into different applications of active, controlled, and intelligent packaging, with CAP/MAP having the largest share of the market estimated to comprise 45.4% in 2008, probably decreasing slightly to approximately 40.5% in 2013 (Restuccia et al., 2010).

Over the past decade, active and intelligent packaging have experienced significant growth and change as new products and technologies have challenged the status quo of the traditional forms of food and beverage packaging (Kotler and Keller, 2006). Firstly introduced in the market of Japan in the mid 1970s, active and intelligent packaging materials and articles, raised the attention of the industry in Europe and in the United States only in the mid 1990s. Although active packaging is used more extensively in Japan, its applications in Europe and North America are starting to increase (Restuccia et al., 2010). Low diffusion in EU countries of active and intelligent packaging is due to cost acceptance (Dainelli et al., 2008). Considering the costs involved, it is anticipated that they will drastically decrease with wider application and thus scaling up of production. Discussions are ongoing as to whether consumers will be ready to incur greater costs for the higher safety/better quality tools (Lahteenmaki and Arvola, 2003).

The importance and technological potential of MAP and vacuum are depicted in Figure 19.1, which demonstrates the number of relevant scientific publications. Figure 19.2 shows the scientific articles published regarding active packaging and

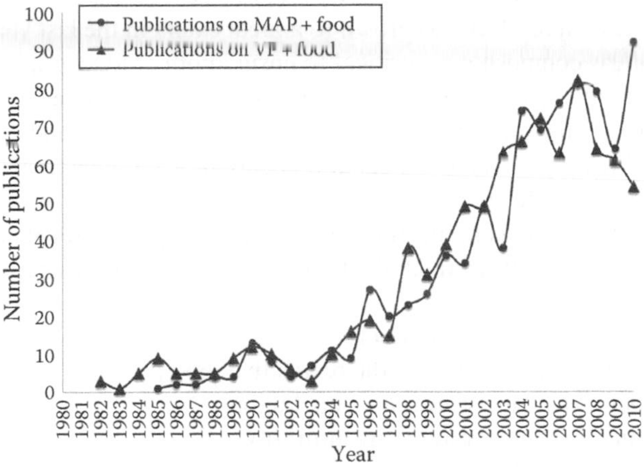

FIGURE 19.1 The number of scientific publications with regard to MAP and vacuum packaging (www.scopus.com).

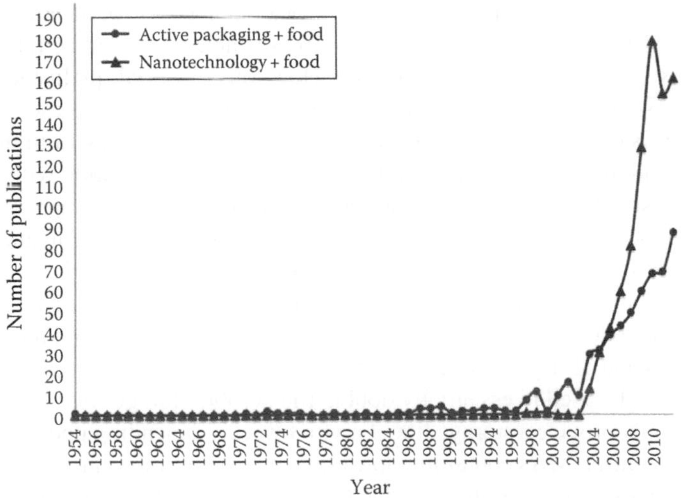

FIGURE 19.2 The number of scientific publications with regard to active packaging and nanotechnology (www.scopus.com).

nanotechnology. Active packaging publications display significant increase in their numbers, after 1999, while according to the number of publications, nanotechnology is anticipated to play potentially a crucial role in food packaging technology in the nearest future.

19.2 ADVANCES IN MAP/CA/UNDER VACUUM

Packaging under MAs is a technology that has the potential to suppress microbial growth and extend the shelf life of foods. However, there are additional advantages of MA processing such as minimization of the use of chemical preservatives and integrated control of both microbial growth and insect infestation (Bhat and Bhat, 2011). The application of MAP has grown to a great extent in the last years; however, optimization of gas composition is critical to ensure both product quality and safety. MAP packaged foods have become increasingly more available as food producers are trying to meet market demands for minimally processed food products with extended shelf lifes. MAP is a practical and economic technique, realized by small technical instruments (Del Nobile et al., 2009).

In general, the following are the most important prerequisites for successful MAP technologies are (Washuettl et al., 2001; Lisinska-Kusnierz and Ucherek, 2003):

1. High initial quality of the product
2. Use of mixtures appropriate to the respective food product
3. Use of appropriate and well-sealed packaging
4. Proper control of temperature and humidity
5. Good hygienic conditions during processing and packaging steps

Del Nobile et al. (2009) studied the possibility of improving the quality of blue fish burgers using thymol (110 ppm), grapefruit seed extract (100 ppm), and lemon extract (120 ppm) in combination with MAP at 4°C. The combined use of the natural preservatives and MAP with a high CO_2 concentration is able to guarantee the microbial acceptability of samples until the 28th day. However, sensorial quality loss limited the samples' shelf life to 22–23 days. In comparison to other mild preservation procedures such as low-dose irradiation, addition of protective cultures, or high-pressure treatments, and the use of natural compounds applied in combination with MAP is a low-cost and simple way to prolong the shelf life of seafood. Rajkumar et al. (2007) investigated the effect of MAP (80% O_2/20% CO_2) packaging, vacuum, and air packaging on turkey meat stored at 4°C. The microbial load was lower in samples stored under MAP compared with the other treatments. However, vacuum-packaged samples had higher odor scores. Turkey meat stored under MAP and vacuum resulted in shelf life of 14 and 21 days, respectively. Although MAs containing high oxygen concentrations can be used to essentially double the aerobic storage life, this packaging system will only provide a storage life of up to 3 weeks, depending upon hygiene and temperature control. MAs containing low oxygen concentrations are not suitable for use with red meats (beef, lamb, etc.), due to metmyoglobin formation at low-oxygen concentrations, and they provide highly variable storage lives with pork and poultry, due to the crudeness of establishment of the package atmosphere. Vacuum packaging is suitable for use with low pH, boneless, beef primal, but is not well suited for packaging irregularly shaped, bone-in or small cuts, or other types of meat. A storage life with vacuum-packaged beef primal of up to 100 days can be expected, and with poultry, depending upon hygiene and temperature control (Singh et al., 2011). Vacuum packaging was considered to be an interesting alternative in the packaging of pork, due to the fact that it increases oxidative stability, which brings about greater color stability and therefore increases the shelf life (Cayuela et al., 2004). MAP does not increase significantly the shelf life of all food products. In some types of food, their shelf life has already been prolonged because of the other treatments, such as salting, smoking, etc. In this case, MAP technologies can be effectively applied to preserve other quality characteristics such as color (Phillips, 1996). The main conclusion is that to optimize shelf life, sensory quality, and microbiological safety, the packaging system applied is product specific. A substantial number of experiments are available, but the presence of contradictions and differences in experimental methodologies often does not promote valid comparisons.

19.3 ADVANCES IN ACTIVE PACKAGING

Active packaging is considered a modern development, which includes a group of techniques in which the package is actively involved with food or works in conjunction with the internal atmosphere to prolong the shelf life and at the same time maintaining safety and quality (Rathore et al., 2010). According to Cooksey (2001), active packaging systems, aimed at quality improvement and shelf-life extension of foods, can be categorized by three concepts: firstly by direct incorporation of active substances into the packaging film; secondly by edible films and coating with bioactive substances; and thirdly by incorporation of the active substances into a sachet, patch, or tablet. Recently, several reviews have been published regarding active packaging of foodstuffs (Joerger, 2007; Coma, 2008). Edible films and surface coatings with bioactive substances are likely to be used to enhance preservation and to add value to products in the future.

There are some very recent published investigations on several aspects of active packaging. Mastromatteo et al. (2010) studied the effectiveness of thymol active coating in peeled shrimp samples and found that there is a slight antimicrobial effect when the coating used was loaded with thymol. Moreover, concentration dependence was also observed. The active coating was also effective in minimizing the sensory quality loss of samples, mainly at the lowest thymol concentration. Mexis et al. (2011b) investigated the effect of active packaging (under an O_2 absorber) in combination with irradiation (1 and 3 kGy) and storage conditions on quality retention of raw, whole, unpeeled almonds. Samples were stored either under fluorescent lighting or in dark at 20°C for 12 months. Results showed that nonirradiated samples under active packaging were considered fresh even after 12 months of storage, while irradiated samples under active packaging were found merely acceptable. Another experiment conducted by Kudachikar et al. (2011) showed that MAP in combination with green keeper (GK) as ethylene absorbent of banana samples results in reduction in physiological loss in weight of 0.7% and 0.8% after 5 and 7 weeks of storage, respectively, compared with a 5% physiological loss in weight in openly stored banana samples after 3 weeks. A steady state of about 8.2% of CO_2 and 2.6% of

O_2 in MAP and GK packages was established after 3 weeks of storage. MAP and GK treatment results in minimal losses in fruit weight, fruit texture, and fruit composition as compared with openly kept control bananas. Results indicated that the shelf life of fruits packed under MAP and GK could be extended up to 7 weeks, as compared with 3 weeks for openly stored control samples. Furthermore, the sensory quality of fully ripe fruits in MAP and GK packages, 5 days after Ethrel dip, was deemed very good. Thus, this treatment under at 12°C ± 1°C and 85%–90% RH can be considered for commercial use to achieve long-term storage and also long-distance transportation of bananas. Mold growth is a significant problem with the shelf life of bakery products. Sanguinetti et al. (2009) demonstrated that active packaging (with an oxygen absorber [OA]) at 20°C can extend the shelf life of cheese tarts up to seven times compared with air packaging. Active packaging was found to be more effective than MAP (70 N_2/30 CO_2 and 20 N_2/80 CO_2) from the point of view of microbiological safety and maintaining original texture and sensory characteristics. Active packaging seems to be a very promising technique to prolong the shelf life of intermediate-moisture bakery products, thereby providing an effective control of *Staphylococci* growth. Mexis et al. (2011a) investigated the effect of active packaging (OA combined with an ethanol emitter [EE]) on Graviera cheese stored at 4°C and 12°C. Yeasts and molds in samples with the OA + EE remained below 2 log CFU/g throughout storage irrespective of temperature. Extension of shelf life was 8–8.5 weeks for samples packed under this packaging regime.

19.4 ADVANCES IN NANOTECHNOLOGY

Nanotechnology is considered the next step in many industries, including food processing and packaging. The applications of nanotechnology in the food industry may include nanoparticulate delivery systems (e.g., micelles, liposomes, nanoemulsion, biopolymeric nanoparticles, and cubosomes), food safety and biosecurity (e.g., nanosensors), and also nanotoxicity (Maynard, 2006; Chau et al., 2007). Nanotechnology in food packaging allows packaging materials to be altered to improve the barrier, mechanical and heat-resistance properties, biodegradability, and flame retarding compared with other typical polymers. It is also considered very promising in active antimicrobial and antifungal surfaces and sensing as well as signaling microbiological and biochemical changes (El Amin, 2005). It is believed that the food packaging industry could potentially attract the largest share of the market for nanotechnology (EFSA, 2009).

Until now, food products like bakery and meat products have attracted the most nanopackaging applications, while with beverages, carbonated drinks and bottled water predominate. However, only a few of these systems have been developed and are currently being applied. The Asia-Pacific region, in particular Japan, is the market leader in active nanopackaging, with 45% of the current market, valued at U.S. $1.86 billions in 2008 and anticipated to increase to U.S. $3.43 billion by 2014, with an annual increase of 13%. In the United States, Japan, and Australia, improved and active packaging is currently successfully applied to prolong the shelf life of products while maintaining nutritional quality and microbiological safety. In Europe, the industrial applications follow a significantly slower pattern. The main reasons for this are (i) legislative restrictions, (ii) lack of knowledge about acceptability to

TABLE 19.1
Advantages of Nanomaterials in Food Packaging

Advantages	
Innovation	The main motivation for applications of nanomaterials in packaging materials is both innovation and new product development. New products offer greater consumer choice and convenience. New products can support social change and lifestyles, can create new markets, and favor economic growth and employment
Light weight	Usage of less packaging material with the same technical performance allows lower material usage. This could give a lower carbon/environmental footprint from the manufacture and transport of the packaging and the packaged food
Greater protection and preservation of food	Improved barrier properties can help maintain food quality and increase shelf life. This can lead potentially to a lower cost and reliable food supply, better nutrition, and less food waste.
Improved performance of biobased materials	The performance of biobased materials can be inferior to synthetic polymers based on oil coal or gas. However, if their performance can be improved by the use of nanomaterials or by using nanotechnology tools to study material properties and improve production methods, then this could allow the synthetic polymers to be eventually substituted with biobased materials

Sources: Bradley, E.L. et al., *Trends Food Sci. Technol.*, in press, 2011; Rhim, J.-W. and Ng, P.K.W., *Crit. Rev. Food Sci. Nutr.*, 47, 411, 2007.

European consumers, and (iii) the efficacy of such systems and the financial and environmental implications these systems could have (Silvestre et al., 2011). The most promising developments introduced in the market to date are likely to significantly improve the quality and shelf life of meat products, by improving barrier properties and incorporating bioactive nanocompounds into or onto the film (nanocomposite). A nanocomposite is a type of multiphase solid material reinforced with nanometer scale particles, fibers, or platelets, which give it better mechanical and chemical properties compared with other conventional composites (Pandey et al., 2005; Lee, 2010). One example are the cellulose–silver nanoparticle hybrid materials that can be potentially used to improve the aseptic conditions during the storage of products of vegetable origin and in absorbent cellulose applied to soak exudates in liners. Thus, especially designed absorbent materials could be optimized to lower the risks during the manipulation of food and storage, resulting in feasible applications of silver-based nanotechnology in food technology (Fernández et al., 2010).

The main benefits offered by nanomaterials and nanotechnology are presented in the following table (Table 19.1).

19.5 RECENT TRENDS IN EU ACTIVE PACKAGING LEGISLATION

Although over the last three to four decades active and intelligent packaging (A & I), occasionally, in conjunction with recently "reinvented" MA packaging have been regarded as promising advances in food packaging, it is of importance to elucidate the purpose of each technology. MA and active packaging aim to extend the shelf life of the product, while intelligent packaging aims to provide indication and, preferably, continuously monitor the freshness of the product.

It is noteworthy, however, that in both packaging cases (A & I), interactions between food and food environment do occur thus posing certain threats to human safety. The aforementioned packaging schemes fall in the category of food contact materials, and the possibility for compound migration from packaging to food must be investigated. EU has been quite active in this field, publishing a number of regulations and directives such as the following:

- Regulation (EC) 1935/2004 (packaging materials coming in contact with foods).
- Regulation (EC) No 1895/2005 (applicable to containers or storage tanks with a capacity greater than 10,000 L or pipelines).
- Regulation (EC) No 2023/2006 shall apply to all sectors and to all stages of manufacture, processing, and distribution of materials and articles, apart from production of starting substances.
- Regulation (EC) No 282/2008, the technical dossier shall contain the information specified in the guidelines for the safety assessment of a recycling process.
- Regulation (EC) No 450/2009 shall apply to active and intelligent materials and articles that are placed on the market within the community.

On the contrary, legislation is considerably less in the United States and Canada, and even the respective changes are not that frequent as in the EU. To be more specific, in the United States it comes under the jurisdiction of the Committee on Energy and Commerce; Food, Drug, and Related Law, as amended through December 31, 2004; and in particular it falls under the Tolerances for Poisonous Ingredients in Food (Sec. 406 [21 USC §346]).

As regards Canada, the corresponding legislation falls under the Food and Drug Regulation from the Canadian Department of Justice, Division 23 associated with food packaging materials.

REFERENCES

Arvanitoyannis, I. 1999. Totally-and-partially biodegradable polymer blends based on natural and synthetic macromolecules: Preparation and physical properties and potential as food packaging materials. *Journal of Macromolecular Science—Reviews in Macromolecular Chemistry and Physics*, C39(2): 205–271.

Bhat, Z.F. and Bhat, H. 2011. Recent trends in poultry packaging: A review. *American Journal of Food Technology*, 6: 531–540.

Bradley, E.L., Castle, L. and Chaudhry, Q. 2011. Applications of nanomaterials in food packaging with a consideration of opportunities for developing countries. *Trends in Food Science & Technology*, in press.

Cayuela, J.M., Gill, M.D., Banon, S. and Garrido, M.D. 2004. Effect of vacuum and modified atmosphere packaging on the quality of pork loin. *European Food Research Technology*, 219: 316–320.

Chau, C.F., Wu, S.H. and Yen, G.C. 2007. The development of regulations for food nanotechnology. *Trends in Food Science & Technology*, 18: 269–280.

Coma, V. 2008. Bioactive packaging technologies for extended shelf life of meat based products. *Meat Science*, 78: 90–103.

Cooksey, K. 2001. Antimicrobial food packaging materials. *Additives for Polymer*, 8: 6–10.

Dainelli, D., Gontard, N., Spyropoulos, D., Zondervan-van den Beuken, E. and Tobback, P. 2008. Active and intelligent food packaging: Legal aspects and safety concerns. *Trends Food Science & Technology*, 19: 99–108.

Day, A. 1996. High oxygen modified atmosphere packaging for fresh prepared produce. *Postharvest News and Information*, 7(3): 31–34.

Del Nobile, M.A., Corbo, M.R., Speranza, B., Sinigaglia, M., Conte, A. and Caroprese, M. 2009. Combined effect of MAP and active compounds on fresh blue fish burger. *International Journal of Food Microbiology*, 135: 281–287.

EFSA, 2009. The potential risks arising from nanoscience and nanotechnologies on food and feed safety. Scientific opinion of the scientific committee (Question No EFSAQ-2007-124a). February 10, 2009.

Eilert, S.J. 2005. New packaging technologies for the 21st century. *Meat Science*, 71: 122–127.

El Amin, A. 2005. Nanotechnology targets new food packaging products. http://www.foodproductiondaily.com/Packaging/Nanotechnology-targets

Fellows, P. 2000. *Food Processing Technology: Principles and Practice*. Cambridge, U.K.: Woodhead Publishing Limited.

Fernández, A., Pierre Picouet, P. and Lloret, E. 2010. Cellulose-silver nanoparticle hybrid materials to control spoilage-related microflora in absorbent pads located in trays of fresh-cut melon. *International Journal of Food Microbiology*, 142: 222–228.

Jacxsens, L., Devlieghere, F. and Debevere, J. 2002. Predictive modelling for packaging design: Equilibrium modified atmosphere packages of fresh-cut vegetables subjected to a simulated distribution chain. *International Journal of Food Microbiology*, 73: 331–341.

Joerger, R.D. 2007. Antimicrobial films for food applications—A quantitative analysis of their effectiveness. *Packaging Technology and Science*, 20: 231–273.

Kotler, P. and Keller, K.L. 2006. *Marketing Managment,* 12th edn., Upper Saddle River, NJ: Pearson.

Kudachikar, V.B., Kulkarni, S.G. and Keshava Prakash, M.N. 2011. Effect of modified atmosphere packaging on quality and shelf life of "Robusta" banana (*Musa sp.*) stored at low temperature. *Journal of Food Science and Technology*, 48(3): 319–324.

Lahteenmaki, L. and Arvola, A. 2003. Testing consumer responses to new packaging concepts, in *Novel Food Packaging Techniques*, Ed. Ahvenainen, R. Cambridge, U.K.: Woodhead Publishing Ltd. pp. 550–562.

Lee, K.T. 2010. Quality and safety aspects of meat products as affected by various physical manipulations of packaging materials. *Meat Science*, 86: 138–150.

Lisinska-Kusnierz, M. and Ucherek, M. 2003. *Modern Packaging*. Cracow, Poland: PTTZ (Polish Food Technologists Society).

Marsh, K. and Bugusu, B. 2007. Food packaging—Roles, materials, and environmental issues. *Food Science*, 72: 39–55.

Mastromatteo, M., Danza, A., Conte, A., Muratore, G. and Del Nobile, M.A. 2010. Shelf life of ready to use peeled shrimps as affected by thymol essential oil and modified atmosphere packaging. *International Journal of Food Microbiology*, 144: 250–256.

Maynard, A.D. 2006. Nanotechnology: Assessing the risks. *Nanotoday*, 1: 22–33.

Mexis, S.F., Chouliara, E. and Kontominas, M.G. 2011a. Quality evaluation of grated Graviera cheese stored at 4 and 12°C using active and modified atmosphere packaging. *Packaging Technology and Science*, 24: 15–29.

Mexis, S.M., Riganakos, K.A. and Kontominas, G.M. 2011b. Effect of irradiation, active and modified atmosphere packaging, container oxygen barrier and storage conditions on the physicochemical and sensory properties of raw unpeeled almond kernels (*Prunus dulcis*). *Journal of the Science of Food and Agriculture*, 91: 634–649.
Pandey, J.K., Kumar, A.P., Misra, M., Mohanty, A.K., Drzal, L.T. and Singh, R.P. 2005. Recent advances in biodegradable nanocomposites. *Journal of NanoScience and Nanotechnology*, 5: 479–526.
Phillips, C.A. 1996. Review: Modified atmosphere packaging and its effects on the microbiological quality and safety of produce. *International Journal of Food Science and Technology*, 31(6): 463–479.
Rajkumar, R., Dushyantha, K., Asha Rajini, R. and Sureshkumar, S. 2007. Effect of modified atmosphere packaging on microbial and physical properties of Turkey meat. *American Journal of Food Technology*, 2: 183–189.
Rathore, H.A., Masud, T., Sammi, S. and Majeed, S. 2010. Innovative approach of active packaging I cardboard carton and its effects on overall quality attributes such as weight loss, total soluble solids, pH, acidity and ascorbic acid contents of Chaunsa white variety of mango at ambient temperature during storage. *Pakistan Journal of Nutrition*, 9: 452–458.
Restuccia, D., Gianfranco Spizzirri, U., Parisi, O.I., Cirillo, G., Curcio, M., Iemma, F., Puoci, F., Vinci, G. and Picci, N. 2010. New EU regulation aspects and global market of active and intelligent packaging for food industry applications. *Food Control*, 21: 1425–1435.
Rhim, J.-W. and Ng, P.K.W. 2007. Natural biopolymer-based nanocomposite films for packaging applications. *Critical Reviews in Food Science and Nutrition*, 47: 411–433.
Sanguinetti, A.M., Secchi, N., Del Caro, A., Stara, G., Roggio, T. and Piga, A. 2009. Effectiveness of active and modified atmosphere packaging on shelf life extension of a cheese tart. *International Journal of Food Science and Technology*, 44: 1192–1198.
Silvestre, C. Duraccio, D. and Cimmino, S. 2011. Food packaging based on polymer nanomaterials. *Progress in Polymer Science*, 36: 1766–1782.
Singh, P., Wani, A.A., Saengerlaub, S. and Langowski, H.-C. 2011. Understanding critical factors for the quality and shelf-life of MAP fresh meat: A review. *Critical Reviews in Food Science and Nutrition*, 51(2): 146–177.
Washuettl, M., Wepner, B. and Tacker, M. 2001. Modified atmosphere packaging development of methods for detection of senescence parameters, respiration rate and changes in gas composition during storage. *Proceedings of 12th IAPRI World Conference on Packaging*, Warsaw, Poland, pp. 18–20.
Yam, K.L., Takhistov, P.T. and Miltz, J. 2005. Intelligent packaging: Concepts and applications. *Journal of Food Science*, 70: 1–10.

Index

C

D

E

F

N

O

P

Q

R

S

T

U

V

W